INTERNATIONAL SERIES OF MONOGRAPHS ON PHYSICS

Series Editors

R. Friend   University of Cambridge
M. Rees   University of Cambridge
D. Sherrington   University of Oxford
G. Veneziano   CERN, Geneva

INTERNATIONAL SERIES OF MONOGRAPHS ON PHYSICS

172. J. Kübler: *Theory of itinerant electron magnetism, Second edition*
171. J. Zinn-Justin: *Quantum field theory and critical phenomena, Fifth edition*
170. V.Z. Kresin, S.G. Ovchinnikov, S.A. Wolf: *Superconducting state - mechanisms and materials*
169. P.T. Chrusciel: *Geometry of black holes*
168. R. Wigmans: *Calorimetry – Energy measurement in particle physics, Second edition*
167. B. Mashhoon: *Nonlocal gravity*
166. N. Horing: *Quantum statistical field theory*
165. T.C. Choy: *Effective medium theory, Second edition*
164. L. Pitaevskii, S. Stringari: *Bose-Einstein condensation and superfluidity*
163. B.J. Dalton, J. Jeffers, S.M. Barnett: *Phase space methods for degenerate quantum gases*
162. W.D. McComb: *Homogeneous, isotropic turbulence - phenomenology, renormalization and statistical closures*
160. C. Barrabès, P.A. Hogan: *Advanced general relativity - gravity waves, spinning particles, and black holes*
159. W. Barford: *Electronic and optical properties of conjugated polymers, Second edition*
158. F. Strocchi: *An introduction to non-perturbative foundations of quantum field theory*
157. K.H. Bennemann, J.B. Ketterson: *Novel superfluids, Volume 2*
156. K.H. Bennemann, J.B. Ketterson: *Novel superfluids, Volume 1*
155. C. Kiefer: *Quantum gravity, Third edition*
154. L. Mestel: *Stellar magnetism, Second edition*
153. R.A. Klemm: *Layered superconductors, Volume 1*
152. E.L. Wolf: *Principles of electron tunneling spectroscopy, Second edition*
151. R. Blinc: *Advanced ferroelectricity*
150. L. Berthier, G. Biroli, J.-P. Bouchaud, W. van Saarloos, L. Cipelletti: *Dynamical heterogeneities in glasses, colloids, and granular media*
149. J. Wesson: *Tokamaks, Fourth edition*
148. H. Asada, T. Futamase, P. Hogan: *Equations of motion in general relativity*
147. A. Yaouanc, P. Dalmas de Réotier: *Muon spin rotation, relaxation, and resonance*
146. B. McCoy: *Advanced statistical mechanics*
145. M. Bordag, G.L. Klimchitskaya, U. Mohideen, V.M. Mostepanenko: *Advances in the Casimir effect*
144. T.R. Field: *Electromagnetic scattering from random media*
143. W. Götze: *Complex dynamics of glass-forming liquids - a mode-coupling theory*
142. V.M. Agranovich: *Excitations in organic solids*
141. W.T. Grandy: *Entropy and the time evolution of macroscopic systems*
140. M. Alcubierre: *Introduction to 3+1 numerical relativity*
139. A.L. Ivanov, S.G. Tikhodeev: *Problems of condensed matter physics - quantum coherence phenomena in electron-hole and coupled matter-light systems*
138. I.M. Vardavas, F.W. Taylor: *Radiation and climate*
137. A.F. Borghesani: *Ions and electrons in liquid helium*
135. V. Fortov, I. Iakubov, A. Khrapak: *Physics of strongly coupled plasma*
134. G. Fredrickson: *The equilibrium theory of inhomogeneous polymers*
133. H. Suhl: *Relaxation processes in micromagnetics*
132. J. Terning: *Modern supersymmetry*
131. M. Mariño: *Chern-Simons theory, matrix models, and topological strings*
130. V. Gantmakher: *Electrons and disorder in solids*
129. W. Barford: *Electronic and optical properties of conjugated polymers*
128. R.E. Raab, O.L. de Lange: *Multipole theory in electromagnetism*
127. A. Larkin, A. Varlamov: *Theory of fluctuations in superconductors*
126. P. Goldbart, N. Goldenfeld, D. Sherrington: *Stealing the gold*
125. S. Atzeni, J. Meyer-ter-Vehn: *The physics of inertial fusion*
123. T. Fujimoto: *Plasma spectroscopy*
122. K. Fujikawa, H. Suzuki: *Path integrals and quantum anomalies*
121. T. Giamarchi: *Quantum physics in one dimension*
120. M. Warner, E. Terentjev: *Liquid crystal elastomers*
119. L. Jacak, P. Sitko, K. Wieczorek, A. Wojs: *Quantum Hall systems*
117. G. Volovik: *The Universe in a helium droplet*
116. L. Pitaevskii, S. Stringari: *Bose-Einstein condensation*
115. G. Dissertori, I.G. Knowles, M. Schmelling: *Quantum chromodynamics*
114. B. DeWitt: *The global approach to quantum field theory*
112. R.M. Mazo: *Brownian motion - fluctuations, dynamics, and applications*
111. H. Nishimori: *Statistical physics of spin glasses and information processing - an introduction*
110. N.B. Kopnin: *Theory of nonequilibrium superconductivity*
109. A. Aharoni: *Introduction to the theory of ferromagnetism, Second edition*
108. R. Dobbs: *Helium three*
105. Y. Kuramoto, Y. Kitaoka: *Dynamics of heavy electrons*
104. D. Bardin, G. Passarino: *The Standard Model in the making*
103. G.C. Branco, L. Lavoura, J.P. Silva: *CP Violation*

101. H. Araki: *Mathematical theory of quantum fields*
100. L.M. Pismen: *Vortices in nonlinear fields*
 99. L. Mestel: *Stellar magnetism*
 98. K.H. Bennemann: *Nonlinear optics in metals*
 94. S. Chikazumi: *Physics of ferromagnetism*
 91. R.A. Bertlmann: *Anomalies in quantum field theory*
 90. P.K. Gosh: *Ion traps*
 87. P.S. Joshi: *Global aspects in gravitation and cosmology*
 86. E.R. Pike, S. Sarkar: *The quantum theory of radiation*
 83. P.G. de Gennes, J. Prost: *The physics of liquid crystals*
 73. M. Doi, S.F. Edwards: *The theory of polymer dynamics*
 69. S. Chandrasekhar: *The mathematical theory of black holes*
 51. C. Møller: *The theory of relativity*
 46. H.E. Stanley: *Introduction to phase transitions and critical phenomena*
 32. A. Abragam: *Principles of nuclear magnetism*
 27. P.A.M. Dirac: *Principles of quantum mechanics*
 23. R.E. Peierls: *Quantum theory of solids*

Theory of Itinerant Electron Magnetism
Second Edition

Jürgen Kübler

Technical University Darmstadt, Germany

OXFORD
UNIVERSITY PRESS

Great Clarendon Street, Oxford, OX2 6DP,
United Kingdom

Oxford University Press is a department of the University of Oxford.
It furthers the University's objective of excellence in research, scholarship,
and education by publishing worldwide. Oxford is a registered trade mark of
Oxford University Press in the UK and in certain other countries

© Jürgen Kübler 2021

The moral rights of the author have been asserted

First Edition published in 2021

Impression: 1

All rights reserved. No part of this publication may be reproduced, stored in
a retrieval system, or transmitted, in any form or by any means, without the
prior permission in writing of Oxford University Press, or as expressly permitted
by law, by licence or under terms agreed with the appropriate reprographics
rights organization. Enquiries concerning reproduction outside the scope of the
above should be sent to the Rights Department, Oxford University Press, at the
address above

You must not circulate this work in any other form
and you must impose this same condition on any acquirer

Published in the United States of America by Oxford University Press
198 Madison Avenue, New York, NY 10016, United States of America

British Library Cataloguing in Publication Data

Data available

Library of Congress Control Number: 2021937964

ISBN 978–0–19–289563–9

DOI: 10.1093/oso/9780192895639.001.0001

Printed and bound by
CPI Group (UK) Ltd, Croydon, CR0 4YY

Links to third party websites are provided by Oxford in good faith and
for information only. Oxford disclaims any responsibility for the materials
contained in any third party website referenced in this work.

Preface

Magnets have been known to mankind for a long time. Historically there is the compass needle; in daily life today magnets are used, for instance, to pin notes to the refrigerator, an admittedly not so very interesting application. Certainly more important are the numerous permanent magnets in electric motors, for instance, in your household, your car, and, in different geometrical shapes, as storage media like the hard disk in your personal computer, as tapes and floppy disks, etc. How does magnetism come about? How much of the multitude of experimental details can we understand? Can we predict certain properties starting with microscopic principles?

To answer these questions we must apply, in the broadest sense, quantum mechanics and statistical mechanics to the field of magnetism. Here, however, one must bear in mind that an immensely large number of electrons move in the solid state of matter; under certain well-described circumstances they produce permanent magnetism.

The quantum mechanical many-electron problem to be solved presents a great challenge that can only be partly mastered using approximations. It is here that this book differs from most others: the local density functional approximation (LDA) is consistently used in an attempt to deal with the entire field of magnetism, including thermodynamic properties. Not all problems are solved; in fact, there remain more questions than answers.

We should eliminate from the very beginning a basic misunderstanding concerning the local density functional approximation: this is not the independent-particle picture that is often associated with the term *band structure*. Let us be a little more precise without going into too much detail.

In a quantum mechanical description of the motion of a particle it is, as we know, a wave function that determines the physics – and part of the wave function of an electron is its spin. The term *spin* is quite descriptive as it acts like a very small magnet. Indeed, it is responsible for magnetism. But since all electrons possess spin, and all metals possess huge numbers of electrons, why is not every metal a magnet? Another ingredient is needed which we now know originates from the *interaction* of the electrons with each other. This interaction leads to a collective effect, i.e. the total macroscopic property is not just the sum of the individual entities. Itinerant-electron magnetism is a collective phenomenon.

Permanent magnets are of fundamental interest, but magnetic materials are also of great practical importance as they provide a large field of technological applications. Therefore, the emphasis in this book is on realistic magnets and, consequently, the equations that describe the many-electron problem in the LDA must be solved using computers. Here the great, recent and continuing improvements of computers are, to a large extent, responsible for the progress made. Thus, in addition to the foundations, consisting of the density functional theory (Chap. 2), representative computational methods are explained (Chap. 3), and a complete computer program is introduced in the appendix of the book and made available through the internet address www.oup.co.uk. This enables the reader to determine the electronic structure of a magnet using a modern PC (or if desired, a more powerful machine; scaling is not so difficult here). Computer-based knowledge of the electronic structure, if properly interpreted, may be the starting point for a deeper understanding of magnetism. This in itself is a great achievement. But one may go even further and use the computer to design magnets that have desired properties needed, say, for some device. Clearly, the computer is not the answer to everything and our understanding has severe limits. But we may at least try – and try we do here. As a reminder it is almost superfluous to say that without the experimentalists progress is not possible.

A large part of the book is devoted to a detailed treatment of the connection between the electronic properties and magnetism and how they differ in the various known magnetic systems (Chap. 4). Trends are discussed and explained for a large class of alloys and compounds. The modern field of artificially layered systems – known as multilayers – which are not only interesting from a fundamental point of view, but are also finding new industrial applications, are dealt with in quite some detail. Finally, an attempt is made to relate the rich thermodynamic properties of magnets with the *ab initio* results originating from the electronic structure (Chap. 5). The old criticism that Stoner theory (with this label temporarily implying band theory) is inadequate to deal with itinerant magnets at finite temperatures is here refuted. Certainly, new degrees of freedom for the electronic motion must be found in order to be successful, but these new degrees of freedom can again be extracted from the electronic structure. We will see, however, that in spite of the progress made this is an open field where a great deal of activity is still expected.

Part of this book grew out of lectures at the physics department of the Darmstadt University of Technology, given to students who had a command of quantum mechanics, statistical mechanics and a basic knowledge of solid state physics. The present treatise in its entirety is certainly too large for a typical course, but selections can easily made so that it may be used as a textbook in a graduate course on the theory of magnetism in the United States or in post-graduate studies in Britain. It is, I hope, also useful for those engaged in basic and applied research in both academic and industrial laboratories.

This book would not have been possible without the help and interest of a great number of people. Very early work was begun with *Dr. A.R. Williams* at the IBM Research Laboratory in Yorktown Heights, New York, who, even after leaving the field of physics, continued to ask those kinds of questions that could not easily be answered and who continued to encourage me in this endeavor up to the present day. Thank you, Art!

Of course, a generation of students and junior scientists was involved in one way or another working with me on the topic of the book. They deserve to be thanked. Of those who left my group very recently or who are still active I would like to express my thanks particularly to *Dr. L.M. Sandratskii* who advanced the issue of noncollinear magnetism enormously. I am grateful to *Dr. M. Uhl* for the active role he played in using spin fluctuations to describe the thermodynamics of itinerant-electron magnets. I also thank *K. Knöpfle* who knew answers to almost all technical problems and

Mrs. Hanna-Daoud whose help with the manuscript is gratefully appreciated. I cordially thank *Prof. R. M. Martin* who, together with *Beverly Martin*, were marvelous hosts during a sabbatical spent at the University of Illinois where I also enjoyed helpful discussions with *Prof. Y.-C. Chang*.

Financial support from the Deutsche Forschungsgemeinschaft (DFG) with its SFB-Program (252) Darmstadt, Frankfurt, Mainz and from the Volkswagen Foundation, who contributed to my stay at the University of Illinois where a large part of this book was written, is acknowledged.

Last but not least, without the never-ending encouragement of *Felicia and Felix Kübler* this book would never have been attempted not to mention completed. Thank you.

Darmstadt
February 2000

Preface to the revised edition

This revised edition contains fewer typos, I hope. A more than moderate effort, however, went into a modernization of the original text. Not that it is really possible to keep up with the rapid progress, but I tried.

Chapter 4 now includes some new work on Heusler compounds and double perovskites. This is where in the recent past much research has been done, especially aiming at materials for the new field of spin electronics, now called spintronics. This required a small excursion into tunnel magnetoresistance, which has been added to the multilayer section. The interaction with my friends in Mainz, especially with Claudia Felser, is gratefully acknowledged.

Special attention I devoted to corrections in Chap. 5, where I hopefully improved the theory of spin-wave spectra and, especially the theory of spin fluctuations. Some of what I discussed in the original missed the point. A complete theory of thermal properties of magnets has, unfortunately, not yet been devised. But some of the approximations have been improved, especially our ability to estimate the Curie temperatures of ferromagnetic compounds.

As before, without the never-ending encouragements of Felicia Kübler these revisions would not have been completed. Thank you.

Fränkisch-Crumbach
February 2009

Preface to the second edition

`Omnia mutantur, nos et mutamur in illis.`- Indeed, so much changed that I can hardly do justice to the topic of this book. I tried anyway and added material that kept me busy in a lively cooperation with the group of Claudia Felser at the Max-Planck Institute for Chemical Physics of Solids in Dresden. Besides magneto-optics, which actually belongs to the previous version of the book where it was simply forgotten, the description of the new development begins with the physics that is connected with the Berry phase. This leads to a greatly improved understanding of the anomalous Hall effect in ferromagnetic and antiferromagnetic materials. The discovery of Weyl fermions in this context adds surface effects to the physics of magnetic materials and provides a novel property of the band structure, which brings us to the field of topology and topological thermodynamics. Of basic interest and important for applications is the discovery of magnetic skyrmions, i.e. swirls of magnetization or vortices that behave like particles. I try to explain some salient features here but leave much to the literature.

The constant and generous support by Claudia Felser is greatly acknowledged, especially, however, her freely sharing of interesting questions. As before, without the never-ending encouragements of Felicia Kübler this second edition would not have been completed. Thank you.

Konstanz
January 2021

Contents

1 **Introduction**   1
   1.1 Basic facts   1
   1.2 Itinerant electrons   5
      1.2.1 Gas of free electrons   5
      1.2.2 Landau levels   7
      1.2.3 The susceptibility of a gas of free electrons   13
      1.2.4 The de Haas–van Alphen effect   15
      1.2.5 The Lifshitz–Kosevich formula   16
      1.2.6 Pauli paramagnetism   22
   1.3 How to proceed   26

2 **Density-Functional Theory**   28
   2.1 Born–Oppenheimer approximation   28
   2.2 Hartree–Fock approximation   31
   2.3 Density-functional theory   39
   2.4 The electron spin: Dirac theory   46
   2.5 Spin-density-functional theory   56
   2.6 The local-density approximation (LDA)   59
   2.7 Nonuniformly magnetized systems   63
   2.8 The generalized gradient approximation (GGA)   70
      2.8.1 Formal properties of density functionals   71
      2.8.2 Scaling relations   75
      2.8.3 The correlation energy of the homogeneous electron gas   78
      2.8.4 Linear response: screening in the electron gas   79
      2.8.5 Analytical expression for GGA   83

3 **Energy-Band Theory**   89
   3.1 Bloch's theorem   89
   3.2 Plane waves, orthogonalized plane waves, and pseudopotentials   96
      3.2.1 Plane waves   97
      3.2.2 OPW method   100
      3.2.3 Pseudopotentials   104

Contents

3.3	Augmented plane waves and Green's functions	108
	3.3.1 APW	108
	3.3.2 Multiple scattering theory	116
3.4	Linear methods	122
	3.4.1 LCAO	123
	3.4.2 Energy derivative of the wave function: ϕ and $\dot{\phi}$	125
	3.4.3 Linear augmented plane waves (LAPW)	129
	3.4.4 Linear combination of muffin-tin orbitals (LMTO)	132
	3.4.5 Augmented spherical waves (ASW)	136
	3.4.6 The Korringa–Kohn–Rostoker atomic sphere approximation (KKR-ASA)	149
	3.4.7 ASW for arbitrary spin configurations	159
	3.4.7.1 Secular equation and density matrix	160
	3.4.7.2 Incommensurate spiral structure	166
	3.4.7.3 Relativistic corrections	171

4 Electronic Structure and Magnetism   173

4.1	Introduction and simple concepts	173
	4.1.1 Stoner theory	180
4.2	The magnetic susceptibility	184
	4.2.1 Linear response	184
	4.2.2 The Stoner condition and other basic facts	189
	4.2.2.1 The Stoner condition	189
	4.2.2.2 Band-structure features of the transition metals	193
	4.2.2.3 Crystal phase stability	197
	4.2.3 The static nonuniform magnetic susceptibility	201
	4.2.3.1 Nonmagnetic V, Cr, and Pd	203
	4.2.3.2 The longitudinal susceptibilities of ferromagnetic Fe, Co, and Ni	205
4.3	Elementary magnetic metals	210
	4.3.1 Ground-state properties of Fe, Co, and Ni	213
	4.3.2 Volume dependence of transition metal magnetism	217
	4.3.3 Band structure of ferromagnetic metals	221
	4.3.3.1 bcc Iron	222
	4.3.3.2 hcp Cobalt	225
	4.3.3.3 Nickel	228
	4.3.4 Electronic structure of antiferromagnetic metals	230
	4.3.4.1 Manganese	231
	4.3.4.2 Chromium	232
	4.3.4.3 Manganese (cont.)	237
	4.3.4.4 Iron	250

4.4	Magnetic compounds		258
	4.4.1 The Slater–Pauling curve		258
		4.4.1.1 Generalized Slater–Pauling curve	264
		4.4.1.2 Constant minority-electron count	267
	4.4.2 Selected case studies		271
		4.4.2.1 CrO_2	271
		4.4.2.2 Heusler compounds	274
		4.4.2.3 Double perovskites	278
		4.4.2.4 Invar	281
4.5	Multilayers		287
	4.5.1 Oscillatory exchange coupling		287
		4.5.1.1 RKKY exchange	288
		4.5.1.2 Free electrons	290
		4.5.1.3 Aliasing	292
		4.5.1.4 Fermi surface effects	295
	4.5.2 Oscillatory exchange coupling in the quantum-well picture		298
		4.5.2.1 Confined states in multilayers	299
		4.5.2.2 A simple model	302
		4.5.2.3 Green's functions	306
	4.5.3 Giant magnetoresistance		312
		4.5.3.1 The Boltzmann equation	313
		4.5.3.2 Approximate evaluation of the conductivity	316
		4.5.3.3 Tunnel magnetoresistance	321
4.6	Relativistic effects		327
	4.6.1 Magneto-optical properties		327
	4.6.2 Symmetry properties of spin–orbit coupling		333
	4.6.3 Noncollinear magnetic structures in uranium compounds		335
		4.6.3.1 The case of U_3X_4	335
		4.6.3.2 The case of UPdSn	339
	4.6.4 Weak ferromagnetism		342
		4.6.4.1 Hematite (α-Fe_2O_3)	343
		4.6.4.2 The case of Mn_3Sn	345
	4.6.5 Magneto-crystalline anisotropy		347
4.7	Berry phase effects in solids		350
	4.7.1 The anomalous Hall effect		352
	4.7.2 The anomalous Hall effect in antiferromagnets		356
4.8	Weyl fermions		361
4.9	Real-case Weyl fermions		366
	4.9.1 Topological effects in magnetic compounds		367
	4.9.2 Topology of antiferromagnetic compounds		371

4.9.3	The Nernst effect	376
4.9.4	Remark about topology	382

5 Magnetism at Finite Temperatures   385

- 5.1 Density-functional theory at $T > 0$ 385
- 5.2 Adiabatic spin dynamics   389
  - 5.2.1 Magnon spectra of bcc Fe, fcc Co, and fcc Ni   393
  - 5.2.2 Magnon spectra for non-primitive lattices and compounds   396
  - 5.2.3 Thermal properties of itinerant-electron ferromagnets at low temperatures   399
- 5.3 Mean-field theories   401
  - 5.3.1 A useful example   402
- 5.4 Spin fluctuations   406
  - 5.4.1 A simple formulation   406
  - 5.4.2 Exploratory results for the elementary ferromagnets   418
    - 5.4.2.1 Nickel   418
    - 5.4.2.2 fcc Cobalt   420
    - 5.4.2.3 Iron   420
    - 5.4.2.4 hcp Cobalt   421
  - 5.4.3 Simple itinerant antiferromagnets   423
    - 5.4.3.1 fcc Mn   424
    - 5.4.3.2 FeRh   425
  - 5.4.4 Previous, semi-empirical spin-fluctuation theories   428
  - 5.4.5 Connection with the fluctuation-dissipation theorem   429
  - 5.4.6 The dynamic approximation   431
    - 5.4.6.1 More on nickel   433
    - 5.4.6.2 The weakly ferromagnetic compound Ni_3Al 436
  - 5.4.7 The spherical approximation   438
    - 5.4.7.1 Exchange in detail   442
  - 5.4.8 Collection of results   446
- 5.5 Magnetic skyrmions   451
  - 5.5.1 Formal properties of magnetic skyrmions   455
  - 5.5.2 The phase transition   457
  - 5.5.3 New developments   460
- 5.6 High-temperature approaches   463
  - 5.6.1 Short-range order   463
  - 5.6.2 The disordered local moment state   467
    - 5.6.2.1 The coherent potential approximation (CPA)   468
    - 5.6.2.2 Lloyd's formula   471
    - 5.6.2.3 The CPA integrated density of states   472

　　　　　5.6.2.4 The spin susceptibility in the paramagnetic state　　　473
　　　　　5.6.2.5 Results　　　476
　　　　　5.6.2.6 Onsager cavity-field approximation　　　477

References　　　481
Index　　　503

1
Introduction

This introductory chapter contains a collection of concepts and facts that can be found in many other textbooks. The reason for dealing with these things here – some are quite elementary, others somewhat more complicated – is an attempt to make this treatise self-contained. The contents of this chapter – in one form or another – center on the magnetic susceptibility, especially the de Haas–van Alphen effect and the Lifshitz–Kosevich formula, which allow the experimentalist to determine the size and shape of the Fermi surface, the latter being so important for itinerant electrons.

1.1 Basic facts

To understand the phenomena of magnetism one might be tempted first to develop a classical description and then complete it with the necessary quantum-mechanical corrections. However, in the 1930s Bohr and van Leeuwen showed that magnetism is incomprehensible within the framework of an exact classical theory based on the magnetism of moving charges (Van Vleck, 1932). This is not hard to see; for, writing the classical Hamiltonian of N particles possessing charges e (e is a negative number for electrons) and masses m in a magnetic field specified by the vector potential $\mathbf{A}$ as

$$\mathcal{H} = \sum_{i=1}^{3N} \frac{1}{2m} \left(p_i - \frac{e}{c} A_i \right)^2 + V(\mathbf{q}_1, \mathbf{q}_2, \ldots, \mathbf{q}_N), \tag{1.1}$$

where $\mathbf{q}_i$ and $\mathbf{p}_i$ are canonical coordinates and *momenta* (c is the velocity of light) and V is an interaction potential, we write the classical partition function as

$$Z = \int e^{-\beta \mathcal{H}} \, dq_1 \ldots dq_{3N} \, dp_1 \ldots dp_{3N}, \tag{1.2}$$

where $\beta = 1/k_\mathrm{B}T$, with k_B being Boltzmann's constant. It is now important to note that the integration over the momenta goes from $-\infty$ to ∞. If we, therefore, substitute

$$\mu_i \doteq p_i - \frac{e}{c} A_i, \tag{1.3}$$

the partition function becomes

$$Z = \int e^{-\beta V} \int \exp\left(-\frac{\beta}{2m}\sum_i \mu_i^2\right) d\mu_1 \ldots d\mu_{3N}\, dq_1 \ldots dq_{3N}, \qquad (1.4)$$

where the μ_i-integration limits are still $-\infty$ to ∞. Hence Z is independent of $\mathbf{A}$, whence the magnetization M, which is given by

$$M = k_B T \frac{\partial}{\partial H} \ln Z, \qquad (1.5)$$

vanishes. Here H is the magnetic field and we have made use of the standard thermodynamic relation

$$M = -\left(\frac{\partial F}{\partial H}\right), \qquad (1.6)$$

where the Helmholtz free energy, F, is related to the partition function by

$$F = -k_B T \ln Z. \qquad (1.7)$$

(Other variants of this argument exist, but they do not seem to show anything different.) We will therefore consistently use quantum mechanics to describe the system of particles that gives rise to the phenomenon of magnetism. To begin with, we want to estimate the magnetic susceptibility for an atom starting with the Hamiltonian which, for a single electron moving in the potential V and the vector potential $\mathbf{A}$, is written as

$$\mathcal{H} = \frac{1}{2m}\left(\mathbf{p} - \frac{e}{c}\mathbf{A}\right)^2 + V + \mu_B\, \boldsymbol{\sigma}\cdot\mathbf{H} + \zeta\, \boldsymbol{\ell}\cdot\boldsymbol{\sigma}. \qquad (1.8)$$

This Hamiltonian can be derived from Dirac's theory with the help of extremely well-justified approximations, see Chap. 2. The magnetic field is $\mathbf{H} = \boldsymbol{\nabla}\times\mathbf{A}$ and the spin–orbit coupling parameter is defined by

$$\zeta = \frac{\hbar}{4m^2 c^2}\frac{1}{r}\frac{dV}{dr} \qquad (1.9)$$

and

$$\mu_B = \frac{|e|\hbar}{2mc} \qquad (1.10)$$

is called the Bohr magneton. $\boldsymbol{\sigma}$ is the Pauli spin matrix vector with components

$$\sigma_z = \begin{pmatrix} 1 & 0 \\ 0 & -1 \end{pmatrix} \quad \sigma_y = \begin{pmatrix} 0 & -i \\ i & 0 \end{pmatrix} \quad \text{and} \quad \sigma_x = \begin{pmatrix} 0 & 1 \\ 1 & 0 \end{pmatrix}.$$

p is the momentum operator, in the real space representation given by

$$\mathbf{p} = -i\hbar \nabla, \tag{1.11}$$

and ∇ is the nabla operator. The quantity $\boldsymbol{\ell}$ is the angular momentum operator, defined by

$$\boldsymbol{\ell} = \mathbf{r} \times \mathbf{p} \tag{1.12}$$

and the term $\mu_B \boldsymbol{\sigma} \cdot \mathbf{H}$ appearing in Eqn (1.8) is usually called the Zeeman term. Now, in elementary quantum mechanics it is shown that, representing a uniform magnetic field **H** by the vector potential $\mathbf{A} = \frac{1}{2} \mathbf{H} \times \mathbf{r}$ (having the gauge $\nabla \cdot \mathbf{A} = 0$), one can rewrite the first term on the right-hand side of Eqn (1.8) as

$$\frac{1}{2m}\left(\mathbf{p} - \frac{e}{c}\mathbf{A}\right)^2 = \frac{p^2}{2m} - \frac{e}{2mc}(\mathbf{r} \times \mathbf{p}) \cdot \mathbf{H} + \frac{e^2}{8mc^2}(\mathbf{H} \times \mathbf{r})^2.$$

The first term is the kinetic energy, and the second is a paramagnetic term, as will be justified in what follows. The last term is simplified to

$$\frac{e^2}{8mc^2} H^2 (x^2 + y^2)$$

if the magnetic field is in the z-direction and calling the z-component simply H. All together we obtain for the Hamiltonian (1.8)

$$\mathcal{H} = \frac{p^2}{2m} + V + \mu_B \left(\frac{\boldsymbol{\ell}}{\hbar} + \boldsymbol{\sigma}\right) \cdot \mathbf{H} + \frac{e^2 H^2}{8mc^2}(x^2 + y^2) + \zeta \boldsymbol{\ell} \cdot \boldsymbol{\sigma}. \tag{1.13}$$

Of course, except for hydrogen, real atoms are not described by a one-electron Hamiltonian; therefore, at least formally, we write out the Hamiltonian for Z electrons, i.e.

$$\mathcal{H} = \sum_{i=1}^{Z} \left[\frac{p_i^2}{2m} + V(\mathbf{r}_i) + \mu_B \left(\frac{\boldsymbol{\ell}_i}{\hbar} + \boldsymbol{\sigma}_i\right) \cdot \mathbf{H} + \frac{e^2 H^2}{8mc^2}(x_i^2 + y_i^2) + \zeta_i \boldsymbol{\ell}_i \cdot \boldsymbol{\sigma}_i\right]$$
$$+ \frac{1}{2} \sum_{i \neq j} \frac{e^2}{|\mathbf{r}_i - \mathbf{r}_j|}. \tag{1.14}$$

We have summed up all one-particle contributions and added the electron–electron Coulomb interaction. The problem of solving the Schrödinger equation with this Hamiltonian is, of course, not tractable. We will see later how, at least to a certain extent, the electron–electron interaction can be taken into account. Here, where we start with drastic approximations, we will simply drop this term. Furthermore, to set the scope,

4 *Introduction*

let us focus our attention first on an atom consisting of closed shells. This may not be very systematic, but is certainly not unreasonable, since the metals we want to consider later possess ions with closed-shell cores besides the itinerant valence electrons. Now, setting out to calculate the static susceptibility, we obtain the magnetization, M, at zero temperature by computing the expectation value of $\mathcal{H}$, and hence, from Eqn (1.6) at $T = 0$

$$M = -\frac{\partial \langle \mathcal{H} \rangle}{\partial H}, \qquad (1.15)$$

from which the molar susceptibility, χ, follows as

$$\chi = N_0 \frac{\partial M}{\partial H}. \qquad (1.16)$$

N_0 is Avogadro's number, equal to $6.002214179 \times 10^{23}\,\text{mol}^{-1}$. Also ignoring spin–orbit coupling(the terms $\zeta_i \, \boldsymbol{\ell}_i \cdot \boldsymbol{\sigma}_i$) we obtain from Eqn (1.14)

$$M = -\frac{\partial \langle \mathcal{H} \rangle}{\partial H} = -\mu_\mathrm{B} \sum_{i=1}^{Z} \left\langle \frac{\ell_{zi}}{\hbar} + \sigma_{zi} \right\rangle - \frac{e^2 H}{4m c^2} \sum_{i=1}^{Z} \langle x_i^2 + y_i^2 \rangle. \qquad (1.17)$$

When the first term on the right-hand side is different from zero, the atom or ion has a permanent magnetic moment that, as we shall see, gives rise to a paramagnetic contribution to the susceptibility. Closed-shell atoms or ions, however, have no permanent magnetic moment (the first term is zero), therefore, using (1.16), we obtain the susceptibility χ_d where we have added the index d to distinguish it from other possible contributions

$$\chi_\mathrm{d} = -\frac{N_0 \, e^2}{4m c^2} \sum_{i=1}^{Z} \langle x_i^2 + y_i^2 \rangle. \qquad (1.18)$$

Because of symmetry, we may write

$$\sum_{i=1}^{Z} \langle x_i^2 \rangle = \sum_{i=1}^{Z} \langle y_i^2 \rangle = \frac{1}{3} \sum_{i=1}^{Z} \langle r_i \rangle^2 = \frac{1}{3} Z \langle r^2 \rangle, \qquad (1.19)$$

where $\langle r^2 \rangle$ is some mean square radius. Hence

$$\chi_\mathrm{d} = -\frac{Z \, e^2 \, N_0}{6m \, c^2} \langle r^2 \rangle. \qquad (1.20)$$

This so-called atomic or diamagnetic susceptibility has been estimated in the past. It is not easy to find any new results, so a reference to Wagner (1972) who compiled a set of representative values should suffice. Following Wagner it is of interest to point out that the He atom is a special case since for this atom the wave function is known quite accurately. He mentions a calculation by Pekeris (1959) and Stewart (1963) from which $\chi_d = -1.8905 \times 10^{-6}$ cm^3 mol^{-1} is obtained, which should be compared with a measured value by Havens (1933) of $\chi_d = -(1.906 \pm 0.006) \times 10^{-6}$ cm^3 mol^{-1}. The agreement is obviously very good. Values for other noble gases or closed-shell ions can be found in Wagner's monograph and have an order of magnitude of $\sim (-2$ to $-40) \times 10^{-6}$ cm^3 mol^{-1}.

1.2 Itinerant electrons

For metals, to which we now turn, we should solve the Schrödinger equation with the Hamiltonian

$$\mathcal{H} = \sum_i \frac{\mathbf{p}_i^2}{2m} + \sum_i V(\mathbf{r}_i) + \frac{1}{2} \sum_{i \neq j} \frac{e^2}{|\mathbf{r}_i - \mathbf{r}_j|}, \qquad (1.21)$$

where, compared with Eqn (1.14), we have dropped spin–orbit coupling and temporarily set the magnetic field equal to zero. The sum on i runs over all electrons present in the system; but we may group them into those originating from the valence states; the latter we assume to be itinerant in extended states. Strictly speaking, the electrons are indistinguishable and the distinction made above may, therefore, appear meaningless. But, under ambient pressure the shell structure of atoms or ions remains intact in the solid state and the distinction is quite physical. We will offer a more quantitative treatment in Sec. 2.1, where we discuss the adiabatic or Born–Oppenheimer approximation, and in Chap. 3 which deals with the energy-band structure of solids.

1.2.1 Gas of free electrons

We now go one step further and assume there are N itinerant electrons whose charge is compensated exactly by the smeared out charges of the ion cores. This sounds like a highly unphysical approximation and in fact it is. But this treatment allows us to discuss some elementary concepts which remain valid even when later on this drastic approximation is dropped in favor of a more realistic treatment of a metal. The potential $V(\mathbf{r})$ appearing in the second term on the right-hand side of Eqn (1.21) is thus assumed to be independent of $\mathbf{r}$. Next, considering the electron–electron interaction, we essentially remove this term by a number of approximations which are treated more systematically in Chap. 2. We are left with the Hamiltonian of a gas of noninteracting (or free) electrons

$$\mathcal{H} = \sum_{i=1}^{N} \frac{\mathbf{p}_i^2}{2m}. \qquad (1.22)$$

Since this simple (and unrealistic) system is extensively covered in many elementary texts, we need not deal with it here in full generality, but state only a few important facts and then look at the magnetic properties of the electron gas.

The gas of electrons obeys Fermi–Dirac statistics; at absolute zero the lowest single-particle states are filled with one spin-up and one spin-down electron up to the Fermi energy which separates the filled from the unfilled states. The single-particle states have energy

$$\varepsilon_k = \frac{\hbar^2 k^2}{2m}, \tag{1.23}$$

where **k** is the wave vector of the single-particle eigenstate (the plane-wave state) and the eigenstates of the electron gas are Slater determinants. At finite temperatures, T, the occupation of states is given by the Fermi function

$$f(\varepsilon_k) = \frac{1}{e^{\beta(\varepsilon_k - \mu)} + 1}, \tag{1.24}$$

where $\beta = 1/k_B T$ and μ is the chemical potential (which is equal to the Fermi energy, ε_F, at $T=0$). Periodic boundary conditions are usually employed which fix the density of states in **k**-space such that a sum over **k** is easily converted to an integral by the prescription (see also Chap. 3)

$$\sum_{\mathbf{k}} \longrightarrow \frac{\Omega}{(2\pi)^3} \int d\mathbf{k}, \tag{1.25}$$

where Ω is the volume of the electron gas. Thus if it consists of N electrons, one has

$$N = 2 \sum_{\mathbf{k}} 1 = 2 \cdot \frac{4\pi \Omega}{(2\pi)^3} \int_0^{k_F} k^2 \, dk, \tag{1.26}$$

or

$$k_F^3 = 3\pi^2 \frac{N}{\Omega} = 3\pi^2 n, \tag{1.27}$$

where k_F is the Fermi radius and we defined the density of the electrons to be $n = N/\Omega$. The Fermi energy is thus

$$\varepsilon_F = \frac{\hbar^2}{2m} k_F^2 = \frac{\hbar^2}{2m} (3\pi^2 n)^{2/3}. \tag{1.28}$$

It is practical to define a temperature by means of

$$\varepsilon_F = k_B T_F. \tag{1.29}$$

T_F is called the Fermi temperature or degeneracy temperature. At finite temperatures (but small compared with T_F) the chemical potential is given by

$$\mu = \varepsilon_\mathrm{F}\left[1 - \frac{\pi^2}{12}\left(\frac{T}{T_\mathrm{F}}\right)^2 + \cdots\right] \qquad (1.30)$$

as is easily seen, using e.g. the Sommerfeld method which is given toward the end of this chapter in Sec. 1.2.6.

After these preliminary remarks we consider the electron gas in the presence of a homogeneous magnetic field $\mathbf{H}$ which is derived from the vector potential $\mathbf{A}$ as usual. The Hamiltonian (1.22) becomes in this case

$$\mathcal{H} = \sum_{i=1}^{N}\frac{1}{2m}\left(\mathbf{p} - \frac{e}{c}\mathbf{A}\right)^2 + \mu_\mathrm{B}\sum_{i=1}^{N}\boldsymbol{\sigma}_i\cdot\mathbf{H}. \qquad (1.31)$$

We will see that the first term on the right-hand side gives rise to Landau levels, a diamagnetic (negative) contribution to the susceptibility and oscillatory contributions to the magnetization as a function of the field strength. The latter phenomenon is called the de Haas–van Alphen effect. The second term on the right-hand side of Eqn (1.31) – the Zeeman term – gives rise to the (positive) Pauli paramagnetic susceptibility. Both contributions will now be treated in detail.

1.2.2 Landau levels

To solve the Schrödinger equation for N noninteracting electrons we turn first to the one-particle properties and consider the Schrödinger equation, temporarily without the Zeeman term

$$\frac{1}{2m^*}\left(\mathbf{p} - \frac{e}{c}\mathbf{A}\right)^2\psi = \varepsilon\psi. \qquad (1.32)$$

Here we used an effective mass, m^*, instead of m, for, as we will see in some more detail in Chap. 3, the influence of the lattice (which we ignored in writing down Eqn (1.31)) can be taken into account in a very rough approximation by replacing the mass of a free electron m by an effective mass, m^*. Further comments on Bloch electrons will made later. Representing the uniform magnetic field now by

$$\mathbf{A} = (0, xH, 0), \qquad (1.33)$$

(this gauge is sometimes called the Landau gauge), we write, instead of Eqn (1.32),

$$\left(\frac{\partial^2}{\partial x^2} + \frac{\partial^2}{\partial z^2}\right)\psi + \left(\frac{\partial}{\partial y} - \mathrm{i}\frac{eHx}{\hbar c}\right)^2\psi + \frac{2m^*\varepsilon}{\hbar^2}\psi = 0. \qquad (1.34)$$

8 *Introduction*

Substituting (p_y and p_z commute with $\mathcal{H}$ of (1.34))

$$\psi = e^{i(k_y y + k_z z)} u(x), \tag{1.35}$$

we transform the Schrödinger equation to

$$\frac{d^2 u}{dx^2} + \left[\frac{2m^*}{\hbar^2} \varepsilon_1 - \left(k_y - \frac{eHx}{\hbar c} \right)^2 \right] u = 0, \tag{1.36}$$

where

$$\varepsilon_1 = \varepsilon - \frac{\hbar^2 k_z^2}{2m^*}. \tag{1.37}$$

Defining

$$x_0 = \frac{\hbar c}{|e| H} k_y, \tag{1.38}$$

and, what is called the cyclotron frequency,

$$\omega_c^* = \frac{|e| H}{m^* c}, \tag{1.39}$$

we obtain the following differential equation for u:

$$-\frac{\hbar^2}{2m^*} \frac{d^2 u}{dx^2} + \frac{m^* \omega_c^2}{2} (x_0 - x)^2 u = \varepsilon_1 u. \tag{1.40}$$

This is the well-known Schrödinger equation for a harmonic oscillator which oscillates with frequency ω_c^* about the point x_0. The eigenvalues are (using Eqn (1.37))

$$\varepsilon = \varepsilon(n, k_z) = \hbar \omega_c^* \left(n + \frac{1}{2} \right) + \frac{\hbar^2 k_z^2}{2m^*}, \tag{1.41}$$

where n is an arbitrary, positive integer. Thus we see that the electron motion in the z-direction, the direction of the magnetic field, is not influenced by the field, whereas the motion perpendicular to the field will be quantized. These discrete energy levels are called *Landau levels*. An elementary treatment of the physics of these energy levels can be found in most introductory texts on solid state physics. Here we want to calculate the free energy of the electrons and thus their magnetization. To be complete we introduce the Zeeman term again (which we temporarily dropped) and write the one-particle energy as

$$\varepsilon_\sigma(n,k_z) = \hbar\omega_c^* \left(n + \frac{1}{2}\right) + \frac{\hbar^2 k_z^2}{2m^*} + \sigma\mu_B H, \tag{1.42}$$

where $\sigma = \pm 1$ denotes the two spin orientations in the magnetic field. The free energy for noninteracting fermions is known from elementary statistical mechanics to be

$$F = N\mu - k_B T \sum_{\sigma=1}^{2} \int \ln\left(1 + e^{-\beta(\varepsilon-\mu)}\right) \mathcal{N}_\sigma(\varepsilon)\,\mathrm{d}\varepsilon, \tag{1.43}$$

where $\mathcal{N}_\sigma(\varepsilon)$ is the density of states with energy ε, which we may write as

$$\mathcal{N}_\sigma(\varepsilon) = \frac{\mathrm{d}Z_\sigma(\varepsilon)}{\mathrm{d}\varepsilon}, \tag{1.44}$$

$Z_\sigma(\varepsilon)$ is the number of states with energy $\varepsilon_\sigma \leq \varepsilon$ and N is the total number of particles in the system. Substituting Eqn (1.44) into Eqn (1.43) and integrating by parts we obtain instead of Eqn (1.43)

$$F = N\mu - \sum_{\sigma=1}^{2} \int f(\varepsilon)\, Z_\sigma(\varepsilon)\,\mathrm{d}\varepsilon, \tag{1.45}$$

where $f(\varepsilon)$ is the Fermi function (1.24). The following treatment, in which we largely follow Wagner (1972), is somewhat complicated and provides results for a highly idealized situation hardly ever encountered in real physical systems. Still, it is important because it paves the way for the more realistic treatment of Sec. 1.2.5 which will allow us to measure real Fermi surfaces.

To do the required integration, we must know the number of states, $\mathcal{N}_\sigma(\varepsilon)$, and, therefore, we first need to count the degeneracies of the Landau levels. For this it is necessary to have appropriate boundary conditions for the Schrödinger equation from which we obtained the Landau levels. We may assume the electron gas is contained in a volume $\Omega = L_x L_y L_z$ centered at the origin and then impose periodic boundary conditions with respect to the y- and z-axes, i.e. from Eqn (1.35)

$$\psi = \psi_{n\,k_y\,k_z}(x,y,z) = \psi_{n\,k_y\,k_z}(x, y + L_y, z + L_z). \tag{1.46}$$

No such boundary condition can be required for the x-axis since the eigenfunctions of the harmonic oscillator have no periodic properties. Just as in the derivation of the rule given by Eqn (1.25), the quantities k_y and k_z are thus integer multiples of $2\pi/L_y$ and $2\pi/L_z$. Since the eigenvalues given by Eqn (1.42) are independent of k_y, there are as many states for each eigenvalue as there are k_y-values. This number is obtained from

10 *Introduction*

the range of k_y-values divided by $2\pi/L_y$, and the range follows from the requirement that x_0 is within the volume Ω, i.e.

$$-\frac{L_x}{2} \leq x_0 \leq \frac{L_x}{2}, \tag{1.47}$$

which implies, with the help of Eqn (1.38),

$$(k_y)_{\min}^{\max} = \pm \frac{|e|\,H\,L_x}{2\hbar c} \tag{1.48}$$

and

$$\frac{(k_y)_{\max} - (k_y)_{\min}}{2\pi/L_y} = \frac{|e|\,H\,L_x\,L_y}{2\pi\,\hbar c} \tag{1.49}$$

is thus the desired degeneracy of each Landau level. To obtain the quantity $Z_\sigma(\varepsilon)$, we multiply the above with the number of possible values of k_z for given n and σ and $\varepsilon \leq \varepsilon_\sigma(n, k_z)$. This is computed, with a glance at Fig. 1.1, as follows. Given a particular ε_n and n (also σ), there is a maximal and minimal k_z: from Eqn (1.42)

$$k_z^{\max}(n, \varepsilon_n) = \frac{\sqrt{2m^*}}{\hbar}\sqrt{\varepsilon_n - \hbar\omega_c^*\left(n + \frac{1}{2}\right) - \mu_B H} \tag{1.50}$$

and

$$k_z^{\min}(n, \varepsilon) = -k_z^{\max}(n, \varepsilon) \tag{1.51}$$

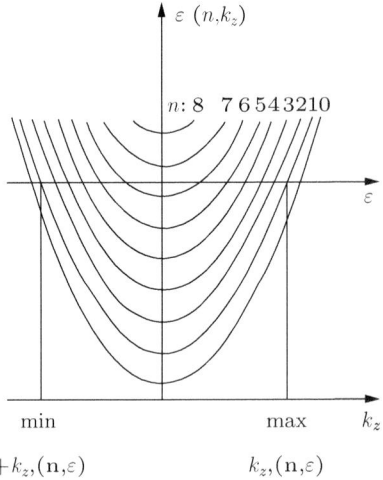

Figure 1.1 *Landau levels (schematic) ignoring the Zeeman term:* $k_z^{\max}(n=1,\ldots\varepsilon)$ *and* $k_z^{\min}(n=1,\ldots\varepsilon)$, *in this example* $n_{\max} = 6$.

1.2 Itinerant electrons

($\sigma = \pm 1$ is assumed). In this range of k_z the number of states with energy $\varepsilon \leq \varepsilon_n$, fixed n, are

$$\frac{k_z^{\max}(n,\varepsilon) - k_z^{\min}(n,\varepsilon)}{2\pi/L_z} = \frac{\sqrt{2m^*}\, L_z}{\pi\hbar}\sqrt{\varepsilon - \hbar\omega_c^*\left(n + \frac{1}{2}\right) - \mu_B H}. \tag{1.52}$$

We thus obtain all possible states with energy $\varepsilon \leq \varepsilon_n$ using Eqn (1.49) and $\Omega = L_x L_y L_z$:

$$Z_\sigma(\varepsilon) = \frac{|e|\,H\,\Omega\,\sqrt{2m^*}}{2\pi^2\hbar^2 c}\sum_{n=0}^{n_{\max}}\sqrt{\varepsilon - \hbar\omega_c^*\left(n + \frac{1}{2}\right) - \sigma\mu_B H}, \tag{1.53}$$

where $n_{\max}$ is the maximal n consistent with the root being real (cf. Fig. 1.1). Using the substitutions

$$\mu_B^* = \frac{|e|\hbar}{2m^*c}, \tag{1.54}$$

$$E_0 = \frac{\mu}{2\mu_B^* H}, \tag{1.55}$$

$$E = \frac{\varepsilon}{2\mu_B^* H}, \tag{1.56}$$

we next evaluate the expression

$$F = N\mu - \frac{4(m^*)^{3/2}\Omega(\mu_B^* H)^{5/2}}{\pi^2\hbar^3}\sum_{\sigma=1}^{2}\int f(E)\sum_{n=0}^{n_{\max}}\left(E - n - \frac{1}{2} - \sigma\frac{1}{2}\frac{m^*}{m}\right)^{1/2} dE, \tag{1.57}$$

where

$$f(E) = \frac{1}{e^{(E-E_0)/\theta} + 1}, \tag{1.58}$$

with

$$\theta = \frac{k_B T}{2\mu_B^* H}. \tag{1.59}$$

The evaluation of the integral in Eqn (1.57) is tedious but straightforward. In the first step one integrates by parts and rearranges terms to obtain

$$F = N\mu + \alpha\int_{-\infty}^{\infty} dE\left[f'\left(E + \frac{m^*}{2m}\right) + f'\left(E - \frac{m^*}{2m}\right)\right]\sum_{n=0}^{n_{\max}}\left(E - n - \frac{1}{2}\right)^{3/2}, \tag{1.60}$$

where

$$\alpha = \frac{8(m^*)^{3/2}\,\Omega(\mu_B^* H)^{5/2}}{3\pi^2\,\hbar^3} = \frac{4N\,(\mu_B^* H)^{5/2}}{\sqrt{2}\,\varepsilon_F^{3/2}}, \tag{1.61}$$

where ε_F is the Fermi energy (μ at $T = 0$). In the second step one uses the identity

$$\sum_{n=0}^{n_{\max}} \left(E - n - \frac{1}{2}\right)^{3/2} = \sum_{\ell=-\infty}^{\infty} (-1)^\ell \int_0^E (E-x)^{3/2}\, e^{2\pi i \ell x}\, dx \tag{1.62}$$

to derive

$$F = N\mu + \alpha \int_{-\infty}^{\infty} dE \left[f'\!\left(E + \frac{m^*}{2m}\right) + f'\!\left(E - \frac{m^*}{2m}\right) \right]$$
$$\cdot \left[\frac{2}{5} E^{5/2} - \frac{1}{16} E^{1/2} - \frac{3}{4\pi^2} \sum_{\ell=1}^{\infty} \frac{(-1)^\ell}{\ell^2} \operatorname{Re} e^{2\pi i \ell E} \int_0^{\sqrt{E}} e^{-2\pi i \ell u^2}\, du \right]. \tag{1.63}$$

Next one best continues with the integrals involving the 5/2- and 1/2-powers of E. Here $f'(E)$ is *approximated* by a negative δ-function and $(E \mp m^*/2m)^{5/2}$ as well as $(E \mp m^*/2m)^{1/2}$ are expanded in a Taylor series about E_0. The terms that result are collected with the first term on the right-hand side of Eqn (1.63) and give for $T \ll T_F$

$$F_0 = N\mu - \frac{N}{\varepsilon_F^{3/2}} \left[\frac{2}{5} \mu^{5/2} - \frac{1}{4} (\mu_B^* H)^2 \mu^{1/2} \left(1 - 3\frac{m^{*2}}{m^2}\right) \right]. \tag{1.64}$$

We will soon see that this part of the free energy gives rise to the *Pauli* and the *Landau susceptibilities*. The remaining part oscillates with the field strength and is responsible for the *de Haas–van Alphen* effect. Thus

$$F = F_0 + F_{\text{OSC}}, \tag{1.65}$$

where

$$F_{\text{OSC}} = -\alpha \int_{-\infty}^{\infty} dE \left[f'\!\left(E + \frac{m^*}{2m}\right) + f'\!\left(E - \frac{m^*}{2m}\right) \right]$$
$$\cdot \frac{3}{4\pi^2} \sum_{\ell=1}^{\infty} \frac{(-1)^\ell}{\ell^2} \operatorname{Re} e^{2\pi i \ell E} \int_0^{\sqrt{E}} e^{-2\pi i \ell u^2}\, du. \tag{1.66}$$

The integral over u is an error integral. But because f' peaks strongly at E_0 one can show by means of an expansion about E_0 that the integration limit, $\sqrt{E}$, may be replaced by ∞. The u-integral then becomes trivial and one obtains

$$F_{\text{OSC}} = -\alpha \frac{3}{4\pi^2 \sqrt{2}} \sum_{\ell=1}^{\infty} \frac{(-1)^\ell}{\ell^{5/2}} \cos\left(\pi \ell \frac{m^*}{m}\right) \text{Re} \int_{-\infty}^{\infty} f'(E)\, e^{2\pi i \ell E - i\pi/4}\, dE. \quad (1.67)$$

In the last step that is needed one cannot replace $f'(E)$ by a δ-function, because the oscillations of the exponential function become too rapid. But one can, after a few simple transformations, calculate the integral by the method of residues and obtain without further approximations

$$F_{\text{OSC}} = \frac{N}{\varepsilon_F^{3/2}} \frac{3}{2} (\mu_B^* H)^{3/2} k_B T \sum_{\ell=1}^{\infty} \frac{(-1)^\ell}{\ell^{3/2}} \cos\left(\pi \ell \frac{m^*}{m}\right)$$
$$\cdot \frac{\cos\left(\frac{\pi}{4} - \frac{\pi \ell \mu}{\mu_B^* H}\right)}{\sinh \frac{\pi^2 \ell k_B T}{\mu_B^* H}}. \quad (1.68)$$

We now discuss the implications of our results.

1.2.3 The susceptibility of a gas of free electrons

First we need to determine the chemical potential which has been, up to now, undetermined. Referring to the free energy as given by Eqn (1.43) we see immediately that

$$\frac{\partial F}{\partial \mu} = 0 \quad (1.69)$$

implies

$$N = \sum_i f(\varepsilon_i), \quad (1.70)$$

where the sum over i extends over all states including the spin. We hence obtain for the chemical potential, using Eqns (1.68), (1.63), and (1.65):

$$\left(\frac{\mu}{\varepsilon_F}\right)^{3/2} = 1 + \gamma_1 \left(1 - 3\frac{m^{*2}}{m^2}\right) + \frac{3}{2} \gamma_1 \gamma_2 \gamma_3 \sum_{\ell=1}^{\infty} \frac{(-1)^\ell}{\ell^{1/2}} \cos\left(\pi \ell \frac{m^*}{m}\right)$$
$$\cdot \frac{\sin\left(\frac{\pi}{4} - \frac{\pi \ell \mu}{\mu_B^* H}\right)}{\sinh \frac{\pi^2 \ell k_B T}{\mu_B^* H}}, \quad (1.71)$$

14 *Introduction*

where we defined temporarily

$$\gamma_1 = \frac{(\mu_B^* H)^2}{8\mu^{1/2} \varepsilon_F^{3/2}}, \tag{1.72}$$

$$\gamma_2 = \pi \frac{k_B T}{\varepsilon_F}, \tag{1.73}$$

$$\gamma_3 = \left(\frac{\mu_B^* H}{\varepsilon_F}\right)^{1/2}. \tag{1.74}$$

These quantities are obviously small compared with 1 so that we can replace the chemical potential μ by the Fermi energy ε_F on the right-hand side of Eqn (1.71) and thus obtain

$$\mu \simeq \varepsilon_F \left[1 + \frac{1}{12} \frac{(\mu_B^* H)^2}{\varepsilon_F^2} \left(1 - 3\frac{m^{*2}}{m^2}\right) \right.$$
$$\left. + \frac{\pi k_B T}{\varepsilon_F} \left(\frac{\mu_B^* H}{\varepsilon_F}\right)^{1/2} \sum_{\ell=1}^{\infty} \frac{(-1)^\ell}{\ell^{1/2}} \cos\left(\pi \ell \frac{m^*}{m}\right) \frac{\sin\left(\frac{\pi}{4} - \frac{\ell \pi \varepsilon_F}{\mu_B^* H}\right)}{\sinh\left(\ell \pi^2 \frac{k_B T}{\mu_B^* H}\right)} \right]. \tag{1.75}$$

Apart from the oscillatory correction to the Fermi energy we obtained above a correction that is proportional to $(\mu_B^* H/\varepsilon_F)^2$. We did not, however, obtain the temperature correction that was found in Eqn (1.30) to be proportional $(T/T_F)^2$. The reason for this is our neglect here of the finite temperature width in the derivative of the Fermi distribution.

The magnetization is now obtained from the thermodynamic relation $M = -\partial F/\partial H$ and we do not need to take into account the dependence of the chemical potential on the magnetic field because $\partial F/\partial \mu = 0$. From the uniform contribution, F_0, given by Eqn (1.64), we obtain at once

$$\chi_0 = \chi_P + \chi_L, \tag{1.76}$$

where

$$\chi_P = \frac{3}{2} N \frac{\mu_B^2}{\varepsilon_F}. \tag{1.77}$$

is the *Pauli* susceptibility and

$$\chi_L = -\frac{1}{2} N \frac{\mu_B^2}{\varepsilon_F} \frac{m^2}{m^{*2}} \approx -\frac{1}{2} N \frac{\mu_B^2}{\varepsilon_F} \tag{1.78}$$

is the *Landau–Peierls* susceptibility. The Pauli susceptibility derives from the electron spin because it contains the Bohr magneton and not the effective magneton. The Landau–Peierls susceptibility is diamagnetic and for free electrons with $m^* = m$ its magnitude is one third of the Pauli susceptibility. It results from the *quantized orbital angular momenta* in the external magnetic field. In some metals, where the effective mass approximation is valid, m^* can be considerably different from the electron mass, m, so that, e.g. in bismuth, the *diamagnetic term exceeds the paramagnetic susceptibility*.

1.2.4 The de Haas–van Alphen effect

Inspecting Eqn (1.68) we see that we obtain *three terms* for the oscillatory part of the magnetization, but only the derivative of the cosine contributes noticeably to the magnetization since the other terms are smaller by a factor of $k_B T/\varepsilon_F$ or $\mu_B^* H/\varepsilon_F$. One obtains

$$M_{\text{OSC}} \cong \frac{3\pi}{2} \frac{N \mu_B^*}{\varepsilon_F} \sqrt{\frac{\varepsilon_F}{\mu_B^* H}} \sum_{\ell=1}^{\infty} \frac{(-1)^\ell}{\ell^{1/2}} \cos\left(\pi \ell \frac{m^*}{m}\right)$$

$$\cdot \frac{\sin\left(\frac{\pi}{4} - \frac{\pi \ell \varepsilon_F}{\mu_B^* H}\right)}{\sinh \frac{\pi^2 \ell k_B T}{\mu_B^* H}}. \tag{1.79}$$

The term $\cos(\pi \ell m^*/m)$ can be traced back to the electron spin. Thus one can see that the spin susceptibility is not just an additive contribution. Hence, it is in principle impossible to calculate the orbital and the spin susceptibility separately. The series in Eqn (1.79) converges quite rapidly even at low temperatures and high magnetic fields. As a rule it is possible in most cases to break off the sum after the $\ell = 1$ term. The period, ΔH^{-1}, of oscillations as a function of the field strength is obtained from

$$\frac{\pi \varepsilon_F}{\mu_B^*} \Delta \frac{1}{H} = 2\pi, \tag{1.80}$$

i.e.

$$\Delta \frac{1}{H} = \frac{2 \mu_B^*}{\varepsilon_F}. \tag{1.81}$$

This characterizes the de Haas–van Alphen effect whose period is then seen to be *independent of the temperature*, a fact always verified by experiments. Apart from having obtained the *de Haas–van Alphen (dHvA) oscillations*, our result shows that the amplitude of the magnetization (and hence the susceptibility) decreases with decreasing field as $\exp(-\pi^2 k_B T/\mu_B^* H)$, which is borne out by experiment, too. But it is not surprising that our result for the gas of free electrons is otherwise not very realistic. So, for instance, it *cannot explain the directional dependence* of the de Haas–van Alphen effect; also the period, Eqn (1.81), only depends on the density since m^* cancels out:

16 *Introduction*

$$\Delta \frac{1}{H} = \frac{2|e|}{e\,\hbar^2\,(3\pi^2 n)^{2/3}}. \tag{1.82}$$

So, after all this elaborate algebra it is disappointing but should not be surprising that we missed experimentally important effects. Thus we must turn to a more complete theory which is connected with the names Onsager (1952) as well as Lifshitz and Kosevich (1955) and which supplies quite a useful description of the dHvA oscillations.

1.2.5 The Lifshitz–Kosevich formula

It would be desirable to start with solutions of the Schrödinger equation for an electron in a periodic potential, $V(\mathbf{r})$, *and* a vector potential, $\mathbf{A}$,

$$\frac{1}{2m}\left(\mathbf{p} - \frac{e}{c}\mathbf{A}\right)^2 \psi + V(\mathbf{r})\psi = \varepsilon\,\psi. \tag{1.83}$$

This problem has been attacked by Zak (1968), as well as by Obermaier and Schellnhuber (1981) and Schellnhuber *et al.* (1981). These theories are fairly complete but not easy to apply to realistic cases. One can, however, avoid some of the difficulties by an application of the *correspondence principle*. It is well known that wave packet solutions of the Schrödinger equation follow the same trajectories as motion derived from the corresponding classical Hamiltonian. This, in fact, is the reason why transport theory based on Boltzmann's equation (Sec. 4.5.3.1) is so successful for electrons in solids. We consider, therefore, a wave packet of momentum $\mathbf{p} = \hbar\,\mathbf{k}$ built up of Bloch functions and postulate that the variation with time of the wave packet momentum in a magnetic field is given by the Lorentz force equation

$$\hbar\,\dot{\mathbf{k}} = \frac{e}{c}\left(\dot{\mathbf{r}} \times \mathbf{H}\right), \tag{1.84}$$

where $\dot{\mathbf{r}} = \mathbf{v}(\mathbf{k})$ is the velocity of the wave packet. This formula is believed to hold quite well up to very high magnetic fields (Ziman, 1969). Equation (1.84) can be integrated if $\mathbf{H}$ is constant in time, and after a suitable choice of the constants of integration one obtains

$$\hbar\,\mathbf{k} = \frac{e}{c}\left(\mathbf{r} \times \mathbf{H}\right). \tag{1.85}$$

We assume the magnetic field is parallel to the z-axis. Then we see that the motion of the wave packet takes place in $\mathbf{k}$-space in planes with $k_z = \text{const}$. Assuming the electron travels along *closed* curves in these planes, we may quantize these orbits according to the Bohr–Sommerfeld quantization rule remembering that it is valid in the limit of large quantum numbers (correspondence principle):

$$\oint \mathbf{p}\cdot d\mathbf{r} = (n+\gamma)\,2\pi\,\hbar, \tag{1.86}$$

where n is an integer, γ is a phase correction (typically, $1/2$), and $\mathbf{p}$ is the momentum conjugate to the position $\mathbf{r}$. In our case the quantization rule will read

$$\oint \left[\hbar \mathbf{k} + \frac{e}{c}\mathbf{A}\right] \cdot d\mathbf{r} = (n+\gamma)\, 2\pi\, \hbar. \tag{1.87}$$

The first integral on the left-hand side can be written by means of Eqn (1.85) as

$$\oint \hbar \mathbf{k} \cdot d\mathbf{r} = \frac{e}{c}\oint (\mathbf{r}\times\mathbf{H})\, d\mathbf{r} = -\frac{e}{c}\oint \mathbf{H}\cdot(\mathbf{r}\times d\mathbf{r}) = -\frac{eH}{c}\, 2S, \tag{1.88}$$

where S is the projection onto the xy-plane of the surface bounded by the electron's orbit (in real space). The second integral on the left-hand side of Eqn (1.87) is (using Stoke's theorem)

$$\oint \mathbf{A}\cdot d\mathbf{r} = \int (\nabla\times\mathbf{A})\cdot d\mathbf{S} = H\cdot S = \Phi \tag{1.89}$$

and represents the *magnetic flux through the surface S*. Thus we obtained the quantization rule

$$-\frac{e}{c}H\, 2S + \frac{e}{c}HS \equiv -\frac{e}{c}HS \equiv \frac{|e|}{c}HS = (n+\gamma)\, 2\pi\, \hbar \tag{1.90}$$

i.e.

$$\Phi = \frac{2\pi\, \hbar c}{|e|}(n+\gamma).$$

From Eqn (1.85) we see that the motion of the electron in k-space and real space is related by the factor $|e|H/\hbar c$, therefore, from Eqn (1.89),

$$S = \frac{2\pi\, \hbar c}{|e|H}(n+\gamma)$$

and hence in reciprocal space

$$\mathcal{A} = S\left(\frac{|e|H}{\hbar c}\right)^2$$

$$\mathcal{A} = \mathcal{A}(n,k_z) = \frac{2\pi\, \hbar c}{|e|H}(n+\gamma)\frac{(|e|H)^2}{(\hbar c)^2},$$

i.e.

$$\mathcal{A}(n,k_z) = \frac{2\pi\, |e|H}{\hbar c}(n+\gamma) \tag{1.91}$$

18 *Introduction*

is the area circumscribed by the orbit in k-space, for a given constant k_z. This is Onsager's (1952) formula. This equation determines implicitly the energy levels, $\varepsilon(n, k_z)$, because for two levels which differ in n by 1, Eqn (1.91) gives

$$\Delta \mathcal{A} = \mathcal{A}(n, k_z) - \mathcal{A}(n-1, k_z) = \Delta\varepsilon \frac{d\mathcal{A}}{d\varepsilon} = \frac{2\pi |e| H}{\hbar c}, \tag{1.92}$$

(remember we assume large quantum numbers n) whence

$$\Delta\varepsilon = \frac{2\pi |e| H}{\hbar c} \bigg/ \frac{d\mathcal{A}}{d\varepsilon}. \tag{1.93}$$

Defining

$$m_c = \frac{\hbar^2}{2\pi} \frac{d\mathcal{A}}{d\varepsilon}, \tag{1.94}$$

we obtain

$$\Delta\varepsilon = \frac{|e| H \hbar}{c \, m_c}$$

$$\Delta\varepsilon = \hbar \omega_c, \tag{1.95}$$

where

$$\omega_c = \frac{|e| H}{m_c \, c}. \tag{1.96}$$

This should be compared with ω_c^*, given by Eqn (1.39),

$$\omega_c^* = \frac{|e| H}{m^* \, c}.$$

The quantity m_c is called the *cyclotron mass*. The energy differences $\Delta\varepsilon$ from Eqn (1.95) and Eqn (1.41) for free electrons agree for a spherical Fermi surface and free electrons, where

$$\mathcal{A} = \pi k^2 \quad \text{and} \quad \varepsilon(k) = \frac{\hbar^2 k^2}{2m^*},$$

whence

$$\mathcal{A} = \frac{\pi \, 2 \, m^*}{\hbar^2} \varepsilon \quad \text{and} \quad \frac{d\mathcal{A}}{d\varepsilon} = \frac{2 \, m^* \, \pi}{\hbar^2}$$

and with (1.94)

$$m_c = m^*.$$

We now rederive the oscillatory part of the free energy using Eqn (1.43) for noninteracting electrons in the form

$$F = N\mu - k_B T \sum_i \ln\left(1 + e^{-\beta(\varepsilon_i - \mu)}\right). \tag{1.97}$$

For simplicity in writing we temporarily ignore the electron spin and use for ε_i the Landau levels, i.e.

$$\varepsilon_i = \varepsilon\left(n + \frac{1}{2}, k_z\right) \tag{1.98}$$

with degeneracy obtained from Eqn (1.49). Then we write

$$F = N\mu - \frac{|e|H\Omega}{2\pi^2 \hbar c} k_B T \int_{-\infty}^{\infty} dk_z \sum_{n=0}^{\infty} \ln\left(1 + e^{-\beta\left(\varepsilon(n+\frac{1}{2}, k_z) - \mu\right)}\right). \tag{1.99}$$

Similar to the identity (1.62) we now use *Poisson's summation* formula. It is obtained for any real function, $G(x)$, which we write as a Fourier series in the interval $n < x < n+1$, i.e.

$$G(x) = \sum_{\ell=-\infty}^{\infty} e^{-2\pi i \ell x} g_\ell \tag{1.100}$$

and

$$g_\ell = \int_n^{n+1} G(x) e^{2\pi i \ell x} \, dx. \tag{1.101}$$

Next we evaluate

$$G\left(n + \frac{1}{2}\right) = \sum_{\ell=-\infty}^{\infty} (-1)^\ell \int_n^{n+1} G(x) e^{2\pi i \ell x} \, dx. \tag{1.102}$$

Summing over all intervals, we have

$$\sum_{n=0}^{\infty} G\left(n + \frac{1}{2}\right) = \int_0^{\infty} G(x) \, dx + 2\operatorname{Re} \sum_{\ell=1}^{\infty} (-1)^\ell \int_0^{\infty} G(x) e^{2\pi i \ell x} \, dx. \tag{1.103}$$

20   *Introduction*

We apply this formula to Eqn (1.99), obtaining

$$F = N\mu - \frac{|e|\,H\,\Omega}{2\pi^2\,\hbar c}\,k_\mathrm{B}T \int_{-\infty}^{\infty} \mathrm{d}k_z \int_0^{\infty} \ln\left(1 + e^{-\beta(\varepsilon(x,k_z)-\mu)}\right)\,\mathrm{d}x$$

$$-\frac{|e|H\Omega}{2\pi^2\,\hbar c}\,k_\mathrm{B}T \cdot 2\mathrm{Re}\sum_{\ell=1}^{\infty}(-1)^\ell \int_{-\infty}^{\infty} \mathrm{d}k_z \int_0^{\infty} \ln\left(1+e^{-\beta(\varepsilon(x,k_z)-\mu)}\right)\cdot e^{2\pi i \ell x}\,\mathrm{d}x.$$

(1.104)

We continue with the last term on the right-hand side only, because we are interested in the oscillatory part of the free energy. Integrating by parts *twice*, we obtain

$$\mathrm{Re}\int_0^{\infty} \ln\left(1+e^{-\beta(\varepsilon(x,k_z)-\mu)}\right) e^{2\pi i \ell x}\,\mathrm{d}x = \frac{1}{k_\mathrm{B}T}\,f\left(\varepsilon(x,k_z)\right)\frac{\partial \varepsilon}{\partial x}\bigg|_{x=0}^{x=\infty}\frac{1}{4\pi^2\,\ell^2}$$

$$+\frac{1}{k_\mathrm{B}T}\frac{1}{4\pi^2\,\ell^2}\,\mathrm{Re}\int_0^{\infty}\left[\frac{\mathrm{d}}{\mathrm{d}x}\,f\left(\varepsilon(x,k_z)\right)\frac{\partial\varepsilon}{\partial x}+f\frac{\partial^2\varepsilon}{\partial x^2}\right] e^{2\pi i \ell x}\,\mathrm{d}x.$$

(1.105)

We drop the first term on the right-hand side because it is not oscillatory; furthermore $\partial^2\varepsilon/\partial x^2 = 0$ because the Landau levels are *linear in n*. Then we notice that $\mathrm{d}f/\mathrm{d}x$ is large where $\varepsilon = \mu$, whence $\partial\varepsilon/\partial x$ can be replaced by $\hbar\omega_c(\mu,k_z)$ (see (1.95)). We thus continue with the evaluation of

$$F_\mathrm{OSC} = -\frac{2|e|\,H\,\Omega}{2\pi^2\,\hbar c}\,k_\mathrm{B}T\sum_{\ell=1}^{\infty}(-1)^\ell \int_{-\infty}^{\infty}\mathrm{d}k_z\,\frac{1}{k_\mathrm{B}T}\frac{\hbar\omega_c(\mu,k_z)}{4\pi^2\,\ell^2}\int_0^{\infty}\frac{\mathrm{d}f}{\mathrm{d}x}\,e^{2\pi i \ell x}\,\mathrm{d}x.$$

(1.106)

Next denote the value of x where $-\mathrm{d}f/\mathrm{d}x$ is maximal by X, i.e.

$$\varepsilon(X,k_z) = \mu.$$

At this value of x the cross-sectional area of the Fermi surface at k_z in k-space, Eqn (1.91), is given by

$$\mathcal{A}(\mu,k_z) = \frac{2\pi\,|e|\,H}{\hbar c}\,(X+\gamma),$$

or

$$X = \frac{c\,\hbar\,\mathcal{A}(\mu,k_z)}{2\pi\,|e|\,H} - \gamma.$$

(1.107)

Expanding the integrand around this value, and using the method of residues to derive (as before)

$$\int_{-\infty}^{\infty} \frac{e^{2\pi i \ell \theta \zeta}}{4\cosh^2 \frac{1}{2}\zeta} d\zeta = -\frac{2\pi^2 \ell \theta}{\sinh 2\pi^2 \ell \theta}, \qquad (1.108)$$

we obtain in this step

$$F_{\text{OSC}} = 2k_{\text{B}}T \frac{|e|H\Omega}{2\pi^2 \hbar c} \sum_{\ell=1}^{\infty} (-1)^\ell \int_{-\infty}^{\infty} dk_z \frac{\cos\left(\frac{\ell c \hbar \mathcal{A}(\mu, k_z)}{|e|H}\right)}{\sinh\left(\frac{2\pi^2 \ell k_{\text{B}} T}{\hbar \omega_c}\right)}. \qquad (1.109)$$

This is a *Fresnel-type integral*; its major contributions come from regions where the *phase is stationary*, i.e. where $\mathcal{A}(\mu, k_z)$ has a maximum or minimum,

$$\left.\frac{\partial \mathcal{A}(\mu, k_z)}{\partial k_z}\right|_{k_z = k_{\text{ext}}} = 0. \qquad (1.110)$$

We expand $\mathcal{A}$ about this point,

$$\mathcal{A} = \mathcal{A}_{\text{ext}} + \frac{1}{2} k^2 \left.\frac{d^2 \mathcal{A}}{dk_z^2}\right|_{k_{\text{ext}}} + \cdots, \qquad (1.111)$$

then, after a simple substitution, using

$$\int_0^\infty \frac{\cos x}{\sqrt{x}} dx = \sqrt{\pi/2},$$

we obtain

$$F_{\text{OSC}} \cong 2k_{\text{B}}T \sum_{\text{ext}} \sum_{\ell=1}^{\infty} (-1)^\ell \Omega \left(\frac{|e|H}{2\pi \ell c \hbar}\right)^{3/2} \frac{1}{\sqrt{\left|\frac{d^2 \mathcal{A}}{dk_z^2}\right|_{\text{ext}}}} \frac{\cos\left(\frac{\ell c \hbar \mathcal{A}_{\text{ext}}}{|e|H} \pm \frac{1}{4}\pi\right)}{\sinh\left(\frac{2\pi^2 \ell k_{\text{B}} T}{\hbar \omega_c}\right)}, \qquad (1.112)$$

where the sum $\sum_{\text{ext}}$ extends over different extremal Fermi surface cross-sections. If we take the electron spin into account, a factor $\cos(\pi \ell m_c/m)$ appears in Eqn (1.112) (Abrikosov, 1988). This is the *Lifshitz–Kosevich* formula for the free energy. The magnetization is obtained by differentiating Eqn (1.112) with respect to the magnetic field as before; keeping the leading terms only, one writes

$$M = k_{\text{B}} T C \frac{1}{\sqrt{\mu_B H}} \sum_{\text{ext}} \sum_{\ell=1}^{\infty} (-1)^\ell \frac{\mathcal{A}_{\text{ext}}}{\sqrt{\ell}} \frac{\sin\left(\frac{\ell c \hbar \mathcal{A}_{\text{ext}}}{|e|H} \pm \frac{1}{4}\pi\right)}{\sinh\left(\frac{2\pi^2 \ell k_{\text{B}} T}{\hbar \omega_c}\right)} \cos\left(\pi \ell \frac{m_c}{m}\right), \qquad (1.113)$$

22 Introduction

where C is a combination of fundamental constants (see e.g. Abrikosov (1988), or Lonzarich in Springford (1980)). We see that the period of the oscillations, when the magnetization is plotted as a function of $1/H$, gives the area of a maximal or minimal cross-section of the Fermi surface normal to the magnetic field, the period being *temperature-independent* given by

$$\Delta\left(\frac{1}{H}\right) = \frac{2\pi |e|}{\hbar c\, \mathcal{A}_{\text{ext}}}. \tag{1.114}$$

When the temperature is not too low it suffices to keep one term, $\ell = 1$, in Eqn (1.113). The amplitude of the oscillations as a function of $1/H$ is then seen to drop off as

$$H^{-1/2} \exp\left(-\frac{2\pi^2 k_\mathrm{B} T\, m_c\, c}{\hbar |e|\, H}\right).$$

At very low temperatures we can no longer ignore the higher harmonics, but the period is still given by Eqn (1.114).

We have, so far, ignored one important effect; this is the role of the *scattering of electrons*. Abrikosov (1988) gives a simple argument by pointing out that collisions have *qualitatively* the same effect as the temperature: they smear out the sharp boundary at the Fermi edge. It is therefore expected that taking account of collisions in Eqn (1.113) will lead to another exponential factor (comparable with that given above). In fact, a rigorous derivation (Dingle, 1952) gives a factor

$$\exp\left(\frac{-2\pi\, c\, m_c}{|e|\, \tau\, H}\right),$$

where τ is an average collision time. Many textbooks deal with the ramifications of the somewhat lengthy derivation given here; see e.g. Ashcroft and Mermin (1976), Springford (1980), Kittel (1966), Abrikosov (1988), Ziman (1964), etc., so we no longer bother with this and continue with Pauli's paramagnetism.

1.2.6 Pauli paramagnetism

According to our derivation in Sec. 1.2.3 the spin susceptibility of an electron gas is to a first approximation given by Eqn (1.77):

$$\chi_\mathrm{P} = \frac{3}{2} N \frac{\mu_\mathrm{B}^2}{\varepsilon_\mathrm{F}},$$

or, introducing for the free-electron gas the density of states for one spin direction given by

$$\mathcal{N}(\varepsilon) = \frac{3}{4} N \frac{\sqrt{\varepsilon}}{\varepsilon_\mathrm{F}^{3/2}}, \tag{1.115}$$

(obviously $\int_0^{\varepsilon_F} \mathcal{N}(\varepsilon) = \frac{1}{2} N$, where N is the number of electrons considered) we write

$$\chi_P = 2\,\mu_B^2 \,\mathcal{N}(\varepsilon_F). \tag{1.116}$$

It is characteristic of this Pauli spin susceptibility that it is independent of the temperature, T, or, if terms of higher order are considered, that it is normally weakly dependent on T. In Fig. 1.2 we show some experimental data for susceptibilities of a selected set of metals. The temperature independence is seen nicely only in Fig. 1.2(a) whereas (b) – (d) show increasing and decreasing behavior.

It is possible to understand these differences *qualitatively* by means of a rather simple consideration. We assume the thermodynamics of the electrons is still describable in

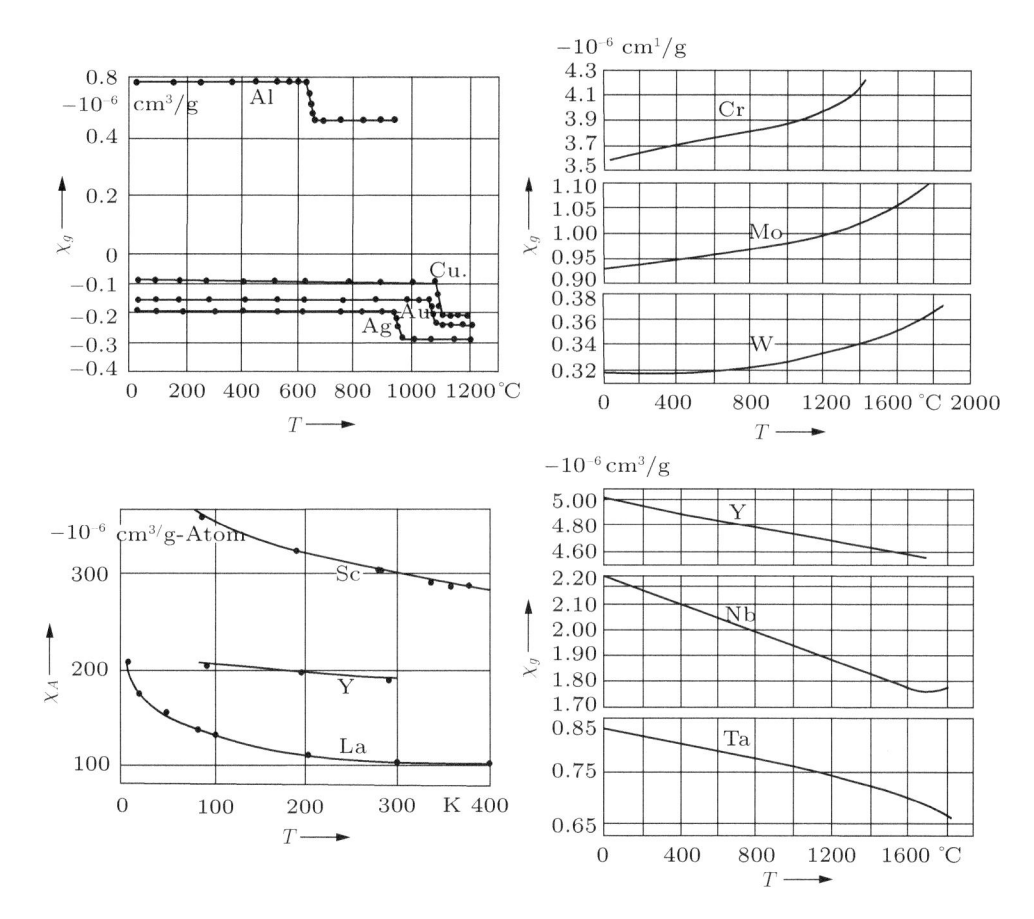

Figure 1.2 *Magnetic susceptibilities of Al, Cu, Au, and Ag (top left); Cr, Mo, and W (top right); Sc, Y, and La (bottom left); V, Nb, and Ta (bottom right). From Landolt and Börnstein (1962), pp. I–7 with kind permission.*

the *independent-particle* picture, but the density of states is not that of free electrons, Eqn (1.115), but is a more complicated function that generally describes *Bloch electrons*, see Chap. 3. Furthermore, we neglect the quantized orbital motion of the electrons. Then the only role of the magnetic field is to redistribute the electrons between the two spin orientations and their energies will be

$$\varepsilon_{\mathbf{k}\sigma} = \varepsilon(\mathbf{k}) \mp \sigma \mu_B H, \qquad (1.117)$$

where $\sigma = \pm 1$ distinguishes the two spin orientations and $\varepsilon(\mathbf{k})$ is the energy (not necessarily proportional to k^2) in the absence of the magnetic field. The total number of up ($\sigma = +1$) and down-spin electrons ($\sigma = -1$) is

$$N_\sigma = \int \mathcal{N}(\varepsilon) f(\varepsilon - \sigma \mu_B H) \, d\varepsilon, \qquad (1.118)$$

whence the magnetic moment follows as

$$M = \mu_B (N_+ - N_-) = \mu_B \int [f(\varepsilon - \mu_B H) - f(\varepsilon + \mu_B H)] \mathcal{N}(\varepsilon) \, d\varepsilon. \qquad (1.119)$$

The function f is, as before, the Fermi–Dirac distribution function which we may expand for small fields H and obtain at once

$$M = 2\mu_B^2 H \int \mathcal{N}(\varepsilon) \left(-\frac{\partial f}{\partial \varepsilon}\right) d\varepsilon \qquad (1.120)$$

to leading order. In the zero-temperature limit, where

$$\lim_{T \to 0} \frac{\partial f}{\partial \varepsilon} = -\delta(\varepsilon - \varepsilon_F),$$

we clearly obtain Eqn (1.116). But we can easily derive the first temperature correction by means of a Sommerfeld expansion of the type that led to Eqn (1.30). The result of this calculation is

$$\chi_p = 2\mu_B^2 \mathcal{N}(\varepsilon_F) \left\{1 + \frac{\pi^2}{6} k_B^2 T^2 \left[\frac{\mathcal{N}''(\varepsilon_F)}{\mathcal{N}(\varepsilon_F)} - \left(\frac{\mathcal{N}'(\varepsilon_F)}{\mathcal{N}(\varepsilon_F)}\right)^2\right]\right\}, \qquad (1.121)$$

where $\mathcal{N}'$ and $\mathcal{N}''$ are the first and second derivates of $\mathcal{N}$ with respect to ε.

Before we sketch the derivation of this formula, we note that this temperature correction is easily evaluated for *free electrons*, Eqn (1.115), to be

$$\chi_p = 2\mu_B^2 \mathcal{N}(\varepsilon_F) \left\{1 - \frac{\pi^2}{12} \left(\frac{k_B T}{\varepsilon_F}\right)^2\right\}. \qquad (1.122)$$

This is a decreasing susceptibility unlike that of Cr, Mo, and W, Fig. 1.2, which therefore cannot be explained using free electrons. However, any realistic density of states is not necessarily an increasing function of the energy, ε, but may be decreasing, or even have a maximum or a minimum at ε_F. Suppose there is a minimum at ε_F. Then the temperature correction in Eqn (1.121) is obviously positive and χ_p is increasing. This connection between χ_p and the density of states is in fact borne out by calculations: the Fermi energy for Cr, Mo, and W is situated in a pronounced minimum (see further details in Chap. 4). Alternatively, the correction is negative when the Fermi energy falls on a maximum of the density of states; this is the case for Sc, as well as V, Nb, and Ta (see Chap. 4). Obviously, other cases are possible. But we must remember that we still have to include interactions between the electrons so that the above-mentioned behavior must be re-evaluated.

Now the derivation of Eqn (1.121) proceeds as follows: the derivative of the Fermi function peaks strongly at the chemical potential, μ; we therefore expand the density of states in the energy region in a Taylor series obtaining in this first step from Eqn (1.120):

$$\chi_p = -2\mu_B^2 \int \frac{\partial f}{\partial \varepsilon} \left\{ \mathcal{N}(\mu) + (\varepsilon - \mu)\mathcal{N}'(\mu) + \frac{1}{2}(\varepsilon - \mu)^2 \mathcal{N}''(\mu) \right\} d\varepsilon. \quad (1.123)$$

Again, because of the above-mentioned property of $\partial f/\partial \varepsilon$ we can extend the lower limit of integration to $-\infty$. The integral over the second member of the right-hand side then vanishes as $\partial f/\partial \varepsilon$ is an even function of $\varepsilon - \mu$. The third integral is, with the substitution $y = \beta(\varepsilon - \mu)$

$$\frac{1}{2}\int_{-\infty}^{\infty}(\varepsilon-\mu)^2 \frac{\partial f}{\partial \varepsilon} d\varepsilon = \frac{1}{2}(k_B T)^2 \int_{-\infty}^{\infty} y^2 \frac{df}{dy} dy$$

$$= -(k_B T)^2 \int_{-\infty}^{\infty} y^2 \frac{e^y}{(e^y+1)^2} dy = -\frac{\pi^2}{6}(k_B T)^2. \quad (1.124)$$

Thus, so far,

$$\chi_p = 2\mu_B \left\{ \mathcal{N}(\mu) + \frac{\pi^2}{6}(k_B T)^2 \mathcal{N}''(\mu) \right\}. \quad (1.125)$$

Next, we take into account the temperature dependence of the chemical potential, which follows from (integrating by parts)

$$N = 2\int_0^\infty f(\varepsilon)\mathcal{N}(\varepsilon) d\varepsilon = -2\int_{-\infty}^\infty Z(\varepsilon) \frac{\partial f}{\partial \varepsilon} d\varepsilon; \quad (1.126)$$

see Eqn (1.44) for the definition of $Z(\varepsilon)$. Now expand about μ:

$$Z(\varepsilon) = Z(\mu) + (\varepsilon - \mu)\mathcal{N}(\mu) + \frac{1}{2}(\varepsilon - \mu)^2 \mathcal{N}'(\mu), \quad (1.127)$$

from which we obtain, from Eqn (1.126),

$$N = 2\,Z(\mu) - \int_{-\infty}^{\infty} (\varepsilon - \mu)^2\,\mathcal{N}'(\mu)\,\frac{\partial f}{\partial \varepsilon}\,\mathrm{d}\varepsilon + \frac{\pi^2}{3}\mathcal{N}'(\mu)\,(k_\mathrm{B}T)^2. \qquad (1.128)$$

Expanding $Z(\mu)$ also about ε_F, i.e.

$$Z(\mu) = 2\,Z(\varepsilon_\mathrm{F}) + (\mu - \varepsilon_\mathrm{F})\,\mathcal{N}(\varepsilon_\mathrm{F}), \qquad (1.129)$$

we get from Eqn (1.128)

$$N = 2\,Z(\varepsilon_\mathrm{F}) + 2\,(\mu - \varepsilon_\mathrm{F})\,\mathcal{N}(\varepsilon_\mathrm{F}) + \frac{\pi^2}{3}\mathcal{N}'(\varepsilon_\mathrm{F})\,(k_\mathrm{B}T)^2.$$

Since $N = 2\,Z(\varepsilon_\mathrm{F})$ (see Eqn (1.44) and thereafter) we see that

$$(\mu - \varepsilon_\mathrm{F}) \cong -\frac{\pi^2}{6}(k_\mathrm{B}T)^2\,\frac{\mathcal{N}'(\varepsilon_\mathrm{F})}{\mathcal{N}(\varepsilon_\mathrm{F})}.$$

So replacing $\mathcal{N}(\mu)$ in Eqn (1.124) by

$$\mathcal{N}(\mu) = \mathcal{N}(\varepsilon_\mathrm{F}) + (\mu - \varepsilon_\mathrm{F})\,\mathcal{N}'(\varepsilon_\mathrm{F}) = \mathcal{N}(\varepsilon_\mathrm{F}) - \frac{\pi^2}{6}(k_\mathrm{B}T)^2\,\frac{\mathcal{N}'(\varepsilon_\mathrm{F})^2}{\mathcal{N}(\varepsilon_\mathrm{F})},$$

we finally obtain Eqn (1.121).

1.3 How to proceed

We eventually want to turn our attention to *spontaneously ordered magnetic metals*. Well known cases are Fe, Co, and Ni or more complicated ferromagnetic compounds and alloys. But antiferromagnetism or even helically ordered magnetic metals are to be considered, too. To explain these phenomena the electrons need to *interact*. Certainly, the free-electron gas will no longer supply an answer, nor will the gas of Bloch electrons considered above. Although the complete Hamiltonian was written down in Eqn (1.21), which contains the full electron–electron Coulomb interaction besides the interaction with the ion cores, we possess no simple and reliable way to extract the dominant interaction that is responsible for the formation of magnetic moments of itinerant electrons and the possible ordering of these moments. One could start with a famous approximation to the many-body Hamiltonian given by Eqn (1.21), the *Hubbard* model. This, in fact, has been done extensively and the literature on this approach cannot be overlooked (see e.g. an early treatise: Marshall (1967) or more modern theories in e.g. Capellmann (1987)). However, we do not use this approach but want to derive the magnetic moments and their ordering within the *energy-band picture*. This means considering Bloch electrons in a way in which interactions between them are *not* omitted. The modern band approach that does the job is derived from *density-functional theory*: Bloch electrons move in an effective potential that is defined in a self-consistent way by the density of all electrons and includes exchange interactions and correlation. Refining this concept to include the *spin density matrix* we will be able to deal with a great number of properties of itinerant-electron systems. This will be the topic of the next chapters.

Before we go on we should comment on the literature in magnetism and the role this treatise plays.

Magnetism is a very large and specialized field with a great many phenomenological facts. This treatise should not be taken to convey a full picture of magnetism. It is rather meant to elucidate the connection of properties inherent in the electronic structure with certain basic facts of itinerant-electron magnetism. It should take its place with many other books (and shorter works) on magnetism of which we may list (in alphabetical order) a few now: Aharoni (2000), Blundell (2001), Bozorth (1951), Fazekas (1999), Gubanov *et al.* (1992), Herring (1966), Lovesey (1984), Mattis (1965, 1981, 1985), Mohn (2003), Moriya (1985), Nolting (1986), Skomski (2008), Staunton (1994), Vonsovski (1974), Wagner (1972), White (1983), and Yosida (1996). This list is not complete; rather it is a selection of works with widely different scopes that perhaps do cover all aspects of interest in magnetism.

2
Density-Functional Theory

In this chapter we attempt a systematic treatment of the many-body problem that we introduced somewhat loosely in Chap. 1. We do this by deriving an effective single-particle Schrödinger equation using density and spin-density-functional theory. Here it is essentially the electron density and somewhat later the spin density that will be the central variables of the theory described in the following sections.

2.1 Born–Oppenheimer approximation

Since we aim at understanding real materials we must drop the notion of a gas of electrons and bring into the field of vision atoms that are condensed to form a solid. Thus we should consider the Hamiltonian of a system containing nuclei of mass M_μ with coordinates $\mathbf{X}_\mu$ and electrons of mass m with coordinates $\mathbf{r}_i$. The corresponding Hamiltonian is easily written out as

$$\mathcal{H} = \sum_\mu \left[-\frac{\hbar^2}{2M_\mu} \nabla_\mu^2 + \sum_{\nu > \mu} V_I(\mathbf{X}_\mu - \mathbf{X}_\nu) \right] \\ + \sum_i \left[-\frac{\hbar^2}{2m} \nabla_i^2 + \sum_{j>i} \frac{e^2}{|\mathbf{r}_i - \mathbf{r}_j|} + \sum_\mu U_{e-I}(\mathbf{r}_i - \mathbf{X}_\mu) \right]. \quad (2.1)$$

The quantity $V_I(\mathbf{X}_\mu - \mathbf{X}_\nu)$ is the interaction potential of the nuclei with each other, while $U_{e-I}(\mathbf{r}_i - \mathbf{X}_\mu)$ represents the interaction between an electron at $\mathbf{r}_i$ and a nucleus at $\mathbf{X}_\mu$. Both are simply appropriate Coulomb potentials.

Now we want to discuss the motion of the electrons separately from the motion of the nuclei and we therefore ask why and how we can achieve this separation. The answer to this question was given long ago by Born and Oppenheimer and can be stated quite simply: Because the electrons are very light compared with the nuclei, they move much more rapidly and can follow the slower motions of the nuclei quite accurately. This, we will see, implies that the electron distribution determines the potential in which the nuclei move. This phenomenon is of considerable conceptual importance, and thus warrants a brief outline of its quantum mechanical justification and the approximations necessary.

The Hamiltonian given in Eqn (2.1) is a sum of a nuclear and an electronic part,

$$\mathcal{H} = \mathcal{H}_n + \mathcal{H}_e, \quad (2.2)$$

in which $\mathcal{H}_n$ contains the first part of Eqn (2.1), i.e.,

$$\mathcal{H}_n = \sum_\mu \left[-\frac{\hbar^2}{2M_\mu} \nabla_\mu^2 + \sum_{\nu > \mu} V_I(\mathbf{X}_\mu - \mathbf{X}_\nu) \right] \quad (2.3)$$

and $\mathcal{H}_e$ contains the remainder, including the interaction of the electrons with the nuclei,

$$\mathcal{H}_e = \sum_i \left[-\frac{\hbar^2}{2m} \nabla_i^2 + \sum_{j>i} \frac{e^2}{|\mathbf{r}_i - \mathbf{r}_j|} + \sum_\mu U_{e-I}(\mathbf{r}_i - \mathbf{X}_\mu) \right]. \quad (2.4)$$

Now we imagine we freeze the positions of the nuclei and solve the Schrödinger equation for the electrons in the presence of the fixed nuclei, i.e. we solve

$$\mathcal{H}_e \, \psi_\mathbf{K}(\mathbf{X}, \mathbf{r}) = E_\mathbf{K}(\mathbf{X}) \, \psi_\mathbf{K}(\mathbf{X}, \mathbf{r}). \quad (2.5)$$

By the single letters $\mathbf{X}$ and $\mathbf{r}$ we here denote the set of all nuclear and electronic coordinates. The energy of the electronic system and the wave function of the electronic state depend on the nuclear positions through the interaction term $U_{e-I}(\mathbf{r}_i - \mathbf{X}_\mu)$. In practice, we are unable to solve Eqn (2.5) exactly and must resort to approximation procedures that are outlined in this and the next chapter. At this point, however, it is practical to proceed as if a complete set of solutions of the many-body states $\{\psi_\mathbf{K}(\mathbf{X}, \mathbf{r})\}$ could be obtained. Then the wave functions for the entire system of electrons and nuclei may be expanded with $\psi_\mathbf{K}$ serving as basis functions. Let Q denote the quantum numbers required to specify the total state of the system. The wave function can then be written as

$$\Phi_Q(\mathbf{X}, \mathbf{r}) = \sum_{\mathbf{K}'} \chi_{\mathbf{K}'}(Q, \mathbf{X}) \, \psi_{\mathbf{K}'}(\mathbf{X}, \mathbf{r}) \quad (2.6)$$

and must satisfy a Schrödinger equation with the full Hamiltonian Eqn (2.1), i.e.

$$\mathcal{H} \Phi_Q(\mathbf{X}, \mathbf{r}) = \varepsilon_Q \, \Phi_Q(\mathbf{X}, \mathbf{r}). \quad (2.7)$$

Let us assume the electronic function $\psi_\mathbf{K}$ can be normalized for all values of $\mathbf{X}$ and are orthogonal with respect to $\mathbf{K}$ for fixed $\mathbf{X}$. We can then substitute Eqn (2.6) into

Eqn (2.7), use Eqn (2.5), premultiply with $\psi_{\mathbf{K}}^*(\mathbf{X},\mathbf{r})$ and integrate over all $\mathbf{r}$'s to obtain a set of coupled equations for the functions $\chi_{\mathbf{K}}$ of the form

$$\sum_{\mathbf{K}'} \{[\mathcal{H}_n + E_{\mathbf{K}}(\mathbf{X})]\,\delta_{\mathbf{K}\mathbf{K}'} + C_{\mathbf{K}\mathbf{K}'}(\mathbf{X})\}\,\chi_{\mathbf{K}'}(Q,\mathbf{X}) = \varepsilon_Q\,\chi_{\mathbf{K}}(Q,\mathbf{X}). \qquad (2.8)$$

The operator $C_{\mathbf{K}\mathbf{K}'}(\mathbf{X})$ has the form

$$C_{\mathbf{K}\mathbf{K}'}(\mathbf{X}) = -\int \psi_{\mathbf{K}}^*(\mathbf{X},\mathbf{r}) \sum_\mu \frac{\hbar^2}{2M_\mu}\left[\nabla_\mu^2\,\psi_{\mathbf{K}'}(\mathbf{X},\mathbf{r}) + 2\,\nabla_\mu\,\psi_{\mathbf{K}'}(\mathbf{X},\mathbf{r})\cdot\nabla_\mu\right]\,d\mathbf{r}. \qquad (2.9)$$

In the lowest approximation, this coupling term is ignored entirely. Then Eqn (2.8) is diagonal, indicating that the energy levels of the system of nuclei are determined by solving the Schrödinger equation

$$[\mathcal{H}_n + E_{\mathbf{K}}(\mathbf{X})]\,\chi_{\mathbf{K}}(Q,\mathbf{X}) = \varepsilon_Q\,\chi_{\mathbf{K}}(Q,\mathbf{X}). \qquad (2.10)$$

We see here that the effective Hamiltonian is obtained by adding to $\mathcal{H}_n$ (as given by Eqn (2.3)) the term $E_{\mathbf{K}}(\mathbf{X})$: The energy eigenvalues of the electronic system contribute to the potential energy of the nuclear system in an important way implying that the latter depends on the states of the electrons. If we can ascertain that the electrons stay in the ground state, $E_{\mathbf{K}_0}(\mathbf{X})$, the effective nuclei Hamiltonian $\mathcal{H}_n + E_{\mathbf{K}_0}(\mathbf{X})$ supplies a well-defined procedure to determine the motion of the nuclei which, in a crystal, gives rise to phonons with their characteristic spectral properties. There are, however, two things to be confirmed: one is the validity of ignoring the coupling terms, $C_{\mathbf{K}\mathbf{K}'}(\mathbf{X})$, in Eqn (2.8); the other is the freezing of the nuclear positions for the electronic problem, Eqn (2.5), when in reality the nuclei move. Concerning the coupling terms, $C_{\mathbf{K}\mathbf{K}'}(\mathbf{X})$, we can give crude arguments which show that they are indeed small. Looking at the second part of $C_{\mathbf{K}\mathbf{K}'}(\mathbf{X})$, Eqn (2.9), we see that it involves the integral

$$\int \psi_{\mathbf{K}}^*(\ldots \mathbf{X}_\mu \ldots,\mathbf{r})\,\nabla_\mu\,\psi_{\mathbf{K}'}(\ldots \mathbf{X}_\mu \ldots,\mathbf{r})\,d\mathbf{r}$$

which can be shown to vanish, if the ψ's are real. To see this we differentiate the normalization integral and obtain

$$\int \psi_{\mathbf{K}}^*(\ldots \mathbf{X}_\mu \ldots,\mathbf{r})\,\nabla_\mu\,\psi_{\mathbf{K}'}(\ldots \mathbf{X}_\mu \ldots,\mathbf{r})\,d\mathbf{r} \\ + \int \psi_{\mathbf{K}'}(\ldots \mathbf{X}_\mu \ldots,\mathbf{r})\,\nabla_\mu\,\psi_{\mathbf{K}}^*(\ldots \mathbf{X}_\mu \ldots,\mathbf{r})\,d\mathbf{r} = 0, \qquad (2.11)$$

which supports our assertion provided the ground state wave function is real. We will in later sections see that this integral is known as the Berry phase. We leave open whether or not it plays an important role in this context. Next, the first part of $C_{\mathbf{KK'}}$ involves the integral

$$\frac{\hbar^2}{2M_\mu} \int \psi_{\mathbf{K}}^*(\ldots \mathbf{X}_\mu \ldots, \mathbf{r}) \nabla_\mu^2 \psi_{\mathbf{K'}}(\ldots \mathbf{X}_\mu \ldots, \mathbf{r}) \, d\mathbf{r}.$$

Here we may assume that the ψ's depend on the difference $\mathbf{X}_\mu - \mathbf{r}_i$, which implies

$$\frac{\hbar^2}{2M_\mu} \int \psi_{\mathbf{K}}^*(\ldots \mathbf{X}_\mu \ldots, \mathbf{r}) \nabla_\mu^2 \psi_{\mathbf{K'}}(\ldots \mathbf{X}_\mu \ldots, \mathbf{r}) \, d\mathbf{r}$$
$$= -\frac{m}{M_\mu} \int \psi_{\mathbf{K}}^*(\mathbf{X}, \ldots \mathbf{r}_i \ldots) \left(-\frac{\hbar^2}{2m} \nabla_i^2\right) \psi_{\mathbf{K'}}(\mathbf{X}, \ldots \mathbf{r}_i \ldots) \, d\mathbf{r}. \quad (2.12)$$

The right-hand side of this equation is of the order of m/M_μ times the electronic kinetic energy. This is small because m/M_μ is of the order 0.5×10^{-3}.

The effect of the nuclear motion concerning the electronic problem can be easily visualized if we assume we solve Eqn (2.5) at each instantaneous position of the nuclei $\mathbf{X}$. The core electrons certainly move with the nucleus as it vibrates, so we may speak of *ionic* vibrations, which is the usual terminology. The effect of the ionic motion on the valence or conduction electrons gives rise to the phenomenon of electron–phonon coupling, which is a more difficult problem (see e.g. Jones and March, 1973). Electron–phonon coupling is beyond the scope of this treatise although it is of great importance not only for normal metals at finite temperatures and superconductors but is also argued to be so for magnetic solids.

We conclude in stating that the electronic Hamiltonian, Eqn (1.21), is quite well justified, i.e., dropping the index e, we write

$$\mathcal{H} = \sum_i \left[-\frac{\hbar^2}{2m} \nabla_i^2 + V(\mathbf{r}_i) + \sum_{j>i} \frac{e^2}{|\mathbf{r}_i - \mathbf{r}_j|} \right], \quad (2.13)$$

where $V(\mathbf{r}_i)$ denotes the electron–nucleon interaction $\sum_\mu U_{e-I}(\mathbf{r}_i - \mathbf{X}_\mu)$ appearing in Eqn (2.4). The approximation we described here is called the Born–Oppenheimer or adiabatic approximation.

2.2 Hartree–Fock approximation

Having established the validity of the electronic Hamiltonian $\mathcal{H}$ (Eqn (2.13)), we now embark on a systematic investigation of the quantum mechanical problem it defines. We begin in this section with the traditional way, viz the Hartree–Fock (HF) approximation. However, we shall not proceed in using the HF approximation as a first step of a systematic perturbation expansion to be followed by summing up appropriate Feynman diagrams. Instead, in the next section we opt for the density-functional theory for which we find the HF approximation sets the stage and defines the basic ingredients.

Referring the reader for a detailed derivation of the Hartree–Fock approximation to the literature (e.g. Slater, 1974; Kittel, 1967; Jones and March, 1973), here we shall describe only the salient features and begin with the variational principle of quantum mechanics. This principle supplies an upper bound to the total energy if we minimize the expression

$$E = \langle \phi | \mathcal{H} | \phi \rangle, \qquad (2.14)$$

where $|\phi\rangle$ is an approximate but normalized state vector (wave function) of a form appropriate to the electron system. Clearly, if $|\phi\rangle$ were the exact ground-state vector, then E would be the ground-state energy. An appropriate choice for $|\phi\rangle$ must reflect the correct symmetry properties of the many-electron state and such a choice is a determinant of single-particle state functions, called a Slater determinant, which we denote by $|\phi_S\rangle$. We then minimize $\langle \phi_S | \mathcal{H} | \phi_S \rangle$ by varying the single-particle functions which specify the Slater determinant. The Euler–Lagrange equations of this minimization procedure are the famous Hartree–Fock equations.

Writing out $\langle \phi_S | \mathcal{H} | \phi_S \rangle$ in terms of the single-particle functions can be done most conveniently in second quantization, where we expand the field operators in terms of a complete orthonormal set (CONS) of single-particle functions $\{\varphi_i(\mathbf{r})\}$, or in the traditional way in which the Slater determinant is specified using the CONS of functions $\{\varphi_i(\mathbf{r})\}$. The result of this analysis in real-space notation is

$$\langle \phi_S | \mathcal{H} | \phi_S \rangle = \sum_{i=1}^{N} \int \varphi_i^*(\mathbf{r}) \left[-\frac{\hbar^2}{2m} \nabla^2 + V(\mathbf{r}) \right] \varphi_i(\mathbf{r}) \, d\mathbf{r}$$

$$+ \frac{1}{2} \sum_{i=1}^{N} \sum_{j=1}^{N} \iint \varphi_i^*(\mathbf{r}) \varphi_j^*(\mathbf{r}') \frac{e^2}{|\mathbf{r} - \mathbf{r}'|} \varphi_i(\mathbf{r}) \varphi_j(\mathbf{r}') \, d\mathbf{r} \, d\mathbf{r}' \qquad (2.15)$$

$$- \frac{1}{2} \sum_{i=1}^{N} \sum_{j=1}^{N} \iint \varphi_i^*(\mathbf{r}) \varphi_j^*(\mathbf{r}') \frac{e^2}{|\mathbf{r} - \mathbf{r}'|} \varphi_j(\mathbf{r}) \varphi_i(\mathbf{r}') \, d\mathbf{r} \, d\mathbf{r}'.$$

Note that the summations extend over the lowest N-state of the N-electron system. This equation is now varied by a functional derivative with respect to $\varphi_n^*(\mathbf{r})$ using a Lagrange parameter to ensure normalization. The Euler–Lagrange equations of this variation are

$$\left(-\frac{\hbar^2}{2m} \nabla^2 + V(\mathbf{r}) \right) \varphi_n(\mathbf{r}) + \sum_{j=1}^{N} \int \varphi_j^*(\mathbf{r}') \varphi_j(\mathbf{r}') \frac{e^2}{|\mathbf{r} - \mathbf{r}'|} d\mathbf{r}' \, \varphi_n(\mathbf{r})$$

$$\qquad (2.16)$$

$$- \sum_{j=1}^{N} \int \varphi_j^*(\mathbf{r}') \varphi_n(\mathbf{r}') \frac{e^2}{|\mathbf{r} - \mathbf{r}'|} d\mathbf{r}' \, \varphi_j(\mathbf{r}) \, \delta_{S_j S_n} = \varepsilon_n \, \varphi_n(\mathbf{r}).$$

These are the Hartree–Fock (HF) equations. We have shortened the derivation by leaving out a spin variable, except in the last step, where the Kronecker delta, $\delta_{S_j S_n}$, appears somewhat abruptly, which ensures that the last term on the left-hand side is nonzero only if the spins of states j and n are parallel. We have also ignored precautions necessary to ensure that the Lagrange parameter is diagonal, leaving these fine points to the literature.

Let us discuss the HF equations in some detail. First one notices that the terms $j = n$ in both sums cancel each other. Hence the electron with quantum number n moves in the field of all other electrons $j \neq n$. The potential in which it moves consists of two terms. One is a local term,

$$\sum_{j=1, j \neq n}^{N} \int |\varphi_j(\mathbf{r}')|^2 \frac{e^2}{|\mathbf{r} - \mathbf{r}'|} \, d\mathbf{r}',$$

which is called the *Hartree* term; it represents physically the Coulomb potential due to the charge distribution

$$\rho(\mathbf{r}) = |e| \sum_{j=1, j \neq n}^{N} |\varphi_j(\mathbf{r})|^2 \tag{2.17}$$

of all electrons other than n. The other term is nonlocal and is called the *exchange* potential. Although Coulombic, it is of quantum mechanical origin and acts only on spin-parallel states.

One may formally define an exchange-charge density by means of

$$\rho_{\text{ex}}^{(n)}(\mathbf{r}, \mathbf{r}') = |e| \sum_{j=1}^{N} \delta_{S_j S_n} \varphi_j^*(\mathbf{r}') \varphi_n(\mathbf{r}') \varphi_n^*(\mathbf{r}) \varphi_j(\mathbf{r}) / |\varphi_n(\mathbf{r})|^2. \tag{2.18}$$

Obviously

$$\int \rho_{\text{ex}}^{(n)}(\mathbf{r}, \mathbf{r}') \, d\mathbf{r}' = |e|, \tag{2.19}$$

thus $\rho_{\text{ex}}^{(n)}(\mathbf{r}, \mathbf{r}')$ contains the charge of one electron at $\mathbf{r}$, called the Fermi hole. Using Eqns (2.17) and (2.18) we can define an effective potential as

$$V_{\text{eff}}^{(n)}(\mathbf{r}) = V(\mathbf{r}) + V_{\text{H}}(\mathbf{r}) + V_{\text{ex}}(\mathbf{r}), \tag{2.20}$$

where V is the electron–nucleon interaction,

$$V_{\text{H}}(\mathbf{r}) = |e| \int \frac{\rho(\mathbf{r})}{|\mathbf{r} - \mathbf{r}'|} \, d\mathbf{r}' \tag{2.21}$$

is the Hartree potential and the exchange potential is

$$V_{\text{ex}}(\mathbf{r}) = -|e| \int \frac{\rho_{\text{ex}}^{(n)}(\mathbf{r},\mathbf{r}')}{|\mathbf{r}-\mathbf{r}'|}\, d\mathbf{r}'. \tag{2.22}$$

Then the HF equations, Eqn (2.16), read

$$\left[-\frac{\hbar^2}{2m}\nabla^2 + V_{\text{eff}}^{(n)}(\mathbf{r})\right]\varphi_n(\mathbf{r}) = \varepsilon_n\,\varphi_n(\mathbf{r}), \tag{2.23}$$

or, with the definition of a HF Hamiltonian,

$$\mathcal{H}_{\text{HF}} = -\frac{\hbar^2}{2m}\nabla^2 + V_{\text{eff}}^{(n)}(\mathbf{r}) = -\frac{\hbar^2}{2m}\nabla^2 + V_{\text{H}}(\mathbf{r}) + V_{\text{ex}}(\mathbf{r}) + V(\mathbf{r}), \tag{2.24}$$

which is formally equivalent to the even simpler

$$\mathcal{H}_{\text{HF}}\,\varphi_n(\mathbf{r}) = \varepsilon_n\,\varphi_n(\mathbf{r}). \tag{2.25}$$

Notice that Eqn (2.25) constitutes what Slater called a "self-consistent field" problem: Proceeding iteratively, we may imagine we determined at some stage a set of functions $\{\varphi_j(\mathbf{r})\}$; this by means of the Eqns (2.17), (2.18), (2.21), (2.22), and (2.24) determines a new HF Hamiltonian, $\mathcal{H}_{\text{HF}}$, and hence by means of Eqn (2.25) a new set of functions $\{\varphi_j(\mathbf{r})\}^{\text{new}}$. When with successive iterations the set $\{\varphi_j(\mathbf{r})\}$ no longer changes we call the result *self-consistent*.

We have tacitly assumed that this process converges and that the $\{\varphi_j(\mathbf{r})\}$ is really a complete orthonormal set of functions. Orthonormality requires that the HF Hamiltonian be Hermitian which it is and which is not hard to show. Convergence is mostly a numerical problem for which suitable algorithms exist.

Suppose now we have solved the HF problem for an atom (which has been done) or for a solid (which has in reality been done very rarely). Then it is quite natural to ask if the Lagrange parameters ε_n possess any physical significance. The question is answered affirmatively by *Koopmans'* theorem which states that the energy of a state where one electron is removed from the HF level ε_n (initially occupied) is

$$E^I = E - \varepsilon_n, \tag{2.26}$$

where E is the ground-state energy given by Eqn (2.15) written out with the HF wave functions. The energy E^I is an ionization energy and is thus measurable. Hence we can state that the HF eigenvalues are excitation energies.

When trying to prove Koopmans' theorem, however, one discovers that it is not exactly true. Instead one must assume that upon removal of an electron all other states remain self-consistent. This is only likely to be true for large systems where such charge rearrangement may be neglected. This theorem also fails for very highly excited states.

With these precautions in mind, Koopmans' theorem is not hard to prove and requires us to take the difference of E as given by Eqn (2.15) with an analogous expression in which, however, the state n is left unoccupied. Since these manipulations are straightforward we leave them to the reader.

Finally, we state a useful formula connecting the total energy, Eqn (2.15), with the HF eigenvalues, ε_i. This is obtained by premultiplying Eqn (2.16) with $\varphi_n^*(\mathbf{r})$, integrating, and using the result to replace the first term on the right-hand side of Eqn (2.15). We obtain

$$E = \sum_{i=1}^{N} \varepsilon_i - \frac{1}{2} \sum_{i=1}^{N} \sum_{j=1}^{N} \iint \varphi_i^*(\mathbf{r}) \varphi_j^*(\mathbf{r}') \frac{e^2}{|\mathbf{r}-\mathbf{r}'|} \varphi_i(\mathbf{r}) \varphi_j(\mathbf{r}') \, d\mathbf{r} \, d\mathbf{r}'$$

$$+ \frac{1}{2} \sum_{i=1}^{N} \sum_{j=1}^{N} \iint \varphi_i^*(\mathbf{r}) \varphi_j^*(\mathbf{r}') \frac{e^2}{|\mathbf{r}-\mathbf{r}'|} \varphi_j(\mathbf{r}) \varphi_i(\mathbf{r}') \, d\mathbf{r} \, d\mathbf{r}'.$$

(2.27)

Thus E is given by the sum of the lowest HF eigenvalues minus Hartree and exchange terms, the latter correcting for double counting.

There is one case for which the HF equations can be solved rigorously (and easily), and for which Koopmans' theorem is exact: the electron gas, again. This is quite an instructive albeit sobering example. More importantly, the electron gas is used in Sec. 2.6 to model an effective potential of great utility to be derived in the density-functional formalism to follow. We therefore close this section by dealing with this example in some detail.

The first step is to specify the potential $V(\mathbf{r})$ in Eqns (2.16) or (2.20). We want the gas to be neutral, therefore we smear out the ionic charges to form a uniform positive background. Let us assume N electrons neutralize this background. Then $V(\mathbf{r})$ is written as

$$V(\mathbf{r}) = -\frac{1}{\Omega} \int \frac{N e^2}{|\mathbf{r}-\mathbf{r}'|} \, d\mathbf{r}', \qquad (2.28)$$

where Ω is the volume with which we define periodic boundary conditions (see also Chap. 3). The translational invariance in the volume Ω suggests we try the single-particle wave functions

$$\varphi_{\mathbf{k}\sigma}(\mathbf{r}) = \frac{1}{\sqrt{\Omega}} e^{i\mathbf{k}\cdot\mathbf{r}} \chi_\sigma, \qquad (2.29)$$

where χ_σ is a spinor, $\chi_+ = \begin{pmatrix} 1 \\ 0 \end{pmatrix}$ and $\chi_- = \begin{pmatrix} 0 \\ 1 \end{pmatrix}$ describing spin-up $(+)$ and spin-down $(-)$ states. The charge distribution ρ, Eqn (2.17), is now

$$\rho = |e| \frac{N-1}{\Omega}. \qquad (2.30)$$

We may write for this $\rho = |e|N/\Omega$ because 1 is canceled by the exchange term. Thus we see that the Hartree potential, Eqn (2.20), cancels the positive background completely and the effective potential, Eqn (2.20), consists of the exchange term $V_{ex}(\mathbf{r})$ only. Hence

$$\mathcal{H}_{HF} \frac{1}{\sqrt{\Omega}} e^{i\mathbf{k}\cdot\mathbf{r}} = \frac{\hbar^2 k^2}{2m} \frac{1}{\sqrt{\Omega}} e^{i\mathbf{k'}\cdot\mathbf{r}}$$

$$-\sum_{\mathbf{k'}} \delta_{S_{\mathbf{k'}} S_{\mathbf{k}}} \frac{1}{\Omega^{3/2}} \int d\mathbf{r'} \frac{e^2}{|\mathbf{r}-\mathbf{r'}|} e^{-i\mathbf{k'}\cdot\mathbf{r'}+i\mathbf{k}\cdot\mathbf{r'}+i\mathbf{k'}\cdot\mathbf{r}}$$

$$= \left(\frac{\hbar^2 k^2}{2m} - \frac{1}{\Omega} \sum_{\mathbf{k'}} \delta_{S_{\mathbf{k'}} S_{\mathbf{k}}} \int \frac{e^2}{|\mathbf{r}-\mathbf{r'}|} e^{-i(\mathbf{k}-\mathbf{k'})(\mathbf{r}-\mathbf{r'})} d\mathbf{r'} \right) \frac{1}{\sqrt{\Omega}} e^{i\mathbf{k}\cdot\mathbf{r}}$$

(2.31)

where the sum on the $\mathbf{k'}$ includes all $|\mathbf{k'}| \leq k_F$, k_F being the Fermi radius. We see that the Fourier transform of the Coulomb potential appears in the exchange term which is easily calculated. Hence, the HF eigenvalue is

$$\varepsilon_{\mathbf{k}} = \frac{\hbar^2 k^2}{2m} - \frac{4\pi e^2}{\Omega} \sum_{\mathbf{k'}} \delta_{S_{\mathbf{k'}} S_{\mathbf{k}}} \frac{1}{|\mathbf{k}-\mathbf{k'}|^2}. \qquad (2.32)$$

Using Eqn (1.25) to transform the sum to an integral, we easily carry out the remaining calculations and obtain for the HF eigenvalue

$$\varepsilon_{\mathbf{k}} = \frac{\hbar^2 k^2}{2m} - \frac{e^2 k_F}{2\pi} F(k/k_F), \qquad (2.33)$$

where the Fermi radius, k_F, is (from Eqn (1.27)) connected with the density $n = N/\Omega$ by

$$k_F = (3\pi^2)^{1/3} n^{1/3} \qquad (2.34)$$

and the function F is

$$F(x) = 2 + \frac{1-x^2}{x} \ln \left| \frac{1+x}{1-x} \right|. \qquad (2.35)$$

The total energy per electron, E/N, is obtained from Eqn (2.27) as

$$E/N = \frac{1}{N} \sum_{\mathbf{k}} \left[2 \frac{\hbar^2 k^2}{2m} - \frac{e^2 k_F}{2\pi} F(k/k_F) \right]. \qquad (2.36)$$

The summation is carried out as before, using Eqn (1.25) and

$$\int_0^1 x^2 F(x)\, dx = 1, \qquad (2.37)$$

resulting in

$$E/N = \frac{3}{5}\frac{\hbar^2 k_F^2}{2m} - \frac{3 e^2 k_F}{4\pi}. \qquad (2.38)$$

It is customary to express these results using different variables and changing to what is called *atomic units*. First the radius of a sphere is defined by the simple relation

$$\frac{4\pi}{3}(r_S a_0)^3 = \frac{1}{n} \qquad (2.39)$$

and the number r_S is used to represent densities. Here $a_0 = \hbar^2/(m e^2) = 0.5292$ Å is the Bohr radius. With this new density variable r_S, the Fermi wave vector, k_F, is calculated to be

$$k_F = \left(\frac{9\pi}{4}\right)^{1/3} \frac{1}{r_S} [a_0]^{-1} = \frac{1.919}{r_S}[a_0]^{-1}. \qquad (2.40)$$

The unit symbol $[a_0]^{-1}$ is usually omitted and the abbreviation a.u., for atomic units, is employed. Numerical values of r_S are easily calculated for real metals from a knowledge of the crystal structure, the valence, and the lattice constant. For instance, for Na one obtains $r_S = 3.982$ and for Be $r_S = 1.867$. Most metals have densities falling within the range represented by roughly $2 \leq r_S \leq 5.5$. Next we use the Rydberg, 1 Ry $= m e^4/2\hbar^2 = 13.606$ eV, to express energies and easily obtain for the total energy per electron

$$E/N = \frac{2.21}{r_S^2} - \frac{0.916}{r_S} \qquad [\text{Ry}]. \qquad (2.41)$$

This equation constitutes the beginning of an expansion of the total energy per particle in terms of the density parameter r_S. The terms to be added to correctly describe the total energy of the interacting electron gas (or liquid) will be discussed in Sec. 2.6 and 2.8.

Counting energies in Rydbergs and lengths in Bohr radii simplifies the HF energy eigenvalues, Eqn (2.33),

$$\varepsilon_k = k^2 - (k_F/\pi) F(k/k_F). \qquad (2.42)$$

We shall make extensive use of atomic units[1] in the following chapters.

---

[1] The recipe for changing to atomic units is $m = 1/2$, $\hbar = 1$, $e^2 = 2$.

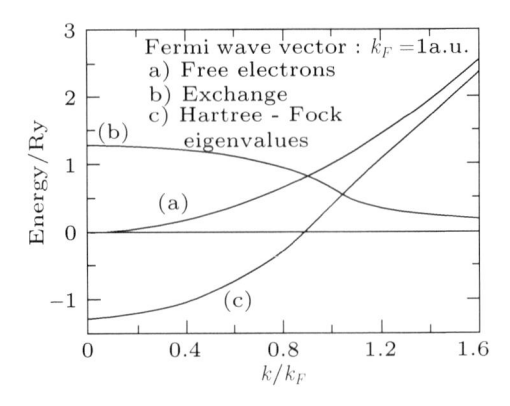

Figure 2.1 (a) The free-electron (Hartree approximation) eigenvalues, k^2, (b) the exchange term $(k_F/\pi) F(k/k_F)$ and (c) the HF eigenvalues.

After this digression into units, we show in Fig. 2.1 the free-electron eigenvalues, k^2, the exchange term $(k_F/\pi) F(k/k_F)$ and the HF eigenvalues for a typical density (roughly that of Na).

A glance at Fig. 2.1 reveals immediately that the bandwidth, Δ, defined as $\Delta = \varepsilon_{k=k_F} - \varepsilon_{k=0}$, is considerably larger in the HF approximation than for free electrons. Furthermore, the weakly visible kink at k_F in the HF eigenvalues leads to a peculiar density of states at the Fermi energy, $\mathcal{N}(\varepsilon_F)$. Using Eqn (3.36) (or the more cumbersome definition in Eqn (1.44)) one easily calculates

$$\mathcal{N}(\varepsilon_F) = \frac{k_F^2/\pi^2}{(d\varepsilon_k/dk)_{k=k_F}} \qquad (2.43)$$

which gives with Eqn (2.33) $(d\varepsilon_k/dk)_{k=k_F} = \infty$, whence $\mathcal{N}(\varepsilon_F) = 0$.

Both of these facts – and other details – are not in accord with experimental observations, where they apply, i.e. in the simple metals. Thus, for instance, the simple Sommerfeld expansion fails since it makes use of

$$\lim_{T \to 0} \frac{\partial f}{\partial \varepsilon} = -\delta(\varepsilon - \varepsilon_F)$$

(see Sec. 1.2.6), where f denotes the Fermi–Dirac distribution function. The well-known expression for the specific heat, $C_V = \gamma T$, with a temperature-independent γ, can no longer be derived. Instead one must calculate how the Fermi distribution changes with temperature and obtains $C_V \propto T/\ln T$ (Bardeen, 1936) which is not at all in accord with experiments. Furthermore, the increased bandwidth in the HF approximation spoils the agreement of the free-electron bandwidth with experiment which was established long ago for the simple metals Li, Na, Be, Hg, and Al (Pines, 1964). The logarithmic singularity in the HF eigenvalues may be traced directly to the long range of the

Coulomb interaction. If, instead, the effective interaction were screened then under certain conditions the logarithmic singularity is removed. Indeed, it is an important feature of density-functional theory that the Coulomb interaction is effectively screened (see Sec. 2.8).

A final point of some importance concerns an approximation for the exchange interaction, $-(e^2 k_F/2\pi) F(k/k_F)$, see Eqn (2.33). By comparing the exchange interaction which was calculated by Hartree for atoms with an *averaged* form of $F(k/k_F)$, Slater found that exchange is quite well represented by the local density. In fact, the average proposed is the exchange contribution per electron to the total energy which is seen from Eqn (2.38) to be $-3 e^2 k_F/4\pi$. Relating the Fermi wave vector, k_F, to the density by means of Eqn (1.27) one may define an effective exchange potential, V_S, due to Slater:

$$V_S = -0.7386 \cdot e^2 \, n(\mathbf{r})^{1/3}. \tag{2.44}$$

This so-called Slater exchange was used extensively, albeit with an adjustable factor, for early electronic structure calculations and we shall find a more rigorous justification for a local-exchange potential of this form later on in this chapter.

2.3 Density-functional theory

A reduction of the complicated many-body problem to an effective single-particle theory which is amenable to numerical calculations and supplies deeper physical insight is the density-functional theory due to Hohenberg and Kohn (1964) and Kohn and Sham (1965). The essential point is that the calculation of the many-body wave function is circumvented and that, instead, knowledge of the ground-state density is sufficient to calculate all physical quantities of interest. This theory has been amazingly successful in a broad spectrum of applications and has, consequently, led to numerous review articles and larger treatises, like, for instance, the work of von Barth (1984), Kohn and Vashishta (1983), Williams and von Barth (1983), Callaway and March (1984), Dreizler and Gross (1990), Eschrig (1996), and others.

For didactic reasons we will initially omit mathematical details. Thus concentrating on the salient features we will be rather brief here. However, in Sec. 2.8 we will investigate in more detail the conceptual, physical, and mathematical foundations of this important theory.

We begin by writing the Hamiltonian for N interacting electrons as

$$\mathcal{H} = T + V + U, \tag{2.45}$$

where T denotes the kinetic energy which we now write as

$$T = -\sum_{i=1}^{N} \nabla_i^2. \tag{2.46}$$

The quantity V denotes the external potential,

$$V = \sum_{i=1}^{N} v_{\text{ext}}(\mathbf{r}_i), \qquad (2.47)$$

which, in the Born–Oppenheimer approximation, consists of the potential due to the ions located at static positions, $\mathbf{R}_\mu$, and of possibly other external fields, $v_{\text{field}}(\mathbf{r})$, i.e.

$$v_{\text{ext}}(\mathbf{r}) = \sum_\mu v_{\text{ion}}(\mathbf{r} - \mathbf{R}_\mu) + v_{\text{field}}(\mathbf{r}). \qquad (2.48)$$

The quantity U denotes the electron–electron Coulomb interaction,

$$U = \sum_{\substack{ij \\ i \neq j}} \frac{1}{|\mathbf{r}_i - \mathbf{r}_j|}. \qquad (2.49)$$

Note that we use atomic units which were explained in the previous section.

The Hamiltonian, Eqn (2.45), has generated a truly immense literature and numerous techniques – most notably field-theoretical ones – to extract physically interesting, albeit approximate solutions. Bypassing here these fascinating topics entirely, we focus our attention on the electron density. This important quantity was already singled out in the previous section. We thus define the electron-density operator by

$$\hat{n}(\mathbf{r}) = \sum_{i=1}^{N} \delta(\mathbf{r} - \mathbf{r}_i) \qquad (2.50)$$

from which we obtain the electron density as

$$n(\mathbf{r}) = \langle \Phi | \hat{n}(\mathbf{r}) | \Phi \rangle \qquad (2.51)$$

where $|\Phi\rangle$ is a many-body state.

Hohenberg and Kohn established formally that the electron density can be assumed to be the crucial variable. We thus cite the following theorems:

1. The total *ground-state energy*, E, of any many-electron system is a functional of the density $n(\mathbf{r})$:

$$E[n] = F[n] + \int n(\mathbf{r})\, v_{\text{ext}}(\mathbf{r})\, d\mathbf{r} \qquad (2.52)$$

where $F[n]$ itself is a functional of the density $n(\mathbf{r})$, but is otherwise independent of the external potential.

2. For any many-electron system the functional $E[n]$ for the total energy has a minimum equal to the ground-state energy at the ground-state density.

This second theorem is of great importance as it leads to a variational principle. To indicate the proof of these theorems we define the functional

$$O[n] = \inf_{|\Phi\rangle \in M(n)} \langle \Phi | O | \Phi \rangle \qquad (2.53)$$

for a physical observable O assuming the infimum to exist, which is quite reasonable for the functional of interest. The set $M(n)$ contains all those many-body wave functions, $|\Phi\rangle$, which yield the density n. It might be further limited by certain conditions, e.g. that this set picks out only wave functions giving the ground state to an external potential or corresponding to an N-particle state. These situations are normally referred to as v-representability or N-representability, respectively.

Using Eqn (2.53) to set up the functionals of the operators H, V, and $F = T + U$ for an N-representable density, we recognize immediately the validity of the first theorem (one uses Eqns (2.50) and (2.51) to obtain the second term on the right-hand side of Eqn (2.53) as the functional $V[n]$). One might hesitate a little when looking at $F = \langle \Phi | T + U | \Phi \rangle$, but a glance at Eqn (2.53) shows that this is indeed a functional of n, i.e. $F = F[n]$.

The proof of the second theorem is carried out in the following way: We denote by $|\Phi_0\rangle$, n_0, and E_0 the ground state, its density, and its energy, respectively. $|\Phi_n\rangle$ may be any state yielding the density $n(\mathbf{r})$ and minimizing the functional $F[n]$ and thus $E[n]$. Now we have to show that in this case $E[n]$ and n are the same as E_0 and n_0. Since E_0 was defined as the ground-state energy, we have

$$E[n] \geq E_0 \qquad (2.54)$$

for any density $n(\mathbf{r})$. On the other hand, if we calculate $F[n]$ with the ground state according to Eqn (2.53), we obtain

$$F[n_0] = \inf_{|\Phi\rangle \in M(n)} \langle \Phi | T + U | \Phi \rangle \leq \langle \Phi_0 | T + U | \Phi_0 \rangle \qquad (2.55)$$

and thus by adding the functional of the external potential

$$E[n_0] \leq E_0. \qquad (2.56)$$

The expressions (2.54) and (2.56) can be fulfilled simultaneously only if $n = n_0$ and thus $E[n] = E[n_0]$. From this it follows that

$$E[n_0] = E_0. \qquad (2.57)$$

Although the basic facts of density-functional theory have now been stated, we still need a key to its application. This was given by Kohn and Sham (1965), who used the variational principle implied by the minimal properties of the energy functional to derive single-particle Schrödinger equations. For this we first split the functional $F[n]$ into three parts:

$$F[n] = T[n] + \iint \frac{n(\mathbf{r})\,n(\mathbf{r}')}{|\mathbf{r}-\mathbf{r}'|}\,\mathrm{d}\mathbf{r}\,\mathrm{d}\mathbf{r}' + E_{xc}[n] \qquad (2.58)$$

which describe the kinetic, the Hartee, and the exchange-correlation energy. In contrast to the Hartree integral, an explicit form of the other functionals, T and E_{xc}, is not known in general. Ignoring this problem for the moment, we use the variational principle and write

$$\frac{\delta E[n]}{\delta n(\mathbf{r})} + \mu \frac{\delta\left(N - \int n(\mathbf{r})\,\mathrm{d}\mathbf{r}\right)}{\delta n(\mathbf{r})} = 0, \qquad (2.59)$$

where μ is a Lagrange multiplier taking care of particle conservation. We now split up the kinetic energy into a term T_0 implying the kinetic energy of noninteracting particles and T_{xc} which stands for the rest, i.e. we write

$$T = T_0 + T_{xc}. \qquad (2.60)$$

This is an important step because we know how to write out the kinetic energy T_0 for noninteracting particles and hence can determine the functional derivative $\delta T_0[n]/\delta n(\mathbf{r})$ by using

$$n(\mathbf{r}) = \sum_{i=1}^{N} |\phi_i(\mathbf{r})|^2 \qquad (2.61)$$

and

$$T_0[n] = \sum_{i=1}^{N} \int \nabla \phi_i^*(\mathbf{r})\,\nabla \phi_i(\mathbf{r}), \qquad (2.62)$$

where the sums extend over the lowest N-occupied states. Here we assumed that we can determine single-particle wave functions $\{\phi_i(\mathbf{r})\}$ which permit us to express the density $n(\mathbf{r})$ with Eqn (2.61). One might indeed question whether the desired densities can be written in this form but we simply accept this step here leaving further intricacies to the specialized literature.

Since the Schrödinger equation is the Euler–Lagrange equation obtained by varying (2.62) plus a potential-energy term (see Merzbacher, 1970), we complete our ansatz by postulating

$$\left(-\nabla^2 + v'(\mathbf{r})\right)\phi_i(\mathbf{r}) = \varepsilon_i\,\phi_i(\mathbf{r}) \qquad (2.63)$$

and attempt to determine the potential energy $v'(\mathbf{r})$ such that the density $n(\mathbf{r})$ obtained from Eqn (2.61) minimizes the total energy. Thus, premultiplying Eqn (2.63) with $\phi_i^*(\mathbf{r})$, requiring the functions $\phi_i(\mathbf{r})$ to be normalized, integrating and summing up, we obtain

$$T_0[n] = \sum_{i=1}^{N} \varepsilon_i - \int v'(\mathbf{r}) \, n(\mathbf{r}) \, \mathrm{d}\mathbf{r}. \tag{2.64}$$

The variation, Eqn (2.59), is now easily carried out. We note that terms containing $\delta\varepsilon_i$ cancel $\delta v'$ (because of (2.63)) and obtain $v'(\mathbf{r})$ which we call the effective potential $v_{\mathrm{eff}}(\mathbf{r})$:

$$v_{\mathrm{eff}}(\mathbf{r}) = v_{\mathrm{ext}}(\mathbf{r}) + 2 \int \frac{n(\mathbf{r}')}{|\mathbf{r} - \mathbf{r}'|} \, \mathrm{d}\mathbf{r}' + v_{xc}(\mathbf{r}) \tag{2.65}$$

with

$$v_{xc}(\mathbf{r}) = \frac{\delta(E_{xc} + T_{xc})}{\delta n(\mathbf{r})}. \tag{2.66}$$

The effective single-particle equation,

$$\left[-\nabla^2 + v_{\mathrm{eff}}(\mathbf{r}) - \varepsilon_i\right] \phi_i(\mathbf{r}) = 0, \tag{2.67}$$

is often called the Kohn–Sham equation. It is a Schrödinger equation with the external potential replaced by the effective potential (Eqn (2.65)) which depends on the density. The density itself, according to Eqn (2.61), depends on the single-particle states ϕ_i. The Kohn–Sham equation thus constitutes a self-consistent field problem.

The Kohn–Sham equation furthermore allows us to derive an alternative expression for the total energy. It follows from Eqn (2.64) that

$$T_0[n] + \int v_{\mathrm{eff}}(\mathbf{r}) \, n(\mathbf{r}) \, \mathrm{d}\mathbf{r} - \sum_{\substack{i=1 \\ \varepsilon_i \leq E_{\mathrm{F}}}}^{N} \varepsilon_i = 0. \tag{2.68}$$

Now combining this with Eqns (2.52), (2.58), and (2.65), we obtain the result

$$E[n] = \sum_{\substack{i=1 \\ \varepsilon_i \leq E_{\mathrm{F}}}}^{N} \varepsilon_i - \iint \frac{n(\mathbf{r}) \, n(\mathbf{r}')}{|\mathbf{r} - \mathbf{r}'|} \, \mathrm{d}\mathbf{r} \, \mathrm{d}\mathbf{r}' - \int v_{xc}(\mathbf{r}) \, n(\mathbf{r}) \, \mathrm{d}\mathbf{r} + \tilde{E}_{xc}[n]. \tag{2.69}$$

The total energy thus consists of the sum over the eigenvalues, ε_i, minus the so-called "double-counting" terms. Note in passing that (at least temporarily) we write $\tilde{E}_{xc}$ instead of E_{xc}, this way taking care of the exchange-correlation kinetic energy, T_{xc}.

Although density-functional theory as outlined above provides a scheme to reduce the entire many-body problem to a Schrödinger-like effective single-particle equation, the physical meaning of the eigenvalues ε_i is controversial. These eigenvalues have, in fact, been used very often and with success to interpret excitation spectra. But there are also cases which are problematic, as we will see. Although we cannot decide whether this failure is inherent to density-functional theory or simply a consequence of the local-density approximation (which will be discussed next), we may still present some results here, which we hope will shed some light on the problem.

Let us start with a relation due to Slater (1972) and Janak (1978) which connects the eigenvalues ε_i to occupation number changes of the orbitals ϕ_i:

$$\varepsilon_i = \frac{\partial E}{\partial n_i}. \qquad (2.70)$$

Here E now denotes a proper generalization of the total energy allowing for fractional occupations. Its construction may readily be achieved by modifying the expressions (2.61) and (2.62) for the electron density and the kinetic energy

$$n(\mathbf{r}) = \sum_{\substack{i=1 \\ \varepsilon_i \leq E_\mathrm{F}}}^N n_i |\phi_i(\mathbf{r})|^2 \qquad (2.71)$$

$$T_0(n) = -\sum_{\substack{i=1 \\ \varepsilon_i \leq E_\mathrm{F}}}^N n_i \int \phi_i^*(\mathbf{r}) \nabla^2 \phi_i(\mathbf{r}) \, \mathrm{d}\mathbf{r} \qquad (2.72)$$

where n_i can vary between 0 and 1. We should note, however, that because of this somewhat loose procedure, the density – besides not being N-representable – also lacks the v-representability if orbitals other than the highest-occupied orbitals are depleted. In such cases the total energy may not be obtained by solving an effective single-particle equation. The problem can be circumvented by fixing the occupation numbers according to the laws of Fermi–Dirac statistics or by turning completely to the finite-temperature extension of density-functional theory as introduced by Mermin (1965). We defer the discussion of this case to Chap. 5.

Next, writing down E in complete analogy with Eqn (2.69), inserting n_i into the first term on the right-hand side and using for the density Eqn (2.71), we differentiate the expression and easily prove Eqn (2.70). Equation (2.70) states that if the occupation of the i-th orbital is changed by an infinitesimal amount δn_i, then the total energy of the system changes by $\varepsilon_i \, \delta n_i$. This still does not fully clarify the connection of the quantities ε_i with the excitation energies of the system for two reasons. First, real excitations involve whole electrons, thus the excitation energy does not correspond simply to the first derivative of the total energy, but to an entire Taylor series (assuming it converges),

$$E(n_i + \Delta n_i) = E(n_i) + (\delta E/\delta n_i)\, \Delta n_i + \frac{1}{2} (\delta^2 E/\delta n_i^2)(\Delta n_i)^2 + \cdots. \qquad (2.73)$$

The importance of second- and higher-order terms, which correspond to relaxation, depends primarily on the localization of the electron density of the states from which the electron was excited and, hence, should not be too important for delocalized states. Second, there are the complications discussed after Eqn (2.72) whose importance is not easily assessed.

Another somewhat different discussion of the same problem stems from von Barth (1984) who considered the homogeneous, interacting electron gas, where the quasiparticle excitation spectrum can formally be obtained from the Dyson equation and can be written in terms of a momentum- and energy-dependent (and in general complex) self-energy as

$$\tilde{\varepsilon}(\mathbf{k}) = \mathbf{k}^2 + \Sigma\left(\mathbf{k}, \tilde{\varepsilon}(\mathbf{k}_F)\right). \tag{2.74}$$

On the other hand, when using density-functional theory to calculate the band energies for the electron gas a $\mathbf{k}$-independent effective potential is involved which supplies a free-electron dispersion curve shifted by a constant. Since density-functional theory correctly describes the Fermi level E_F, as has been shown by Kohn and Sham (1965) (see also Kohn and Vashishta, 1983), we may conclude that

$$\varepsilon(\mathbf{k}) = \mathbf{k}^2 + \Sigma\left(k_F, E(k)\right). \tag{2.75}$$

Comparison of these two dispersion relations yields the result that the density-functional band structure is quite accurate near the Fermi energy provided the variations of the self-energy with $\mathbf{k}$ and energy are small. This condition is presumably fulfilled in most of the conventional metals, but is not, for instance, in the broad class of heavy-fermion systems. The extent of the deviations of the self-energy from its value at k_F and E_F may be characterized by the effective mass m^* which is defined by the relation

$$\frac{m^*}{m} = \frac{1 - \frac{\partial \Sigma(\mathbf{k},\omega)}{\partial \omega}}{1 + \frac{m}{k_F}\frac{\partial \Sigma(\mathbf{k},\omega)}{\partial \mathbf{k}}} \tag{2.76}$$

and which may indeed be several hundred electron masses in heavy-fermion systems (for a review see for instance Fulde *et al.*, 1988).

Summarizing the previous discussion, we can state that we may not be too far off when interpreting the calculated eigenvalues as excitation energies. Nevertheless, caution is necessary. Strictly speaking, this statement applies not only to density-functional theory but also to a greater extent to the local-density approximation, an approximation required in order to obtain a practical computational scheme. Before turning to its discussion, we will extend density-functional theory to magnetic systems. But before we can do this we must introduce the electron spin into our theory in somewhat more detail than in Chap. 1.

2.4 The electron spin: Dirac theory

The preceding discussion in this chapter was exclusively based on the Schrödinger equation and nonrelativistic quantum mechanics. The spin of the electron was largely ignored except when states were counted invoking the Pauli principle, or it was introduced *ad hoc*. However, spin as a dynamical variable (or observable) is of paramount importance in magnetism, as we have indicated in Chap. 1.

A relativistic version of density-functional theory was described some time ago by Rajagopal (1980), see also Eschrig (1996). But, instead of backing up to a field-theoretic formulation to generalize the contents of Sec. 2.3 relativistically, we shorten the discussion by assuming we have done all this and derived the Kohn–Sham equation in relativistic form. This will now be the Dirac equation, which we write as

$$H_D \psi_i = \varepsilon_i \psi_i, \qquad (2.77)$$

where ψ_i is a four-component single-particle wave function and H_D is the Dirac single-particle Hamiltonian,

$$H_D = c\,\boldsymbol{\alpha}\cdot\mathbf{p} + \beta\, m\, c^2 + V. \qquad (2.78)$$

The quantity $\boldsymbol{\alpha}$ is a vector operator whose components may be written using the Pauli spin matrices, σ_k, as

$$\alpha_k = \begin{pmatrix} 0 & \sigma_k \\ \sigma_k & 0 \end{pmatrix}, \qquad (2.79)$$

where in the standard representation

$$\sigma_1 = \begin{pmatrix} 0 & 1 \\ 1 & 0 \end{pmatrix},\ \sigma_2 = \begin{pmatrix} 0 & -i \\ i & 0 \end{pmatrix},\ \sigma_3 = \begin{pmatrix} 1 & 0 \\ 0 & -1 \end{pmatrix} \qquad (2.80)$$

are the x, y, and z Pauli spin matrices. The quantity $\mathbf{p}$ is the momentum operator, $\mathbf{p} = -i\hbar\nabla$ and the matrix β is given by

$$\beta = \begin{pmatrix} 1 & 0 \\ 0 & -1 \end{pmatrix} \qquad (2.81)$$

with

$$\mathbf{1} = \begin{pmatrix} 1 & 0 \\ 0 & 1 \end{pmatrix}. \qquad (2.82)$$

Finally, the quantity V denotes an effective potential which we need not specify here but later want to associate with a potential of the form given by Eqn (2.65) of the previous

2.4 The electron spin: Dirac theory

section. The foundations for the Dirac equation are discussed in many textbooks, as for instance, Messiah (1978), Bjorken and Drell (1964), Sakurai (1967), Baym (1973), and Strange (1999).

Two properties of Dirac's theory for the electron are especially relevant for magnetism and, therefore, deserve attention here; this is, first, a quantum mechanical derivation of the magnetization connecting spin with Maxwell's equations and, second, the role of the spin–orbit interaction as it ultimately leads to magneto-crystalline anisotropy.

The first, quite fundamental topic was treated by Gordon (1928) a long time ago but gained importance again in modern spin-density functional theory and is known as the Gordon decomposition of the current. We begin with the time-dependent Dirac equation for an electron moving in an electromagnetic field that is specified by the scalar potential φ and the vector potential $\mathbf{A}$, this is (in cgs units)

$$i\hbar \frac{\partial \Psi}{\partial t} = \left[c\,\boldsymbol{\alpha} \left(\frac{\hbar}{i} \boldsymbol{\nabla} - \frac{e}{c} \mathbf{A} \right) + \beta\,m\,c^2 + e\,\varphi \right] \Psi, \qquad (2.83)$$

where the notation is as before but Ψ is now the time-dependent four-component wave function. The particle density is, as in Schrödinger's theory (for one electron), given by

$$n(\mathbf{r},t) = \Psi^{+}(\mathbf{r},t)\,\Psi(\mathbf{r},t). \qquad (2.84)$$

The corresponding current must satisfy the continuity equation,

$$\frac{\partial}{\partial t} n(\mathbf{r},t) + \boldsymbol{\nabla} \cdot \mathbf{j}(\mathbf{r},t) = 0, \qquad (2.85)$$

and can easily be constructed to read

$$\mathbf{j}(\mathbf{r},t) = c\,\Psi^{+}(\mathbf{r},t)\,\boldsymbol{\alpha}\,\Psi(\mathbf{r},t). \qquad (2.86)$$

Now, Ψ can be written, from Eqn (2.83) as

$$\Psi(\mathbf{r},t) = \frac{1}{m\,c^2} \left[\beta \left(i\hbar \frac{\partial}{\partial t} - e\,\varphi \right) - c\,\beta\,\boldsymbol{\alpha} \left(\frac{\hbar}{i} \boldsymbol{\nabla} - \frac{e}{c} \mathbf{A} \right) \right] \Psi(\mathbf{r},t), \qquad (2.87)$$

where all we need is $\beta^2 = 1$. We may express Ψ^{+} in an analogous way and substitute Ψ and Ψ^{+} into Eqn (2.86) for component n ($n = 1, 2, 3$) of the vector $\mathbf{j}$. Two relations are needed to simplify the expression for j_n. The first is

$$\beta\,\boldsymbol{\alpha} = -\boldsymbol{\alpha}\,\beta, \qquad (2.88)$$

which is almost obvious. The second is

$$\alpha_n\,\alpha_m \equiv \frac{1}{2}[\alpha_n, \alpha_m] + \frac{1}{2}\{\alpha_n, \alpha_m\} = i\,\varepsilon_{nm\ell}\,\sigma_\ell + \delta_{nm}, \qquad (2.89)$$

where $[,]$ denotes the commutator, $\{,\}$ the anticommutator, and $\varepsilon_{nm\ell}$ is zero if two or more indices are equal and is $+1$ if $nm\ell$ is a positive permutation, -1 otherwise. Collecting the components of $\mathbf{j}$ one can verify after some algebra that

$$\mathbf{j} = \frac{1}{2m} \left[\overline{\Psi} \left(\frac{\hbar}{i} \boldsymbol{\nabla} - \frac{e}{c} \mathbf{A} \right) \Psi + \left(\left(\frac{\hbar}{i} \boldsymbol{\nabla} - \frac{e}{c} \mathbf{A} \right) \overline{\Psi} \right) \Psi \right]$$
$$+ \frac{\hbar}{2m} \left[\boldsymbol{\nabla} \times (\overline{\Psi} \boldsymbol{\sigma} \Psi) - \frac{1}{c} \frac{\partial}{\partial t} (\overline{\Psi} i \boldsymbol{\alpha} \Psi) \right],$$
(2.90)

where $\overline{\Psi} = \Psi^{+} \beta$. Repeating the calculation for the density, one finds

$$n = \frac{1}{2m c^2} \left[\overline{\Psi} \left(i\hbar \frac{\partial}{\partial t} - e\varphi \right) \Psi + \left(\left(-i\hbar \frac{\partial}{\partial t} - e\varphi \right) \overline{\Psi} \right) \Psi \right]$$
$$+ \frac{\hbar}{2m c} \boldsymbol{\nabla} \cdot (\overline{\Psi} i \boldsymbol{\alpha} \Psi).$$
(2.91)

These relations become physically transparent if one defines a *convective* current and density by

$$\mathbf{j}_{\text{conv}} = \frac{\hbar}{2 i m} \left[\overline{\Psi} \boldsymbol{\nabla} \Psi - (\boldsymbol{\nabla} \overline{\Psi}) \Psi \right] - \frac{e \mathbf{A}}{m c} \overline{\Psi} \Psi$$
(2.92)

and

$$n_{\text{conv}} = \frac{i \hbar}{2m c^2} \left(\overline{\Psi} \frac{\partial \Psi}{\partial t} - \frac{\partial \overline{\Psi}}{\partial t} \Psi \right) - \frac{e \varphi}{m c^2} \overline{\Psi} \Psi.$$
(2.93)

Then the current and density may be written as

$$\mathbf{j} = \mathbf{j}_{\text{conv}} + \mathbf{j}_{int}$$
$$n = n_{\text{conv}} + n_{\text{int}},$$
(2.94)

where the *internal* current and density are

$$\mathbf{j}_{\text{int}} = c \boldsymbol{\nabla} \times \mathbf{M} + \frac{\partial \mathbf{P}}{\partial t},$$
(2.95)

$$n_{\text{int}} = -\boldsymbol{\nabla} \cdot \mathbf{P},$$
(2.96)

with

$$\mathbf{M} = \frac{\hbar}{2m c} \overline{\Psi} \boldsymbol{\Sigma} \Psi,$$
(2.97)

2.4 The electron spin: Dirac theory

$$\mathbf{P} = \frac{\hbar}{2mc} \overline{\Psi}(-i\boldsymbol{\alpha}\Psi) \tag{2.98}$$

and

$$\Sigma = \begin{pmatrix} \boldsymbol{\sigma} & 0 \\ 0 & \boldsymbol{\sigma} \end{pmatrix}. \tag{2.99}$$

The convective parts are reminiscent of the current and density of spinless (nonrelativistic) electrons and are determined by the rate of change of Ψ in time and space. Multiplying by the charge of the electron, e, we obtain the internal charge current and the internal charge density as

$$\mathbf{J}_{\text{int}} = c\,\boldsymbol{\nabla} \times \boldsymbol{\mathcal{M}} + \frac{\partial \boldsymbol{\mathcal{P}}}{\partial t},$$

$$\rho_{\text{int}} = -\boldsymbol{\nabla}\boldsymbol{\mathcal{P}}, \tag{2.100}$$

where

$$\boldsymbol{\mathcal{M}} = \frac{e\hbar}{2mc} \overline{\Psi}\boldsymbol{\Sigma}\Psi \tag{2.101}$$

$$\boldsymbol{\mathcal{P}} = \frac{e\hbar}{2mc} \overline{\Psi}(-i\boldsymbol{\alpha}\Psi). \tag{2.102}$$

Now consider Maxwell's equation

$$\boldsymbol{\nabla} \times \mathbf{B} = \frac{4\pi}{c}\mathbf{J} + \frac{1}{c}\frac{\partial \mathbf{E}}{\partial t} \tag{2.103}$$

and split up the current such that $\mathbf{J} = \mathbf{J}_{\text{int}} + \mathbf{J}_{\text{rest}}$, where $\mathbf{J}_{\text{rest}}$ contains all contributions to the charge current other than the internal one. Then with Eqn (2.100)

$$\boldsymbol{\nabla} \times \mathbf{B} = \frac{4\pi}{c}\mathbf{J}_{\text{rest}} + 4\pi\,\boldsymbol{\nabla} \times \boldsymbol{\mathcal{M}} + \frac{1}{c}\frac{\partial \mathbf{E}}{\partial t} + \frac{4\pi}{c}\frac{\partial \boldsymbol{\mathcal{P}}}{\partial t}, \tag{2.104}$$

and collecting, we see that

$$\boldsymbol{\nabla} \times \mathbf{H} = \frac{4\pi}{c}\mathbf{J}_{\text{rest}} + \frac{1}{c}\frac{\partial}{\partial t}\mathbf{D}, \tag{2.105}$$

if we identify

$$\mathbf{H} = \mathbf{B} - 4\pi\boldsymbol{\mathcal{M}}, \tag{2.106}$$

$$\mathbf{D} = \mathbf{E} + 4\pi\boldsymbol{\mathcal{P}}. \tag{2.107}$$

Thus, clearly, the spin observable leads to a magnetic moment given by Eqn (2.101) and an electric moment given by Eqn (2.102). Finally, the energy of a field–particle system is

$$\mathcal{E} = \frac{1}{2c} \int \mathbf{J} \cdot \mathbf{A} \, d\mathbf{x}. \tag{2.108}$$

The internal (or spin) part of the charge current leads to

$$\mathcal{E} = \frac{1}{2} \int (\nabla \times \mathcal{M}) \cdot \mathbf{A} \, d\mathbf{x}$$

$$= \frac{1}{2} \int \mathcal{M} \cdot (\nabla \times \mathbf{A}) \, d\mathbf{x} \tag{2.109}$$

$$= \frac{1}{2} \int \mathcal{M} \cdot \mathbf{B} \, d\mathbf{x},$$

where $\mathbf{B} = \nabla \times \mathbf{A}$ is the magnetic induction.

To describe the second property of the Dirac equation, i.e. the spin–orbit interaction, one proceeds as follows. Consider the time-independent Dirac equation and write the four-component spinor function in terms of two two-component functions ϕ and χ as

$$\psi = \begin{pmatrix} \phi \\ \chi \end{pmatrix} \tag{2.110}$$

where in the case of electrons (positive energy solutions) ϕ is the "large" and χ is the "small" component. The eigenvalue problem,

$$H_D \psi = \varepsilon \psi \tag{2.111}$$

is, with Eqns (2.78) and (2.79), at once seen to lead to a set of coupled equations for ϕ and χ, namely

$$c(\boldsymbol{\sigma} \cdot \mathbf{p}) \chi = (\varepsilon - V - mc^2) \phi$$
$$c(\boldsymbol{\sigma} \cdot \mathbf{p}) \phi = (\varepsilon - V + mc^2) \chi. \tag{2.112}$$

Since it is the radial Schrödinger equation that, as we will see, plays an important role, we next sketch the derivation of the radial Dirac equation. So let the potential be spherically symmetric. Then, in order to obtain the correct classifications (or quantum numbers) of the solutions of the Dirac equation, we must list the constants of motion, i.e., those operators that commute with H_D (and with themselves). They are derived in appropriate textbooks, see e.g. Messiah (1978) or Sakurai (1967), where one finds that the total angular momentum is one such operator:

$$\mathbf{J} = \mathbf{L} + \hbar \mathbf{\Sigma}/2. \tag{2.113}$$

It is a sum of the orbital angular momentum, $\mathbf{L}$, and the spin, see Eqn (2.99). But $\mathbf{L}$ and $\mathbf{\Sigma}$ separately are not constants of motion. Another operator is $\mathbf{K}$ which is explicitly given by

$$\mathbf{K} = \begin{pmatrix} \boldsymbol{\sigma} \cdot \mathbf{L} + \hbar & 0 \\ 0 & -\boldsymbol{\sigma} \cdot \mathbf{L} - \hbar \end{pmatrix}. \tag{2.114}$$

Since $\mathbf{K}$ also commutes with $\mathbf{J}$, the eigenfunctions of H_D in a central potential are eigenfunctions of $\mathbf{K}$, $\mathbf{J}^2$, and $\mathbf{J}_z$. The corresponding eigenvalues are denoted by $-\kappa\hbar$, $j(j+1)\hbar^2$, and $j_z\hbar$, but κ and j are related by

$$\kappa = \pm \left(j + \frac{1}{2}\right) \tag{2.115}$$

as one can easily show. Thus κ is a nonzero integer which can be positive or negative. Pictorially speaking, the sign of κ determines whether the spin is antiparallel ($\kappa > 0$) or parallel ($\kappa < 0$) to the total angular momentum in the nonrelativistic limit. To simplify the notation, instead of attaching all the quantum numbers (κ, j, j_z) to the wave function, we merely write

$$\psi = \begin{pmatrix} \phi \\ \chi \end{pmatrix} = \begin{pmatrix} g(r)\, \mathcal{Y}_{j\ell}^{j_z} \\ i f(r)\, \mathcal{Y}_{j\ell'}^{j_z} \end{pmatrix} \tag{2.116}$$

where g and f are radial functions, and where $\mathcal{Y}_{j\ell}^{j_z}$ stands for a normalized spin angular function (or r-independent eigenfunction of $\mathbf{J}^2$, J_z, $\mathbf{L}^2$, and $\mathbf{S}^2$) formed by the combination of the Pauli spinor with the spherical harmonics, Y_ℓ^m, of order ℓ. Explicitly,

$$\mathcal{Y}_{j\ell}^{j_z} = \pm \sqrt{\frac{\ell + j_z + \frac{1}{2}}{2\ell + 1}}\, Y_\ell^{j_z - 1/2} \begin{pmatrix} 1 \\ 0 \end{pmatrix}$$

$$+ \sqrt{\frac{\ell \mp j_z + \frac{1}{2}}{2\ell + 1}}\, Y_\ell^{j_z + 1/2} \begin{pmatrix} 0 \\ 1 \end{pmatrix} \tag{2.117}$$

where $j = \ell \pm 1/2$. The relations among κ, j, ℓ, and ℓ' are

$$\kappa = j + \frac{1}{2} \quad \text{then} \quad \ell = j + \frac{1}{2}, \ell' = j - \frac{1}{2}$$

$$\kappa = -\left(j + \frac{1}{2}\right) \quad \text{then} \quad \ell = j - \frac{1}{2}, \ell' = j + \frac{1}{2}.$$

(2.118)

Now an easy calculation yields the radial Dirac equations: one starts with Eqn (2.111) using (2.116) and the relation

$$\boldsymbol{\sigma} \cdot \mathbf{p} \equiv \frac{(\boldsymbol{\sigma} \cdot \mathbf{r})}{r^2} (\boldsymbol{\sigma} \cdot \mathbf{r} (\boldsymbol{\sigma} \cdot \mathbf{p}))$$

$$= \frac{(\boldsymbol{\sigma} \cdot \mathbf{r})}{r} \left(-i\hbar \frac{d}{dr} + i\boldsymbol{\sigma} \cdot \mathbf{L}\right)$$

(2.119)

and obtains

$$\hbar c \left(\frac{df}{dr} - \frac{(1+\kappa)}{r} f\right) = -(\varepsilon - V - mc^2) g$$

$$\hbar c \left(\frac{dg}{dr} + \frac{(1+\kappa)}{r} g\right) = (\varepsilon - V - mc^2) f.$$

(2.120)

The band-structure problem (which is the topic of the next chapter) with this full four-component formalism can now be set up and solved numerically following the various procedures to be outlined. A treatment of any of these is beyond the scope of this chapter, but an approximate and physically transparent version is not. It is generally known as the scalar relativistic approximation (Koelling and Harmon, 1977; Takeda, 1978; MacDonald et al., 1980) and it supplies an equation that formally looks like a slightly modified Schrödinger equation. The spin–orbit interaction is separated out explicitly and is normally treated variationally or by perturbation theory, while the new wave equation (often called the scalar relativistic wave equation, which looks quite like the Pauli equation but is not identical with it) simply replaces the Schrödinger equation whenever it appears in any of the procedures to be discussed.

To summarize the salient features of this approximation, we start by defining another energy origin, i.e.

$$\varepsilon' = \varepsilon - mc^2.$$

(2.121)

Then the factor multiplying f on the right-hand side of Eqn (2.120) can formally be written as $2Mc^2$, if we define

$$M = m + \frac{\varepsilon' - V}{2c^2} \qquad (2.122)$$

and the second of Eqns (2.120) gives

$$f = \frac{\hbar}{2Mc}\left(\frac{\mathrm{d}g}{\mathrm{d}r} + \frac{(1+\kappa)}{r}g\right). \qquad (2.123)$$

Substituting this into the first of Eqns (2.120) and carrying out the differentiation, we obtain

$$-\frac{\hbar^2}{2M}\frac{1}{r^2}\frac{\mathrm{d}}{\mathrm{d}r}\left(r^2\frac{\mathrm{d}g}{\mathrm{d}r}\right) + \left[V + \frac{\hbar^2}{2M}\frac{\kappa(\kappa+1)}{r^2}\right]g$$
$$-\frac{\hbar^2}{4M^2c^2}\frac{\mathrm{d}V}{\mathrm{d}r}\frac{\mathrm{d}g}{\mathrm{d}r} - \frac{\hbar^2}{4M^2c^2}\frac{\mathrm{d}V}{\mathrm{d}r}\frac{(1+\kappa)}{r}g = \varepsilon' g. \qquad (2.124)$$

For each of the cases specified in Eqn (2.118), the factor $\kappa(\kappa+1)$ is seen to be

$$\kappa(\kappa+1) = \ell(\ell+1). \qquad (2.125)$$

Thus, apart from the last two terms on the left-hand side, Eqn (2.124) looks like the radial Schrödinger equation. Of course, care must be taken with M, which, as Eqn (2.122) shows, is not a constant. No approximation has yet been made. The formal mass term M is sometimes called the "mass–velocity" term, the term $(\mathrm{d}V/\mathrm{d}r)/(\mathrm{d}g/\mathrm{d}r)$ is known as the Darwin term, and the last term on the left-hand side of Eqn (2.124) is the spin–orbit coupling term; this is true because

$$(\boldsymbol{\sigma}\cdot\mathbf{L} + \hbar)\phi = -\kappa\hbar\phi. \qquad (2.126)$$

We now make the essential approximation and drop the spin–orbit term from the radial equation (2.124) (and from Eqn (2.123)). We may call these approximate functions $\tilde{g}$ and $\tilde{f}$, i.e. we require

$$-\frac{\hbar^2}{2M}\frac{1}{r^2}\frac{\mathrm{d}}{\mathrm{d}r}\left(r^2\frac{\mathrm{d}\tilde{g}}{\mathrm{d}r}\right) + \left[V + \frac{\hbar^2}{2M}\frac{\ell(\ell+1)}{r^2}\right]\tilde{g}$$
$$-\frac{\hbar^2}{4M^2c^2}\frac{\mathrm{d}V}{\mathrm{d}r}\frac{\mathrm{d}\tilde{g}}{\mathrm{d}r} = \varepsilon'\tilde{g} \qquad (2.127)$$

and
$$\tilde{f} = \frac{\hbar}{2Mc} \frac{\mathrm{d}\tilde{g}}{\mathrm{d}r}. \tag{2.128}$$

The latter is needed for proper normalization:
$$\int (\tilde{g}^2 + \tilde{f}^2) \, r^2 \, \mathrm{d}r = 1. \tag{2.129}$$

It may also be used to simplify the numerical procedure to solve Eqn (2.127). But this is more of a technical detail. In passing, we note that by expanding M we easily obtain the Pauli equation from Eqn (2.127).

Equation (2.127) is the scalar relativistic radial equation, and it may be used in place of the Schrödinger equation; no further changes are needed.

It should now be obvious that one needs to formulate the corrections that are brought about when the spin–orbit coupling term is included in the final step of the calculations. The most satisfactory way seems to be the following (MacDonald et al., 1980): Using the radial functions $\tilde{g}$ and $\tilde{f}$, the four-component wave function is first written out as

$$\tilde{\psi} = \begin{pmatrix} \tilde{\phi} \\ \tilde{\chi} \end{pmatrix} \tag{2.130}$$

where
$$\tilde{\phi} = \tilde{g} Y_L \chi_S, \tag{2.131}$$

where Y_L is the spherical harmonic of order $L = (\ell, m)$ and where $\chi_+ = \begin{pmatrix} 1 \\ 0 \end{pmatrix}$ and $\chi_- = \begin{pmatrix} 0 \\ 1 \end{pmatrix}$. Then $\tilde{\chi}$ is obtained using Eqn (2.112),

$$\tilde{\chi} = \mathrm{i} \left(\frac{\boldsymbol{\sigma} \cdot \mathbf{x}}{r} \right) \left(-\tilde{f} + \frac{1}{2Mcr} \tilde{g} \boldsymbol{\sigma} \cdot \mathbf{L} \right) Y_{L-\chi_S}. \tag{2.132}$$

The spin–orbit correction is now obtained by applying the Dirac Hamiltonian H_D, Eqn (2.78) to the wave function $\tilde{\chi}$. The details of this calculation are somewhat tedious but straightforward and give

$$H_D \tilde{\psi} = \varepsilon' \tilde{\psi} + H_{SO} \tilde{\psi} \tag{2.133}$$

where
$$H_{SO} = \frac{\hbar}{(2Mc)^2} \frac{1}{r} \frac{\mathrm{d}V}{\mathrm{d}r} \begin{pmatrix} (\boldsymbol{\sigma} \cdot \mathbf{L}) \mathbf{1} \\ 0 \end{pmatrix}. \tag{2.134}$$

This is the spin–orbit coupling operator that represents a measure of the extent to which the function $\tilde{\psi}$ fails to be a true solution of the spherical potential Dirac equation. In most applications the matrix element of H_{SO} is computed on whatever basis is used in the particular method and is then added to the Hamiltonian matrix in the variational procedure. It is clear that this breaks the degeneracy of the formerly spin-degenerate bands. Details can be found in the paper by Andersen (1975) and a very clear account of spin–orbit coupling in the LAPW method in the paper by MacDonald *et al.* (1980). The latter authors also compare their results with results of a full four-component RAPW calculation and find excellent agreement. Thus it seems that this approximate treatment is satisfactory.

We close this subsection with a short discussion of *spin-polarized* effective potentials to be discussed in Sec. 2.5. This problem appears to be more complex in a relativistic formalism than in the nonrelativistic case.

A desirable approach would start with the Dirac equation for an electron moving in a scalar effective potential, V, and an effective vector potential $\mathbf{A}_{\text{eff}}$:

$$H_D = c\,\boldsymbol{\alpha} \cdot \left(\mathbf{p} - \frac{e}{c}\mathbf{A}_{\text{eff}}\right) + \beta\,m\,c^2 + V. \tag{2.135}$$

There are recent noteworthy attempts with this form of starting point, but only in the framework of *nonrelativistic* density-functional theory has an effective vector potential been derived in a completely satisfactory way (Vignale and Rasolt, 1987, 1988). On the other hand, relativistic theories seem to be noncontradictory only when they work with a spin-polarized scalar effective potential and no effective vector potential (see MacDonald and Vosko, 1979; Doniach and Sommers, 1982; Cortona *et al.*, 1985). Physically, one retains a coupling of the exchange-correlation potential to the electron spin only, discarding a possible coupling to currents and hence to orbital magnetic moments:

$$H_D = c\,\boldsymbol{\alpha} \cdot \mathbf{p} + \beta\,m\,c^2 + V + \beta\,\Sigma_z\,V_1 \tag{2.136}$$

where V is the effective potential that is independent of the spin direction, Σ_z is the z-component of $\boldsymbol{\Sigma}$ (Eqn (2.99)), and V_1 could be called an effective (exchange correlation) "magnetic field".

The consequences of these approximations are not fully known. Strange *et al.* (1984) first used Eqn (2.136) in a fully self-consistent KKR band-structure calculation, but even on this level further approximations must be made which have their origin in the fact that spin is not a good quantum number in the Dirac theory. A thorough discussion of these facts and consequences can be found in a paper by Feder *et al.* (1983). Unfortunately, they also affect the scalar relativistic approximation. A satisfactory approximation is obtained, however, if the spin-dependent potential, V_1, is set to zero in the equation for the small component χ. In this case there are the same steps as before that must be taken to derive the equation for $\tilde{g}$, which now reads

$$-\frac{\hbar^2}{2M}\frac{1}{r^2}\frac{\mathrm{d}}{\mathrm{d}r}\left(r^2\frac{\mathrm{d}\tilde{g}_s}{\mathrm{d}r}\right)+\left[V+sV_1+\frac{\hbar^2}{2M}\frac{\ell(\ell+1)}{r^2}\right]\tilde{g}_s$$

(2.137)

$$-\frac{\hbar^2}{4M^2c^2}\frac{\mathrm{d}V}{\mathrm{d}r}\cdot\frac{\mathrm{d}\tilde{g}_s}{\mathrm{d}r}=\varepsilon'_s\tilde{g}_s$$

where $s = \pm 1$: $+1$ for spin-up and -1 for spin-down electrons. M, Eqn (2.122), $\mathrm{d}V/\mathrm{d}r$, and H_{SO}, Eqn (2.134), do not contain the effect of spin polarization, i.e. the term V_1. This is, of course, a consequence of setting V_1 equal to zero in the small component. Again, the consequences of this approximation are not presently known, but we believe they are unimportant. We see that the scalar relativistic radial equation must be solved for each spin direction separately. The spin–orbit matrix element is then treated as before and results in a coupling of the two spin spaces so that, when everything is done, the bands are no longer labeled by "spin-up" and "spin-down". Later examples will shed more light on the type of solutions.

2.5 Spin-density-functional theory

Apparently, von Barth and Hedin (1972) and Rajagopal and Callaway (1973) were the first to generalize the density-functional theory to include effects of spin polarization. They used a matrix formalism to represent the density and the external potential instead of single variables. We omit the basic proofs here, since they are very similar to those given in Sec. 2.3, and rather concentrate on facts that are typical for spin-density-functional theory. However, we will investigate in more detail the conceptual, physical, and mathematical foundations of spin-density-functional theory in Sec. 2.8.

From the foregoing discussion we recall that in the simplest version of a theory for magnetic materials we attached spinors χ_+ and χ_- to the wave function. Such a two-component spinor function consequently requires a 2×2 matrix for the Hamiltonian, which may again be written in three parts representing the kinetic energy, the external potential, and the Coulomb interaction of the electrons. The corresponding operators are given by

$$T_{\alpha\beta}=-\delta_{\alpha\beta}\sum_{i=1}^{N}\nabla_i^2 \quad (2.138)$$

$$V_{\alpha\beta}=\sum_{i=1}^{N}v_{\alpha\beta}^{\mathrm{ext}}(\mathbf{r}_i) \quad (2.139)$$

$$U_{\alpha\beta}=\sum_{\substack{i=1,j=1\\i\neq j}}^{N}\frac{\delta_{\alpha\beta}}{|\mathbf{r}_i-\mathbf{r}_j|} \quad (2.140)$$

where α and β are spin indices. Comparing this with the corresponding equations of the preceding subsection, we see that it is only the external potential, Eqn (2.139), that needs

2.5 Spin-density-functional theory

special attention. Furthermore, if we write as before the electron density operator, Eqn (2.50), as

$$\hat{n}(\mathbf{r}) = \sum_{i=1}^{N} \delta(\mathbf{r} - \mathbf{r}_i) \quad (2.141)$$

then the electron density is obtained as in Eqn (2.51), where the many-body state is now assumed to be given by a spinor function. But we would like to go a little further and include in our discussion also those cases where the spin state cannot be described by a single, global quantization axis. According to the laws of quantum mechanics such a mixed state is describable by the density matrix ρ which is defined by (Sakurai, 1985; Levy, 1982)

$$\rho(\mathbf{r}'_1 \sigma'_1, \mathbf{r}'_2 \sigma'_2, \ldots, \mathbf{r}'_N \sigma'_N | \mathbf{r}_1 \sigma_1, \mathbf{r}_2 \sigma_2, \ldots, \mathbf{r}_N \sigma_N)$$
$$= \sum_{\ell=1}^{n} c_\ell \, \Phi_\ell(\mathbf{r}'_1 \sigma'_1, \mathbf{r}'_2 \sigma'_2, \ldots, \mathbf{r}'_N \sigma'_N) \, \Phi_\ell^*(\mathbf{r}_1 \sigma_1, \mathbf{r}_2 \sigma_2, \ldots, \mathbf{r}_N \sigma_N) \quad (2.142)$$

where an n-fold degenerate many-particle wave function, Φ_ℓ, is assumed and the c_ℓ are restricted by $0 \leq c_\ell \leq 1$ and $\sum_{l=1}^{n} c_\ell = 1$. For a nondegenerate state, n must equal unity.

The electron density is then given by

$$n(\mathbf{r}) = \mathrm{Tr} \, \langle \rho \hat{n}(\mathbf{r}) \rangle$$
$$= N \sum_{\sigma_1, \ldots, \sigma_N} \int d\mathbf{r}_2, \ldots, d\mathbf{r}_N \, \rho(\mathbf{r}\sigma_1, \mathbf{r}_2\sigma_2, \ldots, \mathbf{r}_N\sigma_N | \mathbf{r}\sigma_1, \mathbf{r}_2\sigma_2, \ldots, \mathbf{r}_N\sigma_N) \quad (2.143)$$

where the symbol Tr denotes the trace. For a complete description of the state of an itinerant magnetic system we need the spin density in addition to the electron density, Eqn (2.143). For this we define the matrix $\tilde{n}(\mathbf{r})$ with elements $n_{\beta\alpha}(\mathbf{r})$ ($\beta = 1, 2$ and $\alpha = 1, 2$) given by

$$n_{\beta\alpha}(\mathbf{r}) = N \sum_{\sigma_2, \ldots, \sigma_N} \int d\mathbf{r}_2 \ldots d\mathbf{r}_N \, \rho(\mathbf{r}\beta, \mathbf{r}_2\sigma_2, \ldots, \mathbf{r}_N\sigma_N | \mathbf{r}\alpha, \mathbf{r}_2\sigma_2, \ldots, \mathbf{r}_N\sigma_N) \quad (2.144)$$

which we call the spin-density matrix.

(It should be noted that the formalism looks much simpler in the notation of second quantization, which, however, we avoid here for didactic reasons.) In general, the quantity $\tilde{n}(\mathbf{r})$ will not be diagonal, but it can be locally diagonalized, and in this case the difference of the eigenvalues supplies the spin density (see Kübler *et al.*, 1988a, 1988b, and Sticht *et al.*, 1989; or Vignale and Rasolt, 1987, 1988 for a somewhat different formulation).

Now,

$$O[\tilde{n}] = \inf_{\rho \in M(\tilde{n})} \mathrm{Tr} \, \langle \rho O \rangle \quad (2.145)$$

is a straightforward extension of the functional given in Eqn (2.53), the only difference being that the constrained search for the infimum now includes all density operators ρ which yield the required density matrix $\tilde{n}$.

We next set up the total energy and subsequently derive the single-particle equation. Thus we proceed by replacing Eqn (2.52) by

$$E[\tilde{n}] = F[\tilde{n}] + V[\tilde{n}] \tag{2.146}$$

where

$$F[\tilde{n}] = T[\tilde{n}] + \iint \frac{n(\mathbf{r}) n(\mathbf{r}')}{|\mathbf{r} - \mathbf{r}'|} \, d\mathbf{r} \, d\mathbf{r}' + E_{xc}[\tilde{n}] \tag{2.147}$$

and

$$V[\tilde{n}] = \inf_{\rho \in M(\tilde{n})} \text{Tr} \langle \rho V \rangle = \inf_{\rho \in M(\tilde{n})} \sum_{\alpha \beta} \int v_{\alpha \beta}^{\text{ext}}(\mathbf{r}) \, n_{\beta \alpha}(\mathbf{r}) \, d\mathbf{r}, \tag{2.148}$$

which can be shown with little effort.

The variation now proceeds as in Sec. 2.3. Again we split up the kinetic energy as $T = T_0 + T_{xc}$, where T_0 is the kinetic energy of noninteracting particles and T_{xc} the remainder.

Next we assume as before that we can determine single-particle functions $\{\phi_{i\alpha}(\mathbf{r})\}$ that permit us to write the elements of the density matrix as

$$n_{\beta \alpha}(\mathbf{r}) = \sum_{\substack{i=1 \\ \varepsilon_{i\alpha}, \varepsilon_{i\beta} \leq E_F}}^{N} \phi_{i\beta}(\mathbf{r}) \phi_{i\alpha}^{*}(\mathbf{r}). \tag{2.149}$$

The indices α and β, due to the electron spin, are $\alpha = 1, 2$ and $\beta = 1, 2$. The kinetic energy of noninteracting particles is

$$T_0[\tilde{n}] = \sum_{\alpha=1}^{2} \sum_{\substack{i=1 \\ \varepsilon_{i\alpha} \leq E_F}}^{N} \int \nabla \phi_{i\alpha}^{*}(\mathbf{r}) \nabla \phi_{i\alpha}(\mathbf{r}) \, d\mathbf{r}. \tag{2.150}$$

The Fermi energy appearing here follows from the density

$$n(\mathbf{r}) = \sum_{\alpha=1}^{2} \sum_{\substack{i=1 \\ \varepsilon_{i\alpha} \leq E_F}}^{N} |\phi_{i\alpha}(\mathbf{r})|^2, \tag{2.151}$$

which is the trace of Eqn (2.149), and the particle number N is $\int n(\mathbf{r}) \, d\mathbf{r} = N$. We complete the ansatz by postulating

$$\sum_{\beta} \left(-\delta_{\alpha \beta} \nabla^2 + v'_{\alpha \beta}(\mathbf{r}) - \varepsilon_i \delta_{\alpha \beta} \right) \phi_{i\beta}(\mathbf{r}) = 0 \tag{2.152}$$

and attempt to determine the potential matrix by minimizing the total energy. Equation (2.152) allows us to express the kinetic energy, T_0, as

$$T_0[\tilde{n}] = \sum_{i=1}^{N} \varepsilon_i - \sum_{\alpha\beta} \int v'_{\alpha\beta}(\mathbf{r})\, n_{\beta\alpha}(\mathbf{r})\, d\mathbf{r}, \qquad (2.153)$$

which may be substituted into Eqn (2.147). The variation of the total energy,

$$\frac{\delta E[\tilde{n}]}{\delta \tilde{n}_{\alpha\beta}(\mathbf{r})} = 0,$$

is now easily determined and yields the effective potential matrix

$$v'_{\alpha\beta}(\mathbf{r}) \equiv v^{\text{eff}}_{\alpha\beta}(\mathbf{r}) = v^{\text{ext}}_{\alpha\beta}(\mathbf{r}) + 2\delta_{\alpha\beta}\int \frac{n(\mathbf{r}')}{|\mathbf{r}-\mathbf{r}'|}\, d\mathbf{r}' + v^{xc}_{\alpha\beta}(\mathbf{r}), \qquad (2.154)$$

where

$$v^{xc}_{\alpha\beta}(\mathbf{r}) = \frac{\delta}{\delta \tilde{n}_{\beta\alpha}(\mathbf{r})}\left(E_{xc}[\tilde{n}] + T_{xc}[\tilde{n}]\right). \qquad (2.155)$$

Equation (2.152) together with (2.154) and (2.155) constitute the Kohn–Sham equations which, in general, are coupled equations that we will discuss in detail in Secs. 2.7 and 3.4.7. In many cases, however, notably for ferromagnets, the exchange-correlation potential, $v^{\text{eff}}_{\alpha\beta}(\mathbf{r})$, is diagonal.

We close this section by writing down an expression for the total energy similar to Eqn (2.69),

$$\begin{aligned} E[\tilde{n}] = &\sum_{\substack{i=1 \\ \varepsilon_i \leq E_F}}^{N} \varepsilon_i - \iint \frac{n(\mathbf{r})\, n(\mathbf{r}')}{|\mathbf{r}-\mathbf{r}'|}\, d\mathbf{r}\, d\mathbf{r}' \\ &- \sum_{\alpha\beta}\int v^{xc}_{\alpha\beta}(\mathbf{r})\, n_{\beta\alpha}(\mathbf{r})\, d\mathbf{r} + E_{xc}[\tilde{n}]. \end{aligned} \qquad (2.156)$$

2.6 The local-density approximation (LDA)

Our discussion in the preceding sections was of a theoretical nature without regard to practical applications. We now turn to an approximation scheme to deal with the exchange-correlation functional which is both accurate and useful. This is the local-density approximation (LDA) where the homogeneous, interacting electron gas serves to model the exchange-correlation energy in the form

$$E_{xc}[n] = \int n(\mathbf{r}) \, \epsilon_{xc}(n(\mathbf{r})) \, d\mathbf{r}, \qquad (2.157)$$

assuming T_{xc} (see Eqn (2.66)) is included in $E_{xc}[n]$.

The symbol ϵ here should not be confused with an eigenenergy; furthermore $\epsilon_{xc}(n)$ is a function of the density instead of a functional and Eqn (2.157) may be viewed as dividing the inhomogeneous electron system into small "boxes," each containing a homogeneous interacting electron gas with a density $n(\mathbf{r})$ appropriate for the "box" at $\mathbf{r}$.

From Eqns (2.66) and (2.157) we derive for the exchange-correlation potential

$$v_{xc}(\mathbf{r}) = \left[\frac{d}{dn} \{ n \, \epsilon_{xc}(n) \} \right]_{n=n(\mathbf{r})}. \qquad (2.158)$$

The local-density approximation may be readily generalized to the spin-polarized case. Assuming the eigenvalues of the density matrix to be $n_\uparrow(\mathbf{r})$ and $n_\downarrow(\mathbf{r})$ the exchange-correlation energy is written as (von Barth and Hedin, 1972)

$$E_{xc}[\tilde{n}] = \int n(\mathbf{r}) \, \epsilon_{xc}(n_\uparrow(\mathbf{r}), n_\downarrow(\mathbf{r})) \, d\mathbf{r}. \qquad (2.159)$$

This results in the following form of the exchange-correlation potential

$$v_\alpha^{xc}(\mathbf{r}) = \left[\frac{\partial}{\partial n_\alpha} \{ n \, \epsilon_{xc}(n_\uparrow, n_\downarrow) \} \right]_{\substack{n_\uparrow = n_\uparrow(\mathbf{r}) \\ n_\downarrow = n_\downarrow(\mathbf{r})}} \qquad (2.160)$$

where $\alpha = \uparrow$ or $\downarrow$. Here we assumed the density matrix and the effective potential to be diagonal. We will discuss this simplification further below.

To obtain an explicit expression for the exchange-correlation-energy density, $\epsilon_{xc}(n_\uparrow, n_\downarrow)$, we initially neglect correlation and begin with the Hartree–Fock approximation for a spin-polarized gas of electrons in atomic units:

$$\epsilon_x(n_\uparrow, n_\downarrow) = -3 \left(\frac{3}{4\pi} \right)^{1/3} \frac{1}{n} \left(n_\uparrow^{4/3} + n_\downarrow^{4/3} \right), \qquad (2.161)$$

where the density is $n = n_\uparrow + n_\downarrow$. This is not hard to derive with the results of Sec. 2.2. Correlation gives rise to another term $\epsilon_{xc}(n_\uparrow, n_\downarrow)$ that must be added to the exchange term to give

$$\epsilon_{xc}(n_\uparrow, n_\downarrow) = \epsilon_x(n_\uparrow, n_\downarrow) + \epsilon_c(n_\uparrow, n_\downarrow). \qquad (2.162)$$

Now, the total energy, and hence, ϵ_c, of a homogeneous but interacting electron gas is known only approximately. But it is known numerically to an accuracy which makes Eqn (2.159) an exceedingly useful approximation. Different parametrizations of the numerical results of different authors exist. In the following we will give the parametrization of Moruzzi *et al.* (1978) which embodies work by Singwi *et al.* (1970), Hedin and Lundqvist (1971), and von Barth and Hedin (1972). However, other parametrizations exist, and are used, for instance, that by Vosko *et al.* (1980), based on quantum Monte Carlo calculations by Ceperley and Alder (1980) and, in addition, the extensions to the relativistic case, which go back to Rajagopal (1980), MacDonald and Vosko (1979), and Ramana and Rajagopal (1983). A particularly important and accurate parametrization is that of Perdew and Wang (1992). It also embodies the quantum Monte Carlo results of Ceperley and Alder and, since it is of great use in gradient corrections to the local-density approximation, we give the details in Sec. 2.8 where we discuss gradient corrections and other properties of exchange and correlation in the electron gas.

Our implementation of the LDA here proceeds as follows: First we write

$$\epsilon_x(n_\uparrow, n_\downarrow) = \epsilon_x^P(r_S) + \left(\epsilon_x^F(r_S) - \epsilon_x^P(r_S)\right) f(n_\uparrow, n_\downarrow) \quad (2.163)$$

$$\epsilon_c(n_\uparrow, n_\downarrow) = \epsilon_c^P(r_S) + \left(\epsilon_c^F(r_S) - \epsilon_c^P(r_S)\right) f(n_\uparrow, n_\downarrow) \quad (2.164)$$

where the function f is given by

$$f(n_\uparrow, n_\downarrow) = \frac{1}{2^{4/3} - 2} \left[\left(\frac{2 n_\uparrow}{n}\right)^{4/3} + \left(\frac{2 n_\downarrow}{n}\right)^{4/3} - 2\right]. \quad (2.165)$$

It interpolates between the fully polarized (F) and nonpolarized (P) cases and hence vanishes when the spin polarization is zero, i.e. $n_\uparrow = n_\downarrow = n/2$. The functions describing exchange in Eqn (2.163) are

$$\epsilon_x^P(r_S) = -\frac{\epsilon_x^0}{r_S}$$
$$\epsilon_x^0 = \frac{3}{2\pi} \left(\frac{9\pi}{4}\right)^{1/3} = 0.91633 \, \text{Ry} \quad (2.166)$$

$$\epsilon_x^F(r_S) = 2^{1/3} \epsilon_x^P(r_S) \quad (2.167)$$

which reproduce the Hartee–Fock results. Here we defined a quantity r_S that is connected with the electron density as (see Sec. 2.2)

$$\frac{4\pi}{3} r_S^3 a_0^3 = \frac{1}{n} \quad (2.168)$$

where a_0 is the Bohr radius. Correlation is described by the functions

$$\epsilon_c^P(r_S) = -c_P \, F\left(\frac{r_S}{r_P}\right) \quad (2.169)$$

$$\epsilon_c^F(r_S) = -c_F \, F\left(\frac{r_S}{r_F}\right) \quad (2.170)$$

where we have defined a function

$$F(z) = (1+z^3) \ln\left(1+\frac{1}{z}\right) + \frac{z}{2} - z^2 - \frac{1}{3}. \tag{2.171}$$

The correlation energy is further modeled by the set of parameters

$$\begin{aligned} c_P &= 0.0504, r_P = 30 \\ c_F &= 0.0254, r_F = 75 \end{aligned} \tag{2.172}$$

as given by von Barth and Hedin (1972) or

$$\begin{aligned} c_P &= 0.045, r_P = 21 \\ c_F &= \frac{c_P}{2} = 0.0225, r_F = 2^{4/3}\, r_P = 52.91668 \end{aligned} \tag{2.173}$$

as given by Moruzzi *et al.* (1978). (The connection between c_F and c_P as well as r_F and r_P given in Eqn (2.173) refers to the RPA scaling, see e.g. Pines (1964).)

From the above equations the exchange-correlation potential can be calculated by means of Eqn (2.160):

$$\begin{aligned} v_\alpha^{xc}(n_\uparrow, n_\downarrow) &= \left[\frac{4}{3}\epsilon_x^P(r_S) + \gamma \cdot \left(\epsilon_c^F(r_S) - \epsilon_c^P(r_S)\right)\right]\left(\frac{2n_\alpha}{n}\right)^{1/3} \\ &+ \mu_c^P(r_S) - \gamma \cdot \left(\epsilon_c^F(r_S) - \epsilon_c^P(r_S)\right) \\ &+ \left[\mu_c^F(r_S) - \mu_c^P(r_S) - \frac{4}{3}\left(\epsilon_c^F(r_S) - \epsilon_c^P(r_S)\right)\right] f(n_\uparrow, n_\downarrow) \end{aligned} \tag{2.174}$$

with

$$\mu_c^P(r_S) = -c_P \ln\left(1 + \frac{r_P}{r_S}\right) \tag{2.175}$$

$$\mu_c^F(r_S) = -c_F \ln\left(1 + \frac{r_F}{r_S}\right) \tag{2.176}$$

and

$$\gamma = \frac{4}{3}\frac{1}{2^{1/3}-1}. \tag{2.177}$$

This completes the specification of the effective potential. The local-density approximation is exact in the limit of a constant density or diagonal and constant-density matrix, respectively. In addition, we would expect it to be quite good for slowly varying densities but not when dealing with realistic systems such as atoms or solids where the density depends strongly on the position. Surprisingly, this is not the case. Instead, experience shows that this approximation does work well in a great number of realistic cases.

Despite its success, there have been numerous attempts to improve the LDA. The earliest attempts came from Kohn and Sham themselves. Like the von Weizsäcker correction to the Thomas–Fermi theory (March, 1983), they proposed to add a gradient correction term to the exchange-correlation energy. However, attempts along these lines were not successful initially, and in many cases the good agreement that is obtained by using the simple approximation above was spoiled.

We will return to improvements of the local spin-density-functional approximation in Sec. 2.8 where we describe some important formal properties of the spin-density functionals and discuss in great detail a generalized gradient approximation that enjoys increasing popularity.

The question remains as to why the LDA does as well as it does. Examples will be found in Chap. 4. This question has received a great deal of attention, especially in connection with investigations of the spatial form of the exchange-correlation potential and the exchange-correlation hole (see Sec. 2.8). Examples include the work of Gunnarsson and Lundqvist (1976), Kohn and Vashishta (1983), Callaway and March (1984) and, most recently, Perdew and Kurth (1998). The theoretical investigation into density-functional theory and refined approximations is still ongoing.

2.7 Nonuniformly magnetized systems

We return now to solutions of the effective single-particle equations (2.152) when it is not sufficient to assume the density matrix is diagonal, i.e. we deal with

$$\sum_{\beta} \left(-\delta_{\alpha\beta} \nabla^2 + v^{\text{eff}}_{\alpha\beta}(\mathbf{r}) - \varepsilon_i \delta_{\alpha\beta}\right) \psi_{i\beta} = 0 \quad \alpha = 1, 2 \tag{2.178}$$

where

$$v^{\text{eff}}_{\alpha\beta}(\mathbf{r}) = v^{\text{ext}}_{\alpha\beta}(\mathbf{r}) + 2\delta_{\alpha\beta} \int \frac{n(\mathbf{r}')}{|\mathbf{r}-\mathbf{r}'|} d\mathbf{r}' + v^{xc}_{\alpha\beta}(\mathbf{r}) \tag{2.179}$$

and

$$v^{xc}_{\alpha\beta}(\mathbf{r}) = \frac{\delta E_{xc}[\tilde{n}]}{\delta n_{\beta\alpha}(\mathbf{r})}, \tag{2.180}$$

with

$$n_{\beta\alpha}(\mathbf{r}) = \sum_{i=1}^{N} \psi_{i\beta}(\mathbf{r}) \psi^*_{i\alpha}(\mathbf{r}), \tag{2.181}$$

and

$$n(\mathbf{r}) = \operatorname{Tr} \tilde{n}(\mathbf{r}). \tag{2.182}$$

The sum in Eqn (2.181) extends over the lowest occupied states.
Here we minimized the total energy which has the form

$$E[\tilde{n}] = T_0[\tilde{n}] + \sum_{\alpha\beta} \int v^{\text{ext}}_{\alpha\beta}(\mathbf{r}) n_{\beta\alpha}(\mathbf{r}) \, d\mathbf{r}$$
$$+ \iint n(\mathbf{r}) \frac{1}{|\mathbf{r}-\mathbf{r}'|} n(\mathbf{r}') \, d\mathbf{r} \, d\mathbf{r}' + E_{xc}[\tilde{n}]. \tag{2.183}$$

Note that we simplified the notation by absorbing T_{xc} into E_{xc}.

Since we know an approximation to E_{xc} only for the case when the density matrix is diagonal, we assume we can *diagonalize* the density matrix *locally*, i.e. we make the unitary transformation

$$\sum_{\alpha\beta} U_{i\alpha}(\mathbf{r}) n_{\alpha\beta}(\mathbf{r}) U^{+}_{\beta j}(\mathbf{r}) = \delta_{ij} \, n_i(\mathbf{r}). \tag{2.184}$$

This implies

$$\frac{\delta n_i}{\delta n_{\alpha\beta}} = U_{i\alpha} \, U^{+}_{\beta i}, \tag{2.185}$$

where the Us and ns depend on the position $\mathbf{r}$. (Since this step may not seem totally convincing it is worthwhile to know that the same result is obtained by calculating the partial derivatives directly from the eigenvalues expressed in terms of $n_{\alpha\beta}$.) For the matrix $\mathbf{U}(\mathbf{r})$ we now choose the well-known spin-1/2 rotation matrix (see e.g. Sakurai, 1985)

$$\mathbf{U} = \begin{pmatrix} \exp\left(\tfrac{1}{2}i\varphi\right) \cos\left(\tfrac{1}{2}\theta\right) & \exp\left(-\tfrac{1}{2}i\varphi\right) \sin\left(\tfrac{1}{2}\theta\right) \\ -\exp\left(\tfrac{1}{2}i\varphi\right) \sin\left(\tfrac{1}{2}\theta\right) & \exp\left(-\tfrac{1}{2}i\varphi\right) \cos\left(\tfrac{1}{2}\theta\right) \end{pmatrix}, \tag{2.186}$$

where φ and θ are polar angles that depend on the position $\mathbf{r}$. A more compact form which brings out the successive rotations is

$$\mathbf{U} = \exp\left(\frac{1}{2} i\theta\, \sigma_y\right) \exp\left(\frac{1}{2} i\varphi\, \sigma_z\right), \tag{2.187}$$

where the σ's are the Pauli spin matrices. Another useful way to write Eqn (2.186) is

$$\mathbf{U} = \begin{pmatrix} \cos\left(\tfrac{1}{2}\theta\right) & \sin\left(\tfrac{1}{2}\theta\right) \\ -\sin\left(\tfrac{1}{2}\theta\right) & \cos\left(\tfrac{1}{2}\theta\right) \end{pmatrix} \begin{pmatrix} \exp\left(\tfrac{1}{2}i\varphi\right) & 0 \\ 0 & \exp\left(-\tfrac{1}{2}i\varphi\right) \end{pmatrix}. \tag{2.188}$$

Of course, the matrix elements of $\mathbf{U}$ can be obtained explicitly and a straightforward calculation gives

$$\tan\varphi = -\frac{\mathrm{Im}\,(n_{12})}{\mathrm{Re}\,(n_{12})}, \tag{2.189}$$

$$\tan\theta = 2\sqrt{(\mathrm{Re}\,(n_{12}))^2 + (\mathrm{Im}\,(n_{12}))^2}/(n_{11} - n_{22}). \tag{2.190}$$

We now use Eqn (2.185) to express the functional derivatives in Eqn (2.180) using the eigenvalues, n_i, of the density matrix, i.e.

$$v_{\alpha\beta}^{xc} = \frac{\delta E_{xc}}{\delta n_{\beta\alpha}} = \sum_{i=1}^{2} \frac{\delta E_{xc}}{\delta n_i} \frac{\delta n_i}{\delta n_{\beta\alpha}} = \sum_{i=1}^{2} \frac{\delta E_{xc}}{\delta n_i} U_{i\beta}\, U_{\alpha i}^{+}. \tag{2.191}$$

The exchange-correlation potential can be expressed in terms of the rotated Pauli spin matrix, i.e.

$$\tilde{\sigma}^z = \mathbf{U}^{+}\,\sigma^z\,\mathbf{U}, \tag{2.192}$$

which we can show in detail as follows. Look at Eqn (2.191) element by element:

$$\begin{aligned} v_{11}^{xc} &= \sum_{i=1}^{2} \frac{\delta E_{xc}}{\delta n_i} U_{i1}\, U_{1i}^{+} = \frac{\delta E_{xc}}{\delta n_1} U_{11}\, U_{11}^{+} + \frac{\delta E_{xc}}{\delta n_2} U_{21}\, U_{12}^{+} \\ &= \frac{\delta E_{xc}}{\delta n_1} \cos^2\frac{\theta}{2} + \frac{\delta E_{xc}}{\delta n_2} \sin^2\frac{\theta}{2}. \end{aligned} \tag{2.193}$$

With

$$\cos^2\frac{\theta}{2} = \frac{1}{2}(\cos\theta + 1) \quad \text{and} \quad \sin^2\frac{\theta}{2} = -\frac{1}{2}(\cos\theta - 1),$$

this becomes

$$v_{11}^{xc} = \frac{1}{2}\sum_{i=1}^{2} \frac{\delta E_{xc}}{\delta n_i} + \frac{1}{2}\left(\frac{\delta E_{xc}}{\delta n_1} - \frac{\delta E_{xc}}{\delta n_2}\right)\cos\theta. \tag{2.194}$$

Next

$$v_{12}^{xc} = \sum_{i=1}^{2} \frac{\delta E_{xc}}{\delta n_1} U_{i2} U_{1i}^+$$

$$= \frac{\delta E_{xc}}{\delta n_2} U_{12} U_{11}^+ + \frac{\delta E_{xc}}{\delta n_1} U_{22} U_{12}^+ \qquad (2.195)$$

$$= \frac{\delta E_{xc}}{\delta n_1} e^{-i\varphi} \cos\frac{\theta}{2} \sin\frac{\theta}{2} - \frac{\delta E_{xc}}{\delta n_2} e^{-i\varphi} \cos\frac{\theta}{2} \sin\frac{\theta}{2}.$$

With

$$\cos\frac{\theta}{2} \sin\frac{\theta}{2} = \frac{1}{2} \sin\theta,$$

this is

$$v_{12}^{xc} = \frac{1}{2} \left(\frac{\delta E_{xc}}{\delta n_1} - \frac{\delta E_{xc}}{\delta n_2} \right) e^{-i\varphi} \sin\theta. \qquad (2.196)$$

Obviously

$$v_{21}^{xc} = \frac{1}{2} \left(\frac{\delta E_{xc}}{\delta n_1} - \frac{\delta E_{xc}}{\delta n_2} \right) e^{i\varphi} \sin\theta, \qquad (2.197)$$

and

$$v_{22}^{xc} = \sum_i \frac{\delta E_{xc}}{\delta n_i} U_{i2} U_{2i}^+ = \frac{\delta E_{xc}}{\delta n_1} U_{12} U_{21}^+ + \frac{\delta E_{xc}}{\delta n_2} U_{22} U_{22}^+$$

$$= \frac{\delta E_{xc}}{\delta n_1} \sin^2\frac{\theta}{2} + \frac{\delta E_{xc}}{\delta n_2} \cos^2\frac{\theta}{2} \qquad (2.198)$$

$$= \frac{1}{2} \sum_{i=1}^{2} \frac{\delta E_{xc}}{\delta n_i} - \frac{1}{2} \left(\frac{\delta E_{xc}}{\delta n_1} - \frac{\delta E_{xc}}{\delta n_2} \right) \cdot \cos\theta.$$

After this exercise we see that we can write

$$\mathbf{v}^{\text{eff}} = v_0 \mathbf{1} + \Delta v \, \tilde{\sigma}^z, \qquad (2.199)$$

where, assuming the external potential, v^{ext}, is diagonal

$$v_0 = v^{\text{ext}} + 2 \int \frac{1}{|\mathbf{r}-\mathbf{r}'|} n(\mathbf{r}')\,\mathrm{d}\mathbf{r}' + \frac{1}{2}\sum_{i=1}^{2} \frac{\delta E_{xc}}{\delta n_i}, \qquad (2.200)$$

$$\Delta v = \frac{1}{2}\left(\frac{\delta E_{xc}}{\delta n_1} - \frac{\delta E_{xc}}{\delta n_2}\right), \qquad (2.201)$$

and

$$\tilde{\sigma}^z = \begin{pmatrix} \cos\theta & e^{-i\varphi}\sin\theta \\ e^{i\varphi}\sin\theta & -\cos\theta \end{pmatrix} = \mathbf{U}^+\sigma_z\,\mathbf{U}. \qquad (2.202)$$

We thus must find a solution of the spinor equation

$$\left(-\mathbf{1}\nabla^2 + \mathbf{1}\,v_0(\mathbf{r}) + \Delta v(\mathbf{r})\,\mathbf{U}^+(\mathbf{r})\,\sigma_z\,\mathbf{U}(\mathbf{r})\right)\psi = \epsilon\,\psi, \qquad (2.203)$$

which is not difficult if we are permitted to make approximations that will be discussed now. To motivate this step we look at Fig. 2.2 in which is shown the computed spin density of ferromagnetic FeCo. This is, of course, a case where the density matrix is diagonal and the spin density is simply the difference of the matrix elements on the diagonal. It should be noticed that the spin density is quite flat between the atoms and that the portions of the nonvanishing spin density can easily be enclosed by spheres whose size may be chosen such that they cover all of space. In the case of FeCo the spin *direction* in each sphere is the same, i.e. at each point within a given sphere the local direction of magnetization is the same and so is the direction of magnetization of each sphere. But now we drop the condition that the direction of magnetization is the same for each sphere. To be more precise, we introduce a local coordinate system for each sphere. This way we replace a fine-grained mesh (the points in space, $\mathbf{r}$) by a coarse-grained mesh (the atomic spheres ν) and consequently use θ^ν and φ^ν instead of $\theta(\mathbf{r})$ and $\varphi(\mathbf{r})$ labeling with ν the type of atomic sphere. The elements of the density matrix we consequently average over each sphere, S_ν, i.e. we calculate

$$n^\nu_{\alpha\beta} = \int_{S_\nu} n_{\alpha\beta}(\mathbf{r})\,\mathrm{d}\mathbf{r} \qquad (2.204)$$

and diagonalize the matrix $n^\nu_{\alpha\beta}$ by means of

$$\sum_{\alpha\beta} U^\nu_{i\alpha}\,n^\nu_{\alpha\beta}\,U^{\nu+}_{\beta j} = n^\nu_i\,\delta_{ij}. \qquad (2.205)$$

We thus speak of a local frame of reference in which the (averaged) density matrix is diagonal. The angles θ^ν and φ^ν give the orientation of the local (or atomic) frame

68 *Density-Functional Theory*

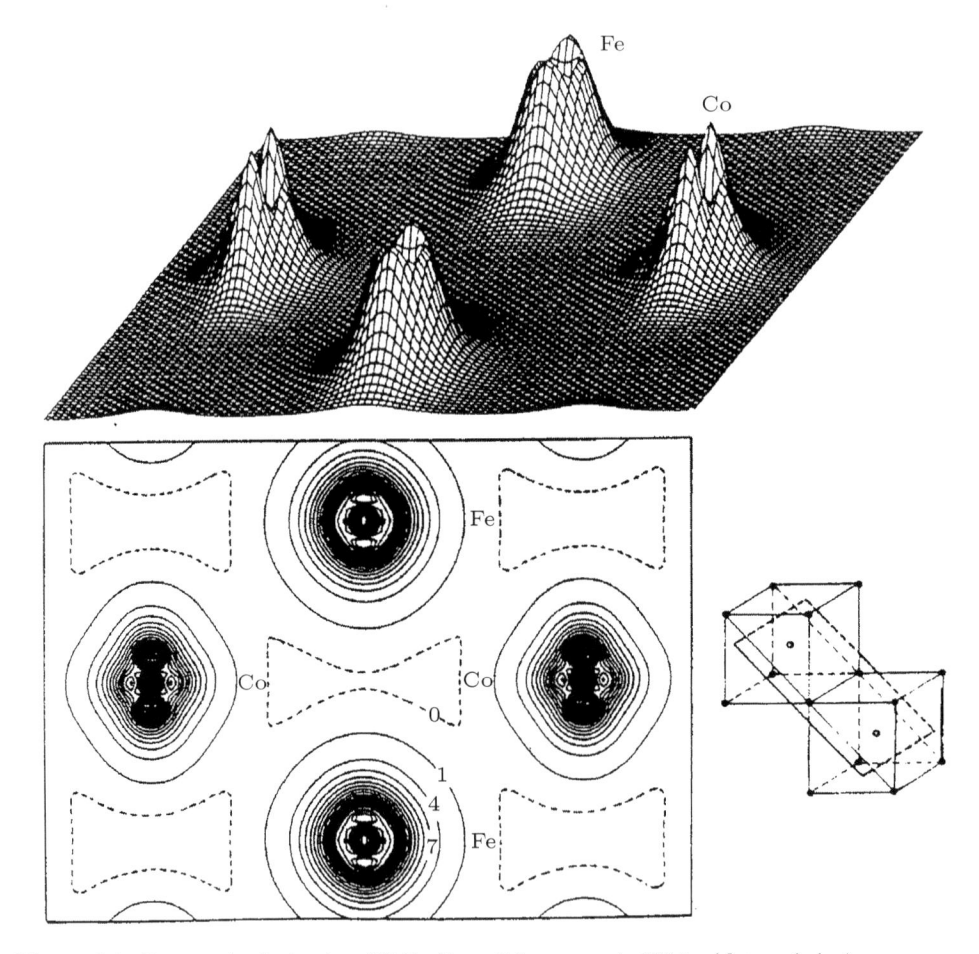

Figure 2.2 *Computed spin density of FeCo (from Schwarz et al., 1984, with permission).*

of reference (or coordinate system) with respect to some global frame of reference (coordinate system), which we may imagine is attached to the crystal lattice and in which the (averaged) density matrix is not in general diagonal.

We continue with the solution of Eqn (2.203) and complete the discussion here by turning to the total energy and expressing it entirely in terms of the diagonalized density matrix, which is an easy exercise that is carried out following Eqn (2.210).

In the local spin-density-functional approximation the result is

$$E = \sum_{i=1}^{N} \varepsilon_i - \iint \frac{n(\mathbf{r})\, n(\mathbf{r}')}{|\mathbf{r} - \mathbf{r}'|} \, d\mathbf{r}\, d\mathbf{r}' - \sum_{\alpha=1}^{2} \int n(\mathbf{r}) \frac{\partial \epsilon_{xc}}{\partial n_\alpha} n_\alpha(\mathbf{r}) \, d\mathbf{r}, \qquad (2.206)$$

where for **r** in the atomic sphere ν, $n_\alpha(\mathbf{r})$ denotes the diagonal element of $\mathbf{U}^\nu \tilde{n}(\mathbf{r})\, \mathbf{U}^{\nu^+}$ and **U** is the matrix defined by Eqn (2.205); the spin density in the ν-th atomic sphere is

$$m(\mathbf{r}) = n_1(\mathbf{r}) - n_2(\mathbf{r}), \tag{2.207}$$

giving a magnetic moment of magnitude m^ν for the ν-th atomic sphere,

$$m^\nu = n_1^\nu - n_2^\nu \qquad (\text{in } \mu_B), \tag{2.208}$$

and the charge density in the ν-th sphere is

$$n(\mathbf{r}) = n_1(\mathbf{r}) + n_2(\mathbf{r}). \tag{2.209}$$

The details of the derivation of Eqn (2.206) are: starting with (see Eqn (2.183))

$$E = T_0 + \sum_{\alpha\beta}\int v^{\text{ext}}_{\alpha\beta} n_{\beta\alpha}\, d\mathbf{r} + \iint \frac{n(\mathbf{r})\, n(\mathbf{r}')}{|\mathbf{r}-\mathbf{r}'|}\, d\mathbf{r}\, d\mathbf{r}' + \int n\, \epsilon_{xc}\, d\mathbf{r}, \tag{2.210}$$

we express the kinetic energy T_0 using

$$\sum_\beta (-\delta_{\alpha\beta} \nabla^2 + v^{\text{ext}}_{\alpha\beta} - \varepsilon_i\, \delta_{\alpha\beta})\, \psi_{i\beta} = 0$$

and assuming the ψ's are normalized, then

$$T_0 = \sum_{i=1}^{N} \varepsilon_i - \sum_{\alpha\beta} \int v^{\text{eff}}_{\alpha\beta}(\mathbf{r})\, n_{\beta\alpha}(\mathbf{r})\, d\mathbf{r}.$$

With v^{eff} in the local-density approximation, i.e.

$$v^{\text{eff}}_{\alpha\beta} = v^{\text{ext}} \delta_{\alpha\beta} + 2\int \frac{n(\mathbf{r}')}{|\mathbf{r}-\mathbf{r}'|}\, d\mathbf{r}'\, \delta_{\alpha\beta}$$

$$+ \frac{1}{2} \delta_{\alpha\beta} \sum_{i=1}^{2} \frac{\partial}{\partial n_i}\, n\, \epsilon_{xc}(n_1, n_2) + \Delta v\, \tilde{\sigma}^z_{\alpha\beta},$$

where

$$\Delta v(\mathbf{r}) = \frac{1}{2}\left(\frac{\partial}{\partial n_1}\, n\, \epsilon_{xc} - \frac{\partial}{\partial n_2}\, n\, \epsilon_{xc}\right),$$

we write

$$E = \sum_{i=1}^{N} \varepsilon_i - \iint \frac{n(\mathbf{r}) n(\mathbf{r}')}{|\mathbf{r} - \mathbf{r}'|} \, d\mathbf{r} \, d\mathbf{r}'$$

$$-\frac{1}{2} \int n^2 \sum_{i=1}^{2} \frac{\partial \epsilon_{xc}}{\partial n_i} \, d\mathbf{r} - \frac{1}{2} \sum_{\alpha\beta} \int n \left(\frac{\partial \epsilon_{xc}}{\partial n_1} - \frac{\partial \epsilon_{xc}}{\partial n_2} \right) \tilde{\sigma}^z_{\alpha\beta} n_{\beta\alpha} \, d\mathbf{r}.$$
(2.211)

We now use $\tilde{\sigma}^z = \mathbf{U}^+ \sigma_z \mathbf{U}$ and $\mathbf{U} \tilde{n} \mathbf{U}^+ = n\,\mathbf{1}$ and obtain for the last two terms of Eqn (2.211):

$$-\sum_{\alpha=1}^{2} \int n(\mathbf{r}) \frac{\partial \epsilon_{xc}}{\partial n_\alpha} n_\alpha \, d\mathbf{r}$$

which proves Eqn (2.206).

A method of how to solve Eqn (2.203) will be described in Chap. 3.

2.8 The generalized gradient approximation (GGA)

The local-density approximation for spin-polarized systems, the so-called LSD, was described in Sec. 2.6. In this approximation the exchange-correlation energy functional, denoted by $E_{xc}[\tilde{n}]$ in Eqn (2.159), is written in terms of $\epsilon_{xc}(n_\uparrow, n_\downarrow)$, the exchange-correlation energy per particle of an electron gas with uniform spin densities $n_\uparrow, n_\downarrow$ and $n = n_\uparrow + n_\downarrow$. We assume throughout this section that the spin-density matrix is diagonal and rewrite Eqn (2.159) in the form

$$E_{xc}^{\text{LSD}}[n_\uparrow, n_\downarrow] = \int n(\mathbf{r}) \epsilon_{xc}(n_\uparrow(\mathbf{r}), n_\downarrow(\mathbf{r})) \, d\mathbf{r}, \qquad (2.212)$$

where we added the abbreviation LSD to distinguish the exchange-correlation energy functional from other approximations to be discussed in this section. Equation (2.212) is clearly valid when the spin densities vary slowly over space. However, this condition does not really seem appropriate for real atoms, molecules, and solids, temporarily setting aside the surprisingly successful applications of the LSD approximation. Indeed, the next systematic correction in the slowly varying density limit would seem to be a gradient approximation (GA) of the form

$$E_{xc}^{\text{GA}}[n_\uparrow, n_\downarrow] = E_{xc}^{\text{LSD}}[n_\uparrow, n_\downarrow] + \sum_{\sigma,\sigma'} \int C_{xc}^{\sigma,\sigma'}(n_\uparrow, n_\downarrow) \, \boldsymbol{\nabla} n_\sigma \cdot \boldsymbol{\nabla} n_{\sigma'} \, n_\sigma^{-2/3} n_{\sigma'}^{-2/3} \, d\mathbf{r}.$$
(2.213)

But this approximation was found to be less accurate than the LSD because it violates certain conditions to be discussed in this section in an attempt to justify a gradient correction that has been quite successfully applied.

The next step is a generalization of Eqn (2.212); it is called the generalized gradient approximation (GGA) and is written as

$$E_{xc}^{GGA}[n_\uparrow, n_\downarrow] = \int f(n_\uparrow, n_\downarrow, \nabla n_\uparrow, \nabla n_\downarrow) \, d\mathbf{r}. \quad (2.214)$$

While the input quantity $\epsilon_{xc}(n_\uparrow, n_\downarrow)$ to LSD is in principle unique – there is the electron gas in which $n_\uparrow$ and $n_\downarrow$ are constant and for which LSD is exact – there is *no* unique input quantity $f(n_\uparrow, n_\downarrow, \nabla n_\uparrow, \nabla n_\downarrow)$ to construct the GGA. However, thanks to the work of Perdew and his collaborators (1986–1998), Burke (1997), etc. a "conservative philosophy of approximations" is known that allows us to construct a nearly unique GGA possessing all the known correct formal features of LSD. Before we can justify the GGA formulas we need to determine some important formal properties of density functionals that were employed in arriving at the expressions to be given. In the following we make free use of an excellent review article by Perdew and Kurth (1998).

2.8.1 Formal properties of density functionals

We consider for simplicity a nondegenerate N-particle wave function $\Phi(\mathbf{r}_1 \sigma_1, \mathbf{r}_2 \sigma_2, \ldots, \mathbf{r}_N \sigma_N)$ that depends on the electron coordinates, $\mathbf{r}_i$, and spins, σ_i ($i = 1, \ldots, N$). Then the expectation value of the total Hamiltonian, $\langle \Phi | \mathcal{H} | \Phi \rangle$, is most easily written down if one defines the so-called "one-electron *reduced* density matrix,"

$$\rho_1(\mathbf{r}' \sigma, \mathbf{r}\sigma) \\ = N \sum_{\sigma_2, \ldots, \sigma_N} \int d\mathbf{r}_2 \ldots \int d\mathbf{r}_N \, \Phi^*(\mathbf{r}'\sigma, \mathbf{r}_2\sigma_2, \ldots, \mathbf{r}_N\sigma_N) \, \Phi(\mathbf{r}\sigma, \mathbf{r}_2\sigma_2, \ldots, \mathbf{r}_N\sigma_N), \quad (2.215)$$

and the "two electron-reduced density matrix"

$$\rho_2(\mathbf{r}', \mathbf{r}) = N(N-1) \sum_{\sigma_1, \ldots, \sigma_N} \int d\mathbf{r}_3 \ldots \int d\mathbf{r}_N \, |\Phi(\mathbf{r}' \sigma_1, \mathbf{r}\sigma_2, \ldots, \mathbf{r}_N \sigma_N)|^2. \quad (2.216)$$

These two density matrices allow us to write out the expectation value of the kinetic energy, Eqn (2.138), and the Coulomb interaction, Eqn (2.140), in the form

$$\langle T \rangle = -\sum_\sigma \int d\mathbf{r} \, \nabla_\mathbf{r}^2 \rho_1(\mathbf{r}' \sigma, \mathbf{r}\sigma)|_{\mathbf{r}'=\mathbf{r}} \quad (2.217)$$

and

$$\langle U \rangle = \int d\mathbf{r} \int d\mathbf{r}' \frac{\rho_2(\mathbf{r}', \mathbf{r})}{|\mathbf{r}' - \mathbf{r}|}, \qquad (2.218)$$

which are easily verified. The particle density is obviously

$$n(\mathbf{r}) = \sum_\sigma \rho_1(\mathbf{r}\,\sigma, \mathbf{r}\,\sigma) = \sum_\sigma n_\sigma(\mathbf{r}). \qquad (2.219)$$

The positive numbers $\rho_2(\mathbf{r}', \mathbf{r})\, d\mathbf{r}'\, d\mathbf{r}$ are interpreted as the joint probability of finding an electron in the volume element $d\mathbf{r}'$ at $\mathbf{r}'$, *and* an electron in $d\mathbf{r}$ at $\mathbf{r}$. This can be written formally as

$$\rho_2(\mathbf{r}', \mathbf{r}) = n(\mathbf{r})\, n_2(\mathbf{r}, \mathbf{r}'), \qquad (2.220)$$

thereby defining the *conditional probability* $n_2(\mathbf{r}, \mathbf{r}')$, which is the probability of finding an electron in $d\mathbf{r}'$, given that there is one at $\mathbf{r}$. Integrating this equation over $\mathbf{r}'$ and using the definition, Eqn (2.216) we get immediately

$$\int n_2(\mathbf{r}, \mathbf{r}')\, d\mathbf{r}' = N - 1. \qquad (2.221)$$

One next separates out the exchange-correlation part by writing

$$n_2(\mathbf{r}, \mathbf{r}') = n(\mathbf{r}') + n_{xc}(\mathbf{r}, \mathbf{r}'), \qquad (2.222)$$

thus defining $n_{xc}(\mathbf{r}, \mathbf{r}')$, the density at $\mathbf{r}'$ of the *exchange-correlation hole* about an electron at $\mathbf{r}$. This becomes even more transparent if we integrate this equation over $\mathbf{r}'$; using Eqn (2.221), we see that

$$\int n_{xc}(\mathbf{r}, \mathbf{r}')\, d\mathbf{r}' = -1, \qquad (2.223)$$

which says that, if an electron is definitely at $\mathbf{r}$, it is missing from the rest of the system (therefore the use of the term "hole"). The density of the exchange-correlation hole gives a simple and important expression for the exchange-correlation energy functional. To obtain this we rewrite $\langle U \rangle$, Eqn (2.218), by substituting Eqn (2.220) and Eqn (2.222):

$$\langle U \rangle = \iint \frac{n(\mathbf{r})\, n(\mathbf{r}')}{|\mathbf{r} - \mathbf{r}'|}\, d\mathbf{r}\, d\mathbf{r}' + \iint \frac{n(\mathbf{r})\, n_{xc}(\mathbf{r}, \mathbf{r}')}{|\mathbf{r} - \mathbf{r}'|}\, d\mathbf{r}\, d\mathbf{r}'. \qquad (2.224)$$

2.8 The generalized gradient approximation (GGA)

Noting that the first term on the right-hand side is the Hartree energy, we see that the exchange-correlation energy is

$$E_{xc} = \iint \frac{n(\mathbf{r})\, n_{xc}(\mathbf{r}, \mathbf{r}')}{|\mathbf{r} - \mathbf{r}'|}\, d\mathbf{r}\, d\mathbf{r}'. \qquad (2.225)$$

In the next, rather obvious step, we split up $n_{xc}(\mathbf{r}, \mathbf{r}')$ into exchange and correlation parts, i.e.

$$n_{xc}(\mathbf{r}, \mathbf{r}') = n_x(\mathbf{r}, \mathbf{r}') + n_c(\mathbf{r}, \mathbf{r}'), \qquad (2.226)$$

and inquire how the sum rule, Eqn (2.223), carries over to the exchange and correlation densities. Physical insight is gained by switching off the Coulomb interaction by means of a coupling constant over which we integrate to obtain the full interaction again. This is also called the *adiabatic connection* which describes how a hypothetical noninteracting system goes over to the real interacting one. For this purpose we first rewrite Eqn (2.147) as

$$F[\tilde{n}] = \min_{\Phi \to \tilde{n}} \langle \Phi | T + U | \Phi \rangle, \qquad (2.227)$$

where the minimization is over all wave functions that give the spin-density matrix $\tilde{n}$. A coupling constant is now introduced by generalizing this expression to:

$$F_\lambda[\tilde{n}] = \min_{\Phi_\lambda \to \tilde{n}} \langle \Phi_\lambda | T + \lambda U | \Phi_\lambda \rangle. \qquad (2.228)$$

In this minimization the spin-density matrix is held constant at its physical value by making the external potential λ-dependent, i.e. $V[\tilde{n}]$ (see Eqn (2.148)) is replaced by $V_\lambda[\tilde{n}]$ and the coupling constant λ is nonnegative being $\lambda = 1$ for the real system. The exchange-correlation energy as a function of the coupling constant is now defined by

$$E_{xc}^\lambda = F_\lambda - T_0 - \lambda E^H, \qquad (2.229)$$

where

$$E^H = \iint \frac{n(\mathbf{r})\, n(\mathbf{r}')}{|\mathbf{r} - \mathbf{r}'|}\, d\mathbf{r}\, d\mathbf{r}' \qquad (2.230)$$

is the Hartree energy. Now the physical exchange-correlation energy is seen to be given by the coupling-constant integral

$$E_{xc} = \int_0^1 d\lambda\, \frac{dE_{xc}^\lambda}{d\lambda} = \int_0^1 d\lambda\, \left(\langle \Phi_\lambda | U | \Phi_\lambda \rangle - E^H \right). \qquad (2.231)$$

74 Density-Functional Theory

In deriving this equation one should bear in mind that the λ-dependence of the wave function in Eqn (2.228) does not contribute to the first derivative because of the variational principle.

The coupling-constant dependence of the exchange-correlation energy is next used to define the appropriate exchange-correlation hole density $n_{xc}(\mathbf{r}, \mathbf{r}')$ by rewriting Eqn (2.225):

$$E_{xc}^\lambda = \iint \frac{n(\mathbf{r})\, n_{xc}^\lambda(\mathbf{r}, \mathbf{r}')}{|\mathbf{r} - \mathbf{r}'|}\, d\mathbf{r}\, d\mathbf{r}'. \quad (2.232)$$

Using Φ_λ to define $\rho_2^\lambda(\mathbf{r}', \mathbf{r})$, Eqn (2.216), we obtain Eqn (2.222) for the λ-dependent densities, and since in the minimization the spin-density matrix is held constant at its physical value, we obtain

$$\int n_{xc}^\lambda(\mathbf{r}, \mathbf{r}')\, d\mathbf{r}' = -1. \quad (2.233)$$

Since for $\lambda = 0$ the exchange-correlation hole density becomes the exchange density (by definition there is no correlation energy for noninteracting electrons) we conclude from Eqn (2.233) that

$$\int n_x(\mathbf{r}, \mathbf{r}')\, d\mathbf{r}' = -1, \quad (2.234)$$

whence, from Eqn (2.226),

$$\int n_c(\mathbf{r}, \mathbf{r}')\, d\mathbf{r}' = 0. \quad (2.235)$$

These last two equations state that exchange exhausts the exchange-correlation sum rule completely and that the Coulomb repulsion changes the shape of the hole but not its integral.

In the $\lambda = 0$ limit the wave function $\Phi^{\lambda=0}$ is a single Slater determinant and the minimization gives Kohn–Sham equations with single-particle solutions, $\{\phi_{i\sigma}\}$, that determine the one-electron reduced density matrix, Eqn (2.215),

$$\rho_1^{\lambda=0}(\mathbf{r}'\sigma, \mathbf{r}\sigma) = \sum_\sigma \sum_{i=1}^N \phi^*_{i\sigma}(\mathbf{r}')\, \phi_{i\sigma}(\mathbf{r}). \quad (2.236)$$

Comparing with the Hartree–Fock exchange term, i.e. the last term on the right-hand side of Eqn (2.15) (after making appropriate changes to atomic units) and using

$$E_x = \iint \frac{n(\mathbf{r})\, n_x(\mathbf{r}, \mathbf{r}')}{|\mathbf{r} - \mathbf{r}'|}\, d\mathbf{r}\, d\mathbf{r}', \quad (2.237)$$

which is Eqn (2.232) for $\lambda = 0$, we read off with (2.236)

$$n_x(\mathbf{r}, \mathbf{r}') = -\sum_\sigma \frac{|\rho_1^{\lambda=0}(\mathbf{r}\,\sigma, \mathbf{r}'\,\sigma)|^2}{n(\mathbf{r})}. \tag{2.238}$$

An immediate consequence of this is that

$$n_x(\mathbf{r}, \mathbf{r}') \leq 0, \tag{2.239}$$

the exchange-hole density and obviously also the exchange energy, (2.237), are negative. Furthermore, because of Eqn (2.219) and (2.237) we can write

$$E_x = E_x^\uparrow + E_x^\downarrow. \tag{2.240}$$

Equations (2.219) and (2.238) finally determine the "on-top" value of the exchange-hole density as

$$n_x(\mathbf{r}, \mathbf{r}) = -\frac{n_\uparrow^2(\mathbf{r}) + n_\downarrow^2(\mathbf{r})}{n(\mathbf{r})}. \tag{2.241}$$

2.8.2 Scaling relations

Another set of very useful relations that constitute constraints on the density functionals is obtained by scaling uniformly the coordinates or the spins. We begin with the wave function and continue with the density.

Let $\Phi(\mathbf{r}_1\,\sigma_1, \ldots, \mathbf{r}_N\,\sigma_N)$ be a normalized wave function which we might, for instance, assume to be an optimized restricted trial wave function. We define the uniformly scaled wave function using a parameter $\gamma > 0$ by

$$\Phi_\gamma(\mathbf{r}_1\,\sigma_1, \ldots, \mathbf{r}_N\,\sigma_N) = \gamma^{3N/2}\,\Phi(\gamma\mathbf{r}_1\,\sigma_1, \ldots, \gamma\mathbf{r}_N\,\sigma_N). \tag{2.242}$$

It is easy to check that the norms obey

$$\langle \Phi_\gamma | \Phi_\gamma \rangle = \langle \Phi | \Phi \rangle = 1 \tag{2.243}$$

and the density corresponding to the scaled wave function is, from Eqn (2.215) and (2.219), seen to be

$$n_\gamma(\mathbf{r}) = \gamma^3\, n(\gamma\mathbf{r}). \tag{2.244}$$

This equation clearly conserves the electron number:

$$\int d\mathbf{r}\, n_\gamma(\mathbf{r}) = \int d\mathbf{r}\, n(\mathbf{r}) = N. \tag{2.245}$$

Physically, values of $\gamma > 1$ describe densities $n_\gamma(\mathbf{r})$ that are on average higher and more contracted than $n(\mathbf{r})$, while values of γ in the range $0 < \gamma < 1$ describe densities that are on average lower and more expanded.

We are now in the position to describe the coordinate scaling of other physical quantities. Thus the Hartree energy, Eqn (2.230), scales as

$$E^{\mathrm{H}}[n_\gamma] = \int d\mathbf{r} \int d\mathbf{r}' \frac{n_\gamma(\mathbf{r})\, n_\gamma(\mathbf{r}')}{|\mathbf{r}-\mathbf{r}'|} = \gamma \int d\mathbf{r}_1 \int d\mathbf{r}_2 \frac{n(\mathbf{r}_1)\, n(\mathbf{r}_2)}{|\mathbf{r}_1-\mathbf{r}_2|} = \gamma E^{\mathrm{H}}[n], \quad (2.246)$$

where we used Eqn (2.244) and an obvious transformation of the integration variables.

The kinetic energy, T_0, of noninteracting particles, Eqn (2.150), scales as

$$T_0[n_\gamma] = \gamma^2\, T_0[n]. \quad (2.247)$$

In deriving and checking this equation one should remember to scale the single-particle functions, $\phi(\mathbf{r})$, to be consistent with Eqn (2.242) and conserve the single-particle norm, i.e. $\phi_\gamma(\mathbf{r}) = \gamma^{3/2}\, \phi(\gamma \mathbf{r})$. Using this relation again and combining Eqn (2.236) with Eqn (2.238) and (2.244), we obtain from Eqn (2.237) the scaled exchange energy:

$$E_x[n_\gamma] = \gamma\, E_x[n]. \quad (2.248)$$

These scaling relations show that in the high-density limit ($\gamma \to \infty$) the kinetic energy, $T_0[n_\gamma]$, dominates the Hartree and exchange energies, $E^{\mathrm{H}}[n_\gamma]$ and $E_x[n_\gamma]$, whereas in the low-density limit ($\gamma \to 0$) $E^{\mathrm{H}}[n_\gamma]$ and $E_x[n_\gamma]$ dominate $T_0[n_\gamma]$.

Unfortunately, in contrast to $T_0[n]$, $E^{\mathrm{H}}[n]$ and $E_x[n]$, the correlation energy, $E_c[n] = E_{xc}[n] - E_x[n]$, does not possess a simple scaling relation. This is not hard to see: denote the wave function that minimizes Eqn (2.227) and yields the density $n_\gamma(\mathbf{r})$ by Φ_{n_γ}. A little reflection reveals that this is *not* the scaled wave function Φ_γ defined in Eqn (2.242). The latter, although it yields the density $n_\gamma(\mathbf{r})$, minimizes the expectation value of $T + \gamma U$, and it is this expectation value which scales like γ^2 under wave function scaling. This way we can see that

$$E_c[n_\gamma] = \gamma^2\, E_c^{1/\gamma}[n], \quad (2.249)$$

where $E_c^{1/\gamma}[n]$ is the density functional for the correlation energy in a system for which the electron–electron interaction is not U but $\gamma^{-1} U$.

We complete this discussion by giving results that are important for the construction of the GGA but are somewhat too involved to derive here. Such a relation concerns the limit $\gamma \to \infty$ of the correlation energy,

$$\lim_{\gamma \to \infty} E_c[n_\gamma] = \mathrm{const.}, \quad (2.250)$$

where const. is a negative constant. Perdew and Kurth (1998) sketch a proof of this relation that uses the dominating behavior of the kinetic energy as $\gamma \to \infty$ to enable the other terms in the total energy to be amenable to perturbation theory. For finite values of γ Levy and Perdew (1985) gave *scaling inequalities*:

$$E_c[n_\gamma] > \gamma E_c[n] \quad (\gamma > 1), \tag{2.251}$$

$$E_c[n_\gamma] < \gamma E_c[n] \quad (\gamma < 1). \tag{2.252}$$

Furthermore, it should be clear from the variational principle that $E_c[n] \leq 0$.

There is also an important local lower bound due to Lieb and Oxford (1981):

$$E_x[n] \geq E_{xc}[n] \geq 2.273 \, E_x^{\text{LDA}}[n], \tag{2.253}$$

where

$$E_x^{\text{LDA}}[n] = -3 \left(\frac{3}{8\pi} \right)^{1/3} \int d\mathbf{r} \, n(\mathbf{r})^{4/3}, \tag{2.254}$$

is the exchange energy in the local-density approximation, see also Eqn (2.161). It is readily seen that $E_x^{\text{LDA}}[n]$ satisfies the scaling relation (2.248).

Finally, spin-scaling relations are useful if density functionals are known for spin-unpolarized systems that one wants to convert into spin-density functionals. A relation of this type that is of interest can be obtained from Eqn (2.240) which we rewrite as

$$E_x[n^\uparrow, n^\downarrow] = E_x[n^\uparrow, 0] + E_x[0, n^\downarrow]. \tag{2.255}$$

The corresponding density functional, appropriate to a spin-unpolarized system, is

$$E_x[n] = E_x[n/2, n/2] = 2 \, E_x[n/2, 0], \tag{2.256}$$

from which we get

$$E_x[n/2, 0] = \frac{1}{2} E_x[n], \tag{2.257}$$

whence Eqn (2.255) becomes

$$E_x[n^\uparrow, n^\downarrow] = \frac{1}{2} E_x[2n^\uparrow] + \frac{1}{2} E_x[2n^\downarrow]. \tag{2.258}$$

Applying this relation to Eqn (2.254), we at once find the local-density expression for a spin-polarized system:

$$E_x^{\text{LDA}}[n^\uparrow, n^\downarrow] = \int d\mathbf{r} \, n(\mathbf{r}) \, \epsilon_x \left(n(\mathbf{r})^\uparrow, n(\mathbf{r})^\downarrow \right), \tag{2.259}$$

where in agreement with Eqn (2.161)

$$\epsilon_x(n^\uparrow, n^\downarrow) = -3\left(\frac{3}{4\pi}\right)^{1/3} \frac{1}{n}\left((n^\uparrow)^{4/3} + (n^\downarrow)^{4/3}\right). \quad (2.260)$$

Although this is not a new result, the above derivation lends support to Eqn (2.161).

Since two electrons of antiparallel spin repel one another by the Coulomb force, thereby contributing to the correlation energy, there is no simple spin-scaling relation for E_c. We next turn to the correlation energy of the homogeneous electron gas.

2.8.3 The correlation energy of the homogeneous electron gas

For an accurate construction of the GGA a form of the correlation energy is employed that is more precise than the relations given in Sec. 2.6. The parametrization of the correlation energy per electron, $\epsilon_c(r_s)$, of the uniform electron gas of density given by r_s is described as follows (for the connection of r_s with the density see Eqn (2.168)).

An exact analytical expression for $\epsilon_c(r_s)$ is known only in extreme limits. The high density ($r_s \longrightarrow 0$) is known from many-body perturbation theory by Gell-Mann and Brueckner (1957) and by Onsager *et al.* (1966) and is given by

$$\epsilon_c(r_s) = 2\left(c_0 \ln r_s - c_1 + c_2 r_s \ln r_s - c_3 r_s + \cdots\right) \quad (r_s \longrightarrow 0). \quad (2.261)$$

The positive constants c_0 and c_1 are known and are given in Table 2.1. In the low-density ($r_s \longrightarrow \infty$) limit the uniform fluid phase is unstable against the formation of a lattice of localized electrons which is close-packed and is known as the Wigner lattice. Since the energies of these two phases are nearly degenerate as $r_s \longrightarrow \infty$, they have the same kind of dependence upon r_s which is written as (Coldwell-Horsfall and Maradudin, 1960)

$$\epsilon_c(r_s) = -d_0/r_s + d_1/r_s^{3/2} + \cdots \quad (r_s \longrightarrow \infty). \quad (2.262)$$

Estimating the coefficients d_0 and d_1, Perdew and Wang (1992) gave an expression that encompasses both limits above:

$$\epsilon_c(r_s) = -4c_0(1 + \alpha_1 r_s)\ln\left[1 + \frac{1}{2c_0(\beta_1 r_s^{1/2} + \beta_2 r_s + \beta_3 r_s^{3/2} + \beta_4 r_s^2)}\right], \quad (2.263)$$

where

$$\beta_1 = \frac{1}{2c_0}\exp\left(-\frac{c_1}{2c_0}\right), \beta_2 = 2c_0 \beta_1^2. \quad (2.264)$$

The coefficients α_1, β_3, and β_4 are found by fitting to accurate quantum Monte Carlo correlation energies (Ceperley and Alder, 1980) for $r_s = 2, 5, 10, 20, 50$, and 100 and are listed in Table 2.1.

Table 2.1 *Coefficients for $\epsilon_c(r_s)$ defined by Eqn (2.263), for $\epsilon_c(r_s, 1)$, the fully polarized version of $\epsilon_c(r_s)$, and for $-\alpha_c(r_s)$ all parametrized in the form of Eqn (2.263).*

	c_0	c_1	α_1	β_3	β_4
$\epsilon_c(r_s)$	0.031091	0.046644	0.21370	1.6382	0.49294
$\epsilon_c(r_s,1)$	0.015545	0.025599	0.20548	3.3662	0.62517
$-\alpha_c(r_s)$	0.016887	0.035475	0.11125	0.88026	0.49671

Equation (2.263) with the parametrization above provides a representation of the correlation energy of the unpolarized system. For the polarized system we denote the correlation energy by $\epsilon_c(r_s, \zeta)$, where ζ is defined as the relative spin polarization, i.e.

$$\zeta = (n_\uparrow - n_\downarrow)/(n_\uparrow + n_\downarrow). \tag{2.265}$$

For the fully polarized system, for which $n_\uparrow = n$ and $n_\downarrow = 0$, i.e. $\zeta = 1$, Eqn (2.263) with different parameters represents $\epsilon_c(r_s, \zeta = 1)$. These are given in the second row of Table 2.1. A useful interpolation formula for arbitrary relative spin polarization based upon a study of the random-phase approximation is (Vosko *et al.*, 1980):

$$\epsilon_c(r_s, \zeta) = \epsilon_c(r_s) + \alpha_c(r_s)\left(1 - \zeta^4\right) f(\zeta)/f''(0) + [\epsilon_c(r_s, 1) - \epsilon_c(r_s)] f(\zeta) \zeta^4, \tag{2.266}$$

with

$$f(\zeta) = \left[(1+\zeta)^{4/3} + (1-\zeta)^{4/3} - 2\right]/(2^{4/3} - 2). \tag{2.267}$$

The functions $\epsilon_c(r_s, 1)$ and $-\alpha_c(r_s)$ are parametrized as $\epsilon_c(r_s)$ given by Eqn (2.263), the coefficients being collected in Table 2.1. This completes the description of the correlation energy for the homogeneous electron gas.

2.8.4 Linear response: screening in the electron gas

Since we are ultimately interested in inhomogeneous systems we must introduce a local-density variation into the homogeneous electron gas. We do this by perturbing the latter with a small static external potential at the point **r** in space, $\delta w(\mathbf{r})$, which induces a modulation of the density that we write as

$$\delta n(\mathbf{r}) = \int d\mathbf{r}' \, \chi(\mathbf{r} - \mathbf{r}') \, \delta w(\mathbf{r}'), \tag{2.268}$$

where χ is a linear density response function. In contrast to Sec. 4.2.1 where we discuss linear response theory in great detail and where we deal predominantly with the response of an inhomogeneous system, we may here assume a wave-like perturbation and, furthermore, a non-spin-polarized electron gas,

$$\delta w(\mathbf{r}) = \delta w(\mathbf{q}) \exp(i\mathbf{q} \cdot \mathbf{r}). \tag{2.269}$$

Then Eqn (2.268) becomes

$$\delta n(\mathbf{q}) = \chi(\mathbf{q})\, \delta w(\mathbf{q}), \tag{2.270}$$

where, as is easy to see, $\delta n(\mathbf{q}) = \delta n(\mathbf{r}) \exp(-i\mathbf{q} \cdot \mathbf{r})$ and $\chi(\mathbf{q})$ is the Fourier transform of $\chi(\mathbf{x})$, $\mathbf{x} = \mathbf{r} - \mathbf{r}'$:

$$\chi(\mathbf{q}) = \int \chi(\mathbf{x}) \exp(-i\mathbf{q} \cdot \mathbf{x})\, d\mathbf{x}. \tag{2.271}$$

Now a small external perturbation changes the Kohn–Sham effective potential from v^{eff} to $v^{\text{eff}} + \delta v^{\text{eff}}$ and we see from Eqn (2.65) that the change in the effective potential linear in $\delta n(\mathbf{r})$ is

$$\delta v^{\text{eff}}(\mathbf{r}) = \delta w(\mathbf{r}) + \int d\mathbf{r}' \left\{ \frac{2}{|\mathbf{r} - \mathbf{r}'|} + \frac{\delta^2 E_{xc}}{\delta n(\mathbf{r})\, \delta n(\mathbf{r}')} \right\} \delta n(\mathbf{r}'). \tag{2.272}$$

We connect the change in δn resulting from the change in the effective potential with $\delta v^{\text{eff}}(\mathbf{r})$ through

$$\delta n(\mathbf{r}) = \int d\mathbf{r}'\, \chi_0(\mathbf{r} - \mathbf{r}')\, \delta v^{\text{eff}}(\mathbf{r}'), \tag{2.273}$$

where we defined χ_0, the density response function of the noninteracting uniform electron gas. The function χ_0 can be evaluated using first-order perturbation theory. In fact this step is carried out in detail in Sec. 4.2.1 for a general, inhomogeneous system, resulting in an answer given by Eqn (4.50). Thus we can take this formula and reduce it to our needs here. This means we assume plane wave for the wave functions occurring in Eqn (4.50), as given in Eqn (2.29). Applying the Fourier transform, implied by Eqn (2.271), to the resulting $\chi_0(\mathbf{r} - \mathbf{r}')$, and converting the sums to integrals by means of Eqn (1.25), we obtain

$$\chi_0(\mathbf{q}) = \frac{1}{\pi^3} \int \frac{f_\mathbf{k} - f_{\mathbf{k}-\mathbf{q}}}{\varepsilon_\mathbf{k} - \varepsilon_{\mathbf{k}-\mathbf{q}}}\, d\mathbf{k}, \tag{2.274}$$

where $f_\mathbf{k}$ is the Fermi function at $T = 0$, resulting from the unit step functions that occur in Eqn (4.50). The remaining integration in Eqn (2.274) can be carried out easily if the eigenvalues in the denominator are assumed to be those for the electron gas, i.e. $\varepsilon_\mathbf{k} = k^2$, $\varepsilon_{\mathbf{k-q}} = (\mathbf{k}-\mathbf{q})^2$. The result is known as the Lindhard susceptibility (see e.g. Jones and March, 1973) and is

$$\chi_0(\mathbf{q}) = -\frac{k_F}{2\pi^2} F_L\left(\frac{q}{2k_F}\right), \qquad (2.275)$$

where the Lindhard function is given by

$$F_L(x) = \frac{1}{2} + \frac{1-x^2}{4x}\ln\left|\frac{1+x}{1-x}\right|. \qquad (2.276)$$

It equals $1 - x^2/3 - x^4/15$ as $x \to 0$, has the value $1/2$ at $x = 1$, approaches $1/(3x^2) + 1/(15x^4)$ as $x \to \infty$, and dF/dx diverges logarithmically as $x \to 1$.

We next replace the $\mathbf{r}$-dependent infinitesimals in Eqn (2.272) by their Fourier coefficients and read off:

$$\delta v^{\text{eff}}(\mathbf{q}) = \delta\omega(\mathbf{q}) + \left(\frac{8\pi}{q^2} - \frac{2\pi}{k_F^2}\gamma_{xc}(\mathbf{q})\right)\delta n(\mathbf{q}). \qquad (2.277)$$

Here $8\pi/q^2$ is the Fourier transform of the Coulomb potential and the second term in parentheses is the Fourier transform of the exchange-correlation term $\delta^2 E_{xc}/\delta n(\mathbf{r})\delta n(\mathbf{r}')$, which is a function of the difference $(\mathbf{r}-\mathbf{r}')$ since we deal with a homogeneous system. Now using the Fourier component of Eqn (2.273) on the left-hand side of Eqn (2.277) and Eqn (2.270) on the right-hand side and cancelling $\delta n(\mathbf{q})$, we obtain the desired expression of the density response function:

$$\chi(\mathbf{q}) = \frac{\chi_0(\mathbf{q})}{1 - (8\pi/q^2 - 2\pi\gamma_{xc}(\mathbf{q})/k_F^2)\chi_0(\mathbf{q})}. \qquad (2.278)$$

We may use this result to see how the applied perturbation $\delta\omega(\mathbf{q})$ is screened in the electron gas, i.e. we use the concept of the dielectric function writing

$$\delta v^{\text{eff}}(\mathbf{q}) = \frac{\delta\omega(\mathbf{q})}{\epsilon_0(\mathbf{q})} \qquad (2.279)$$

and read off from Eqn (2.277), using Eqn (2.270) and (2.278), an expression for the dielectric function:

$$\epsilon_0(\mathbf{q}) = 1 - \left(\frac{8\pi}{q^2} - \frac{2\pi}{k_F^2}\gamma_{xc}(\mathbf{q})\right)\chi_0(\mathbf{q}). \qquad (2.280)$$

Comparing with Eqn (2.278) we see furthermore that

$$\chi(\mathbf{q}) = \frac{\chi_0(\mathbf{q})}{\epsilon_0(\mathbf{q})}. \qquad (2.281)$$

We emphasize that $\epsilon_0(\mathbf{q})$ is not the standard dielectric function $\epsilon(\mathbf{q})$; the latter describes the response of the electrostatic potential alone to a variation of the external potential, thus

$$\delta w(\mathbf{q}) + \frac{8\pi}{q^2} \delta n(\mathbf{q}) = \frac{\delta w(\mathbf{q})}{\epsilon(\mathbf{q})}, \tag{2.282}$$

where ϵ follows easily as

$$\frac{1}{\epsilon(\mathbf{q})} = 1 + \frac{8\pi}{q^2} \chi(\mathbf{q}). \tag{2.283}$$

The results are particularly simple in the long wavelength limit ($q \to 0$). We will see subsequently that in this limit $\gamma_{xc}(\mathbf{q})$ becomes constant, hence

$$\epsilon_0(\mathbf{q}) = 1 + \frac{4\pi}{q^2} \frac{k_F}{\pi^2} - \frac{2\pi}{k_F^2} \frac{k_F}{\pi^2} \gamma_{xc}(q \to 0) = 1 + \frac{k_s^2}{q^2} - C_{xc}, \tag{2.284}$$

where we collected the effects of exchange and correlation in the constant C_{xc} and $k_s = \sqrt{4k_F/\pi}$ is called the inverse *Thomas–Fermi screening length*. The physics described by k_s becomes apparent when we neglect exchange and correlation, i.e. set $C_{xc} = 0$, and Fourier transform Eqn (2.279) using Eqn (2.284) and the Fourier transform of the Coulomb potential for $\delta w(\mathbf{q})$ to get

$$\delta v^{\text{eff}}(\mathbf{r}) = \frac{1}{\pi^2} \int \frac{\exp(i\mathbf{q} \cdot \mathbf{r})}{q^2 + k_s^2} d\mathbf{q} = \frac{2}{r} \exp(-k_s r). \tag{2.285}$$

This result is known as the *Thomas–Fermi screened Coulomb potential*: the inverse of k_s is thus seen to set the characteristic distance over which an external potential is screened out. We see that the uniform electron gas leaves only a very weak effective potential, Eqn (2.279), if the external potential is slowly varying; furthermore, the response function $\chi(\mathbf{q})$ is weaker than $\chi_0(\mathbf{q})$, Eqn (2.281). But we must emphasize the two rough approximations made: there is first the long wavelength limit, $q \to 0$, in the dielectric function. This approximation can be overcome with some effort and the improvements that result are mainly visible at distances far from the perturbing potential where characteristic oscillations are found (known as Friedel oscillations, see e.g. Kittel, 1966 or Jones and March, 1973). Second, there is the neglect of exchange and correlation, the effects of which are much more subtle.

Turning now to exchange and correlation we point out that Perdew and Kurth (1998) discuss a small q-expansion of the exchange contribution to $\gamma_{xc}(\mathbf{q})$, that is, neglecting correlation, they give

$$\gamma_x(\mathbf{q}) = 1 + \frac{5}{9}\left(\frac{q}{2k_F}\right)^2 + \frac{73}{225}\left(\frac{q}{2k_F}\right)^4 \quad (q \to 0), \tag{2.286}$$

which is obtained from a numerically tabulated function. When correlation is included the dependence of $\gamma_{xc}(\mathbf{q})$ on the density is not only through $(q/2k_F)$ and is so far known only from quantum Monte Carlo studies of the weakly perturbed uniform electron gas (Moroni *et al.* 1995). However, in the local density functional approximation (LDA) one obtains γ_{xc} which is independent of q. In this case

$$\frac{\delta^2 E_{xc}^{\mathrm{LDA}}}{\delta n(\mathbf{r})\delta n(\mathbf{r'})} = \delta(\mathbf{r}-\mathbf{r'})\frac{\delta^2 \left[n\epsilon_{xc}(n)\right]}{\delta n^2}, \qquad (2.287)$$

where $\delta(\mathbf{r}-\mathbf{r'})$ is the Dirac delta function. The Fourier transform is thus trivial and splitting up the exchange-correlation energy into Eqn (2.254), the exchange contribution, and into a correlation term, we readily obtain

$$\gamma_{xc}^{\mathrm{LDA}}(\mathbf{q}) = 1 - \frac{k_F^2}{2\pi}\frac{\partial^2}{\partial n^2}\left[n\epsilon_c(n)\right]. \qquad (2.288)$$

Thus $\gamma_{xc}^{\mathrm{LDA}}(\mathbf{q})$ is independent of $\mathbf{q}$, which must be the $q \to 0$ limit of the exact $\gamma_{xc}(\mathbf{q})$. The quantum Monte Carlo calculations of Moroni *et al.* (1995) for the exact (or nearly exact) $\gamma_{xc}(\mathbf{q})$ show that, indeed, the LDA result is remarkably close for $q \to 0$ and the finite q-behavior is fairly well represented by the terms given in Eqn (2.286) valid approximately for $q \simeq 2k_\mathrm{F}$ and $q \leq 2k_\mathrm{F}$.

We are now ready to discuss the generalized gradient approximation as devised by Perdew *et al.* (1996a,b).

2.8.5 Analytical expression for GGA

Following Perdew *et al.* (1996a,b) we write the exchange-correlation functional in the form

$$E_{xc}^{\mathrm{GGA}}[n_\uparrow, n_\downarrow] = \int n\epsilon_x^{\mathrm{P}}(r_S) F_{xc}(r_S, \zeta, s)\,d\mathbf{r}, \qquad (2.289)$$

where $n = n(\mathbf{r})$ is the electron density, $\epsilon_x^{\mathrm{P}}(r_S)$ the exchange energy per particle of the unpolarized uniform electron gas, see Eqn (2.166), and $F_{xc}(r_S, \zeta, s)$ is called the enhancement factor which depends on the density parameter r_S (Eqn (2.168), in atomic units),

$$r_S = \left(\frac{3}{4\pi n}\right)^{1/3}, \qquad (2.290)$$

the local relative spin polarization $\zeta = \zeta(\mathbf{r})$,

$$\zeta(\mathbf{r}) = \frac{n_\uparrow(\mathbf{r}) - n_\downarrow(\mathbf{r})}{n(\mathbf{r})} \qquad (2.291)$$

and the local inhomogeneity parameter, $s = s(\mathbf{r})$, or reduced density gradient,

$$s(\mathbf{r}) = \frac{|\boldsymbol{\nabla} n|}{2k_\mathrm{F} n} = \frac{3}{2\alpha}|\boldsymbol{\nabla} r_S|, \quad (2.292)$$

where $\alpha = (9\pi/4)^{1/3}$ and the local Fermi radius k_F is $k_\mathrm{F} = \alpha/r_S$ (see Eqn (2.40)). The inhomogeneity parameter s measures how fast and by how much the density varies on the scale of the local Fermi wavelength, $2\pi/k_\mathrm{F}$. Under uniform density scaling, Eqn (2.244), r_S obviously scales as $r_S \to \frac{1}{\gamma} r_S(\gamma \mathbf{r})$, ζ is invariant, and so is $s(\mathbf{r}) \to s_\gamma(\mathbf{r}) = s(\gamma \mathbf{r})$.

The enhancement factor F_{xc} is specified in two parts by decomposing it into an exchange and a correlation contribution,

$$F_{xc}(r_S, \zeta, s) = F_x(\zeta, s) + F_c(r_S, \zeta, s). \quad (2.293)$$

The exchange contribution is obtained by scaling the unpolarized case to the spin-polarized case by using the spin-scaling relation given in Eqn (2.258). This is achieved by first defining

$$F_x(s) = 1 + \kappa - \frac{\kappa}{1 + \mu s^2/\kappa}, \quad (2.294)$$

where $\kappa = 0.804$ and $\mu = 0.21952$. Then $F_x(\zeta, s)$ is obtained by spin-scaling $F_x(s)$:

$$F_x(\zeta, s) = \frac{1}{2}\left[(1+\zeta)^{4/3} F_x\left(s/(1+\zeta)^{1/3}\right) + (1-\zeta)^{4/3} F_x\left(s/(1-\zeta)^{1/3}\right)\right]. \quad (2.295)$$

The details in this step are the following: we observe uniform density scaling and expand the exchange energy, Eqn (2.254), for very small s as

$$E_x[n] = A_x \int d\mathbf{r}\, n^{4/3}\left(1 + \mu s^2 + \cdots\right). \quad (2.296)$$

Because of the uniformity of the electron gas, there can be no term linear in s and terms linear in $\nabla^2 n$ can be recast as s^2-terms by means of integration by parts. Furthermore, using linear response theory, in particular starting with the screened Coulomb interaction, Eqn (2.285), one can obtain the value for μ given above (for more details concerning μ, see however Perdew and Kurth (1998) and references given therein). For small s Eqn (2.294) approaches the factor $(1 + \mu s^2)$ in the integrand of Eqn (2.296). Finally, the functional form of $F_x(s)$, Eqn (2.294), was chosen to obey the Lieb–Oxford bound, Eqn (2.253). This bound is assumed to be an equality and the maximum value of Eqn (2.295), $F_x(\zeta = 1, s \to \infty) = 2^{1/3}(1+\kappa)$, is equated to (2.273) which results in the value of κ given above.

2.8 The generalized gradient approximation (GGA)

For the correlation contribution one defines a function

$$H = 2 c_0 \phi^3 \ln\left\{1 + \frac{\beta}{c_0}\left[\frac{1+At^2}{1+At^2+A^2t^4}\right]t^2\right\}, \quad (2.297)$$

with

$$A = \frac{\beta}{c_0}\left[\exp\left(-\frac{\epsilon_c(r_s,\zeta)}{2c_0\phi^3}\right) - 1\right]^{-1}, \quad (2.298)$$

where

$$\phi = \phi(\zeta) = \frac{(1+\zeta)^{2/3} + (1-\zeta)^{2/3}}{2}, \quad (2.299)$$

$$t = \frac{|\nabla n|}{2\phi k_s n} = \frac{(3\pi^2/16)^{1/3} s}{\sqrt{r_s}\,\phi}. \quad (2.300)$$

The quantity t is another dimensionless density gradient, $k_s = 2\sqrt{k_F/\pi}$ is the Thomas–Fermi screening wave number, $\beta = 0.066725$, and $c_0 = (1-\ln 2)/\pi^2 = 0.031091$ (see Table 2.1). In contrast to the scale-invariant $s(\mathbf{r})$ the dimensionless density gradient t scales as $t(\mathbf{r}) \to \sqrt{\gamma}\,t(\gamma\mathbf{r})$. The quantity $\epsilon_c(r_s,\zeta)$ appearing in Eqn (2.298) is the correlation energy as parametrized in Eqn (2.266).

The expression for the enhancement factor is now given as

$$F_{xc}(r_s,\zeta,s) = F_x(\zeta,s) + \frac{\epsilon_c(r_s,\zeta) + H[r_s,\zeta,t=t(s)]}{\epsilon_x^P(r_s)}. \quad (2.301)$$

Perdew *et al.* (1996a,b) (see also Perdew and Kurth, 1998) justify the functional form of the correlation contribution to $F_{xc}(r_s,\zeta,s)$ and the choice of coefficients as follows.

It is first observed that in the high-density ($r_s \to 0$) limit, the screening length $\sim 1/k_s \sim \sqrt{r_s}$ is the important length scale for the correlation hole; the latter is therefore expanded in terms of t as

$$E_c[n] = \int d\mathbf{r}\, n\left[\epsilon_c(n) + 2\beta(n)\,t^2 \cdots\right]. \quad (2.302)$$

The high-density limit of the function $\beta(n)$ in the integrand was obtained using many-body techniques by Ma and Brueckner (1968) who showed that $\beta = 0.066725$ for $r_s \to 0$. The limit $t \to 0$ of Eqn (2.297) is easily found to be

$$H \to 2\beta\phi^3 t^2 \quad (t \to 0), \quad (2.303)$$

where the spin scaling $\phi^3 t^2 \sim \phi s^2$ is that appropriate for exchange, thus justifying H in this limit. In the opposite limit, $t \to \infty$, which describes rapidly varying densities, one verifies:

$$H \to 2c_0 \phi^3 \ln\left(1 + \frac{\beta}{c_0 A}\right) \to 2c_0 \phi^3 \ln\left(\exp\left[-\frac{\epsilon_c(r_s,\zeta)}{2c_0 \phi^3}\right]\right) \to -\epsilon_c(r_s,\zeta). \quad (2.304)$$

Thus correlation is seen to vanish in Eqn (2.301). To support this limit it is argued that the sum rule, Eqn (2.235), on the correlation-hole density is in the limit $t \to \infty$ only satisfied by $n_c = 0$ resulting in vanishing correlation, see the correlation part of Eqn (2.225).

Finally, uniform density scaling to the high-density limit ($\gamma \to \infty$) is used in which case the correlation energy must scale to a negative constant, Eqn (2.250). To show this we write for Eqn (2.261) in good approximation $\epsilon_c(r_s,\zeta) = \phi^3 (c_0 \ln r_s - c_1)$ and subsequently obtain for the limit $r_s \to 0$ the value of At^2 occurring in Eqn (2.297):

$$At^2 = \xi s^2/\phi^2, \quad (2.305)$$

where s and ϕ are defined in Eqn (2.292) and (2.299), and ξ is obtained as

$$\xi = (3\pi^2/16)^{2/3} \frac{\beta}{c_0} \exp(-c_1/c_0) = 0.72162. \quad (2.306)$$

For $r_s \to 0$ one then finds that the logarithmic singularity cancels out in Eqn (2.301) and one obtains after some algebra

$$E_c[n_\gamma] \to -2c_0 \int d\mathbf{r}\, n\, \phi^3 \ln\left[1 + \frac{1}{\xi s^2/\phi^2 + (\xi s^2/\phi^2)^2}\right] \quad (\gamma \to \infty), \quad (2.307)$$

which is a negative constant as required.

In Fig. 2.3(a) we graph the enhancement factor for a vanishing spin polarization, $\zeta = 0$, as a function of the reduced density gradient, s, in the physical range $0 \le s \le 3$, and in Fig. 2.3(b) we show the difference between the enhancement factors for full spin polarization, $\zeta = 1$, and that for $\zeta = 0$, also as a function of s. Notice that in both break Figs. 2.3(a) and (b) the LSD curves would be horizontal straight lines coinciding with the GGA curves at $s = 0$. The asymptote for $s \to \infty$ is $1 + \kappa = 1.804$ in Fig. 2.3(a) and is $(2^{1/3} - 1)(1 + \kappa) = 0.469$ in Fig. 2.3(b). Notice, furthermore, that an earlier numerical version of GGA, the so-called PW91-GGA (Perdew, 1991), yields enhancement factors that are virtually indistinguishable from those shown in Fig. 2.3(a) and (b) (Perdew et al., 1996a,b). The work on refinements of the GGA is continuing as a recent reference shows: Perdew et al. (2008).

Since, as was said before, the construction of the GGA is not unique, one must in the end compare results of calculations for realistic systems with experimental data, such as the atomization energies of molecules. This was done with great success by Perdew et al. (1996a,b) for a large selection of different molecules. In Chap. 4 we will encounter examples of metallic magnets where the GGA is of importance, too. One should remember, however, that the GGA is not a remedy for all ills. Different cases require different corrections, a list of those being SIC, orbital-polarization corrections, "LDA+U," and

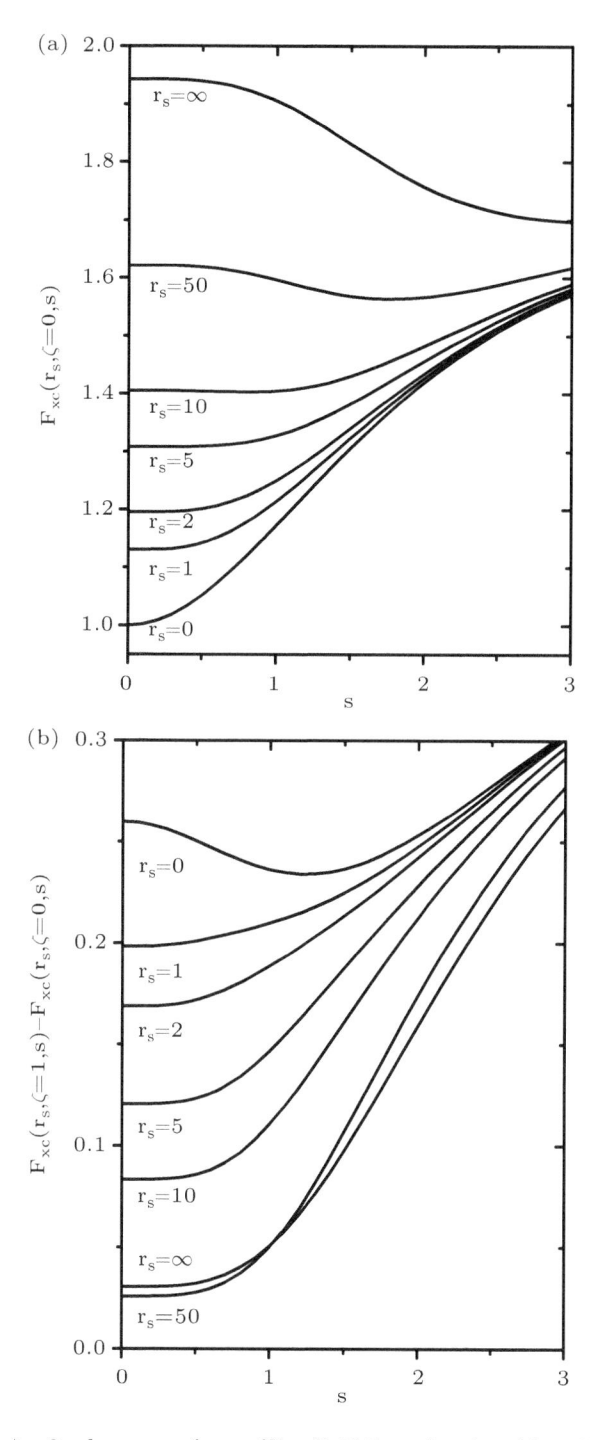

Figure 2.3 *(a) The $\zeta = 0$ enhancement factor of Eqn (2.301) as a function of the reduced density gradient, $s(\mathbf{r}) = \frac{|\nabla n|}{2k_F n}$. The asymptote for $s \longrightarrow \infty$ is $1 + \kappa = 1.804$. (b) Difference of the fully spin-polarized and unpolarized enhancement factors of Eqn (2.301) as a function of s. The asymptote for $s \longrightarrow \infty$ is 0.469.*

exact exchange. SIC, the self-energy correction, is important for localized electron states (Perdew and Zunger, 1981; Temmerman et al., 1998); orbital-polarization correction becomes a necessity mainly in relativistic systems (Sandratskii, 1998). "LDA+U" is of importance whenever correlation effects play a dominant role (Anisimov et al., 1988, 1991), and an exact treatment of exchange is asked for in insulators when the band gap is to be treated correctly (Städele et al., 1997).

A very powerful and recent approach to treat the correlation problem on the basis of realistic band structures is emerging in what is called the *dynamical mean field theory* (DMFT). The literature on DMFT is growing extremely fast, so it must suffice to illustrate the key idea by mentioning but two publications, one by Katsnelson and Lichtenstein (2000) and another by Held et al. (2001).

3
Energy-Band Theory

In this chapter we want to discuss in some detail the methodology for calculating energy bands. Although we focus our attention mainly on linear methods which have become well-known for their transparency and high numerical speed, we will also deal with more traditional methods to supply some historical background and to illustrate the problems we face.

3.1 Bloch's theorem

Let us start by considering a crystal of some symmetry and a single electron moving in an effective potential which we here simply denote by $V(\mathbf{r})$, i.e. we look for solutions of the Schrödinger equation

$$\left(-\frac{\hbar^2}{2m}\nabla^2 + V(\mathbf{r})\right)\psi_i(\mathbf{r}) = \varepsilon_i\,\psi_i(\mathbf{r}) \tag{3.1}$$

where the wave function we seek is $\psi_i(\mathbf{r})$, the energy eigenvalue ε_i, and i is a label for quantum numbers to be specified below. The symmetry of the potential is the same as that of the crystal lattice, thus its most prominent aspect is translational periodicity. It is now over 70 years since Bloch (1929) first discussed the consequences of this symmetry and we now call the appropriate wave function a Bloch function. Its properties are of fundamental importance for solid state physics; we therefore take the time to establish them here.

First notice that if the potential acting on the electron at position $\mathbf{r}$ in the crystal is $V(\mathbf{r})$, then it is unchanged if the position is displaced by any lattice translation vector $\mathbf{R}_j$, i.e.

$$V(\mathbf{r} + \mathbf{R}_j) = V(\mathbf{r}). \tag{3.2}$$

The vectors $\mathbf{R}_j$ are customarily expressed by three independent translation vectors $\mathbf{a}$, $\mathbf{b}$, $\mathbf{c}$,

$$\mathbf{R}_j = \ell_j\,\mathbf{a} + m_j\,\mathbf{b} + n_j\,\mathbf{c}, \tag{3.3}$$

where the quantities ℓ_j, m_j, and n_j are integers. The vectors **a**, **b**, **c** are defined in a standard way by the crystal lattice and are to be found among one of the possible 14 Bravais lattices in three dimensions, see e.g. Kittel (1986). It is convenient to define a set of translation operators T_j having the property

$$T_j f(\mathbf{r}) = f(\mathbf{r} + \mathbf{R}_j), \tag{3.4}$$

where $f(\mathbf{r})$ is any function of position. The operators T_j obviously form a group and they commute with each other and, because of (3.2), with the Hamiltonian

$$H = -\frac{\hbar^2}{2m}\nabla^2 + V(\mathbf{r}), \tag{3.5}$$

thus

$$[T_j, H] = 0. \tag{3.6}$$

As a consequence, the wave functions $\psi_i(\mathbf{r})$ can be chosen to be eigenfunctions of the energy *and* all the translations, i.e. besides Eqn (3.1) we have

$$T_j \psi_i(\mathbf{r}) = \psi_i(\mathbf{r} + \mathbf{R}_j) = \lambda_j \psi_i(\mathbf{r}), \tag{3.7}$$

where λ_j is the eigenvalue that describes the effect of the operator T_j on the function ψ_i. Next we require that the group representation be unitary or, in other words, that the norm of ψ_i is unchanged by translations. Thus λ_j must be a complex number of modulus unity and we write

$$\lambda_j = \exp(i\theta_j), \tag{3.8}$$

where θ_j is real.

Let us now consider two translation operators, T_j and T_ℓ, that act in succession. Obviously

$$T_j T_\ell \psi_i(\mathbf{r}) = \psi_i(\mathbf{r} + \mathbf{R}_j + \mathbf{R}_\ell) = \lambda_j \lambda_\ell \psi_i(\mathbf{r}) = \lambda_{j+\ell} \psi_i(\mathbf{r}), \tag{3.9}$$

and hence from Eqn (3.8)

$$\theta_j + \theta_\ell = \theta_{j+\ell}. \tag{3.10}$$

This is satisfied if

$$\theta_j = \mathbf{k} \cdot \mathbf{R}_j, \tag{3.11}$$

where **k** is an arbitrary vector that is the same for each operation. It characterizes the particular wave function $\psi_i(\mathbf{r})$ and is thus part of the subscript i. Allowing for other quantum numbers, n, besides **k** we may therefore replace $\psi_i(\mathbf{r})$ by $\psi_{n\mathbf{k}}(\mathbf{r})$ and the energy eigenvalue ε_i by $\varepsilon_{n\mathbf{k}}$. The Bloch function now satisfies

$$\psi_{n\mathbf{k}}(\mathbf{r} + \mathbf{R}_j) = e^{i\mathbf{k}\cdot\mathbf{R}_j}\,\psi_{n\mathbf{k}}(\mathbf{r}). \tag{3.12}$$

This is Bloch's theorem.

It is now apparent that the vector **k** is of great importance in describing the Bloch function. In fact, we see from Eqn (3.12) that it has the dimension of an inverse length. It is thus a vector in reciprocal space which is conveniently described using the *reciprocal lattice*. The latter is defined by vectors $\mathbf{K}_s$ obeying

$$\mathbf{K}_s \cdot \mathbf{R}_j = 2\pi\, n_{sj}, \tag{3.13}$$

where the quantities n_{sj} are integers (positive, negative, or zero). Equation (3.13) is to hold for all translations $\mathbf{R}_j$; the end points of all the vectors $\mathbf{K}_s$ define a lattice of points, which is called the reciprocal lattice. Just as Eqn (3.3) expresses the translation $\mathbf{R}_j$ as a linear combination of the basic translations **a, b, c**, one can express the reciprocal lattice vectors as linear combinations of fundamental translation vectors in reciprocal space **A, B, C**, i.e.

$$\mathbf{K}_s = g_{s1}\,\mathbf{A} + g_{s2}\,\mathbf{B} + g_{s3}\,\mathbf{C}, \tag{3.14}$$

where the quantities g_{s1}, g_{s2}, g_{s3} are again integers. Equation (3.13) requires that $\mathbf{A}\cdot\mathbf{a} = 2\pi$, $\mathbf{B}\cdot\mathbf{b} = 2\pi$, $\mathbf{C}\cdot\mathbf{c} = 2\pi$, all other scalar products vanishing. These equations are satisfied if

$$\mathbf{A} = 2\pi\frac{\mathbf{b}\times\mathbf{c}}{\mathbf{a}\cdot(\mathbf{b}\times\mathbf{c})}, \quad \mathbf{B} = 2\pi\frac{\mathbf{c}\times\mathbf{a}}{\mathbf{a}\cdot(\mathbf{b}\times\mathbf{c})}, \quad \mathbf{C} = 2\pi\frac{\mathbf{a}\times\mathbf{b}}{\mathbf{a}\cdot(\mathbf{b}\times\mathbf{c})}. \tag{3.15}$$

We thus have a prescription for constructing the reciprocal lattice from the direct lattice vectors. No doubt the reader recognizes these last equations and is familiar with them from the theory of diffraction by crystals.

An immediate consequence of the above definitions follows for the wave vector **k** labeling the Bloch functions. To see this consider two vectors **k** and **k'** which satisfy

$$\mathbf{k}' = \mathbf{k} + \mathbf{K}_s, \tag{3.16}$$

where $\mathbf{K}_s$ is an arbitrary reciprocal lattice vector. Evidently, because of Eqn (3.13), one has

$$e^{i\mathbf{k}\cdot\mathbf{R}_j} = e^{i\mathbf{k}'\cdot\mathbf{R}_j} \tag{3.17}$$

for all vectors $\mathbf{R}_j$. Consequently, the wave functions $\psi_{n\mathbf{k}}(\mathbf{r})$ and $\psi_{n\mathbf{k}'}(\mathbf{r})$ possess the same translation eigenvalue. The wave vectors $\mathbf{k}$ and $\mathbf{k}'$ satisfying (3.16) are said to be equivalent and we adopt the convention that

$$\psi_{n\mathbf{k}}(\mathbf{r}) = \psi_{n\,\mathbf{k}+\mathbf{K}_s}(\mathbf{r}) \tag{3.18}$$

for any $\mathbf{K}_s$. This convention implies for the eigenvalue $\varepsilon_{n\mathbf{k}}$ of the Schrödinger equation

$$H\,\psi_{n\mathbf{k}}(\mathbf{r}) = \varepsilon_{n\mathbf{k}}\,\psi_{n\mathbf{k}}(\mathbf{r}), \tag{3.19}$$

that

$$\varepsilon_{n\mathbf{k}} = \varepsilon_{n\mathbf{k}+\mathbf{K}_s}, \tag{3.20}$$

and for $\mathbf{k}$ itself that its domain is restricted to some unit cell in reciprocal space. It has become customary to choose a cell that has the full symmetry of the reciprocal lattice; it is called the Brillouin zone and is constructed as follows. Choose as origin an arbitrary lattice point, then draw the vectors connecting the origin with other lattice points. Next the planes that are perpendicular bisectors of these vectors are constructed. The Brillouin zone (BZ) is the smallest volume containing the origin bounded by these planes. Some of the most common BZs are useful for later examples and we therefore illustrate them here.

There are three Bravais lattices in the *cubic system*, the simple cubic (sc), the body centered cubic (bcc), and the face centered cubic (fcc) lattices. The sc-Brillouin zone is just a cube. The other cubic cases are more interesting; the *primitive translation* vectors of a bcc lattice are given by

$$\mathbf{a} = \frac{a}{2}(\hat{\mathbf{x}}+\hat{\mathbf{y}}-\hat{\mathbf{z}}),\ \mathbf{b} = \frac{a}{2}(-\hat{\mathbf{x}}+\hat{\mathbf{y}}+\hat{\mathbf{z}}),\ \mathbf{c} = \frac{a}{2}(\hat{\mathbf{x}}-\hat{\mathbf{y}}+\hat{\mathbf{z}}), \tag{3.21}$$

and those of an fcc lattice by

$$\mathbf{a} = \frac{a}{2}(\hat{\mathbf{x}}+\hat{\mathbf{y}}),\ \mathbf{b} = \frac{a}{2}(\hat{\mathbf{y}}+\hat{\mathbf{z}}),\ \mathbf{c} = \frac{a}{2}(\hat{\mathbf{z}}+\hat{\mathbf{x}}), \tag{3.22}$$

where a is the lattice constant and $\hat{\mathbf{x}}$, $\hat{\mathbf{y}}$, $\hat{\mathbf{z}}$ are unit Cartesian vectors. Using Eqn (3.15) we obtain for the primitive translations in reciprocal space:

$$\mathbf{A} = \frac{2\pi}{a}(\hat{\mathbf{x}}+\hat{\mathbf{y}}),\ \mathbf{B} = \frac{2\pi}{a}(\hat{\mathbf{y}}+\hat{\mathbf{z}}),\ \mathbf{C} = \frac{2\pi}{a}(\hat{\mathbf{x}}+\hat{\mathbf{z}}) \tag{3.23}$$

for the bcc lattice and

$$\mathbf{A} = \frac{2\pi}{a}(\hat{\mathbf{x}}+\hat{\mathbf{y}}-\hat{\mathbf{z}}),\ \mathbf{B} = \frac{2\pi}{a}(-\hat{\mathbf{x}}+\hat{\mathbf{y}}+\hat{\mathbf{z}}),\ \mathbf{C} = \frac{2\pi}{a}(\hat{\mathbf{x}}-\hat{\mathbf{y}}+\hat{\mathbf{z}}) \tag{3.24}$$

for the fcc lattice. We see that the reciprocal lattice of bcc is the fcc lattice and vice versa.

The BZs are now easily drawn and are depicted in Fig. 3.1 for the simple cubic (sc) lattice, in Fig. 3.2 for the fcc lattice, and in Fig. 3.3 for the bcc lattice. For future reference we add the BZ of a hexagonal lattice in Fig. 3.4. Some special **k**-points are labeled following the convention first introduced by Bouckaert *et al.* (1936), see also Bradley and Cracknell (1972). These labels are used in plots of the band structure and in classifications of further symmetries of the electron states. This is so because most interesting crystals possess more symmetry than the translational invariance which gives rise to Bloch functions. The additional symmetry operations, like rotations and

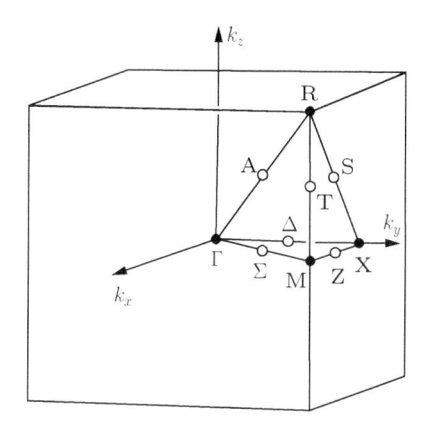

Figure 3.1 *Brillouin zone for the simple cubic (sc) lattice. Cartesian coordinates of the points shown are:* $\Gamma = (0\,0\,0), X = \left(0\,\frac{1}{2}\,0\right), M = \left(\frac{1}{2}\,\frac{1}{2}\,0\right), R = \left(\frac{1}{2}\,\frac{1}{2}\,\frac{1}{2}\right)$ *(in units of $2\pi/a$).*

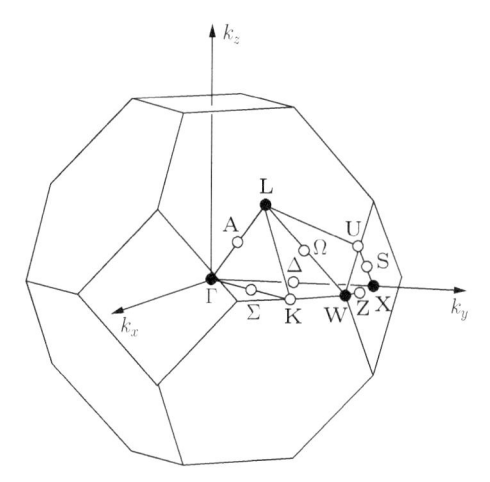

Figure 3.2 *Brillouin zone for the fcc lattice. Cartesian coordinates of the points shown are:* $\Gamma = (0\,0\,0)$, $W = \left(1\,\frac{1}{2}\,0\right), L = \left(\frac{1}{2}\,\frac{1}{2}\,\frac{1}{2}\right), X = (0\,1\,0), K = \left(\frac{3}{4}\,\frac{3}{4}\,0\right)$ *(in units of $2\pi/a$).*

94 *Energy-Band Theory*

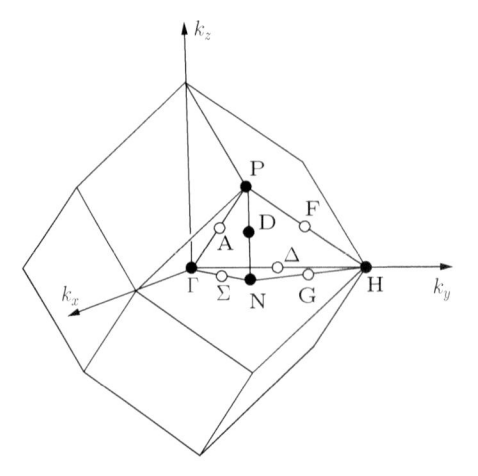

Figure 3.3 *Brillouin zone for the bcc lattice. Cartesian coordinates of the points shown are:* $\Gamma = (0\,0\,0), H = (1\,0\,0), N = \left(\frac{1}{2}\,\frac{1}{2}\,0\right), P = \left(\frac{1}{2}\,\frac{1}{2}\,\frac{1}{2}\right)$ *(in units of $2\pi/a$).*

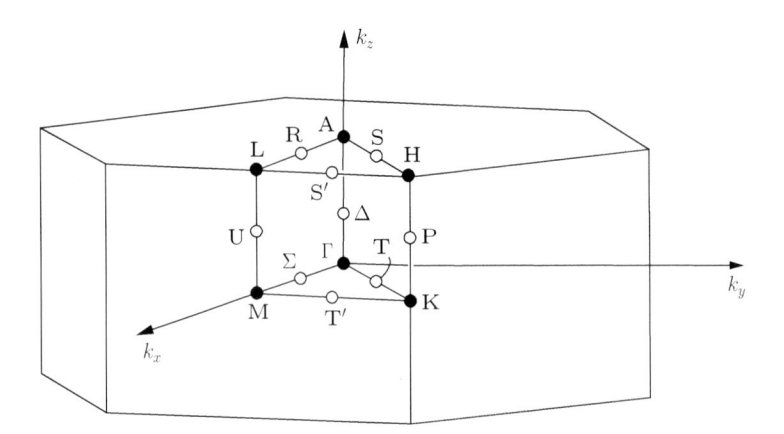

Figure 3.4 *Brillouin zone for the hexagonal lattice. Cartesian coordinates of the points shown are:* $\Gamma = (0\,0\,0), L = \left(\frac{1}{\sqrt{3}}\,0\,\frac{a}{2c}\right), A = \left(0\,0\,\frac{a}{2c}\right), K = \left(\frac{1}{\sqrt{3}}\,\frac{1}{3}\,0\right), M = \left(\frac{1}{\sqrt{3}}\,0\,0\right)$ *(in units of $2\pi/a$).*

reflections, again form groups, the so-called point groups. These, taken together with translations, form the space group of a crystal. The corresponding operators again commute with the Hamiltonian and, therefore, lead to classifications of the electron states which, in general, are different for different **k**-vectors. There is insufficient space to deal with this interesting but specialized subject here. But if later examples require it, we will give the necessary explanations there. The reader interested in more detail is referred to Koster (1957), Callaway (1964), Bradley and Cracknell (1972) or Ludwig and Falter (1988), to name just a few.

We close this section by determining the number of points in the BZ and whence the density of states. This is done by imposing the so-called Born–von Kármán (or periodic) boundary conditions on the Bloch functions by requiring

$$\psi_{n\mathbf{k}}(\mathbf{r}+N_1\,\mathbf{a}) = \psi_{n\mathbf{k}}(\mathbf{r}) = \psi_{n\mathbf{k}}(\mathbf{r}+N_2\,\mathbf{b}) = \psi_{n\mathbf{k}}(\mathbf{r}+N_3\,\mathbf{c}), \quad (3.25)$$

where $N = N_1 \cdot N_2 \cdot N_3$ is the total number of primitive cells in the crystal. These rather artificial boundary conditions are adopted under the assumption that the bulk properties of the solid will not depend on this particular choice, which can therefore be made for reasons of analytical convenience. Equations (3.25) and Bloch's theorem together with Eqn (3.13) now imply

$$\exp{(\mathrm{i}\,2\pi\,N_1\,k_1)} = \exp{(\mathrm{i}\,2\pi\,N_2\,k_2)} = \exp{(\mathrm{i}\,2\pi\,N_3\,k_3)} = 1, \quad (3.26)$$

where k_1, k_2, k_3 are the components of $\mathbf{k}$ expressed as

$$\mathbf{k} = k_1\,\mathbf{A} + k_2\,\mathbf{B} + k_3\,\mathbf{C}, \quad (3.27)$$

and hence the general form of the allowed $\mathbf{k}$-vectors is

$$\mathbf{k} = \frac{m_1}{N_1}\,\mathbf{A} + \frac{m_2}{N_2}\,\mathbf{B} + \frac{m_3}{N_3}\,\mathbf{C}, \quad (3.28)$$

where m_1, m_2, m_3 are integers (positive, negative, or zero). It follows that the volume, $\mathrm{d}\mathbf{k}$, of $\mathbf{k}$-space per allowed value of $\mathbf{k}$ is the volume of the parallelepiped with edges $\mathbf{A}/N_1$, $\mathbf{B}/N_2$, and $\mathbf{C}/N_3$, i.e.

$$\mathrm{d}\mathbf{k} = \frac{\Omega_{\mathrm{BZ}}}{N}, \quad (3.29)$$

where Ω_{BZ} is the volume of the BZ,

$$\Omega_{\mathrm{BZ}} = \mathbf{A}\cdot(\mathbf{B}\times\mathbf{C}). \quad (3.30)$$

It is easy to see with the basic relations (3.15) that

$$\Omega_{\mathrm{BZ}} = (2\pi)^3/\Omega_{\mathrm{UC}}, \quad (3.31)$$

where Ω_{UC} is the volume of the unit cell,

$$\Omega_{\mathrm{UC}} = \mathbf{a}\cdot(\mathbf{b}\times\mathbf{c}). \quad (3.32)$$

Writing for the volume of the crystal,

$$\Omega = N\,\Omega_{\mathrm{UC}}, \quad (3.33)$$

we see from Eqns (3.29) and (3.31) that

$$\mathrm{d}\mathbf{k} = \frac{(2\pi)^3}{\Omega}. \tag{3.34}$$

This relation gives a simple prescription to convert a sum over **k**-vectors to an integral:

$$\sum_{k} \to \frac{\Omega}{(2\pi)^3} \int \mathrm{d}\mathbf{k} = N \frac{\Omega_{\mathrm{UC}}}{(2\pi)^3} \int \mathrm{d}\mathbf{k}. \tag{3.35}$$

Equation (3.35) states that the number of allowed wave vectors in the BZ is equal to the number of primitive cells in the crystal. A further application of Eqn (3.35) is an important formula for the *density of states*, $\mathcal{N}(\varepsilon)$, which is written as

$$\mathcal{N}(\varepsilon) = \frac{1}{\Omega_{\mathrm{BZ}}} \sum_{n} \int_{\mathrm{BZ}} \mathrm{d}\mathbf{k}\, \delta(\varepsilon - \varepsilon_{n\mathbf{k}}), \tag{3.36}$$

where $\delta(x)$ is the Dirac δ-function and the integral is taken over the BZ. For an actual computation of the density of states certain numerical techniques must be used, like for instance the tetrahedron method. Since this is a rather specialized subject, we will not go into any details, but see Coleridge *et al.* (1982) or Skriver (1984), to give but two examples. The relation for $\mathcal{N}(\varepsilon)$ itself, however, is quite plausible because the δ-function just does the counting of states, giving unity for the integral whenever $\varepsilon = \varepsilon_{n\mathbf{k}}$ and zero otherwise, the constant of proportionality ensuring that

$$\int_{\varepsilon_a}^{\varepsilon_b} \mathcal{N}(\varepsilon)\, \mathrm{d}\varepsilon = n_{ab}, \tag{3.37}$$

where n_{ab} is the number of bands in the interval $(\varepsilon_a, \varepsilon_b)$. Using well-known properties of the δ-function, one can easily show that Eqn (3.36) is equivalent to the form given in Eqn (1.44).

3.2 Plane waves, orthogonalized plane waves, and pseudopotentials

We now begin with the description of some important methods of solutions of the band-structure problem embodied in Eqn (3.1), focusing mainly on the underlying physics while leaving aside details which can be found in more specialized texts. To enable the reader to follow the literature in this field and for our own convenience in the following sections we will adopt atomic units, see Sec. 2.2.

The solution of the band-structure problem will throughout be attempted by an expansion of the Bloch function in terms of a suitable set of functions, $\{\chi_j(\mathbf{r})\}$, i.e.

$$\psi_{\mathbf{k}}(\mathbf{r}) = \sum_j a_j(\mathbf{k})\,\chi_j(\mathbf{r}). \tag{3.38}$$

The sum on j is *finite* in practice and the set of functions, $\{\chi_j(\mathbf{r})\}$, is in general not orthonormal; completeness is another problem. The expansion coefficients, $a_j(\mathbf{k})$, are obtained by a variational procedure: one substitutes Eqn (3.38) into (3.1) multiplying from the left with $\psi_{\mathbf{k}}^*(\mathbf{r})$ and integrating; the resulting expression is then varied, i.e. with H defined by Eqn (3.5) we can write

$$\delta \sum_{ij} a_i^*(\mathbf{k})\,a_j(\mathbf{k}) \left\{ \int \chi_i^*(\mathbf{r})\,H\,\chi_j(\mathbf{r})\,\mathrm{d}\mathbf{r} - \varepsilon_{\mathbf{k}} \int \chi_i^*(\mathbf{r})\,\chi_j(\mathbf{r})\,\mathrm{d}\mathbf{r} \right\} = 0 \tag{3.39}$$

which gives

$$\sum_j \left\{ \int \chi_i^*(\mathbf{r})\,H\,\chi_j(\mathbf{r})\,\mathrm{d}\mathbf{r} - \varepsilon_{\mathbf{k}} \int \chi_i^*(\mathbf{r})\,\chi_j(\mathbf{r})\,\mathrm{d}\mathbf{r} \right\} a_j(\mathbf{k}) = 0. \tag{3.40}$$

This equation is called the secular equation. Its detailed solution depends on the type of functions $\chi_j(\mathbf{r})$ chosen (they may be energy-dependent or they may not) but the eigenvalues, $\varepsilon_{\mathbf{k}}$, always follow by equating the determinant of $\{\ldots\}$ to zero and finding the roots.

3.2.1 Plane waves

Returning to the basic expansion, we begin by making a seemingly simple choice and select a Fourier series. In the language of band theory this is called a plane-wave expansion; each plane wave is written as a Bloch function, $\exp\left[\mathrm{i}\,(\mathbf{k} + \mathbf{K}_s) \cdot \mathbf{r}\right]$, where $\mathbf{K}_s$ is a reciprocal lattice vector. It is easy to see that this is indeed a Bloch function by applying a translation and observing Eqn (3.13). Thus a valid expansion is a sum over reciprocal lattice vectors and has the form

$$\psi_{\mathbf{k}}(\mathbf{r}) = \frac{1}{\sqrt{\Omega}} \sum_s a_s(\mathbf{k})\,\mathrm{e}^{\mathrm{i}(\mathbf{k}+\mathbf{K}_s)\cdot\mathbf{r}}, \tag{3.41}$$

where Ω is the volume of the crystal and we have temporarily dropped the band index in our notation. The factor $\Omega^{-1/2}$ ensures that $\psi_{\mathbf{k}}$ is normalized to unity in the volume Ω provided that

98 *Energy-Band Theory*

$$\sum_s |a_s(\mathbf{k})|^2 = 1. \tag{3.42}$$

To obtain the expansion coefficients $a_s(\mathbf{k})$ and the eigenvalues $\varepsilon_\mathbf{k}$ we set up the matrix of the Hamiltonian on the plane-wave basis. The resulting secular equation can be written as

$$\sum_s \{[(\mathbf{k}+\mathbf{K}_s)^2 - \varepsilon_\mathbf{k}]\delta_{st} + V(\mathbf{K}_t - \mathbf{K}_s)\} a_s(\mathbf{k}) = 0. \tag{3.43}$$

The solutions for the eigenvalues, $\varepsilon_\mathbf{k}$, and the eigenvectors, $a_s(\mathbf{k})$, thus constitute a standard linear algebra problem and are obtained by diagonalizing the matrix $\{[(\mathbf{k}+\mathbf{K}_s)^2 - \varepsilon_\mathbf{k}]\delta_{st} + V(\mathbf{K}_t - \mathbf{K}_s)\}$. The quantity $V(\mathbf{K})$ is the Fourier coefficient of the effective crystal potential. It is obtained by expressing the latter as a sum of identical terms, $V_c(\mathbf{r})$, centered on each unit cell of the crystal,

$$V(\mathbf{r}) = \sum_\nu V_c(\mathbf{r}-\mathbf{R}_\nu). \tag{3.44}$$

A straightforward calculation gives

$$V(\mathbf{K}) = \Omega_{\mathrm{UC}}^{-1} \int_{\Omega_{\mathrm{UC}}} e^{i\mathbf{K}\cdot\mathbf{r}} V_c(\mathbf{r})\, d\mathbf{r}, \tag{3.45}$$

where again use was made of Eqn (3.13).

Physically we can view the potential $V_c(\mathbf{r})$ as something very closely related to an atomic potential, or a sum of atomic potentials if the unit cell contains more than one kind of atom. Thinking of a valence electron moving through the crystal it seems clear that the potential cannot be weak in the intra-atomic regions although it may well be so in the region between the atoms. In fact for a solid made up of atoms with atomic number Z (for Al, e.g., $Z = 13$) and valence Z_V ($Z_V = 3$ for Al) the leading contribution to $V(\mathbf{r})$ very close to the nucleus is the Coulomb potential, $-Ze^2/r$, changing to $-Z_V e^2/r$ outside the core region because of electrostatic screening by the core electrons, and further out, because of periodicity, becoming rather flat between two atoms. As a consequence of this a great number of large Fourier coefficients is obtained from Eqn (3.45) that must be included in the secular Eqn (3.43), of very small wavelengths (large $|\mathbf{K}_t - \mathbf{K}_s|$) near the nuclei and large wavelengths in the interstitial region. In a lucid description of the problem Heine (1970) has indeed estimated the rank of the secular equation (3.43) to be about $10^6 \times 10^6$.

In view of this it is a paradoxical fact that experimentally the electronic properties of at least the simple metals (Na, K, Rb, Ca, Al, etc.) seem to be well described by the free-electron model if one is willing to accept a small change in the apparent electron mass (see e.g. Ashcroft and Mermin, 1976). Following the experimental facts let us, therefore, accept free-electron behavior and set $V(\mathbf{K}) = 0$ for all reciprocal lattice vectors $\mathbf{K}$, but assume the latter are still defined. This gives a rather artificial solid where there is no potential, but still a lattice; this is called the "empty lattice". The secular equation (3.43) is now trivial; it gives solutions, $\varepsilon_{\mathbf{k}}$, for a given $\mathbf{k}$ for each reciprocal lattice vector $\mathbf{K}_s$. The resulting ε versus $\mathbf{k}$ curves are plotted for an fcc lattice in Fig. 3.5 along some typical directions in the BZ and in Fig. 3.6 for the bcc lattice.

These ε versus $\mathbf{k}$ curves are called "energy bands". Most states are highly degenerate and we indicate some of the degeneracies by the numbers appearing near the "bands".

Now to describe the band structure of the simple metals—at least in the zeroth approximation—we must first ignore all core electrons. The *valence* electrons are then placed in the lowest energy states according to Fermi–Dirac statistics, filling these up to

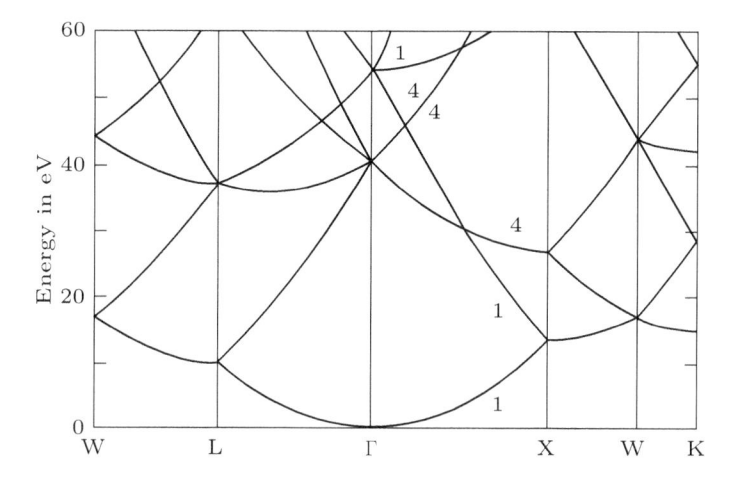

Figure 3.5 *Free-electron dispersion curves for the fcc lattice along certain symmetry lines of the Brillouin zone. The points indicated are the same as in Fig. 3.2.*

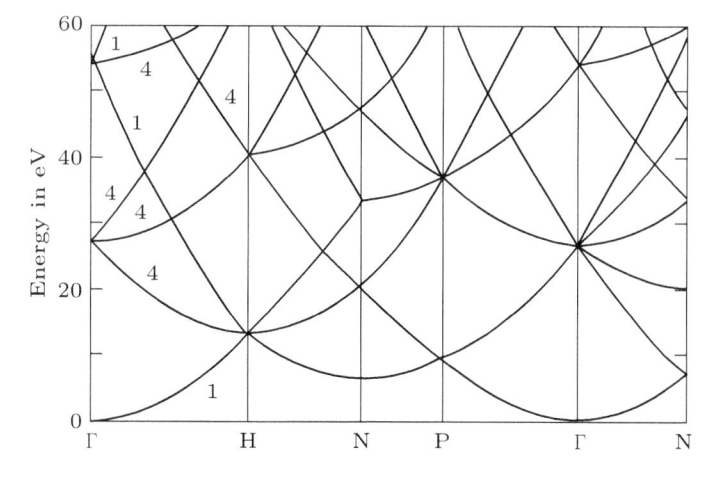

Figure 3.6 *Same as Fig. 3.5 but for the bcc lattice. The points now refer to Fig. 3.3.*

the Fermi energy. For monovalent metals the lowest band in Figs. 3.5 and 3.6 becomes half-filled giving a Fermi energy of about 5 eV for the arbitrary but convenient choice of the lattice constant of $a = 2\pi$ Bohr. This band-filling ratio results from our counting of states following Eqn (3.35) when we include spin degeneracy, i.e. occupy each possible **k**-state twice, once with a spin-up and once with a spin-down electron.

Proceeding one step further we can empirically include a weak crystal potential in Eqn (3.43) and arrive at the nearly free-electron model. Here the crystal potential in Eqn (3.43) is conveniently treated by perturbation theory (Ashcroft and Mermin, 1976, for example) and without doing any calculations we know that in lowest order it will lift the degeneracies of the bands. Since the lowest band is nondegenerate, the effect of the perturbation in the half-filled band case will be strong above the Fermi energy, in Fig. 3.5, for instance, at the **k**-points W, L, X, etc. where the degeneracies appear first. As a result the lowest *occupied* band will be deformed only slightly, giving rise to small changes in the apparent electron mass, as is observed. But how can all this be justified?

To gain an idea about the solutions of the crystal Schrödinger equation we observe that these are so difficult to obtain because one simultaneously copes with a very strong potential near the nuclei and a rather flat and weak one in between them (Heine, 1970). This is the reason why a plane-wave ansatz for a real crystal potential fails. Historically, two methods had been devised to deal with this problem by modifying the plane waves used in the expansion of Eqn (3.41) near the nuclei. One was Slater's (1937) augmented plane wave (APW) method, the other Herring's (1940) orthogonalized plane wave (OPW) method. We will deal with Slater's APW method in some detail in Sec. 3.3 and, mainly for didactic reasons, give a brief description of the OPW method in the following section (see also Callaway 1964, 1974).

3.2.2 OPW method

Using Dirac's notation to denote a plane wave by $|\mathbf{K}\rangle$ and a core state by $|c\rangle$, we write an OPW, $|\chi_\mathbf{K}\rangle$, in the form

$$|\chi_\mathbf{K}\rangle = |\mathbf{K}\rangle - \sum_c |c\rangle \langle c|\mathbf{K}\rangle. \qquad (3.46)$$

The sum extends over the core states and we note as a reminder that the real-space notation is obtained from a scalar product with the bra-vector $\langle \mathbf{r}|$, for instance,

$$\langle \mathbf{r}|\mathbf{K}\rangle = \frac{1}{\sqrt{\Omega}} e^{i\mathbf{K}\cdot\mathbf{r}}. \qquad (3.47)$$

Obviously, because of the orthogonality of different core states, we see that $\langle c|\chi_\mathbf{K}\rangle = 0$, thus $|\chi_\mathbf{K}\rangle$ is orthogonal to the core states. The OPW, Eqn (3.46), consists of a smooth part, $|\mathbf{K}\rangle$ (for not too large $\mathbf{K}$) and a part that possesses small wavelength oscillations over the core region, $|c\rangle \langle c|\mathbf{K}\rangle$; one expects, therefore, that an expansion of the Bloch function of the form

$$|\psi_{\mathbf{k}}\rangle = \sum_s a_s(\mathbf{k}) |\chi_{\mathbf{k}+\mathbf{K}_s}\rangle \qquad (3.48)$$

will converge quite rapidly. This is also borne out by actual calculations. For Eqn (3.48) to be valid as stated there, the OPW, $|\chi_{\mathbf{K}}\rangle$, must be a Bloch function and, strictly speaking, this is not so unless the state $|c\rangle$ is constructed to have Bloch symmetry. This can formally be achieved by a linear combination of core states, i.e. by interpreting the ket $|c\rangle$ as

$$|c\rangle = \frac{1}{\sqrt{N}} \sum_\mu e^{i\mathbf{k}\cdot\mathbf{R}_\mu} |c_\mu\rangle \qquad (3.49)$$

the sum extending over the lattice and $|c_\mu\rangle$ representing the core state c-centered at $\mathbf{R}_\mu$. The **k**-vector can be suppressed in the notation—as we have done—and no problems arise provided the core states centered on different atoms do not overlap, a condition that is generally fulfilled.

Let us look at a half-way realistic example that demonstrates the idea behind an OPW and first write out Eqn (3.46) in real space:

$$\langle\mathbf{r}|\chi_{\mathbf{K}}\rangle = \langle\mathbf{r}|\mathbf{K}\rangle - \sum_c \langle\mathbf{r}|c\rangle \langle c|K\rangle, \qquad (3.50)$$

where

$$\langle\mathbf{r}|\chi_{\mathbf{K}}\rangle = \chi_K(\mathbf{r}) \qquad (3.51)$$

is an OPW and (see Eqn (3.47)):

$$\langle\mathbf{r}|\mathbf{K}\rangle = \frac{1}{\sqrt{\Omega}} e^{i\mathbf{K}\cdot\mathbf{r}}$$

is a PW. Furthermore

$$\langle\mathbf{r}|c\rangle = \frac{1}{\sqrt{\Omega N}} \sum_\mu e^{i\mathbf{K}\cdot\mathbf{R}_\mu} u_c(\mathbf{r}-\mathbf{R}_\mu) \qquad (3.52)$$

is a Bloch-symmetrized core wave function and

$$\langle c|\mathbf{K}\rangle = \frac{1}{\sqrt{\Omega N}} \sum_\mu \int e^{-i\mathbf{K}\cdot\mathbf{R}_\mu} u_c^*(\mathbf{r}-\mathbf{R}_\mu) e^{i\mathbf{K}\cdot\mathbf{r}} \, d\mathbf{r}$$

$$= \frac{1}{\sqrt{\Omega_{\mathrm{UC}}}} \int u_c^*(\mathbf{r}) e^{i\mathbf{K}\cdot\mathbf{r}} \, d\mathbf{r}. \qquad (3.53)$$

This is still quite general. But now assume we want to determine the band structure of Li, which possesses one 1s core state and in the metallic state has the bcc crystal structure. Thus $\Omega_{\text{UC}} = a^3/2$, where a is the lattice constant. The 1s core function may be approximated by

$$u_{1s}(\mathbf{r}) = \left(\frac{\alpha^3}{\pi}\right)^{1/2} e^{-\alpha r}, \tag{3.54}$$

assuming α has been chosen appropriately. Then we determine $\langle c|\mathbf{K}\rangle$ as

$$\langle c|\mathbf{K}\rangle = \sqrt{\frac{2}{a^3}} \left(\frac{\alpha^3}{\pi}\right)^{1/2} \int e^{i\mathbf{K}\cdot\mathbf{r} - \alpha r} \, d\mathbf{r}$$

$$= 8\sqrt{\frac{2\pi\alpha^3}{a^3}} \frac{\alpha}{(\alpha^2 + K^2)^2}. \tag{3.55}$$

Now put the pieces together in Eqn (3.50)

$$\chi_\mathbf{K}(\mathbf{r}) = \frac{1}{\sqrt{\Omega}} e^{i\mathbf{K}\cdot\mathbf{r}} - \frac{1}{\sqrt{N}} \sum_\mu e^{i\mathbf{K}\cdot\mathbf{R}_\mu} u_{1s}(\mathbf{r} - \mathbf{R}_\mu) 8 \sqrt{\frac{2\pi\alpha^3}{a^3}} \frac{\alpha}{(\alpha^2 + K^2)^2}$$

$$= \frac{1}{\sqrt{N}} \left\{ \sqrt{\frac{2}{a^3}} e^{i\mathbf{K}\cdot\mathbf{r}} - 8\sqrt{\frac{2\pi\alpha^3}{a^3}} \left(\frac{\alpha^3}{\pi}\right)^{1/2} \sum_\mu e^{i\mathbf{K}\cdot\mathbf{R}_\mu} e^{-\alpha|\mathbf{r}-\mathbf{R}_\mu|} \frac{\alpha}{(\alpha^2 + K^2)^2} \right\}$$

$$= \frac{1}{\sqrt{N}} \sqrt{\frac{2}{a^3}} \left[e^{i\mathbf{K}\cdot\mathbf{r}} - \frac{8\alpha^4}{(\alpha^2 + K^2)^2} \sum_\mu e^{i\mathbf{K}\cdot\mathbf{R}_\mu} e^{-\alpha|\mathbf{r}-\mathbf{R}_\mu|} \right]$$

i.e.

$$\chi_\mathbf{K}(\mathbf{r}) = \frac{1}{\sqrt{\Omega}} \left[e^{i\mathbf{K}\cdot\mathbf{r}} - \frac{8\alpha^4}{(\alpha^2 + K^2)^2} \sum_\mu e^{i\mathbf{K}\cdot\mathbf{R}_\mu} e^{-\alpha|\mathbf{r}-\mathbf{R}_\mu|} \right]. \tag{3.56}$$

Since it is reasonable to assume that the cores do not overlap we approximate this basis function by

$$\chi_\mathbf{K}(\mathbf{r}) \simeq \frac{1}{\sqrt{\Omega}} \left[e^{i\mathbf{K}\cdot\mathbf{r}} - \frac{8\alpha^4}{(\alpha^2 + K^2)^2} e^{-\alpha r} \right]. \tag{3.57}$$

The Bloch function is now expanded as in Eqn (3.41), i.e.

$$\psi_\mathbf{k}(\mathbf{r}) = \sum_s a_s(\mathbf{k}) \chi_{\mathbf{k}+\mathbf{K}_s}(\mathbf{r}). \tag{3.58}$$

It is worth looking at the first term in the expansion, in particular for $\mathbf{k} = 0$ (the Γ-point). Here we get

$$\chi_0(\mathbf{r}) = \frac{1}{\sqrt{\Omega}}[1 - 8\,e^{-\alpha r}]$$

which is plotted in Fig. 3.7.

We see that χ_0 has one node like a 2s function. This is obviously produced by the orthogonalization to the 1s core and indeed we expect the electron at the bottom of the valence band to be 2s-like. But we also notice that the OPW does *not* behave like an atomic function for $\alpha r \to \infty$, rather it becomes flat between the atoms. (We may assume the Brillouin zone boundary to be at roughly $\alpha r = 3\text{--}6$.) Thus the OPW is qualitatively similar to the lowest-order wave function in the Wigner–Seitz approximation. The latter wave function has zero-slope boundary conditions at the Wigner–Seitz sphere radius. It is elementary to see that the energy of such a state is lower than that having zero value at infinity (the atomic case). This is the central finding of the simplified treatment of metallic binding by Wigner and Seitz (Ashcroft and Mermin, 1976).

Although Li is not an interesting case for magnetism, we see that it supplies an illustrative wave function for the one k-point, $\mathbf{k} = 0$. Its energy, $\varepsilon_{\mathbf{k}} = 0$, is the off-set for the nearly parabolic valence band that is observed experimentally. To calculate the complete band structure we would be required to set up the secular equation, which can be written as

$$\sum_s \{[(\mathbf{k} + \mathbf{K}_s)^2 - \varepsilon_{\mathbf{k}}]\delta_{st} + V^{\mathrm{OPW}}(\mathbf{K}_t - \mathbf{K}_s)\} a_s(\mathbf{k}) = 0, \qquad (3.59)$$

where the term $V^{\mathrm{OPW}}(\mathbf{K}_t - \mathbf{K}_c)$ is given by

$$V^{\mathrm{OPW}}(\mathbf{K}_t - \mathbf{K}_s) = V(\mathbf{K}_t - \mathbf{K}_s) + \sum_c (\varepsilon_{\mathbf{k}} - \varepsilon_c) \langle \mathbf{k} + \mathbf{K}_t | c \rangle \langle c | \mathbf{k} + \mathbf{K}_s \rangle. \qquad (3.60)$$

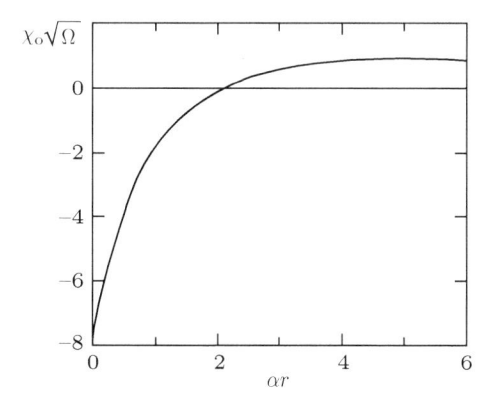

Figure 3.7 *The OPW $\chi_0(\mathbf{r})$ for Li.*

The quantity $V(\mathbf{K}_t - \mathbf{K}_s)$ is the Fourier coefficient given by Eqn (3.45) and in the derivation of Eqn (3.60) the core states were assumed to be eigenstates of H with energy ε_c.

We do not continue with the OPW method, but use it as convenient starting point for the pseudopotential theory.

3.2.3 Pseudopotentials

One of the worldwide most frequently used computer programs is VASP, a modern method to determine the electronic structure of solids and surfaces. It is thoroughly described by Hafner (2008) and it is based on pseudopotentials of a modern kind (see also Kresse und Furthmüller (1996)). Here we therefore give a short historical introduction, beginning with the early empirical versions and continuing with the modern so-called norm-conserving pseudopotentials.

First, it was observed by Heine (1970) that the OPW secular equation has the same form as that for plane waves. The latter is simply the Schrödinger equation $(T+V)|\psi_\mathbf{k}\rangle = \varepsilon_\mathbf{k}|\psi_\mathbf{k}\rangle$, where T is the kinetic energy in the plane-wave representation and V is the full potential. The aim is to construct a Schrödinger equation $(T+V_{ps})|\phi_\mathbf{k}\rangle = \varepsilon_\mathbf{k}|\phi_\mathbf{k}\rangle$ with a 'simpler' pseudopotential V_{ps} a 'smoother' ket $|\phi_\mathbf{k}\rangle$, the same energy eigenvalues $\varepsilon_\mathbf{k}$ but eliminating the core states. There were many successful attempts—a clever one was that by Austin et al. (1962), who suggested a non-Hermitian potential V_A defined by

$$(V+V_A)|\phi\rangle = V|\phi\rangle + \sum_c |c\rangle\langle F_c|\phi\rangle , \qquad (3.61)$$

where c labels the core states, for which $\langle\psi|c\rangle = 0$ is required, and $F_c = F_c(\mathbf{r})$ is quite an arbitrary function chosen suitably. The ket $|\phi\rangle$ is supposed to be subject to the same boundary conditions as $|\psi\rangle$ and they should not be orthogonal, $\langle\psi|\phi\rangle \neq 0$. Then it is seen at once that eigenvalues of the problem

$$(T+V)|\psi\rangle = \varepsilon|\psi\rangle \qquad (3.62)$$

and that of

$$(T+V+V_A)|\phi\rangle = \varepsilon'|\phi\rangle , \qquad (3.63)$$

are the same, $\varepsilon = \varepsilon'$. Simply form the scalar product of $\langle\psi|$ with eqn (3.63). Concentrating first on the simpler atomic case, we may require that V_A cancels the nuclear potential as best as possible. Then we may transfer this potential to the crystal, as suggested by Heine (1970). We now construct a simple pseudopotential.

Considering the expression

$$V_{ps} = V - \sum_c |c\rangle\langle c|V \qquad (3.64)$$

we notice that the cancellation of V would be complete if the core states formed a complete set, i.e. $\sum_c |c\rangle\langle c| = 1$; of course they are not, but the cancellation could still be considerable. Now assume V_{ps} acts on a smooth function ϕ which is assumed not to vary over the core region, then we obtain from eqn (3.64) an expression for F_c, see eqn (3.61), $\langle F_c|\phi\rangle = \langle c|V|\phi\rangle$, which we can approximate by taking ϕ out of the integral, obtaining in real space

$$V_{ps}(\mathbf{r}) \cong V(\mathbf{r}) - \sum_c \langle c|V\rangle \psi_c(\mathbf{r}), \qquad (3.65)$$

where $\psi_c(\mathbf{r}) = \langle \mathbf{r}|c\rangle$ and the sum on c only contains the angular-momentum components occurring in ψ. This is easy to calculate for the atomic case and shown in Fig. 3.8. Outside the core $Z = -1$, the valence of Na$^+$; inside the core the cancellation is easily seen, but a singularity remains at the origin. Many empirical or model potentials of this kind were introduced in the 1960s and 1970s with great success by, for instance, Cohen and Heine (1970), Heine and Weaire (1970), and Harrison (1966). Once the pseudopotential of the bare ion had been obtained, it was transferred to the electron gas, which then screens it to give the total pseudopotential of the solid. This screening was achieved by using a linear dielectric function method.

When one attempts to obtain the charge density the empirical pseudopotentials turned out to be unsuitable because of a problem with the normalization of the pseudo- and the "exact" wave function. In the development that followed one therefore abandoned any notion of orthogonalizing to the core states. As before, however, it is the isolated atom or ion that is used to construct the pseudopotential. Following the work of Hamann

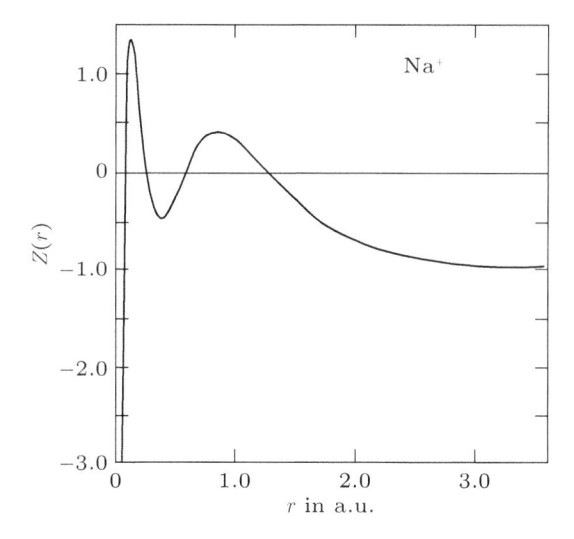

Figure 3.8 *A simple pseudopotential generated for Na$^+$. The pseudopotential is given by $V_{ps} = \frac{Z(r)}{r}$. (Constructed using wave functions of Hartree and Hartree (1948)).*

(1979), Bachelet et al. (1982), and Hamann (1989), we require that the pseudo-wave function for the atom be nodeless, and when normalized, that it becomes identical to the true valence-wave function beyond some "core-radius", R. The real atom and the pseudo-atom should, furthermore, have the same valence eigenvalues for some chosen configuration and the pseudo-charge contained in the sphere with radius R should be identical to the real charge in the sphere. Hamann (1979) coined the term "norm conserving" to describe the pseudopotentials constructed in this fashion (NCPP).

For the pseudopotential to be useful, its core portion must be transferable to other situations where the external potential has changed, such as in molecules, solids, or surfaces. Topp and Hopfield (1973) were the first to use an identity derived from the Wronskian of the radial Schrödinger equation for constructing a pseudopotential and Hamann (1979) used it to ensure the transferability of the NCPP. For deriving this relation (Messiah, 1976), we consider the Schrödinger equation for the radial function $u_l(\varepsilon, r)$ for some energy ε, not necessarily an eigenvalue

$$[-\frac{d^2}{dr^2} + V_l(r)]u_l(\varepsilon, r) = \varepsilon\, u_l(\varepsilon, r) \tag{3.66}$$

and similarly for $u_l(\varepsilon + d\varepsilon, r)$. The potential V_l is arbitrary and includes $l(l+1)/r^2$; it must not, however, depend on ε, as did most of the older pseudopotentials. Now one easily derives

$$u_l^2(\varepsilon, r) \frac{d}{d\varepsilon}\frac{d}{dr} \ln u_l(\varepsilon, r)|_{r=R} = -\int_0^R u_l^2(\varepsilon, r) dr. \tag{3.67}$$

The consequence of this equation is that for two potentials $V_l^{(1)}$ and $V_l^{(2)}$ the linear energy variation around ε of their scattering phase shifts at R, $\frac{d}{dr}\ln u_l(\varepsilon, r)|_{r=R}$, is identical provided the solutions $u_l^{(1)}$ and $u_l^{(2)}$ with $u_l^{(1)}(R) = u_l^{(2)}(R)$ possess the same charge density inside the sphere of radius R. The requirement that the pseudo-wave function agrees with the full wave function for $r > R$ guaranties that the charge is identical for $r > R$, and that the scattering properties of the NCPP and the full potential to first order have the same energy variation when transferred to other systems. In the following we go through a specific example following Hamann (1989). In a first step one calculates the self-consistent local-density potential $V(r)$ for the *atom* of interest. This gives the potential $V_l(r)$ in eqn (3.66) which in this step is supposed to be an eigenvalue equation giving ε and u_l. Next a "core radius" r_{cl} is to chosen to be about 1/2 of the radius at which $u_l(r)$ has its outermost maximum. It is now possible to pick an outer radius at which the pseudo-wave function can be accurately converged to the full-potential wave function and $R_l = 2.5 r_{cl}$ is found to be practical. Equation (3.66) is then integrated with $\varepsilon = \varepsilon_l$ from the origin and stopped at R_l normalizing u_l such that $4\pi \int_0^{R_l} u_l^2(r) dr = 1$. The normalized values of $u_l(R_l)$ and the first derivatives $\frac{d}{dr} u_l(r)|_{r=R_l}$ are recorded. Next, a so-called intermediate pseudopotential is defined by

$$V_{1l} = [1 - f(r/r_{cl})]V(r) + c_l f(r/r_{cl}), \tag{3.68}$$

where as a cut-off function the use of $f(r/r_{cl}) = \exp[-(r/r_{cl})^{3.5}]$ was found to be practical. An intermediate nodeless wave function $w_{1l}(r)$ is obtained that is iteratively used to allow c_l to be such that the scattering phase shifts agree, $\frac{d}{dr}\ln w_{1l}(r)|_{R_l} = \frac{d}{dr}\ln u_l(r)|_{R_l}$. A scale factor, furthermore, is found to make the two wave functions identical at R_l, $\gamma_l = u_l(R_l)/w_{1l}(R_l)$, and a final pseudo-wave function is constructed by adding a norm-conserving term,

$$w_{2l}(r) = \gamma_l[w_{1l}(r) + \delta_l g_l(r)], \tag{3.69}$$

with δ_l such that $\int_0^{R_l} w_{2l}^2(r)dr = 1$ and $g_l(r) = r^{l+1}f(r/r_{cl})$. The final pseudopotential is found by inverting the Schrödinger equation

$$V_{2l}(r) = \varepsilon_l + \frac{1}{w_{2l}(r)}\frac{d^2}{dr^2}w_{2l}(r) \tag{3.70}$$

which, with the parameters given above, can be done analytically (Bachelet *et al.*, 1982).

For either the function $u_l(r)$ or $w_{2l}(r)$ the right-hand side of the key equation (3.67) are the same by construction, therefore, the logarithmic derivatives at energy ε are the same to first order in $(\varepsilon - \varepsilon_l)$ by a Taylor series expansion, so the scattering power of V_{2l} for l partial waves is the same as the full potential $V(r)$ to a good approximation at energies away from the selected set of ε_l.

Finally, in an important step, the pseudopotentials $V_{2l}(r)$ is "unscreened" to give the ionic pseudopotential defined by

$$V_l^{\text{ion}}(r) = V_{2l}(r) - V^{\text{H}}(n,r) - V_{xc}(n,r), \tag{3.71}$$

where V^{H} is the Hartree- and V_{xc} is the exchange-correlation potential corresponding to the atom density obtained from $n(r) = \sum_l n_l[w_{2l}(r)/r]^2$, where n_l are the occupancies of the valence states. As an example the ionic pseudopotential for aluminium obtained by the above procedure is shown in Fig. 3.9. As in the older pseudopotentials the cancellation in the core region is apparent, but no singularity remains at the origin. For the band calculation the NCPP is transferred to the solid and screened again by the Hartree- and exchange correlation potentials corresponding to the self-consistent solid state charge densities obtained from the plane-wave expansion of the Bloch functions. Note that dielectric screening is no longer needed as in the case of the empirical pseudopotentials. The paper by Bachelet *et al.* (1982) contains pseudopotentials for the entire periodic table of the elements from hydrogen to plutonium.

The pseudopotential construction is somewhat more involved in the VASP method thus the interested reader is referred to Kresse and Furthmüller (1996) or the review article by Hafner (2008).

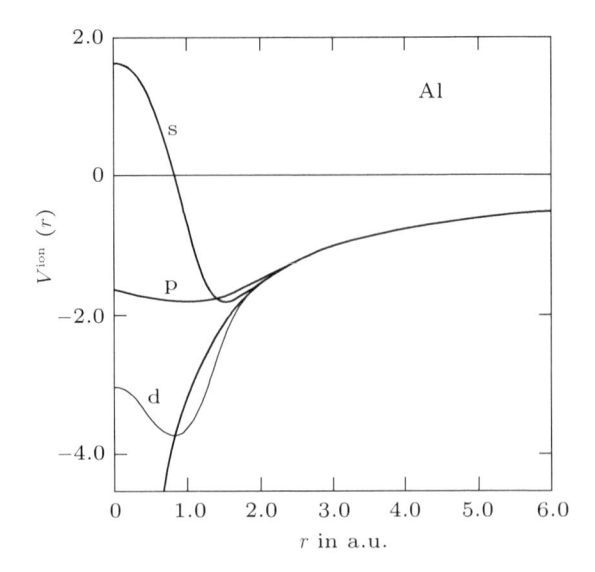

Figure 3.9 *The norm-conserving pseudopotential for s, p, and d states of Al; after Bachelet et al. (1982). The unlabeled curve represents the bare ion potential* $-\frac{Z_{val}}{r}$.

3.3 Augmented plane waves and Green's functions

At the beginning of Slater's (1937, 1965) augmented plane wave (APW) method there is an assumption about the form of the crystal potential. It is assumed to be spherically symmetric inside nonoverlapping spheres centered at the atom sites and constant outside. Because of the particular shape of the potential, it is said to be a "muffin-tin" potential and the spheres (chosen as large as possible) are called "muffin-tin" spheres. We shall call the space outside the spheres interstitial space (IS) denoting the radius of the sphere centered at atom ν in the unit cell by S_ν. Although this is not a perfect representation of the potential actually found in crystals, it enables one to set up a rather rigorous solution of the Schrödinger equation. Having found these solutions, the departures of the potential from the assumed form can be treated as a perturbation.

3.3.1 APW

The basic expansion for the Bloch function is again written as

$$|\psi_\mathbf{k}\rangle = \sum_s a_s(\mathbf{k}) |\chi_{\mathbf{k}+\mathbf{K}_s}\rangle, \tag{3.72}$$

the (finite) sum on s extending over reciprocal lattice vectors as before, but the basis functions are constructed in a rather elaborate way. In the interstitial region of space, i.e. for **r** in IS, they are plane waves

$$\langle \mathbf{r} | \chi_{\mathbf{k}+\mathbf{K}_s} \rangle = e^{i(\mathbf{k}+\mathbf{K}_s)\cdot \mathbf{r}} \tag{3.73}$$

because the potential is constant here. This, of course, satisfies the Bloch condition. Inside the sphere centered at $\boldsymbol{\tau}_\nu$ in the unit cell one writes

$$\langle \mathbf{r} | \chi_{\mathbf{k}+\mathbf{K}_s} \rangle = \sum_L C_{L\nu}(\mathbf{k}+\mathbf{K}_s)\, R_{\ell\nu}(\varepsilon,r)\, Y_L(\hat{\mathbf{r}}). \tag{3.74}$$

The quantity L denotes both the angular momentum and magnetic quantum numbers ℓ and m, i.e. $\sum_L = \sum_\ell \sum_{m=-\ell}^{\ell}$, $Y_L(\hat{\mathbf{r}})$ are spherical harmonics, $Y_{\ell m}(\theta,\varphi)$, and the functions $R_{\ell\nu}(\varepsilon,r)$ are (numerical) solutions of the radial Schrödinger equation regular at the origin

$$-\frac{1}{r^2}\frac{d}{dr}\left(r^2 \frac{dR_{\ell\nu}}{dr}\right) + \left[\frac{\ell(\ell+1)}{r^2} + V_\nu(r) - \varepsilon\right] R_{\ell\nu}(\varepsilon,r) = 0. \tag{3.75}$$

The energy, ε, is assumed to be a variable here and not an eigenvalue. The coefficients $C_{L\nu}(\mathbf{k}+\mathbf{K}_s)$ in Eqn (3.74) are obtained by requiring that the values of $\langle \mathbf{r}|\chi_{\mathbf{k}+\mathbf{K}_s}\rangle$ match continuously the plane waves given by Eqn (3.73) at the surface of the muffin-tin spheres. This is what is meant by augmentation, and is achieved mathematically by using the well-known identity

$$e^{i\mathbf{k}\cdot\mathbf{r}} = 4\pi \sum_L i^\ell\, j_\ell(kr)\, Y_L^*(\hat{\mathbf{k}})\, Y_L(\hat{\mathbf{r}}), \tag{3.76}$$

where the quantities $j_\ell(kr)$ are spherical Bessel functions of order ℓ which, we remind the reader, are admissible solutions of the radial Schrödinger equation (3.75) for a vanishing potential and $\varepsilon = k^2$. For the coefficients $C_{L\nu}(\mathbf{k}+\mathbf{K}_s)$ one easily obtains with Eqn (3.76)

$$C_{L\nu}(\mathbf{k}+\mathbf{K}_s) = 4\pi\, i^\ell\, e^{i(\mathbf{k}+\mathbf{K}_s)\cdot\boldsymbol{\tau}_\nu}\, j_\ell(|\mathbf{k}+\mathbf{K}_s|S_\nu)\, Y_L^*(\widehat{\mathbf{k}+\mathbf{K}_s})/R_{\ell\nu}(\varepsilon,S_\nu). \tag{3.77}$$

This, together with Eqns (3.73) and (3.74), defines an augmented plane wave (APW), which, by construction, is in general discontinuous in *slope* on the muffin-tin sphere boundary. Therefore, the variational expression that yields the expansion coefficients $a_s(\mathbf{k})$ must be treated with some care.

Let us denote by H_0 the Hamiltonian underlying the APW method, i.e.

$$H_0 = -\nabla^2 + V_{\mathrm{mt}}, \tag{3.78}$$

where the muffin-tin potential V_{mt} is constant in interstitial space and has the value $V_\nu(r)$ inside the muffin-tin sphere centered at $\boldsymbol{\tau}_\nu$. The integral over the unit cell, $\int_{\mathrm{UC}} \psi_\mathbf{k}^*(\mathbf{r})\, H_0\, \psi_\mathbf{k}(\mathbf{r})\, d\mathbf{r}$, can now be written as the sum of an integral over the spherical volumes defined by the muffin-tin spheres (MT) and an integral extending over interstitial space, hence

$$\int_{UC} \psi_{\mathbf{k}}^*(\mathbf{r}) H_0 \psi_{\mathbf{k}}(\mathbf{r}) \, d\mathbf{r} = -\int_{MT} \psi_{\mathbf{k}}^*(\mathbf{r}) \nabla^2 \psi_{\mathbf{k}}(\mathbf{r}) \, d\mathbf{r} - \int_{IS} \psi_{\mathbf{k}}^*(\mathbf{r}) \nabla^2 \psi_{\mathbf{k}}(\mathbf{r}) \, d\mathbf{r}$$

$$+ \int_{UC} \psi_{\mathbf{k}}^*(\mathbf{r}) V_{mt} \psi_{\mathbf{k}}(\mathbf{r}) \, d\mathbf{r}.$$
(3.79)

Applying Green's identity to the first two terms on the right-hand side we derive

$$\int_{UC} \{[\nabla \psi_{\mathbf{k}}^*(\mathbf{r})] [\nabla \psi_{\mathbf{k}}(\mathbf{r})] + \psi_{\mathbf{k}}^*(\mathbf{r}) V_{mt} \psi_{\mathbf{k}}(\mathbf{r}) - \varepsilon \psi_{\mathbf{k}}^*(\mathbf{r}) \psi_{\mathbf{k}}(\mathbf{r})\} \, d\mathbf{r}$$

$$= \int_{UC} \psi_{\mathbf{k}}^*(\mathbf{r}) (H_0 - \varepsilon) \psi_{\mathbf{k}}(\mathbf{r}) \, d\mathbf{r} + \int_S \psi_{\mathbf{k}}^*(\mathbf{r}^{(-)}) \left(\frac{\partial \psi_{\mathbf{k}}(\mathbf{r}^{(-)})}{\partial r} - \frac{\partial \psi_{\mathbf{k}}(\mathbf{r}^{(+)})}{\partial r} \right) dS,$$
(3.80)

where the last term on the right-hand side is a surface integral over the muffin-tin spheres, $\mathbf{r}^{(-)}$ and $\mathbf{r}^{(+)}$ denoting positions just inside and outside, respectively, the muffin-tin sphere boundaries. This results from the above-mentioned discontinuity in slope.

Now, the important point to notice is that the Euler–Lagrange equation obtained from varying the left-hand side of Eqn (3.80) is just the Schrödinger equation (see e.g. Merzbacher, 1970):

$$(H_0 - \varepsilon) \psi_{\mathbf{k}}(\mathbf{r}) = 0.$$
(3.81)

It follows that the entire right-hand side of Eqn (3.80) must be varied to obtain the coefficients $a_s(\mathbf{k})$ of Eqn (3.72) and not only the integral $\int_{UC} \psi_{\mathbf{k}}^*(\mathbf{r})(H_0 - \varepsilon) \psi_{\mathbf{k}}(\mathbf{r}) \, d\mathbf{r}$. Thus substituting Eqn (3.72) into the right-hand side of Eqn (3.80) and varying the coefficients $a_s^*(\mathbf{k})$, we obtain the algebraic equation

$$\sum_s \{\langle t|H_0 - \varepsilon_{\mathbf{k}}|s\rangle + \langle t|S|s\rangle\} a_s(\mathbf{k}) = 0,$$
(3.82)

where we defined

$$\langle t|H_0 - \varepsilon_{\mathbf{k}}|s\rangle = \int_{UC} \chi_{\mathbf{k}+\mathbf{K}_t}^*(\mathbf{r}) (H_0 - \varepsilon_{\mathbf{k}}) \chi_{\mathbf{k}+\mathbf{K}_s}(\mathbf{r}) \, d\mathbf{r}$$
(3.83)

and

$$\langle t|S|s\rangle = \int_S \chi_{\mathbf{k}+\mathbf{K}_t}^*(\mathbf{r}) \left[\frac{\partial}{\partial r} \chi_{\mathbf{k}+\mathbf{K}_s}(\mathbf{r}^{(-)}) - \frac{\partial}{\partial r} \chi_{\mathbf{k}+\mathbf{K}_s}(\mathbf{r}^{(+)}) \right] dS.$$
(3.84)

Let us temporarily simplify the discussion by assuming there is only one atom per unit cell. Inside the muffin-tin sphere an APW is an eigenfunction of H_0 and hence, in evaluating Eqn (3.83), only that part of the integral remains that extends over the interstitial region, i.e.

$$\langle t|H_0 - \varepsilon_{\mathbf{k}}|s\rangle = \left[(\mathbf{k}+\mathbf{K}_s)^2 - \varepsilon_{\mathbf{k}}\right] \int_{\mathrm{IS}} e^{i(\mathbf{K}_s - \mathbf{K}_t)\cdot\mathbf{r}} \, d\mathbf{r} \qquad (3.85)$$

(here $\varepsilon_{\mathbf{k}}$ is counted from the constant interstitial potential). The remaining integral can be shown to be

$$\int_{\mathrm{IS}} e^{i(\mathbf{K}_s - \mathbf{K}_t)\cdot\mathbf{r}} \, d\mathbf{r} = \Omega_{UC}\, \delta_{st} - 4\pi\, S^2 \, \frac{j_1(|\mathbf{K}_s - \mathbf{K}_t|S)}{|\mathbf{K}_s - \mathbf{K}_t|}. \qquad (3.86)$$

The calculation of the surface integral, Eqn (3.84), is straightforward. Using the addition theorem for spherical harmonics, dividing through by the volume of the unit cell, and collecting terms, we can write Eqn (3.82) in the form

$$\sum_s \left\{ \left[(\mathbf{k}+\mathbf{K}_s)^2 - \varepsilon_{\mathbf{k}}\right]\delta_{ts} + V^{\mathrm{APW}}(\mathbf{K}_t, \mathbf{K}_s)\right\} a_s(\mathbf{k}) = 0, \qquad (3.87)$$

where

$$V^{\mathrm{APW}}(\mathbf{K}_t, \mathbf{K}_s) = \frac{4\pi S^2}{\Omega_{\mathrm{UC}}} \sum_\ell (2\ell+1)\, P_\ell(\cos\theta_{ts})\, j_\ell(|\mathbf{k}+\mathbf{K}_t|S)\, j_\ell(|\mathbf{k}+\mathbf{K}_s|S)$$

$$\left[\frac{d}{dr}\ln R_\ell(\varepsilon,r) - \frac{d}{dr}\ln j_\ell(|\mathbf{k}+\mathbf{K}_s|r)\right]_S - \left[(\mathbf{k}+\mathbf{K}_s)^2 - \varepsilon_{\mathbf{k}}\right]$$

$$\cdot \frac{4\pi S^2}{\Omega_{\mathrm{UC}}} \frac{j_1(|\mathbf{K}_s - \mathbf{K}_t|S)}{|\mathbf{K}_s - \mathbf{K}_t|}. \qquad (3.88)$$

$P_\ell(x)$ is a Legendre polynomial of order ℓ and θ_{ts} is the angle between the vectors $\mathbf{k}+\mathbf{K}_t$ and $\mathbf{k}+\mathbf{K}_s$. This form of the APW equation is quite convenient for a discussion of the limit of nearly free-electron behavior. But there exists an alternative form of the APW equation that is more practical for actual calculations; details can be found in the book of Loucks (1967).

Armed with the APW equations we can repeat the discussion of nearly free-electron behavior. But we can broaden our understanding now by dealing semi-quantitatively with an important example where both nearly free-electron behavior and strong deviations from it occur, i. e. the case of copper (the same is true for all transition metals).

So we begin by following Slater (1965) and examine Eqns (3.87) and (3.88) looking at the diagonal elements first. Since for small arguments, x, the asymptotic form of the spherical Bessel function is

$$j_\ell(x) \xrightarrow[x \to 0]{} x^\ell/(2\ell+1)!!, \tag{3.89}$$

and $P_\ell(1) = 1$, we can write Eqn (3.88) for $\mathbf{K}_s = \mathbf{K}_t$ as

$$V^{\mathrm{APW}}(\mathbf{K}_s, \mathbf{K}_s)$$

$$= \frac{4\pi S^2}{\Omega_{\mathrm{UC}}} \sum_\ell (2\ell+1) j_\ell^2(|\mathbf{k}+\mathbf{K}_s|S) \cdot \Delta D_\ell(\varepsilon) - [(\mathbf{k}+\mathbf{K}_s)^2 - \varepsilon_\mathbf{k}] \frac{4\pi S^3}{3\Omega_{\mathrm{UC}}}, \tag{3.90}$$

where ΔD_ℓ is the difference of the logarithmic derivatives

$$\Delta D_\ell(\varepsilon) = \frac{\mathrm{d}}{\mathrm{d}r} \ln R_\ell(\varepsilon, r)|_S - \frac{\mathrm{d}}{\mathrm{d}r} \ln j_\ell(|\mathbf{k}+\mathbf{K}_s|r)|_S. \tag{3.91}$$

We use perturbation theory (justifying this below) and replace the secular determinant by the product of its diagonal terms. If we set this equal to zero, we can set one such term equal to zero and read off from Eqns (3.87) and (3.90)

$$\varepsilon_\mathbf{k} \cong (\mathbf{k}+\mathbf{K}_s)^2 + \frac{4\pi S^2}{\Omega_{\mathrm{UC}} - \Omega_{\mathrm{MT}}} \sum_\ell (2\ell+1) j_\ell^2(|\mathbf{k}+\mathbf{K}_s|S) \Delta D_\ell(\varepsilon_\mathbf{k}), \tag{3.92}$$

where Ω_{MT} is the volume of the muffin-tin sphere. The situation $\Delta D_\ell(\varepsilon) \simeq 0$ is that of nearly free electrons because, as was stated after Eqn (3.76), the spherical Bessel functions are the solutions of the radial Schrödinger equation for the empty lattice, i.e. free electrons. Thus, in this case, Eqn (3.76) gives $\varepsilon_\mathbf{k} \cong (\mathbf{k}+\mathbf{K}_s)^2$ which may be substituted into the nondiagonal matrix elements, $V^{\mathrm{APW}}(\mathbf{K}_t, \mathbf{K}_s)$, this way removing the last term on the right-hand side of Eqn (3.90), giving $V^{\mathrm{APW}}(\mathbf{K}_t, \mathbf{K}_s) \simeq 0$ for $\mathbf{K}_t \neq \mathbf{K}_s$, so justifying our use of perturbation theory.

The next step is to examine calculated values of the difference of the logarithmic derivatives, $\Delta D_\ell(\varepsilon)$, and make their functional form plausible. The dashed curves in Fig. 3.10 are the logarithmic derivatives ($\times S$) of free electrons, $S\,\mathrm{d}/\mathrm{d}r \ln j_\ell(kr)|_{r=S}$, with $k^2 = \varepsilon$ and $S = 2.415$ Bohr appropriate for Cu. The solid lines are estimated values compiled from the literature (Slater, 1965; Mattheiss et al., 1968) of calculated logarithmic derivatives, $S\,\mathrm{d}/\mathrm{d}r \ln R_\ell(\varepsilon, r)|_S$ which, except for $\ell = 2$, are rather similar to those of free electrons. This remark applies only to the logarithmic derivatives at S and not the radial functions themselves which are completely different from the spherical Bessel functions further inside the atom where they show strong core oscillations because of orthogonality requirements. For $\ell = 2$, however, we cannot really ignore the off-diagonal elements of $V^{\mathrm{APW}}(\mathbf{K}_t, \mathbf{K}_s)$. Proceeding thus with the numerical results, we examine the calculation by Burdick (1963), which is a rather famous case. It was obtained by the APW method using an empirical potential constructed by Chodorow (1939) on the basis of Hartree and Hartree–Fock calculations for Cu^+.

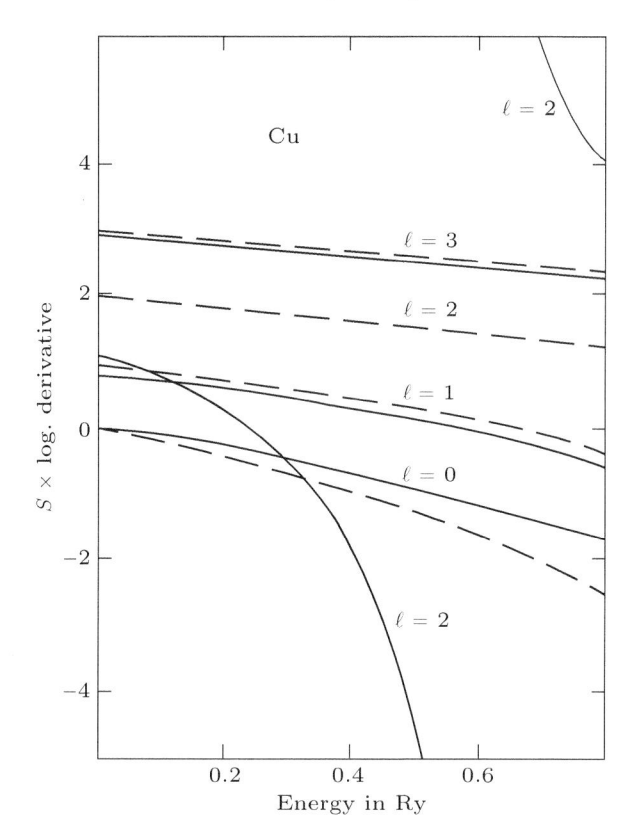

3.3 Augmented plane waves and Green's functions 113

Figure 3.10 *Logarithmic derivatives as a function of energy for copper (solid lines) and for free electrons (dashed lines).*

Burdick's results are shown in Fig. 3.11 where we also compare them with measurements from angle-resolved photoemission experiments by Thiry *et al.* (1979). The agreement between the measured and calculated energies is quite astonishing. Therefore we want to look at the calculated band structure in some detail. It begins with $\varepsilon = 0$ Ry at the Γ-point in the BZ (corresponding to $\mathbf{k} = (0, 0, 0)$); the label Γ_1, designates the same symmetry as an *s*-wave function ($\ell = 0$). The parabolic $\varepsilon_{\mathbf{k}}$ curve is interrupted (we call it hybridized) by flatter states starting at Γ with labels $\Gamma_{25'}$ and Γ_{12}; these denote d state symmetries, $\Gamma_{25'}$ being threefold degenerate having transformation properties like xy, yz, and zx and Γ_{12} being twofold degenerate transforming like $x^2 - y^2$ and $2z^2 - x^2 - y^2$. (These products are an abbreviated notation of real $\ell = 2$ spherical harmonics.) The bands labeled Δ_5 and Λ_3 are twofold degenerate d states. We furthermore call attention to the states $X_{4'}$ at $\varepsilon \simeq 0.804$ Ry and $L_{2'}$ at $\varepsilon \simeq 0.608$ Ry where the band structure appears roughly parabolic again. These labels designate p symmetries ($\ell = 1$) whose logarithmic derivative is very close to the free-electron value (see Fig. 3.10). Indeed, with the lattice

114 *Energy-Band Theory*

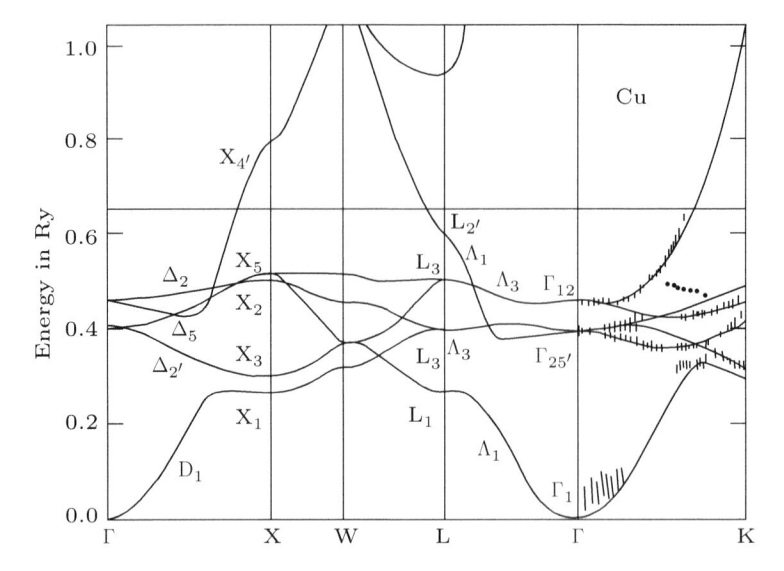

Figure 3.11 *Band structure for Cu as calculated by Burdick (1963). Experimental results (vertical bars between Γ and K) are angle-resolved photoemission data by Thiry et al. (1979). Black dots are presumably surface states.*

constant of Cu, $a = 6.8309$ Bohr, the free-electron value at X is $(2\pi/a)^2 = 0.846$ Ry and at L it is $(\pi\sqrt{3}/a)^2 = 0.635$ Ry reasonably close to the $X_{4'}$ and $L_{2'}$ values.

We will learn more about the typical "fingerprints" of d functions in metals when discussing the canonical band concept in Sec. 3.4.5. Concerning the non-d states we emphasize that although some of the energies are close to free-electron values, the wave functions are far from the plane waves representing free electrons. Although they are described with good accuracy by a single plane wave between the atomic sphere, inside such a sphere the wave functions go through all the oscillations characteristic of atomic wave functions. The point is that a single plane wave joins smoothly onto the solution of the Schrödinger equation inside the sphere, the energy of this solution being the same as that of the plane wave (Slater, 1965).

We want to close this subsection on the APW method with some technical remarks. First there is the number of basis functions that are required to obtain converged results. This number is *not* small, indeed, roughly 40 APW basis functions are needed for each atom in the unit cell; for compounds the rank of the secular equation (3.87) can thus become rather large. Its eigenvalues, $\varepsilon_\mathbf{k}$, are usually found by tracing the roots of the secular determinant. Since this is often done by triangularization, it is not hard to obtain the eigenvectors $a_s(\mathbf{k})$ belonging to each eigenvalue $\varepsilon_\mathbf{k}$ by back-substitution and hence the corresponding wave function $\psi_\mathbf{k}(\mathbf{r})$ from Eqn (3.72). To normalize it one needs matrix elements of the form $\langle \chi_{\mathbf{k}+\mathbf{K}_t} | \chi_{\mathbf{k}+\mathbf{K}_s} \rangle$ which are readily obtained using the Eqns (3.73), (3.74), and (3.77); one finds

$$\langle \chi_{\mathbf{k}+\mathbf{K}_t} | \chi_{\mathbf{k}+\mathbf{K}_s} \rangle = \Omega_{\text{UC}}\, \delta_{ts} + 4\pi \sum_{\nu=1}^{n} S_\nu^2 \, e^{i(\mathbf{K}_s - \mathbf{K}_t)\cdot \boldsymbol{\tau}_\nu}$$

$$\cdot \left\{ \sum_\ell (2\ell+1)\, P_\ell(\cos\theta_{ts})\, j_\ell(|\mathbf{k}+\mathbf{K}_t|S_\nu)\, j_\ell(|\mathbf{k}+\mathbf{K}_s|S_\nu)\, I_{\ell\nu}(\varepsilon_\mathbf{k}) \right.$$

$$\left. - \frac{j_1(|\mathbf{K}_t - \mathbf{K}_s| S_\nu)}{|\mathbf{K}_t - \mathbf{K}_s|} \right\}, \qquad (3.93)$$

where the symbols are defined as before and the quantity $I_{\ell\nu}(\varepsilon_\mathbf{k})$ denotes the integral

$$I_{\ell\nu}(\varepsilon_\mathbf{k}) = \frac{1}{S_\nu^2\, R_{\ell\nu}^2(\varepsilon_\mathbf{k}, S_\nu)} \int_0^{S_\nu} r^2\, R_{\ell\nu}^2(\varepsilon_\mathbf{k}, r)\, dr. \qquad (3.94)$$

It is not hard to see (and quite useful to know) that this quantity is also given by

$$I_{\ell\nu}(\varepsilon) = -\frac{d}{d\varepsilon}\frac{d}{dr} \ln R_{\ell\nu}(\varepsilon, r)\big|_{S_\nu}. \qquad (3.95)$$

(For the derivation of Eqn (3.95) one writes down the Schrödinger equation for r inside the ν-th muffin-tin as

$$\left[-\frac{d^2}{dr^2} + V_0(r) \right] u_{\ell\nu}(\varepsilon, r) = \varepsilon\, u_{\ell\nu}(\varepsilon, r),$$

where $u_{\ell\nu}(\varepsilon, r) = r R_{\ell\nu}(\varepsilon, r)$. Doing the same for the function $u_{\ell\nu}(\varepsilon + d\varepsilon, r)$, premultiplying with $u_{\ell\nu}(\varepsilon + d\varepsilon, r)$ or $u_{\ell\nu}(\varepsilon, r)$, respectively, integrating over the ν-th muffin-tin, subtracting and letting $d\varepsilon \to 0$, we find

$$u_{\ell\nu}^2(\varepsilon, r)\, \frac{d}{d\varepsilon}\frac{d}{dr} \ln u_{\ell\nu}(\varepsilon, r)\big|_{r=S_\nu} = -\int_0^{S_\nu} u_{\ell\nu}^2(\varepsilon, r)\, dr .)$$

One sees that the sum on ℓ in Eqn (3.93) because of Eqn (3.94) is related to the probability of finding the electron inside the muffin-tin sphere, ν, the dominating term being recognizable by means of Eqn (3.95) as that with a large negative energy derivative of the logarithmic derivative at the sphere boundary, as, for instance, the d state in Cu, see Fig. 3.10.

Finally, if the self-consistent potential deviates from the assumed muffin-tin form in the interstitial region one can define a potential $\Delta V(\mathbf{r})$, which describes this deviation, and add it to the Hamiltonian H_0 in Eqn (3.83). As a result one obtains an additional matrix element in V^{APW}, Eqn (3.88), that is the plane-wave matrix element of ΔV. This is so because $\Delta V(\mathbf{r})$ is zero (or constant) inside the muffin-tin sphere. However, deviations from spherical symmetry inside the muffin-tin sphere are much harder to handle and are (if at all) treated by perturbation theory using as unperturbed basis functions the muffin-tin APWs. Although for most metals and intermetallic compounds this so-called shape approximation for the potential is seldom really serious there are prominent exceptions and problems occur, for example, if shear distortions are considered. In these cases one must drop the muffin-tin approximation. Calculations that do are called "full potential" and in many modern studies are conducted using a linearized version of the APW method to be discussed in Sec. 3.4.3.

3.3.2 Multiple scattering theory

We now turn to a description of multiple scattering theory (MST) which is a different and rather refined method to solve the single-particle Schrödinger equation (3.1). It is also sometimes called the Green's function or KKR method where KKR stands for the names Korringa (1947), Kohn and Rostoker (1954) who first introduced this method into the electronic-structure theory of crystalline solids. Besides crystalline solids it has been successfully applied to studies of the electronic structure of molecules and surfaces, and it is the basis of most theories of liquids, disordered systems, and impurities in various metallic and nonmetallic hosts. In all applications its content is the same and easy to state (Weinberger, 1990): it is a prescription for obtaining the transition matrix T describing the scattering of the entire system in terms of the corresponding matrix t for individual scatterers. The central result of the theory can be expressed in a form independent of the details of the problem and is obtained as the result of summing all possible trajectories of the particle through the system. Thus if G describes free particle propagation from any individual scattering event to the next, we may describe multiple scattering by means of

$$T = t + tGt + tGtGt + \cdots = t + tG(t + tGt + tGtGt + \cdots) = t + tGT, \quad (3.96)$$

or

$$T = (1 - tG)^{-1} t. \quad (3.97)$$

Stationary states (Messiah, 1978) of a system are given by the singularities of T and hence are obtained from the relation

$$\det\left(t^{-1} - G\right) = 0. \quad (3.98)$$

Application of this simple formula requires us to find a suitable representation; this is where the real work starts.

It begins with the definition of the free-particle Green's function, $G_0(\varepsilon, \mathbf{r} - \mathbf{r}')$, which as usual is required to solve the equation

$$(\nabla^2 + \varepsilon) G_0(\varepsilon, \mathbf{r} - \mathbf{r}') = \delta(\mathbf{r} - \mathbf{r}'), \quad (3.99)$$

where $\delta(\mathbf{r}-\mathbf{r}')$ is the Dirac δ-function. A standing wave free-particle solution of this equation is easily constructed (Messiah, 1976) to be

$$G_0(\varepsilon, \mathbf{r}-\mathbf{r}') = -\frac{1}{4\pi} \frac{\cos(\kappa|\mathbf{r}-\mathbf{r}'|)}{|\mathbf{r}-\mathbf{r}'|}, \tag{3.100}$$

where $\kappa^2 = \varepsilon$. The use of the standing wave Green's function is appropriate to the determination of the stationary states for which the Schrödinger equation is written in integral form as

$$\psi_\mathbf{k}(\mathbf{r}) = \int d\mathbf{r}' \, G_0(\varepsilon_\mathbf{k}, \mathbf{r}-\mathbf{r}') V(\mathbf{r}') \psi_\mathbf{k}(\mathbf{r}'). \tag{3.101}$$

Applying the operator $(-\nabla^2 - \varepsilon_\mathbf{k})$ to this equation and using Eqn (3.99) one verifies directly that $\psi_\mathbf{k}(\mathbf{r})$ satisfies the Schrödinger equation.

There are many different ways to proceed from here. A modern and very complete account can be found in the book by Weinberger (1990). Since we are only interested in a rather special aspect of the method, we may choose the simplest way to solve the problem and assume, first, there is one atom per unit cell and, second, the potential is of muffin-tin (MT) form as in Sec. 3.3.1, i.e. it is spherically symmetric inside the MT sphere and zero outside. Thus we write

$$V(\mathbf{r}) = \sum_\nu V_c(\mathbf{r}-\mathbf{R}_\nu), \tag{3.102}$$

where V_c is zero outside the MT sphere. We next write down the Bloch function for some energy, ε, not yet an eigenvalue, as

$$\psi_\mathbf{k}(\varepsilon, \mathbf{r}) = \int d\mathbf{r}' \, G_0(\varepsilon, \mathbf{r}-\mathbf{r}') \sum_c V_c(\mathbf{r}'-\mathbf{R}_\nu) \psi_\mathbf{k}(\varepsilon, \mathbf{r}'). \tag{3.103}$$

Substituting for the integration variable $\mathbf{r}' = \mathbf{R}_\nu + \mathbf{x}$ and using the Bloch property, Eqn (3.12), we obtain

$$\psi_\mathbf{k}(\varepsilon, \mathbf{r}) = \int_{\Omega_\mathrm{UC}} d\mathbf{x} \, G(\varepsilon, \mathbf{k}, \mathbf{r}-\mathbf{x}) V_c(\mathbf{x}) \psi_\mathbf{k}(\varepsilon, \mathbf{x}), \tag{3.104}$$

where the integration now extends over the volume, Ω_UC, of the unit cell and the Green's function is Bloch transformed according to

$$G(\varepsilon, \mathbf{k}, \mathbf{r}-\mathbf{x}) = \sum_{\mathbf{R}_\nu} G_0(\varepsilon, \mathbf{r}-\mathbf{x}-\mathbf{R}_\nu) e^{i\mathbf{k}\cdot\mathbf{R}_\nu}. \tag{3.105}$$

We next need the identity (Messiah, 1976)

$$G_0(\varepsilon, \mathbf{r} - \mathbf{r}') = -\kappa \sum_L j_\ell(\kappa r_<) \, n_\ell(\kappa r_>) \, Y_L(\hat{\mathbf{r}}) \, Y_L(\hat{\mathbf{r}}'), \qquad (3.106)$$

where $r_>$ is the larger, $r_<$ the smaller of $\mathbf{r}$ and $\mathbf{r}'$ and the spherical harmonics may be chosen to be real here. The quantities $n_\ell(\kappa r)$ are spherical Neumann functions of order ℓ which are also solutions of the radial Schrödinger equation (3.75), but are, in contrast to the spherical Bessel functions, $j_\ell(\kappa r)$, irregular at the origin. By separating the sum in Eqn (3.105) into the term $\mathbf{R}_\nu = 0$ and $\mathbf{R}_\nu \neq 0$, one derives

$$\begin{aligned} G(\varepsilon, \mathbf{k}, \mathbf{r} - \mathbf{r}') = &- \kappa \sum_L j_\ell(\kappa r_<) \, n_\ell(\kappa r_>) \, Y_L(\hat{\mathbf{r}}) \, Y_L(\hat{\mathbf{r}}') \\ &- \sum_{LL'} j_\ell(\kappa r) \, B_{LL'}(\varepsilon, \mathbf{k}) \, j_{\ell'}(\kappa r') \, Y_L(\hat{\mathbf{r}}) \, Y_{L'}(\hat{\mathbf{r}}'). \end{aligned} \qquad (3.107)$$

An explicit expression for $B_{LL'}(\varepsilon, \mathbf{k})$ is given following Eqn (3.123).

We next replace $V_c(\mathbf{x}) \, \psi_{\mathbf{k}}(\varepsilon, \mathbf{k})$ by $(\nabla_x^2 + \varepsilon) \, \psi_{\mathbf{k}}(\varepsilon, \mathbf{x})$ in Eqn (3.104), which gives

$$\psi_{\mathbf{k}}(\varepsilon, \mathbf{r}) = \int_{|\mathbf{x}| < S - \delta} d\mathbf{x} \, G(\varepsilon, \mathbf{k}, \mathbf{r} - \mathbf{x}) \, (\nabla_x^2 + \varepsilon) \, \psi_{\mathbf{k}}(\varepsilon, \mathbf{x}), \qquad (3.108)$$

where δ is an infinitesimal and then use Green's identity,

$$\begin{aligned} \int d\mathbf{r} \, [f(\mathbf{r}) \, \nabla^2 \, g(\mathbf{r}) - g(\mathbf{r}) \, \nabla^2 \, f(\mathbf{r})] \\ = \int_S [f(\mathbf{r}) \, \frac{\partial}{\partial \mathbf{r}} \, g(\mathbf{r}) - g(\mathbf{r}) \, \frac{\partial}{\partial \mathbf{r}} \, f(\mathbf{r})] \, dS \end{aligned} \qquad (3.109)$$

to obtain

$$\int_{S-\delta} [G(\varepsilon, \mathbf{k}, \mathbf{r} - \mathbf{x}) \, \frac{\partial}{\partial \mathbf{x}} \, \psi_{\mathbf{k}}(\varepsilon, \mathbf{x}) - \psi_{\mathbf{k}}(\varepsilon, \mathbf{x}) \, \frac{\partial}{\partial \mathbf{x}} \, G(\varepsilon, \mathbf{k}, \mathbf{r} - \mathbf{x})] \, dS = 0. \qquad (3.110)$$

In the final step we require a solution of the Schrödinger equation containing only the *single-site* potential $V_c(\mathbf{r})$. This is obtained from the radial equation (3.75) as in the APW case and is denoted as $R_\ell(\varepsilon, \mathbf{r}) \, Y_L(\hat{\mathbf{r}})$ which allows us to expand the Bloch function as

$$\psi_{\mathbf{k}}(\varepsilon, \mathbf{r}) = \sum_L \gamma_L(\mathbf{k}) \, R_\ell(\varepsilon, \mathbf{r}) \, Y_L(\hat{\mathbf{r}}). \qquad (3.111)$$

Substituting this into Eqn (3.110) and using Eqn (3.107) for $G(\varepsilon, \mathbf{k}, \mathbf{r} - \mathbf{x})$ we get after integrating out the angular variables and cancelling a factor of $j_\ell(\kappa r)$

$$\sum_{L'} [B_{LL'}(\varepsilon, \mathbf{k}) \{R_{\ell'}, j_{\ell'}\} + \kappa \, \delta_{LL'} \{R_\ell, n_\ell\}] \gamma_{L'} = 0. \qquad (3.112)$$

Here we used so-called Wronskians which, for two functions $f(r)$, and $g(r)$, are defined as

$$\{f, g\} = \left[f(r) \frac{\mathrm{d}g(r)}{\mathrm{d}r} - g(r) \frac{\mathrm{d}f(r)}{\mathrm{d}r} \right]_{r=S}. \qquad (3.113)$$

The result, Eqn (3.112), is conveniently expressed using the scattering phase shift, $\eta_\ell(\varepsilon)$, of a single scattering center. For this we match at the MT-sphere radius, S, the inside solutions, $R_\ell(\varepsilon, r)$, to proper scattering functions which are the spherical Bessel and Neumann functions $j_\ell(\kappa r)$ and $n_\ell(\kappa r)$:

$$R_\ell(\varepsilon, S) = C_\ell \left[j_\ell(\kappa S) - \tan \eta_\ell \, n_\ell(\kappa S) \right] \qquad (3.114)$$

and

$$\frac{\mathrm{d}}{\mathrm{d}r} R_\ell(\varepsilon, r)|_S = C_\ell \left[\frac{\mathrm{d}}{\mathrm{d}r} j_\ell(\kappa r) - \tan \eta_\ell \, \frac{\mathrm{d}}{\mathrm{d}r} n_\ell(\kappa r) \right]_S. \qquad (3.115)$$

Here $\kappa^2 = \varepsilon$ and $\eta_\ell = \eta_\ell(\varepsilon)$ is the scattering phase shift of angular momentum ℓ (Messiah, 1978). It can be simply expressed with the logarithmic derivative

$$D_\ell(\varepsilon) = \frac{\mathrm{d}}{\mathrm{d}r} \ln R_\ell(\varepsilon, r)|_S \qquad (3.116)$$

and Eqns (3.114) and (3.115) as

$$\tan \eta_\ell = \left[\frac{\mathrm{d}}{\mathrm{d}r} j_\ell(\kappa r)|_S - D_\ell(\varepsilon) j_\ell(\kappa S) \right] \bigg/ \left[\frac{\mathrm{d}}{\mathrm{d}r} n_\ell(\kappa r)|_S - D_\ell(\varepsilon) n_\ell(\kappa S) \right]. \qquad (3.117)$$

A simple and useful alternative to express $\tan \eta_\ell$ is obtained by using the Wronskians. Multiplying Eqn (3.114) by $\mathrm{d}j_\ell/\mathrm{d}r$, Eqn (3.115) by j_ℓ and repeating this procedure using $\mathrm{d}n_\ell/\mathrm{d}r$ and n_ℓ, we get

$$\tan \eta_\ell = \frac{\{R_\ell, j_\ell\}}{\{R_\ell, n_\ell\}}. \qquad (3.118)$$

Thus, if $\{R_{\ell'}, j_{\ell'}\} \neq 0$ we can write Eqn (3.112) in the form

$$\sum_{L'} [B_{LL'}(\varepsilon, \mathbf{k}) + \kappa \, \delta_{LL'} \cot \eta_\ell] \, \alpha_{L'}(\mathbf{k}) = 0, \qquad (3.119)$$

with

$$\alpha_L(\mathbf{k}) = \{R_\ell, j_\ell\} \, \gamma_L(\mathbf{k}). \qquad (3.120)$$

Equation (3.119) is the standard Korringa–Kohn–Rostoker (KKR) form. We may compare this equation with Eqn (3.98): since Eqn (3.119) has solutions when the determinant of the bracketed part vanishes, the inverse of the single-site scattering matrix, t^{-1}, is seen to be proportional to $\delta_{LL'} \cot \eta_\ell$.

It remains to work out the so-called structure constants, $B_{LL'}(\varepsilon, \mathbf{k})$. To indicate the steps we refer back to Eqns (3.105) and (3.106) and compare with Eqn (3.107). Obviously

$$\sum_{LL'} j_\ell(\kappa r) \, B_{LL'}(\varepsilon, \mathbf{k}) \, j_\ell(\kappa r') \, Y_L(\hat{\mathbf{r}}) \, Y_L(\hat{\mathbf{r}}') $$
$$= \kappa \sum_{\mathbf{R}_\nu \neq 0} \sum_L j_\ell(\kappa|r - r'|) \, n_\ell(\kappa|\mathbf{R}_\nu|) \, Y_L(\widehat{r - r'}) \, Y_L(\hat{\mathbf{R}}_\nu) \cdot e^{i\mathbf{k} \cdot \mathbf{R}_\nu} \qquad (3.121)$$

and we see that we must carry out a lattice sum, which is not a trivial task. Although we will need the structure constants for the linearized methods that follow, we omit a detailed derivation but indicate the starting point (for complete details see Ham and Segall, 1961). This is a relation due to Ewald:

$$\frac{e^{i\kappa|\mathbf{r}-\mathbf{R}|}}{|\mathbf{r}-\mathbf{R}|} = \frac{2}{\sqrt{\pi}} \int_C \exp\left[-(\mathbf{r}-\mathbf{R})^2 \zeta + \frac{\kappa^2}{4\zeta^2}\right] d\zeta. \qquad (3.122)$$

For a description of the integration contour C see Ham and Segall (1961). This expression is used to express the Green's function, Eqn (3.100), and the lattice sum is then carried out using the Ewald method. The result of this procedure can be written for any number, N, of atoms per unit cell which are specified by the basis vector $\boldsymbol{\tau}_\alpha$, $\alpha = 1, \ldots, N$:

$$B_{LL'}(\boldsymbol{\tau}_\alpha - \boldsymbol{\tau}_\beta, \varepsilon, \mathbf{k}) = 4\pi \sum_{L''} C_{LL'L''} \, D_{L''}(\boldsymbol{\tau}_\alpha - \boldsymbol{\tau}_\beta, \varepsilon, \mathbf{k}) \, e^{-i\mathbf{k} \cdot (\boldsymbol{\tau}_\beta - \boldsymbol{\tau}_\alpha)}, \qquad (3.123)$$

where

$$C_{LL'L''} = \int d\Omega \, Y_L(\hat{\mathbf{r}}) \, Y_{L'}(\hat{\mathbf{r}}) \, Y_{L''}(\hat{\mathbf{r}}), \qquad (3.124)$$

and

$$D_L(\boldsymbol{\tau}_\alpha - \boldsymbol{\tau}_\beta, \varepsilon, \mathbf{k}) = D_L^{(1)}(\boldsymbol{\tau}_\alpha - \boldsymbol{\tau}_\beta, \varepsilon, \mathbf{k}) + D_L^{(2)}(\boldsymbol{\tau}_\alpha - \boldsymbol{\tau}_\beta, \varepsilon, \mathbf{k}) + \delta_{L,0}\, \delta_{\boldsymbol{\tau}_\alpha - \boldsymbol{\tau}_\beta, 0}\, D^{(3)}. \quad (3.125)$$

Here

$$D_L^{(1)}(\boldsymbol{\tau}_\alpha - \boldsymbol{\tau}_\beta, \varepsilon, \mathbf{k}) = (4\pi/\Omega_{UC})\, \kappa^{-\ell} \exp(\varepsilon/q_0)$$
$$\cdot \sum_s e^{i\mathbf{K}_s \cdot (\boldsymbol{\tau}_\alpha - \boldsymbol{\tau}_\beta)}\, Y_L(\widehat{\mathbf{K}_s + \mathbf{k}})\, |\mathbf{K}_s + \mathbf{k}|^\ell\, \frac{\exp[-(\mathbf{K}_s + \mathbf{k})^2/q_0]}{\varepsilon - (\mathbf{K}_s + \mathbf{k})^2}, \quad (3.126)$$

$$D_L^{(2)}(\boldsymbol{\tau}_\alpha - \boldsymbol{\tau}_\beta, \varepsilon, \mathbf{k}) = \pi^{-1/2}\, (-2)^{\ell+1}\, i^\ell\, \kappa^{-\ell} \sum_\nu{}'\, \exp[i\mathbf{k} \cdot (\mathbf{R}_\nu - \boldsymbol{\tau}_\alpha + \boldsymbol{\tau}_\beta)]$$
$$\cdot Y_L(\widehat{\mathbf{R}_\nu + \boldsymbol{\tau}_\beta - \boldsymbol{\tau}_\alpha}) \cdot |\mathbf{R}_\nu - \boldsymbol{\tau}_\alpha + \boldsymbol{\tau}_\beta|^\ell \int_{\sqrt{q_0}}^\infty z^{2\ell} \quad (3.127)$$
$$\cdot \exp\left[-\frac{1}{4}(\boldsymbol{\tau}_\beta - \boldsymbol{\tau}_\alpha + \mathbf{R}_\nu)^2 z^2 + \varepsilon/z^2\right] dz,$$

and

$$D^{(3)} = -\frac{\sqrt{q_0}}{2\pi} \sum_{n=0}^\infty \frac{(\varepsilon/q_0)^n}{n!\,(2n-1)}. \quad (3.128)$$

The symbol $\sum_\nu'$ is supposed to exclude from the summation the term $\mathbf{R}_\nu = \boldsymbol{\tau}_\alpha - \boldsymbol{\tau}_\beta$. The quantity q_0 is the Ewald parameter which may be chosen to achieve optimal convergence of the terms $D_L^{(1)}$ and $D_L^{(2)}$.

The corresponding KKR equation appropriate to the case of N atoms per unit cell is

$$\sum_{L'} \sum_{\beta=1}^N \{B_{LL'}(\boldsymbol{\tau}_\alpha - \boldsymbol{\tau}_\beta, \varepsilon, \mathbf{k}) + \kappa \cot \eta_{\ell\beta}(\varepsilon)\, \delta_{LL'}\, \delta_{\alpha\beta}\}\, \alpha_{L'\beta}(\mathbf{k}) = 0, \quad (3.129)$$

where $\eta_{\ell\beta}(\varepsilon)$ is the scattering phase shift of the atom numbered β and $\alpha = 1, \ldots, N$.

Concerning the size of the secular equation it is not *a priori* clear how many angular momentum terms must be included in Eqns (3.119) or (3.129). But we expect that for higher angular momenta the logarithmic derivatives approach those of free electrons very quickly. This is, in fact, borne out by experience and can be seen to be true for $\ell \geq 3$, for the example of Cu in Fig. 3.10. This, in turn, means that as

$$D_\ell(\varepsilon) \longrightarrow \frac{\mathrm{d}}{\mathrm{d}r} \ln j_\ell(\kappa r)|_S, \qquad (3.130)$$

or

$$\{R_\ell, j_\ell\} \longrightarrow 0 \qquad (3.131)$$

the scattering phase shift becomes

$$\tan \eta_\ell \longrightarrow 0, \qquad (3.132)$$

and therefore, because of Eqns (3.120),

$$\alpha_L(\mathbf{k}) \longrightarrow 0. \qquad (3.133)$$

This obviously limits the size of the secular equation to rather small values. For instance, for elementary transition metals the maximal value of ℓ can be chosen to be $\ell_{\max} = 3$, i.e. the rank of the secular equation is 16.

The eigenvalues of the KKR equations [Eqns (3.119) or (3.129)] must be obtained by root tracing, as in the case of the APW method. With the eigenvalues so obtained one determines the expansion coefficient α from the KKR equations and hence the wave function ψ. Some technical questions concerning details of the construction of the wave function are addressed in the work of Segall and Ham (1968).

We mentioned earlier that relaxing the muffin-tin approximation for the shape of the potential in the KKR method is possible but not easy. In its simplest incarnation it leads to a nondiagonal scattering matrix $t_{LL'}$, and, consequently, the KKR equation is nondiagonal not only in the structure-constant part, B, but also in the t-matrix part. The structure constants themselves, however, remain unchanged since they are defined by properties of the crystal lattice and not by the scattering strengths of the atoms.

Except for the difficulties arising when the shape approximation of the potential is relaxed, the APW and KKR methods can be made arbitrarily accurate; but they are slow calculationally because the secular equations are nonlinear requiring time-consuming root tracing. Linear methods, however, do exist and are exceedingly important; therefore we will deal with them next.

3.4 Linear methods

The introduction of linear methods into band theory was an important and far-reaching step first taken some 25 years ago by Andersen (1975). It lead to a large variety of calculations for realistic and complex systems allowing considerable insight into their chemical and physical properties at the same time shifting the burden away from technical problems. Linear methods also supplied tools necessary to analyze and interpret the results of numerical calculations and are in this respect superior to—let us call them—the classical methods discussed in the previous two sections. A price, however, has to be paid: slight inaccuracies. Being proponents of linear methods ourselves, we are perhaps not quite unprejudiced and believe the price is not too high.

3.4.1 LCAO

Before we can deal with linear methods in detail, we must return to the general expansion introduced earlier and discuss what is called in rather general terms "linear combination of atomic orbitals" (LCAO). In this section we particularly mean energy-independent orbitals that are associated with a specific atom or ion at a well-defined position in the crystal but not necessarily related to eigenfunctions of the atom in question. Given such orbitals, $\chi_j(\mathbf{r} - \mathbf{R}_\nu - \boldsymbol{\tau})$, a Bloch function is written as

$$\chi_{j\mathbf{k}}(\mathbf{r}) = \sum_\nu e^{i\mathbf{k}\cdot\mathbf{R}_\nu} \chi_j(\mathbf{r} - \mathbf{R}_\nu - \boldsymbol{\tau}), \tag{3.134}$$

where the summation extends over the Bravais lattice $\{\mathbf{R}_\nu\}$, $\boldsymbol{\tau}$ is a basis vector when the crystal is not primitive, and j is a combination of quantum numbers characterizing the orbital χ. The Bloch function defined by Eqn (3.134) is, of course, not an eigenfunction of the crystal Schrödinger equation. To obtain a variational estimate we use the expansion

$$\psi_\mathbf{k}(\mathbf{r}) = \sum_j a_j(\mathbf{k})\, \chi_{j\mathbf{k}}(\mathbf{r}), \tag{3.135}$$

and determine the coefficients variationally as before. Let us here denote the matrix elements of the Hamiltonian by

$$\int \chi^*_{i\mathbf{k}}(\mathbf{r})\, H\, \chi_{j\mathbf{k}}(\mathbf{r})\, d\mathbf{r} \equiv \langle i|H|j\rangle \tag{3.136}$$

and the overlap matrix by

$$\int \chi^*_{i\mathbf{k}}(\mathbf{r})\, \chi_{j\mathbf{k}}(\mathbf{r})\, d\mathbf{r} \equiv \langle i|j\rangle. \tag{3.137}$$

Then the secular equation is

$$\sum_j \{\langle i|H|j\rangle - \varepsilon \langle i|j\rangle\} a_j = 0 \tag{3.138}$$

for each i. The distinguishing feature of this secular equation is its linearity in energy, ε, because the orbitals χ, and hence the matrix elements, are chosen to be independent of energy. Note that the dependence on $\mathbf{k}$ of the matrix elements does not mean energy dependence in this context. The convergence properties of Eqn (3.138), of course, depend strongly on the choice of the set of orbitals $\{\chi_j(\mathbf{r} - \mathbf{R})\}$. Here a number of possibilities exists. Roughly speaking, we may distinguish two: one is *augmented* in the vicinity of the atoms. This is similar to the APW or KKR methods, but the augmentation is energy-independent. We will discuss this in detail later on in this section. The other type simply consists of a judicious choice of functions without augmentation. In both cases, however, the efficiency is largely determined by the ease with which multicenter integrals occurring in Eqn (3.138) can be evaluated. What we mean by this is the following: an integral of the type $\langle i|j \rangle$ contains terms like

$$\int \chi_i^*(\mathbf{r} - \mathbf{R}_\nu - \boldsymbol{\tau}) \, \chi_j(\mathbf{r} - \mathbf{R}_\mu - \boldsymbol{\tau}') \, d\mathbf{r},$$

so when the integration variable $\mathbf{r}$ ranges near the site $\mathbf{R}_\lambda + \boldsymbol{\tau}''$, but $\lambda \neq \nu$, $\lambda \neq \mu$, and $\mu \neq \nu$, then this integral is called a three-center term; the meaning of analogous two- and one-center integrals should be clear now. A determination of these multicenter terms can always be done numerically, but this is inefficient and time-consuming. It becomes much more efficient when there exists an expansion theorem which allows us to express the orbital centered at some site $\mathbf{R}$ in the crystal in terms of orbitals centered at some other site, $\mathbf{R}'$, i.e. for $\mathbf{R} \neq \mathbf{R}'$ and $\mathbf{r}$ in a sphere at $\mathbf{R}$ that does not overlap the same size sphere at $\mathbf{R}'$, we desire a relation

$$\chi_j(\mathbf{r} + \mathbf{R}') = \sum_i T_{ji}(\mathbf{R}', \mathbf{R}) \, \chi_i(\mathbf{r} + \mathbf{R}) \tag{3.139}$$

with mathematically well-defined coefficients T_{ji}. Examples are the spherical Bessel, Neumann, and Hankel functions. Other sets of functions possessing an expansion theorem are Slater-type orbitals ($r^\beta e^{-\alpha r}$) and Gaussian-type orbitals ($r^\beta e^{-\lambda r^2}$) (GTO).

Eschrig (1989) has devised an LCAO scheme using Slater-type orbitals (STO) which yields a rather small secular equation. This is achieved by choosing the STO such that they are accurate, variational solutions of the atomic-like problem in the vicinity of the nuclei. In effect this is similar to augmentation, but the details are quite different. Since this method is rather new, more experience with it is needed before an impartial judgement can be made.

In closing our short discussion of LCAO methods we point out that they also constitute an important tool for *empirical* descriptions of band structures. In this case the detailed nature of the orbitals is left open, except for their angular dependence which is formulated with spherical harmonics, as before, see e.g. Slater and Koster (1954). The actual matrix elements occurring in Eqn (3.138) are simplified, by neglecting, for instance, three-center terms and all interactions beyond next-nearest neighbors, and are then treated as adjustable parameters to describe some measured or otherwise calculated features of the band structure (Harrison, 1980; Papaconstantopoulos, 1986). These techniques are often called "empirical tight binding."

We next turn to a discussion of linear methods and begin by looking at the energy dependence of the single-site wave function, ϕ.

3.4.2 Energy derivative of the wave function: ϕ and $\dot{\phi}$

At the center of linearized methods is the concept of augmentation which, as we learned in Sec. 3.3.1, consists of replacing a wave function in the vicinity of the nucleus by a numerical solution of the single-site Schrödinger equation. Following the treatment by Andersen (1984), we assume the potential is spherically symmetric within some sphere of radius S but has a rather general form in the interstitial region, i.e. we write

$$V(\mathbf{r}) = \sum_{\nu\tau} V_{c\tau}(|\mathbf{r} - \mathbf{R}_\nu - \boldsymbol{\tau}|) + V_i(\mathbf{r}), \qquad (3.140)$$

where $V_{c\tau}(\mathbf{r})$ is assumed to vanish *outside* the sphere of radius S_τ and the potential in the interstitial, $V_i(\mathbf{r})$, is assumed to vanish *inside* these spheres; the vectors $\{\mathbf{R}_\nu\}$ define the Bravais lattice, $\{\boldsymbol{\tau}\}$ the basis.

The energy dependence of the APW or KKR basis functions clearly stems from the solutions as a function of energy, ε, of the Schrödinger equation (3.75) for APWs, or multiple scattering theories. There are different possibilities to sidestep this energy dependence. Andersen achieves it by expanding the solution of the single-sphere Schrödinger equation in terms of $\phi(\mathbf{r})$ belonging to one, initially arbitrarily chosen energy, $\varepsilon = E_\nu$, i.e.

$$(-\nabla^2 + V_c - E_\nu)\phi(\mathbf{r}) = 0 \qquad (3.141)$$

and its energy derivative,

$$\frac{\partial}{\partial \varepsilon} \phi(\varepsilon, \mathbf{r})|_{\varepsilon = E_\nu},$$

which is denoted by

$$\dot{\phi}(\mathbf{r}) = \frac{\partial}{\partial \varepsilon} \phi(\varepsilon, \mathbf{r})|_{\varepsilon = E_\nu}. \qquad (3.142)$$

To simplify the notation we have dropped the index τ on $V_{c\tau}$. The motivation for this choice to expand the energy-dependent functions can superficially be argued to be the Taylor series, but a more physically compelling reason will be given later on in the ASW subsection.

Only a few facts and definitions are needed to make Andersen's scheme transparent. First we see from the Schrödinger equation for $\phi(\varepsilon, \mathbf{r})$ that the equation for $\dot{\phi}$ is

$$(-\nabla^2 + V_c - E_\nu)\dot{\phi}(\mathbf{r}) = \phi(\mathbf{r}). \qquad (3.143)$$

If the wave function ϕ is normalized to unity in the sphere, i.e.

$$\langle \phi | \phi \rangle = 1, \qquad (3.144)$$

defining the brackets appropriately, one sees at once that

$$\langle \phi | \dot\phi \rangle = 0, \qquad (3.145)$$

hence ϕ and $\dot\phi$ are orthogonal. We next define the radial function belonging to $\phi(\varepsilon, \mathbf{r})$ by the symbol $\phi(\varepsilon, r)$. Note that here we distinguish these two functions only by their arguments. This differs from our notation in the previous section and is done to keep the notation as close as possible to that of Andersen. Then the logarithmic derivative is defined by

$$D(\varepsilon) = \left[S \frac{d}{dr} \ln \phi(\varepsilon, r) \right]_S, \qquad (3.146)$$

where, compared with Eqn (3.116), an extra factor of S appears. We can now expand any function, Φ, having the logarithmic derivative D in terms of ϕ and $\dot\phi$ by

$$\Phi(r) = \phi(r) + w\, \dot\phi(r). \qquad (3.147)$$

Denoting the logarithmic derivatives of ϕ and $\dot\phi$ by $D\{\phi\}$ and $D\{\dot\phi\}$, respectively, we obtain by a simple calculation

$$w = w(D) = -\frac{\phi(S)}{\dot\phi(S)} \frac{D - D\{\phi\}}{D - D\{\dot\phi\}}, \qquad (3.148)$$

where, we remind the reader, D is the logarithmic derivative of the function Φ we want to expand.

The four parameters occurring here, i.e. $\phi(S)$, $\dot\phi(S)$, $D\{\phi\}$, and $D\{\dot\phi\}$, are not independent. Substituting into the norm, Eqn (3.144), the equation (3.143) for ϕ and using (3.141) as well as Green's identity, one can derive

$$1 = S\,\phi(S)\,\dot\phi(S)[D\{\phi\} - D\{\dot\phi\}]. \qquad (3.149)$$

Using this relation and Eqn (3.148) for w in Eqn (3.147) we obtain for the value of $\Phi(r)$ at the sphere the relation

$$\Phi(S) = [S\dot\phi(S)\,(D - D\{\dot\phi\})]^{-1}. \qquad (3.150)$$

With the help of these relations a useful parametrization of the logarithmic derivative function $D(\varepsilon)$ may at last be found (Andersen, 1984) by using $\Phi(r)$ expanded above by ϕ and $\dot\phi$, Eqn (3.147) as a variational estimate to $\varepsilon(D)$, the inverse to $D(\varepsilon)$; a straightforward calculation gives

$$\varepsilon(D) = \langle \Phi | -\nabla^2 + V_c | \Phi \rangle / \langle \Phi | \Phi \rangle = E_\nu + w(D)/(1 + w^2(D)\,\langle \dot\phi^2 \rangle), \qquad (3.151)$$

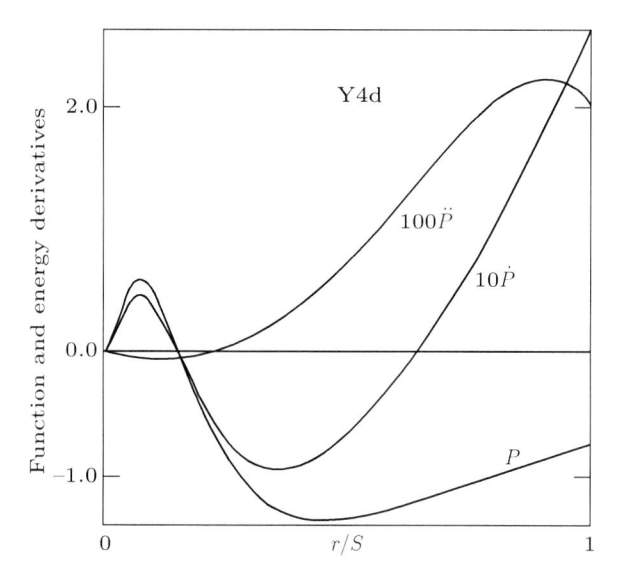

Figure 3.12 Radial d function, P, and energy derivatives, $\dot{P}$, $\ddot{P}$, of Y, given by Andersen (1975).

or the less precise energy

$$\tilde{\varepsilon}(D) = E_\nu + \omega(D), \tag{3.152}$$

where

$$\langle \dot{\phi}^2 \rangle = \langle \dot{\phi} | \dot{\phi} \rangle, \tag{3.153}$$

for which Andersen (1975) derives

$$\langle \dot{\phi}^2 \rangle = -\frac{1}{3} \ddot{\phi}(S)/\phi(S). \tag{3.154}$$

The value of these relations stems from the observation that each successive energy derivative of the function $\phi(r)$ decreases by an order of magnitude. This is illustrated in Fig. 3.12 by means of the d function of Y (taken from Andersen (1975)).

An explicit example for the parametrization given in Eqn (3.152) sheds light on the quality of these approximations. It is again obtained from free electrons and the simple case of $\ell = 0$, for which the zero-order spherical Bessel function is

$$\phi(\varepsilon, r) = \alpha \frac{\sin(\sqrt{\varepsilon}\, r)}{\sqrt{\varepsilon}\, r}, \tag{3.155}$$

where α is obtained by normalizing in the sphere of radius S and

$$D(\varepsilon) = [x \cot x - 1]_{x = S\sqrt{\varepsilon}}. \tag{3.156}$$

In Fig. 3.13 we compare the exact logarithmic derivative, Eqn (3.156), over a large energy range with the approximations (3.151) and (3.152) evaluated for an energy E_ν satisfying $D(E_\nu) = -1$.

After these preliminaries we are now in a position to define the energy-independent orbital $\chi_{j\mathbf{k}}(\mathbf{r})$, Eqn (3.134), for the whole crystal given the functions ϕ and $\dot{\phi}$ in the spheres. Following Andersen (1984) we write

$$\chi_{j\mathbf{k}}(\mathbf{r}) = \chi^e_{j\mathbf{k}}(\mathbf{r}) + \sum_{L\tau} \{\phi_{L\tau}(\mathbf{r}_\tau)\, \Pi_{L\tau j}(\mathbf{k}) + \dot{\phi}_{L\tau}(\mathbf{r}_\tau)\, \Omega_{L\tau j}(\mathbf{k})\}, \tag{3.157}$$

where $\chi^e_{j\mathbf{k}}(\mathbf{r})$ is a so-called envelope function that is defined to be nonzero only outside the spheres and we have appended the angular momentum index, L, and the site index, τ, to ϕ and $\dot{\phi}$,

$$\phi_{L\tau}(\mathbf{r}_\tau) = \phi_{\ell\tau}(r_\tau)\, Y_L(\hat{\mathbf{r}}_\tau), \tag{3.158}$$

with $\mathbf{r}_\tau = \mathbf{r} - \boldsymbol{\tau}$. ϕ and $\dot{\phi}$ are assumed to be zero outside their respective spheres. The matrices Π and Ω are constructed to have Bloch symmetry, i.e.

$$\Pi_{L\boldsymbol{\tau}+\mathbf{R}_\nu j}(\mathbf{k}) = e^{i\mathbf{k}\cdot\mathbf{R}_\nu}\, \Pi_{L\tau j}(\mathbf{k}); \tag{3.159}$$

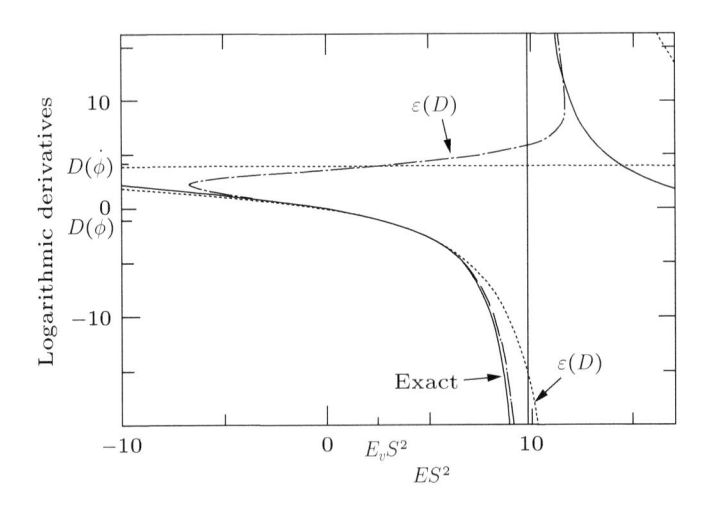

Figure 3.13 *Exact logarithmic derivative $D(E)$ and the parametrizations $\varepsilon(D)$ and $\tilde{\varepsilon}(D)$ for free electrons with $\ell = 0$ as given by Andersen (1975).*

similarly for Ω, and are calculated from the requirement that $\chi_{j\mathbf{k}}(\mathbf{r})$ is everywhere continuous and differentiable. Their determination becomes possible once the envelope function is specified. The procedure is thus quite general and we will explicitly complete the construction of $\chi_{j\mathbf{k}}(\mathbf{r})$ for the case of linear augmented plane waves (LAPW) and linear augmented muffin-tin orbitals (LMTO). Obviously, other schemes are possible; for instance, the choice of a Slater orbital for the envelope function supplies a formalism to handle multicenter integrals and results in a linear method developed by Davenport (1984; Fernando *et al.*, 1989a,b).

Before specifying the envelope function we may determine the secular equation in terms of Π and Ω. Using the definitions for the integrals supplied by Eqns (3.136) and (3.137), where we now imply integration over the unit cell, we easily derive by means of the Schrödinger equation for ϕ and $\dot{\phi}$, Eqns (3.141) and (3.143), the orthogonality relation, Eqn (3.145):

$$\langle i|H - E_\nu|j\rangle = \langle i|H - E_\nu|j\rangle^e + \sum_{L\tau} \Pi^+_{L\tau i}(\mathbf{k})\, \Omega_{L\tau j}(\mathbf{k}) \qquad (3.160)$$

and

$$\langle i|j\rangle = \langle i|j\rangle^e + \sum_{L\tau} \Pi^+_{L\tau i}(\mathbf{k})\, \Pi_{L\tau j}(\mathbf{k}) + \sum_{L\tau} \Omega^+_{L\tau i}(\mathbf{k})\, \langle \dot{\phi}^2_{L\tau}\rangle\, \Omega_{L\tau j}(\mathbf{k}), \qquad (3.161)$$

where the notation $\langle \cdots \rangle^e$ denotes integration over the interstitial, the integrand containing the envelope function. The secular equation now becomes

$$\sum_j \{\langle i|H|j\rangle - \varepsilon\, \langle i|j\rangle\}\, a_j(\mathbf{k})$$

$$= \sum_j \{\langle i|H - E_\nu|j\rangle^e + \sum_{L\tau} \Pi^+_{L\tau i}(\mathbf{k})\, \Omega_{L\tau j}(\mathbf{k}) - \varepsilon'\, \langle i|j\rangle\}\, a_j(\mathbf{k}) \qquad (3.162)$$

$$= 0,$$

where ε' is the energy counted from E_ν (which, in general, will depend on angular momentum).

3.4.3 Linear augmented plane waves (LAPW)

With a choice of a plane wave for the envelope function we obtain the LAPW method. The quantum label j becomes a reciprocal lattice vector $\mathbf{K}_s$ and Bloch symmetrization is not necessary. We write for $\mathbf{r}$ outside the spheres

$$\chi^e_{\mathbf{K}_s \mathbf{k}}(\mathbf{r}) = e^{i(\mathbf{k}+\mathbf{K}_s)\cdot \mathbf{r}} \qquad (3.163)$$

and augment this plane wave inside the spheres using the angular momentum expansion about the site τ

$$e^{i\mathbf{K}\cdot\mathbf{r}_\tau} = e^{i\mathbf{K}\cdot\tau} 4\pi \sum_L i^\ell j_\ell(Kr) Y_L^*(\hat{K}) Y_L(\hat{\mathbf{r}}), \qquad (3.164)$$

where $\mathbf{r}_\tau = \mathbf{r} + \tau$. Inside the sphere at τ we use a function $\Phi_{\ell\tau}(r)$ having the logarithmic derivative

$$D_{\ell\tau K} = [xj_\ell'(x)/j_\ell(x)]_{x=KS_\tau}, \qquad (3.165)$$

i.e. using Eqn (3.147),

$$\Phi_{\ell\tau} \equiv \Phi_{\ell\tau}(D_{\ell\tau K}, r) = \phi_{\ell\tau}(r) + \omega_{\ell\tau K}\, \dot{\phi}_{\ell\tau}(r) \qquad (3.166)$$

and replace $j_\ell(Kr)$ in Eqn (3.164) by

$$j_\ell(Kr) \longrightarrow j_\ell(KS_\tau)\Phi_{\ell\tau}(D_{\ell\tau K}, r)/\Phi_{\ell\tau}(D_{\ell\tau K}), \qquad (3.167)$$

where we have abbreviated

$$\Phi_{\ell\tau}(D_{\ell\tau K}) = \Phi_{\ell\tau}(D_{\ell\tau K}, S_\tau), \qquad (3.168)$$

which is given by Eqn (3.150).

The quantities Π and Ω follow at once and are

$$\Pi_{L\tau\mathbf{K}_s}(\mathbf{k}) = 4\pi\, i^\ell\, e^{i\mathbf{G}_s\cdot\tau}\, j_\ell(G_s S_\tau)\, Y_L(\hat{G}_s)/\Phi_{\ell\tau}(D_{\ell\tau G_s}), \qquad (3.169)$$

and

$$\Omega_{L\tau\mathbf{K}_s}(\mathbf{k}) = \Pi_{L\tau\mathbf{K}_s}(\mathbf{k}) \cdot \omega_{\ell\tau G_s}, \qquad (3.170)$$

where

$$\mathbf{G}_s = \mathbf{k} + \mathbf{K}_s \qquad (3.171)$$

and $\omega_{\ell\tau G_s}$ is given by Eqn (3.148).

Now it is not hard to write out the secular equation in full generality and the interested reader may do this as an exercise. For simplicity in writing, however, we restrict the treatment and assume an elementary solid, i.e. $\tau = 0$ (we drop this symbol from the notation); furthermore we use the muffin-tin approximation for the potential. The integral in the secular equation (3.162) containing the envelope function is then given by Eqn (3.85) and (3.86) in the APW section. For the remaining terms in Eqn (3.162) we use Eqn (3.170) and write

$$\sum_L \Pi^+_{L\mathbf{K}_s} \Omega_{L\mathbf{K}_t} - \varepsilon' \sum_L \Pi^+_{L\mathbf{K}_s} \Pi_{L\mathbf{K}_t} - \varepsilon' \sum_L \Omega^+_{L\mathbf{K}_s} \langle \dot{\phi}_L^2 \rangle \Omega_{L\mathbf{K}_t}$$
$$= \sum_L \Pi^+_{L\mathbf{K}_s} (\omega_{\ell G_s} - \varepsilon' \beta_{\ell G_s G_t}) \Pi_{L\mathbf{K}_t}, \qquad (3.172)$$

where

$$\beta_{\ell G_s G_t} = 1 + \langle \dot{\phi}_L^2 \rangle \omega_{\ell G_s} \omega_{\ell G_t}. \qquad (3.173)$$

Next we use the identity

$$\frac{1}{\Phi_\ell(D_{\ell K_s})} \frac{1}{\Phi_\ell(D_{\ell K_t})} = S \frac{D_{\ell K_s} - D_{\ell K_t}}{\omega_{\ell K_t} - \omega_{\ell K_s}}, \qquad (3.174)$$

which may be proved directly with the help of Eqns (3.148) – (3.150). If we now define the quantity

$$W^\ell_{st} = \frac{1}{\Omega_{\text{UC}}} \frac{\omega_{\ell K_t} - \omega_{\ell K_s}}{G_t^2 - G_s^2} \sum_m \Pi_{L\mathbf{K}_s} \Pi_{L\mathbf{K}_t} \qquad (3.175)$$

and insert the expression for Π, Eqn (3.169), we obtain

$$W^\ell_{st} = \frac{4\pi S}{\Omega_{\text{UC}}} (2\ell+1) P_\ell(\cos\theta_{st}) j_\ell(G_s S) j_\ell(G_t S) \cdot \frac{D_{\ell K_s} - D_{\ell K_t}}{G_t^2 - G_s^2} \qquad (3.176)$$

(remember $L = (\ell, m)$). With the help of Eqn (3.86) one finds

$$\sum_\ell W^\ell_{st} = \frac{4\pi S^3}{\Omega_{\text{UC}}} \frac{j_1(|\mathbf{K}_s - \mathbf{K}_t|S)}{|\mathbf{K}_s - \mathbf{K}_t|S}. \qquad (3.177)$$

Defining two more abbreviations by

$$\Gamma^\ell_{st} = -(\mathbf{G}_t^2 - E_\nu) + \frac{G_t^2 - G_s^2}{\omega_{\ell K_t} - \omega_{\ell K_s}} \cdot \omega_{\ell K_t} \qquad (3.178)$$

and

$$\Delta^\ell_{st} = \beta_{\ell G_s G_t} \frac{G_t^2 - G_s^2}{\omega_{\ell K_t} - \omega_{\ell K_s}} - 1 \qquad (3.179)$$

we finally obtain

$$\langle s|H|t\rangle - \varepsilon \langle s|t\rangle = \Omega_{\text{UC}} \left\{ [(\mathbf{k}+\mathbf{K}_t)^2 - \varepsilon]\delta_{st} + \sum_\ell W_{st}^\ell (\Gamma_{st}^\ell - \varepsilon' \Delta_{st}^\ell) \right\}. \quad (3.180)$$

This is the secular equation for the LAPW method; see also Andersen (1975) and (1984).

The major advantage of the LAPW method is its high accuracy and relative ease with which it can treat a general potential. Another similar LAPW technique has been devised by Krakauer *et al.* (1979).

One of the frequently used and highly successful linearized APW programs is that developed by Schwarz *et al.* (2002) called Wien2K. The modern version of this program is full-potential, even though it requires the choice of muffin-tin spheres. For its detailed description we refer the reader to the literature. A critical discussion of the precision of these and other modern methods to tackle the density-functional method can be found in a publication by Lejaeghere *et al.* (2016).

3.4.4 Linear combination of muffin-tin orbitals (LMTO)

We next turn to the "linear muffin-tin orbital method" (LMTO). It is obtained by requiring the envelope function to satisfy Laplace's equation (Andersen, 1975, 1984), i.e.

$$\nabla^2 \chi_L^e(\mathbf{r}) = 0. \quad (3.181)$$

In spherical geometry there are two well-known types of solutions, one type is regular at infinity and is given by

$$\chi_{L-}^e(\mathbf{r}) = \left(\frac{r}{S}\right)^{-\ell-1} Y_L(\hat{\mathbf{r}}), \quad (3.182)$$

the other types is regular at the origin and is given by

$$\chi_{L+}^e(\mathbf{r}) = \left(\frac{r}{S}\right)^\ell Y_L(\hat{\mathbf{r}}). \quad (3.183)$$

$S > 0$ is arbitrary at this point. The expansion theorem, Eqn (3.139), is the well-known series of a static multipole potential:

$$\left(\left|\frac{\mathbf{r}-\mathbf{R}}{S}\right|\right)^{-\ell-1} Y_L(\widehat{\mathbf{r}-\mathbf{R}}) = -\sum_{L'} \left(\frac{r}{S}\right)^{\ell'} Y_{L'}(\hat{\mathbf{r}}) \frac{1}{2(2\ell'+1)} \mathcal{S}_{L'L}(\mathbf{R}) \quad (3.184)$$

connecting the two types of solutions, χ_{L-}^e and χ_{L+}^e. $\mathcal{S}$ is given by

$$\mathcal{S}_{L'L}(\mathbf{R}) = (4\pi)^{1/2} g_{L'L} \left(\frac{S}{R}\right)^{\ell'+\ell+1} Y_{\ell'+\ell,m'-m}^*(\hat{\mathbf{R}}), \quad (3.185)$$

with $L = (\ell, m)$, $L' = (\ell', m')$ and

$$g_{L'L} = (-1)^{\ell+m+1} 2[(2\ell'+1)(2\ell+1)(\ell'+\ell-m'+m)!(\ell'+\ell+m'-m)!]^{1/2}$$
$$\cdot [(2\ell'+2\ell+1)(\ell'+m')!(\ell'-m')!(\ell+m)!(\ell-m)!]^{-1/2}. \qquad (3.186)$$

Restricting for simplicity our discussion to the case of one atom per unit cell again, we now express the energy-independent orbital in the form of Eqn (3.157) noting that the angular momentum, L, replaces the index j:

$$\chi_{L\mathbf{k}}(\mathbf{r}) = \chi^e_{L-}(\mathbf{r}) + \sum_{L'} [\phi_{L'}(\mathbf{r}) \Pi_{L'L}(\mathbf{k}) + \dot{\phi}_{L'}(\mathbf{r}) \Omega_{L'L}(\mathbf{k})]. \qquad (3.187)$$

The matrices Π and Ω are determined as follows.
At $r = S$ the logarithmic derivative of χ^e_{L-} is $D = -\ell - 1$. We therefore match the function χ^e_{L-} at S with the function $\Phi_\ell(D = -\ell - 1, r)/\Phi_\ell(-)$, where we have abbreviated

$$\Phi_\ell(-) = \Phi_\ell(D = -\ell - 1, S). \qquad (3.188)$$

In other words, for $r < S$ we replace

$$\left(\frac{r}{S}\right)^{-\ell-1} \longrightarrow \Phi_\ell(-\ell-1, r)/\Phi_\ell(-), \qquad (3.189)$$

and use suitable E_ν's to expand Φ in terms of ϕ and $\dot{\phi}$ as given in Eqn (3.147). The sphere radius S will be specified a little later. If we place one orbital like Eqn (3.189) at the sphere centered at the origin, translational symmetry requires the same orbital to be placed at all other sites in such a way that they obey Bloch's theorem. Since the "tail" of one orbital centered at $R \neq 0$ is given in the central sphere by the expansion, Eqn (3.184), the tails of *all* orbitals superimpose with correct Bloch symmetry and are given by

$$-\sum_{L'} (r/S)^{\ell'} Y_{L'}(\hat{\mathbf{r}}) \frac{1}{2(2\ell'+1)} \mathcal{S}^{\mathbf{k}}_{L'L},$$

where we define

$$\mathcal{S}^{\mathbf{k}}_{L'L} = \sum_{\nu \neq 0} e^{i\mathbf{k}\cdot\mathbf{R}_\nu} \mathcal{S}_{L'L}(\mathbf{R}_\nu). \qquad (3.190)$$

The quantities Π and Ω are now obtained by matching the "heads," Eqn (3.189), and the tails above. The logarithmic derivatives of the functions χ^e_{L+} occurring in the tails are obviously $D = \ell$ and we abbreviate

134 *Energy-Band Theory*

$$\Phi_\ell(D=\ell, S) = \Phi_\ell(+). \tag{3.191}$$

Then for $r < S$ the desired orbital may be written as

$$\chi_{L\mathbf{k}}(\mathbf{r}) = \Phi_\ell(-\ell-1, \mathbf{r})/\Phi_\ell(-) - \sum_{L'} \Phi_{L'}(\ell', \mathbf{r})/\Phi_\ell(+) \frac{1}{2(2\ell'+1)} \mathcal{S}^{\mathbf{k}}_{L'L}. \tag{3.192}$$

Using the relation between Φ and $\phi, \dot{\phi}$, Eqn (3.147), and comparing Eqn (3.187) with Eqn (3.192), we can read off Π and Ω. The results are most compactly written by defining

$$\omega_{\ell-} = \omega_\ell(-\ell-1) \, , \, \omega_{\ell+} = \omega_\ell(\ell), \tag{3.193}$$

$$\Delta_\ell = \omega_{\ell+} - \omega_{\ell-}, \tag{3.194}$$

$$\tilde{\Phi}_\ell = \Phi_\ell(-)\sqrt{\frac{S}{2}}, \tag{3.195}$$

and by using the identity

$$\omega_{\ell-} - \omega_{\ell+} = S(2\ell+1)\Phi_\ell(+)\Phi_\ell(-), \tag{3.196}$$

which is easily derived from Eqns (3.148) and (3.150). One finds

$$\Pi_{L''L}(\mathbf{k}) = \tilde{\Phi}_\ell^{-1}\delta_{LL''} + (\tilde{\Phi}_{\ell''}/\Delta_{\ell''})\mathcal{S}^{\mathbf{k}}_{L''L}, \tag{3.197}$$

and

$$\Omega_{L''L}(\mathbf{k}) = \omega_{\ell''-}\tilde{\Phi}_{\ell''}^{-1}\delta_{LL''} + (\tilde{\Phi}_{\ell''}\omega_{\ell''+}/\Delta_{\ell''})\mathcal{S}^{\mathbf{k}}_{L''L}. \tag{3.198}$$

These quantities Π and Ω define a so-called linear muffin-tin orbital (LMTO), and, apart from slight differences in notation, they agree with Skriver's (1984) expressions. Their specification is, however, still incomplete because the partitioning of space into interstitial and spherical regions has yet to be defined, in particular, the sphere radius S. We do this now.

In closely packed solids each atom has 8–12 nearest neighbors and the assumed muffin-tin (MT) form of the potential is a good approximation. Far more convenient and sometimes even more accurate is an approximation in which the interstitial region is eliminated by enlarging the MT spheres and neglecting the slight overlap. The MT spheres in this approximation thus become the Wigner–Seitz atomic spheres (S is the Wigner–Seitz radius) which are supposed to fill space completely (Wigner and Seitz, 1955). This is called the atomic sphere approximation (ASA) and, for example, in the case of a body centered cubic (bcc) crystal the radius S is related to the lattice constant, a, by

$$\frac{4\pi}{3}S^3 = \frac{1}{2}a^3 \qquad (3.199)$$

(or $1/4\,a^3$ in the case of fcc crystals).

In open structures the interstitial positions often have such high symmetry that both the atomic and the repulsive interstitial potential can be approximated by spherically symmetric ones and the atomic and interstitial spheres together form a close packing. We will now use the ASA and continue by stating that the terms referring to the interstitial region and the non-muffin-tin part of the potential drop out, but, of course, the boundary conditions incorporated in the quantities Π and Ω remain. The secular matrix given in Eqn (3.162) (with Eqn (3.161)) then consists solely of products involving Π and Ω given above and can be written as

$$\langle L|H|L'\rangle - \varepsilon\langle L|L'\rangle = \{[\omega_- + E_\nu(1+\omega_-^2\langle\dot{\phi}^2\rangle)]/\tilde{\Phi}^2\}_\ell\,\delta_{LL'}$$

$$+ \{\{[\omega_+ + E_\nu(1+\omega_-\omega_+\langle\dot{\phi}^2\rangle)]/\Delta\}_\ell + \{\cdots\}_{\ell'} - 1\}\,\mathcal{S}^{\mathbf{k}}_{LL'}$$

$$+ \sum_{L''}\mathcal{S}^{\mathbf{k}}_{LL''}\{\tilde{\Phi}^2[\omega_+ + E_\nu(1+\omega_+^2\langle\dot{\phi}^2\rangle)]/\Delta^2\}_{\ell''}\,\mathcal{S}^{\mathbf{k}}_{L''L'}$$

$$-\varepsilon\Big\{\{[1+\omega_-^2\langle\dot{\phi}^2\rangle]/\tilde{\Phi}^2\}_\ell\,\delta_{LL'} + \{\{[1+\omega_+\omega_-\langle\dot{\phi}^2\rangle]/\Delta\}_\ell$$

$$+\{\cdots\}_{\ell'}\}\mathcal{S}^{\mathbf{k}}_{LL'} + \sum_{L''}\mathcal{S}^{\mathbf{k}}_{LL''}[\tilde{\Phi}^2(1+\omega_+^2\langle\dot{\phi}^2\rangle)/\Delta^2]_{\ell''}\,\mathcal{S}^{\mathbf{k}}_{L''L'}\Big\}.$$

$$(3.200)$$

The terms without a factor of $\mathcal{S}$ are one-center, those with one factor of $\mathcal{S}$ are two-center, and those with two factors of $\mathcal{S}$ are three-center terms. The dots are meant to repeat the contents of the preceding curly bracket.

Although the errors introduced by the atomic sphere approximation are unimportant for many applications, there are cases where energy bands of high accuracy are needed and where one should include the non muffin-tin perturbations in some form. Skriver (1984) gives expressions which account to first order for the differences between the atomic spheres and the atomic polyhedron and correct for the neglect of higher partial waves. These extra terms which are added to the LMTO matrices, Eqn (3.200), are called the "combined correction terms." We will omit their treatment here but will discuss this problem in some detail in connection with the augmented spherical wave (ASW) method which we turn to now.

3.4.5 Augmented spherical waves (ASW)

The augmented spherical wave method (ASW) was developed by Williams *et al.* (1979) and without Andersen's LMTO technique it would not have been conceived. The differences are essentially two. There is a different choice of the envelope function and a different and simpler method for augmenting the basis functions. For the former, solutions of the wave equation,

$$(\nabla^2 + \varepsilon_0) \chi_L^e(\mathbf{r}) = 0, \qquad (3.201)$$

are used which, of course, coincide with the Laplace equation if $\varepsilon_0 = 0$. In spherical geometry there are again two types of solutions. For negative energies, the solutions regular at infinity are given by spherical Hankel functions $h_\ell^+(\kappa r)$, $\kappa = \sqrt{\varepsilon_0}$, and those regular at the origin are spherical Bessel functions $j_\ell(\kappa r)$. For positive energies the spherical Hankel functions must be replaced by spherical Neumann functions and the envelope functions in this case become identical with the basis used in multiple-scattering theories (see Sec. 3.3.2). The energy parameter, ε_0, obviously controls the degree of localization and it could, in principle, be used as a variational parameter. This was indeed done initially, but in all practical cases it was found that a choice of a fixed negative and small value ($\varepsilon_0 = -0.01$ Ry) gave best results for the electronic properties of solids at or near ambient pressure. We will, therefore, limit our discussion to the negative energy case and, whenever possible, suppress the parameter $\varepsilon_0 = \kappa^2$ in the notation.

For mathematical reasons it is convenient to define the envelope function in the following way

$$\chi_L^e(\mathbf{r}) = H_L(\mathbf{r}) = i^\ell \, \kappa^{\ell+1} \, h_\ell^+(\kappa r) \, Y_L(\hat{\mathbf{r}}), \qquad (3.202)$$

the factor $\kappa^{\ell+1}$ being introduced to cancel out the leading energy dependence for small values of κr, because here

$$h_\ell^+(x) \underset{x \to 0}{=} x^{-\ell-1} \frac{(2\ell+1)!!}{(2\ell+1)}. \qquad (3.203)$$

The expansion theorem, Eqn (3.139), that allows an easy treatment of multicenter integrals—and augmentation—connects the two types of functions in the following way

$$H_L(\mathbf{r} + \mathbf{R}) = \sum_{L'} J_{L'}(\mathbf{r}) \, B_{L'L}(\mathbf{R}). \qquad (3.204)$$

Here we assume $\mathbf{R} \neq 0$ and

$$J_L(\mathbf{r}) = i^\ell \, \kappa^{-\ell} \, j_\ell(\kappa r) \, Y_L(\hat{\mathbf{r}}). \qquad (3.205)$$

The reason for the factor $\kappa^{-\ell}$ is again a cancellation of the leading energy dependence, see Eqn (3.89), and we will show the quantity $B_{L'L}(\mathbf{R})$ is given by

$$B_{LL'}(\mathbf{R}) = 4\pi \sum_{L''} C_{LL'L''} \kappa^{\ell+\ell'-\ell''} H^*_{L''}(\mathbf{R}), \tag{3.206}$$

where, as before, $C_{LL'L''}$ is a Gaunt coefficient

$$C_{LL'L''} = \int d\hat{\mathbf{r}}\, Y^*_L(\hat{\mathbf{r}})\, Y_{L'}(\hat{\mathbf{r}})\, Y_{L''}(\hat{\mathbf{r}}). \tag{3.207}$$

To derive the expansion theorem we first express a plane wave, $e^{i\mathbf{k}\cdot(\mathbf{r}-\mathbf{r}')}$, in terms of spherical waves using Eqn (3.76), twice: once to write

$$e^{i\mathbf{k}\cdot(\mathbf{r}-\mathbf{r}')} = 4\pi \sum_L i^\ell\, j_\ell(k|\mathbf{r}-\mathbf{r}'|)\, Y^*_L(\hat{\mathbf{k}})\, Y_L(\widehat{\mathbf{r}-\mathbf{r}'}) \tag{3.208}$$

and then

$$e^{i\mathbf{k}\cdot\mathbf{r}} \left(e^{i\mathbf{k}\cdot\mathbf{r}'}\right)^* = (4\pi)^2 \sum_{LL'} i^{\ell-\ell'}\, j_\ell(kr)\, j_{\ell'}(kr')\, Y^*_L(\hat{\mathbf{k}})\, Y_{L'}(\hat{\mathbf{k}})\, Y^*_{L'}(\hat{\mathbf{r}}')\, Y_L(\hat{\mathbf{r}}) \tag{3.209}$$

which gives, in the notation defined by Eqns (3.202) and (3.205),

$$J_L(\mathbf{r}-\mathbf{r}') = 4\pi \sum_{L'L''} \kappa^{\ell'+\ell''-\ell}\, C_{L'L''L}\, J_{L'}(\mathbf{r})\, J^*_{L''}(\mathbf{r}'). \tag{3.210}$$

Next we need the Green's function (Messiah, 1976) (also compare with Eqns (3.100) and (3.106))

$$\frac{e^{ik|\mathbf{r}-\mathbf{r}'|}}{|\mathbf{r}-\mathbf{r}'|} = 4\pi k \sum_L j_\ell(kr_<)\, h^+_\ell(kr_>)\, Y_L(\hat{\mathbf{r}})\, Y_L(\hat{\mathbf{r}}'), \tag{3.211}$$

which gives for $|\mathbf{x}| < |\mathbf{R}+\mathbf{r}|$

$$\frac{e^{ik|\mathbf{r}+\mathbf{R}-\mathbf{x}|}}{|\mathbf{r}+\mathbf{R}-\mathbf{x}|} = 4\pi \sum_L J^*_L(\mathbf{x})\, H_L(\mathbf{r}+\mathbf{R}) \tag{3.212}$$

and for $|\mathbf{r}-\mathbf{x}| < |\mathbf{R}|$

$$\frac{e^{ik|\mathbf{r}+\mathbf{R}-\mathbf{x}|}}{|\mathbf{r}+\mathbf{R}-\mathbf{x}|} = 4\pi \sum_L J_L(\mathbf{r}-\mathbf{x})\, H^*_L(\mathbf{R}). \tag{3.213}$$

With the help of Eqn (3.210) one obtains the desired result. Note that these formulas are also valid for complex **k** (Messiah, 1976). Furthermore, the LMTO structure constants $\mathcal{S}_{LL'}(\mathbf{R})$ can be seen to follow from $B_{LL'}(\mathbf{R})$ when the energy parameter, κ, which is suppressed in the notation, is allowed to go to zero.

The mathematics of the augmentation process is straightforward. The spherical wave $H_L(\mathbf{r})$ is continued into the atomic sphere in which it is centered by a numerical solution of the Schrödinger equation which joins smoothly to $H_L(\mathbf{r})$ at the sphere radius. We may follow Andersen and call this "head augmentation." At atomic spheres where $H_L(\mathbf{r})$ is *not* centered we use Eqn (3.204) and continue the spherical wave into the sphere by a *linear combination* of numerical solutions of the Schrödinger equation which joins smoothly to $H_L(\mathbf{r} - \mathbf{R})$ at the sphere radius. This we may call "tail augmentation." If we use the tilde to denote augmentation, we replace the head inside the sphere $|\mathbf{r} - \mathbf{R}_\nu| \leq S_\nu$ by the function

$$\tilde{H}_{L\nu}(\mathbf{r} - \mathbf{R}_\nu) = i^\ell \, \tilde{h}_{\ell\nu}(|\mathbf{r} - \mathbf{R}_\nu|) \, Y_L(\widehat{\mathbf{r} - \mathbf{R}_\nu}) \qquad (3.214)$$

where $\tilde{h}_{\ell\nu}(|\mathbf{r} - \mathbf{R}_\nu|)$ is a regular solution of the radial Schrödinger equation inside the atomic sphere labeled ν satisfying boundary conditions

$$\frac{d^n}{dr^n} \left[\tilde{h}_{\ell\nu}(|\mathbf{r} - \mathbf{R}_\nu|) - \kappa^{\ell+1} \, h_\ell^+ \left(\kappa |\mathbf{r} - \mathbf{R}_\nu|\right) \right]_{|\mathbf{r} - \mathbf{R}_\nu| = S_\nu} = 0 \, , \; n = 0, 1, \qquad (3.215)$$

which makes the function continuous and differentiable across the spherical surface $|\mathbf{r} - \mathbf{R}_\nu| = S_\nu$. The eigenvalues resulting from the head augmentation we denote by $\varepsilon_{\ell\nu}^{(H)}$. The tail of the spherical wave centered at $\mathbf{R}_\nu$ is replaced in the atomic sphere at $\mathbf{R}_\mu$ ($\mathbf{R}_\mu \neq \mathbf{R}_\nu$), $|\mathbf{r} - \mathbf{R}_\mu| \leq S_\mu$, by

$$\tilde{H}_{L\nu}(\mathbf{r} - \mathbf{R}_\nu) = \sum_{L'} \tilde{J}_{L'\mu}(\mathbf{r} - \mathbf{R}_\mu) \, B_{L'L}(\mathbf{R}_\mu - \mathbf{R}_\nu) \qquad (3.216)$$

where

$$\tilde{J}_{L'\mu}(\mathbf{r} - \mathbf{R}_\mu) = i^\ell \, \tilde{j}_{\ell'\mu}(|\mathbf{r} - \mathbf{R}_\mu|) \, Y_{L'}(\widehat{\mathbf{r} - \mathbf{R}_\mu}), \qquad (3.217)$$

and $\tilde{j}_{\ell\mu}(|\mathbf{r} - \mathbf{R}_\mu|)$ is the regular solution of the radial Schrödinger equation appropriate to the atomic sphere centered at $\mathbf{R}_\mu$ satisfying boundary conditions

$$\frac{d^n}{dr^n} \left[\tilde{j}_{\ell\mu}(|\mathbf{r} - \mathbf{R}_\mu|) - \kappa^{-\ell} \, j_\ell(\kappa |\mathbf{r} - \mathbf{R}_\mu|) \right]_{|\mathbf{r} - \mathbf{R}_\mu| = S_\mu} = 0 \, , \; n = 0, 1, \qquad (3.218)$$

which makes the function continuous and differentiable across the spherical surface $|\mathbf{r} - \mathbf{R}_\mu| = S_\mu$. The eigenvalues resulting from the tail augmentation we denote by $\varepsilon_{\ell\mu}^{(J)}$. Note that $\tilde{h}_{\ell\nu}(r)$ and $\tilde{j}_{\ell\nu}(r)$ are solutions of the same radial Schrödinger equation; they

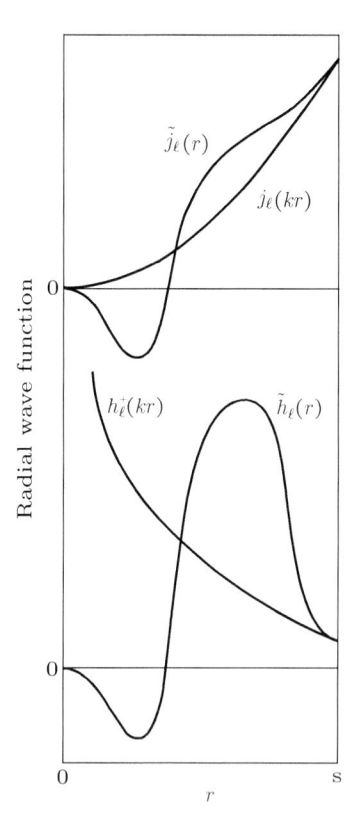

Figure 3.14 *Schematic representation of the spherical Hankel and Bessel functions, $h_\ell^+(\kappa,r)$ and $j_\ell(\kappa,r)$, and the corresponding augmented functions, $\tilde{h}_\ell(r)$ and $\tilde{j}_\ell(r)$, all for $\ell = 2$ as given by Williams et al. (1979).*

differ only in normalization and energy. The four functions $h_\ell^+(\kappa r)$, $\tilde{h}_\ell(r)$, $j_\ell(\kappa r)$, $\tilde{j}_{\ell\nu}(r)$ are compared in Fig. 3.14.

We may pause here and consider augmentation in further detail by furnishing a connection between the ASWs and ϕ and $\dot{\phi}$. Figure 3.15 illustrates qualitatively two augmented orbitals centered on two different sites 0 and **R**. It is at once apparent that the tail augmentation achieved by the functions $\tilde{J}_{L\mu}$ is a feature not possessed by an otherwise perfectly chosen basis orbital that is a solution of Schrödinger's equation for one given atom or atomic sphere. Imagine now that we want to describe the state of a homonuclear, diatomic molecule with the two functions depicted in Fig. 3.15. To do this we form two different linear combinations, bonding and antibonding. Ignoring normalization and suppressing angular momentum indices in the notation, we express the bonding orbital as

$$\Phi_B(\mathbf{r}) = \tilde{\chi}(\mathbf{r}) + \tilde{\chi}(\mathbf{r} - \mathbf{R}), \tag{3.219}$$

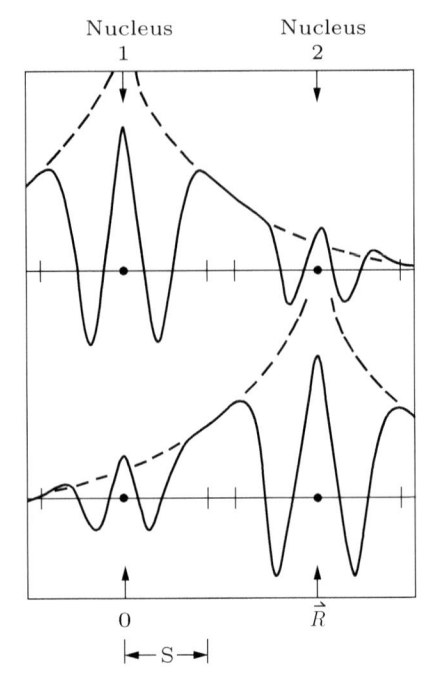

Figure 3.15 *Schematic diagram of two ASWs centered at different sites. Solid lines: augmented functions; dashed lines: envelope functions.*

where the parts on the right-hand side are meant to be those shown in Fig. 3.15. Φ_B has zero slope between the two nuclei and hence a finite charge density there which supplies the bonding. The antibonding orbital is

$$\Phi_A(\mathbf{r}) = \tilde{\chi}(\mathbf{r}) - \tilde{\chi}(\mathbf{r} - \mathbf{R}) \qquad (3.220)$$

having a node between the two nuclei and hence zero charge density there. If the energy of the bonding orbital is ε_B and that of the antibonding is ε_A, then $\varepsilon_B < \varepsilon_A$. Let us next express the head and the tail, at the left nucleus in Fig. 3.15, say, by Φ_B and Φ_A; we see, trivially, that

$$\tilde{\chi}(\mathbf{r}) = \frac{1}{2}\left[\Phi_B(\mathbf{r}) + \Phi_A(\mathbf{r})\right] \qquad (3.221)$$

and

$$\tilde{\chi}(\mathbf{r} - \mathbf{R}) = \frac{1}{2}\left[\Phi_B(\mathbf{r}) - \Phi_A(\mathbf{r})\right]. \qquad (3.222)$$

Then we may use a single-site expansion for Φ_B and Φ_A in terms of ϕ and $\dot\phi$, Eqn (3.147), and simplify by means of the approximate relation between energy and ω, Eqn (3.152); the result is

$$\tilde\chi(\mathbf{r}) = \phi(\mathbf{r}) + \frac{1}{2}\left(\varepsilon_B + \varepsilon_A\right)\dot\phi(\mathbf{r}) \tag{3.223}$$

and

$$\tilde\chi(\mathbf{r}-\mathbf{R}) = \frac{1}{2}\left(\varepsilon_B - \varepsilon_A\right)\dot\phi(\mathbf{r}), \tag{3.224}$$

where we count ε_B and ε_A from E_ν. This is the result of the digression: although the augmentation procedures underlying the ASW and LMTO methods differ in an essential way, the augmented tail is proportional to $\dot\phi(\mathbf{r})$ whereas the head is proportional to a linear combination of $\phi(\mathbf{r})$ and $\dot\phi(\mathbf{r})$. We now complete the description of the ASW method.

The augmented spherical waves are now defined in all regions of the polyatomic system; they are energy independent, continuous, and continuously differentiable. As before we form a Bloch function, $\chi_{L\nu\mathbf{k}}(\mathbf{r})$, which we use in the variational procedure writing an arbitrary translation as a Bravais vector, $\mathbf{R}$, plus a basis vector, $\boldsymbol{\tau}_\nu$,

$$\mathbf{R}_\nu = \mathbf{R} + \boldsymbol{\tau}_\nu. \tag{3.225}$$

We define

$$\chi_{L\nu\mathbf{k}}(\mathbf{r}) = \sum_{\mathbf{R}} e^{i\mathbf{k}\cdot\mathbf{R}}\,\tilde{H}_{L\nu}(\mathbf{r}-\boldsymbol{\tau}_\nu-\mathbf{R}). \tag{3.226}$$

We are now in a position to determine the elements of the secular matrix which we denote as

$$\sum_{L'\nu'}\left\{\langle\nu\tilde{L}|\mathcal{H}|\tilde{L}'\nu'\rangle - \varepsilon\,\langle\nu\tilde{L}|\tilde{L}'\nu'\rangle\right\}a_{L'\nu'}(\mathbf{k}) = 0 \tag{3.227}$$

where $a_{L\nu}(\mathbf{k})$ are the expansion coefficients that determine

$$\psi_\mathbf{k}(\mathbf{r}) = \sum_{L\nu} a_{L\nu}(\mathbf{k})\,\chi_{L\nu\mathbf{k}}(\mathbf{r}); \tag{3.228}$$

furthermore $\mathcal{H} = -\nabla^2 + V(\mathbf{r})$. The brackets $\langle\cdots\rangle$ indicate integration over all space, e.g.

$$\langle\nu\tilde{L}|\tilde{L}'\nu'\rangle = \int d\mathbf{r}\,\tilde{H}_L^*(\mathbf{r}-\mathbf{R}_\nu)\,\tilde{H}_{L'}(\mathbf{r}-\mathbf{R}_{\nu'}) \tag{3.229}$$

and the tilde denotes augmentation.

Taking the effective potential to be zero in the interstitial region (which we are free to do if we later eliminate it by means of the atomic sphere approximation) allows us to write the matrix element as follows

$$\langle \nu \tilde{L} | \mathcal{H} | \tilde{L}' \nu' \rangle = \langle \nu L | \mathcal{H}_0 | L' \nu' \rangle + \sum_{\nu''} \{ \langle \nu \tilde{L} | \mathcal{H} | \tilde{L}' \nu' \rangle_{\nu''} - \langle \nu L | \mathcal{H}_0 | L' \nu' \rangle_{\nu''} \}, \quad (3.230)$$

where $\mathcal{H}_0 = -\nabla^2$ denotes the free-particle Hamiltonian. In other words, we first construct matrix elements of $\mathcal{H}_0$ using unaugmented spherical waves everywhere; then we replace the contributions from the intra-atomic regions by integrations over the full Hamiltonian and ASWs. The reason for this manipulation is that the integrations over the atomic spheres are performed using spherical harmonic expansions and the expansion of the differences in $\{\cdots\}$ in Eqn (3.230) converges much more rapidly than does the corresponding expansion of $\langle \nu \tilde{L} | \mathcal{H} | \tilde{L}' \nu' \rangle_{\nu''}$ alone. This improved ℓ convergence represents a gain both in efficiency and in accuracy.

In the remaining details, which we give here for completeness, we exploit the fact that, in all integrals in Eqn (3.230), the states occurring are eigenfunctions of the relevant Hamiltonian, e.g.

$$\langle \nu L | \mathcal{H}_0 | L' \nu' \rangle = \kappa^2 \langle \nu L | L' \nu' \rangle. \quad (3.231)$$

Furthermore, the integrals over the atomic spheres are of the three basic types discussed above: one-center, two-center, or three-center. One-center contributions are those in which in Eqn (3.230) $\nu = \nu' = \nu''$; in this case the integral involving augmented functions is

$$\langle \nu'' \tilde{L} | \mathcal{H} | \tilde{L}' \nu' \rangle_{\nu''} = \varepsilon_{\ell\nu}^{(H)} \langle \tilde{H}_{L\nu} | \tilde{H}_{L\nu} \rangle_\nu \delta_{LL'}. \quad (3.232)$$

Two-center contributions are those in which $\nu = \nu'' \neq \nu'$ or $\nu \neq \nu'' = \nu'$; in this case

$$\langle \nu'' \tilde{L} | \mathcal{H} | \tilde{L}' \nu' \rangle_{\nu''} = \varepsilon_{\ell\nu}^{(J)} \langle \tilde{H}_{L\nu''} | \tilde{J}_{L\nu''} \rangle_{\nu''} B_{LL'}(\mathbf{R}_{\nu''} - \mathbf{R}_{\nu'}), \quad (3.233)$$

and

$$\langle \nu \tilde{L} | \mathcal{H} | \tilde{L}' \nu'' \rangle_{\nu''} = B_{LL'}^+ (\mathbf{R}_\nu - \mathbf{R}_{\nu''}) \langle \tilde{J}_{L'\nu''} | \tilde{H}_{L'\nu''} \rangle_{\nu''} \varepsilon_{\ell'\nu''}^{(H)}, \quad (3.234)$$

where

$$B_{LL'}^+ (\mathbf{R}_\nu - \mathbf{R}_{\nu''}) \equiv B_{L'L}^* (\mathbf{R}_{\nu''} - \mathbf{R}_\nu). \quad (3.235)$$

Three-center contributions are those in which $\nu \neq \nu'' \neq \nu'$; now one obtains

$$\langle \nu \tilde{L} | \mathcal{H} | \tilde{L}' \nu' \rangle_{\nu''} = \sum_{L''} B^{+}_{LL''}(\mathbf{R}_\nu - \mathbf{R}_{\nu''}) \varepsilon^{(J)}_{\ell'' \nu''} \langle \tilde{J}_{L'' \nu''} | \tilde{J}_{L'' \nu''} \rangle_{\nu''} B_{L''L'}(\mathbf{R}_{\nu''} - \mathbf{R}_{\nu'}). \tag{3.236}$$

The evaluation of the integrals in Eqn (3.230) involving unaugmented functions proceeds in the same way, except for the one-center term which combines with the integral over all of space to give

$$\langle \nu L | L' \nu \rangle - \langle \nu L | L' \nu \rangle_\nu = \delta_{LL'} \langle H_L | H_L \rangle'_\nu, \tag{3.237}$$

where

$$\langle H_L | H_L \rangle'_\nu = \kappa^{2\ell+2} \int_{S_\nu}^{\infty} r^2 \, \mathrm{d}r \, |h^{+}_\ell(\kappa r)|^2, \tag{3.238}$$

i.e. the integral excludes the (diverging) part in the interior of the sphere. Finally, the integral over all of space in Eqn (3.230) involving spherical waves centered at different nuclei is given by an energy derivative of the corresponding structure constant, i.e. for $\nu \neq \nu'$

$$\langle \nu L | L' \nu' \rangle = \dot{B}_{LL'}(\mathbf{R}_\nu - \mathbf{R}_{\nu'}) \equiv \frac{\mathrm{d}}{\mathrm{d}\kappa^2} B_{LL'}(\mathbf{R}_\nu - \mathbf{R}_{\nu'}). \tag{3.239}$$

To derive this equation one first differentiates the free-particle Schrödinger equation with respect to energy (like Eqn (3.143)):

$$\left(\nabla^2 + \kappa^2 \right) \dot{H}_L(\mathbf{r} - \mathbf{R}_{\nu'}) + H_L(\mathbf{r} - \mathbf{R}_{\nu'}) = 0. \tag{3.240}$$

Then multiplying this equation by $H^*_L(\mathbf{r} - \mathbf{R}_\nu)$, subtracting the result from $\dot{H}_{L'}(\mathbf{r} - \mathbf{R}_{\nu'}) (\nabla^2 + \kappa^2) H^*_L(\mathbf{r} - \mathbf{R}_\nu) = 0$, and integrating over all space provides a relationship between the desired integral and

$$\int \mathrm{d}\mathbf{r} \left\{ \dot{H}_{L'}(\mathbf{r} - \mathbf{R}_{\nu'}) \nabla^2 H^*_L(\mathbf{r} - \mathbf{R}_\nu) - H^*_L(\mathbf{r} - \mathbf{R}_\nu) \nabla^2 \dot{H}_{L'}(\mathbf{r} - \mathbf{R}_{\nu'}) \right\}.$$

After integrating by parts, the remaining quantity becomes an integral over small spherical surfaces containing the points of singularity ($\mathbf{r} = \mathbf{R}_\nu$ and $\mathbf{r} = \mathbf{R}_{\nu'}$). The introduction of the structure constant expansion (3.204) permits the surface integrals to be easily evaluated, giving us Eqn (3.238). The expression for the individual integrals which enter our representation for the Hamiltonian matrix elements can now be combined to complete the specification of the secular matrix:

$$\langle \nu \tilde{L} | \mathcal{H} | \tilde{L}' \nu' \rangle = [\varepsilon_{\ell\nu}^{(H)} \langle \tilde{H}_{L\nu} | \tilde{H}_{L\nu} \rangle_\nu + \kappa^2 \langle H_L | H_L \rangle_\nu'] \delta_{\nu\nu'} \delta_{LL'}$$

$$+ \kappa^2 \dot{\hat{B}}_{LL'}(\boldsymbol{\tau}_\nu - \boldsymbol{\tau}_{\nu'}, \mathbf{k}) + \hat{B}_{LL'}^+(\boldsymbol{\tau}_\nu - \boldsymbol{\tau}_{\nu'}, \mathbf{k}) [\varepsilon_{\ell'\nu'}^{(H)} \langle \tilde{J}_{L'\nu'} | \tilde{H}_{L'\nu'} \rangle_{\nu'}$$

$$- \kappa^2 \langle H_{L'} | J_{L'} \rangle_{\nu'}] + [\varepsilon_{\ell\nu}^{(J)} \langle \tilde{H}_{L\nu} | \tilde{J}_{L\nu} \rangle_\nu - \kappa^2 \langle H_L | J_L \rangle_\nu] \hat{B}_{LL'}(\boldsymbol{\tau}_\nu - \boldsymbol{\tau}_{\nu'}, \mathbf{k})$$

$$+ \sum_{\nu''} \sum_{L''} \hat{B}_{LL''}^+(\boldsymbol{\tau}_\nu - \boldsymbol{\tau}_{\nu''}, \mathbf{k}) [\varepsilon_{\ell''\nu''}^{(J)} \langle \tilde{J}_{L''\nu''} | \tilde{J}_{L''\nu''} \rangle_{\nu''}$$

$$- \kappa^2 \langle J_{L''} | J_{L''} \rangle_{\nu''}] \hat{B}_{L''L'}(\boldsymbol{\tau}_{\nu''} - \boldsymbol{\tau}_{\nu'}, \mathbf{k}), \tag{3.241}$$

where translational symmetry has been exploited, i.e. the sum indicated in Eqn (3.226) has been taken to yield the structure constants appropriate to energy-band theory,

$$\hat{B}_{LL'}(\boldsymbol{\tau}_\nu - \boldsymbol{\tau}_{\nu'}, \mathbf{k}) = \sum_{\mathbf{R}} e^{i\mathbf{k}\cdot\mathbf{R}} B_{LL'}(\boldsymbol{\tau}_\nu - \boldsymbol{\tau}_{\nu'} - \mathbf{R}). \tag{3.242}$$

Here $B_{LL'}(\mathbf{x})$ is defined to be zero when $\mathbf{x}$ vanishes. The required structure constants are those given by Eqns (3.123) – (3.128) of the KKR theory, except for a factor of $\kappa^{\ell+\ell'}$, with $\varepsilon_0 = \kappa^2$, and with one more small exception: when the energy parameter κ^2 is negative the cosine in Eqn (3.100) together with the Neumann function in Eqn (3.106) must be replaced by an exponential and a Hankel function, respectively. The net result of this difference is that $D^{(3)}$, Eqn (3.128), is given by

$$D^{(3)} = -2\sqrt{q_0} \left[e^{-\kappa^2/q_0} + \left(\frac{-\pi \kappa^2}{q_0} \right)^{1/2} + \frac{2\kappa^2}{q_0} \sum_{n=0}^{\infty} \frac{\kappa^2/q_0}{n!(2n+1)} \right]. \tag{3.243}$$

The normalization matrix $\langle \nu \tilde{L} | \tilde{L}' \nu' \rangle$ is obtained from Eqn (3.241) by setting the energies κ^2, $\varepsilon_{L\nu}^{(H)}$, and $\varepsilon_{L\nu}^{(J)}$ equal to unity. Furthermore, a simple manipulation of the Schrödinger equation for the augmented Bessel and Hankel functions, $\tilde{j}_\ell$ and $\tilde{h}_\ell^+$, shows that

$$\langle \tilde{H}_{L\nu} | \tilde{J}_{L\nu} \rangle_\nu = \langle \tilde{J}_{L\nu} | \tilde{H}_{L\nu} \rangle_\nu = \left(\varepsilon_{\ell\nu}^{(H)} - \varepsilon_{\ell\nu}^{(J)} \right)^{-1}, \tag{3.244}$$

so that like LMTOs the ASW secular matrix involves only four potential-dependent quantities (for each ℓ and ν): $\varepsilon_{\ell\nu}^{(H)}$, $\varepsilon_{\ell\nu}^{(J)}$, $\langle \tilde{H}_{L\nu} | \tilde{H}_{L\nu} \rangle_\nu$, and $\langle \tilde{J}_{L\nu} | \tilde{J}_{L\nu} \rangle_\nu$.

The integrals involving augmented functions require one-dimensional (radial) numerical integrations and those involving unaugmented spherical Bessel and Hankel functions can be found in standard mathematical texts. The form of the secular equation is similar to the LMTO equation (3.200), but the more precise integration embodied in Eqn (3.230) supplies extra terms which are comparable to the "combined correction terms" mentioned earlier and lead to fast convergence because the difference converges faster than the separate terms in the brackets of the sum in Eqn (3.241).

We close this section with a brief discussion of the construction of the charge density within the ASW and LMTO methods (Williams et al., 1979; Andersen, 1984). It is well known that the Rayleigh–Ritz variational procedure gives rather good estimates of the energies but does poorly on the wave functions. We therefore do not use the coefficients $a_{L\nu}$ that are eigenvectors of the secular equation together with the basis functions $\chi_{L\nu\mathbf{k}}$, Eqn (3.226), directly in constructing the electron density, but rather emphasize the role of the most reliable aspect of the Rayleigh—Ritz procedure, the variationally determined eigenenergies. To do this we take the electron density to have the form it would have in an accurate (KKR or APW) calculation, i.e.

$$n_\nu(r_\nu) = n_\nu^{(c)}(r_\nu) + \sum_\ell \int^{\varepsilon_F} d\varepsilon \, \mathcal{N}_{\ell\nu}(\varepsilon) R_{\ell\nu}^2(\varepsilon, r_\nu), \qquad (3.245)$$

where $n_\nu(r_\nu)$ is the spherical average of the electron density in the ν-th atomic sphere, $r_\nu = |\mathbf{r} - \mathbf{R}_\nu|$; $n_\nu^{(c)}(r_\nu)$ is the contribution of the core levels; $R_{\ell\nu}(\varepsilon, r_\nu)$ is a normalized solution of the Schrödinger equation appropriate to the ν-th sphere; $\mathcal{N}_{\ell\nu}(\varepsilon)$ is the valence electron state density decomposed according to angular momentum and atomic site, and the energy integration extends over the occupied part of the valence band. Now, although it is clear that the *total* valence state density, $\mathcal{N}(\varepsilon)$, Eqn (3.36), indeed requires only a knowledge of the eigenenergies, the *partial* state densities, $\mathcal{N}_{\ell\nu}(\varepsilon)$, appearing in our representation of the electron density require the decomposition of the norm of each eigenstate, i.e.

$$\mathcal{N}_{\ell\nu}(\varepsilon) = \frac{1}{\Omega_{\mathrm{BZ}}} \sum_n \int_{\mathrm{BZ}} d\mathbf{k} \, \delta(\varepsilon - \varepsilon_{n\mathbf{k}}) \, q_{\ell\nu n}(\mathbf{k}), \qquad (3.246)$$

where the quantities $q_{\ell\nu n}(\mathbf{k})$ (to be determined in due course) are the angular momentum ℓ and site ν decompositions of the single-electron norm associated with each eigenstate,

$$\sum_{\ell\nu} q_{\ell\nu n}(\mathbf{k}) = 1, \qquad (3.247)$$

so that

$$\mathcal{N}(\varepsilon) = \sum_{\ell\nu} \mathcal{N}_{\ell\nu}(\varepsilon). \qquad (3.248)$$

(The notation is that introduced in connection with Eqn (3.36).) It might seem therefore that the representation of the electron density in terms of the partial state densities merely exchanges one problem for another, i.e. the ℓ and ν decomposition of each eigenstate requires wave function information which we wanted to avoid because it is given somewhat unreliably by the Rayleigh–Ritz procedure. Still, the present procedure is warranted because it exploits the fact that the normalization of a radial wave function varies more slowly with energy than the wave function itself. This can be seen if the relation, which connects the norm with the energy derivative of the logarithmic derivative, Eqns (3.94) and (3.95), is differentiated with respect to energy; the result is found to involve second energy derivatives of the wave function whose size we discussed before and found to drop off rather fast; see Fig. 3.12.

The required decomposition of the normalization of each state, $q_{\ell\nu n}(\mathbf{k})$, is obtained straightforwardly and involves only a small correction that stems from the overlap of the atomic spheres and the associated problem of the convergence of the spherical harmonics expansions. With the coefficients $a_{L\nu n}(\mathbf{k})$ that diagonalize the secular problem and overlap matrix we know that

$$\sum_{L\nu}\sum_{L'\nu'} a^*_{L\nu n}(\mathbf{k}) \langle \nu \tilde{L}|\tilde{L}'\nu'\rangle a_{L'\nu' n}(\mathbf{k}) = 1. \qquad (3.249)$$

Setting all the energies, $\varepsilon^{(H)}_{\ell\nu}$, $\varepsilon^{(J)}_{\ell\nu}$, and κ^2, appearing in Eqn (3.241) to unity provides an explicit expression for the normalization matrix $\langle \nu \tilde{L}|\tilde{L}'\nu'\rangle$, which when substituted into Eqn (3.247) allows the normalization to be written as a single summation over atomic sites and angular momenta, plus a small quantity $\delta_n(\mathbf{k})$:

$$1 = \sum_{L\nu} \Big\{ a^*_{L\nu n}(\mathbf{k}) \langle \tilde{H}_{L\nu}|\tilde{H}_{L\nu}\rangle_\nu a_{L\nu n}(\mathbf{k}) + a^*_{L\nu n}(\mathbf{k}) \langle \tilde{H}_{L\nu}|\tilde{J}_{L\nu}\rangle_\nu A_{L\nu n}(\mathbf{k})$$
$$+ A^*_{L\nu n}(\mathbf{k}) \langle \tilde{J}_{L\nu}|\tilde{H}_{L\nu}\rangle_\nu a_{L\nu n}(\mathbf{k}) + A^*_{L\nu n}(\mathbf{k}) \langle \tilde{J}_{L\nu}|\tilde{J}_{L\nu}\rangle_\nu A_{L\nu n}(\mathbf{k}) \Big\} + \delta_n(\mathbf{k}), \qquad (3.250)$$

where the coefficients $A_{L\nu n}(\mathbf{k})$ are given by

$$A_{L\nu n}(\mathbf{k}) = \sum_{L'\nu'} B_{LL'}(\boldsymbol{\tau}_\nu - \boldsymbol{\tau}_{\nu'}, \mathbf{k}) a_{L'\nu' n}(\mathbf{k}). \qquad (3.251)$$

Thus, if it were not for the quantity $\delta_n(\mathbf{k})$, Eqn (3.250) would constitute the required ν and ℓ decomposition of the norm; $\delta_n(\mathbf{k})$ is the exactly computed contribution to the normalization due to unaugmented spherical waves minus the integral over the atomic spheres of the spherical harmonic expansion of the same quantity. In the ASA and for closely packed systems it is thus rather small, of the order of a few percent, generally. There is no unique way to distribute this quantity over the various atomic sites and the angular momenta; an acceptable approximation, however, is to omit it altogether but renormalize the sum over the curly bracket in Eqn (3.250) to unity.

The final step in the construction of the charge density is rather self-evident if we remember that the energy dependence of the solutions to the radial Schrödinger equation is approximately linear. In the *LMTO method*, $R_{\ell\nu}(\varepsilon, r_\nu)$ in Eqn (3.245) is therefore expanded in a Taylor series in $\varepsilon - E_\nu$. Defining energy moments as

$$\overline{\lambda}_{\ell\nu}^{(q)} = \int^{\varepsilon_F} (\varepsilon - E_{\nu\ell})^q \, \mathcal{N}_{\ell\nu}(\varepsilon) \, d\varepsilon, \qquad (3.252)$$

one easily sees that the valence-charge density (the second term on the right-hand side of Eqn (3.245)) can be written as

$$n_\nu^{\text{val}}(r_\nu) = \sum_\ell \Big\{ n_{\ell\nu} R_{\ell\nu}^2(E_{\nu\ell}, r_\nu) + 2\overline{\lambda}_{\ell\nu}^{(1)} R_{\ell\nu}(E_{\nu\ell}, r_\nu) \dot{R}_{\ell\nu}(E_{\nu\ell}, r_\nu)$$

$$+ \overline{\lambda}_{\ell\nu}^{(2)} \left[\dot{R}_{\ell\nu}^2(E_{\nu\ell}, r_\nu) + R_{\ell\nu}(E_{\nu\ell}, r_\nu) \ddot{R}_{\ell\nu}(E_{\nu\ell}, r_\nu) \right] \Big\}, \qquad (3.253)$$

where

$$n_{\ell\nu} = \overline{\lambda}_{\ell\nu}^{(0)} = \int^{\varepsilon_F} \mathcal{N}_{\ell\nu}(\varepsilon) \, d\varepsilon \qquad (3.254)$$

is the partial charge in state ℓ of the ν-th atomic sphere. Only the first term on the right-hand side of Eqn (3.253) contributes net charge to the sphere, the remaining terms merely redistributing the spectral weight of the charge within that sphere. This is so because integration of Eqn (3.253) over the sphere gives

$$\int_0^{S_\nu} n_\nu^{\text{val}}(r_\nu) r_\nu^2 \, dr_\nu \equiv \langle n_\nu^{\text{val}} \rangle$$

$$= \sum_\ell \Big\{ n_{\ell\nu} + 2\overline{\lambda}_{\ell\nu}^{(1)} \langle R_{\ell\nu} \dot{R}_{\ell\nu} \rangle + \overline{\lambda}_{\ell\nu}^{(2)} \left[\langle \dot{R}_{\ell\nu}^2 \rangle + \langle R_{\ell\nu} \ddot{R}_{\ell\nu} \rangle \right] \Big\}$$

$$= \sum_\ell n_{\ell\nu}, \qquad (3.255)$$

due to the orthogonality of $R_{\ell\nu}$ and $\dot{R}_{\ell\nu}$, see Eqn (3.145), and because of

$$0 = \frac{\partial}{\partial \varepsilon} \langle R_{\ell\nu}(\varepsilon, r_\nu) \dot{R}_{\ell\nu}(\varepsilon, r_\nu) \rangle = \langle \dot{R}_{\ell\nu}^2 \rangle + \langle R_{\ell\nu} \ddot{R}_{\ell\nu} \rangle. \qquad (3.256)$$

In the ASW method (where energy derivatives are not used directly) each partial state density is sampled with *two* characteristic energies and corresponding weights. In particular, energy moments are defined similar to Eqn (3.252):

$$\lambda_{\ell\nu}^{(q)} = \int^{\varepsilon_F} \varepsilon^q \, \mathcal{N}_{\ell\nu}(\varepsilon) \, d\varepsilon. \qquad (3.257)$$

148 Energy-Band Theory

Then, for each site ν and angular momentum ℓ, two energies $E^{(1)}_{\ell\nu}$, $E^{(2)}_{\ell\nu}$, and two weights $W^{(1)}_{\ell\nu}$, $W^{(2)}_{\ell\nu}$ are required to satisfy the relation

$$\sum_{i=1}^{2} W^{(i)}_{\ell\nu} [E^{(i)}_{\ell\nu}]^q = \lambda^{(q)}_{\ell\nu}, \qquad (3.258)$$

where $q = 0, 1, 2, 3$. For the two pairs of unknown weights and energies there are four equations that can be solved analytically. In Fig. 3.16 we illustrate the relation of the weights and energies to the s and d state densities corresponding to some band structure whose further details are not relevant here. The valence-charge density is now given by the compact representation

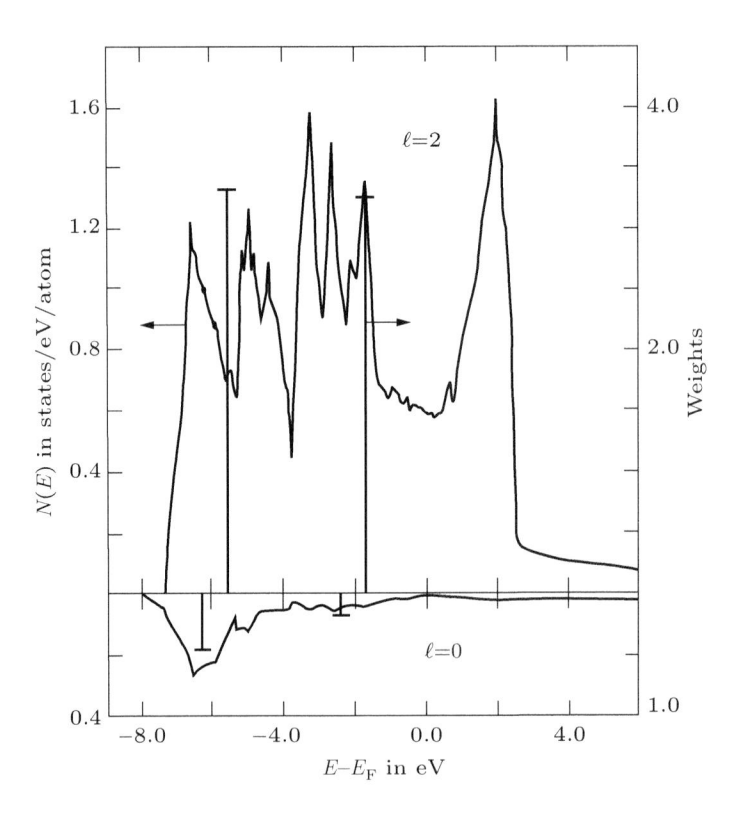

$$n^{\mathrm{val}}_{\nu}(r_\nu) = \sum_{\ell} \sum_{i=1}^{2} W^{(i)}_{\ell\nu} R^2_{\ell\nu} (E^{(i)}_{\ell\nu}, r_\nu), \qquad (3.259)$$

Figure 3.16 *An example of moment analysis of s and d state densities.*

where the radial functions $R_{\ell\nu}$ are numerical solutions to the Schrödinger equation with energies

$$\varepsilon = E_{\ell\nu}^{(i)} \qquad (3.260)$$

obtained from the set of equations (3.258). It is conceptually preferable to think of the environment of a given atom as characterized by *boundary conditions* which the intra-atomic wave functions are obliged to satisfy, rather than by sampling energies. The effective one-electron potential and the radial Schrödinger equation containing it indeed imply an unambiguous relationship between the energies $E_{\ell\nu}^{(i)}$ and the logarithmic derivative $D_{\ell\nu}^{(i)}$ of the corresponding solution at the surface of the atomic sphere, i.e.

$$D_{\ell\nu}^{(i)} = \left[\frac{r \frac{\mathrm{d}}{\mathrm{d}r} R_{\ell\nu}(\varepsilon, r)}{R_{\ell\nu}(\varepsilon, r)} \right]_{r=S_\nu, \varepsilon=E_{\ell\nu}^{(i)}}. \qquad (3.261)$$

This characterization of the environment in terms of electronic configurations, $\{W_{\ell\nu}^{(i)}\}$, and boundary conditions, $\{D_{\ell\nu}^{(i)}\}$, imposed at finite radii, $\{S_\nu\}$, provides a link with intuitive theories of metallic bonding and crystal structure. For instance, the basis of the theory of metallic cohesion due to Wigner and Seitz (1955) is the difference between atomic boundary conditions and those characteristic of partially filled energy bands; see also the discussion in Sec. 3.2.2. Furthermore, it also has a practical virtue: it decouples intra- and interatomic self-consistency. In other words, once the $W_{\ell\nu}^{(i)}$ and $D_{\ell\nu}^{(i)}$ have been produced by a band calculation, the atomic calculations they specify can be iterated to self-consistency before another band calculation is performed. In this way every band calculation performed in the course of obtaining interatomic self-consistency is based on atomic potentials which are internally self-consistent. This fact substantially reduces the number of band calculations required for total self-consistency. We add that these virtues unfortunately disappear to a large extent if the atomic sphere approximation and the shape approximation for the potential are abandoned.

A modern treatment of the ASW method has been published by Eyert (2012). He especially deals with the full potential aspects of the method.

3.4.6 The Korringa–Kohn–Rostoker atomic sphere approximation (KKR-ASA)

By now it should have become apparent to the reader that there is a deep connection between the LMTO and ASW methods on the one hand and the Green's function or KKR method on the other hand. It is worthwhile to pursue this connection somewhat further because it will shed considerably more light on the band-structure problem. Let us therefore return to the envelope function given in Eqn (3.202) and change the notation slightly to emphasize the energy dependence. Thus we may write instead of Eqn (3.202)

150 *Energy-Band Theory*

$$H_L(\varepsilon,\mathbf{r}) = \hat{h}_\ell(\kappa r) Y_L(\hat{\mathbf{r}}), \qquad (3.262)$$

where

$$\hat{h}_\ell(\kappa r) = \kappa^{\ell+1} h_\ell^+(\kappa r) \qquad (3.263)$$

is the spherical Hankel function with the leading κ-dependence removed and $\kappa^2 = \varepsilon$ is the energy. We similarly define

$$J_L(\varepsilon,\mathbf{r}) = \hat{j}_\ell(\kappa r) Y_L(\hat{\mathbf{r}}), \qquad (3.264)$$

where

$$\hat{j}_\ell(\kappa r) = \kappa^{-\ell} j_\ell(\kappa r) \qquad (3.265)$$

is the analogous spherical Bessel function; see also Eqn (3.205).

The expansion theorem, Eqn (3.204), then reads

$$H_L(\varepsilon,\mathbf{r}+\mathbf{R}) = \sum_{L'} J_{L'}(\varepsilon,\mathbf{r}) B_{L'L}(\varepsilon,\mathbf{R}), \qquad (3.266)$$

where $\mathbf{R} \neq 0$ and the set $\{\mathbf{R}\}$ may be assumed to define a crystal lattice. If we assume that $\phi_\ell(\varepsilon,r)$ is a solution of energy ε to the radial Schrödinger equation inside the atomic sphere of radius S, we can construct a solution of the crystal Schrödinger equation by matching in each unit cell value and slope of $\sum_L a_L(\mathbf{k}) \phi_\ell(\varepsilon,r) Y_L(\hat{\mathbf{r}})$ at $r = S$ to the Bloch function

$$\sum_L A_L \left[H_L(\varepsilon,\mathbf{r}) + \sum_{\mathbf{R}\neq 0} H_L(\varepsilon,\mathbf{r}+\mathbf{R}) \mathrm{e}^{\mathrm{i}\mathbf{k}\cdot\mathbf{R}} \right]$$

$$= \sum_L A_L \left[H_L(\varepsilon,\mathbf{r}) + \sum_{L'} J_{L'}(\varepsilon,\mathbf{r}) \sum_{\mathbf{R}\neq 0} B_{L'L}(\varepsilon,\mathbf{R}) \mathrm{e}^{\mathrm{i}\mathbf{k}\cdot\mathbf{R}} \right] \qquad (3.267)$$

$$= \sum_L A_L \left[H_L(\varepsilon,\mathbf{r}) + \sum_{L'} J_{L'}(\varepsilon,\mathbf{r}) \hat{B}_{L'L}(\varepsilon,\mathbf{k}) \right].$$

Here $\hat{B}_{L'L}(\varepsilon,\mathbf{k})$ are the Bloch-transformed expansion coefficients $B_{LL'}(\varepsilon,\mathbf{R})$ and are thus the KKR structure constants discussed following Eqn (3.241). Multiplying the matching conditions with $Y_L^*(\hat{\mathbf{r}})$ and integrating over the solid angle we obtain

$$a_L\,\phi_\ell(\varepsilon,S) = A_L\,\hat{h}_\ell(\kappa S) + \sum_{L'} A_{L'}\,B_{LL'}(\varepsilon,\mathbf{k})\,\hat{j}_\ell(\kappa S),$$

$$a_\ell \left[\frac{\mathrm{d}}{\mathrm{d}r}\phi_\ell(\varepsilon,r)\right]_{r=S} = A_L \left[\frac{\mathrm{d}}{\mathrm{d}r}\hat{h}_\ell(\kappa r)\right]_{r=S} + \sum_{L'} A_{L'}\,B_{LL'}(\varepsilon,\mathbf{k})\left[\frac{\mathrm{d}}{\mathrm{d}r}\hat{j}_\ell(\kappa r)\right]_{r=S} \tag{3.268}$$

which gives, using Wronskians again, Eqn (3.113):

$$\sum_{L'}\left[\hat{B}_{LL'}(\varepsilon,\mathbf{k})\,\{\phi_{\ell'},\hat{j}_{\ell'}\} + \{\phi_\ell,\hat{h}_\ell\}\,\delta_{LL'}\right] a_{L'}(\mathbf{k}) = 0. \tag{3.269}$$

Except for the different treatment of the leading κ-dependence in the spherical Bessel and Hankel functions these are the KKR equations. The occurrence of Hankel instead of Neumann functions can be traced back to the use of running—or standing—wave Green's functions, respectively. The KKR equations (3.269) are valid for all energies, positive and negative.

Although it is interesting to see another derivation of the KKR equations, the purpose of this section is an approximation that we now want to make. Looking at the matching condition, Eqn (3.268), we see that the orbital is still continuous and differentiable even if we use different energy values on the left- and right-hand sides. Thus we may keep the full-energy dependence of the orbital ϕ_ℓ on the left-hand side—this way it is still a solution in the sphere of radius S—but let the energy on the right-hand side go to zero. The Bloch function ceases to be a solution of the Schrödinger equation in the interstitial, but if we practically eliminate it by going over to the atomic sphere approximation, we expect the error to be small. For $\kappa \to 0$ the functions $\hat{j}_\ell$ and $\hat{h}_\ell$ become (Messiah, 1976)

$$\hat{j}_\ell = r^\ell/(2\ell+1)!! \tag{3.270}$$

$$\hat{h}_\ell = \frac{(2\ell+1)!!}{(2\ell+1)}\frac{1}{r^{\ell+1}}, \tag{3.271}$$

with which the Wronskians are easily determined. Cancelling common factors and identifying, except for factors of S and double factorials of ℓ, $B_{LL'}(0,\mathbf{k})$ with $\mathcal{S}^{\mathbf{k}}_{LL'}$ of the LMTO method, we finally write, instead of (3.269),

$$\sum_{L'}\left[\mathcal{S}^{\mathbf{k}}_{LL'} - P_\ell(\varepsilon)\,\delta_{LL'}\right] M_{\ell'}(\varepsilon)\,a_{L'}(\mathbf{k}) = 0, \tag{3.272}$$

where

$$P_\ell(\varepsilon) = 2(2\ell+1)\,(D_\ell(\varepsilon)+\ell+1)/(D_\ell(\varepsilon)-\ell), \tag{3.273}$$

$$M_\ell(\varepsilon) = \phi_\ell(\varepsilon, S)\,(D_\ell(\varepsilon) - \ell)/(2\ell+1), \tag{3.274}$$

and $D_\ell(\varepsilon)$ are the logarithmic derivatives defined by Eqn (3.146).

This set of linear, homogeneous equations has nontrivial solutions for the eigenvectors $a_L(\mathbf{k})$ at the energies $\varepsilon = \varepsilon_\mathbf{k}$ for which the determinant of the coefficient matrix vanishes, i.e.

$$\det\left(\mathcal{S}^\mathbf{k}_{LL'} - P_\ell(\varepsilon)\delta_{LL'}\right) = 0. \tag{3.275}$$

Andersen, who derived this equation slightly differently, calls Eqn (3.272) or Eqn (3.275) the KKR-ASA equations. However, the matrix $\mathcal{S}^\mathbf{k}_{LL'}$ does not depend on energy, in contrast to the corresponding KKR structure constants $\hat{B}_{LL'}(\varepsilon, \mathbf{k})$. The entire energy dependence is contained in the term $P_\ell(\varepsilon)$ which is called the *potential function* since it involves the logarithmic derivatives and through them the characteristics of the potential. The KKR-ASA equations therefore establish the link between the potential and the structure-dependent parts of the energy-band problem and provide the connection between ε and $\mathbf{k}$ which is the energy-band structure. In this approximation we deal with a boundary value problem in which the lattice through the structure constants, $\mathcal{S}^\mathbf{k}_{LL'}$, imposes a $\mathbf{k}$-dependent and nonspherically symmetric boundary condition on the solutions $P_\ell(\varepsilon)$ inside the atomic Wigner–Seitz sphere.

To obtain more insight into the band-structure problem we notice that the potential function, $P_\ell(\varepsilon)$, does not depend on the magnetic quantum number, m. Therefore, one may transform the structure matrix

$$\mathcal{S}^\mathbf{k}_{LL'} \equiv \mathcal{S}^\mathbf{k}_{\ell m, \ell' m'} \tag{3.276}$$

from the given representation to another one where the subblocks to each ℓ are diagonalized and denote the $2\ell+1$ diagonal elements by $\mathcal{S}^\mathbf{k}_{\ell i}$, $i = 1, \ldots, 2\ell+1$. This unitary transformation is independent of potential and atomic volume and leaves the form of the KKR-ASA equations invariant. If one, therefore, neglects hybridization, i.e. sets the elements of $\mathcal{S}^\mathbf{k}_{LL'}$, with $\ell \neq \ell'$ equal to zero, the "unhybridized bands," $\varepsilon_{n\ell i}(\mathbf{k})$, are simply found as the n-th solution of

$$P_\ell(\varepsilon) = \mathcal{S}^\mathbf{k}_{\ell i}. \tag{3.277}$$

For each value of ℓ, the right-hand side of this equation is $2\ell+1$ functions of $\mathbf{k}$ which are called *canonical bands* and are illustrated for $\ell = 2$, i.e. d states, in Fig. 3.17 along a few lines of high symmetry in the Brillouin zone for fcc, bcc, and hcp lattices. We see that at the center of the Brillouin zone Γ, in the case of fcc and bcc lattices there are two energy levels, the lower of which is *triply* degenerate, and the upper *doubly* degenerate. The former (as in the case of Cu, Fig. 3.11) comprise the xy-, yz-, and xz-orbitals which are equivalent to one another in a cubic environment and are called T_{2g} (or $\Gamma_{25'}$) orbitals. The latter comprise the $x^2 - y^2$ and $3z^2 - r^2$ orbitals called E_g (or Γ_{12}) which by pointing

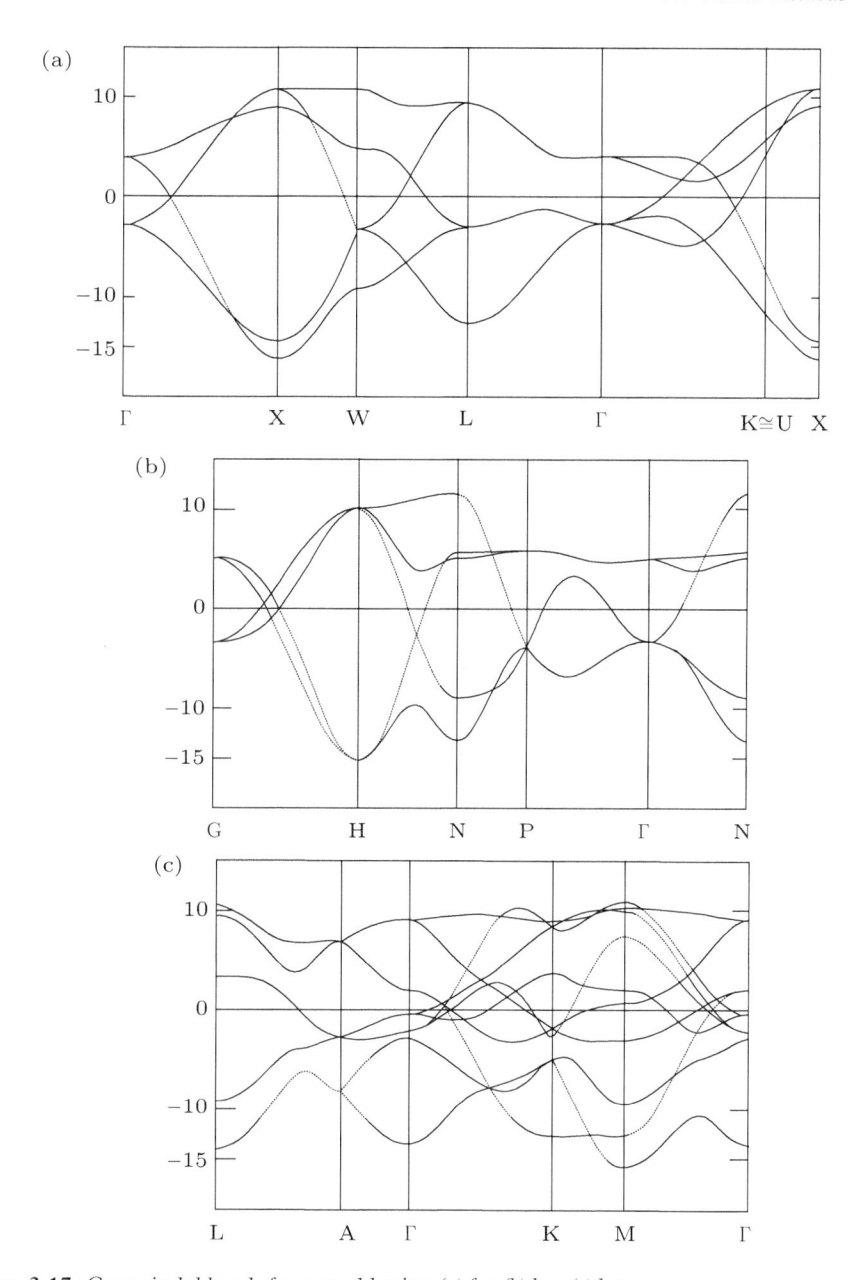

Figure 3.17 *Canonical d bands for several lattices (a) fcc; (b) bcc; (c) hcp.*

154 *Energy-Band Theory*

along the cubic axis are not equivalent to the T_{2g} orbitals. The degeneracy is partially lifted when going into the zone because eigenfunctions which are equivalent at $\mathbf{k} = 0$ may become nonequivalent for $\mathbf{k} \neq 0$ due to the translational phase factor $e^{i\mathbf{k}\cdot\mathbf{R}}$. The situation is somewhat more complicated for hcp lattices (Fig. 3.17(c)) because there are two atoms in the unit cell and hence twice as many bands. We see six energy levels at Γ, the two lower ones being nondegenerate and the others doubly degenerate.

Equation (3.277) is a monotonic mapping of the canonical bands onto an energy scale specified by the n-th branch of the potential function. In Fig. 3.18(a) and (b), which were computed for Fe by the ASW method, we easily identify the canonical d bands of the bcc lattice. The two nearly identical but shifted band structures describe the eigenstates of (a) the majority-, and (b) minority-spin electrons. As we will see in more detail later on, these electrons move independently of each other in spin-polarized potentials. Comparing Fig. 3.18 with the canonical bands in Fig. 3.17(b) we recognize

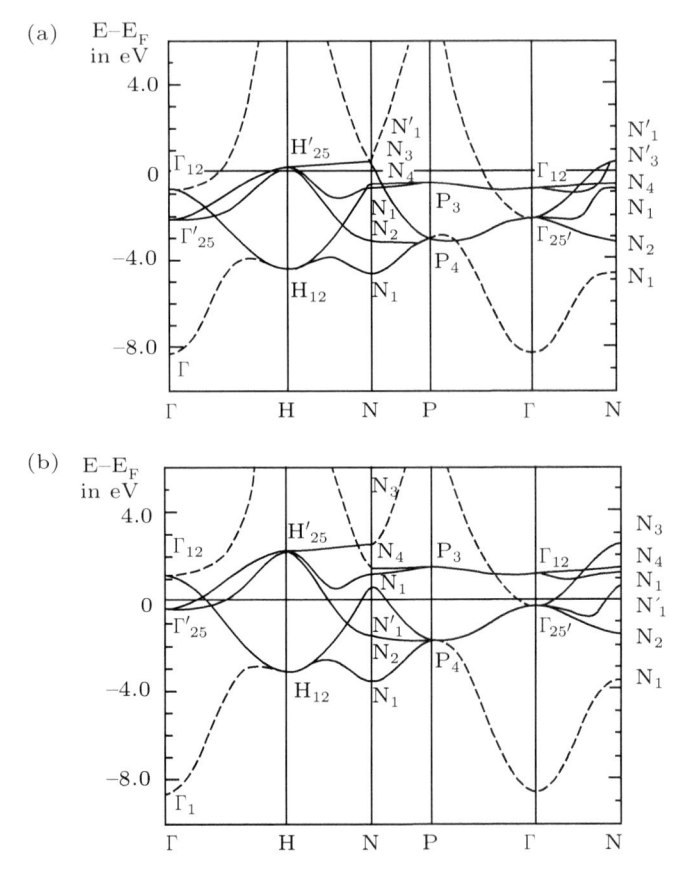

Figure 3.18 *Band structure of bcc Fe: (a) majority-, (b) minority-spin electrons. Dashed lines: s band; solid lines: d bands.*

the s band (dashed lines) hybridizing in a characteristic fashion with the d bands (solid lines). In Fig. 3.19(a) and (b) we show the band structure of fcc Ni, which should be compared with the canonical bands in Fig. 3.17(a). Note again the hybridization of the s band (dashed) with the d bands (solid lines).

We close this subsection by linearizing the potential function, in this way obtaining characteristic *potential parameters* and establishing the mathematical connection between the LMTO and the KKR-ASA formulations.

The qualitative features of the potential function defined by Eqn (3.272) are shown in Fig. 3.20: $P_\ell(\varepsilon)$ has a singularity for an energy ε_+ for which $(D_\ell(\varepsilon_+) = \ell$ and is zero at ε_-, where $D_\ell(\varepsilon_-) = -\ell - 1$. With the approximate relationship between energy and $\omega(D)$, Eqn (3.152), the corresponding quantities ω are $\omega_{\ell+}$ and $\omega_{\ell-}$ defined by Eqn (3.193). Thus the simple expression

$$P_\ell(\varepsilon) \simeq \frac{1}{\gamma_\ell}(\varepsilon - \omega_{\ell-})/(\varepsilon - \omega_{\ell+}) \tag{3.278}$$

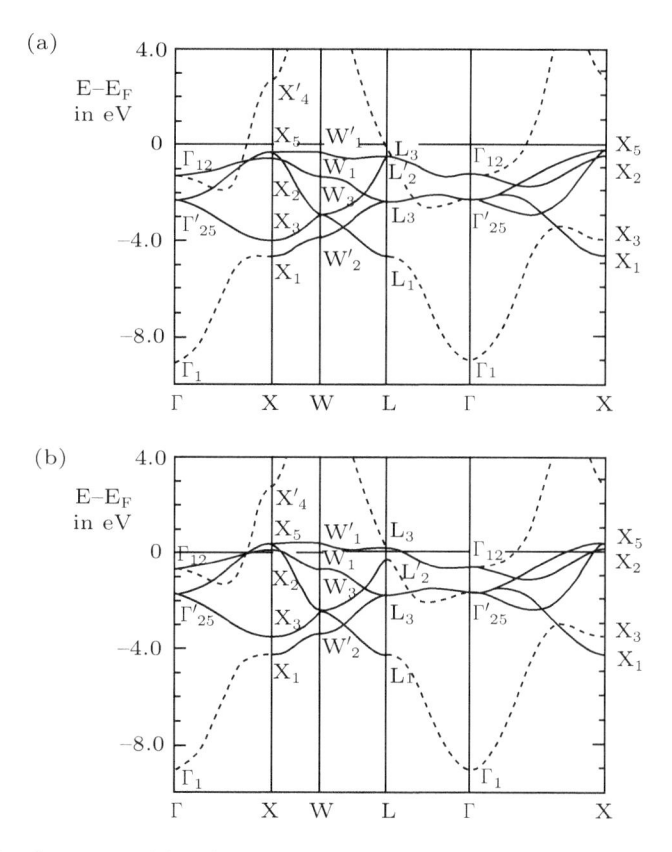

Figure 3.19 Band structure of fcc Ni: (a) majority-, (b) minority-spin electrons. Dashed lines: s band; solid lines: d bands.

156 *Energy-Band Theory*

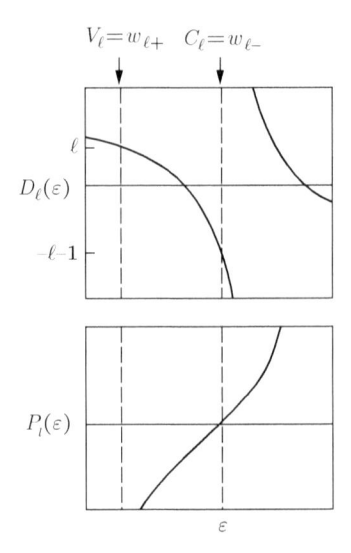

Figure 3.20 *Qualitative features of the logarithmic derivative $D_\ell(\varepsilon)$ and the potential function $P_\ell(\varepsilon)$ and their connection to $\omega_{\ell+}$ and $\omega_{\ell-}$.*

has the correct analytical form and γ_ℓ is a scale factor which is related with typical LMTO parameters. To see this, first manipulate Eqns (3.147) and (3.148) to eliminate $D\{\phi\}$ and $D\{\dot\phi\}$ in favor of $\Phi_\ell(+)$ and $\Phi_\ell(-)$ defined by Eqns (3.191) and (3.188) and derive the identity

$$\frac{\omega(D_\ell) - \omega_{\ell-}}{\omega(D_\ell) - \omega_{\ell+}} = \frac{\Phi_\ell(-)}{\Phi_\ell(+)} \cdot \frac{D_\ell + \ell + 1}{D_\ell - \ell}. \tag{3.279}$$

Using Eqn (3.152) and comparing with Eqn (3.278), we find

$$\gamma_\ell^{-1} = 2(2\ell+1)\,\Phi_\ell(+)/\Phi_\ell(-). \tag{3.280}$$

The potential parameter $\omega_{\ell-}$ is commonly denoted by C_ℓ and it describes, for $\ell > 0$, the center of the band corresponding to angular momentum ℓ. This is so because one can establish the fact that for $\ell > 0$

$$\sum_{i=1}^{2\ell+1} \mathcal{S}_{\ell i}^{\mathbf{k}=0} = 0, \tag{3.281}$$

i.e. the center of gravity of the canonical bands is zero for $\ell > 0$. This coincides with the value of $\varepsilon = C_\ell$, where $P_\ell = 0$ because of Eqn (3.277). The parameter $\omega_{\ell+}$ is often denoted by V_ℓ and the difference $C_\ell - V_\ell$ is related to the bandwidth. Instead of listing this value to characterize a band it is more common to define the so-called band-mass parameter through the relation

$$\mu_\ell^{-1} = \gamma_\ell (C_\ell - V_\ell) S^2, \quad (3.282)$$

and use this number as a potential parameter. It can be seen to be unity for free electrons.

With the help of these parameters, a simple picture emerges of the structure of unhybridized bands. For, inverting Eqn (3.278) to express $\varepsilon(P_\ell)$, we obtain easily

$$\varepsilon(P_\ell) = C_\ell + \frac{1}{\mu_\ell S^2} \frac{P_\ell}{1 - \gamma_\ell P_\ell}, \quad (3.283)$$

where μ_ℓ is defined by Eqn (3.282). Inserting now Eqn (3.277) we obtain an explicit expression for unhybridized bands:

$$\varepsilon_{\ell i}(\mathbf{k}) = C_\ell + \frac{1}{\mu_\ell S^2} \frac{\mathcal{S}_{\ell i}^{\mathbf{k}}}{1 - \gamma_\ell \mathcal{S}_{\ell i}^{\mathbf{k}}}. \quad (3.284)$$

The interpretation is that the pure ℓ band is obtained from the canonical ℓ band structure by fixing the position through the parameter C_ℓ, scaling it by $\mu_\ell S^2$, and distorting it by γ_ℓ. This distortion is generally small in the case of d bands.

We complete this discussion by finally establishing the connection between the LMTO and KKR-ASA equation. It is, however, not an exact one, because we must drop the terms containing $\langle \dot{\phi}_L^2 \rangle$. Then, referring to Eqn (3.160) the Hamiltonian matrix, $\mathbf{H}$, can be written as

$$\mathbf{H} = \mathbf{\Pi}^+ \, \mathbf{\Omega}, \quad (3.285)$$

and the overlap matrix, $\mathbf{O}$, as

$$\mathbf{O} = \mathbf{\Pi}^+ \, \mathbf{\Pi}, \quad (3.286)$$

where we imply matrix multiplications, and the elements of Π and Ω are given by Eqns (3.197) and (3.198). The LMTO equations are then simply

$$\mathbf{H} - \varepsilon \, \mathbf{O} = \mathbf{\Pi}^+ \left(\mathbf{\Omega} - \varepsilon' \, \mathbf{\Pi} \right), \quad (3.287)$$

where ε' is counted from E_ν.

If we now define the KKR-ASA matrix as $\mathbf{\Lambda}$ with elements from Eqns (3.272), (3.278), and (3.280):

$$\Lambda_{LL'} = \mathcal{S}_{LL'}^{\mathbf{k}} - 2(2\ell+1) \frac{\Phi_\ell(+)}{\Phi_\ell(-)} \frac{\varepsilon' - \omega_{\ell-}}{\varepsilon' - \omega_{\ell+}} \delta_{LL'}, \quad (3.288)$$

and another matrix γ with elements

$$\gamma_{LL'} = \tilde{\Phi}_\ell(\omega_{\ell+} - \varepsilon') \Delta_\ell^{-1} \delta_{LL'} \qquad (3.289)$$

we obtain by means of a simple multiplication

$$\gamma \Lambda = \Pi^+ (\Omega - \varepsilon' \Pi) = \mathbf{H} - \varepsilon \mathbf{O}. \qquad (3.290)$$

Hence, the KKR-ASA matrix Λ is a factor of the LMTO matrix if the terms containing $\langle \dot{\phi}_L^2 \rangle$ are neglected as well as the combined correction terms. The potential parameters and the canonical bands are therefore useful tools for interpreting and documenting numerical results that are obtained by solving the more precise LMTO equations.

We can now be very brief in discussing similarly the ASW method. Comparing the functions $\hat{h}_\ell$ and $\hat{j}_\ell$, Eqns (3.263) and (3.264), with those used in the augmentation conditions, Eqns (3.225) and (3.228), and rewriting the KKR-ASA equation now as

$$\sum_{L'} \left[\hat{B}_{LL'}(\varepsilon_0, \mathbf{k}) + P_\ell(\varepsilon) \delta_{LL'} \right] M_{\ell'} a_{L'}(\mathbf{k}) = 0, \qquad (3.291)$$

where ε_0 is the ASW energy parameter and

$$P_\ell(\varepsilon) = \frac{\{\phi_\ell, \hat{h}_\ell\}}{\{\phi_\ell, \hat{j}_\ell\}}, \qquad (3.292)$$

$$M_\ell = \{\phi_\ell, j_\ell\}, \qquad (3.293)$$

we see that the potential function must be

$$P_\ell(\varepsilon) \propto \frac{\varepsilon - \varepsilon_\ell^{(H)}}{\varepsilon - \varepsilon_\ell^{(J)}}, \qquad (3.294)$$

with $\varepsilon_\ell^{(H)}$ and $\varepsilon_\ell^{(J)}$ being the "Hankel" and "Bessel" energies of ASWs. Thus the potential parameter C_ℓ of the LMTO method becomes the Hankel energy $\varepsilon_\ell^{(H)}$ which consequently gives the center of a band corresponding to angular momentum, ℓ, the parameter V_ℓ becomes the Bessel energy, $\varepsilon_\ell^{(J)}$, and the distortion parameter can easily be seen to be

$$\gamma_\ell = \langle \tilde{j}_\ell | \tilde{h}_\ell \rangle / \langle \tilde{h}_\ell | \tilde{h}_\ell \rangle \qquad (3.295)$$

the scaling parameter being given by (3.282). The structure constants $\hat{B}_{LL'}(\varepsilon_0, \mathbf{k})$ are exactly those of the ASW method which, hence, define for small ε_0 canonical bands (in fact those given in Fig. 3.17).

3.4.7 ASW for arbitrary spin configurations

None of the methods described in the previous sections were in any way specifically tailored for magnetic systems, which, however, are the topic of this treatise. The time has thus come to turn our attention to spin polarization.

We begin by referring back to Chap. 2 where we started Sec. 2.7 with the effective single-particle equation (2.178)

$$\sum_{\beta} \left(-\delta_{\alpha\beta} \nabla^2 + v_{\alpha\beta}^{\text{eff}}(\mathbf{r}) - \varepsilon_i \, \delta_{\alpha\beta} \right) \psi_{i\beta} = 0. \tag{3.296}$$

If there is a common quantization axis for all electrons, the effective potential is diagonal and is given by

$$v_{\alpha\beta}^{\text{eff}}(\mathbf{r}) = \delta_{\alpha\beta} \left(2 \int \frac{n(\mathbf{r}')}{|\mathbf{r} - \mathbf{r}'|} \, d\mathbf{r}' + \frac{\delta E_{xc}}{\delta n_\alpha(\mathbf{r})} - \sum_{n\nu} \frac{2 Z_\nu}{|\mathbf{r}_{n\nu}|} \right), \tag{3.297}$$

where $\sum_{n\nu} 2 Z_\nu / |\mathbf{r}_{n\nu}|$ originates from the nuclei of atomic number Z_ν located at $\boldsymbol{\tau}_\nu + \mathbf{R}_\mu$, $\mathbf{r}_{n\nu} = \mathbf{r} - \boldsymbol{\tau}_\nu - \mathbf{R}_\mu$, and where the elements of the density matrix are for $\alpha = 1, 2$

$$n_\alpha(\mathbf{r}) = \sum_{i=1}^{N} |\psi_{i\alpha}|^2 \tag{3.298}$$

and

$$\tilde{n}_{\alpha\beta}(\mathbf{r}) = 0 \quad \text{if} \quad \alpha \neq \beta, \tag{3.299}$$

see Eqn (2.181). The charge density is

$$n(\mathbf{r}) = \sum_{\alpha=1}^{2} n_\alpha(\mathbf{r}) \tag{3.300}$$

and the spin density follows

$$m(\mathbf{r}) = n_1(\mathbf{r}) - n_2(\mathbf{r}), \tag{3.301}$$

see Eqn (2.207).

The result can also be expressed in the language of Sec. 2.7 using the spin-1/2 rotation matrix which for the case at hand is

$$\mathbf{U} = 1 \tag{3.302}$$

and Eqn (2.203) can be written with the Pauli spin matrix σ_z as

$$\left(-\mathbf{1}\nabla^2 + \mathbf{1}\cdot v_0(\mathbf{r}) + \Delta v(\mathbf{r})\,\sigma_z\right)\psi_i = \varepsilon_i\,\psi_i,$$

where ψ_i is a spinor function having components ψ_{i1} and ψ_{i2}.

The potentials $v_0(\mathbf{r})$ and $\Delta v(\mathbf{r})$ are given by (2.200) and (2.201) or can be read off from Eqn (3.297). Thus we may use a scalar wave function for spin up, ↑, corresponding to $\alpha = 1$ for which the Schrödinger equation reads

$$\left(-\nabla^2 + v_0(\mathbf{r}) + \Delta v(\mathbf{r})\right)\psi_{i\uparrow} = \varepsilon_{i\uparrow}\,\psi_{i\uparrow}, \qquad (3.303)$$

and another one for spin down, ↓, ($\alpha = 2$)

$$\left(-\nabla^2 + v_0(\mathbf{r}) - \Delta v(\mathbf{r})\right)\psi_{i\downarrow} = \varepsilon_{i\downarrow}\,\psi_{i\downarrow}. \qquad (3.304)$$

Except for a common origin the ↑ electrons move independently of the ↓ electrons and the spin degeneracy is lifted, i.e. $\varepsilon_{i\uparrow} \neq \varepsilon_{i\downarrow}$. The energy difference $\varepsilon_{i\uparrow} - \varepsilon_{i\downarrow}$ is called exchange splitting. This is clearly seen in Figs. 3.18 and 3.19 which were obtained for the ferromagnets iron and nickel using the ASW method. For obvious reasons we use the term *collinear* for the ferromagnetic spin arrangement, but, in a more general context we speak of collinear spin arrangements when there is one common quantization axis for all electrons.

Although simple versions of antiferromagnetic order are still collinear and can be dealt with again by considering the crystal as a compound consisting of ↑ and ↓ atoms it is nevertheless advantageous to deal with this case together with the more general noncollinear spin arrangements.

In this section we therefore discuss band-structure theory (in particular the ASW method) for noncollinear magnetic order and somewhat later include the treatment of spin–orbit coupling. Furthermore, a special case is that of spiral magnetic order which is very useful, among other things, for the calculation of low-lying excitations. These we will later need for the discussion of finite temperature effects.

Noncollinear order is characterized by the absence of a spin quantization axis common for the whole crystal. As a result, each one-electron eigenstate now *must* be treated as a two-component spinor function and central to the theory is the density matrix, $\tilde{n}_{\alpha\beta}(\mathbf{r})$, defined in Eqn (2.181).

3.4.7.1 Secular equation and density matrix

We discussed in Sec. 2.7 an approximation that renders the problem manageable. This is a local coordinate system not for every point $\mathbf{r}$ in space but for each atomic sphere, which amounts to replacing a fine-grained mesh by a coarse-grained mesh. The local coordinate system is chosen such that the density matrix integrated over the atomic sphere is diagonal, i.e. we calculate

$$n_{\alpha\beta}^{\nu} = \int_{S_\nu} \tilde{n}_{\alpha\beta}(\mathbf{r})\,d\mathbf{r} \qquad (3.305)$$

for each atomic sphere S_ν and diagonalize the matrix $n^\nu_{\alpha\beta}$ using the spin-1/2 rotation matrix **U** whose elements are given in Eqn (2.186):

$$\sum_{\alpha\beta} U_{i\alpha}(\theta_\nu, \varphi_\nu) \, n^\nu_{\alpha\beta} \, U^+_{\beta j}(\theta_\nu, \varphi_\nu) = n^\nu_i \, \delta_{ij}. \tag{3.306}$$

The angles θ_ν and φ_ν are given by Eqns (2.189) and (2.190) where the quantities on the right-hand side are now understood to stand for the integrated elements of the density matrix. Each angle θ_ν and φ_ν then acquires an index ν that labels the appropriate atomic sphere S_ν.

The band-structure problem may now be defined by the simple two-component Schrödinger equation

$$\left(-\nabla^2 \, \mathbf{1} + \mathbf{v}^{eff}(\mathbf{r})\right) \psi_i = \varepsilon_i \, \psi_i \tag{3.307}$$

but the effective potential is written in such a way that the local coordinate system is explicitly apparent. From the foregoing arguments this is easily seen to be

$$\mathbf{v}^{\text{eff}}(\mathbf{r}) = \sum_{n\nu} \mathbf{U}^+(\theta_\nu, \varphi_\nu) \begin{pmatrix} v^{\text{eff}}_{+\nu}(|\mathbf{r}_{n\nu}|) & 0 \\ 0 & v^{\text{eff}}_{-\nu}(|\mathbf{r}_{n\nu}|) \end{pmatrix} \mathbf{U}(\theta_\nu, \varphi_\nu) \, \Theta_\nu(|\mathbf{r}_{n\nu}|), \tag{3.308}$$

where the diagonal parts of the effective potential are given by

$$v^{\text{eff}}_{\pm\nu}(|\mathbf{r}_{n\nu}|) = v_{0\nu}(|\mathbf{r}_{n\nu}|) \pm \Delta v_\nu(|\mathbf{r}_{n\nu}|) \tag{3.309}$$

and, from Eqn (2.200) and (2.201),

$$v_{0\nu}(|\mathbf{r}_{n\nu}|) = v^{\text{ext}}_\nu(|\mathbf{r}_{n\nu}|) + 2 \int \frac{n(\mathbf{r}')}{|\mathbf{r}_{n\nu} - \mathbf{r}'|} d\mathbf{r}' + \frac{1}{2} \sum_{i=1}^{2} \left.\frac{\delta E_{xc}}{\delta n_i}\right|_{\mathbf{r}=\mathbf{r}_{n\nu}} \tag{3.310}$$

$$\Delta v_\nu(|\mathbf{r}_{n\nu}|) = \frac{1}{2}\left(\frac{\delta E_{xc}}{\delta n_1} - \frac{\delta E_{xc}}{\delta n_2}\right)\bigg|_{\mathbf{r}=\mathbf{r}_{n\nu}}. \tag{3.311}$$

The quantity $\Theta_\nu(r)$ is the step function, that is unity for $r < S_\nu$ and zero for $r > S_\nu$; $\mathbf{r}_{n\nu}$ abbreviates $\mathbf{r} - \mathbf{R}_n - \boldsymbol{\tau}_\nu$ and E_{xc} is the exchange-correlation potential that in the local-density approximation is given by Eqn (2.159). Note that the effective potential is spin-dependent only by virtue of the exchange-correlation potential, E_{xc}. We remind the reader that the set of vectors $\{\mathbf{R}_\mu\}$ describes the crystal lattice, and $\boldsymbol{\tau}_\nu$, $\nu = 1, \ldots, N$, are the basis vectors. Finally, the external potential, $v^{\text{ext}}_\nu(|\mathbf{r}_{n\nu}|)$, consists of the Coulomb potential due to the nucleus at the center of the ν-th atomic sphere plus the Madelung potential due to the Coulomb potential of all other atoms.

Now, if we are to use any of the linearized methods described above, we must begin by selecting an envelope function, but in contrast to the previous subsections, this choice now also includes an appropriate spin function. Let us therefore first denote by

$$|1\rangle = \begin{pmatrix} 1 \\ 0 \end{pmatrix}, \qquad (3.312)$$

and

$$|2\rangle = \begin{pmatrix} 0 \\ 1 \end{pmatrix} \qquad (3.313)$$

the unit spinors. If there is a common quantization axis for all electrons, i.e. $\mathbf{U}$ in (3.308) is independent of ν, in which case we may as well choose the z-axis such that $\mathbf{U} = \mathbf{1}$, then envelope functions are written as $\chi_L^e(\mathbf{r})|1\rangle$ or $\chi_L^e(\mathbf{r})|2\rangle$, where, for the ASW method, $\chi_L^e(\mathbf{r})$ is the spherical Hankel function as defined in Eqn (3.202). When the quantization varies from site to site, it is clear that we must define a site-dependent spinor which we obtain by rotating the unit spinor $|\sigma\rangle$:

$$|\nu\sigma\rangle = \sum_{\sigma'} U^+_{\sigma\sigma'}(\nu)|\sigma'\rangle \qquad (3.314)$$

where

$$\mathbf{U}(\nu) \doteq \mathbf{U}(\theta_\nu, \varphi_\nu) \qquad (3.315)$$

is the spin-1/2 rotation matrix, Eqn (2.186). Then an ASW envelope function centered at $\mathbf{R}_n + \boldsymbol{\tau}_\nu \doteq \mathbf{R}_{n\nu}$ is

$$\chi^e_{Ln\nu\sigma}(\mathbf{r}_{n\nu}) = H_L(\mathbf{r}_{n\nu})|\nu\sigma\rangle, \qquad (3.316)$$

i.e. a spin-dependent Hankel function, which we may call a "Hankel spinor," $H_L(\mathbf{r})$ being given by Eqn (3.202).

The "head augmentation," i.e. the replacement of the Hankel spinor at $\mathbf{R}_n + \boldsymbol{\tau}_\nu$ inside the $n\nu$-th atomic sphere now involves an *augmented* Hankel spinor

$$\tilde{H}_{L\nu\sigma}(\mathbf{r}_{n\nu}) = i^\ell \, \tilde{h}_{\ell\nu\sigma}(|\mathbf{r}_{n\nu}|) \, Y_L(\widehat{\mathbf{r}_{n\nu}}) \, |\nu\sigma\rangle \qquad (3.317)$$

and it is easily seen that the function $\tilde{h}_{\ell\nu\sigma}(r)$ satisfies the radial Schrödinger equation at $r = |\mathbf{r}_{n\nu}|$ in the local atomic (spin-diagonal) frame of reference

$$\left[-\frac{1}{r^2}\frac{d}{dr}\left(r^2 \frac{d}{dr}\right) + \frac{\ell(\ell+1)}{r^2} + v^{\mathrm{eff}}_{\sigma\nu}(r) - \varepsilon \right] \tilde{h}_{\ell\nu\sigma}(r) = 0, \qquad (3.318)$$

3.4 Linear methods

with boundary conditions given by Eqn (3.215). The effective potential, $v_{\sigma\nu}^{\text{eff}}(r)$, is given by Eqns (3.309) to (3.311) and the eigenvalues, ε, we denote by $\varepsilon_{\ell\nu\sigma}^{(H)}$.

Next, the expansion theorem, Eqn (3.204), must be generalized to hold for Hankel spinors, too. Obviously by writing out Eqn (3.204) for $|\mathbf{r}_{m\mu}| < S_\mu$ and $\mathbf{R}_{m\mu} \neq \mathbf{R}_{n\nu}$ and multiplying both sides by $|\nu\sigma\rangle$ defined in (3.314) we get

$$H_L(\mathbf{r}_{n\nu})|\nu\sigma\rangle = \sum_{L'} J_{L'}(\mathbf{r}_{m\mu}) B_{L'L}(\mathbf{R}_{m\mu} - \mathbf{R}_{n\nu}) |\nu\sigma\rangle. \qquad (3.319)$$

A "Bessel spinor" is obtained on the right-hand side by inserting the decomposition of $\mathbf{1}$, i.e.

$$\mathbf{1} := \begin{pmatrix} 1 & 0 \\ 0 & 1 \end{pmatrix} = \sum_{\sigma'} |\mu\sigma'\rangle \langle\mu\sigma'|, \qquad (3.320)$$

which is valid at any site μ. Thus

$$H_L(\mathbf{r}_{n\nu})|\nu\sigma\rangle = \sum_{L'\sigma'} J_{L'}(\mathbf{r}_{m\mu})|\mu\sigma'\rangle \langle\mu\sigma'|B_{L'L}(\mathbf{R}_{m\mu} - \mathbf{R}_{n\nu})|\nu\sigma\rangle \qquad (3.321)$$

is the desired expansion theorem. It is now clear that an augmented Bessel spinor may be defined by analogy to (3.317) as

$$\tilde{J}_{L\nu\sigma}(\mathbf{r}_{n\nu}) = i^\ell \tilde{j}_{\ell\nu\sigma}(|\mathbf{r}_{n\nu}|) Y_L(\hat{\mathbf{r}}_{n\nu}) |\nu\sigma\rangle \qquad (3.322)$$

where the function $\tilde{j}_{\ell\nu\sigma}(r)$ satisfies the same radial Schrödinger equation as $\tilde{h}_{\ell\nu\sigma}(r)$, viz. (3.318) where $\tilde{h} \to \tilde{j}$, but the boundary conditions are given by Eqn (3.218) leading to different solutions with eigenvalues ε that we denote by $\varepsilon_{\ell\nu\sigma}^{(J)}$.

The augmented spinor functions are now defined in all of space: for $\mathbf{r}$ such that $|\mathbf{r}_{n\nu}| < S_\nu$ by Eqn (3.317) and for $\mathbf{r}$ with $|\mathbf{r}_{m\mu}| < S_\mu$, but $\mathbf{R}_{m\mu} \neq \mathbf{R}_{n\nu}$, by

$$\tilde{H}_{Ln\nu\sigma}(\mathbf{r}_{n\nu}) = \sum_{L'\sigma'} \tilde{J}_{L'm\mu\sigma'}(\mathbf{r}_{m\mu}) \langle\mu\sigma'|\nu\sigma\rangle B_{L'L}(\mathbf{R}_{m\mu} - \mathbf{R}_{n\nu}). \qquad (3.323)$$

The matrix element $\langle\mu\sigma'|\nu\sigma\rangle$ can easily be expressed by the $\mathbf{U}$'s using Eqn (3.314) and may be combined with the B's to give new, spin-dependent structure constants which we denote by

$$\begin{aligned} B_{L'L}^{\sigma'\sigma}(\mathbf{R}_{m\mu} - \mathbf{R}_{n\nu}) &= \langle\mu\sigma'|\nu\sigma\rangle B_{L'L}(\mathbf{R}_{m\mu} - \mathbf{R}_{n\nu}) \\ &= \left(\mathbf{U}^+(\mu)\,\mathbf{U}(\nu)\right)^+_{\sigma'\sigma} B_{L'L}(\mathbf{R}_{m\mu} - \mathbf{R}_{n\nu}). \end{aligned} \qquad (3.324)$$

Now, except for some added indices and an additional sum on σ', Eqn (3.323) looks like (3.316), indicating that the further procedure will largely follow the development in Sec. 3.4.5. In particular, defining as in Eqn (3.226) a Bloch function by

$$\chi_{L\nu\mathbf{k}\sigma}(\mathbf{r}) = \sum_{\mathbf{R}_n} e^{i\mathbf{k}\cdot\mathbf{R}_n}\,\tilde{H}_{L\nu\sigma}(\mathbf{r}-\boldsymbol{\tau}_\nu-\mathbf{R}_n) \qquad (3.325)$$

and subsequently expanding the solution, $\psi_\mathbf{k}$, of the two-component Schrödinger equation (3.307) as

$$\psi_\mathbf{k}(\mathbf{r}) = \sum_{L\nu\sigma} a_{L\nu\sigma}(\mathbf{k})\,\chi_{L\nu\mathbf{k}\sigma}(\mathbf{r}), \qquad (3.326)$$

we may reproduce the steps following Eqn (3.228) and obtain the elements of the secular equation,

$$\sum_{L'\nu'\sigma'} \left(\langle \nu\tilde{L}\sigma|\mathcal{H}|\nu'\tilde{L}'\sigma'\rangle - \varepsilon\,\langle \nu\tilde{L}\sigma|\nu'\tilde{L}'\sigma'\rangle\right) a_{L'\nu'\sigma'}(\mathbf{k}) = 0, \qquad (3.327)$$

in the form

$$\langle \nu\tilde{L}\sigma|\mathcal{H}|\nu'\tilde{L}'\sigma'\rangle = \left[\varepsilon^{(H)}_{L\nu\sigma}\langle\tilde{H}_{L\nu\sigma}|\tilde{H}_{L\nu\sigma}\rangle_\nu + \kappa^2\,\langle H_L|H_L\rangle'_\nu\right]\delta_{\nu\nu'}\delta_{LL'}\delta_{\sigma\sigma'}$$

$$+ \kappa^2\,\dot{\hat{B}}^{\sigma\sigma'}_{LL'}(\boldsymbol{\tau}_\nu-\boldsymbol{\tau}_{\nu'},\mathbf{k}) + \hat{B}^{+\sigma\sigma'}_{LL'}(\boldsymbol{\tau}_\nu-\boldsymbol{\tau}_{\nu'},\mathbf{k})$$

$$\cdot \left[\varepsilon^{(H)}_{\ell'\nu'\sigma'}\langle\tilde{J}_{L'\nu'\sigma'}|\tilde{H}_{L'\nu'\sigma'}\rangle_{\nu'} - \kappa^2\,\langle J_{L'}|H_{L'}\rangle_{\nu'}\right]$$

$$+ \left[\varepsilon^{(J)}_{\ell\nu\sigma}\langle\tilde{H}_{L\nu\sigma}|\tilde{J}_{L\nu\sigma}\rangle_\nu - \kappa^2\,\langle H_L|J_L\rangle_\nu\right]\cdot \hat{B}^{\sigma\sigma'}_{LL'}(\boldsymbol{\tau}_\nu-\boldsymbol{\tau}_{\nu'},\mathbf{k})$$

$$+ \sum_{\nu''}\sum_{\sigma''}\sum_{L''} \hat{B}^{+\sigma\sigma''}_{LL''}(\boldsymbol{\tau}_\nu-\boldsymbol{\tau}_{\nu''},\mathbf{k})\left[\varepsilon^{(J)}_{\ell''\nu''\sigma''}\langle\tilde{J}_{L''\nu''\sigma''}|\tilde{J}_{L''\nu''\sigma''}\rangle_{\nu''}\right.$$

$$\left. - \kappa^2\,\langle J_{L''}|J_{L''}\rangle_{\nu''}\right]\hat{B}^{\sigma''\sigma'}_{L''L'}(\boldsymbol{\tau}_{\nu''}-\boldsymbol{\tau}_{\nu'},\mathbf{k}). \qquad (3.328)$$

Here the Bloch sum apparent in Eqn (3.325) has been carried out to yield the spinor structure constants

$$\hat{B}^{\sigma\sigma'}_{LL'}(\boldsymbol{\tau}_\nu-\boldsymbol{\tau}_{\nu'},\mathbf{k}) = \sum_{\mathbf{R}_n} e^{i\mathbf{k}\cdot\mathbf{R}_n}\,B^{\sigma\sigma'}_{LL'}(\boldsymbol{\tau}_\mu-\boldsymbol{\tau}_\nu-\mathbf{R}_n), \qquad (3.329)$$

where $B_{LL'}^{\sigma\sigma'}(\mathbf{x})$ is defined to be zero when $\mathbf{x}$ vanishes and is given by (3.324), as well as (3.123)–(3.128), and (3.243). The normalization matrix $\langle\nu\tilde{L}\sigma|\nu'\tilde{L}'\sigma'\rangle$ is obtained from Eqn (3.328) by setting the energies, κ^2, $\varepsilon_{\ell\nu\sigma}^{(H)}$, and $\varepsilon_{\ell\nu\sigma}^{(J)}$ equal to unity and the brackets on the right-hand side of Eqn (3.328) are defined as in Sec. 3.4.5. Note that the size of the secular equation, Eqn (3.327), is twice as large as that of the spin-independent equation, Eqn (3.227).

Suppose now we have solved the secular equation for eigenvalues and eigenvectors. The next and important step is then to make the calculation self-consistent, i.e. we must determine the density matrix which supplies all we need, in particular the new quantities that are characteristic of magnetism: these are the spin densities and the orientations of the atomic, local frames of reference with respect to the global coordinate system. In short, we must calculate both the magnitude and the direction of the magnetic moment. The self-consistency procedure follows the steps after Eqn (3.245) in Sec. 3.4.5 with the added spin-degree of freedom. This leads instead of Eqn (3.246) to a density-of-states matrix, $\mathcal{N}_{L\nu}^{\sigma\sigma'}(\varepsilon)$, given by

$$\mathcal{N}_{L\nu}^{\sigma\sigma'}(\varepsilon) = \frac{1}{\Omega_{BZ}} \sum_n \int_{BZ} d\mathbf{k}\, \delta(\varepsilon - \varepsilon_{n\mathbf{k}})\, q_{L\nu n}^{\sigma\sigma'}(\mathbf{k}) \qquad (3.330)$$

with

$$\begin{aligned}q_{L\nu n}^{\sigma\sigma'}(\mathbf{k}) &= a_{L\nu\sigma}^*(k)\, \langle\tilde{H}_{L\nu\sigma}|\tilde{H}_{L\nu\sigma'}\rangle_\nu\, a_{L\nu\sigma'}(k) \\ &+ a_{L\nu\sigma}^*(k)\, \langle\tilde{H}_{L\nu\sigma}|\tilde{J}_{L\nu\sigma'}\rangle_\nu\, b_{L\nu\sigma'}(k) \\ &+ b_{L\nu\sigma}^*(k)\, \langle\tilde{J}_{L\nu\sigma}|\tilde{H}_{L\nu\sigma'}\rangle_\nu\, a_{L\nu\sigma'}(k) \\ &+ b_{L\nu\sigma}^*(k)\, \langle\tilde{J}_{L\nu\sigma}|\tilde{J}_{L\nu\sigma'}(k)\rangle_\nu\, b_{L\nu\sigma'}(k),\end{aligned} \qquad (3.331)$$

where

$$b_{L\nu\sigma}(k) = \sum_{L'\nu'\sigma'} \hat{B}_{LL'}^{\sigma\sigma'}(\boldsymbol{\tau}_\nu - \boldsymbol{\tau}_{\nu'}, \mathbf{k})\, a_{L'\nu'\sigma'}(k). \qquad (3.332)$$

Since the secular equation (3.327) gives an eigenvector for each eigenvalue $\varepsilon = \varepsilon_{n\mathbf{k}}$, the quantities a (and therefore also b) above should carry another index n. We have, however, condensed the notation somewhat by using the symbol k for both the band index n and the wave vector $\mathbf{k}$. A small renormalization discussed in connection with Eqn (3.250) appears here also but is assumed to be absorbed in Eqn (3.331).

The elements of the integrated density matrix, Eqn (3.305), are now obtained as

$$n^\nu_{\alpha\beta} = \sum_L \int_{-\infty}^{E_F} \mathcal{N}^{\alpha\beta}_{L\nu}(\varepsilon)\,d\varepsilon. \qquad (3.333)$$

Diagonalizing $n^\nu_{\alpha\beta}$ we obtain the angles θ_ν and φ_ν by means of Eqn (3.306). These are the new degrees of freedom that must be made self-consistent. The charge and spin densities pose no further problem since they are now calculated in the local, atomic frame of reference using the method of moments outlined following Eqn (3.252), where we simply add a single-spin index at obvious places.

From the very beginning of this section we have tacitly assumed we know the crystallographic symmetry of the problem. This is indicated by our denoting by $\{\mathbf{R}_n\}$ the set of translations (or the Bravais lattice) and by $\{\boldsymbol{\tau}_\nu\}$, $\nu = 1, \ldots, N$, the N basis vectors. However, the symmetry given by the locations of the chemical constituents of the crystal must be reconsidered when we add the spin degrees of freedom to the constituents, which frequently "breaks" the chemical symmetry. This means that not only may the point-group symmetry be reduced, but also the translational periodicity may be altered, resulting, under certain circumstances, in a larger unit cell than that required by the chemical properties of the crystal. In principle this poses no difficulties once the new symmetry is defined, but in practice this can lead to enormously increased computer times and memory requirements. There is one important case where an apparent loss of translational symmetry is deceiving and in reality does not lead to an enlarged unit cell. We deal with this case next

3.4.7.2 Incommensurate spiral structure

A spiral magnetic structure is sketched in Fig. 3.21 and is defined by giving the Cartesian coordinates of the magnetization vector, $\mathbf{m}_{n\nu}$, as

$$\mathbf{m}_{n\nu} = m_\nu \left[\cos(\mathbf{q}\cdot\mathbf{R}_n + \varphi_\nu)\sin\theta_\nu, \sin(\mathbf{q}\cdot\mathbf{R}_n + \varphi_\nu)\sin\theta_\nu, \cos\theta_\nu\right]. \qquad (3.334)$$

Here m_ν is the magnitude of the magnetic moment at site $\boldsymbol{\tau}_\nu$, n labels the atom at $\mathbf{R}_n$, and $(\mathbf{q}\cdot\mathbf{R}_n + \varphi_\nu)$ as well as θ_ν are polar angles, as depicted in Fig. 3.21. On first sight it appears that the periodicity is lost with respect to lattice translations nonorthogonal to $\mathbf{q}$. One should notice, however, that all atoms of the spiral structure separated by a translation $\mathbf{R}_n$ are equivalent, possessing magnetic moments of equal magnitude. This equivalence leads to an interesting property for the single-particle spinor functions that can be deduced if we succeed in constructing the operator describing this symmetry transformation.

Indeed, as was first pointed out by Herring (1966) and later by Sandratskii (1986a,b), transformations combining a lattice translation $\mathbf{R}_n$ and a spin rotation about the z-axis by an angle $\mathbf{q}\cdot\mathbf{R}_n$ leave the spiral structure invariant, provided spin-orbit coupling is neglected. The symmetry operators describing this transformation are members of a group that Brinkman and Elliott (1966) called a spin-space group (SSG). Quite

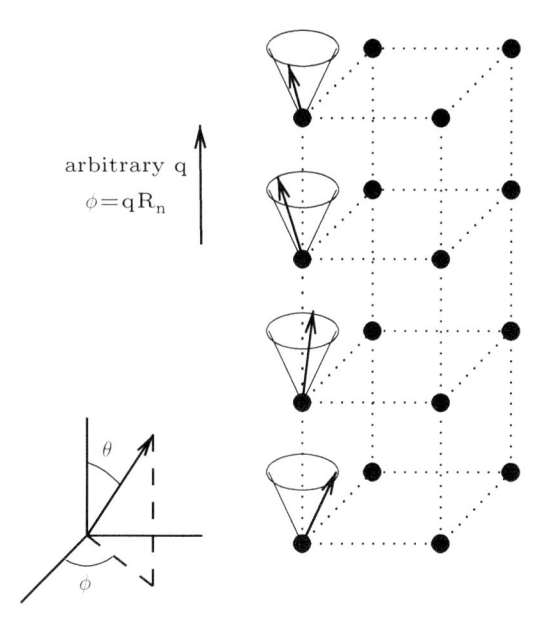

Figure 3.21 *Spiral magnetic structure having the* **q**-*vector parallel to the z-axis. All dots should carry an arrow which designates the magnetic moment. For the sake of clarity we omit most arrows.*

generally, the elements of the SSG are denoted by $\{\alpha_S|\alpha_R|\mathbf{t}\}$ and defined by operating on a two-component spinor, $\psi(\mathbf{r})$, in the following way

$$\{\alpha_S|\alpha_R|\mathbf{t}\}\,\psi(\mathbf{r}) = U(\alpha_S)\,\psi(\alpha_R^{-1}\mathbf{r} - \alpha_R^{-1}\mathbf{t}). \tag{3.335}$$

Here $U(\alpha_S)$ is the spin-1/2 rotation matrix for a spin rotation through α_S, α_R denotes a space rotation, and $\mathbf{t}$ a space translation.[1] The operators we seek are now obtained by specifying the spin rotation as

$$\alpha_S = -\mathbf{q}\cdot\mathbf{R}_n, \tag{3.336}$$

and unity, ε, for the space rotation α_R, giving $\{-\mathbf{q}\cdot\mathbf{R}_n|\varepsilon|\mathbf{R}_n\}$ as generalized translations. They have the following properties:

(i) A spinor is transformed according to

$$\{-\mathbf{q}\cdot\mathbf{R}_n|\varepsilon|\mathbf{R}_n\}\,\psi(\mathbf{r}) = \begin{pmatrix} \exp(-i\,\mathbf{q}\cdot\mathbf{R}_n/2) & 0 \\ 0 & \exp(i\,\mathbf{q}\cdot\mathbf{R}_n/2) \end{pmatrix} \psi(\mathbf{r} - \mathbf{R}_n). \tag{3.337}$$

---

[1] The quantities $U(\alpha_S)$ form a special unitary group called SU(2), see e.g. Sakurai (1985), pp 168

(ii) They commute with the Kohn–Sham Hamiltonian, $\mathcal{H}_\mathbf{q}$, of a spiral structure, where we define $\mathcal{H}_\mathbf{q}$ by Eqns (3.307) and (3.308) as

$$\mathcal{H}_\mathbf{q} = -1\nabla^2 + \sum_{n\nu} \Theta(|\mathbf{r}_{n\nu}|) U^+(\theta_\nu, \varphi_\nu, \mathbf{q}) \begin{pmatrix} v^{\text{eff}}_{+\nu}(|\mathbf{r}_{n\nu}|) & 0 \\ 0 & v^{\text{eff}}_{-\nu}(|\mathbf{r}_{n\nu}|) \end{pmatrix} U(\theta_\nu, \varphi_\nu, \mathbf{q}). \tag{3.338}$$

Here, from Eqn (2.188),

$$U(\theta_\nu, \varphi_\nu, \mathbf{q}) = \begin{pmatrix} \cos\frac{\theta_\nu}{2} & \sin\frac{\theta_\nu}{2} \\ -\sin\frac{\theta_\nu}{2} & \cos\frac{\theta_\nu}{2} \end{pmatrix} \begin{pmatrix} \exp\left(\frac{i\varphi_\nu}{2} + \frac{i\mathbf{q}\cdot\mathbf{R}_n}{2}\right) & 0 \\ 0 & \exp\left(\frac{-i\varphi_\nu}{2} - \frac{i\mathbf{q}\cdot\mathbf{R}_n}{2}\right) \end{pmatrix}. \tag{3.339}$$

By an easy calculation we see at once that

$$[\mathcal{H}_\mathbf{q}, \{-\mathbf{q}\cdot\mathbf{R}_n|\varepsilon|\mathbf{R}_n\}] = 0, \tag{3.340}$$

which is the desired commutator property.

(iii) The generalized translations obviously obey the multiplication law

$$\{-\mathbf{q}\cdot\mathbf{R}_n|\varepsilon|\mathbf{R}_n\}\{-\mathbf{q}\cdot\mathbf{R}_m|\varepsilon|\mathbf{R}_m\} = \{-\mathbf{q}(\mathbf{R}_n + \mathbf{R}_m)|\varepsilon|\mathbf{R}_n + \mathbf{R}_m\}, \tag{3.341}$$

which clearly commute thus forming an Abelian group isomorphic to the group of ordinary space translations by vectors $\mathbf{R}_n$. Now, if two groups are isomorphic they have the same irreducible representation (IR). But the IR of ordinary space translations constitutes Bloch's theorem, which we stated in Eqn (3.12).

Thus we obtain the *generalized* Bloch theorem for a spiral structure:

$$\{-\mathbf{q}\cdot\mathbf{R}_n|\varepsilon|\mathbf{R}_n\}\psi_\mathbf{k}(\mathbf{r}) = e^{-i\mathbf{k}\cdot\mathbf{R}_n}\psi_\mathbf{k}(\mathbf{r}), \tag{3.342}$$

where because of property (3.340) the $\psi_\mathbf{k}(\mathbf{r})$ are also eigenspinors of $\mathcal{H}_\mathbf{q}$. The vectors $\mathbf{k}$ lie in the first Brillouin zone which is defined in the usual way (Sec. 3.1) by the set of vectors $\mathbf{R}_n$. This is the central result of this subsection and we may summarize that the spin spiral $\mathbf{m}_{n\nu}$ defined by Eqn (3.334) does not break the translational symmetry of the lattice, although, in general, the point-group symmetry may be reduced. This statement is independent of the choice of **q** which, therefore, need not be commensurate with the lattice. A practical consequence is that we need no supercells to solve the Kohn–Sham–Schrödinger equation in the presence of spin spirals. Indeed, from the definition of the generalized translation, $\{-\mathbf{q}\cdot\mathbf{R}_n|\varepsilon|\mathbf{R}_n\}$, we see that it suffices to pick any **q** inside the first Brillouin zone of the crystal; **q**-vectors outside give nothing new.

It is instructive to find the compatibility relations between the states characterized by the spiral vector **q** and those of the nonmagnetic crystals obtained in the limit $v_{+\nu}^{\mathrm{eff}} \to v_{-\nu}^{\mathrm{eff}}$, see Eqn (3.338).

Let the Bloch function in the limit $v_{+\nu}^{\mathrm{eff}} \to v_{-\nu}^{\mathrm{eff}}$ be labeled by the **k**-vector $\mathbf{k}_0$. Then, since the symmetry property expressed by Eqn (3.337) is also valid in this limit, Eqn (3.337) gives, with Bloch's law

$$\{-\mathbf{q}\cdot\mathbf{R}_n|\varepsilon|\mathbf{R}_n\}\psi(\mathbf{r})$$

$$= \begin{pmatrix} \exp\left[-\mathrm{i}\left(\tfrac{1}{2}\mathbf{q}+\mathbf{k}_0\right)\cdot\mathbf{R}_n\right] & 0 \\ 0 & \exp\left[-\mathrm{i}\left(-\tfrac{1}{2}\mathbf{q}+\mathbf{k}_0\right)\cdot\mathbf{R}_n\right] \end{pmatrix}\psi(\mathbf{r}). \quad (3.343)$$

Comparing this with Eqn (3.342) we read off

$$\mathbf{k} = \mathbf{k}_0 + \frac{1}{2}\sigma\,\mathbf{q}, \quad (3.344)$$

where $\sigma = \pm 1$, i.e. a state of the nonmagnetic crystal with Bloch vector $\mathbf{k}_0$ and spin projection σ is transformed into a spiral state with wave vector

$$\mathbf{k} = \mathbf{k}_0 + \frac{1}{2}\mathbf{q} \quad \text{or} \quad \mathbf{k} = \mathbf{k}_0 - \frac{1}{2}\mathbf{q}$$

depending on the spin projection σ. Expressing the same result differently, we can say that states of a nonmagnetic crystal belonging to different representations

$$\left(\mathbf{k}-\frac{1}{2}\mathbf{q},+\right) \quad \text{and} \quad \left(\mathbf{k}+\frac{1}{2}\mathbf{q},-\right)$$

treated as a limiting case of a spiral structure will belong to the same representation characterized by a vector **k**. These states will hybridize when $v_{+\nu}^{\mathrm{eff}} \neq v_{-\nu}^{\mathrm{eff}}$. This interaction of states, shifted in **k**-space by the vector **q** in the traditional description of the nonmagnetic crystal, is important in the formation of the spiral structure. We will see an explicit example in Sec. 4.3.

We are now in a position to return to the solution of the band-structure problem. For a spiral structure that is defined by the vectors $\mathbf{m}_{n\nu}$ given in Eqn (3.334) the site-dependent spinor is, with Eqns (3.314) and (3.339), written as

$$|\nu n \sigma\rangle = \begin{pmatrix} \exp(-i\mathbf{q}\cdot\mathbf{R}_n/2) & 0 \\ 0 & \exp(i\mathbf{q}\cdot\mathbf{R}_n/2) \end{pmatrix} |\nu \sigma\rangle \qquad (3.345)$$

which allows us to write the envelope function centered at $\mathbf{R}_{n\nu}$ as

$$\chi^e_{Ln\nu\sigma}(\mathbf{r}_{n\nu}) = \begin{pmatrix} \exp(-i\mathbf{q}\cdot\mathbf{R}_n/2) & 0 \\ 0 & \exp(i\mathbf{q}\cdot\mathbf{R}_n/2) \end{pmatrix} H_L(\mathbf{r}_{n\nu})|\nu\sigma\rangle; \qquad (3.346)$$

compare with (3.316). A generalized Bloch function is now easily constructed as

$$\chi_{L\nu\mathbf{k}\sigma}(\mathbf{r}) =$$

$$\sum_n e^{i\mathbf{k}\cdot\mathbf{R}_n} \begin{pmatrix} \exp(-i\mathbf{q}\cdot\mathbf{R}_n/2) & 0 \\ 0 & \exp(i\mathbf{q}\cdot\mathbf{R}_n/2) \end{pmatrix} \tilde{H}_{L\nu\sigma}(\mathbf{r}-\boldsymbol{\tau}_L-\mathbf{R}_n), \qquad (3.347)$$

where the augmented Hankel spinor is defined by Eqn (3.317), (3.322), and (3.323). To show that Eqn (3.347) indeed possesses generalized Bloch symmetry, one applies the operator $\{-\mathbf{q}\cdot\mathbf{R}_m|\varepsilon|\mathbf{R}_m\}$ using the rule embodied in Eqn (3.337). Finally, the secular equation is derived and has the form of Eqn (3.328), except that the Fourier-transformed structure constants must be replaced by the elements of the matrix

$$G_{LL'}(\boldsymbol{\tau}_\nu - \boldsymbol{\tau}_{\nu'}, \mathbf{k})$$

$$\doteq U(\nu) \begin{pmatrix} \hat{B}_{LL'}(\boldsymbol{\tau}_\nu - \boldsymbol{\tau}_{\nu'}, \mathbf{k}-\tfrac{1}{2}\mathbf{q}) & 0 \\ 0 & \hat{B}_{LL'}(\boldsymbol{\tau}_\nu - \boldsymbol{\tau}_{\nu'}, \mathbf{k}+\tfrac{1}{2}\mathbf{q}) \end{pmatrix} U^+(\nu'), \qquad (3.348)$$

where $\hat{B}_{LL'}(\boldsymbol{\tau}, \mathbf{k})$ are obtained from Eqn (3.242) and

$$U(\nu) = U(\theta_\nu, \varphi_\nu) = \begin{pmatrix} \cos\tfrac{\theta_\nu}{2} & \sin\tfrac{\theta_\nu}{2} \\ -\sin\tfrac{\theta_\nu}{2} & \cos\tfrac{\theta_\nu}{2} \end{pmatrix} \begin{pmatrix} \exp(i\varphi_\nu/2) & 0 \\ 0 & \exp(-i\varphi_\nu/2) \end{pmatrix}. \qquad (3.349)$$

3.4.7.3 Relativistic corrections

Our treatment of the effects of spin polarization is not complete without considering relativistic corrections, the most prominent of which being spin–orbit coupling. We thus return to Sec. 2.4 where we discussed an approximation that allows us to reformulate the Kohn–Sham Hamiltonian in a way easy to incorporate in the present treatment, rather than dealing with the full, four-component Dirac formalism. This is the scalar relativistic wave equation given in Eqn (2.127) or (2.137) where we add the spin–orbit coupling defined in Eqn (2.134) to be modified to apply to spin-polarized situations.

We thus focus our attention on the radial Schrödinger equation in the spin diagonal, atomic frame of reference that must be solved to obtain the augmented Hankel and Bessel functions, depending on the boundary conditions supplied by the envelope function. Hence, rewriting Eqn (3.318) by using Eqn (2.137) in atomic units to replace the bracket we write

$$\mathcal{H}_{SC}\,\varphi(r) = 0, \qquad (3.350)$$

where the scalar relativistic operator $\mathcal{H}_{SC}$ is obtained as

$$\begin{aligned}\mathcal{H}_{SC} = &-\frac{1}{r^2}\frac{\mathrm{d}}{\mathrm{d}r}\left(r^2\frac{\mathrm{d}}{\mathrm{d}r}\right) + \frac{\ell(\ell+1)}{r^2} + (V(r)-\varepsilon) \\ &\cdot\left(1-\frac{V(r)-\varepsilon}{c^2}\right) + \left(V(r)-\varepsilon-c^2\right)^{-1}\frac{\mathrm{d}V}{\mathrm{d}r}\frac{\mathrm{d}}{\mathrm{d}r}.\end{aligned} \qquad (3.351)$$

Here $V(r)$ stands for the effective potential $v_{\sigma\nu}^{\mathrm{eff}}$ and $\varphi(r)$ for any of the augmented functions. The physical meaning of the third and fourth terms on the right-hand side was explained preceding Eqn (2.126). Solving Eqn (3.350) gives eigenvalues and wave functions that differ numerically from those obtained in the Schrödinger formalism, but otherwise the form of Eqns (3.328)–(3.333) remains unchanged.

Next we must add spin–orbit coupling for which the Hamiltonian is given by Eqn (2.134). Refining the approximations that were discussed following Eqn (2.137) somewhat and using atomic units we write

$$\mathcal{H}_{SO} = \frac{1}{(2c)^2}\frac{1}{r}\left[\begin{pmatrix} M_+^{-2}\frac{\mathrm{d}V_+}{\mathrm{d}r} & 0 \\ 0 & M_-^{-2}\frac{\mathrm{d}V_-}{\mathrm{d}r} \end{pmatrix}\sigma_z\hat{L}_z + M_{av}^{-2}\frac{\mathrm{d}V_{av}}{\mathrm{d}r}\left(\sigma_x\hat{L}_x + \sigma_y\hat{L}_y\right)\right], \qquad (3.352)$$

where

$$V_{av} \equiv V_{av}(r) = \frac{1}{2}\left(V_+(r) + V_-(r)\right) \qquad (3.353)$$

is an average of the spin-polarized effective potentials and

$$M_\alpha = \frac{1}{2}\left(1 - c^2 V_\alpha\right) \tag{3.354}$$

with $\alpha = +, -$, and av. The quantities $\hat{L}_x$, $\hat{L}_y$, and $\hat{L}_z$ are the standard angular momentum operators whence matrix elements $\langle\nu\tilde{L}\sigma|\mathcal{H}_{\text{SO}}|\nu\tilde{L}'\sigma'\rangle$ are easily computed in the basis used to determine the secular equation. This term can be treated by perturbation theory or, more consistently, as an added part to Eqn (3.328), i.e. as part of the variational determination of the eigenstates with the trial functions being determined scalar relativistically using Eqn (3.351). The variational treatment is more precise than perturbation theory since $\mathcal{H}_{\text{SO}}$ is then part of the self-consistency cycle—it is, therefore, to be preferred.

There is another relativistic correction term that is quite important for heavy elements, in particular for uranium and uranium compounds. Physically, the spin–orbit coupling term induces an orbital polarization so that matrix elements of the angular momentum operator no longer vanish but, especially for the heavy elements, this polarization is found to be too weak. What is missing is a feedback effect of this polarization on the states and hence on the charge and spin densities. This was studied by, among others, Eriksson *et al.* (1989) who used the Hartree–Fock approximation and suggested the added Hamiltonian

$$\mathcal{H}_{\text{orb}} = I_{\text{orb}} L_z \hat{L}_z, \tag{3.355}$$

where L_z is the projection of the atomic orbital moment on to the local atomic z-axis and the parameter I_{orb} has a numerical value in the millirydberg range depending on the states on which $\mathcal{H}_{\text{orb}}$ operates. Physically, $\mathcal{H}_{\text{orb}}$ takes into account interactions responsible for Hund's second rule that requires the orbital moments to be maximal. In practice, matrix elements of $\mathcal{H}_{\text{orb}}$ are computed and added to the variational treatment just like spin–orbit coupling. But in contrast to the latter no completely satisfactory derivation exists within density-functional theory, so this correction has an empirical nature.

At this point of the development we have described those methodologies that are necessary for the study of the electronic structure of magnets. There are new and interesting developments that aim at either a more accurate description of the electronic structure, so-called *full potential schemes*, or at the ability to handle larger and larger systems, mostly having in mind employing computers that make use of a parallel architecture. To continue with a description of these modern developments would lead to excessive length. We therefore stop here.

4
Electronic Structure and Magnetism

4.1 Introduction and simple concepts

We begin this chapter by selecting first the three ferromagnetic transition metals, iron, cobalt, and nickel, and state some basic experimental facts. As we progress, we will expand our view including, among others, the antiferromagnetic metals, chromium and manganese, finally turning to the large class of transition metal compounds and alloys. It should be clear from the outset that it will be impossible to deal with all known systems here so that we must make a selection of typical cases.

Wohlfarth, in an important review (1980), stated that the three ferromagnetic transition metals are of importance to the whole subject of metallic magnetism and their properties should be thoroughly understood before attempting to understand those of the transition metal alloys. Historically the important question was whether the magnetic carriers were localized or itinerant and each of these two viewpoints had eminent proponents. The deeper meaning of the terms "localized" and "itinerant" should become clear as we proceed. However, even though not all cases are fully understood presently, it seems clear that the $T = 0$ properties of a majority of metallic magnets, *excluding* those involving rare earths, can be explained in the itinerant-electron picture.

In what follows we will often use the terms "band picture" or "band theory" instead of "itinerant-electron picture or theory." Although one may be able to state that these terms define different concepts, we will nevertheless use them interchangeably. It should be understood, however, that we do not take the term "band electrons" to mean *independent* electrons, as exchange and correlation effects are contained in the effective potential that we imply to be constructed from the electron density or density matrix as explained in Chap. 2. In some frameworks of definition, for instance that of Mott (1964), the electrons we talk about are itinerant since they participate in the Fermi surface. In spite of this they do not necessarily form delocalized magnetic moments, but they conspire to such an extent that the moments are localized on some scale. What exactly they look like we will see later on.

The band picture was originally proposed by Stoner in 1938 and 1939. However, when we use the term "Stoner theory" in the following, we do not imply the original theory in its entirety, especially not those aspects that deal with thermodynamic ($T > 0$)

properties of itinerant-electron magnets. However, the theory we will describe applies to ground-state properties ($T = 0$), for which it supplies a criterion for the ferromagnetic instability based on *intra-atomic* exchange, much in the way that Stoner originally postulated.

Before turning to the discussion of experimental data, we briefly address the question of units. Experimentally, the magnetic moment is often given per unit of mass in 1 A m²/kg (in the International System of Units) corresponding to 1 emu/g (in cgs units). A magnetic moment per unit of volume is deduced from the magnetic moment per unit of mass by multiplying by the density (expressed in kg/m³). In theoretical work it is more common to express the moment as the number of unpaired spins per atom or per formula unit, leading to a magnetic moment M in units of μ_B (Bohr magnetons per atom or per formula unit (f.u.)) where $1\ \mu_B = |e|\hbar/2m = 0.578 \cdot 10^{-4}\ \text{eV/T} = 0.927 \cdot 10^{-23}\ \text{J/T}$.

In Table 4.1 we summarize the fundamental magnetic properties of bcc iron, hcp cobalt, and fcc nickel.

A theory of magnetism must supply the value of the saturation magnetization, q_s, which is a zero-temperature property, but also the dependence of the magnetization on the temperature should find an explanation. Here it is especially the temperature where the macroscopic magnetization vanishes, the Curie temperature, T_c, which one would like to calculate from first principles. Finally, we want to understand the susceptibility for temperatures above T_c, for which the entries q_c are relevant, as we will see. It is the phenomenology of these facts that we want to discuss in this section together with the problems that naive theories face. In Fig. 4.1 we show the measured reduced magnetization of bcc Fe, fcc Co, and fcc Ni as a function of the reduced temperature, T/T_c. The reduced magnetization is denoted by M/M_s, where $M_s = q_s$ of Table 4.1, except for fcc Co for which Crangle and Goodman (1971) used an extrapolated value of $q_s = 166.1$ emu/g since fcc Co is stable only at high temperatures, hcp Co being the low-temperature form.

For a first attempt to explain the values of the saturation magnetization given in Table 4.1, we are tempted to look at free Fe, Co, and Ni ions for which we determine

Table 4.1 *Fundamental magnetic properties of bcc Fe, hcp Co, and fcc Ni (from Wohlfarth (1980)).*

	Fe	Co	Ni
Saturation magnetization [emu/g]	221.71 ± .08	162.55	58.57 ± .03
Saturation magnetization (q_s) [μ_B per atom]	2.216	1.715	0.616
Curie temperature (T_c) [K]	1044 ± 2	1388 ± 2	627.4 ± .3
kT_c [meV]	90.0	119.6	54.1
Number of magnetic carriers from Curie–Weiss law (q_c)	2.29	2.29	0.90
q_c/q_s	1.03	1.34	1.46

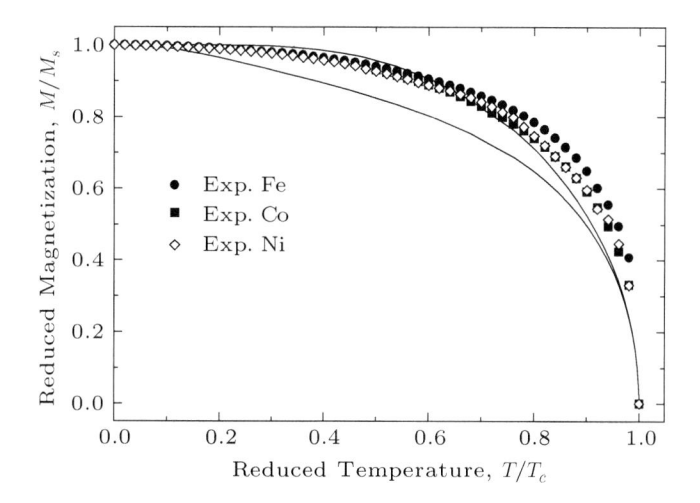

Figure 4.1 *Reduced magnetization, M/M_s, for bcc Fe, fcc Co, and fcc Ni as a function of the reduced temperature, T/T_c. Solid lines give the reduced magnetization calculated in the mean-field approximation from the spin-1/2 and from the spin-10/2 (lower curve) Brillouin function.*

the magnetic moments from Hund's rules well known from the quantum mechanics of atoms and ions. These rules give the sequence of occupation of the electronic states by requiring first that for a given configuration the term with maximum spin, S (called the maximum multiplicity) possesses the lowest energy; second, for a given configuration and multiplicity the term with the largest value of the angular momentum, L, possesses the lowest energy, and, third, for the given configuration, multiplicity, and angular momentum, the value of the total angular momentum J is a minimum if the configuration represents a less than half-filled shell, but is a maximum if the shell is more than half filled. The effective magnetic moment of the ion is then, to a good approximation, given by

$$\mu_{\text{eff}} = g\,\mu_B\,\sqrt{J(J+1)}, \tag{4.1}$$

where g is the Landé g-factor.

The orbital angular momentum is nearly completely quenched in the transition metals, but we will see that spin–orbit coupling leads to a very small contribution here. Therefore, to a good approximation, the magnitude of the magnetic moments is given by Hund's first rule and is due only to the spin of the electrons. Depending on the assumed configuration, we obtain for Fe, Co, and Ni magnetic moments (in μ_B) of 5, 4, and 3 (or 4, 3, and 2), respectively, which cannot even explain the trend in the measured saturation moments.

We therefore abandon this simple-minded approach and temporarily assume for the magnetic moment some adjustable values, for which we determine in the simplest possible way the magnetization, M, as a function of the temperature, T and the magnetic field, H. Thus, assuming initially N independent (i.e. noninteracting) magnetic moments of particles having the angular momentum quantum number j and denoting by m the magnetic quantum number, we begin by evaluating

$$M = Z^{-1} \sum_{m=-j}^{j} mg\,\mu_\mathrm{B}\, e^{mg\,\mu_\mathrm{B}\,H\,\beta} \tag{4.2}$$

to obtain the average moment per particle, M. Here (as usual) $\beta = 1/k_\mathrm{B} T$, g is the Landé factor and

$$Z = \sum_{m=-j}^{j} e^{mg\,\mu_\mathrm{B}\,H\,\beta}. \tag{4.3}$$

Using the abbreviation

$$a = g\,\mu_\mathrm{B}\, H\,\beta, \tag{4.4}$$

we easily determine Z by summing a geometric series to obtain

$$Z = \frac{e^{ja} - e^{-(j+1)a}}{1 - e^{-a}} = \frac{\sinh\left(j + \tfrac{1}{2}\right) a}{\sinh \tfrac{a}{2}}. \tag{4.5}$$

Equation (4.2) is now evaluated by calculating

$$M = g\,\mu_\mathrm{B}\, Z^{-1} \frac{\mathrm{d}Z}{\mathrm{d}a} \tag{4.6}$$

which gives

$$\frac{M}{M_s} = B_j\,(j\,g\,\mu_\mathrm{B}\,H\,\beta), \tag{4.7}$$

where $M_s = g\,\mu_\mathrm{B}\, j$ and

$$B_j(x) = \frac{2j+1}{2j} \coth \frac{2j+1}{2j} x - \frac{1}{2j} \coth \frac{1}{2j} x \tag{4.8}$$

is known as the Brillouin function. The relation (4.7) describes the experimental results perfectly for those systems for which the assumptions above are justifiable, as for instance for Gd^{3+} in $Gd_2(SO_4)_3 \cdot 8H_2O$ and for Fe^{3+} in $NH_4Fe(SO_4)_2 \cdot 12H_2O$ (Henry, 1952). We demonstrate this by showing in Fig. 4.2 interesting experimental data for these paramagnetic salts.

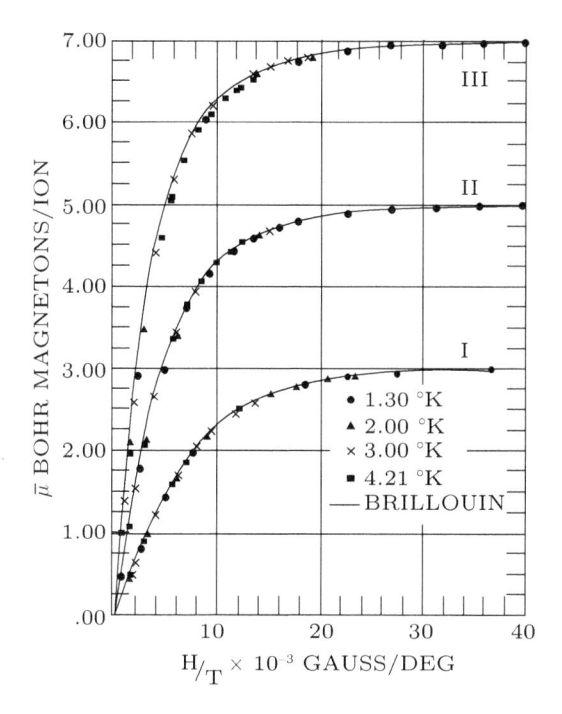

Figure 4.2 *Comparison of the Brillouin function with measured magnetic moments for spherical samples of Gd^{3+} in $Gd_3SO_4 \cdot 8H_2O$, Fe^{3+} in $NH_4Fe(SO_4)_2 \cdot 12H_2O$, and Cr^{3+} in $KCr(SO)_4 \cdot 12H_2O$ (from Henry, 1952).*

Now, returning to our problem and following Weiss, we postulate the existence of an internal magnetic field, H_m, also known as a molecular field, that is proportional to the magnetization, i.e.

$$H_m = WM. \qquad (4.9)$$

Historically, nothing was known about the physical origin of this field which, of course, is different now, and the constant W can (and will be) related to interatomic exchange interactions. Here we eliminate W in favor of the Curie temperature by substituting H_m into Eqn (4.7) and expanding the Brillouin function for small x:

$$B_j(x) \underset{x \to 0}{=} \frac{j+1}{3j} x, \qquad (4.10)$$

which gives, for finite β and $M \to 0$,

$$k_B T_c = \frac{1}{3} j(j+1) \left(g \, \mu_B\right)^2 W, \qquad (4.11)$$

i.e. T_c is the temperature at which the magnetization M vanishes. We next substitute Eqn (4.9) into Eqn (4.7); eliminating W with Eqn (4.11), we obtain

$$\frac{M}{M_s} = B_j\left(\frac{3j}{j+1}\cdot\frac{M/M_s}{T/T_c}\right). \tag{4.12}$$

This implicit equation for M/M_s is easily solved by defining

$$\zeta = \frac{M}{M_s}, \tag{4.13}$$

$$t = \frac{T}{T_c}, \tag{4.14}$$

$$\zeta = x\,t, \tag{4.15}$$

whence from Eqn (4.12)

$$t = \frac{1}{x} B_j\left(\frac{3j}{j+1} x\right). \tag{4.16}$$

Assuming for x any value in $0 < x < \infty$ we obtain t from (4.16) and ζ from (4.15). The results are shown in Fig. 4.1, the solid curves being calculated for $j = 1/2$ and lower curve for $j = 10/2$. Except for some small temperature range neither curve fits the data, although qualitatively "they capture the spirit."

In spite of the rather meager success, we may continue with the Weiss molecular field model and determine the magnetic susceptibility for temperatures above the Curie temperature. Using the expansion (4.10) and adding an external field, H, to the molecular field, we obtain with Eqn (4.7)

$$\frac{M}{M_s} = \frac{j+1}{3}\cdot g\,\mu_B\,\beta\,(H + W\,M) \tag{4.17}$$

from which the magnetic susceptibility, $\chi = M/H$, follows as

$$\chi = \frac{C}{T - T_c}, \tag{4.18}$$

where

$$C = \frac{1}{3} j(j+1)\,(g\,\mu_B)^2/k_B. \tag{4.19}$$

This is the famous Curie–Weiss law.

In Fig. 4.3 we show the measured high-temperature susceptibilities of Fe, Co, and Ni and see that the statement "the inverse susceptibility is linear for temperatures larger than

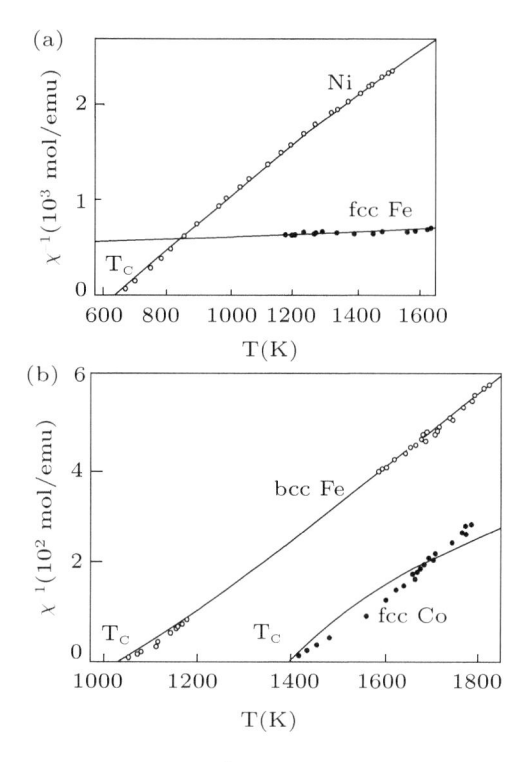

Figure 4.3 *Inverse magnetic susceptibility, χ^{-1}, for Fe, Co, and Ni (after Shimizu, 1981).*

the Curie temperature, T_c," holds only approximately. One can still obtain C [Eqn (4.19)] from the slope, and rewriting it as

$$C = \frac{1}{3} q_c (q_c + 2) \mu_B^2 / k_B \qquad (4.20)$$

we collect the values of q_c, which are called the "number of magnetic carriers." These are given in Table 4.1 where the ratio q_c/q_s is also listed. The trend in the number of magnetic carriers, q_c/q_s, allows a systematic discussion of the magnetic properties of ferromagnets provided we look at a large number of different systems. This was done by Rhodes and Wohlfarth (1963) who took the ratio q_c/q_s from experiment and plotted it as a function of the Curie temperature.

The results are shown in Fig. 4.4 which is called the Rhodes–Wohlfarth plot.

This phenomenological curve gives intuitive insight into the different mechanisms responsible for magnetic order. For, considering a system with localized moments, we do not expect the magnitude of the moment to change much when it is measured below and above T_c, giving ratios q_c/q_s of order unity. We attribute the rather systematic deviation of q_c/q_s from unity that is the dominant feature of the Rhodes–Wohlfarth plot to magnetic moments due to itinerant, i.e. essentially delocalized, magnetic moments.

180 *Electronic Structure and Magnetism*

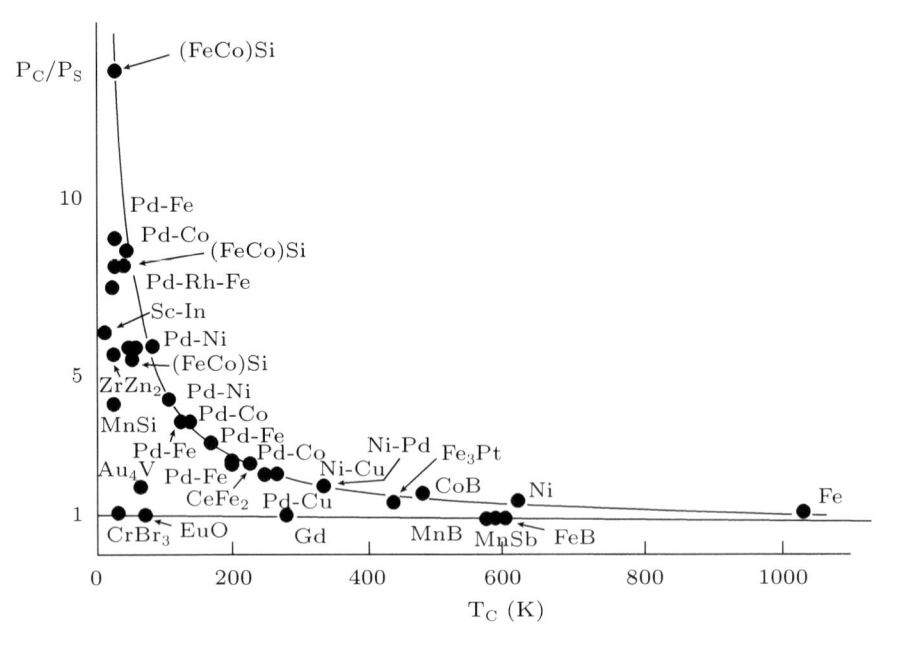

Figure 4.4 *Rhodes–Wohlfarth plot (Rhodes and Wohlfarth, 1963): ratio q_c/q_s, where q_c is obtained from the experimental Curie constant, Eqn (4.20), and q_s is the saturation magnetization, versus the Curie temperature. In the original graph, which we show here, the quantities q are denoted by P.*

The challenge is that we do not deal with two disjunct cases, localized or itinerant, but with a distribution which requires a unified treatment that interpolates between the two limits. Having tried out a primitive theory based on localized moments, we therefore now turn to a simple version of an itinerant-electron theory, presenting, except for some changes in the notation, essentially Stoner's original derivation.

4.1.1 Stoner theory

We have seen in Chapter 3 that electrons moving in the periodic potential of the solid are characterized by the quantum numbers n and $\mathbf{k}$, i.e. the band index and wave vector, and a spin projection for which, in the simplest case, we can choose up and down. We thus abandon the quantum numbers of the free atom states and try to work out a magnetic solution in the band picture. The physical picture is that the periodic potential breaks up the atom states and redistributes the valence electrons in Bloch states. In this introductory section we will again use a molecular field to represent exchange, postponing its justification to a later section. To render the calculation transparent we assume the electrons move in a simple s band characterized by the highly idealized density of states per particle and spin

$$\mathcal{N}(\varepsilon) = \frac{3}{4} \frac{\sqrt{\varepsilon}}{\varepsilon_F^{3/2}}, \qquad (4.21)$$

see Eqn (1.115), where ε_F is the Fermi energy. It should be clear that this choice of $\mathcal{N}(\varepsilon)$ is made for computational convenience only. In reality the electron states leading to magnetism originate from d rather than s bands. In its simplest incarnation they could, for instance, be described by the canonical d bands shown in Fig. 3.17. But since with these bands analytical calculations are not easily carried out, we resort to Eqn (4.21).

The aim is now to probe the stability of the nonmagnetic case by introducing a molecular field

$$H_m = I\zeta, \qquad (4.22)$$

where I denotes the molecular field constant and ζ is the magnetization given by the difference of the number of up $(+)$ and down $(-)$ electrons that H_m produces, i.e.

$$\zeta = n_+ - n_- \qquad (4.23)$$

and

$$n_\sigma = \int_0^{\varepsilon_{F\sigma}} \mathcal{N}(\varepsilon)\, d\varepsilon. \qquad (4.24)$$

Here $\varepsilon_{F\sigma}$ $(\sigma = \pm)$ are shifted Fermi energies connected with ζ by integrating Eqn (4.24) and comparing with (4.23):

$$\frac{\varepsilon_{F\pm}}{\varepsilon_F} = (1 \pm \zeta)^{2/3}. \qquad (4.25)$$

We next calculate the total energy of the electrons moving in the molecular field H_m. There are two contributions, the kinetic energy,

$$E_K = \int_0^{\varepsilon_{F+}} \varepsilon \mathcal{N}(\varepsilon)\, d\varepsilon + \int_0^{\varepsilon_{F-}} \varepsilon \mathcal{N}(\varepsilon)\, d\varepsilon, \qquad (4.26)$$

and the field energy

$$E_m = -I \int_0^\zeta \zeta'\, d\zeta', \qquad (4.27)$$

giving, after integrating,

$$E_K + E_m = E(\zeta) = \frac{9}{20} \frac{1}{\mathcal{N}_0} \left[(1+\zeta)^{5/3} + (1-\zeta)^{5/3}\right] - \frac{I}{2}\zeta^2. \qquad (4.28)$$

182 *Electronic Structure and Magnetism*

Here we have, for convenience in comparing with later, more general results, expressed the total energy using the spin-added density of states at the Fermi energy, which from (4.21) is

$$\mathcal{N}_0 = \frac{3}{2} \frac{1}{\varepsilon_F}. \tag{4.29}$$

This now allows us to find in a simple way the condition for a magnetic instability that is signaled by a *decreasing* function $E(\zeta)$ when ζ becomes finite, the stable case being an increasing $E(\zeta)$ for $\zeta > 0$. Therefore, we consider

$$\frac{dE}{d\zeta} = \frac{3}{4} \frac{1}{\mathcal{N}_0} \left[(1+\zeta)^{2/3} - (1-\zeta)^{2/3} \right] - I\zeta \tag{4.30}$$

and

$$\frac{d^2 E}{d\zeta^2} = \frac{1}{2\mathcal{N}_0} \left[(1+\zeta)^{-1/3} + (1-\zeta)^{-1/3} \right] - I. \tag{4.31}$$

The condition for an extremum, $dE/d\zeta = 0$, gives

$$\frac{(1+\zeta_0)^{2/3} - (1-\zeta_0)^{2/3}}{\zeta_0} = \frac{4}{3} I \mathcal{N}_0, \tag{4.32}$$

and Eqn (4.31) for the second derivative of the total energy can be rewritten as

$$\frac{d^2 E}{d\zeta^2} = \frac{1}{4} \left(\frac{1}{\mathcal{N}_+} + \frac{1}{\mathcal{N}_-} \right) - I, \tag{4.33}$$

where we have defined spin-projected densities of states at the Fermi energy using Eqn (4.21) and (4.25) as

$$\mathcal{N}_\pm = \frac{3}{4} \frac{1}{\varepsilon_F^{3/2}} \sqrt{\varepsilon_{F\pm}} = \frac{1}{2} \mathcal{N}_0 (1 \pm \zeta)^{1/3}. \tag{4.34}$$

The nonmagnetic state is unstable when the second derivative is negative, which for $\zeta = 0$ is seen from Eqn (4.33) to occur when

$$\mathcal{N}_0 I > 1. \tag{4.35}$$

This relation is called the *Stoner condition* which is much more general than this simple theory suggests. In Fig. 4.5 we graph the total energy, $2(E(\zeta) - E(0))/I$, Eqn (4.28) with $I\mathcal{N}_0$ as parameter and in Fig. 4.6 the stable solution, ζ_0, from (4.32) as a function of $I\mathcal{N}_0$ which is also called the Stoner parameter. If we assume that the molecular field constant

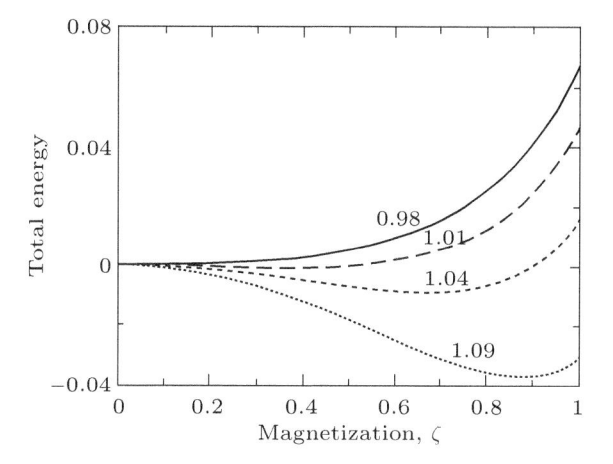

Figure 4.5 *Total energy in units of $I/2$ as a function of the magnetization, ζ, for different value of $I\mathcal{N}_0$ indicated in the figure.*

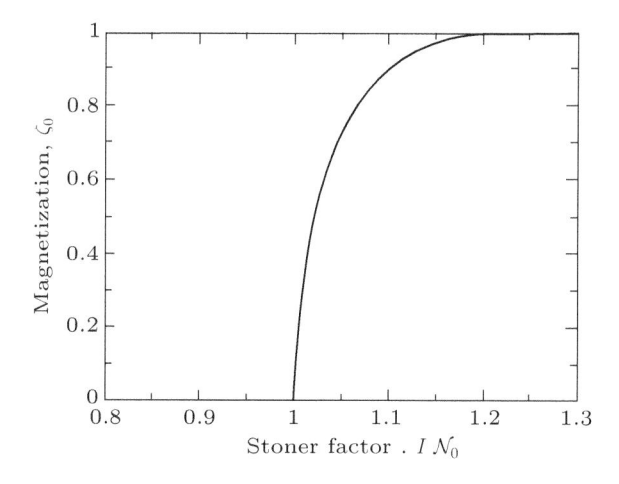

Figure 4.6 *Magnetization, ζ_0, from Eqn (4.32) as a function of the Stoner parameter, $I\mathcal{N}_0$.*

is approximately of the same order of magnitude for the different transition metals (an assumption that is borne out by calculations and experiments discussed later) we see that with increasing values of the density of states at ε_F, $\mathcal{N}_0$, or, what amounts to the same thing, with decreasing bandwidth, the nonmagnetic solution becomes unstable at some critical value of $\mathcal{N}_0$ that is fixed by the Stoner condition, Eqn (4.35). The magnetization that sets in is seen in Fig. 4.6 to saturate quickly with increasing $I\mathcal{N}_0$ tending toward unity which is, of course, a feature of the simple s band assumed at the outset (one should remember that a single band can hold only one electron per spin direction).

In the original work of Stoner the constant I was denoted as $k_B \theta$ and θ was interpreted as a characteristic temperature which Stoner speculated was related to the Curie temperature. This point of view is nowadays clarified and has no physical basis; with the use of realistic band structures and *ab initio* values of I the Curie temperatures on this basis are found to be unacceptably large; see Chap. 5, where we return to temperature properties. We close this subsection by writing down the susceptibility which is obtained from Eqn (4.33) as

$$\chi = 1 \bigg/ \frac{d^2 E}{d\zeta^2} = \left[\frac{1}{4} \left(\frac{1}{\mathcal{N}_+} + \frac{1}{\mathcal{N}_-} \right) - I \right]^{-1} \quad (4.36)$$

which for $\zeta \to 0$ is seen with Eqn (4.34) to become

$$\chi = \frac{\mathcal{N}_0}{1 - I \mathcal{N}_0} \quad (4.37)$$

(compare with Chap. 1).

Again, the Stoner condition is clearly seen to signal the magnetic instability. χ is called the *enhanced susceptibility* because for metals near a magnetic instability the denominator $1 - I \mathcal{N}_0$ can become very small.

The formulation appearing in this section is highly idealized; still Derlet and Dudarev (2007) have recently perfected this approach to such an extent that it constitutes the basis for realistic materials modeling.

We will see in the following section how the enhanced susceptibility, Eqn (4.37), and whence the Stoner condition, Eqn (4.35), together with a formula for the Stoner factor I are obtained from density-functional theory thus supplying a way to calculate these quantities *ab initio* in a realistic description of metallic magnets.

4.2 The magnetic susceptibility

4.2.1 Linear response

The magnetic susceptibility is but one example of a more general concept, namely that of the linear response functions. We have already seen an application of this theory in Sec. 2.8, where we discussed—among other things—the phenomenon of *screening*. However, the discussion there was restricted to the density response of the homogeneous, non-spin-polarized electron gas, whereas here we focus our attention on the magnetic response of the inhomogeneous system.

It is the density-functional approach that presently seems to offer the only practical way to incorporate effects of exchange and correlation into the response functions of inhomogeneous systems with pronounced band-structure properties. We therefore present an exposition of this theory which, after some approximations, gives an expression of the Stoner parameter that is amenable to *ab initio* calculations. Furthermore, our results are useful because they also apply to nonuniformly magnetized systems like antiferromagnets and others.

As we have seen in Chap. 3 we can calculate the spin-density matrix $\tilde{n}_{\alpha\beta}(\mathbf{r})$, defined in Eqn (2.149) in the local density-functional approximation for the ground state of any electronic system subject to a local spin-dependent external potential which we here denote by $\omega_{\alpha\beta}(\mathbf{r})$. It describes the coupling of the charge and the spin of the electrons to external electric and magnetic fields assumed to be time-independent. The *static* response functions are thus obtained by calculating the density matrices in the presence of two external potentials differing by an infinitesimal quantity $\delta\omega_{\alpha\beta}(\mathbf{r})$. By definition, the static linear spin-density response function $\chi_{\alpha\beta,\alpha'\beta'}(\mathbf{r},\mathbf{r}')$ measures the proportionality between the perturbation $\delta\omega$ and the resulting change δn in the density matrix (Williams and von Barth, 1983):

$$\delta\tilde{n}_{\alpha\beta}(\mathbf{r}) = \sum_{\alpha'\beta'} \int d\mathbf{r}'\, \chi_{\alpha\beta,\alpha'\beta'}(\mathbf{r},\mathbf{r}')\, \delta\omega_{\alpha'\beta'}(\mathbf{r}'). \tag{4.38}$$

Referring back to Chap. 2, Eqn (2.154), we see that a small external perturbation $\delta\omega$ changes the effective one-electron potential from $v^{\text{eff}}_{\alpha\beta}$ to $v^{\text{eff}}_{\alpha\beta} + \delta v^{\text{eff}}_{\alpha\beta}$ thereby giving rise to new one-electron orbitals $\{\psi_{i\alpha}\}$ for $\alpha = 1$ and 2.

The derivation is considerably more transparent if we assume we can find a single quantization axis (see e.g. the discussion in Sec. 3.4.7) and restrict the external potential to be diagonal. In this case the density matrix remains diagonal and the response function depends on two-spin indices only, so that we can rewrite Eqn (4.38) as

$$\delta n_\alpha(\mathbf{r}) = \sum_{\beta=1}^{2} \int d\mathbf{r}'\, \chi_{\alpha\beta}(\mathbf{r},\mathbf{r}')\, \delta\omega_\beta(\mathbf{r}'). \tag{4.39}$$

The change in the effective potential linear in $\delta n_\alpha(\mathbf{r})$ is, from Eqn (2.154),

$$\delta v^{\text{eff}}_\alpha(\mathbf{r}) = \delta\omega_\alpha(\mathbf{r}) + \sum_{\beta=1}^{2} \int d\mathbf{r}' \left\{ \frac{2}{|\mathbf{r}-\mathbf{r}'|} + \frac{\delta^2 E_{xc}}{\delta n_\alpha(\mathbf{r})\, \delta n_\beta(\mathbf{r}')} \right\} \delta n_\beta(\mathbf{r}'), \tag{4.40}$$

where E_{xc} includes as before the exchange-correlation kinetic energy T_{xc}. The change in δn_α resulting from the change of the effective potential we connect with $\delta v^{\text{eff}}_\alpha(\mathbf{r})$ by means of

$$\delta n_\alpha(\mathbf{r}) = \sum_{\beta=1}^{2} \int d\mathbf{r}'\, \chi^{0}_{\alpha\beta}(\mathbf{r},\mathbf{r}')\, \delta v^{\text{eff}}_\beta(\mathbf{r}'). \tag{4.41}$$

$\chi^0_{\alpha\beta}$ is easily calculated by first-order perturbation theory (see e.g. Merzbacher, 1970) which gives for the first-order change of the wave function

$$\delta\psi_{i\alpha}(\mathbf{r}) = \sum_{j \neq i} \frac{\psi_{j\alpha}(\mathbf{r}) \langle \delta v^{\text{eff}}_\alpha \rangle_{ji}}{\varepsilon_{i\alpha} - \varepsilon_{j\alpha}}, \qquad (4.42)$$

whence

$$\delta n_\alpha(\mathbf{r}) = \sum_{i=1}^{N} \sum_{j \neq i} \frac{\psi^*_{j\alpha}(\mathbf{r}) \psi_{i\alpha}(\mathbf{r}) \langle \delta v^{\text{eff}}_\alpha \rangle^*_{ji}}{\varepsilon_{i\alpha} - \varepsilon_{j\alpha}} + \text{c.c..} \qquad (4.43)$$

Since

$$\langle \delta v^{\text{eff}}_\alpha \rangle_{ji} = \int d\mathbf{r}' \, \psi^*_{j\alpha}(\mathbf{r}') \, \delta v^{\text{eff}}_\alpha(\mathbf{r}') \, \psi_{i\alpha}(\mathbf{r}'), \qquad (4.44)$$

we obtain by comparing with Eqn (4.41)

$$\chi^0_{\alpha\beta}(\mathbf{r}, \mathbf{r}') = \delta_{\alpha\beta} \sum_{i=1}^{N} \sum_{j \neq i} \frac{\psi_{i\alpha}(\mathbf{r}) \psi^*_{i\alpha}(\mathbf{r}') \psi^*_{j\alpha}(\mathbf{r}) \psi_{j\alpha}(\mathbf{r}')}{\varepsilon_{i\alpha} - \varepsilon_{j\alpha}} + \text{c.c..} \qquad (4.45)$$

It is an easy exercise to show that this expression for χ^0 can be written in the more symmetric form

$$\chi^0_{\alpha\beta}(\mathbf{r}, \mathbf{r}') = \delta_{\alpha\beta} \sum_{\substack{ij \\ (i \neq j)}} \frac{\theta(\varepsilon_F - \varepsilon_{i\alpha}) - \theta(\varepsilon_F - \varepsilon_{j\alpha})}{\varepsilon_{i\alpha} - \varepsilon_{j\alpha}} \cdot \psi_{i\alpha}(\mathbf{r}) \psi^*_{i\alpha}(\mathbf{r}') \psi^*_{j\alpha}(\mathbf{r}) \psi_{j\alpha}(\mathbf{r}'), \qquad (4.46)$$

where $\theta(x)$ is the unit step function which is 1 for $x > 0$ and 0 for $x < 0$. There is, however, another contribution to χ^0 that results from the first-order shift of the eigenvalues which leads to

$$\delta n(\mathbf{r}) = \sum_i \left[\theta(\varepsilon_F - \varepsilon_{i\alpha} - \langle \delta v^{\text{eff}}_\alpha \rangle_{ii}) - \theta(\varepsilon_F - \varepsilon_{i\alpha}) \right] |\psi_{i\alpha}(\mathbf{r})|^2. \qquad (4.47)$$

For small $\langle \delta v^{\text{eff}}_\alpha \rangle_{ii}$ this becomes

$$\delta n(\mathbf{r}) = -\sum_i \int d\mathbf{r}' \, \delta(\varepsilon_F - \varepsilon_{i\alpha}) \, \delta v^{\text{eff}}_\alpha \, |\psi_{i\alpha}(\mathbf{r}')|^2 \, |\psi_{i\alpha}(\mathbf{r})|^2, \qquad (4.48)$$

whence with Eqn (4.41) and (4.46)

$$\chi^0_{\alpha\beta}(\mathbf{r},\mathbf{r}') = \delta_{\alpha\beta}\,\chi^0_\alpha(\mathbf{r},\mathbf{r}'), \qquad (4.49)$$

where

$$\chi^0_\alpha(\mathbf{r},\mathbf{r}') = -\sum_i \delta(\varepsilon_F - \varepsilon_{i\alpha})\,|\psi_{i\alpha}(\mathbf{r}')|^2\,|\psi_{i\alpha}(\mathbf{r})|^2$$

$$+ \sum_{\substack{ij\\(i\neq j)}} \frac{\theta(\varepsilon_F - \varepsilon_{i\alpha}) - \theta(\varepsilon_F - \varepsilon_{j\alpha})}{\varepsilon_{i\alpha} - \varepsilon_{j\alpha}} \qquad (4.50)$$

$$\cdot \psi_{i\alpha}(\mathbf{r})\,\psi^*_{i\alpha}(\mathbf{r}')\,\psi_{j\alpha}(\mathbf{r}')\,\psi^*_{j\alpha}(\mathbf{r}).$$

This equation is the response function for noninteracting electrons, i.e. the Lindhard expression, but the space dependence is that of the true wave functions rather then the usually assumed plane waves, see Sec. 2.8. The first part on the right-hand side is the *intra-band* term which can be regarded as the limit as $i \to j$ of the term excluded from the second part on the right-hand side, the latter being the *interband* part of the Lindhard susceptibility (1954).

Having determined χ^0 we now derive an integral equation for the susceptibility χ of the interacting system. This is obtained from Eqn (4.40) by multiplying by χ^0, integrating, summing, using Eqn (4.39) on the left-hand side and observing that the change of the external potential is arbitrary. Hence

$$\chi_{\alpha\beta}(\mathbf{r},\mathbf{r}') = \delta_{\alpha\beta}\chi^0_\alpha(\mathbf{r},\mathbf{r}') + \sum_{\beta'} \int d\mathbf{r}_1\,d\mathbf{r}_2\,\chi^0_\alpha(\mathbf{r},\mathbf{r}_1)$$

$$\cdot \left[\frac{2}{|\mathbf{r}_1 - \mathbf{r}_2|} + K^{xc}_{\alpha\beta'}(\mathbf{r}_1,\mathbf{r}_2)\right]\chi_{\beta'\beta}(\mathbf{r}_2,\mathbf{r}'), \qquad (4.51)$$

where

$$K^{xc}_{\alpha\beta}(\mathbf{r},\mathbf{r}') = \frac{\delta^2 E_{xc}}{\delta n_\alpha(\mathbf{r})\,\delta n_\beta(\mathbf{r}')}. \qquad (4.52)$$

The result demonstrates how the full response of the system is obtained by letting the electrons respond as free particles to an effective field which consists of the external field, the induced Coulomb field, and an additional field that is related to the total exchange-correlation energy and describes the action of exchange and correlation.

Since we want to probe the magnetic instability of the nonmagnetic system, we now specialize the discussion to a non-spin-polarized system in which case there are two independent response functions, χ_n and χ_m, describing, respectively, the charge-density response to an external potential and the spin-density response to an external applied magnetic field.

From the possible response functions we here single out the *spin susceptibility*, χ_m, by writing with Eqn (4.39) the change in the magnetization that results from an external magnetic field which interacts with the spins as $-\boldsymbol{\sigma}\,\delta\mathbf{B}$ giving for $\delta\mathbf{B}$ in the z-direction

$$\delta m(\mathbf{r}) = \delta n_1(\mathbf{r}) - \delta n_2(\mathbf{r})$$

$$= -\sum_{\beta=1}^{2}\int d\mathbf{r}'\,[\chi_{1\beta}(\mathbf{r},\mathbf{r}') - \chi_{2\beta}(\mathbf{r},\mathbf{r}')]\,\delta B_\beta(\mathbf{r}'). \tag{4.53}$$

Hence, with $\delta B = \delta B_1 = -\delta B_2$,

$$\delta m(\mathbf{r}) = \int d\mathbf{r}'\,\chi_m(\mathbf{r},\mathbf{r}')\,\delta B(\mathbf{r}'), \tag{4.54}$$

where

$$\chi_m(\mathbf{r},\mathbf{r}') = \chi_{12}(\mathbf{r},\mathbf{r}') + \chi_{21}(\mathbf{r},\mathbf{r}') - \chi_{11}(\mathbf{r},\mathbf{r}') - \chi_{22}(\mathbf{r},\mathbf{r}') \tag{4.55}$$

is the desired spin susceptibility expressed with the quantities $\chi_{\alpha\beta}(\mathbf{r},\mathbf{r}')$. We now write out this combination of terms using Eqn (4.51). Since for non-spin-polarized systems χ^0, Eqn (4.50), is independent of the spin direction, we find that the Coulomb term in Eqn (4.51) drops out and derive

$$\chi_m(\mathbf{r},\mathbf{r}') = \chi^0(\mathbf{r},\mathbf{r}') + \int d\mathbf{r}_1\,d\mathbf{r}_2\,\chi^0(\mathbf{r},\mathbf{r}_1)$$
$$\cdot I_{xc}(\mathbf{r}_1,\mathbf{r}_2)\,\chi_m(\mathbf{r}_2,\mathbf{r}'). \tag{4.56}$$

Here we have used

$$\chi^0(\mathbf{r},\mathbf{r}') = -\chi^0_1(\mathbf{r},\mathbf{r}') - \chi^0_2(\mathbf{r},\mathbf{r}') = -2\,\chi^0_1(\mathbf{r},\mathbf{r}') \tag{4.57}$$

(observing a sign change due to $\delta v_\alpha^{\text{eff}} = -\delta B_\alpha^{\text{eff}}$) and

$$I_{xc}(\mathbf{r},\mathbf{r}') = -\frac{\delta^2 E_{xc}}{\delta m(\mathbf{r})\,\delta m(\mathbf{r}')} \tag{4.58}$$

which is obtained from $K^{xc}_{\alpha\beta}$ using $m(\mathbf{r}) = n_1(\mathbf{r}) - n_2(\mathbf{r})$.

In the local density-functional approximation where

$$E_{xc}[\tilde{n}] = \int n(\mathbf{r})\,\varepsilon_{xc}\,[n_1(\mathbf{r}), n_2(\mathbf{r})]\,d\mathbf{r}, \tag{2.158}$$

the local field interaction, Eqn (4.58), becomes a δ-function interaction:

$$I_{xc}(\mathbf{r}_1, \mathbf{r}_2) = \nu_{xc}[n(\mathbf{r}_1)]\,\delta(\mathbf{r}_1 - \mathbf{r}_2), \tag{4.59}$$

where

$$\nu_{xc}[n(\mathbf{r})] = \left[-\frac{d^2}{dm^2}\,n\,\varepsilon_{xc}(n,m)\right]_{\substack{n=n(\mathbf{r})\\m\to 0}} \tag{4.60}$$

which can be seen to be positive, and the susceptibility is

$$\chi_m(\mathbf{r},\mathbf{r}') = \chi^0(\mathbf{r},\mathbf{r}') + \int d\mathbf{r}_1\,\chi^0(\mathbf{r},\mathbf{r}_1)\,\nu_{xc}[n(\mathbf{r}_1)]\,\chi_m(\mathbf{r}_1,\mathbf{r}'). \tag{4.61}$$

This is a complicated integral equation that is not easily solved, but we will make some more comments on this problem later on. Here we want to extract the Stoner condition for the ferromagnetic instability.

4.2.2 The Stoner condition and other basic facts

4.2.2.1 The Stoner condition

An approximate solution for the susceptibility is obtained by specifying a homogeneous, uniform external magnetic field for which Eqn (4.54) gives with (4.61)

$$\begin{aligned}\delta m(\mathbf{r}) &= \int \chi_m(\mathbf{r},\mathbf{r}')\,d\mathbf{r}'\,\delta B \\ &= \int \chi^0(\mathbf{r},\mathbf{r}')\,d\mathbf{r}'\,\delta B + \iint d\mathbf{r}'\,d\mathbf{r}_1\,\chi^0(\mathbf{r},\mathbf{r}_1)\,\nu_{xc}[n(\mathbf{r}_1)]\,\chi_m(\mathbf{r}_1,\mathbf{r}')\,\delta B \\ &= \chi_0(\mathbf{r})\,\delta B + \int d\mathbf{r}_1\,\chi^0(\mathbf{r},\mathbf{r}_1)\,\nu_{xc}[n(\mathbf{r}_1)]\,\delta m(\mathbf{r}_1),\end{aligned} \tag{4.62}$$

where we obtain from Eqn (4.50) and (4.57)

$$\chi_0(\mathbf{r}) = \int d\mathbf{r}'\,\chi^0(\mathbf{r},\mathbf{r}') = \sum_i \delta(\varepsilon_F - \varepsilon_i)\,|\psi_i(\mathbf{r})|^2, \tag{4.63}$$

and notice that the interband part does not contribute. Janak (1977) defines a quantity

$$\gamma(\mathbf{r}) = \sum_i \frac{\delta(\varepsilon_F - \varepsilon_i)\,|\psi_i(\mathbf{r})|^2}{\mathcal{N}_0}, \tag{4.64}$$

where $\mathcal{N}_0 = \sum_i \delta(\varepsilon_F - \varepsilon_i)$ is the density of states at the Fermi energy. This gives

$$\chi_0(\mathbf{r}) = \int \chi^0(\mathbf{r},\mathbf{r}')\,d\mathbf{r}' = \int \chi^0(\mathbf{r}',\mathbf{r})\,d\mathbf{r}' = \mathcal{N}_0\,\gamma(\mathbf{r}), \tag{4.65}$$

with the further obvious property

$$\int \gamma(\mathbf{r})\,d\mathbf{r} = 1 \qquad \text{or} \qquad \int \chi_0(\mathbf{r})\,d\mathbf{r} = \mathcal{N}_0. \tag{4.66}$$

A solution of Eqn (4.62) is now obtained by trying the approximation (Janak, 1977)

$$\delta m(\mathbf{r}) = C\,\gamma(\mathbf{r}). \tag{4.67}$$

Substituting this into Eqn (4.62) and integrating over $\mathbf{r}$ we obtain

$$C = \mathcal{N}_0\,\delta B + C\,\mathcal{N}_0 \int \gamma^2(\mathbf{r}')\,\nu_{xc}[n(\mathbf{r}')]\,d\mathbf{r}'. \tag{4.68}$$

With the C that results Eqn (4.61) becomes

$$\delta m(\mathbf{r}) = \frac{\mathcal{N}_0\,\gamma(\mathbf{r})\,\delta B}{1 - \mathcal{N}_0 \int \gamma^2(\mathbf{r})\,\nu_{xc}[n(\mathbf{r})]\,d\mathbf{r}}. \tag{4.69}$$

Thus, after integrating over $\mathbf{r}$, we obtain the uniform susceptibility as

$$\chi = \frac{\mathcal{N}_0}{1 - I\,\mathcal{N}_0}, \tag{4.70}$$

where

$$I = \int \gamma^2(\mathbf{r})\,\nu_{xc}[n(\mathbf{r})]\,d\mathbf{r}. \tag{4.71}$$

We thus succeeded in rederiving the enhanced susceptibility, Eqn (4.37), without Stoner's simplifying assumptions. The Stoner parameter I is seen to be an "exchange-correlation" integral since ν_{xc} contains effects of both exchange and correlation. It has become customary, however, to call I simply the "Stoner exchange constant."

Since all quantities occurring can be clearly identified knowing the band structure (the $\{\varepsilon_i\}$ and $\{\psi_i(\mathbf{r})\}$) and the exchange correlation E_{xc}, I can be calculated using state

4.2 The magnetic susceptibility

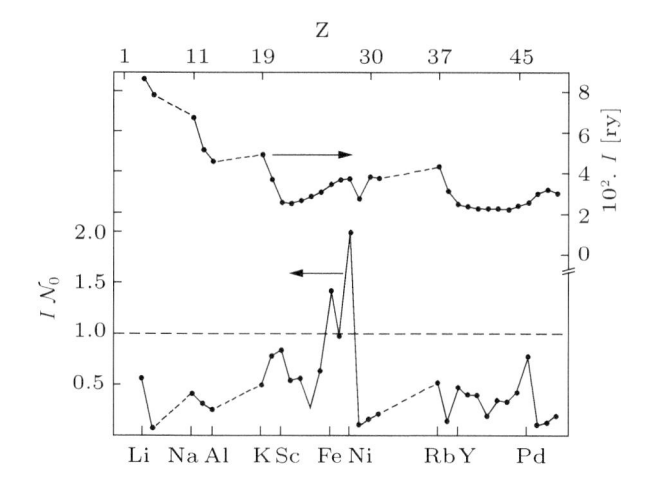

Figure 4.7 *Calculated (Janak 1977) values of the Stoner product $I\mathcal{N}_0$ and the Stoner exchange constant, I, for 32 elemental metals.*

of the art local density-functional methods. This was first done for 32 elemental metals by Janak (1977) who obtained the density of states at ε_F, $\mathcal{N}_0$, and the Stoner exchange constant from nonmagnetic ground-state calculations employing the KKR method. His results are collected in Fig. 4.7, in which both I and $I\mathcal{N}_0$ are plotted as a function of the atomic number, Z.

In these calculations, the crystal structure was taken to be either fcc or bcc, and materials with hexagonal or more complicated structures were usually treated as fcc. Of the two quantities calculated, the exchange constant I turned out to be rather insensitive to the crystal structure, the density of states, $\mathcal{N}_0$, however, being more sensitive. This can lead to large changes in the susceptibility enhancement with crystal structure. The lattice constants used were within 0.5% of those values where the total energy is minimum. This leads to atomic volumes that are slightly smaller than those observed. Many other calculations carried out later established that this underestimate is a defect of the local density-functional approximation that leads to a weak but peculiar chemical overbinding.

The most fundamental property of the results shown in Fig. 4.7 is the occurrence of ferromagnetism: from Eqn (4.70), any material for which $I\mathcal{N}_0 > 1$ should be ferromagnetic; this happens in Fe, Co, and Ni. Since these are experimentally the only ferromagnetic metals in this set, the result is indeed quite significant, supplying very strong support here for the correctness of the itinerant-electron picture. The calculated trend in the exchange constants deserves further notice. There is a gradual overall decrease in I with increasing atomic number Z that is due to the dependence of $\nu_{xc}[n(\mathbf{r})]$ on the density; ν_{xc} varies roughly as $n^{-2/3}$ and, as $n(\mathbf{r})$ gradually increases with increasing Z, ν_{xc} and thus I gradually decrease. Also of some interest is the roughly parabolic behavior of I within each transition series. This finds an explanation in the fact that the quantity $\gamma(\mathbf{r})$ reflects two characteristic forms, free-electron-like (roughly constant)

192 Electronic Structure and Magnetism

and d-like (relatively peaked and localized within the unit cell). Proceeding through a transition series, $\gamma(\mathbf{r})$ is initially free-electron-like, becoming a sum of free-electron and d contributions in the middle of the series, and is primarily d-like toward the end of the series. Thus, in going through the transition series, I is largest at either end (fully free-electron-like or fully d-like $\gamma(\mathbf{r})$) passing through a minimum near the middle of the series.

It is apparent from Fig. 4.7 that ferromagnetism occurs basically because of the spatial localization of the d orbitals near the top of the d band. This localization produces both a large density of states and a relative maximum in the exchange integral I. The product $I\mathcal{N}_0$ is large enough for ferromagnetism at the end of the 3d series, but not the 4d series because the 4d wave function extends further out from the nucleus (it possesses one more node than the 3d function) which implies a larger interaction between the neighbors, a larger bandwidth, and thus a smaller density of states. Even though the exchange integral is considerably larger in the light elements than in the 3d series, there is no possibility of ferromagnetism because of the much smaller density of states. To give further support to these statements we show the calculated densities of states $\mathcal{N}_0$ at the Fermi energy in Fig. 4.8.

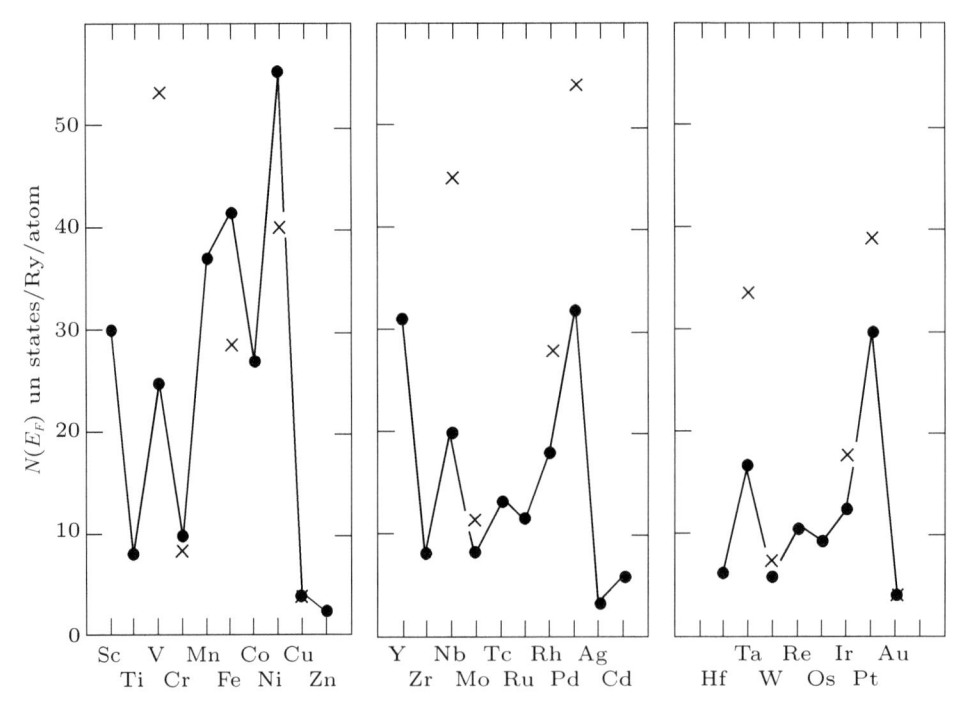

Figure 4.8 *Density of states at the Fermi energy, $\mathcal{N}_0$, for the three transition series based on data from Moruzzi et al. (1978) and Papaconstantopoulos (1986). Experimental data are marked by crosses and were obtained from the Sommerfeld specific-heat values.*

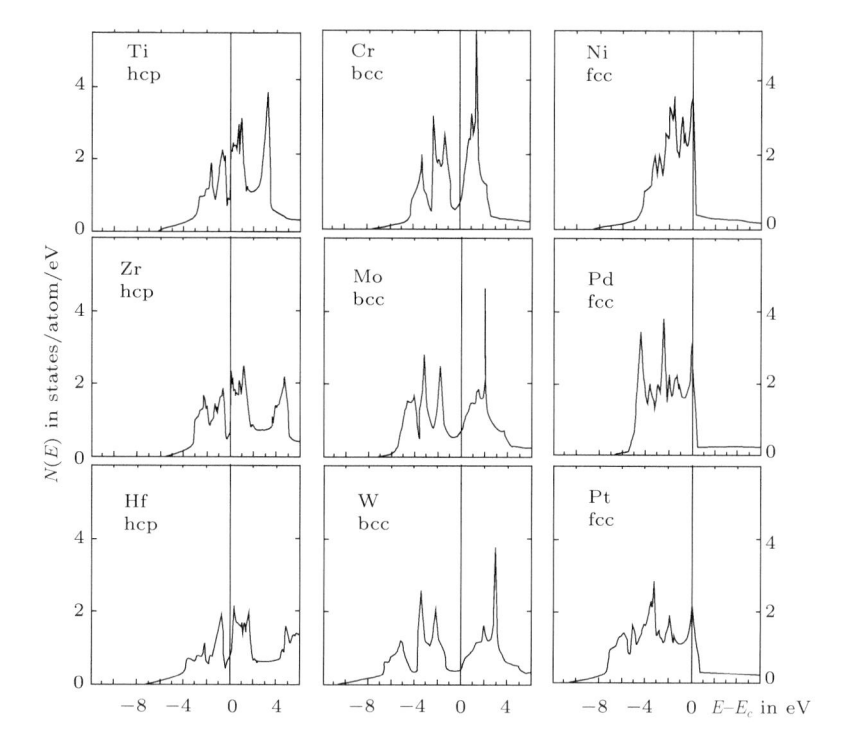

Figure 4.9 *Densities of states of selected transition metals (3d to 5d) from top to bottom and in the sequence hcp, bcc, fcc from left to right. ASW method used.*

The striking and huge variations of $\mathcal{N}_0$ reflect the peak structure of the density of states shown for nine typical transition metals in Fig. 4.9. As the bands are successively filled the Fermi energy moves across the bands thus causing the large variations in Fig. 4.8.

At this point it is natural to inquire into some typical and quite conspicuous features that can be seen in Fig. 4.9.

To prepare for this discussion we must have a deeper look into band-structure properties.

4.2.2.2 Band-structure features of the transition metals

The gross feature of the band structure of the transition metals is easily stated: as we move across the series, the d states are filled, and the bands first widen slightly moving up in energy, then they move down and become narrower.

Figure 4.10 shows the trend of the most important band parameters of the 4d transition metals.

It is inspired by a figure of Pettifor (1977a,b) to which it bears great similarity, but the details are different since we used parameters from fully hybridized, self-consistent scalar relativistic ASW calculations and slightly different definitions of the band parameters.

194   *Electronic Structure and Magnetism*

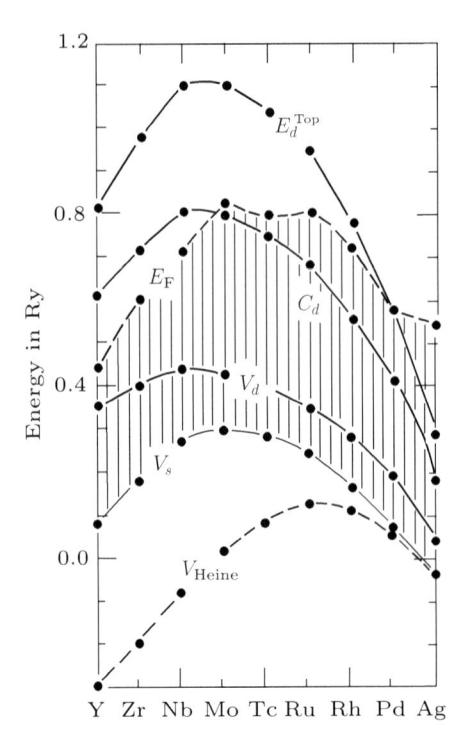

Figure 4.10 *Band parameters of the 4d transition metals: Fermi energy E_F; top, center, and bottom E_d^{TOP}, C_d, and V_d of the d band; bottom of the s band V_s and the Heine power law V_{Heine} (see text).*

The bottom of the s band is denoted by V_s; it is the Bessel function energy as explained in Sec. 3.4.5 and agrees closely with the lowest Γ_1 state in the band structure. The quantity V_d is the bottom of the d band which is obtained as the energy of the bonding state (zero logarithmic derivative) in the self-consistent potential. C_d is the center of the d band which is very close to the equivalent quantity in LMTO theory, but here it is the Hankel function energy. E_F denotes the Fermi energy and E_d^{TOP} is the top of the d band which is obtained as the energy of the antibonding state (infinite logarithmic derivative) in the self-consistent potential and is slightly higher in energy than the corresponding band parameter in Pettifor's (1977a,b) analysis. The quantities V_d and E_d^{TOP} are thus the energies obtained from so-called Wigner–Seitz (1955) boundary conditions. The trends in the 3d and 5d series are approximately the same as in the 4d series. The bands are, however, narrower for 3d electrons, as their radial wave function is more contracted and has one fewer nodes. On the other hand the bands are slightly wider for the 5d electrons, since their radial wave functions overlap more and have one more node than the 4d electrons. The parabolic trend that is so clearly visible in Fig. 4.10 is explained as follows: on the left-hand side, bonding states are occupied that accumulate charge mostly in between two atoms, thus keeping the charge *away* from the nuclei. As a consequence the lattice shrinks and the bands move up. Obviously, the equilibrium Wigner–Seitz (or atomic) radii, S, must reflect this behavior—and they do, as the curve V_{Heine} shows, which because of Heine's (1967) power law is an energy given by

$$V_{\text{Heine}} = \text{const.} + C/S^5 \qquad (4.72)$$

where the constant and C are chosen arbitrarily so that the curve neatly fits into Fig. 4.10. In the middle of the series nonbonding and further to the right antibonding states are occupied. Since the nuclear charge is only poorly screened it pulls electrons *toward* the nuclei, thus lowering the energy of the states and decreasing the overlap and thereby narrowing the bands. At the same time the lattice expands. Finally, the Fermi energy, E_F, displays a parabolic trend because of the rising *and* the filling of the d band and then decreases on account of the lowering in the center of gravity of the d band. Figures 4.11 and 4.12 depict small sections of the band structure of all transition metals. They were obtained using self-consistent scalar relativistic ASW calculations with experimental lattice constants and structure information (Landolt and Börnstein, 1973). The d bands can be clearly recognized in all cases if one compares them with the canonical bands in Fig. 3.17 and remembers the role of hybridization.

We return now to Fig. 4.9 where we show densities of states. We have made this selection to allow the same structure type to occur down a group, but different ones to occur across the period. It is seen that each structure type has a characteristic "fingerprint" which is similar to unhybridized canonical d state densities: this can be seen if Fig. 4.9 is compared with a canonical d state density. But the features are generally sharper in Fig. 4.9 than in canonical densities of states because hybridization with sp bands repels bands of the same symmetry, thereby creating local gaps and hence a more pronounced structure. An early and interesting explanation of the different structure types in the density of states of transition metal d electrons was attempted by Mott (1964). What we are concerned about is the pronounced minimum in the case of the bcc density of states which is in contrast to fcc and hcp. This is an important feature which shows up in phase stabilities and, as Mott (1964) pointed out, in bcc metallic alloys, but not in fcc or hcp alloys. Let us concentrate on the difference between the bcc and fcc density of states and consult the canonical d band structure shown in Figs. 3.17(a) and (b). Mott's original argument can now be made with confidence: he started out by considering unhybridized d bands, just as they occur in Fig. 3.17(b) along Γ to H. Along other directions, however, the bands hybridize, resulting in two narrow bands at the bottom and two narrow bands at the top being crossed by *one* band as in, for instance, the direction Γ to N, or even Γ to P, where for reasons of symmetry the lower and upper states are doubly degenerate. This is different in the case of the fcc bands shown in Fig. 3.17(a). Although the bands hybridize as well, there are *two* bands that connect the bottom and the top, for example, along the direction Γ to K and X, in such a way that the bottom and top states are more mixed up and not separated in two groups as in bcc.

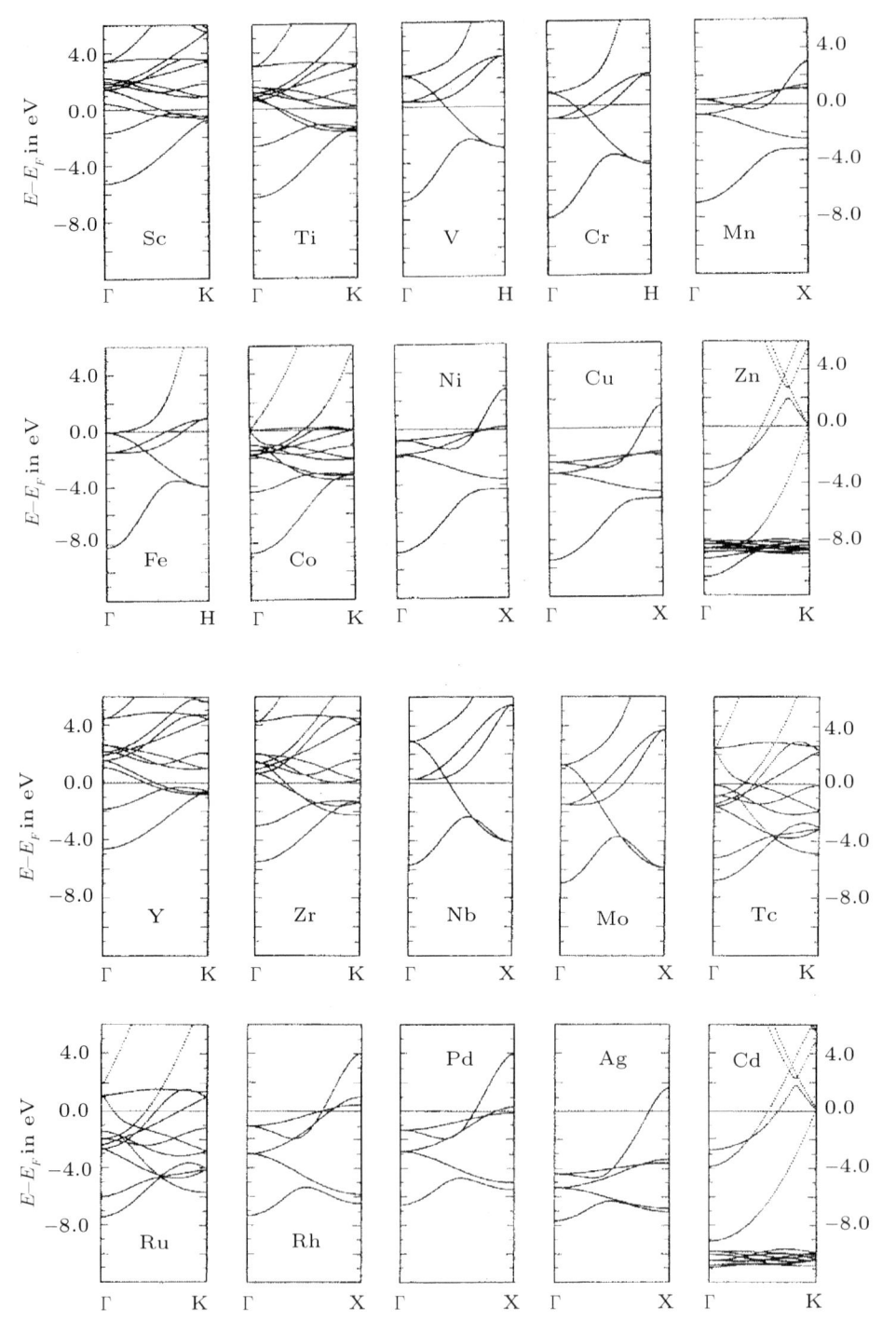

Figure 4.11 *Parts of the band structures for the transition metals calculated with the ASW method at the experimental lattice constants taken from Landolt and Börnstein (1973), (a) 3d metals, (b) 4d metals.*

4.2.2.3 Crystal phase stability

We are now in a position to discuss the physical origin of the occurrence of the various crystal structures observed in the transition metals. These are hcp → bcc → hcp → fcc as the d bands are progressively filled upon moving from left to right across the series of the 4d and 5d transition metals. Thus with a glance at Fig. 4.11(b), Y and Zr are hcp, Nb and Mo are bcc, Tc and Ru are hcp again, and Rh, Pd, and Ag are fcc. This is repeated in Fig. 4.12 where La and Hf are hcp, Ta and Wo are bcc, Re and Os are hcp again and Ir, Pt, and Au are fcc. However, we must exclude and treat separately the magnetic 3d metals to which we turn in Sec. 4.3.1.

At low temperatures the crystal structure of a metal is determined by the total energy E; in addition, there is a small contribution E_0, from the zero-point motion which may be neglected for the following reason: the zero-point energy in the Debye model is given by

$$E_0 = \frac{9}{8} k_B \theta_D$$

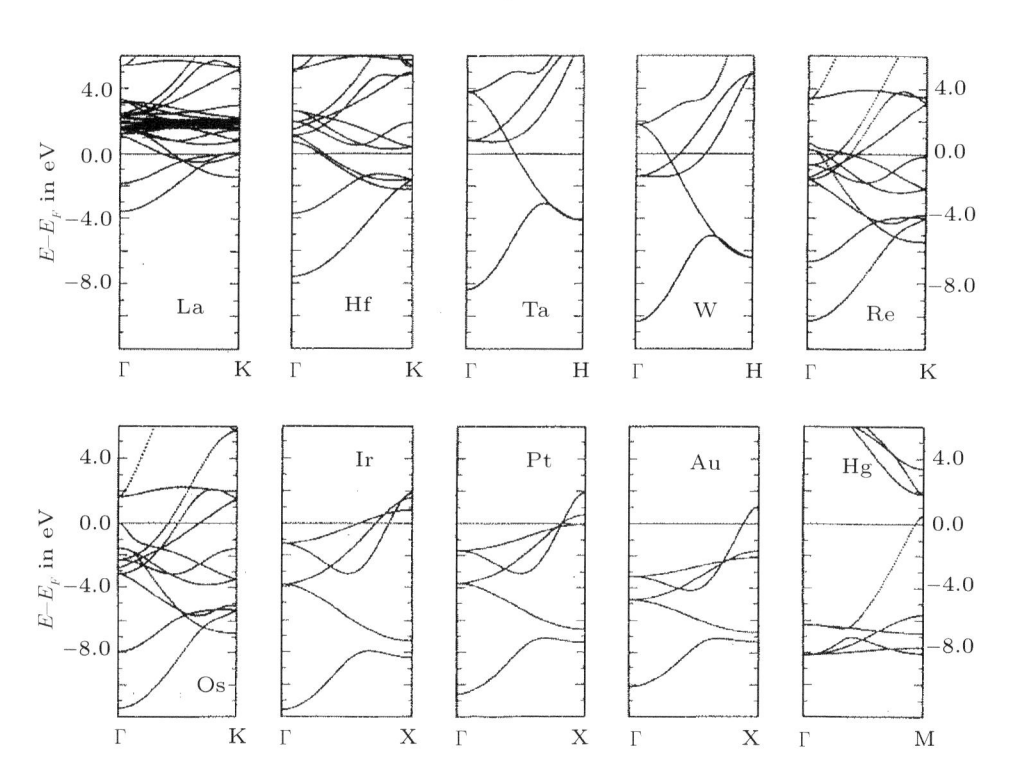

Figure 4.12 *Parts of the band structures for the transition metals calculated with the ASW method at the experimental lattice constants taken from Landolt and Börnstein (1973): 5d metals (La is assumed to have the hcp structure and Hg is calculated within an sc lattice).*

where k_B is Boltzmann's constant and θ_D is the Debye temperature. The latter is observed to vary at most by 10 K (Gschneidner, 1964) between different structures of the same metal and thus the corresponding change of zero-point energy is about 0.1 mRy which is more than an order of magnitude smaller than the energy difference we will be concerned with here. Hence, if the stability of some structure is to be determined with respect to some reference structure, which we may, for instance, take to be fcc, the total energy per atom is calculated for both phases and the energy difference is formed,

$$\Delta E = E_\mathrm{bcc} - E_\mathrm{fcc}$$

or

$$\Delta E = E_\mathrm{hcp} - E_\mathrm{fcc}.$$

The stable crystal phase is distinguished by having the lowest value of ΔE. Physical insight, however, is gained not by subtracting two big numbers as indicated by ΔE but by using the density of states discussed above; this is possible because of what has become to be known as *Andersen's force theorem* which we, therefore, briefly discuss here.

To do this we write the total energy as

$$E = \int^{E_\mathrm{F}} \varepsilon \mathcal{N}(\varepsilon)\, \mathrm{d}\varepsilon - E_1, \tag{4.73}$$

where $\mathcal{N}(\varepsilon)$ is the density of states, and E_1 is the double-counting term defined in Chap. 2 by, e.g. Eqn (2.69).

Next we calculate the first-order change in E when the system is subjected to some perturbation; for example, a perturbation of the boundary condition, of the occupation counts, of the external potential, or of the shape and size of the volume of the unit cell. We assume these changes to be parametrized by some quantities $X_1, X_2, \ldots$ called $\{X_i\}$; the self-consistent total energy is then a function of these parameters and the aim is to calculate

$$\delta E = \sum_i \frac{\delta E}{\delta X_i}\, \delta X_i. \tag{4.74}$$

For a given perturbation, specified by some set of $\{\Delta X_i\}$, we constrain the system to move to new self-consistency in two steps. In the first step, the effective potential is held fixed while the Schrödinger equation, appropriate for the new parameter set $\{X_i + \Delta X_i\}$, is solved giving rise to a new density and new energy eigenvalues. Denoting this restricted variation by the symbol δ^*, we obtain the complete variation by adding δ^SS which consists of a second step in which the potential is allowed to relax to self-consistency while keeping the parameter set fixed at $\{X_i + \Delta X_i\}$, i.e.

$$\delta = \delta^* + \delta^\mathrm{SS}. \tag{4.75}$$

Thus, for instance, the change of the single-particle (or *band*) term in Eqn (4.73),

$$E_{\rm B} = \int^{E_{\rm F}} \varepsilon \mathcal{N}(\varepsilon)\, {\rm d}\varepsilon, \qquad (4.76)$$

is

$$\delta E_{\rm B} = \delta^* E_{\rm B} + \delta^{\rm SS} E_{\rm B}. \qquad (4.77)$$

Hence the change of the entire total energy is

$$\delta E = \delta^* E_{\rm B} + \delta^{\rm SS} E_{\rm B} - \delta E_1. \qquad (4.78)$$

The essential observation now is that the last two terms on the right-hand side cancel in first order leaving only a surface term when the volume is changed, i.e. in the local-density approximation

$$\delta E = \delta^* E_{\rm B} - \int_{\delta\Omega} n^2 \frac{{\rm d}\epsilon_{xc}}{{\rm d}n} \delta {\bf S} \cdot {\rm d}{\bf S}, \qquad (4.79)$$

where d**S** is the surface element and δ**S** describes the change in the volume Ω in the sense that a point **S** is taken to **S** + δ**S** by the perturbation. The proof of the cancellation is not difficult; one writes out the term δE_1 and uses first-order perturbation theory for the term $\delta^{\rm SS} E_{\rm B}$ (see e.g. Methfessel and Kübler, 1982).

The basic idea behind Eqn (4.79), *Andersen's force theorem*, is that in a linear approximation the interaction drops out of the total energy change. In particular, the self-consistency step can be ignored, except for the surface term in the case of a volume change. We thus have a mathematical justification for thinking in terms of independent electrons and interpreting band structures and density of states. We also have a simple method to obtain forces from a non-self-consistent calculation.

Returning now to the discussion of the *crystal-phase stability*, we first point out that self-consistent calculations of the total energy differences $\Delta E = E_{\rm bcc} - E_{\rm fcc}$ and $\Delta E = E_{\rm hcp} - E_{\rm fcc}$ showed that use of the force theorem is exceedingly well justified. So we may calculate

$$\delta E = \delta^* \int^{E_{\rm F}} \varepsilon \mathcal{N}(\varepsilon)\, {\rm d}\varepsilon \qquad (4.80)$$

at equal volumes using the self-consistent fcc potential in the bcc or hcp geometry and thus obtain the desired value of $\Delta E \cong \delta E$. This was first done by Skriver (1985) whose results reproduced the observed sequence of structures for the transition metals. The most conspicuous feature of his results (which can easily be reproduced) is the pronounced bcc stability in the middle of the series which finds a ready explanation by the density of states, $\mathcal{N}(\varepsilon)$, that determines δE above.

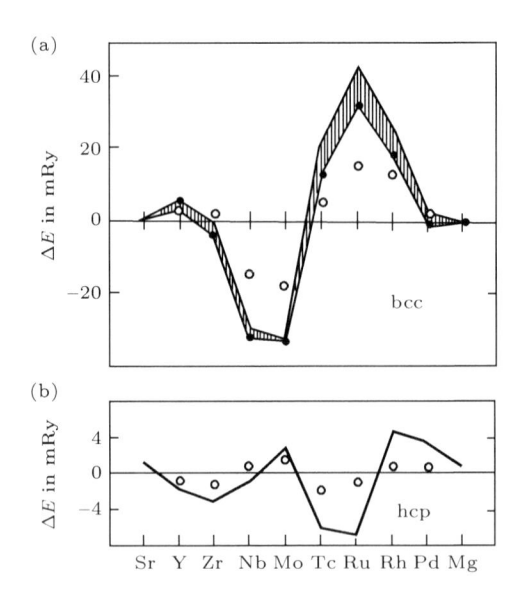

Figure 4.13 *Total energy of the bcc (a) and hcp structures (b), relative to the total energy of the fcc structure for the 4d transition metals. After Skriver (1985) and Kübler and Eyert (1992).*

The canonical d bands described earlier give an explanation for the gross features of the density of states that, within each crystal structure, is to a good approximation independent of the element involved, except, of course, for the width of the bands; recall Fig. 4.9. We draw attention in particular to the deep minimum marking the center of the bcc bands. So if we imagine we determine the value of the integral in Eqn (4.80) using separately the hcp, bcc, and fcc densities of states for an element in the middle of the transition series (where the d band is approximately half filled) and compare the numerical results, it is very plausible that we find the value for the bcc density of states to be lowest. Indeed, the total energy difference, $\Delta E = E_{\text{bcc}} - E_{\text{fcc}}$, for Mo is ~ -37 mRy and so is the value obtained with Eqn (4.80). Thus it is the degree of band filling and the crystal structure that determines the phase stability. The hcp-fcc energy differences are much smaller but find the same explanation.

In Fig. 4.13 we show the bcc-fcc (upper part) and hcp-fcc (lower part) structural energy differences for the 4d transition metal series. Here the open circles are enthalpy differences derived by Miedema and Niessen (1983) from phase-diagram studies; they are usually accepted as the experimental counterpart to the calculated $T = 0$ total energy differences. The latter are obtained from self-consistent ASW calculations and are shown as solid dots, whereas the other border of the shaded area is obtained by the force theorem which also supplied the results shown for the hcp-fcc energy differences. In Fig. 4.14, taken from the work of Skriver (1985), we show the structural total energy differences for the 3d transition metal series. These results are particularly interesting because of their failure to explain the crystal structures of Fe and Co; we take up this topic in Sec. 4.3.1.

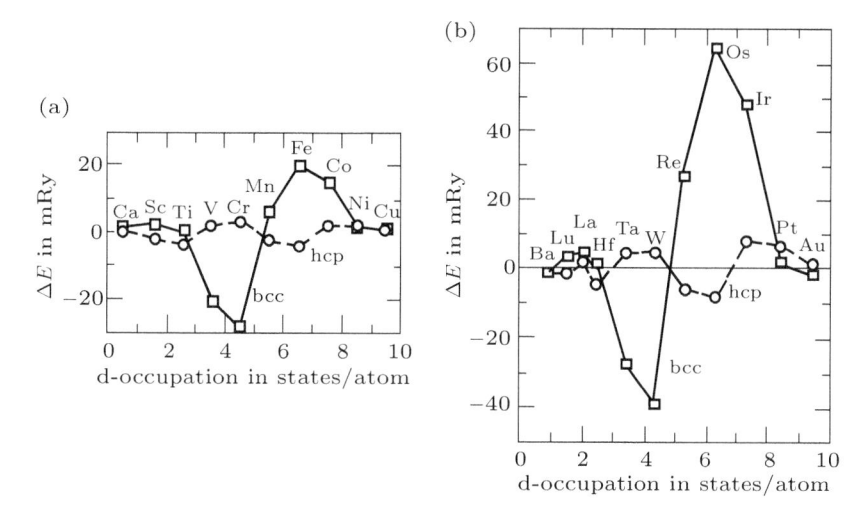

Figure 4.14 *Total energy of the bcc and hcp structures, relative to the total energy of the fcc structure for the 3d transition metals. After Skriver (1985).*

After this overview we return to the susceptibility and discuss calculations of the nonuniform magnetic susceptibility that throws more light on the magnetic properties of a selection of transition metals.

4.2.3 The static nonuniform magnetic susceptibility

In contrast to the estimate in Eqn (4.70) of the enhanced uniform susceptibility, we now want to go one step further and probe the response of the electron system to a magnetic field that is spatially nonuniform in such a way that its variation in space can be characterized by a wave vector **q**. For this purpose one could, in principle, attempt to solve the integral equation (4.61). Indeed, such an approach was suggested by Vosko *et al.* (1978), but no particular application was performed. We (Sandratskii and Kübler, 1992) here circumvent the integral equation and proceed differently by specifying a staggered magnetic field of the form

$$\Delta \mathbf{B}(\mathbf{r}) = \Delta B \cdot \sum_n \Big(\cos(\mathbf{q} \cdot \mathbf{R}_n), \sin(\mathbf{q} \cdot \mathbf{R}_n), 0 \Big) \Theta(|\mathbf{r} - \mathbf{R}_n|), \qquad (4.81)$$

where $\Theta(r)$ is the step function that equals unity for r smaller than the atomic sphere radius and zero otherwise, and the large parentheses contain Cartesian coordinates. The staggered field thus describes a spiral with polar angle $\theta = 90°$; compare with Sec. 3.4.7. To simplify the calculations further we only deal with cases where the magnetic moment of a given atom is parallel to the field. Therefore, in all calculations the magnetic structure possesses the form given by Eqn (4.81), i.e. the magnetic moment of the n-th atom is parallel to the vector $(\cos(\mathbf{q} \cdot \mathbf{R}_n), \sin(\mathbf{q} \cdot \mathbf{R}_n), 0)$. For nonmagnetic crystals this condition is always fulfilled because the symmetry of the unperturbed state (with no magnetic moments) guarantees that the field given by Eqn (4.81) and the induced moment are parallel.

For magnetic crystals one must ensure that this condition is fulfilled. To do this, the calculation is performed in two steps for every value of $\mathbf{q}$. In the first step the calculations are carried to self-consistency constraining the symmetry of the magnetic structure such that it coincides with the symmetry of the magnetic field chosen. This field is then turned on to be finite and the response of the electrons is calculated. The symmetry of the crystal, finally, guarantees that the magnetic response has the configuration of the field, as desired. The detailed prescription for incorporating the magnetic field in both cases is the replacement of the unperturbed potentials Δv in (3.309) by $\Delta v \mp \Delta B$. The response after the first iteration yields, because of the symmetry requirements, in the local (spin-diagonal) atomic frame of reference (FOR) $\Delta m = \chi_0 \Delta B$, or in the global FOR

$$\Delta m(\mathbf{q}) = \chi_0(\mathbf{q})\, \Delta B(\mathbf{q}). \tag{4.82}$$

A full self-consistent calculation in the presence of ΔB allows the system to relax the spin and charge densities and hence yields the enhanced susceptibility, $\chi(\mathbf{q})$, for which we now give a simple derivation that is valid for both nonmagnetic and magnetic systems; it is inspired by the work of Heine (Small and Heine, 1984) and approximates the theory given in Sec. 4.2.1 and 4.2.2.

The induced magnetic moment, Δm, changes the exchange-correlation potential leading to an effective increase in the applied field. The physics is in principle given by Eqn (4.62) for which we write in the local FOR

$$\Delta m = \chi_0 \Delta B + \chi_0 I \Delta m \tag{4.83}$$

which again is possible because of the imposed symmetry restrictions. In the global FOR (4.83) reads

$$\Delta m(\mathbf{q}) = \chi_0(\mathbf{q})\, \Delta B(\mathbf{q}) + \chi_0(\mathbf{q})\, I(\mathbf{q})\, \Delta m(\mathbf{q}). \tag{4.84}$$

The self-consistent response is now written like Eqn (4.54) for which we assume in the local FOR

$$\Delta m = \chi \Delta B. \tag{4.85}$$

In the global FOR we thus write

$$\Delta m(\mathbf{q}) = \chi(\mathbf{q})\, \Delta B(\mathbf{q}) \tag{4.86}$$

and obtain for the enhanced susceptibility with Eqn (4.84)

$$\chi(\mathbf{q}) = \frac{\chi_0(\mathbf{q})}{1 - I(\mathbf{q})\, \chi_0(\mathbf{q})}, \tag{4.87}$$

and hence

$$I(\mathbf{q}) = \chi_0(\mathbf{q})^{-1} - \chi(\mathbf{q})^{-1}. \tag{4.88}$$

In the calculations at hand a small magnetic field corresponding to a spin splitting of 2 mRy was employed.

We next focus our attention on three metals, two of which are nonmagnetic, viz. V and Pd and one, Cr, possesses a spin-density wave in the ground state. We will return to the latter property in Sec. 4.3.4, assuming here and making certain that Cr is nonmagnetic.

4.2.3.1 Nonmagnetic V, Cr, and Pd

The calculated susceptibilities of V, Cr, and Pd are shown in Figs. 4.15–4.17 and appear to be very different. The susceptibility enhancement is strong for the cases of Cr and Pd. For Cr the maximum of the enhancement corresponds to an antiferromagnetic configuration and reaches the value of 12.9 at $\mathbf{q} = (0,0,1)$ (in units of $2\pi/a$ throughout). In the case of Pd, the enhancement is maximal for a ferromagnetic configuration ($\mathbf{q} = 0$) and reaches the value of 6.85. For other values of $\mathbf{q}$ the enhanced χ quickly approaches the unenhanced χ_0.

The situation shown in Fig. 4.15 for V is different. The enhancement is much weaker and χ is a nonmonotonic function of $\mathbf{q}$. An obvious explanation of the strong enhancement obtained for antiferromagnetic configurations of Cr and a ferromagnetic configuration of Pd is the fact that here magnetic states "lie nearby"; in fact, a rather small increase of the lattice constants in both cases leads to transitions to magnetic states, antiferromagnetic for Cr and ferromagnetic for Pd. The total energy, correspondingly, has a rather flat minimum as a function of the amplitude of the magnetic moment m at $m = 0$, which leads to large values of the magnetic susceptibility. This is different in

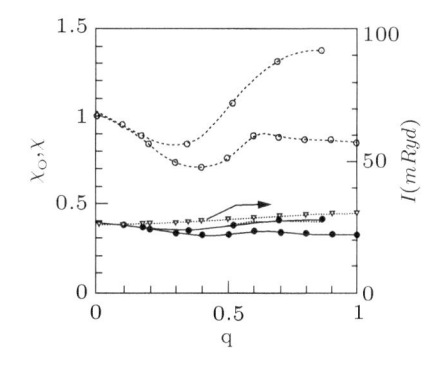

Figure 4.15 *The **q**-dependence of unenhanced and enhanced susceptibilities and Stoner parameters of V for two directions in reciprocal space. Susceptibilities are given in units of $\chi(0)$, see Table 4.2. Length of the **q**-vector is given in units of $2\pi/a$. • $-\chi_0$; ○ $-\chi$; ∇ $-I$. The curves for different directions may be distinguished by the last point of the curve: for the (0,0,1)-direction we have $q = 1$, for the (1,1,1)-direction the last point is close to 0.9 (Sandratskii and Kübler, 1992).*

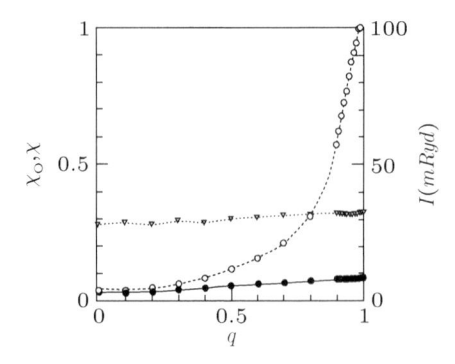

Figure 4.16 *The **q**-dependence of unenhanced and enhanced susceptibilities and Stoner parameters of Cr for the (0,0,1)-direction. The same symbols as in Fig. 4.15 are used. Susceptibilities are given in units of the maximal value of enhanced susceptibility $\chi_{max} = 8.03 \cdot 10^{-4}$ emu/mol (Sandratskii and Kübler, 1992).*

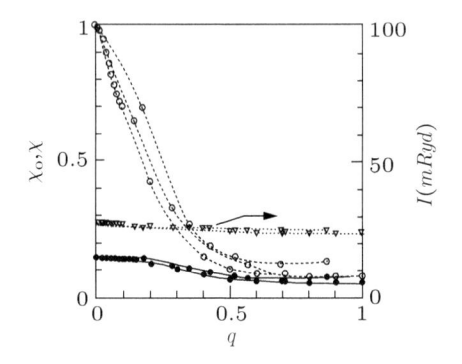

Figure 4.17 *The **q**-dependence of unenhanced and enhanced susceptibilities and Stoner parameter of Pd for three directions in reciprocal space. The same units and the same symbols as in Fig. 4.15 are used. The curves for different directions may be distinguished by the last point of the curve: for the (0,0,1)-direction this is $q = 1$, for the (1,1,1)-direction the last point is close to $q = 0.9$, for the (1,1,0)-direction the last point is close to 0.7 (Sandratskii and Kübler, 1992).*

the case of V where ferro- and antiferromagnetic states seem to coexist (Moruzzi and Marcus (1990a, b) for larger volumes. This property is obviously correlated with the type of **q**-dependence of the susceptibilities of these metals. To estimate the extent of the anisotropy of the susceptibility and exchange parameter, calculations for V and Pd were carried out for different directions in reciprocal space. For both metals we found the parameter I to be almost isotropic, whereas the anisotropy of the susceptibility is quite pronounced. However, in the case of Pd, the overall behavior of the susceptibility, i.e. its sharp maximum at $\mathbf{q} = 0$ and its monotonic decrease with increasing **q** is common for all directions of **q**. This is different for V where in the (0,0,1)-direction the susceptibility possesses a maximum at **q** = 0 in contrast to the (1,1,1)-direction where the susceptibility is maximal at the Brillouin-zone boundary.

It is seen that the Stoner parameter depends on **q** only weakly. This suggests that the exchange parameter can indeed be treated as a local atomic property, independent of **q**, having values in the range of 25–30 mRy for all atoms. This is in good agreement with Himpsel (1991) who reached this conclusion on the basis of experimental information on magnetic 3d crystals. From the calculations it appears that this is also so for the nonmagnetic 3d and 4d crystals (see also Fig. 4.7).

The weak dependence of I on **q** allows us to trace back the strong **q**-dependence of the enhanced susceptibility to the **q**-dependence of the unenhanced susceptibility, essentially. For **q** = 0 we saw (see Eqn (4.70)) that χ_0 is connected with a property of the unperturbed state, namely the density of states at the Fermi energy, because in this case the applied field lifts the spin degeneracy of the bands leading to a repopulation of the shifted spin-up and spin-down bands at the Fermi energy. For nonvanishing **q** the situation is substantially more complicated because the change of the electronic structure cannot be reduced to merely shifted states. In the case of Pd we see that the high density of states at the Fermi energy does not lead to a high unenhanced susceptibility for large **q**. As pointed out in Sec. 3.4.7 for **q** $\neq$ 0, states of opposite spin projection are separated by the vector **q** in reciprocal space and they, in general, hybridize. Bonding states have lower energy and possess a positive spin projection on the local direction of the magnetic field; antibonding states have higher energy and possess a negative spin projection. However, those states close to the Fermi energy in zero field that become antibonding states will be emptied in a finite field. If the number of such states is substantial, they will lead to a noticeable increase of the local magnetic moment and to a large unenhanced susceptibility. An example of this is the case of Cr (see Sec. 4.3.4) where, as we will see, there are large ("nesting") pieces of the Fermi surface separated by the vector **q** close to (0,0,1) and, as a result, the susceptibility χ_0 for **q** close to (0,0,1) is of the order of three times higher than for **q** = 0. Some numerical details of these calculations are collected in Table 4.2 which also contains results for ferromagnetic Fe, Co, and Ni, to be discussed next.

4.2.3.2 The longitudinal susceptibilities of ferromagnetic Fe, Co, and Ni

We have seen in Sec. 4.2.2 that the ground states of Fe, Co, and Ni, at least for the crystal structures investigated, cannot be nonmagnetic. We start, therefore, in ferromagnetic states and calculate the nonuniform longitudinal susceptibilities beginning with the total energies as a function of the spiral vector **q**. The results are summarized in Fig. 4.18 where the total energy differences, $E(\mathbf{q}) - E_{\text{ferro}}$, are shown as a function of **q**. We see that the ferromagnetic states (**q** = 0) are stable except for fcc Fe which will be discussed in Sec. 4.3 where we will take up the electronic structure of the elementary magnetic metals in some detail. Figure 4.19 shows that for bcc Fe the magnetic moment is approximately unchanged up to **q** = 0.3 and then decreases for increasing **q**, the larger **q**, the faster its decrease.

For **q** = 1, different methods give different results for the value of the self-consistent atomic magnetic moment, a fact that is explained by detailed investigations which showed

Table 4.2 *Parameters of calculation and computational results for* $\mathbf{q} = 0$.

Metal	Crystal structure	Lattice param. (au)	$\chi(0)$ (emu/mol $\times 10^{-4}$)		$\chi(0)/\chi_0(0)$		$I(0)$ (mRy)	
			Present calcul.	Others' results	Present calcul.	Other's results	Present calcul.	Others' results
V	bcc	5.54	1.15		2.26	2.34^d	26	26^d 29^g
		5.7	1.53	$\sim 1.6^a$	2.60	2.73^a	25	
Cr	bcc	5.3	0.3		1.36	1.36^d	28	28^d
Pd	fcc	7.42	5.18	$\sim 7.1^a$	6.85	4.46^d	27	25^d
				11.2^b		5.0^e		24.5^b
						$4.6-9.4^c$		26^g
Fe	bcc	5.27	0.21	0.37^c	1.24		28	34^d 32.5^b
								33.5^f 34^g
								30^c
Co	fcc	6.448^h	0.25	0.22^c	1.00		0	36^d 30^c
								36^g
Co	hcp	4.738	0.14		1.23		32	
		7.690						
Ni	fcc	6.55	0.31	0.13^b	1.43		33	37^d 34.5^b
				0.26^c				33^c 37^g

^{a}Stenzel and Winter (1986), ^{b}Poulson *et al.* (1976), ^{c}Yamada *et al.* (1980), ^{d}Janak (1977), ^{e}Jarlborg (1986), ^{f}Luchini *et al.* (1991), ^{g}Gunnarsson (1976). ^{h}This lattice constant has been chosen somewhat too small.

that for values $\mathbf{q}$ close to 1 there are two states, magnetic and nonmagnetic, for which the difference in the total energy is very small. It reacts sensitively to the details of the computational procedure, the choice of the lattice constant, the basis set, etc., the calculations thus showing a pronounced instability at $\mathbf{q} = 1$. In the case of fcc Co and Ni, Figs. 4.20 and 4.22, the important result emerges that starting at $\mathbf{q} = 0.6$ for Co and $\mathbf{q} = 0.5$ for Ni the magnetic moment vanishes, i.e. for large values of $\mathbf{q}$ the self-consistent state is nonmagnetic. Since the function $m(\mathbf{q})$ contains information about the dependence of the local magnetic moments on the angles between the spins of adjacent atoms, our finding is important for finite temperature properties of itinerant-electron magnets to be discussed in Chapter 5.

4.2 The magnetic susceptibility

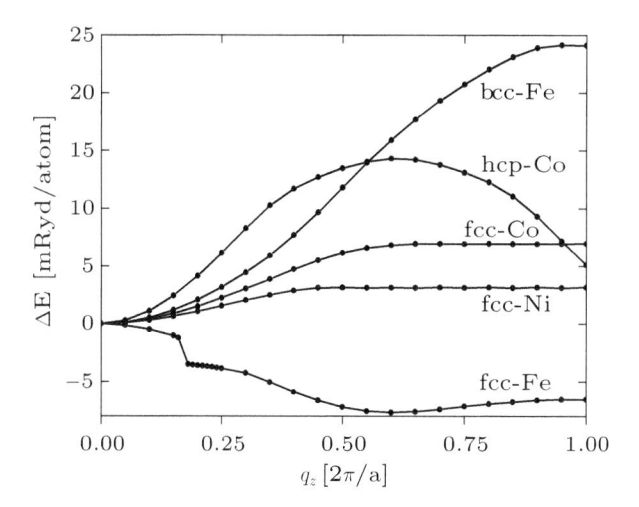

Figure 4.18 *The total energy difference, $E(\mathbf{q}) - E_{\text{ferro}}$, as a function of $\mathbf{q} = (0,0,q_z)$ for Fe, Co, and Ni. Concerning hcp Co note the differences in the value of the A-point in the hexagonal Brillouin zone compared with the X- and H-points of the cubic zones. The kink in the curve for fcc Fe will find a natural explanation later on.*

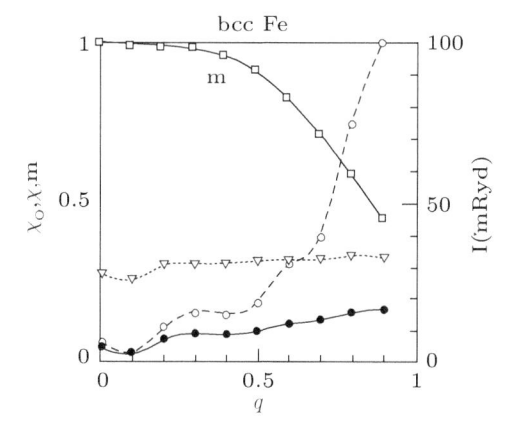

Figure 4.19 *The $\mathbf{q}$-dependence of unenhanced and enhanced susceptibilities, Stoner parameter and local magnetic moment of bcc Fe for the (0,0,1)-direction. The same units and the same symbols as in Fig. 4.15 are used for susceptibilities and the Stoner parameters. □ – magnetic moment. Magnetic moment is given in units of $m_0(0) = 2.13\,\mu_B$ (Sandratskii and Kübler, 1992).*

In the case of hcp Co, Fig. 4.21, the magnetic moment decreases slowly with increasing $\mathbf{q}$ keeping 60% of its ferromagnetic value at $\mathbf{q} = 0.9$. However, at $\mathbf{q} = 1.0$ the state of lowest energy is again nonmagnetic.

The non-self-consistent, $\chi_0(\mathbf{q})$, and self-consistent response, $\chi(\mathbf{q})$, of the local atomic moments to a small applied field, as well as the values of the exchange parameters, I,

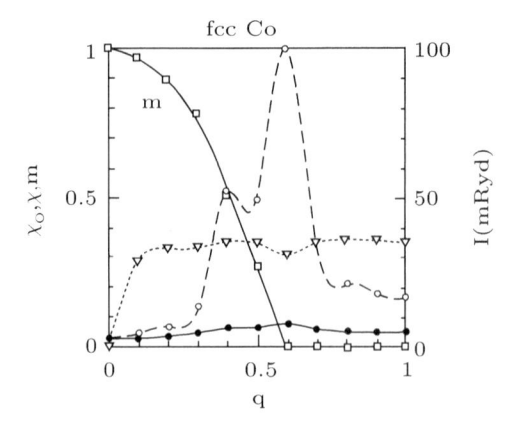

Figure 4.20 *The q-dependence of unenhanced and enhanced susceptibilities, Stoner parameter and local magnetic moment of fcc Co for the (0,0,1)-direction. The same units and the same symbols as in Fig. 4.18 are used.* $m_0(0) = 1.54\,\mu_B$ *(Sandratskii and Kübler, 1992).*

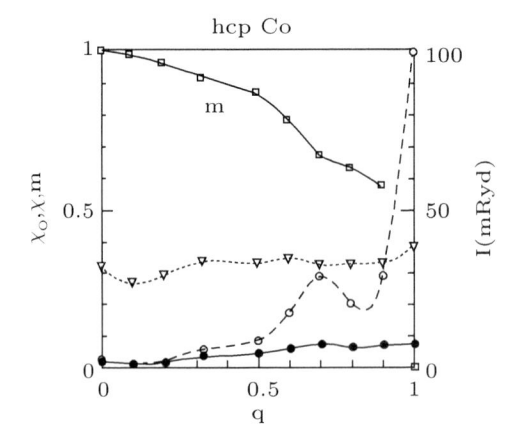

Figure 4.21 *The q-dependence of unenhanced and enhanced susceptibilities, Stoner parameter and local magnetic moment of hcp Co for the (0,0,1)-direction. The same units and the same symbols as in Fig. 4.19 are used for susceptibilities, the parameter I, and the magnetic moment. The value of* $\mathbf{q}$ *is given in units of* $2\pi/a$. $m_0(0) = 1.54\,\mu_B$ *(Sandratskii and Kübler, 1992).*

are also shown in Figs. 4.19–4.20 as functions of $\mathbf{q}$. In all four cases the susceptibility is minimal for the ferromagnetic configuration or for configurations close to it and the values of the enhancements are very small here. This shows that a given change of the amplitude of the magnetic moment costs more energy in the ferromagnetic case than for configurations having large $\mathbf{q}$. To clarify this last statement it is useful to write for the total energy the following approximate relation

$$E_\mathbf{q}(m) = E_\mathbf{q}(m_0(\mathbf{q})) + \frac{(m - m_0(\mathbf{q}))^2}{2\,\chi(\mathbf{q})}, \qquad (4.89)$$

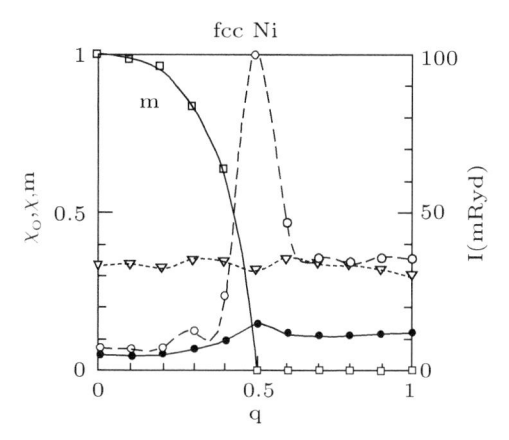

Figure 4.22 *The* **q**-*dependence of unenhanced and enhanced susceptibilities, Stoner parameter and local magnetic moment of fcc Ni for the (0,0,1)-direction. The same units and the same symbols as in Fig. 4.19 are used.* $m_0(0) = 0.59\,\mu_B$ *(Sandratskii and Kübler, 1992).*

where $m_0(\mathbf{q})$ is the self-consistent magnetic moment in the absence of a magnetic field. This relation follows from the connection of the second derivative of the total energy with the inverse susceptibility.

In the case of bcc Fe the susceptibility has a clear tendency to increase with increasing **q**. This is quite different for Ni and fcc Co, where the enhanced susceptibility has a sharp peak (Figs. 4.20 and 4.22) near those values of **q** which mark the boundary between the magnetic and nonmagnetic states. For hcp Co (Fig. 4.21) the borderline between magnetic and nonmagnetic states is close to the Brillouin zone boundary and again we see a sharp increase of the enhanced susceptibility. Thus, near these **q**-values, there are two different self-consistent states with nearly the same total energy giving rise to a very flat minimum of the total energy as a function of the magnetic moment length, m. Any consistent theory of finite-temperature properties of itinerant-electron magnets must take into account the variation of the energy of the self-consistent state with **q** as well as the corresponding variation with the size of the local moment. The large peak in the enhanced susceptibility for Ni and Co close to the boundary between the magnetic and nonmagnetic states will result in an important contribution to the partition function of the crystal due to larger statistical weights of these states. We will come back to the susceptibility in Chap. 5, where it will be analyzed in more detail. We return now to a discussion of the **q**-dependence of the Stoner parameter, I. It is quite remarkable that in view of the pronounced **q**-dependence of the susceptibilities the quantity I is a constant as it is, except for the case of fcc Co. On the basis of these results we repeat that the Stoner parameter can, for many practical purposes, be treated as a constant characterizing a particular atom. Substantially lower values of I obtained for ferromagnetic configurations of fcc Co show, however, that Eqn (4.89) fails to describe the connection of the unenhanced and enhanced susceptibilities in this case. This difficulty may be related to the fact that, in contrast to Sec. 4.2.1, the effects of a charge-density disturbance caused by the magnetic field is not explicitly singled out here. The following explanation may be suggested for the fact that this difficulty occurs for configurations close to ferromagnetic and for magnetic crystals only. In this case applying a field leads to transitions of some electrons from spin-down states to spin-up states. If there is a large difference between the initial and the final states the charge distribution may be noticeably disturbed. For Fe and Co the difference of the spin-up and spin-down states at the Fermi energy can be substantial because of the large magnetic moment of the ferromagnetic state and, consequently, the large exchange splitting of the spin-up and spin-down bands. Note that for a nonmagnetic crystal this splitting is zero and for Ni it is relatively small. None of these difficulties occur for large $\mathbf{q}$, where there is a strong hybridization of the spin-up and spin-down states. In this case the field will not cause transitions from occupied to empty states, rather it leads to an increase of the positive spin projection in the bonding states and of the negative projection in the antibonding states, and a sizable charge density rearrangement does not take place.

4.3 Elementary magnetic metals

We have seen that the uniform magnetic susceptibility, Eqn (4.70) together with (4.71), predicts that Fe, Co, and Ni must be ferromagnetic since the Stoner condition is $I\mathcal{N}_0 > 1$, see Fig. 4.7. Thus, if in a band-structure calculation we disturb an initially nonmagnetic spin density slightly, these metals will converge into ferromagnetic ground states provided we specify the symmetry accordingly. It is instructive and worthwhile to take a small detour to follow the path "down the total energy" to the magnetic ground state. To do this we *constrain* the calculation to yield a desired value of the magnetization, say M, by adding an appropriate Lagrange multiplier to the total energy. For a ferromagnet, in which the spin-up and spin-down electrons move essentially independently (see Sec. 3.4.7), we may simply use a scalar for the Lagrange parameter and write the total energy as (Dederichs et al., 1984)

$$\tilde{E}[n, M] = E[n_+, n_-] - B \left[\int_\Omega m(\mathbf{r}) \, d\mathbf{r} - M \right], \qquad (4.90)$$

where M is the desired magnetization per volume, Ω. Here $E[n_+, n_-]$ on the right-hand side is the total energy in the absence of the constraining B and, as usual,

$$m(\mathbf{r}) = n_+(\mathbf{r}) - n_-(\mathbf{r}), \qquad (4.91)$$

where n_+, n_- denote the spin densities. The Euler–Lagrange equations from Eqn (4.90) are the Kohn–Sham equations (see Sec. 2.5)

$$\left[-\nabla^2 + v_0(\mathbf{r}) \pm [\Delta v(\mathbf{r}) - B] \right] \psi_{i\pm} = \varepsilon_{i\pm} \psi_{i\pm}, \qquad (4.92)$$

where $v_0(\mathbf{r})$ and $\Delta v(\mathbf{r})$ are given by Eqn (2.200) and (2.201) and i denotes both the band index and the k-vector. We see that the Lagrange parameter B acts like a magnetic field. Technically, it is not easy to find the appropriate B that results in the desired magnetization because B may be multivalued. Therefore it is more practical to proceed from Eqn (4.90) and evaluate

$$\frac{\delta \tilde{E}}{\delta B} = 0$$

which gives, of course,

$$M = \int_\Omega m(\mathbf{r}) \, d\mathbf{r} = \int_\Omega n_+(\mathbf{r}) \, d\mathbf{r} - \int_\Omega n_-(\mathbf{r}) \, d\mathbf{r}. \qquad (4.93)$$

The right-hand side, however, can be obtained from densities of states $\mathcal{N}_\pm(\varepsilon)$ to be calculated from (4.92) setting $B = 0$ but adjusting the Fermi energies $\varepsilon_{F\pm}$ such that

$$\int^{\varepsilon_{F\pm}} \mathcal{N}_\pm(\varepsilon) \, d\varepsilon = \frac{1}{2} (Z_{\text{val}} \pm M), \qquad (4.94)$$

where Z_{val} is the number of valence electrons necessary to neutralize the volume Ω. Obviously, the input quantity is now M which gives ε_{F+} and ε_{F-} conserving the charge appropriately, i.e.

$$\int^{\varepsilon_{F+}} \mathcal{N}_+(\varepsilon) \, d\varepsilon + \int^{\varepsilon_{F-}} \mathcal{N}_-(\varepsilon) \, d\varepsilon = Z_{\text{val}} \qquad (4.95)$$

and $B = B(M)$ follows as

$$\frac{\partial \tilde{E}}{\partial M} = B, \qquad (4.96)$$

which implies that $B \to 0$ as the equilibrium is approached, where consequently ε_{F+} becomes equal to ε_{F-}. Excitation energies may be obtained by integrating Eqn (4.96)

$$\Delta \tilde{E} = \int_0^M B(M') \, dM'. \qquad (4.97)$$

We show at the center of Fig. 4.23 the relative total energy $\Delta E / \Delta E_s$ as a function of the relative magnetization M/M_s obtained from constrained calculations for bcc Fe, hcp Co, and fcc Ni at fixed volumes. The total energy origin is the energy of the nonmagnetic metal and the values of ΔE_s, the energy gained in the ferromagnetic equilibrium, and M_s, the saturation magnetization, are collected in Table 4.3 together with the atomic sphere radii used.

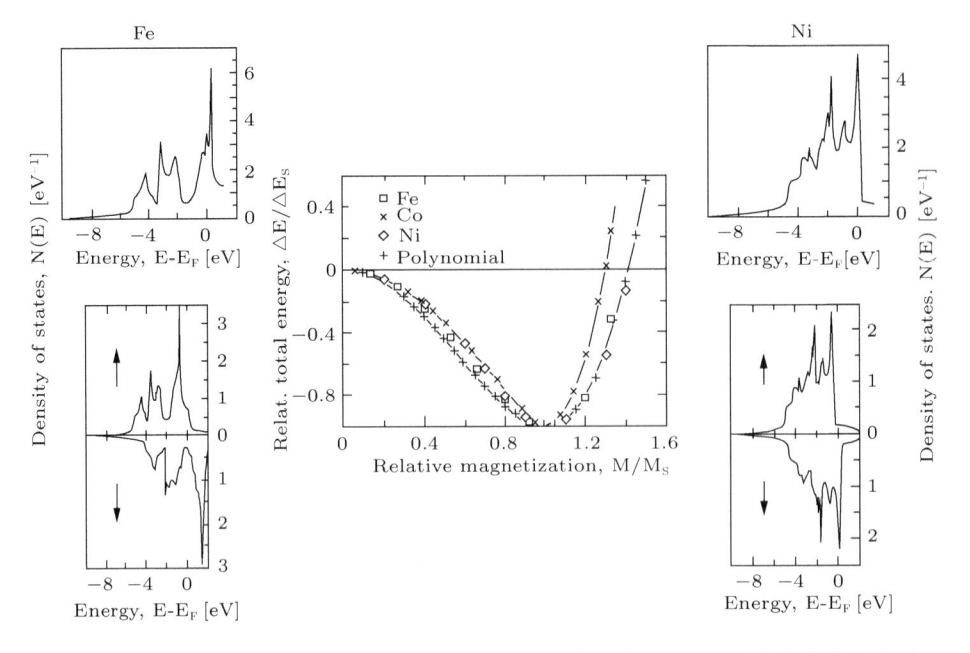

Figure 4.23 *Center portion: relative total energy changes for bcc Fe, hcp Co, and fcc Ni, $\Delta E/\Delta E_s$ as a function of the relative magnetization M/M_s. ΔE_s and M_s are given in Table 4.3. Densities of states at left (right) for Fe (Ni), nonmagnetic on top, in ferromagnetic equilibrium on bottom.*

Table 4.3 *Atomic sphere radius, S (a.u.), total energy gain, ΔE_s(mRy/atom), and calculated saturation magnetization, M_s (μ_B/atom) of Fe, Co, and Ni.*

Metal	Crystal structure	S	ΔE_s	M_s
Fe	bcc	2.62	25	2.20
Co	hcp	2.57	11	1.52
Ni	fcc	2.57	3	0.61

Note that these calculations ignore the change of volume that is brought about by the magnetic pressure to be discussed subsequently.

At the left and right of Fig. 4.23 we show the density of states of Fe (left) and Ni (right) in the nonmagnetic states (top) and in the equilibrium ferromagnetic states (bottom). The total energy change at the center of Fig. 4.23 is (for later purposes) compared with the simple polynomial

$$\frac{\Delta E}{\Delta E_s} = -2\left(\frac{M}{M_s}\right)^2 + \left(\frac{M}{M_s}\right)^4 \tag{4.98}$$

which mimics the trend in the total energy change but does not provide a fit. This is not surprising and indicates that the total energy change should be approximated by a higher-order polynomial.

4.3.1 Ground-state properties of Fe, Co, and Ni

In connection with the calculations of Janak (1977) discussed after Fig. 4.7 we pointed out that a determination of atomic volumes from the minimum of the total energy yields results that are systematically somewhat smaller than the experimental values. However, when the magnetic moment develops it gives rise to magnetic pressure which expands the lattice noticeably. The magnetic pressure may be estimated from the total energy change with magnetization which we approximate using Eqns (4.70) and (4.73) as

$$E(m) = E(0) + \frac{m^2}{2\mathcal{N}_0} - \frac{m^2}{2} I. \qquad (4.99)$$

Since we may assume I to be only weakly dependent on the volume, we evaluate the magnetic pressure from the kinetic energy by differentiating with respect to the volume, Ω:

$$P_M = -\frac{\partial}{\partial \Omega}\left(\frac{m^2}{2\mathcal{N}_0}\right)_{m=M}. \qquad (4.100)$$

Assuming Heine's scaling law, $\mathcal{N}_0 \propto S^5$, we obtain

$$P_M = \frac{M^2}{2}\frac{1}{\mathcal{N}_0^2}\frac{\partial \mathcal{N}_0}{\partial \Omega} = \frac{5}{3} E_K/\Omega, \qquad (4.101)$$

where $E_K = M^2/2\mathcal{N}_0$ is the kinetic energy. This leads to a magnetic pressure of roughly 210 kbar, 180 kbar, and 10 kbar for Fe, (fcc) Co, and Ni, respectively. We show the change of the atomic sphere radii (Wigner–Seitz radii) in Fig. 4.24 and the concomitant change of the bulk moduli in Fig. 4.25 for the metals Cr, Mn, Fe, Co, and Ni. (The values shown for Cr and Mn are rough estimates using Eqn (4.101) with $M = 0.45$ and 2.4 for Cr and Mn, respectively.)

We comment that the magnetic pressure acts to improve the theoretical results but is not strong enough to overcome the overbinding inherent in the local density-functional approximation.

The description of other important ground-state properties is summarized in Fig. 4.26, which shows that the magnetic moments at the calculated equilibrium volumes reproduce the measured moments almost exactly. The chemical trend is well described in each case and is put into a broader perspective when we discuss the Slater–Pauling curve in Sec. 4.4.1. The trend in the pressure dependence of the magnetic moments, the hyperfine field and its pressure dependence is also well described by the calculations. It was correctly pointed out (Eastman et al., 1979) that the ability of such calculations

214   *Electronic Structure and Magnetism*

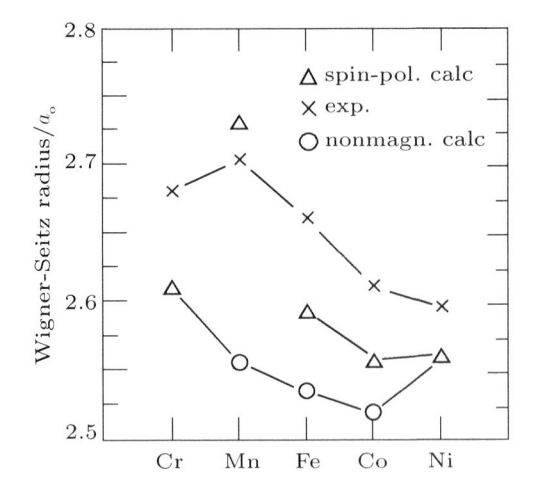

Figure 4.24 *Wigner–Seitz radii for the metals Cr, Mn, Fe, Co, and Ni calculated with data published by Moruzzi et al. (1978); crosses (×), circles (○), and triangles (△) correspond to experimental values (×), nonmagnetic calculations (○), and spin-polarized calculations (△), respectively.*

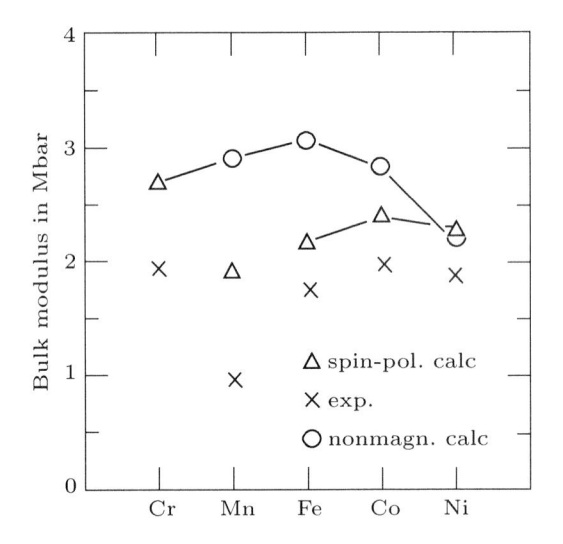

Figure 4.25 *Bulk moduli for the metals Cr, Mn, Fe, Co, and Ni calculated with data given by Moruzzi et al. (1978); symbols as in Fig. 4.24.*

to describe the hyperfine field and its variation with pressure is particularly noteworthy because the hyperfine field (in T: 52.4 times the spin density at the nucleus in atomic units) would seem to be an intrinsically nonlocal effect. That is, the spin density at the nucleus is due to all (core and valence) s electrons (aside from relativistic effects neglected here) which have been polarized by the magnetization in the valence 3d shell.

4.3 Elementary magnetic metals

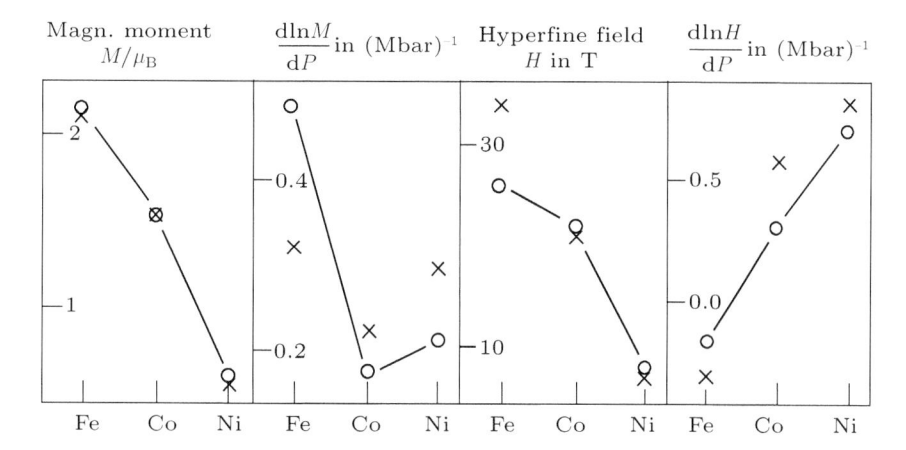

Figure 4.26 *Ground-state magnetic properties of Fe, Co, and Ni: crosses and circles indicate experimental and theoretical values, respectively, as given by Eastman et al. (1979).*

Table 4.4 *Occupation number of the spin-up and spin-down d band for bcc Fe, hcp Co, and fcc Ni.*

	Spin-up	Spin-down	Total
Fe	4.4	2.2	6.6
Co	4.6	2.9	7.5
Ni	4.6	4.0	8.6

One might think that the nonlocality of the exchange interaction is of importance in this context; it is in this sense that the success of the *local* spin-density treatment was felt to be somewhat surprising.

We finally discuss the *sequence of crystal structures* for Mn, Fe, Co, and Ni which in Sec. 4.2.2 were erroneously found to assume the same structures as the analogous 4d transition metals, i.e. in Fig. 4.13 Mn and Fe are predicted to be hcp, whereas Co and Ni are predicted to be fcc. We remind the reader that the observed ground states are bcc for Fe, hcp for Co, and fcc for Ni. We exclude Mn which, as we will see in greater detail in Sec. 4.3.4, has an unusually complex crystal structure. We argue now that the incorrect predictions are due to the appearance of magnetism in these metals and find for the magnetic crystals a pleasantly simple explanation for the observed structures (Söderlind et al., 1994).

We begin with the important observation that for Fe, Co, and Ni the majority (or spin-up) d band is almost completely filled, containing nearly five electrons. This fact follows from experimental reasoning and is brought out by all self-consistent electronic structure calculations. To support this statement we give in Table 4.4 the spin-decomposed d occupation for these metals.

Since we know that in a ferromagnetic state the spin-up (↑) and spin-down (↓) electrons move independently in their respective ↑ or ↓ bands (we here ignore the small role of spin–orbit coupling), we can assume the canonical band picture to hold separately for the ↑ and ↓ electrons. Now we recall from Sec. 4.2.2 that the structural energy differences for the nearly filled d bands are very small, see Figs. 4.13 and 4.14. Because the ↑ bands are nearly filled, we hence neglect their contribution to the structural energy differences and consider only the contribution from the ↓ electrons. Now, to compare the fractional filling of the ↓ band (containing a maximum of five electrons) with the fractional filling of the spin-degenerate d band (containing a maximum of ten electrons) considered in Sec. 4.2.2, we multiply the spin ↓ band occupation number by 2. We thus argue that as regards the fractional filling, the spin ↓ d occupation of spin-polarized Fe having 2, $n_d^\downarrow = 4.4$, corresponds closely to the d occupation of nonmagnetic Mo having a 4d occupation count of 4.5 and W with a 5d count of 4.3. Since Mo and W are bcc we conclude that the spin ↓ electrons of Fe favor the bcc crystal structure, as observed. Similarly, the spin ↓ d occupation of spin-polarized Co, having 2, $n_d^\downarrow = 5.9$, places it in between the 4d metals Tc and Ru (the latter having $n_d = 6.5$) and the 5d metals Re and Os (the latter having $n_d = 6.2$), all of them being hcp metals. Finally, ferromagnetic Ni, for which 2, $n_d^\downarrow = 8$, corresponds to somewhere between nonmagnetic Rh ($n_{4d} = 7.6$) and Pd ($n_{4d} = 8.7$) as well as somewhere between Ir ($n_{5d} = 7.2$) and Pt ($n_{5d} = 8.4$), all of them being fcc metals. This relation between the saturated ferromagnetic 3d metals and the 4d and 5d metals is shown schematically for a selected part of the periodic table of the elements in Table 4.5.

We see in Table 4.5 that the observed crystal structure sequence bcc → hcp → fcc for Fe, Co, and Ni is obtained correctly and finds the same explanation as the crystal structure sequence of the nonmagnetic metals in the middle of the 4d and 5d rows, Nb to Pd, and Ta to Pt, respectively.

However, this explanation depends on the assumption that Fe, Co, and Ni are all saturated ferromagnets; this is not really true for Fe with a spin ↑ d count noticeably lower than that of Co and Ni, see Table 4.4. Therefore, a more accurate analysis is required, at least for Fe. To do this one calculates the total energy difference $\Delta E = E_{\text{bcc}} - E_{\text{fcc}}$,

Table 4.5 *Section of the periodic table of the elements connecting the fractional minority (↓) d band filling in Fe, Co, and Ni (upper part) with the d band filling of the 4d and 5d transition metals (lower part).*

	bcc		hcp		fcc	
3d	Fe		Co		Ni	
4d	Nb	Mo	Tc	Ru	Rh	Pd
5d	Ta	W	Re	Os	Ir	Pt

$\underbrace{\qquad\qquad}_{\text{bcc}}$ $\underbrace{\qquad\qquad}_{\text{hcp}}$ $\underbrace{\qquad\qquad}_{\text{fcc}}$

obtaining $E_{\rm bcc}$ and $E_{\rm fcc}$ from separate self-consistent LDA calculations in which the volume is varied to locate the proper total energy minimum. If this is done one discovers that the total energy of the fcc state is lower by about 1 to 2 mRy than the bcc state. Thus an LDA analysis fails. This failure is now known to be due to the local-density approximation, the correct total energy differences requiring better approximations to the exchange correlation energy functional. An approximation that works is the gradient correction discussed in Sec.2.8. We will return to this problem in Sec. 4.3.4.

Other peculiarities in the ground-state properties of Fe, Co, and Ni are observed in the elastic constants which appear anomalous for these ferromagnets when they are compared with the 4d metals Ru, Rh, and Pd as well as the 5d metals Os, Ir, and Pt. These anomalies have been reduced to effects comparable to those that lead to the sequence of crystal structures. The interested reader can find a detailed discussion of these findings in the paper by Söderlind et al. (1994).

4.3.2 Volume dependence of transition metal magnetism

Up to this point we were concerned with magnetic properties of transition metals near their equilibrium atomic volumes which are found to be somewhat smaller than the corresponding experimental values. Disregarding this small discrepancy now we want to look at the magnetic properties of transition metals assuming atomic volumes which span a rather large range about the equilibrium values. This is easily done in calculations where we may assume any volume we please. Experimentally, however, this is much more difficult to do, especially when expanded volumes are to be considered. Although at this point our investigations are theoretically motivated, we note that recent advances of experimental techniques open the way to studying nonequilibrium structures of expanded volumes, as was first demonstrated by Prinz (1985) who fabricated epitaxially grown nonequilibrium films on suitably chosen substrates. We have seen that the presence or absence of magnetism in transition metals is determined by a competition between intra-atomic exchange and interatomic electron motion measured by the kinetic energy. Since the latter depends strongly on the interatomic separation, and because the d bands are partially filled, transition metals are necessarily magnetic at sufficiently large volumes and necessarily nonmagnetic at sufficiently small volumes. Thus, a normally magnetic transition metal, like iron, becomes nonmagnetic when compressed. Conversely, a normally nonmagnetic transition metal, like Pd (see Fig. 4.17), becomes magnetic when expanded, a fact which, however, is not easily established experimentally. The question we address here can be put in simple words: is the transition from nonmagnetic to magnetic behavior a continuous one or is it discontinuous?

At a given volume, the possible magnetic states of a system are determined by the variation of the total energy with magnetic moment, $E(M)$, examples of which are shown in the center portion of Fig. 4.23. We saw that the relation given in Eqn (4.98) roughly mimics the calculated $E(M)$ curves. We now write the same relation as

$$E(M) - E_0 = A(V) M^2 + B(V) M^4 \qquad (4.102)$$

where E_0 is the total energy of the nonmagnetic state. It would be misleading to imply that there is some universal connection between E and M, in fact near the volume where the magnetic state evolves there are a number of distinct possibilities as was first pointed out by Moruzzi (1986) (also Moruzzi and Marcus, 1993). We now outline these possibilities and subsequently show that they are indeed found by actual total energy calculations. The simplest possible evolution from nonmagnetic to magnetic behavior is obtained by assuming that Eqn (4.102) is valid near the volume where the magnetic state becomes stable. The minimum energy requirement is from Eqn (4.102)

$$\frac{\mathrm{d}E}{\mathrm{d}M} = 2\,A(V)\,M + 4\,B(V)\,M^3 = 0 \qquad (4.103)$$

from which

$$M = \sqrt{\frac{-A(V)}{2\,B(V)}} \qquad (4.104)$$

provided $M \neq 0$. If we assume

$$A(V) = -a(V - V_c) \qquad (4.105)$$

we obtain for $V \geq V_c$

$$M \propto (V - V_c)^{1/2}, \qquad (4.106)$$

and $M = 0$ for $V < V_c$.

Thus, M exhibits a square-root singularity at $V = V_c$ where $\mathrm{d}M/\mathrm{d}V = \infty$ and V_c is a "critical" volume. This simple connection between E and M as well as M and V is sketched schematically in Fig. 4.27(a).

Note that the total energy is always counted from the energy of the nonmagnetic state, $E(M = 0)$. Moruzzi classified this evolution as a type-I transition.

However, the simple Landau expansion, Eqn (4.102), does not necessarily describe the total energy correctly. Thus more complex behavior occurs if a second minimum develops at finite M. In this case, at some intermediate volume, the system exhibits two local minima in $E(M)$, one at $M = 0$ and another one at finite M. At this volume the system shows a behavior that was described by Rhodes and Wohlfarth (1963) as "metamagnetic." It can be driven from the nonmagnetic to the magnetic state by overcoming the potential barrier through, for instance, the application of a magnetic field. This case is shown schematically in Fig. 4.27(b). An immediate implication of the existence of two minima at this intermediate volume is the existence of two critical volumes, V_{c_1} and V_{c_2}, marking the termination of each minimum. The corresponding $M(V)$ behavior is also shown in Fig. 4.27(b). The two branches are now separated and there is an overlapping region where the behavior is discontinuous and multivalued. Moruzzi classified this type of evolution as a type-II transition.

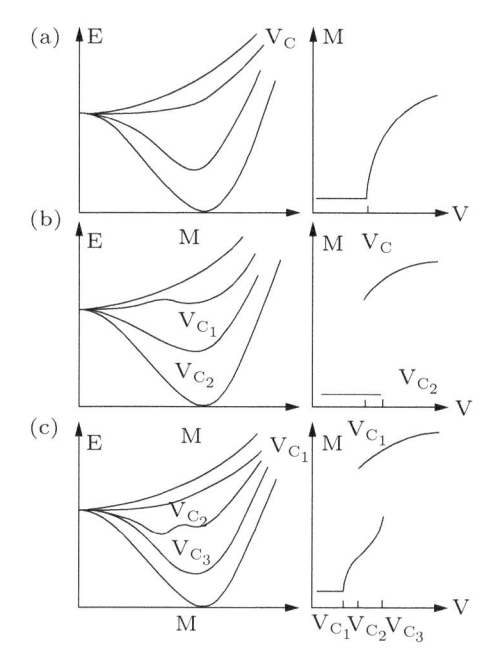

Figure 4.27 *Schematic representation of the three simplest types of evolution from nonmagnetic to magnetic behavior. The lowermost and uppermost $E(M)$ curves in each left-hand panel correspond to the limiting nonmagnetic (low volume) and magnetic (high volume) behaviors. The panels (a), (b), and (c) represent type-I, type-II, and type-III transitions, respectively (see text). The expected $M(V)$ behavior is shown by the right-hand panels. After Moruzzi (1986).*

A third possibility arises if the second minimum appears at a larger volume than the onset of the type-I transition resulting in the complex evolution shown in Fig. 4.27(c). At an intermediate volume, the system possesses two local minima in $E(M)$, both at finite M, giving three critical volumes. Here the low-volume, nonmagnetic minimum terminates at a third critical volume labeled V_{c_1}, where it joins with the lower termination of the low-moment minimum in the same manner as the type-I transition shown in Fig. 4.27(a). The low-moment minimum must also terminate at a second critical volume labeled V_{c_3} at a finite M value. Furthermore, in this termination, the singularity is approached from below. The high-moment minimum terminates at the critical volume labeled V_{c_2} in the same manner as the type-II transition shown in Fig. 4.27(b). The corresponding $M(V)$ behavior is shown in Fig. 4.27(c) and is classified as a type-III transition. Examples of each of these three types of transitions can be identified in real total energy calculations, by using, for instance, the ASW method. In Fig. 4.28 we show some examples for Fe, Co, Ni, and V (Moruzzi, 1986) giving the calculated magnetic moments, M, as a function of V/V_0, where V_0 is the calculated equilibrium volume (corresponding to zero pressure). A much larger collection of similar results can be found in the review by Moruzzi and Marcus (1993). Note that the singular behavior occurs at a 1.5% volume expansion for

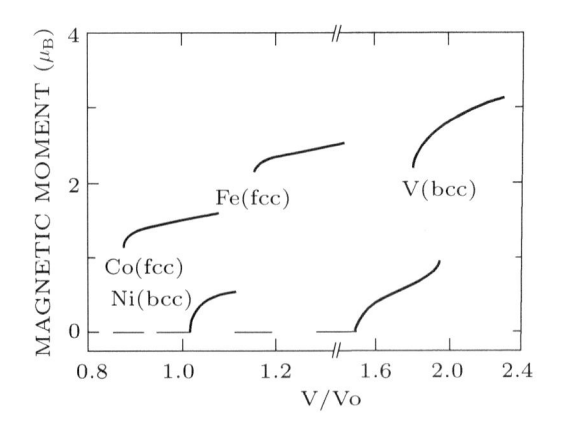

Figure 4.28 *Calculated $M(V)$ curves for bcc nickel, fcc cobalt, bcc vanadium, and fcc iron showing critical behavior in the transition region. The horizontal axis is the reduced volume, V/V_0, where V_0 is the equilibrium volume. The horizontal scale for vanadium is different from that for iron, cobalt, and nickel. Note that bcc nickel, fcc cobalt, bcc vanadium, and fcc iron exhibit type-I, type-II, and type-III transitions, respectively (see text). All results are based on nonrelativistic augmented spherical wave electronic structure calculations. After Moruzzi (1986).*

bcc nickel, at approximately a 10% volume compression for fcc cobalt, in the vicinity of a 20% volume expansion for fcc iron, and at an unphysically large volume expansion for bcc vanadium. Moruzzi and Marcus demonstrate that all transition metals undergo similar transitions. Normally occurring bcc iron and fcc nickel and nonequilibrium bcc cobalt undergo transitions and exhibit singularities at large volume compression. For fcc vanadium, a type-I transition is found at approximately the same expanded volume as bcc vanadium. A type-I transition is also found for fcc palladium at a small volume expansion, with moments comparable to those of nickel.

The origin of the singular behavior lies in the details of the densities of states (DOS) which, as we have seen, determine the kinetic energy price required to exploit intra-atomic exchange. The DOS is a volume- and structure-dependent quantity dominated by the d bands, see Fig. 4.9. At small volumes, the kinetic energy price is so high that exchange cannot be exploited, implying that the nonmagnetic state is stable. At large volume where the bands become narrow, the kinetic energy price is low and the system takes advantage of exchange by becoming magnetic. The different maxima in the DOS presumably control the kinetic energy price in different ways for incremental changes of the magnetic moment. We close this subsection by remarking that, so far, we have only discussed transitions from nonmagnetic to ferromagnetic states; but many more possibilities exist, like transitions to antiferromagnetic and noncollinear configurations which considerably enrich phase diagrams such as that shown in Fig. 4.28, as we will show in Sec. 4.3.4 in some detail. It should thus be clear, therefore, that a comparison with experimental results, if at all tractable, must allow for more possibilities than those pictured in Fig. 4.28.

4.3.3 Band structure of ferromagnetic metals

A great number of band-structure calculations for magnetic metals has appeared in the past, like the early work of Asano and Yamashita (1973) and that of Callaway and co-workers (also Moruzzi *et al.*, 1978; Hathaway *et al.*, 1985; Johnson *et al.*, 1984, and many others). They all show features like the band structure of *ferromagnetic Fe* reproduced in Fig. 3.18 which was obtained with an LDA calculation using the ASW method as explained in Chap. 3. The band structure of ferromagnetic Fe shown in Fig. 4.29 is due to Callaway and Wang (1977) and is given as another example. Here the spin-up bands (majority) are drawn as solid lines, and the spin-down bands (minority) as dashed lines. The exchange splitting is obvious, but is not the same for all bands and all **k**-points, so that the average $\langle \Delta \varepsilon_\mathbf{k} \rangle$ must be suitably defined and is ≈ 1.5 eV for states near the Fermi energy. Hence with a calculated magnetic moment of $m = 2.2$ (μ_B), the Stoner parameter becomes $I \approx 0.68$ eV using

$$\langle \Delta \varepsilon_\mathbf{k} \rangle = I \cdot m. \tag{4.107}$$

This is somewhat larger than Janak's (1977) value (0.46 eV), and, indeed, full agreement should not be expected since the latter originated from a formula, Eqn (4.71), which was derived assuming an infinitesimally small magnetization. After seeing in Sec. 4.3.1

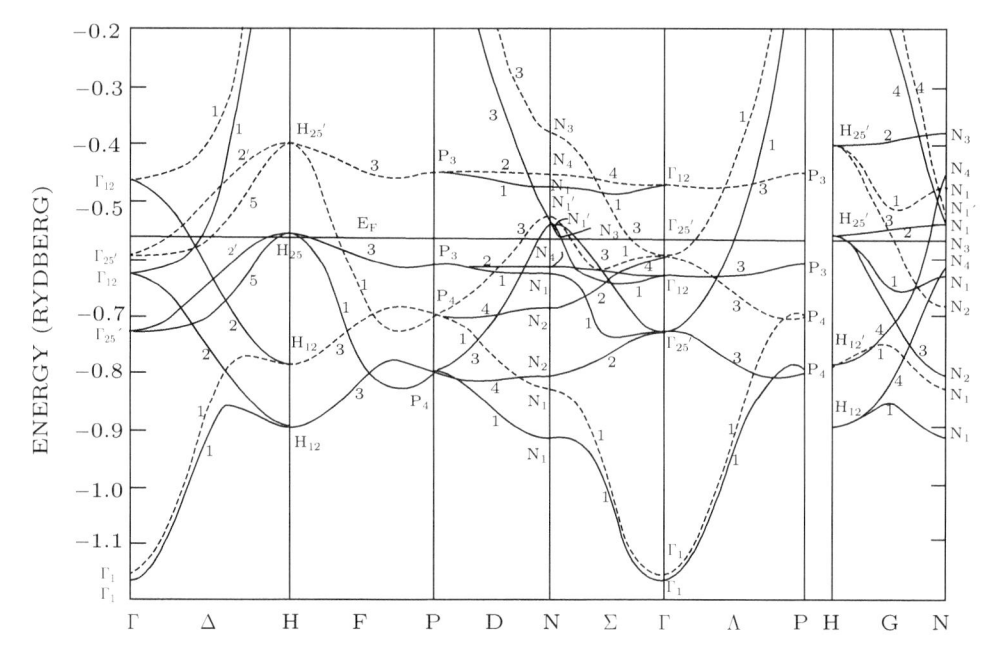

Figure 4.29 *Energy bands of ferromagnetic Fe. Reprinted with permission from Callaway and Wang (1977).*

that the ground-state properties are well described by self-consistent band theory, we will now inquire into the validity of the band structure itself, i.e. compare it directly with experimental information. This comparison will include measurements of Fermi surfaces (via the de Haas–van Alphen effect, see Sec. 1.2.4) and energy-band dispersions (via photoemission methods). We remind the reader of our word of warning in Sec. 2.3., where excited states were discussed. Nevertheless, we will make this comparison and will come back to additional remarks along similar lines shortly.

Angle-resolved photoemission data reveal that the band structure obtained on the basis of the local spin-density functional approximation (LDA) supplies a description of the single-particle excitations that is reasonable for Fe, less so for Co, and much less so for Ni (Eastman et al., 1979).

4.3.3.1 bcc Iron

We first look at the case of Fe in some detail and compare experimental results with calculated bands in Fig. 4.30 taken from Eastman et al. (1979). The experimental band dispersions are from photoemission measurements of Eastman et al. (1979) and Himpsel and Eastman (1980) (circles and crosses) and crossings at the Fermi energy (solid and hollow triangles for majority and minority bands, respectively) are from de Haas–van Alphen experiments. The calculated bands (solid and dashed lines) are the self-consistent LDA bands of Callaway and Wang (1977) shown before in Fig. 4.29. It is interesting that the exchange splitting at the P-point has been observed to decrease from $\Delta = 1.5$ eV at 293 K to about 1.2 eV at 973 K which is just a little short of the Curie temperature of $T_c = 1043$ K. It is thus most likely that the exchange splitting does not disappear (at least not locally) at T_c. Quite extensive photoemission measurements by Turner et al. (1984) allow for comparison with the data of Himpsel et al. (1979) and in addition furnish new

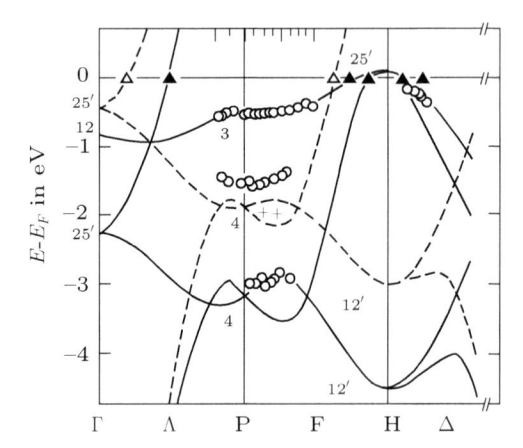

Figure 4.30 *Experimental band structure for Fe as given by Eastman et al. (1979). Solid and open triangles denote band crossings at the Fermi surface from dHvA data, and solid and dashed lines indicate the bands calculated by Callaway and Wang (1977) for majority- and minority spin-electrons, respectively.*

Table 4.6 *Negative binding energies of Fe and exchange splitting determined by experiment and compared with four different LDA calculations. All values are in eV measured from the Fermi energy.*

k-point	M^a	C^b	H^c_1	H^c_2	T^d (exp.)
$\langle \Gamma_{1\uparrow\downarrow} \rangle$	8.42	8.12	8.83	8.23	8.15 ± 0.20
$\Gamma_{25'\uparrow}$	2.48	2.25	2.51	2.26	2.35 ± 0.10
$\Gamma_{12\uparrow}$	0.97	0.86	0.99	0.94	0.78 ± 0.10
$\Gamma_{25'\downarrow}$	0.45	0.43	0.69	0.34	0.27 ± 0.05
$H_{12\uparrow}$	5.17	4.50	5.28	4.64	3.80 ± 0.30
$H_{12\downarrow}$	3.71	2.99	3.71	2.99	2.50 ± 0.30
$P_{4\uparrow}$	3.50	3.17	3.57	3.18	3.20 ± 0.10
$P_{3\uparrow}$	0.68	0.53	0.72	0.72	0.60 ± 0.08
$P_{4\downarrow}$	1.95	1.83	2.17	1.71	1.85 ± 0.10
$N_{1\uparrow}$	5.24	4.75	5.41	4.82	4.50 ± 0.23
$N_{2\uparrow}$	3.65	3.27	3.74	3.32	3.00 ± 0.15
$N_{1\uparrow}$	0.94	0.86	0.96	0.93	0.70 ± 0.08
$N_{4\uparrow}$	0.72	0.69	0.76	0.76	0.70 ± 0.08
$N_{1\downarrow}$	3.92	3.60	4.16	3.52	3.60 ± 0.20
$N_{2\downarrow}$	1.82	1.62	2.09	1.58	1.40 ± 0.10
Exchange splitting					
$\Gamma_{25'}$	2.03	1.82	1.82	1.92	2.08 ± 0.10
H_{12}	1.46	1.51	1.57	1.65	1.30 ± 0.30
P_4	1.55	1.34	1.39	1.47	1.35 ± 0.10
N_2	1.83	1.65	1.64	1.74	1.60 ± 0.15
Simple splitting					
	1.72	1.58	1.60	1.69	1.58 ± 0.3

[a] Moruzzi *et al.* (1978); [b] Callaway and Wang (1977); [c] Hathaway *et al.* (1985); [d] Turner *et al.* (1984).

data at the N-point in the BZ. We can thus undertake a comprehensive comparison of a number of different band-structure results both with themselves and with the experimental data; this is done in Table 4.6.

The columns denoted by M and H_1 contain the band structure obtained at the same, theoretical equilibrium-lattice constant ($a = 5.32$ a.u.), but in M (Moruzzi *et al.*, 1978) the KKR method was used, and in H_1 (Hathaway *et al.*, 1985) the full potential (i.e., *not* shape-approximated, see Sec. 3.4.3) LAPW method was employed. The columns denoted by C and H_2, on the other hand, contain the band structure obtained at the observed experimental lattice constant ($a = 5.406$ a.u.). The method of H_2 (Hathaway *et al.*, 1985) is that of H_1, and C (Callaway and Wang, 1977) the LCGO (linear combination of Gaussian orbitals) method; the latter makes no shape approximation in the potential either. It is not clear why the columns C and H_2 are as different as they are, and the agreement of C with the experimental data T (Turner *et al.*, 1984) is to be considered excellent. This is the reason we chose to show this particular band structure in Fig. 4.29. The error, however, that the local spin-density-functional approximation makes in the determination of the equilibrium volume is amplified in the band structure which is seen to react very sensitively to changes in the lattice constant. It must be taken as an empirical fact that the calculations at the experimental lattice constant describe the photoemission data best.

The Fermi surface of bcc iron has been studied in great detail in the past; an early and thorough account can be found in the paper by Eastman *et al.* (1979). Here we want to point out the gross features only, using Fe as an example.

The Fermi surface of Fe shows substantial complexity with three sheets belonging to the minority-spin electrons and three sheets belonging to the majority-spin electrons. The band-structure calculations mentioned above and experimental measurements give general confirmation for the existence of all of these, but minor questions still remain about the precise topology.

Choosing a theoretical lattice constant ($a = 5.32$ a.u.) and spin–orbit coupling we can visualize all sheets, but will not get the precise topology unless a great deal of fine tuning is done. With a moderate amount of adjustments the minority Fermi surfaces one obtains look like Fig. 4.31.

This consists of a central electron sheet, electron balls located along δ that join the center piece, and a large hole sheet (octahedron-like) centered at H. Here we made very small adjustments of the Fermi energy to get a topology that is close to that published in earlier work (Eastman *et al.*, 1979). What is important to us here is the wave function character of the states at the Fermi surface. The band structure of iron shown in Fig. 3.18 allows us to distinguish s,p electrons from the d electrons, the latter again originating either from t_{2g}, or in different notation, $\Gamma_{25'}$ states or from e_g, or Γ_{12} states. It can be seen that all "types of electrons" s,p,d make up the *minority* Fermi surfaces, although to a different degree in different directions. We stress this point because in the older literature one finds the notion that one type of electrons supplies the conducting states at the Fermi surface and the other type is localized giving rise to magnetism. This notion must clearly be revised.

The majority Fermi surfaces are shown in Fig. 4.32. This consists of hole arms with varying t_{2g}-e_g character, a large electron surface centered at Γ, and—not so easy to see—a hole pocket centered on H. It is again only at the small sections pointing toward N, where two arms come close to each other, that an appreciable number of sp electrons can be found. (The length of these arms is extremely sensitive to details of the band structure;

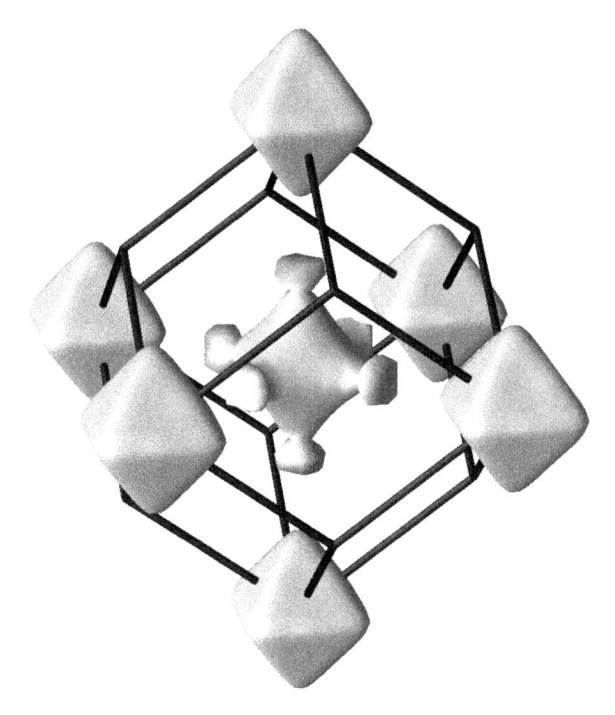

Figure 4.31 *Minority-spin electron Fermi surface of bcc Fe: Large electron sheet centered at Γ, electron balls along δ joining the center piece, and a large hole sheet (octahedron-like) centered at H.*

furthermore, a fourth sheet may appear in both spin directions with a very small change of the Fermi energy.)

4.3.3.2 hcp Cobalt

Turning now to the case of cobalt we find the situation less satisfactory. On the experimental side, there are the photoemission data by Himpsel and Eastman (1980), but they are limited since they give the band structure only along the hexagonal z-axis. Other measurements exist, but unfortunately we know of none (except for those of Himpsel and Eastman) that have been analyzed to allow a comparison with calculated band structures. Himpsel and Eastman compared their results with the calculations of Moruzzi et al. (1978) for fcc Co. Although, as they comment, this did lead to a successful analysis, the procedure is not quite convincing in all respects. It is true that even in 1980 a self-consistent calculation for hcp Co did not exist. Thus Himpsel and Eastman had no other choice but to compare measurements of an hcp crystal with calculations for an assumed fcc system, but their statements that the hcp band structure along the direction Γ to A (see Fig. 3.4 for BZ) is the same as the fcc band structure along the direction Γ to L provided the latter is folded about the midpoint between Γ and L is not quite true. To shed light on this problem, we carry out a self-consistent ASW calculation for hcp Co at

226 *Electronic Structure and Magnetism*

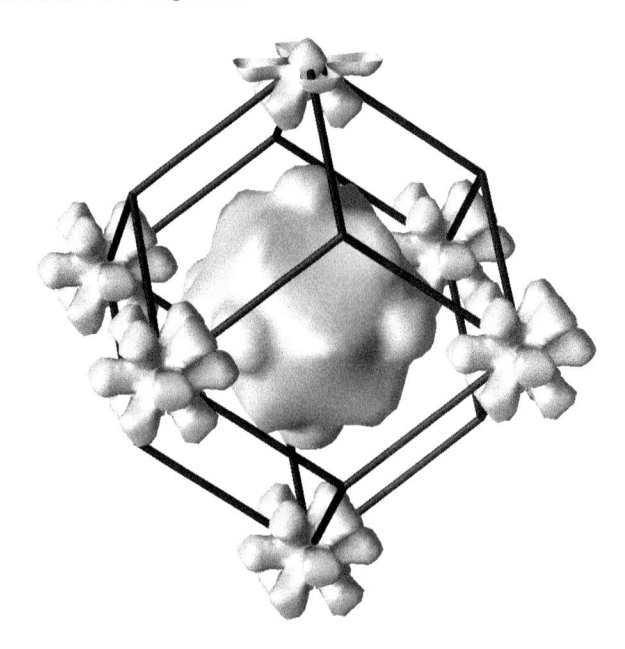

Figure 4.32 *Majority-spin electron Fermi surface of bcc Fe. Seen are: hole arms with varying t_{2g}-e_g character, a large electron surface centered at Γ, and—not so easy to see—a hole pocket centered at H.*

the experimental lattice constant and obtain both the hcp and fcc band structure using for the latter the potentials converged in the hcp geometry.

Our results are depicted in Fig. 4.33 and are shown separately for spin-up and spin-down—for simplicity only showing the section of the band structure that is relevant for a comparison with the experimental results which are also provided in Fig. 4.33. Visual inspection shows to what extent the assumption of Himpsel and Eastman is justified: it holds for all states except those within a range of 2 eV directly below the Fermi energy, E_F. In particular, the state labeled Γ_{12} in cubic symmetry is shifted considerably in hexagonal symmetry, and for the down-spin electrons just below E_F, we find a state that we labeled Γ_{6+} which fits nicely with the state labeled $\Gamma_{12\downarrow}$ by Himpsel and Eastman. The results of our numerical comparison are collected in Table 4.7 which, in addition to the results of Moruzzi *et al.* (1978), also contains the experimental values. The labels given are those of Himpsel and Eastman, and in the 2 eV range below E_F they are not necessarily the labels appropriate for our ASW results. The differences of our calculations compared to those of Moruzzi *et al.*, in addition to the problems in the 2 eV range below E_F, arise because they used a different theoretical equilibrium lattice constant. Our results are in very good agreement with LMTO results of Jarlborg and Peter (1984) and, were it not for the limited amount of material to compare, the agreement of theory with experiment is to be considered satisfactory.

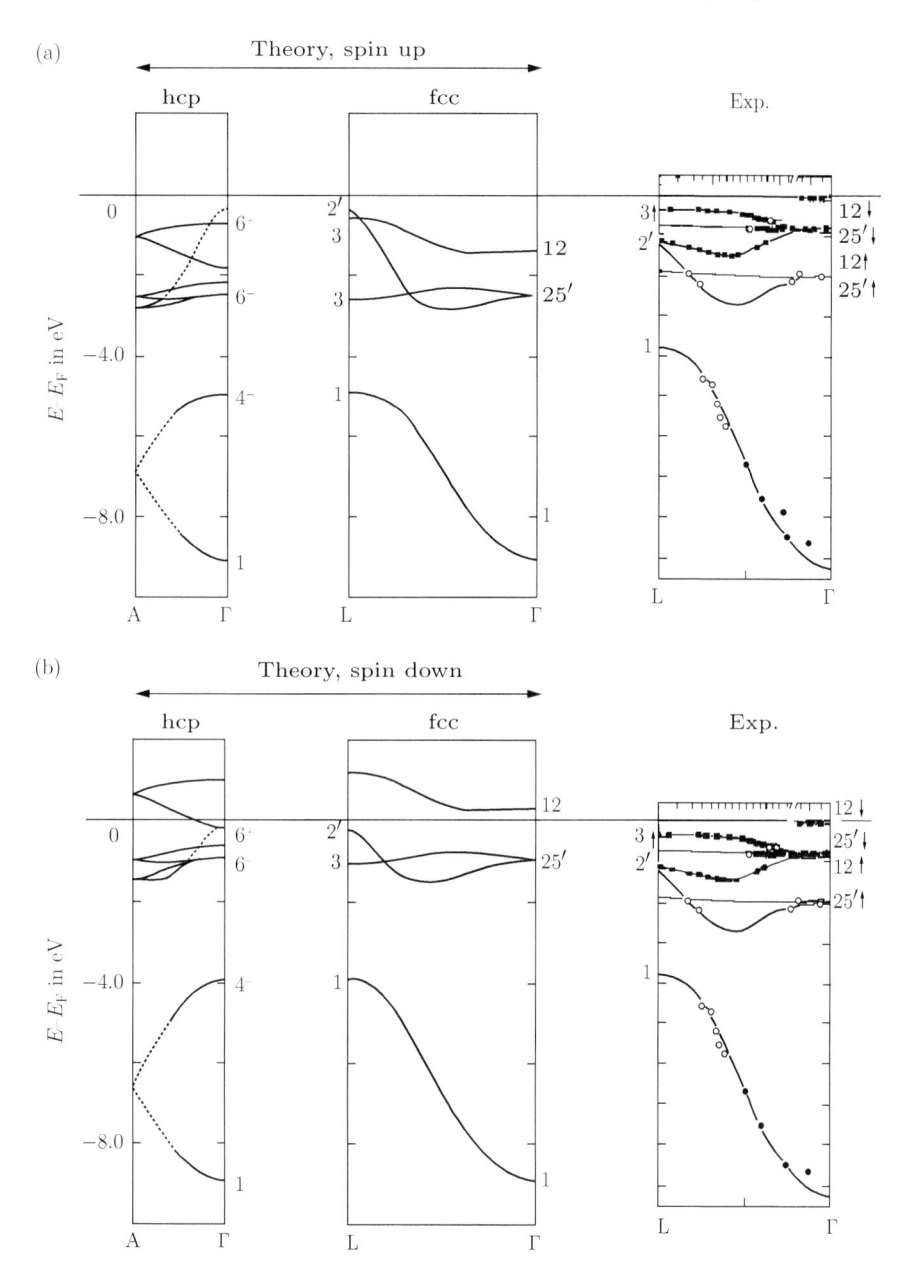

Figure 4.33 *Energy bands for Co, calculated with the ASW method for the hcp and fcc structure for majority (a) and minority (b) spin; the experimental band structure as given by Himpsel and Eastman (1980).*

Table 4.7 *Binding energies of Co and exchange splitting determined by experiments and compared with two different LDA calculations. All values are in eV measured from the Fermi energy.*

k-point fcc	k-point hcp	M^a	ASW	H^b (exp.)
$\langle\Gamma_1\rangle$	Γ_1^+	−8.90	−9.0	−8.7 ± 1.0
$\Gamma'_{25\uparrow}$	Γ_1^+, Γ_6^-	−2.65	−2.46	−2.0 ± 0.3
$\Gamma_{12\uparrow}$	Γ_6^+	−1.43	−0.72	−0.9 ± 0.2
$\Gamma'_{25\downarrow}$	Γ_1^+, Γ_6^-	−1.14	−0.94	−0.8 ± 0.1
$\Gamma_{12\downarrow}$	Γ_6^+	+0.22	+0.17	−0.05 ± 0.05
$\langle L_1\rangle$	Γ_4^-	−4.75	−4.47	−3.8 ± 0.5
$L_{3\uparrow}$	Γ_5^+	−2.76	−2.15	−1.9 ± 0.4
$\langle L'_2\rangle$	Γ_3^+	−0.23	−1.0	−1.5 ± 0.2
$L_{3\downarrow}$	Γ_5^+	−1.28	−0.58	−0.75 ± 0.2
$L_{3\uparrow}$	Γ_5^-	−0.53	−0.31	−0.35 ± 0.05
Exchange splitting				
$\Gamma_{25'}$		1.51	1.52	1.2 ± 0.3
Γ_{12}		1.65	0.55	0.85 ± 0.2
L_3		1.48	1.57	1.15 ± 0.4
Simple averages				
		1.55	1.21	1.07 ± 0.4

^{a} Moruzzi *et al.* (1978); ^{b} Himpsel and Eastman (1980).

4.3.3.3 Nickel

The amount of literature on the electronic structure of Ni is overwhelming. Without trying to be complete, we mention on the theoretical side the work of Callaway (1981 and ref. therein), Moruzzi *et al.* (1978), Andersen (1984), and on the experimental side Eastman *et al.* (1979), Dietz *et al.* (1978), Maetz *et al.* (1982), Kisker (1983), and Mårtensson and Nilsson (1984).

The situation is summarized by Fig. 4.34, where parts (a) and (b) are taken from Mårtensson and Nilsson, and (c) is from the LSDFA-KKR calculation of Moruzzi *et al.* (1978). This time it is rather immaterial whether we take a band structure calculated at the theoretical equilibrium lattice constant (as that of Moruzzi *et al.*) or one calculated at the experimental volume and compare with the experimental data; the situation remains the same and is as follows. In Fig. 4.34(a) we see in the upper left-hand corner two bands marked with arrows that designate the spin directions. The splitting seen is

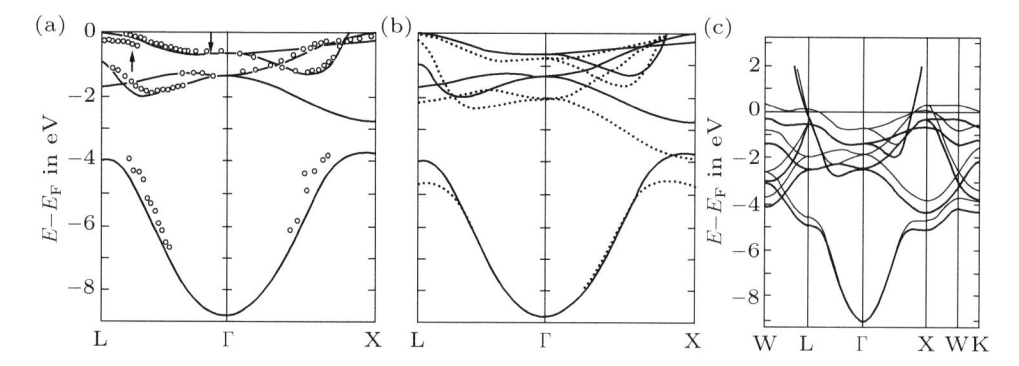

Figure 4.34 *Energy bands for Ni: comparison of the semi-empirical band structure given by Mårtensson and Nilsson (1984) (a) to the experimental values (circles) given by Himpsel et al. (1979) and (b) to the calculated bands by Moruzzi et al. (1978). (c) Spin-polarized band structure as given by Moruzzi et al. (1978) (thick and thin lines for majority and minority spin).*

thus the exchange splitting, and it is ≈ 0.3 eV, a value verified by all other experimental work cited above; this should be compared with the calculated value of ≈ 0.5 eV that, likewise, is verified by all other calculations cited above.

Mårtensson and Nilsson (1984) fitted their extensive photoemission data with bands that they call semi-empirical. They are shown as solid lines in Fig. 4.34(a) and (b). In the former they are compared with photoemission data by Himpsel *et al.* (1979) (circles), in the latter with minority bands of Moruzzi *et al.* (1978) (Fig. 4.34(c)) that are slightly shifted; one clearly sees that the measured d bandwidth is smaller by about 30% compared with the calculated width. This is also in good agreement with other work. There is no universally accepted explanation of this discrepancy, but certain observations made as early as 1979 still seem plausible (Eastman *et al.*, 1979): In the local density-functional approximation the dynamical behavior of an electron can be viewed as being that of a neutral quasiparticle; i.e. an analysis of the treatment of exchange and correlation in this theory (Kohn and Vashishta, 1983) shows that as an electron moves through the system, it carries with it a perfectly neutralizing "hole" in the distribution of other electrons. This hole is allowed by the theory to breath as it moves; i.e. it is spatially small where the local electron density is large and expands when the electron to which it is pinned moves to regions of low average electron density. But the description of the hole is very approximate; the hole is rigidly pinned to the electron, and its form is approximated by homogeneous electron gas results. This implies in particular that when the correlated motion of the electron is atomic in character—as is, for example, the correlated motion responsible for multiplet structure in free atoms—the local density theory becomes inappropriate. The boundary of the adequacy of this simple description of correlation appears to be in the 3d transition metals. Nickel, from several points of view, appears to be a case in which two-particle effects cannot be ignored. So the self-energy of the hole, left as the final state in the photoemission process, seems to be responsible for the small exchange splitting (Liebsch, 1979), and it is this effect, in addition to the narrower

230 Electronic Structure and Magnetism

d bands, that self-consistent band-structure calculations based on the local spin-density approximation fail to reproduce.

4.3.4 Electronic structure of antiferromagnetic metals

We turn to the subject of antiferromagnetism in the pure metals after first posing the following question: why is antiferromagnetism confined to the elements near the middle of the 3d transition series (Cr, Mn, and fcc Fe)? This question has been answered in a variety of ways in recent years (Hamada, 1981; Heine and Samson, 1983) and earlier by Moriya (1964). In its simplest form the physics of itinerant-electron antiferromagnetism can be described as follows: An electron traveling through an antiferromagnetic crystal experiences exchange-correlation forces that point in opposite directions on two sublattices, and that polarize each sublattice in such a way that on one sublattice the magnetization (Eqn 4.91) is positive, and on the other it is negative.

The situation can therefore be compared with the states of a diatomic molecule (Williams *et al.*, 1982). Leaving aside all the wiggles of wave functions, we show in Fig. 4.35 the asymmetry introduced into the states of a diatomic molecule by the energy difference between the atomic states which interact to form bonding and antibonding hybrids. An important aspect of this physical effect is the ionic component of covalent bonding. It reflects the energy gained by concentrating the charge associated with the bonding molecular orbital on the atom possessing the more attractive atomic level. The

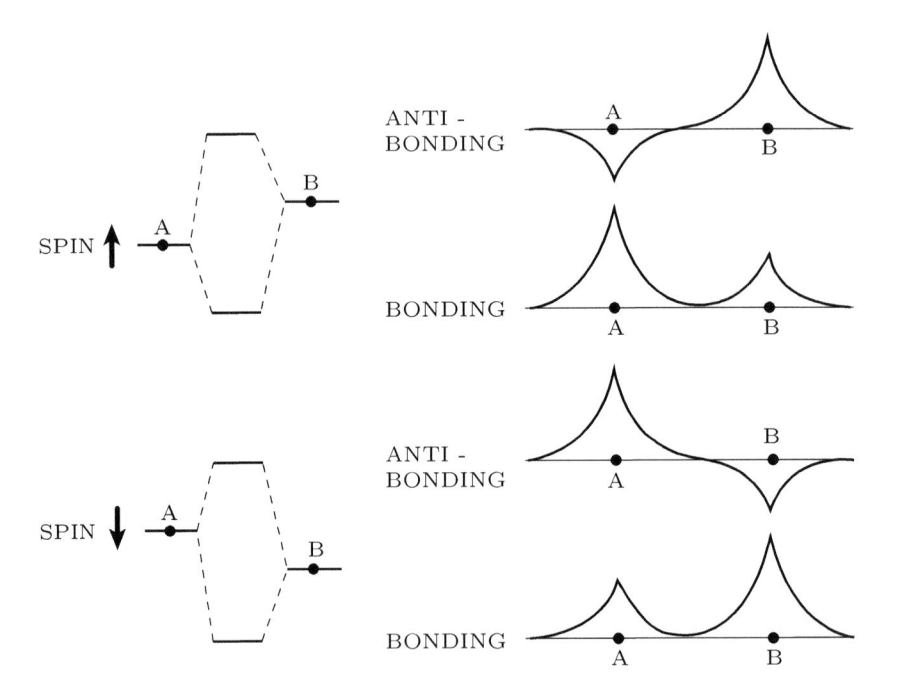

Figure 4.35 *Schematic drawing of bonding and antibonding orbitals.*

connection between ionicity and antiferromagnetism is that the exchange interaction can lower the energy of the states of a given spin on one of the sublattices. The bonding molecular orbitals (or band states) exploit these exchange-energy differences by placing more charge there. In the same way, the bonding molecular orbital (or band states) for the *other* spin concentrates charge on the *other* atom or sublattice. (It is clear that the atom as a whole remains neutral.) The concentration of the different spins of different atoms or sublattices is what we mean by antiferromagnetism. It should be noted, however, that the antibonding molecular orbitals (or band states) concentrate charge in precisely the reverse manner. Therefore, to exploit the exchange interaction in this way, the system must preferentially occupy the bonding levels, and this is the connection between antiferromagnetism and the half-filled d band. We substantiate our explanation by means of the density of states from a self-consistent ASW calculation for bcc Mn.

4.3.4.1 Manganese

Manganese exists in four allotropic forms, but only one has the simple bcc crystal structure. This is δ-Mn which exists just below the melting point, between 1406 K and 1517 K. Unfortunately, its magnetic moment is not known. The calculation at the experimental lattice constant ($a = 3.081$ Å) gives a magnetic moment of 3.18 μ_B per atom. The sublattices are interpenetrating simple cubic, as in the CsCl structure, and the density of states is shown in Fig. 4.36. Being guided by Fig. 4.35 we identify part

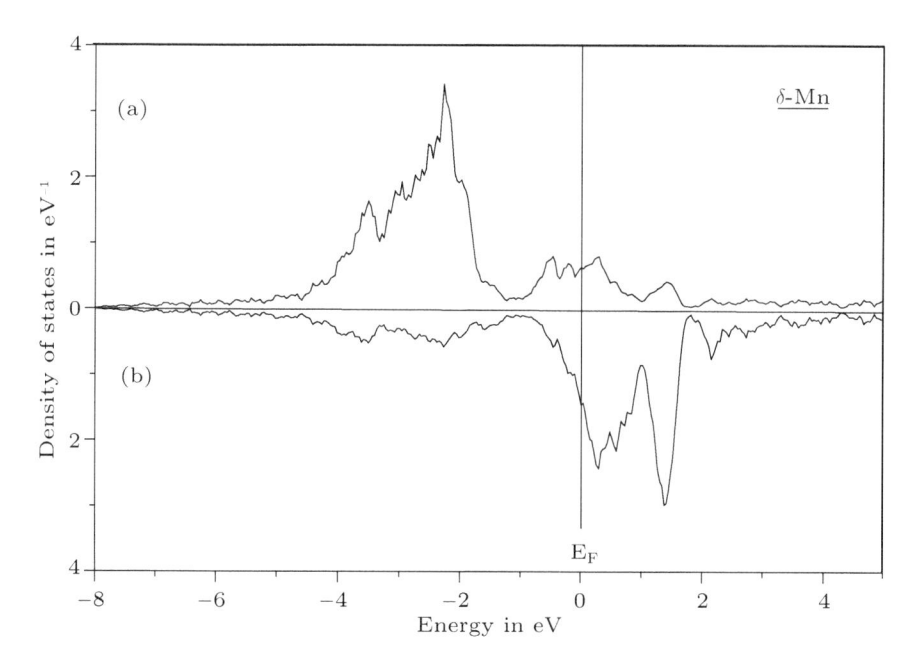

Figure 4.36 *Density of states for antiferromagnetic δ-Mn* ($\mathbf{q} = (0, 0, 1)\, 2\pi/a$), *separated for the spin-up (a) and spin-down (b) sublattices. Lattice constant: $a = 5.8207$ a.u., magnetic moment: $\mu = 3.18\,\mu_B$.*

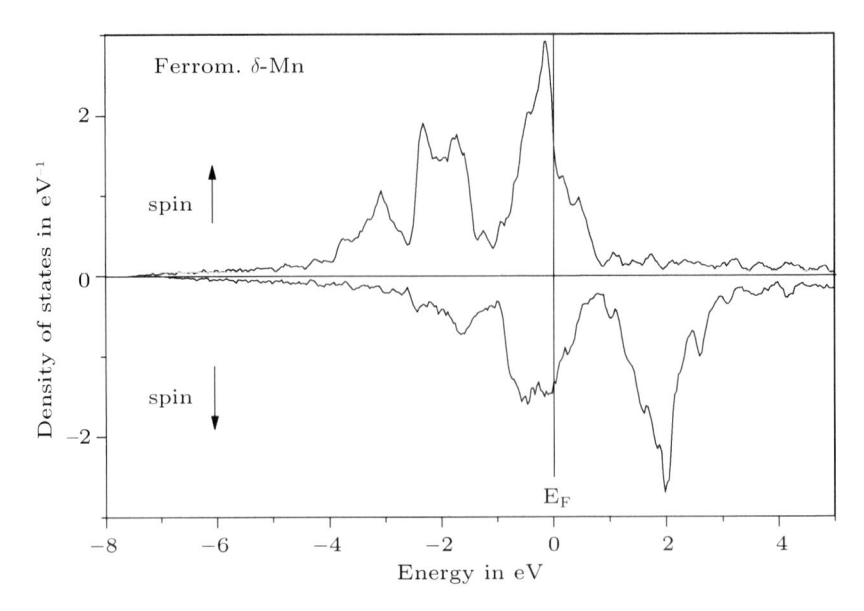

Figure 4.37 *Density of states for assumed ferromagnetic δ-Mn. Lattice constant:* $a = 5.8207$ *a.u., magnetic moment:* $\mu = 2.24\,\mu_B$.

(a) as the spin-up density of states which are the majority-spin electrons on sublattice A and the minority-spin electrons on sublattice B and part (b) as the spin-down density of states which are the majority-spin electrons on sublattice B and the minority-spin electrons on sublattice A. This symmetry leads to a *spin-degenerate* band structure in contrast to ferromagnetic bands. The difference between the antiferromagnetic and the ferromagnetic cases are further stressed by comparing Fig. 4.36 with Fig. 4.37 which shows the density of states of assumed ferromagnetic Mn. It is seen that Fig. 4.37 consists of essentially one curve which is shifted by the exchange splitting in opposite directions for the two kinds of spins, whereas the curves of Fig. 4.36 are distinct because of shifted *spectral weight*. We will come back to the case of Mn again after discussing Cr, but mention here that computationally Fig. 4.36 is obtained using the special spin-spiral state defined by $\mathbf{q} = (0,0,1)\,2\pi/a$, whereas Fig. 4.37 has $\mathbf{q} = 0$ with a magnetic moment that has dropped to $2.24\,\mu_B$.

4.3.4.2 Chromium

The case of chromium is of special interest and an excellent review of the experimental situation is that of Fawcett (1988). As early as 1962 Lomer (1962) proposed Cr to be antiferromagnetic on account of a particular band-structure feature, namely "nesting." But before going into any detail and describing this mechanism, we want to stress an argument by Heine and Samson (1983) that brings us back to the beginning of this subsection: If one has Fermi surfaces with good nesting, then the states near the Fermi energy are particularly significant and will affect the fine tuning, as we will see, or even

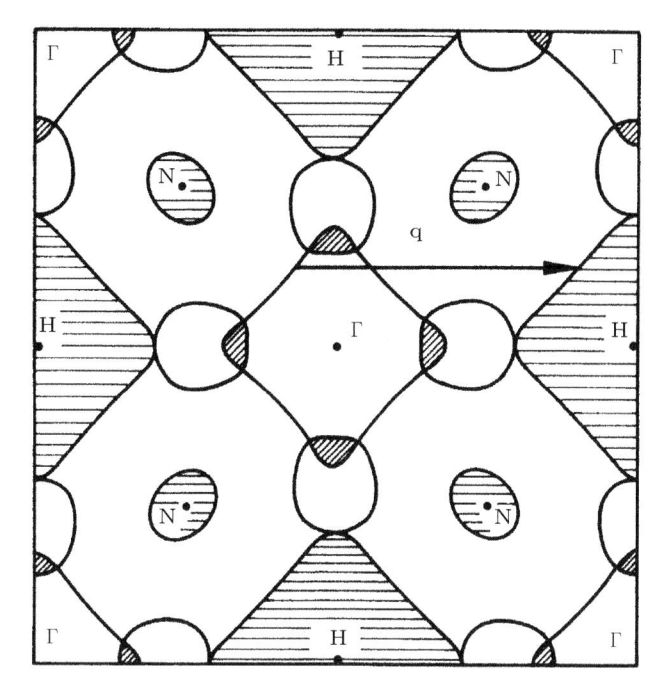

Figure 4.38 *Fermi surface cross-section in the (001)-plane for Cr, reprinted with permission from Fry et al. (1981).*

provide extra help that makes the difference in Cr either being magnetic or nonmagnetic. However, this effect does not alter the overriding general point that Cr, having a half-filled d band, will have an inherent tendency toward changes with a twofold superlattice, i.e. antiferromagnetism.

Beginning now with the concept of Fermi surface nesting, we use moderately new band-structure information and show in Fig. 4.38 the results by Fry *et al.* (1981), which were obtained using self-consistent LCGO calculations for nonmagnetic Cr. The Fermi surface cross-sections in the (001)-plane shown here consist of an electron surface centered at the point Γ and a hole surface centered at H. (We may ignore the other details.) Since the band structure obtained by Fry *et al.* is in rather good agreement with an ASW calculation that we will use for further considerations, we do not show their band-structure results here but the reader may glance at Fig. 4.40(a) to verify our Fermi surface assignments. It is seen from Fig. 4.38 that there are parallel portions of electron and hole surfaces that can be brought to coincidence by a rigid shift of one part onto the other; the necessary translation may, for instance, be defined by the vector **q** drawn in Fig. 4.38. This Fermi surface feature is not uncommon; it exists, for instance, in Mo and other, more complicated systems, and it is called *nesting*. It is important physically because it can have a large effect on the spin susceptibility, χ, or any generalized susceptibility.

To see this we return to the linear response Eqn (4.50) which we write in a nonmagnetic state as

$$\chi^0(\mathbf{r},\mathbf{r}') = \sum_{\mathbf{k}\mu}\sum_{\mathbf{k}'\nu} \frac{\theta(\varepsilon_F - \varepsilon_{\mathbf{k}\mu}) - \theta(\varepsilon_F - \varepsilon_{\mathbf{k}'\nu})}{\varepsilon_{\mathbf{k}\mu} - \varepsilon_{\mathbf{k}'\nu}} \quad (4.108)$$
$$\cdot \psi_{\mathbf{k}\mu}(\mathbf{r})\,\psi^*_{\mathbf{k}\mu}(\mathbf{r}')\,\psi_{\mathbf{k}'\nu}(\mathbf{r}')\,\psi^*_{\mathbf{k}'\nu}(\mathbf{r}),$$

where we have used appropriate Bloch states labeled by the wave vectors $\mathbf{k}$, $\mathbf{k}'$ and the band indices μ, ν. We double Fourier transform this equation (see also Jones and March, 1973) and obtain, using Bloch symmetry (Eqn (3.12) together with (3.20)), the result

$$\chi^0(\mathbf{q},\mathbf{q}') = \iint d\mathbf{r}\,d\mathbf{r}'\, e^{-i\mathbf{q}\cdot\mathbf{r}}\, e^{i\mathbf{q}'\cdot\mathbf{r}'}\, \chi^0(\mathbf{r},\mathbf{r}') \quad (4.109)$$
$$= \delta_{\mathbf{q},\mathbf{q}'}\,\chi_0(\mathbf{q}),$$

where

$$\chi_0(\mathbf{q}) = \sum_{\mathbf{k}\nu\mu} \frac{[f(\varepsilon_{\mathbf{k}\nu}) - f(\varepsilon_{\mathbf{k}-\mathbf{q}\mu})]}{\varepsilon_{\mathbf{k}-\mathbf{q}\mu} - \varepsilon_{\mathbf{k}\nu} + i\delta} \cdot |\langle \mathbf{k}\nu|\,e^{i\mathbf{q}\cdot\mathbf{r}}\,|\mathbf{k}-\mathbf{q}\mu\rangle|^2. \quad (4.110)$$

Here we have generalized the expression slightly by using the Fermi distribution $f(\varepsilon)$ instead of the θ function and including a suitable infinitesimal $i\delta$ (see also Jones and March, 1973). Being guided by the treatment of Sec. 4.2.3 we now write down an approximate equation for the enhanced susceptibility of the form

$$\chi(\mathbf{q}) = \frac{\chi_0(\mathbf{q})}{1 - I\,\chi_0(\mathbf{q})}, \quad (4.111)$$

where $\chi_0(\mathbf{q})$ is given by (4.110). It should be clear that for very low temperatures and for the case of nesting Fermi surfaces the susceptibility $\chi_0(\mathbf{q})$ can be very large for $\mathbf{q}$-vectors that define a "nesting translation" of the type indicated in Fig. 4.38. This is true because the denominator of Eqn (4.110) is near zero for many values of $\mathbf{k}$ in the sum, and the numerator is near unity. Obviously, because of Eqn (4.111), the spin susceptibility $\chi(\mathbf{q})$ can become very large for these values of $\mathbf{q}$ and we obtain a generalized Stoner condition

$$I\,\chi_0(\mathbf{q}) \geq 1 \quad (4.112)$$

that signals a magnetic instability. Indeed, a glance at Fig. 4.38 reveals that the nesting wave vector is nearly of the same length as the distance Γ to H, i.e., half a reciprocal lattice vector. Hence the magnetic instability is likely to occur when the magnetization of ferromagnetic planes alternates up and down as we go along one of the cubic axes; this is

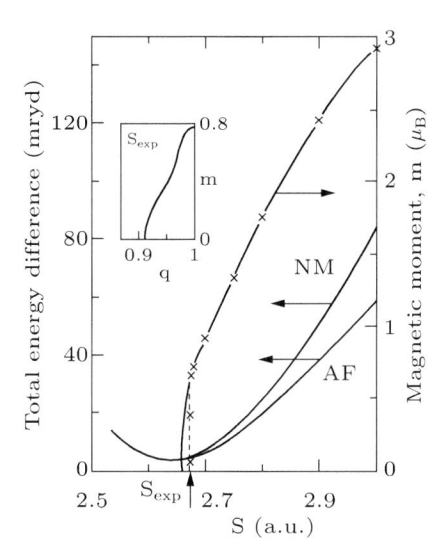

Figure 4.39 *Magnetic moment and total energy difference of assumed nonmagnetic (NM) and antiferromagnetic (AF) (q = $(2\pi/a)$ (0, 0, 1.0)) chromium as a function of the atomic-sphere radius S. The arrows indicate experimental volumes having atomic sphere radius S_{exp}. Crosses: calculations include spin–orbit coupling in a CsCl supercell. Inset: Magnetic moment as a function of spin spiral q = $(2\pi/a)$ (0, 0, q) at experimental volume.*

an antiferromagnetic arrangement with which we began our discussion in this subsection. Of course, these arguments must be augmented with real calculations to prove that the instability in fact occurs. If it does, then general experience is that the symmetry of the ground state formed conforms with the instability (but this need not necessarily be so).

It is extremely interesting that the value of q is not exactly one-half of a reciprocal lattice vector but is somewhat less. Therefore, in the observed structure, although it is as stated, the *magnitude* of the magnetization of adjacent planes varies sinusoidally with a long period that is *incommensurate* with the lattice: a spin-density wave exists either in the form of a transverse or a longitudinal standing wave, the periodicity being given by a q-vector that is q = $(2\pi/a)$ (0, 0, 0.952).

A simple approximation to the magnetic state of Cr thus consists of ignoring the spin-density wave and assuming an antiferromagnet with q = $(2\pi/a)$ (0, 0, 1.0), i.e. ferromagnetic planes that alternate up and down as one moves along the z-axis. Figure 4.39, which was obtained by means of scalar relativistic ASW calculations, shows the strong dependence of the magnetic moment on the volume, denoted by the atomic sphere radius S. In fact, at the volume where the total energy possesses its minimum, Cr is nonmagnetic, at least in this assumed type of order. At the experimental atomic volume, marked by an arrow in the figure, the magnetic moment is nonvanishing. The situation does not change very much if spin–orbit coupling (SOC) is included in the calculation. In this case a CsCl type supercell was used. The total energy obtained is somewhat lower than before but cannot be distinguished in the graph in Fig. 4.39 and the critical volume seems to be slightly larger when SOC is included; see the crosses that mark the values obtained using SOC. The value of the moment calculated at the experimental lattice constant in this case is 0.71 μ_B which should be compared with the experimental amplitude of the spin-density wave of 0.62 μ_B.

Although one should not confuse the observed spin-density wave with a spin spiral of the type discussed in Sec. 3.4.7, it is of some interest to calculate the magnetic state with such a spiral. As the inset in Fig. 4.39 shows, a nonvanishing magnetic moment results at the experimental volume for incommensurate values of q in the range $0.91 < q \leq 1.0$ for a spiral state defined by $\mathbf{q} = (2\pi/a)\,(0, 0, q)$, and magnetization given by the vectors

$$\mathbf{M}_n = M\,(\cos(\mathbf{q}\cdot\mathbf{R}_n), \sin(\mathbf{q}\cdot\mathbf{R}_n), 0) \tag{4.113}$$

where $\mathbf{M}_n$ is the magnetic moment of magnitude M of the n-th atom.

A realistic calculation of the magnetic state in Cr was attempted by Hirai (1997) who used the KKR method and very long supercells to simulate commensurate spin-density waves. The total energy he determined levels off at values of q of about 17/18 to 21/22 (as before $\mathbf{q} = (2\pi/a)\,(0, 0, q)$) attaining an amplitude of approximately 0.7 μ_B, but the total energy gain is very small (about 0.10 mRy).

It is quite remarkable that Hirai succeeded in analyzing his results such that he obtained higher harmonics of the spin-density wave state. Writing for the magnetic moment of the atom at site n

$$M_n = M_1\cos(\mathbf{q}\cdot\mathbf{R}_n) + M_3\cos(3\mathbf{q}\cdot\mathbf{R}_n) + \cdots \tag{4.114}$$

he determined M_1 and M_3/M_1 to be about 0.69 μ_B and -0.03, respectively, which compares quite favorably with the experimental values (Mori and Tsunoda, 1993) of 0.62 μ_B and -0.02. A similar analysis of even harmonics in the charge density resulted in a value of the coefficient of $\cos(2\mathbf{q}\cdot\mathbf{R}_n)$ that was much smaller than the experimental estimate. A recent thorough treatment along the same lines is that by Hafner et al. (2002).

One can obtain considerable insight into the factors that stabilize the simple antiferromagnetic state of chromium by examining its band structure in detail. Beginning, therefore, with the nonmagnetic state we show its band structure in Fig. 4.40(a) in the bcc Brillouin zone. The same bands are shown again in Fig. 4.40(b) but are now folded into the simple cubic Brillouin zone appropriate for the CsCl structure, the twofold superlattice necessary to describe the antiferromagnetic spin arrangement. The evolution of the band structure of the latter, shown finally in Fig. 4.40(c), is now readily apparent; through the mechanism of hybridization a gap opens up along the directions Σ and Λ and degeneracies are lifted at the zone boundaries, most notably at R and X. As a consequence, bands below the Fermi edge are lowered in energy and thus conceivably lead to stabilizing the antiferromagnetic state. Furthermore, the lifted degeneracies at R and X are of interest since a comparison of Fig. 4.40(b) and (c) shows that a Fermi surface piece centered at X exists in the nonmagnetic state but has disappeared (or become very small) in the antiferromagnetic band structure (see the state

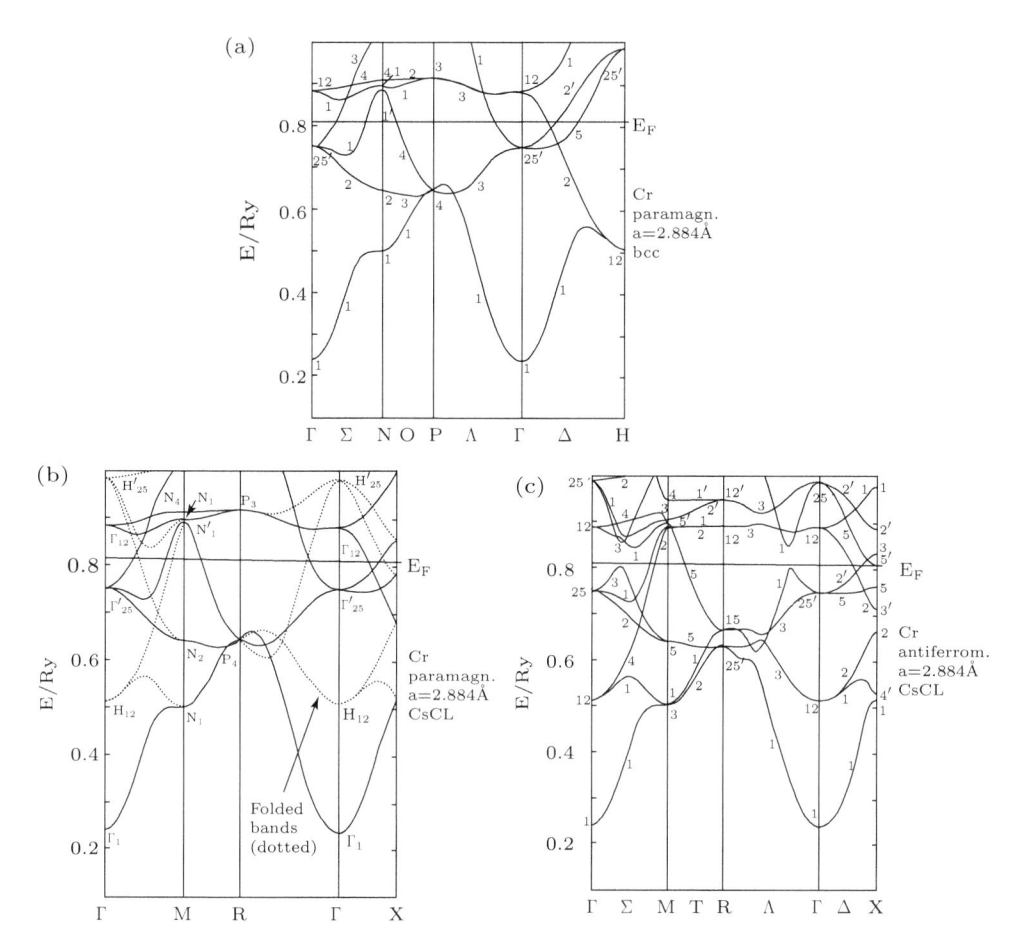

Figure 4.40 *Energy bands for chromium: (a) nonmagnetic in bcc structure, (b) nonmagnetic folded into the Brillouin zone of the CsCl structure, (c) antiferromagnetic in CsCl supercell. For labels see Fig. 3.1.*

labeled $X_{5'}$ at E_F). Singh *et al.* (1988) measured two-dimensional angular correlations of the positron-annihilation radiation from Cr in its paramagnetic and antiferromagnetic phases, as well as from paramagnetic $Cr_{0.95}V_{0.05}$ alloys, and found a clear difference between Fermi surface topologies in the two phases. They attributed this difference convincingly to the above-noted band-structure changes. Furthermore, angle resolved photoemission measurements by Johansson *et al.* (1980) are in good agreement with the occupied band structure along the symmetry line Σ, except near Σ_3 where they seem to be much less dispersive.

4.3.4.3 Manganese (cont.)

We now return to the case of manganese and continue first with the discussion of δ-Mn. Although this high-temperature form does not have much in common with the complicated low-temperature phase to be discussed below, it seems that δ-Mn embodies much of the essential physics of Mn. Thus, beginning with simple collinear spin arrangements, we calculate, using the ASW method, the ground-state electronic structure as a function of the volume and obtain different magnetic states for which the total energy and the magnetic moments are graphed in Fig. 4.41. Starting with a small volume corresponding to an atomic radius of $S = 2.50$ a.u., we find two states of low-magnetic moments called low-spin ferromagnetic (LS-FM) states; their total energy differences are too small to be resolved in the figure. At $S = 2.7$ a.u. antiferromagnetism (AFM) sets in whose total energy becomes lower than the total energy of the LS-FM state at about $S = 2.74$ a.u., see Fig. 4.41, upper panel. At about $S = 2.87$ a.u. the unstable LS-FM state gives way to an unstable high-spin ferromagnetic (HS-FM) state; see the sequence of total energies in the upper panel and the corresponding magnetic moments in the lower panel of Fig. 4.41. The common tangent to the AFM- and LS-FM total energy curves, denoted by SPM, gives the approximate energy of a spiral magnetic state to be discussed below.

It is of interest to have a closer look at the crossing of the two total energy curves at about $S = 2.74$ a.u. and inquire how at this point of degeneracy the low-spin ferromagnetic state goes over into the antiferromagnetic state having an atomic magnetic moment that is more than twice as large as that of the LS-FM state. The result of this inquiry is surprising as was first observed by Mohn et al. (1997).

Choosing an atomic volume corresponding to $S = 2.75$ a.u., i.e. slightly larger than the cross-over value, we continue the calculations with noncollinear spin arrangements and determine spin-spiral states with a polar angle of $\theta = \pi/2$ as a function of the **q**-vector varying from $(0,0,0)\, 2\pi/a$, the ferromagnet, to $(0,0,1)\, 2\pi/a$, the antiferromagnet, or from $(0,0,0)\, 2\pi/a$ via $(\frac{1}{2},\frac{1}{2},\frac{1}{2})\, 2\pi/a$ to $(1,1,1)\, 2\pi/a$, the latter being equivalent to $(0,0,1)\, 2\pi/a$ again. The ensuing total energies and the corresponding magnetic moments are shown in Fig. 4.42.

We see that a number of stable and metastable states (marked by arrows) exist in an energy range of about 3 mRy. Furthermore, the total energy curve consists of parabola-like segments which form cusps (also marked by arrows) at each intersection where the corresponding magnetic moment varies strongly.

Comparison of Fig. 4.41 with Fig. 4.42 shows that the magnetic moments span the entire range defined by the LS-FM and AFM values at the volume under consideration. If we now imagine we have prepared δ-Mn at the volume given by $S = 2.75$ a.u. in the ferromagnetic state and quench it down to a state of lower total energy, then it is apparent that the system may end up in either of the stable or metastable states seen in Fig. 4.42, most likely being "frustrated," but with some likelihood of obtaining the state of lowest total energy (the minimum marked by an X in Fig. 4.42) which is certainly noncollinear. We will devote some more attention to the concept of frustration in connection with β-Mn below. The situation was compared by Mohn et al. (1997) with the model for a spin glass where very many local energy minima are separated by barriers through which the system can tunnel.

The behavior shown in Fig. 4.42 begins at a smaller volume, roughly given by $S \simeq 2.69$ a.u., with smaller amplitudes but very similar shape and continues to larger volumes until

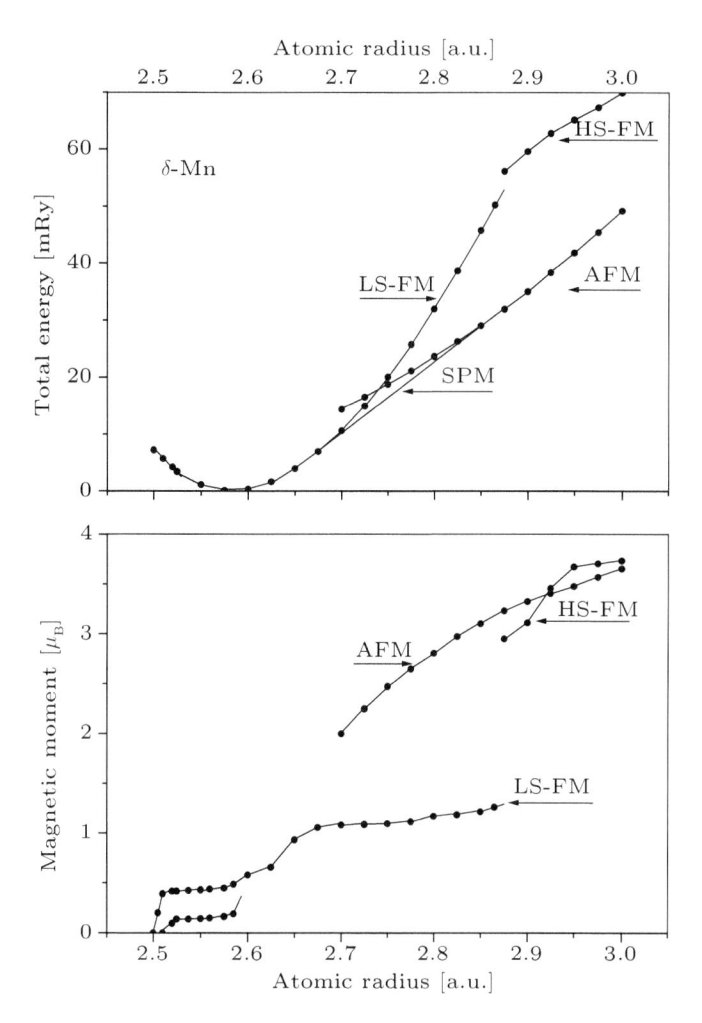

Figure 4.41 *Total energy (upper panel) of δ-Mn as a function of the volume which is labeled by the atomic sphere radius, S, in a.u. (multiples of the Bohr radius). LS-FM abbreviates the low-spin and HS-FM the high-spin ferromagnetic phases; AFM denotes the collinear antiferromagnetic phase. The common tangent, denoted by SPM, gives the approximate total energy of a spiral magnetic state. In the lower panel the corresponding magnetic moments are given, except for those of the SPM states, whose values can be inferred from Fig. 4.42.*

about $S \simeq 2.8$ a.u. where the collinear antiferromagnetic state defined by $(0,0,1)\,2\pi/a$ finally becomes stable. In all the cases examined the total energy minimum occurs at $\mathbf{q} = (0.3, 0.3, 0.3)\,2\pi/a$, see the value marked X in Fig. 4.42, and lies approximately on the common tangent drawn in the upper panel of Fig. 4.41 and denoted by SPM, the spiral magnetic state.

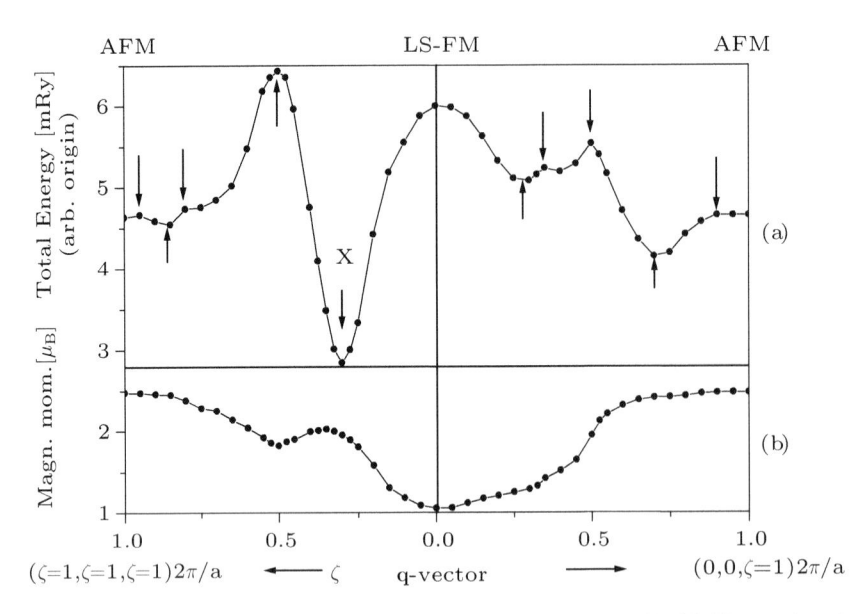

Figure 4.42 *Dependence of the total energy (a) and the magnetic moment (b) of δ-Mn as a function of the spin-spiral* **q**-*vector for a volume corresponding to an atomic sphere radius of $S = 2.75$ a.u. Right-hand side* $\mathbf{q} = (0,0,\zeta)\, 2\pi/a$, *left-hand side* $\mathbf{q} = (\zeta,\zeta,\zeta)\, 2\pi/a$.

The reason for the stabilization of these different possibilities can in our case be traced back to the respective band structures, where, as in the case of Cr, only states near the Fermi energy E_F matter. For noncollinear magnetic structures we will show in some detail the evolution of bands when we discuss γ-Fe; here it suffices to say that the nondiagonal part of the spin-density matrix leads to hybridizing bands that in collinear spin arrangements would simply cross. If hybridization leads to gaps straddling E_F it assists in stabilizing that particular structure (see the discussion below of γ-Fe). This is the case in δ-Mn for the local minima seen in the upper portion of Fig. 4.42. We show examples of band structures in Fig. 4.43 for three different wave vectors: (a) $\mathbf{q} = (0,0,0.35)\, 2\pi/a$, (b) $\mathbf{q} = (0,0,0.7)\, 2\pi/a$, and (c) $\mathbf{q} = (0,0,0.875)\, 2\pi/a$. In (a) the gap opens near the Γ-point, in (b) it is an indirect gap in which E_F is pinned, and in (c), which applies to a cusp, E_F lies just above a gap destabilizing this spin arrangement.

Mohn *et al.* (1997) drew attention to these peculiar states of δ-Mn mainly in view of a possible origin for spin glasses formed by Mn. Although of great fundamental interest, this is not the issue here. However, the existence of these states could play a role in explaining the complicated real ground state of Mn which we will turn to after a discussion of two other phases of Mn.

In the temperature range 1368–1406 K an fcc phase of Mn stabilizes which is called γ-Mn. This high-temperature phase can be retained at low temperatures by rapid quenching, provided it is doped with a sufficient concentration of impurities such as carbon, nickel, or iron (Honda *et al.*, 1976), or its properties can be inferred from

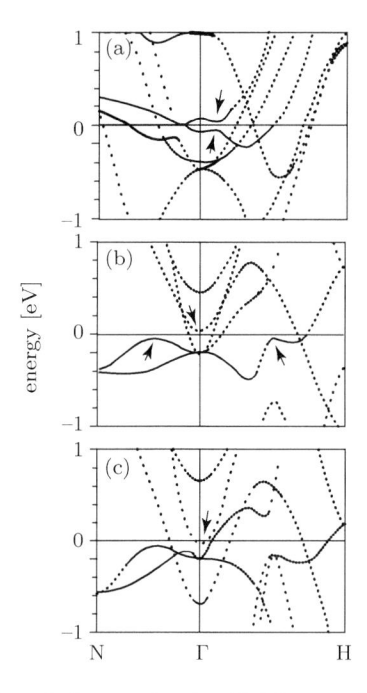

Figure 4.43 *Section of the spin-spiral energy bands of δ-Mn near the Fermi edge chosen to be at the energy origin for three* **q**-*vectors* $(0,0,\zeta)\,2\pi/a$ *with* $\zeta = 0.35$ *(a),* $\zeta = 0.7$ *(b), and* $\zeta = 0.875$ *(c). The arrows indicate the position of the band gaps discussed in the text (after Mohn et al., 1997).*

extrapolations to $x \longrightarrow 0$ of fcc alloys, like $Mn_{1-x}Fe_x$ (Yamaoka et al., 1974). One obtains from these extrapolations values for the magnetic moment that lie between 1.7 and 2.4 μ_B and, furthermore, small tetragonal crystallographic distortions ($c/a \cong 0.94$). A great number of band-structure calculations can be found in the literature starting as early as 1971 with work by Asano and Yamashita (1971), followed by work of Cade (1980, 1981), Duschanek et al. (1989) and many others. Of interest to us here are calculations of the magnetic states as a function of the volume and inquiries into the stability of collinear or noncollinear states.

The simplest antiferromagnetic state in an fcc lattice consists of ferromagnetic planes whose spin projection alternates ↑ and ↓ as one moves along the z-axis. This is called antiferromagnetic order of type AFI. The corresponding twofold superlattice is the CuAu structure which is tetragonal with a lattice constant of $a_{fcc}/\sqrt{2}$, where a_{fcc} is the lattice constant of the fcc phase. Another possible ordering consists of ferromagnetic planes whose spin projection alternates ↑ and ↓ as one moves along one of the body diagonals of the cubic cell. We disregard this type of ordering because its total energy was found to be higher than that of the AFI state (Kübler and Eyert, 1992). For the same reason we also disregard a noncollinear ordering where the magnetic moments point along the four different body diagonals forming a tetrahedral order that must be described with *four* interpenetrating simple cubic lattices.

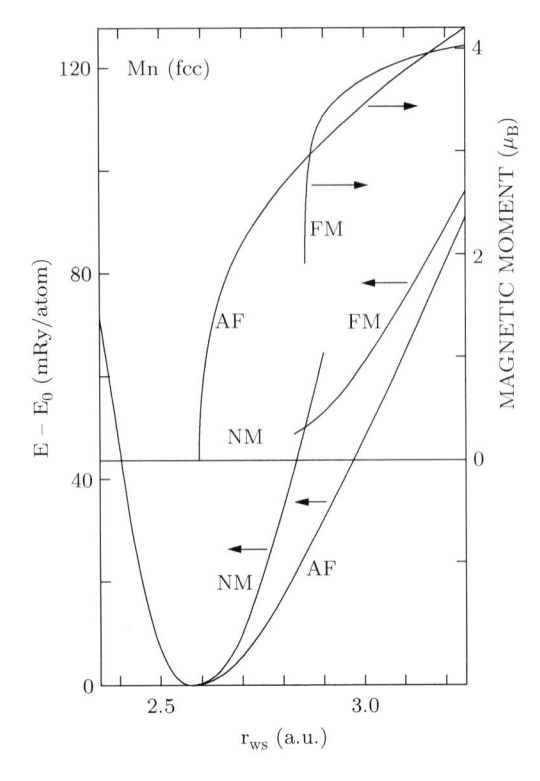

Figure 4.44 *Total energy and magnetic moment of γ-Mn as a function of the volume given by the atomic sphere radius which is denoted here by r_{WS}. AF: collinear antiferromagnetic type AFI (see text), NM: nonmagnetic; FM: ferromagnetic. Reprinted with permission from Moruzzi and Marcus (1993).*

The total energy from ASW calculations by Moruzzi and Marcus (1993) for nonmagnetic (NM), ferromagnetic (FM), and antiferromagnetic (AF) states—the latter being collinear of type AFI—are shown together with the corresponding magnetic moments in Fig. 4.44. An important result here is the existence of a stable antiferromagnetic state that extends down to the total energy minimum. A ferromagnetic state sets in at a larger volume, its total energy, however, being above that of the ferromagnetic state. Investigations of spiral magnetic states by Mryasov *et al.* (1991) and this author revealed that none have low-lying total energies, in contrast to the case of δ-Mn. Furthermore, the small observed tetragonal distortions give rise to only very small changes to these results. In view of the strong rate of change of the calculated magnetic moment near the total energy minimum, we conclude that the observed values in the range 1.7–2.4 μ_B are not in disagreement with the theoretical results.

Another phase of manganese is found in the temperature range 1000–1368 K; it is cubic with 20 atoms per unit cell and is called β-Mn. This phase can be quenched down to liquid helium temperatures where it is found to be nonmagnetic (Kasper and Roberts, 1956). This low-temperature state, however, is found to be quite unusual. A recent and

4.3 Elementary magnetic metals

Table 4.8 *Atomic positions of the β-Mn structure. The parameters x and y are defined in the text.*

Site I	x,x,x	$\frac{3}{4}-x,\frac{3}{4}-x,\frac{3}{4}-x$	$\frac{1}{2}+x,\frac{1}{2}-x,-x$
	$\frac{1}{4}-x,\frac{3}{4}+x,\frac{1}{4}+x$	$-x,\frac{1}{2}+x,\frac{1}{2}-x$	$\frac{1}{4}+x,\frac{1}{4}-x,\frac{3}{4}+x$
	$\frac{1}{2}-x,-x,\frac{1}{2}+x$	$\frac{3}{4}+x,\frac{1}{4}+x,\frac{1}{4}-x$	
Site II	$\frac{1}{8},y,\frac{1}{4}+y$	$\frac{1}{4}+y,\frac{1}{8},y$	$y,\frac{1}{4}+y,\frac{1}{8}$
	$\frac{3}{8},-y,\frac{3}{4}+y$	$\frac{3}{4}+y,\frac{3}{8},-y$	$-y,\frac{3}{4}+y,\frac{3}{8}$
	$\frac{5}{8},\frac{1}{2}-y,\frac{3}{4}-y$	$\frac{3}{4}-y,\frac{5}{8},\frac{1}{2}-y$	$\frac{1}{2}-y,\frac{3}{4}-y,\frac{5}{8}$
	$\frac{7}{8},\frac{1}{2}+y,\frac{1}{4}-y$	$\frac{1}{4}-y,\frac{7}{8},\frac{1}{2}+y$	$\frac{1}{2}+y,\frac{1}{4}-y,\frac{7}{8}$

thorough investigation is that of Nakamura *et al.* (1997) who measured thermodynamic and transport properties as well as nuclear magnetic resonance and neutron scattering spectra. From this combination of measurements it appears that one portion of the Mn atoms (those located at sites I, see below) are nonmagnetic or nearly so and another portion, located at sites II, are magnetic, having local moments that are hard to estimate but are placed at roughly 1 μ_B. The noteworthy property is that there is no sign of magnetic order down to a temperature of 1.4 K. There is at present no unambiguous explanation of this finding, although there is a conjecture by Nakamura *et al.* that we find worthwhile to discuss. Of special interest in this context is a density-functional calculation of Asada (1995) who obtained a nearly vanishing magnetic moment for the site-I atoms while finding the site-II atoms to order antiferromagnetically with a local moment of about 2 μ_B. ASW calculations support these findings but result in somewhat smaller magnetic moments for the site-II atoms (about 1.3 μ_B).

For a better understanding of the problem one must begin with the crystal structure of β-Mn. It consists of a simple cubic lattice and is described by giving the atomic positions as listed in Table 4.8. The parameters x and y are obtained experimentally and have been reported by different authors to be $x = 0.061$ or $x = 0.0636$ and $y = 0.206$ or $y = 0.2022$. A schematic diagram of the unit cell is shown in Fig. 4.45(a) where the site-II atoms that are nearest neighbors have been connected by straight lines.

The numbering sequence corresponds to the sequence of columns in Table 4.8 beginning with $\frac{1}{8}, \ldots$ and the permutations in row 1, etc. The atom numbered 14 is equivalent to atom 4, and 13 is equivalent to 6. The triangles that emerge are separately perpendicular to one of the four body diagonals of the cubic cell and each site-II atom bonds with six neighboring site-II atoms, as three triangles share those corners. These twisted triangles are shown in Fig. 4.45(b).

With the value of y given above the triangles are not of equal size, but a closer consideration of the geometry reveals that they become identical if $y = (9 - \sqrt{33})/16 = 0.20346$. In this case the sides of the equilateral triangles measure 0.4207 in units of the lattice constant. The location of the site-I atoms are given by the unnumbered dots in Fig. 4.45(a). A slight change of x from the experimental values given above to

244 *Electronic Structure and Magnetism*

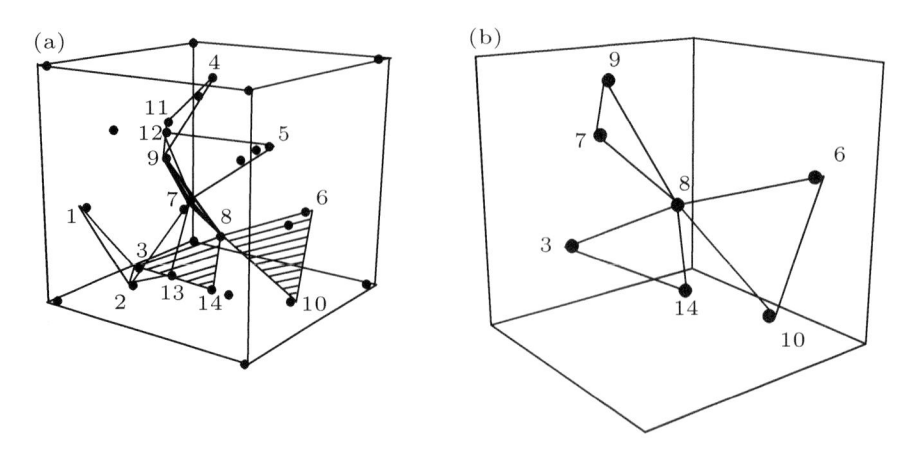

Figure 4.45 *(a) Schematic diagram of the crystal structure of β-Mn. The site-I atoms are given by the unnumbered dots. The site-II atoms are numbered and nearest neighbors are connected by straight lines. The numbering sequence corresponds to the sequence of columns in Table 4.8 beginning with $\frac{1}{8}$, ... and the permutations in row 1, etc. The atom numbered 14 is equivalent to atom 4, and 13 is equivalent to 6. (b) An element of (a) showing three twisted, corner-sharing triangles that are separately perpendicular to three of the four-body diagonals of the cube.*

$x = 1/(9 + \sqrt{33}) = 0.06782$ will place them such that each site-I atom has the three atoms of a site-II triangle at a distance of 0.41278 in units of the lattice constant. Thus one site-I atom and three site-II atoms form a nearly regular tetrahedron and in this somewhat idealized structure the nearest neighbors of the site-I atoms are the site-II atoms.

Concerning the value of the magnetic moment of the site-I atoms, the calculation of Asada (1995) and the measurements of Nakamura *et al.* (1997) agree, thus allowing us to neglect their magnetic moments. Accepting this as a basically unexplained fact, we face the still enormous but simpler problem of antiferromagnetism on a network of triangles formed by the site-II atoms as shown in Fig. 4.45. This problem warrants a brief excursion into properties of *frustrated systems*. The simplest example to consider is a system consisting of three spins each at the corner of an equilateral triangle; we assume, furthermore, the spins are classical and can only point up or down. These objects are called Ising spins and denoted by s_i at site i. A model Hamiltonian for this system is usually written as

$$\mathcal{H} = -J s_1 s_2 - J s_2 s_3 - J s_3 s_1, \qquad (4.115)$$

where J is called the exchange constant measuring the strength of the interaction between two given spins. It is obvious how this Hamiltonian is generalized to an arbitrary network. Assuming only nearest-neighbor interactions, one writes in this case

$$\mathcal{H} = -J \sum_{\langle i,j \rangle} s_i s_j, \qquad (4.116)$$

where $\sum_{\langle i,j \rangle}$ denotes pairwise summation. For the case of negative J we next wish to write down that configuration of spins which results in the lowest energy. Thus attempting to arrange the spins on the corners of the triangle in an antiparallel way, we see quickly that this is not possible; in fact, one of the interactions is always "broken," i.e. the spins are parallel and the interactions are obviously competing. In the context of investigations of spin glasses Toulouse (1977) has coined the term *frustration* for systems with competing interactions. This not being at the center of the present treatise, we refer the reader interested in this topic to the clear exposition by Liebmann (1986). Here it suffices to say that in the infinite triangular lattice the ground-state entropy remains finite as $T \to 0$ and the ordering temperature is $T_c = 0$. Furthermore, the number of lattices studied that exhibit frustration is quite numerous, yet the site-II lattice of β-Mn (Fig. 4.45) does not seem to have attracted any attention by model theories.

Although the frustrated Ising model is a good starting point for the discussion of a possible mechanism to understand the missing order in β-Mn, it clearly cannot be applied without appreciable modifications. The most important shortcoming is the Hamiltonian, Eqn (4.116), itself. For one thing, the exchange constant J will certainly not be short ranged, but, if applicable at all, in metals will couple more than nearest neighbors. This is plausible from our results for γ-Mn, because the fcc lattice is also frustrated as is easily seen by considering the tetrahedron formed by the nearest neighbors, but there is no indication of competing interactions. Asada (1995) remarks, however, that there are 26 nonequivalent spin alignments, $\{s_i\}$, that presumably give a compensated magnetic state with a vanishing cell sum, $\sum_i s_i = 0$. However, it has yet to be shown that these different configurations result in nearly degenerate total energies, or give rise to an energy landscape comparable to that obtained for δ-Mn (Fig. 4.42). Another problem with the model Hamiltonian, Eqn (4.116), is, besides the Ising nature of the spins, the variable magnitude of the moments in a metal when their relative orientation is changed; this is quite apparent from our determination of the nonuniform susceptibility in Sec. 4.2.3. To sum up, the magnetism of itinerant-electron systems is not likely to be described correctly by Eqn (4.116) or its quantum version, the Heisenberg (1926) Hamiltonian. We will return to questions like this in Chap. 5. Still, the conjecture of Nakamura *et al.* is interesting: frustration is the likely cause for the vanishing order in the ground state, but the precise mechanism, especially the nature of the fluctuations that quench the order, is still to be understood.

We now complete our discussion with the low-temperature ($T \leq 1000$ K) form of manganese which is called α-Mn. Even though it is cubic, it is surprisingly complex, consisting of 58 atoms per cubic cell (or 29 per primitive unit cell). If we ignore the magnetic moments the lattice is bcc with which we may start showing three-dimensional sketches in Fig. 4.46 together with site labels I–III. For ease in understanding the crystal structure, the two simple cubic sublattices are shown separately in (a) and (b) which, taken together, define the body-centered lattice.

Site I possesses point-group symmetry $\bar{4}3m$, site II $3m$, and site III m. There are sites IV (also of symmetry m) which are not labeled in the figure. The atomic positions for the bcc unit cell are collected in Table 4.9.

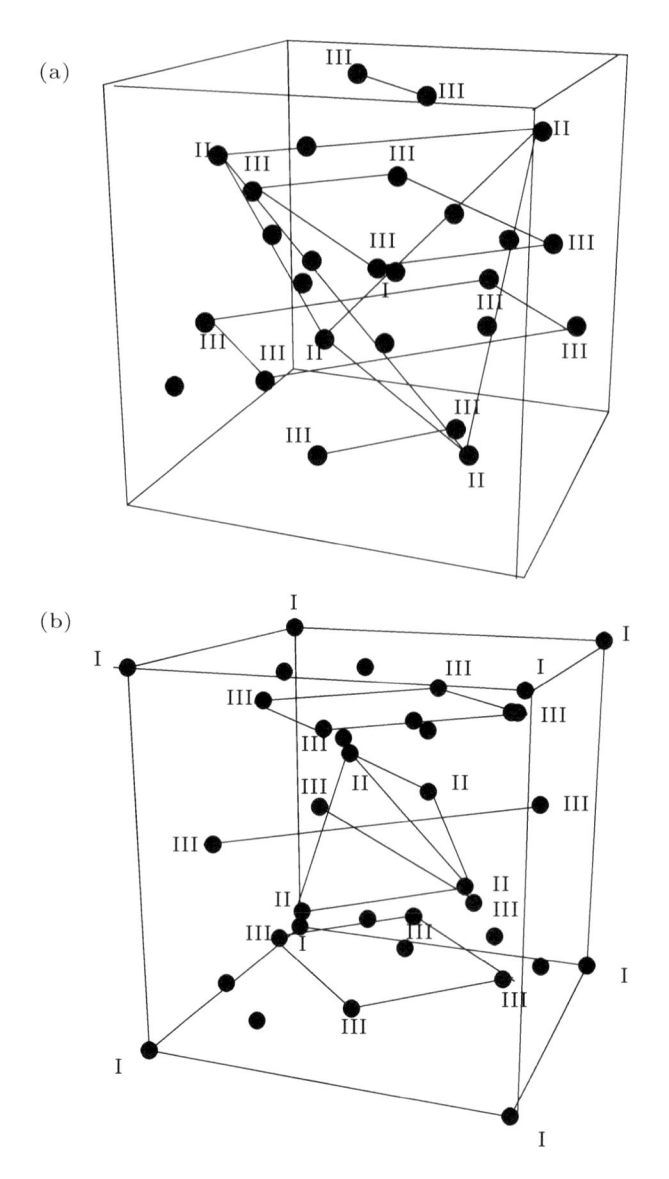

Figure 4.46 *Three-dimensional sketches of the bcc unit cell of α-Mn: (a) pictures the simple cubic sublattice that contains the body center whereas (b) contains the corner atoms. The full bcc structure is obtained by fitting (a) into (b). Site labels and connections are explained in the text.*

Table 4.9 *Atomic positions for α-Mn (x and z: room-temperature values from Lawson et al., 1994).*

	Site I:	0,0,0		

	Site II:	x,x,x	$\bar{x},\bar{x},x$	
	$x = 0.318$	$\bar{x},x,\bar{x}$	$x,\bar{x},\bar{x}$	

Site III:	x,x,z	$\bar{x},\bar{x},z$	$\bar{x},x,\bar{z}$	$x,\bar{x},\bar{z}$
$x = 0.350$	x,z,x	$\bar{x},\bar{z},x$	$\bar{x},z,\bar{x}$	$x,\bar{z},\bar{x}$
$z = 0.035$	z,x,x	$\bar{z},\bar{x},x$	$\bar{z},x,\bar{x}$	$z,\bar{x},\bar{x}$

Site IV:		
$x = 0.091$	$x,x,z...$ as site III	
$z = 0.283$		

The essential feature to be noticed is the simple cubic cell, clearly to be seen by the atoms labeled I in 4.46(b). There are eight atoms labeled II located on two tetrahedra, one inside the other, as seen in (a) and (b). Next moving up from the bottom of the figures one sees atoms labeled III: two in one row, two rows of four atoms each, and again two in one row, the situation being repeated above the body center. The remaining, unlabeled atoms are located on sites IV. Site-III Mn atoms in the same plane are connected by lines; however, for sites II the two tetrahedra are outlined completely.

The complexity of the crystal structure is surprising since most elements have simple structures. It is believed that it arises from an instability of the half-filled 3d electron shell which leads to the formation of a "self-intermetallic" compound, i.e. elemental Mn is understood to be an intermetallic compound between Mn atoms in different electronic configurations.

Below 95 K α-Mn orders antiferromagnetically. Neutron diffraction measurements by Yamada *et al.* (1980) and, more recently by Lawson *et al.* (1994), yield magnetic moments whose values are collected in Table 4.10. The type of order is noncollinear and can be described with a polar angle, θ, that is given in parentheses in Table 4.10. The differences in the experimental values originate mainly from uncertainties in the form factor, a problem that is briefly described as follows.

A detailed theory of neutron diffraction in magnetic crystals can be found in the book by Lovesey (1984); here it suffices to say that the neutrons probe almost entirely the valence electron density as the core polarization is, relatively, very small. It is shown by, for instance, Lovesey that the scattering intensity contains the atomic magnetic scattering amplitude, $F(\mathbf{q})$, which is given by the Fourier transform of the magnetization, $m(\mathbf{r})$, i.e.

$$F(\mathbf{q}) = \int m(\mathbf{r}) \exp(i\mathbf{q}\cdot\mathbf{r}) \, d\mathbf{r}. \tag{4.117}$$

248 Electronic Structure and Magnetism

Table 4.10 Measured and calculated magnetic moments (in μ_B) of α-Mn. Polar angle θ in parentheses.

	Sites I	Sites II	Sites III$_1$	Sites III$_2$	Sites IV$_1$	Sites IV$_2$
Experiment Yamada et al. (1970)	2.32 (0)	1.99 (28)	0.72 (141)	0.62 (136)	0.17 (131)	0.36 (67)
Experiment Lawson et al. (1994)	2.83 (0)	1.83 (6)	0.74 (125)	0.48 (132)	0.59 (140)	0.66 (44)
Theory* (collinear cubic) Asada (1995)	2.67	2.29		0.59	0.52	
Theory* ASW (collinear cubic)	2.76	2.34		1.42	0.04	
Theory* ASW (collinear tetragonal)	2.71	2.31	1.17	1.15	0.24	~ 0

* calculated at experimental lattice constants

The *magnetic form factor* is the scattering amplitude divided by the total moment. Assuming that the magnetization, $m(\mathbf{r})$, is solely due to the electron spin, i.e. ignoring here possible orbital contributions, and using the fact that to a good approximation it is spherically symmetric in the atomic sphere (see for instance Fig. 2.2), we obtain with the standard expansion for the exponential function (see Eqn (3.76))

$$F(q) = \langle j_0 \rangle \doteq \frac{4\pi}{m_s} j_0(qr) \, r^2 \, m(r) \, dr, \tag{4.118}$$

where j_0 is the spherical Bessel function of order 0, $m(r)$ is the radial spin density in the local atomic frame of reference, S is the atomic sphere radius and

$$m_s = 4\pi \int_0^S r^2 \, m(r) \, dr \tag{4.119}$$

is the atomic magnetic moment in units of μ_B.

Now the uncertainty in the form factor originates from the fact that $m(r)$ is given by $m(r) = n_1(r) - n_2(r)$, see Eqn (3.301) and the separate spin-up and spin-down contributions are spherical averages of

$$n_\alpha(\mathbf{r}) = \sum_{i=1}^{N} |\psi_{i\alpha}(\mathbf{r})|^2, \tag{4.120}$$

($\alpha = 1, 2$) and thus involve the wave functions for which various assumptions were made by Yamada *et al.* (1980).

It is not hard to compute and compare the form factors using self-consistent wave functions for, say, δ- or γ-Mn. Very small differences are found among them even when the magnetic moments differ greatly. The same is true if one compares with the atomic form factors for the $3d^5$ $4s^2$ state of neutral Mn determined by Freeman and Watson (1961). The latter was used by Lawson *et al.* (1994). Since the paper of Yamada *et al.* is quite explicit one can easily select values of magnetic moments that follow from a given choice of form factors, hence the values given in the first row of Table 4.10 are those that one can approximately read off the paper using the Freeman–Watson values.

The observed magnetic order is easily described using Fig. 4.46. The moments of the site-I atoms point up at the corners of the cube and down at the center. The moments of the site-II atoms are tilted away from the *c*-axis but the two experiments give different values for the tilt angle θ, the projections, however, onto the *c*-axis being down for all four atoms of the larger tetrahedron and up for the smaller that it encloses. The sites III and IV split into pairs as indicated in the table because of a small tetragonal lattice distortion. Here the measured tilt angles are in better agreement with each other and the projections onto the *c*-axis are for the eight site-III$_1$ moments pair-wise down, up, up, down moving up the *c*-axis. For the 16 site-III$_2$ moments, four in a plane, the projections are up, down, down, up, again moving up the *c*-axis. There are also eight site-IV$_1$ and 16 site-IV$_2$ atoms that are ordered opposite to the site-III atoms.

A few attempts have been made at calculations. The third and fourth rows of Table 4.10 contain results by Asada (1995) and Uhl (unpublished), respectively. Here, in contrast to the last row that is also due to Uhl (unpublished), a cubic lattice was assumed in the calculations. The results for the large moments on site I and II are in good agreement with each other and in reasonable agreement with experiment, but the smaller ones on sites III and IV differ considerably. This may be due to the gradient correction used by Asada, but there are also ambiguities in the ordering of the moments that may have caused the discrepancies. Only calculations using collinear order are listed because attempts to verify the noncollinear moment arrangements failed. These latter calculations were carried out without spin–orbit coupling using as input values the experimental information for the polar angles θ. All calculations attempted so far converged to collinear order. We cannot exclude effects due to frustration that conceivably lead to finding metastable total energy minima, but the present results may also be due to the neglect of spin–orbit coupling.

A new and very complete *ab initio* study of α-Mn was undertaken by Hobbs *et al.* (2003). This paper excels by its handsome illustrations, which nicely show the non-collinear magnetic order and demonstrate the subtle dependence on the volume of the crystal. The authors also compare their results exhaustively with other theoretical work and with experimental data. Furthermore, all phases of manganese, especially β-Mn, were put into proper perspective in a companion paper by Hafner and Hobbs (2003).

4.3.4.4 Iron

The high-temperature form of iron has the fcc crystal structure and is called γ-Fe. When stabilized at low temperatures it is at a cross-over point for ferromagnetic and antiferromagnetic states, its magnetic properties depending sensitively on the atomic volume. Fabrication of γ-Fe was done in the past by appropriate thermal treatment of supersaturated CuFe alloys resulting at room temperatures and below in precipitates (small clusters of the order of 500 Å diameter) in the Cu matrix. The precipitates can be diluted with Co to become γ-$Fe_{97}Co_3$ which increases their structural stability. The magnetic structure of these γ-Fe and γ-$Fe_{97}Co_3$ precipitates was reported to be an incommensurate spin-density wave with Néel temperatures of about 50 and 30 K, respectively (Tsunoda, 1989; Tsunoda *et al.*, 1993). In more recent studies thin Fe films were grown epitaxially on Cu(001) and several different structures were observed depending on the growth conditions and film thickness (Li *et al.*, 1994; Müller *et al.*, 1995; Ellerbrock *et al.*, 1995). Although the matter is still under debate, it seems clear that in the thickness range of 6–10 monolayers Fe the fcc phase is stable and shows an antiferromagnetic structure at low temperatures.

It was in connection with alloying problems, in particular concerning Fe-Ni Invar, that a truly immense amount of effort was devoted to γ-Fe. We will deal with this topic briefly in the next section. Here we want to draw attention to an early microscopic model for γ-Fe by Weiss (1963) and Kaufman *et al.* (1963) which these authors arrived at by extrapolations from various alloys. They could explain a number of alloy properties by postulating the existence of two states in γ-Fe, one being a high-spin state having a large volume and the other a low-spin state with a small volume. In fact, in an early ASW calculation (Kübler, 1981) of the total energy and magnetic moment of iron, a scenario similar to the two-state model of Weiss *et al.* was found which was later derived again and refined by a number of authors (Moruzzi *et al.*, 1986; Krasko, 1987; Marcus and Moruzzi, 1988a,b; Moruzzi *et al.* 1989, Podgorny, 1989). But a problem soon emerged which concerned the relative values of the total energy of bcc and fcc Fe (Wang *et al.*, 1985; Hathaway *et al.*, 1985; Krasko, 1989): depending somewhat on the computational method employed the total energy of the fcc state was found to be lower than that of the bcc state by a couple of milliRydbergs. We pointed this out before in Sec. 4.3.1. Since the ground-state crystal structure of iron is manifestly bcc, this finding signals a failure of the density-functional theory in the approximation used, the culprit being most likely the local density-functional approximation (LDA). The source of this failure was indeed located (Asada and Terakura, 1992; Stixrude *et al.*, 1994) by using corrections to the LDA that were proposed by Becke (1992) and Perdew and that are now called the generalized gradient correction (GGA), see Sec. 2.8. Applying the

GGA one finds it compensates for the overbinding inherent in the LDA, shifting up the total energy minimum relative to that of bcc Fe and increasing the theoretical volume slightly. However, the GGA is not in all respects a standardized remedy, especially since experience shows that it has to be used with full potential methods, over-correcting in methods that employ the atomic sphere approximation. Such a full-potential GGA calculation was used by Knöpfle *et al.* (2000) who showed with a modified ASW method that, indeed, the correct ground state of iron is reproduced. However, for fast estimates we resort to the LDA again. The following LDA results for γ-Fe should, however, be taken with a grain of salt. The necessary corrections will be dealt with subsequently.

Using the LDA in the ASW method we calculate the total energy and the magnetic moment as a function of the volume assuming different magnetic states. The results are collected in Fig. 4.47 and demonstrate that γ-Fe is nonmagnetic (NM) at the theoretical equilibrium volume. However, increasing the volume an antiferromagnetic state defined by a q-vector of $\mathbf{q} = (0,0,1)$ in units of $2\pi/a$ sets in at $S \cong 2.56$ a.u. (corresponding to a lattice constant of $a = 3.466$ Å). This antiferromagnetic state is denoted by AFM in Fig. 4.47, but is usually called AFI and was explained in detail before in connection with γ-Mn. A high-spin ferromagnetic state, denoted by HS-FM in Fig. 4.47, sets in at $S \cong 2.65$ a.u. (corresponding to a lattice constant of $a = 3.588$ Å). The pressure at the onset of the HS-FM state nearly vanishes and the curvature of the HS-FM total energy implies a reduced bulk modulus compared with the NM and AFM states. As in the case of γ-Mn the common tangent in the upper panel of Fig. 4.47 is the approximate locus of a spiral magnetic state denoted by SPM. Thus, as was pointed out already in Sec. 4.2.3 (Fig. 4.18), at expanded volumes γ-Fe orders in a spiral. For energetic reasons we disregard the noncollinear ordering where the magnetic moments point along the edges of the tetrahedron formed by the nearest-neighbor Fe atoms; this order must be described with *four* interpenetrating simple cubic lattices and was discussed by Antropov *et al.* (1996). These authors, using the atomic sphere approximation and gradient corrections, found even more complicated ground states which we will not discuss here any further.

In Fig. 4.48 we look at the ground states of γ-Fe in more detail by scanning $\mathbf{q}$-space and choosing two different volumes, *viz.* $S = 2.642$ a.u., corresponding to a lattice constant of $a = 3.577$ Å as well as $S = 2.661$ a.u., corresponding to a lattice constant of $a = 3.603$ Å. The former is the reported volume of γ-Fe and the latter is the volume of Cu in which we envision the precipitates of γ-Fe to occur (Tsunoda, 1989; Tsunoda *et al.*, 1993). As in the case of δ-Mn (compare with Fig. 4.41 and 4.42), one sees besides a number of metastable states a stable spiral state with $\mathbf{q}_s = (0,0,0.6)\,2\pi/a$ that is prominent at both volumes. Metastable states are seen to occur between the points L and X at $\mathbf{q}_1 = (0.35, 0.35, 0.35)2\pi/a$, at W for the larger volume, between the points X and W at $\mathbf{q}_2 = (0.25, 0, 1)2\pi/a$, and between the point W and Γ at $\mathbf{q}_3 = (0.25, 0, 0.5)\,2\pi/a$.

In an attempt to relate these theoretical findings to the experimental data of Tsunoda (1989) and Tsunoda *et al.* (1993) we face the overbinding problem inherent in the LDA. However, the fact is noteworthy that both the measured ground state and the calculated ground state at the experimental volume are noncollinear. One could argue that the Cu host with the larger lattice constant stabilizes γ-Fe in a volume larger than the

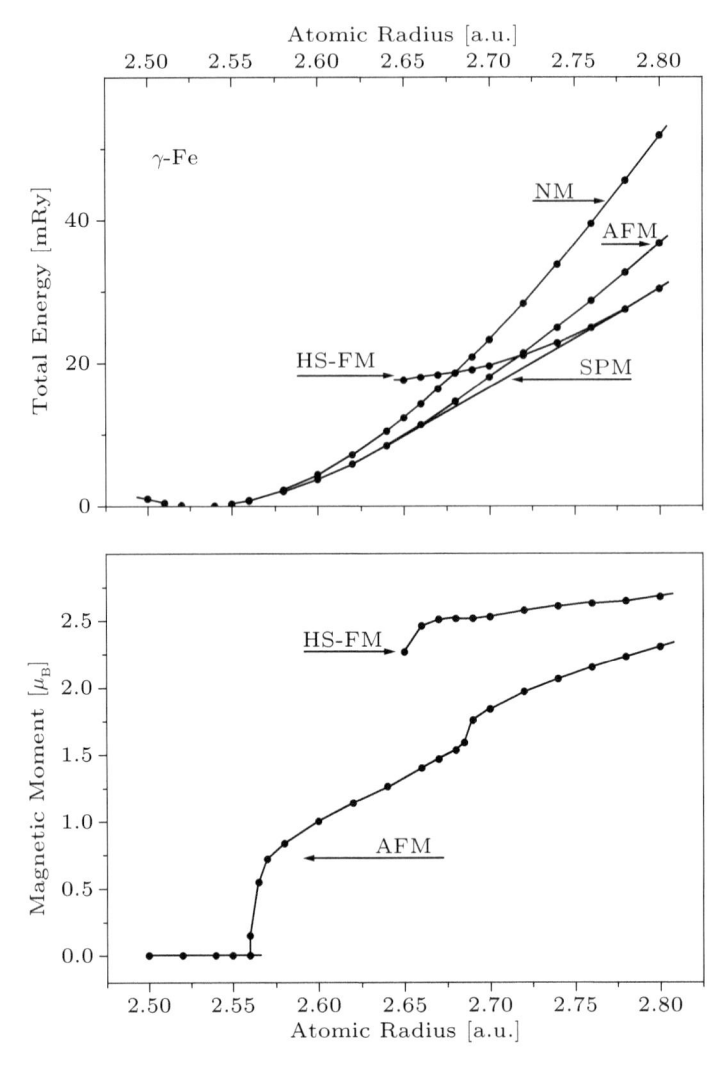

Figure 4.47 *Total energy (upper panel) of γ-Fe as a function of the volume which is labeled by the atomic sphere radius, S, in a.u. (multiples of the Bohr radius). HS-FM abbreviates the high-spin ferromagnetic phases; AFM denotes the collinear antiferromagnetic phase. The common tangent, denoted by SPM, gives the approximate total energy of a spiral magnetic state. In the lower panel the corresponding magnetic moments are given, except for those of the SPM states, whose values can be inferred from Fig. 4.48. A low-spin ferromagnetic state that is found in a very small volume range about $S = 2.67$ a.u. is not shown in either the upper and lower panels.*

4.3 Elementary magnetic metals 253

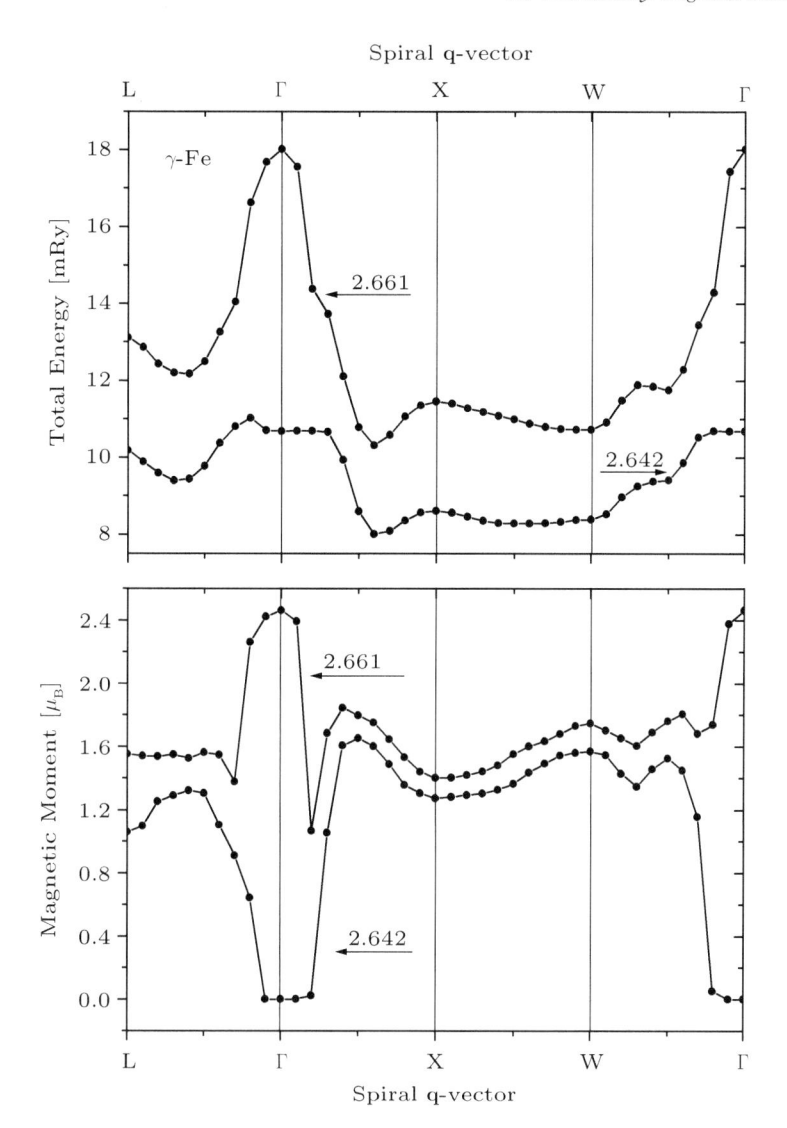

Figure 4.48 *Total energy, upper panel, and the magnetic moment, lower panel, of γ-Fe as a function of the spin-spiral **q**-vector for two volumes defined by the atomic sphere radii $S = 2.642$ a.u., corresponding to a lattice constant of $a = 3.577$ Å, and $S = 2.661$ a.u., corresponding to a lattice constant of $a = 3.603$ Å. The labels used (L, Γ, X and W) are those common for the fcc Brillouin zone (see Sec. 3.1, Fig. 3.2).*

nonmagnetic equilibrium. Next one must compare the experimental with the calculated ordering **q**-vectors which are $\mathbf{q}_{\text{exp}} = (0.1, 0, 1)\, 2\pi/a$ versus $\mathbf{q}_{calc} = \mathbf{q}_s = (0, 0, 0.6)\, 2\pi/a$. These values, obviously, do not agree, but one should notice the metastable state at $\mathbf{q}_2 = (0.25, 0, 1)\, 2\pi/a$ near the experimental $\mathbf{q}_{\text{exp}}$. Again one might question the accuracy

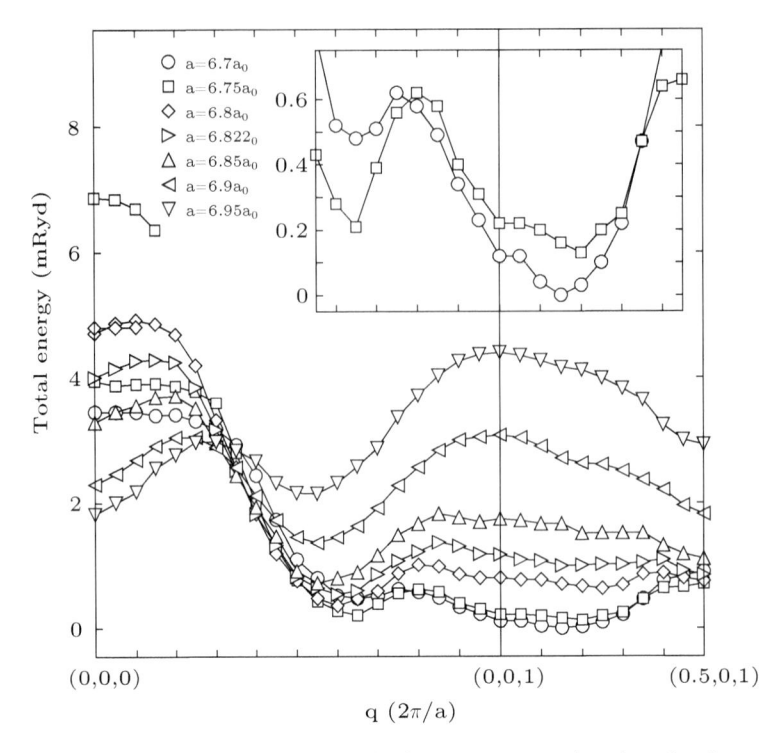

Figure 4.49 *Total energy of γ-iron for different lattice constants as a function of* **q** *along two directions in the Brillouin zone. In this case the GGA was used, after Knöpfle et al. (2000).*

of the LDA. Indeed, Körling and Ergon (1996) repeated the search for the ground state of γ-Fe using the aforementioned GGA in the atomic sphere approximation and found essentially the same scenario as that obtained by the LDA here, except for some important shifts in the relative values of the total energy, but again no agreement with the measured value of the ordering **q**-vector.

It is the ASW-GGA calculation by Knöpfle *et al.* (2000) that finally brings the theory in agreement with the experimental data. We show in Fig. 4.49 the total energy for different lattice constants as a function of the spiral vector **q** in the relevant parts of the Brillouin zone. For a lattice constant $a \leq 6.75$ a.u. we observe the formation of a total-energy minimum at $\mathbf{q} \approx (0.15, 0, 1)$ (in units of $2\pi/a$) which is in perfect agreement with experiment. The inset in Fig. 4.49 exposes the details in the region of this minimum. At $a = 6.75$ a.u. the minimum near the experimental **q** value is slightly lower than that at $\mathbf{q} = (0, 0, 0.65)$. The trend becomes much more pronounced for $a = 6.7$ a.u. Knöpfle *et al.* also show that in spite of a large interatomic angle the intra-atomic noncollinearity within the atom increases only slowly. The leading variation of the magnetization direction occurs at the border of the atomic sphere where the magnetic density is small. In the

4.3 Elementary magnetic metals 255

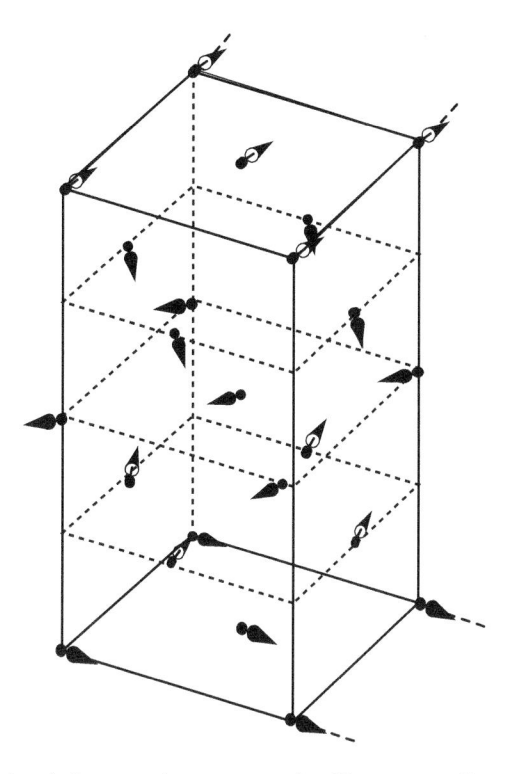

Figure 4.50 *Calculated spiral magnetic structure of γ-Fe corresponding to $\mathbf{q}_s = (0,0,0.6)2\pi/a$, showing two unit cells (after Uhl et al., 1992).*

spatial region where the 3d states have a large probability density, determining the direction of the atomic magnetic moment, the deviation of the intra-atomic magnetization from the direction of the atomic moment is small compared with the interatomic angle. These authors conclude that in the case of γ-Fe the magnetic moment, being formed by the itinerant 3d electrons, appears atomic in nature and is well defined locally.

It is worthwhile to complete the theoretical picture by showing a sketch of the magnetic structure in Fig. 4.50 in which the directions of the moments corresponding to the ordering vector $\mathbf{q} = (0,0,0.6)$ are depicted. A physical, although qualitative, reason for the occurrence of the minimum of the total energy at this particular $\mathbf{q}_s$ may, furthermore, be found in the energy-band structure of γ-Fe. The arguments are as follows.

First, to define the notation we repeat the expression for the spiral magnetization, Eqn (3.334), omitting the basis index ν which is not needed in this case,

$$\mathbf{M}_n = m\left[\cos(\mathbf{q}\cdot\mathbf{R}_n+\varphi)\sin\theta\,\hat{e}_x + \sin(\mathbf{q}\cdot\mathbf{R}_n+\varphi)\sin\theta\,\hat{e}_y + \cos\theta\,\hat{e}_z\right], \quad (4.121)$$

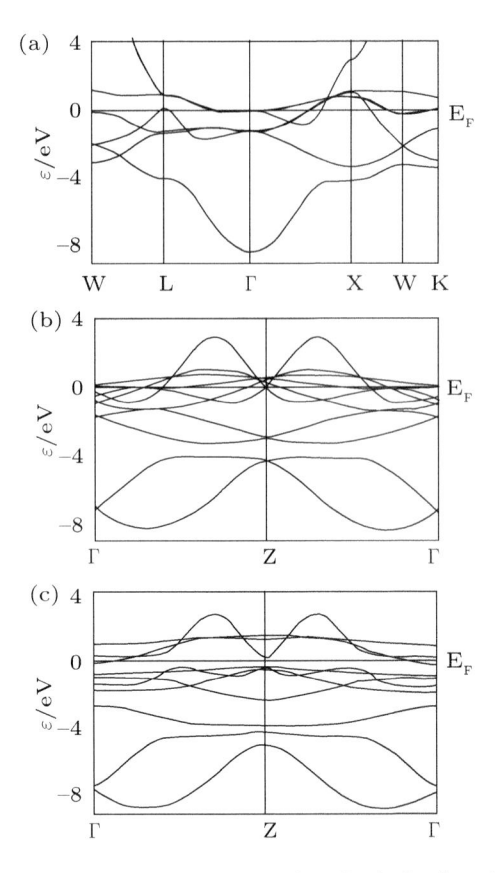

Figure 4.51 *(a) Band structure of assumed nonmagnetic γ-Fe. (b) Section of the band structure shown in (a) between Γ and X, but spin-up bands shifted relative to spin-down bands by $\mathbf{q}_s = (0,0,0.6)\, 2\pi/a$. (c) Band structure shown in (b) but now calculated with self-consistent magnetic moments (after Uhl et al., 1992).*

where m is the magnitude of the local magnetization vector, $\mathbf{M}_n$, $\mathbf{R}_n$ is a lattice translation vector, and φ is an arbitrary phase angle. Referring to Eqns (3.323) and (3.347) we see that the wave function for this spiral state can be written in the following form

$$|\psi_{\mathbf{k},\mathbf{q},\theta}\rangle = |u_{\mathbf{k},\mathbf{q},\theta}\rangle_\uparrow \begin{pmatrix} \cos(\tfrac{\theta}{2})\exp(-i\varphi/2) \\ \sin(\tfrac{\theta}{2})\exp(+i\varphi/2) \end{pmatrix} + |u_{\mathbf{k},\mathbf{q},\theta}\rangle_\downarrow \begin{pmatrix} -\sin(\tfrac{\theta}{2})\exp(-i\varphi/2) \\ \cos(\tfrac{\theta}{2})\exp(+i\varphi/2) \end{pmatrix}, \qquad (4.122)$$

where

$$\langle \mathbf{r}|u_{\mathbf{k},\mathbf{q},\theta}\rangle_\sigma = \sum_L C_{L\sigma}\tilde{H}_{L\sigma}(\mathbf{r}) + \sum_{LL'\sigma'} C_{L\sigma} G_{L\sigma L'\sigma'}(\mathbf{k},\mathbf{q},\theta)\tilde{J}_{L'\sigma'}(\mathbf{r}). \qquad (4.123)$$

4.3 Elementary magnetic metals 257

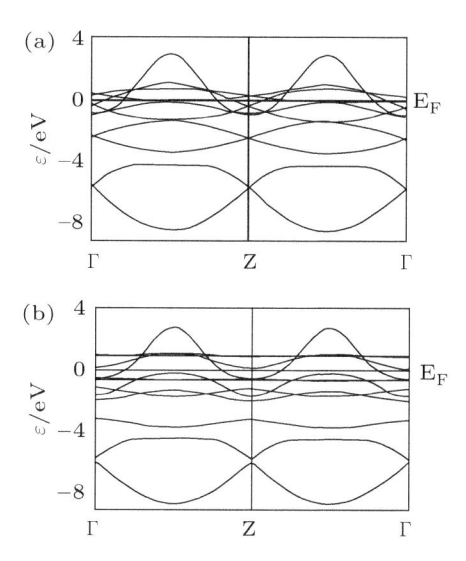

Figure 4.52 *(a) As Fig. 4.51(b), but spin-up band shifted relative to spin-down bands by* $\mathbf{q}_{AF} = (0,0,1)2\pi/a$. *(b) As Fig. 4.51(c), but moment and exchange splitting corresponding to the antiferromagnetic state defined by* $\mathbf{q}_{AF}$ *(after Uhl et al., 1992).*

Here $\tilde{H}_{L\sigma}(\mathbf{r})$ is the augmented Hankel function and $\tilde{J}_{L'\sigma}(\mathbf{r})$ the augmented Bessel function (see Sec. 3.4.7). The coefficients $C_{L\sigma}$ are obtained variationally from the secular equation and the quantity $G_{L'\sigma L\sigma'}(\mathbf{k},\mathbf{q},\theta)$ is defined by

$$G_{L'L}(\mathbf{k},\mathbf{q},\theta) = U \begin{pmatrix} B_{L'L}(\mathbf{k}-\frac{1}{2}\mathbf{q}) & 0 \\ 0 & B_{L'L}(\mathbf{k}+\frac{1}{2}\mathbf{q}) \end{pmatrix} U^+, \qquad (4.124)$$

where the $B_{L'L}(\mathbf{k})$ are the KKR structure constants defined by Eqn (3.123) and U, U^+ are the spin-1/2 rotation matrices, Eqn (2.186).

We are now ready to show in Fig. 4.51(a) the band structure of γ-Fe in the standard representation, but assuming for the moment a nonmagnetic state. Let us focus our attention on the Fermi energy, E_F, and the typical upturn of the e_g-band above E_F at the X-point. This feature is repeated in Fig. 4.51(b) but now shifting the degenerate spin-up band relative to the spin-down bands by the vector $\mathbf{q}_s$. This representation of the band structure is in accord with the wave function above. In Fig. 4.51(b) we see that after the shift of the nonmagnetic bands we get an intersection of bands at the point $X \equiv Z$ just at the Fermi level. Both bands are characterized by opposite spin projections. The situation changes when the spin polarization is taken into account; the augmented Bessel and Hankel functions in Eqn (4.123) become dependent on the spin projection when the magnetic moment is determined self-consistently and the intersecting bands hybridize and whence are repelled. Figure 4.51(c) demonstrates this and shows that the states below the Fermi energy, E_F, shift to lower energies and the states above E_F to higher energies. Thus this particular portion of the bands is seen to lead to a lowering of the total energy.

We repeat in Fig.4.52(a) and (b) the same steps for the antiferromagnetic state described by the vector $\mathbf{q}_{AF} = (0,0,1)2\pi/a$ and see by comparing with Fig. 4.51(b) and (c) that the energy gain mechanism is less effective for $\mathbf{q}_{AF}$ than for $\mathbf{q}_s$.

4.4 Magnetic compounds

When two or more elementary metals are brought together to form intermetallic compounds we are faced with a vast amount of possible topics. We might begin by asking what is new and different compared to the elementary constituents. Thus the evolution of the electronic structure of the compounds from that of the constituents is of primary interest and there is a concomitant connection with the magnetic properties. We might as well ask what are the *positions* of the nuclei in an alloy of a given composition and how much—if at all—they are influenced by the magnetic properties, thus inquiring into the connection of the crystallographic structure with magnetism. Another modern topic that is not only of interest in basic magnetism but also begins to acquire technological importance is concerned with systems that are artificially made by sputtering, molecular beam epitaxy or other experimental techniques. These "tailored magnets" do not essentially depend on chemistry as alloys or compounds and show unusual physical properties that we will briefly discuss in Sec. 4.5.

To begin with, we limit our attention to binary compounds made of the elements A and B and simply assume the compound $A_{1-x}B_x$ to exist in some ordered crystal structure or as a disordered alloy. We then turn our attention to the magnetic properties of $A_{1-x}B_x$. We later come back to a short discussion of the phase stability and close this section by looking at some particular magnetic compounds in detail.

4.4.1 The Slater–Pauling curve

Basic ground-state properties of alloys or compounds are the values of their saturation magnetization. The collection of these data for a great number of transition metal alloys are displayed by what is traditionally called the Slater–Pauling curve, where the average saturation magnetization of ferromagnets is plotted as a function of the electron concentration. The regularities and trends in the Slater–Pauling curve are perhaps the most prominent properties of itinerant-electron ferromagnets; they have from the very beginning been ascribed to broad band-structure features, albeit initially in misleading ways.

In Fig. 4.53 we show this curve taken from the book of Bozorth (1951). The Slater–Pauling curve has been reprinted in many other books and can be found for example in Kittel's *Introduction to Solid State Physics* (1966). A somewhat richer version appeared

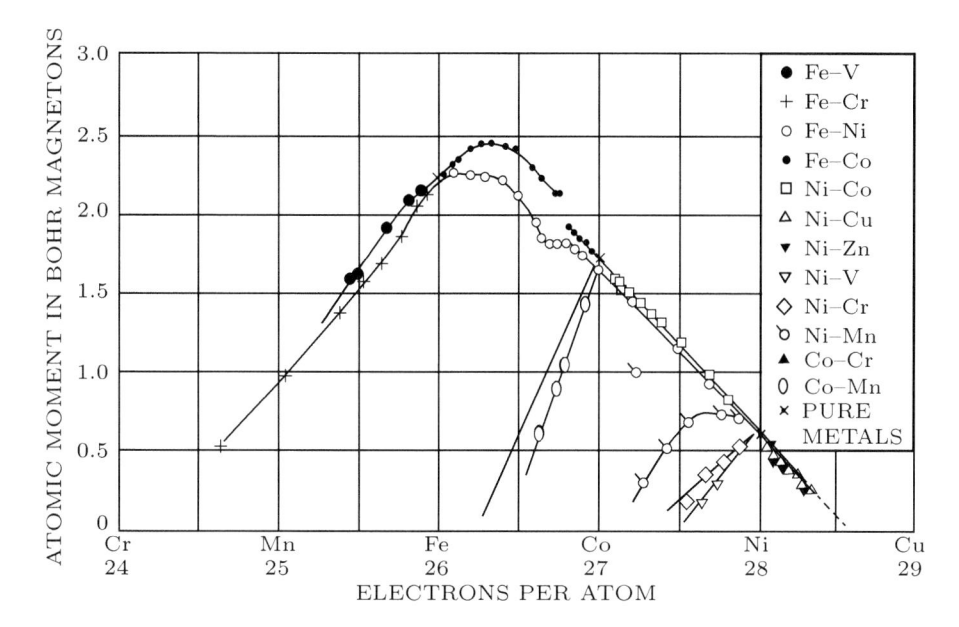

Figure 4.53 *Slater–Pauling curve: average saturation magnetization of Fe, Co, and Ni based alloys (or compounds) as a function of the average number of electrons per atom; from Bozorth (1951).*

in a paper by Dederichs *et al.* (1991) containing newer but unreferenced experimental data. What we can easily identify is quite regular behavior, like the magnetization of NiCu, NiCo, and FeNi alloys on the right side of Fig. 4.53 and that of FeV and FeCr alloys on the left, the latter being bcc coordinated and the former fcc coordinated. But striking departures from this regular behavior occur like, e.g. the magnetization of CoCr alloys, etc.

A short *historical review* of interpretations of the Slater–Pauling curve will clarify the physics here and set into proper perspective a simple and modern approach due to Williams *et al.* (1983b, 1984) and Malozemoff *et al.* (1983, 1984a,b) that results in a transparent generalization of the theory and gives its connection with the underlying electronic band structure. Our discussion here generally follows Malozemoff *et al.* (1984a).

Considering the alloy $A_{1-x}B_x$ we may call A the host (for instance Fe, Co, or Ni) and B the solute (either another transition metal or a metalloid). We will use the terms "host" and "solute" quite indiscriminately although they are plausible only when x is small. Quite generally, its *atom-averaged* moment expressed in multiples of the Bohr magneton (μ_B) is the difference of the *atom-averaged* number of spin-up and spin-down electrons, i.e.

$$M = (1-x)\, M_A + x\, M_B = N_\uparrow - N_\downarrow. \tag{4.125}$$

Here we denote by M the atom-averaged moment and by M_A, M_B the moments per atom of the host and solute, respectively, $N_\uparrow$ and $N_\downarrow$ being the atom-averaged number of spin-up and spin-down electrons. If we denote the average electronic valence, that is, the number of electrons outside the last filled shell, by Z then, clearly,

$$Z = (1-x)Z_A + x Z_B = N_\uparrow + N_\downarrow, \tag{4.126}$$

where Z_A and Z_B are the valences of the host and solute, respectively. One can trivially eliminate either $N_\uparrow$ or $N_\downarrow$ from these equations to obtain

$$M = 2 N_\uparrow - Z \tag{4.127}$$

or

$$M = Z - 2 N_\downarrow. \tag{4.128}$$

Now let us consider some special cases. Assuming $N_\uparrow$ is constant under alloying and identifying from Eqn (4.127) the pure host moment as

$$M_A^0 = 2 N_\uparrow - Z_A \tag{4.129}$$

we find from Eqn (4.127)

$$M = M_A^0 - x(Z_B - Z_A). \tag{4.130}$$

Alternatively, assuming $N_\downarrow$ is constant and redefining M_A^0 appropriately, we find

$$M = M_A^0 + x(Z_B - Z_A). \tag{4.131}$$

Of course, the assumptions concerning $N_\uparrow$ and $N_\downarrow$ are crucial and must be discussed, but what we can read off from Eqns (4.130) and (4.131) is the change of sign in the concentration dependence. This sign change is the origin of the triangular shape of the Slater–Pauling curve (Fig. 4.53). In other words, the two sides (but obviously not the distinct deviations) correspond to either spin-up or spin-down electron numbers remaining constant. Take for example as host Co and as solute Ni for which $Z_B - Z_A = 1$. Then Eqn (4.130) describes the Ni-Co data in Fig. 4.53 decreasing from approximately 1.6. Alternatively, for Fe as host and Cr as solute we have $Z_B - Z_A = -2$ and Eqn (4.131) describes the Fe-Cr data decreasing from approximately 2.2 on the left side of Fig. 4.53. We are now in the position to discuss some earlier theories of ferromagnetism, one of which was the rigid-band theory due to Mott (1935), Slater (1936), and Pauling (1938). It presumes that the d and sp electrons of the alloy form a common band invariant under alloying. This band is filled by electrons according to the average chemical valence. Ferromagnets were called *strong* when the spin-up d band is full, as in Co and Ni. This situation is depicted for Co in Fig. 4.54 which is the result of an LDA-ASW

4.4 Magnetic compounds

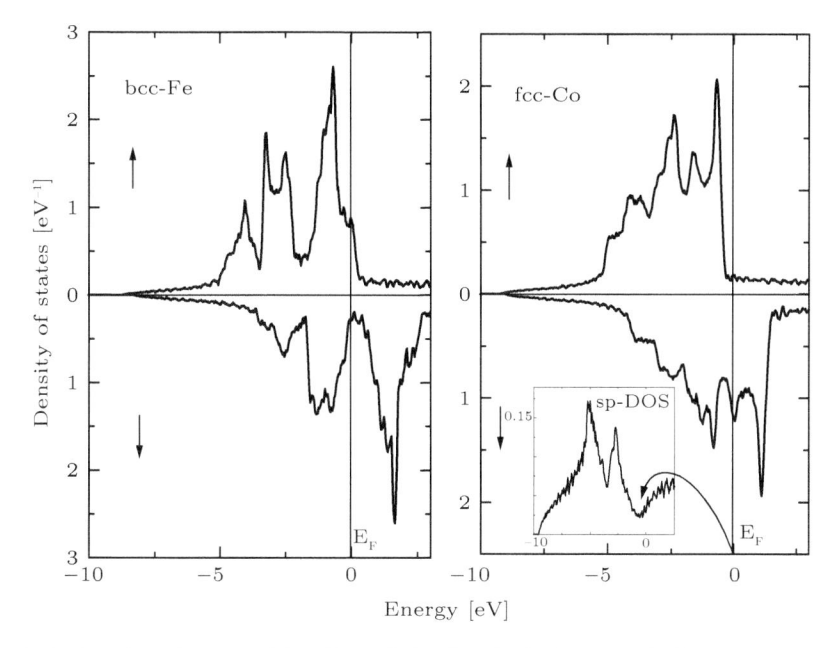

Figure 4.54 *Density of states of bcc Fe and fcc Co obtained by LDA-ASW calculations using experimental values for the atomic volumes. Inset in the Co figure shows the density of states of the s and p electrons.*

calculation. The filled spin-up d band can be easily recognized by its steep upper band edge approximately 0.5 eV below E_F. Writing then $N_\uparrow = N_{d\uparrow} + N_{sp\uparrow}$, we see that the part due to the d electrons, $N_{d\uparrow}$, does not change with alloying and $2N_{d\uparrow} = 10$. Thus Eqn (4.127) becomes

$$M = 10 - Z + 2N_{sp\uparrow}. \quad (4.132)$$

In his original work, Mott (1935) also assumed $N_{sp\uparrow}$ is constant, although he had more difficulty justifying this assumption. Thus he arrived at Eqn (4.130), which became known as the *rigid-band formula*. We hasten to add that this is not the only way to derive the result as we shall see. The simple Eqn (4.130) works well for Ni-Cu, Ni-Zn, Ni-Co, Co-rich Fe-Co, and Ni-rich Fe-Ni forming the right-hand side portion of the Slater–Pauling curve.

An equally simple theory has been applied to bcc Fe based alloys (Mott, 1964; Berger, 1965). Alloys such as Fe-Cr and Fe-rich bcc Fe-Co are assumed to form rigid bands having the Fermi energy pinned in a deep valley in the bcc spin-down band. The situation was pictured very much like in bcc Fe, for which an LDA-ASW band structure easily supplies density of states as those shown for convenience here in Fig. 4.54.

Splitting up the density of states again as $N_\downarrow = N_{d\downarrow} + N_{sp\downarrow}$ we may assume that the pinned Fermi energy assures that $N_{d\downarrow}$ remains constant under alloying. If, for some reason, $N_{sp\downarrow}$ also remains constant, we get Eqn (4.131). This theory explains the bcc Fe-rich Fe-Co, the Fe-Cr, and the Fe-V alloys on the left side of Fig. 4.53.

In spite of the success of rigid-band theory, improved theoretical and experimental techniques showed that the bands were simply *not* rigid, even for the simplest cases. A thorough review of the basic experimental facts, especially those concerning photoemission, and of the theory of the electronic structure of alloys, is that by Ehrenreich and Schwartz (1976).

The first explanation of the Slater–Pauling curve that did not use the rigid-band theory came from the work of Friedel (1958) who exploited the concept of *screening* (see also Sec. 2.8.4) to explain Eqn (4.130). The argument is that the system being metallic cannot sustain an internal electric field of long range, thus the excess charge of an impurity must be screened out on an atomic scale. A glance at the Co density of states in Fig. 4.54 reveals that the charge available for this process will come from the states of high density of the minority-spin electrons thus changing $N_\downarrow$ and leaving the up-spin electrons unchanged. This obviously leads to Eqn (4.130).

Another crucial observation of Friedel concerns the physics of impurities consisting of early transition metal atoms. When they are introduced into the host they will reduce the number of up-spin d-electrons, $N_{d\uparrow}$, but this reduction in $N_{d\uparrow}$ is not by a general amount, but by precisely five. The reason is the strongly repulsive potential of the impurity which pushes the up-spin d electrons to a position above the Fermi energy. None of the early works, however, succeeded completely in explaining the constancy of the spin-up sp electrons when *metalloids* were added to the transition metal ferromagnets.

It is through the work of Williams *et al.* (1983b, 1984) and Malozemoff *et al.* (1983, 1984a,b) that a unified theory of the Slater–Pauling curve emerged. They incorporated the findings of Terakura and Kanamori (1971) (see also Terakura, 1976, 1977) who showed how it is possible that adding a large number of metalloid sp electrons to a transition metal does not increase the value of $N_{sp\uparrow}$ appearing in the expression of the magnetization, Eqn (4.132). Their theory shows that, first, there are two mechanisms by which the valence s and p states can be created. One is that states in addition to those originally below the Fermi level are simply pulled down below the Fermi level, E_F, by the attractive metalloid potential thus increasing $N_{sp\uparrow}$. The other is polarization of neighboring occupied states of the host, i.e. states centered on the solute are created of linear combinations of states originally centered on nearest neighbor sites; this will not change $N_{sp\uparrow}$ since states below E_F are simply rearranged (note, nothing is said about the spin-$\downarrow$-electrons). Second, and more importantly, Terakura and Kanamori showed that the effect of polarization largely dominates the creation of new states from above E_F. The reason is that the system can only respond to a perturbation if states of the correct symmetry and spatial distribution are energetically available. Thus, if there is a gap in the spectrum of such states, then the perturbation must be very strong to displace states across the gap. An important property of the host electronic structure is hybridization of sp with d states that results in a depression near the Fermi level of the sp density of states, thus creating an approximate gap in the sp spectrum. It is this gap that holds $N_{sp\uparrow}$ nearly constant, by forcing the transition metal host to screen the solute sp potential by means of the polarization mechanism. Terakura and Kanamori referred to this gap as a "Fano-antiresonance." Once it is known what to look for, one indeed easily finds this gap in an electronic-structure calculation. This is demonstrated by the inset in Fig. 4.54 where the density of states of the sp electrons (sp DOS) near the Fermi level is plotted. Since, however, this gap is not perfect, the value of $N_{\text{sp}\uparrow}$ may still change to some extent.

Having thus explained the constancy of $N_{\text{sp}\uparrow}$ and keeping in mind the abrupt change of $N_{\text{d}\uparrow}$ when passing from the late to the early transition metals, we now define the *magnetic valence*, Z_m, as the contribution of the two integers $N_{\text{d}\uparrow}$ and Z to the magnetization; that is, Eqn (4.127) gives

$$M = Z_m + 2 N_{\text{sp}\uparrow}, \tag{4.133}$$

if we define (Williams et al., 1983b; Malozemoff et al., 1984a,b)

$$Z_m = 2 N_{\text{d}\uparrow} - Z. \tag{4.134}$$

Here the value of $N_{\text{d}\uparrow}$ is 0 for increasing Z up to iron where it becomes $N_{\text{d}\uparrow} = 5$. For easy use we display a portion of the periodic table of the elements in Table 4.11 giving the values of the chemical valence, Z, and the magnetic valence, Z_m. The magnetic valences for Ru and Rh are put in parentheses because their values are ambiguous. For more detail see Williams et al. (1984). In the formula for the magnetization, Eqn (4.133), one then puts $N_{sp\uparrow} \simeq 0.3$ for the late transition metals and calculates the average magnetic valence of a compound or alloy consisting of n elements with atom fraction x_i, $i = 1, \ldots, n$ by means of

$$Z_m = \sum_{i=1}^{n} x_i Z_m^i \tag{4.135}$$

taking Z_m^i from Table 4.11.

This prescription gives for the pure metals Co and Ni a magnetic moment of $\sim 1.6\,\mu_{\text{B}}$ and $\sim 0.6\,\mu_{\text{B}}$, respectively, and for Fe $\sim 2.6\,\mu_{\text{B}}$. The fact that experimentally the magnetic moment of Fe is not 2.6 μ_{B} but rather 2.2 μ_{B} reveals the weakness of Fe:

Table 4.11 *Section of the periodic table of the elements containing the number of valence electrons (maximum chemical valence, number on left) and magnetic valence (number on right)*

										Al	Si	P
										3 −3	4 −4	5 −5
Sc	Ti	V	Cr	Mn	Fe	Co	Ni	Cu	Zn	Ga	Ge	As
3 −3	4 −4	5 −5	6 −6	7 −7	8 2	9 1	10 0	1 −1	2 −2	3 −3	4 −4	5 −5
Y	Zr	Nb	Mo	Tc	Ru	Rh	Pd	Ag	Cd	In	Sn	Sb
3 −3	4 −4	5 −5	6 −6	7 −7	8 (2)	9 (1)	10 0	1 −1	2 −2	3 −3	4 −4	5 −5

the penetration of the up-spin d band by the Fermi level, see Fig. 4.54. We now may plot the magnetization of compounds or alloys as a function of the magnetic valence and call the result the generalized Slater–Pauling curve.

4.4.1.1 Generalized Slater–Pauling curve

We show a portion of the generalized Slater–Pauling curve on the left side of Fig. 4.55.

The data that appear here contain as a subset the data on the right of the conventional Slater–Pauling curve, Fig. 4.53. It is seen at once that the deviations from regular behavior have disappeared except for the Fe-Ni systems, for which alloys in the Invar composition range (~25% Ni) are anomalous. This departure is caused by the onset of magnetic weakness as the Fermi level moves out of the state-density peak responsible for magnetism in the fcc coordinated alloys. It is noted, furthermore, that the data points parallel to but above the 45° line indicate values of $N_{sp\uparrow}$ that vary reaching values of 0.45 and in some cases being even larger, especially when the magnetization becomes small.

To support the physical picture Williams *et al.* (1983b, 1984) and Malozemoff *et al.* (1983, 1984a,b) carried out numerous band-structure calculations (see especially Malozemoff *et al.*, 1984b). These showed clearly that there is no rigid-band behavior; the band structure of any combination of these elements is very complicated, due to hybridization of the various d bands with one another and with the sp band. All this

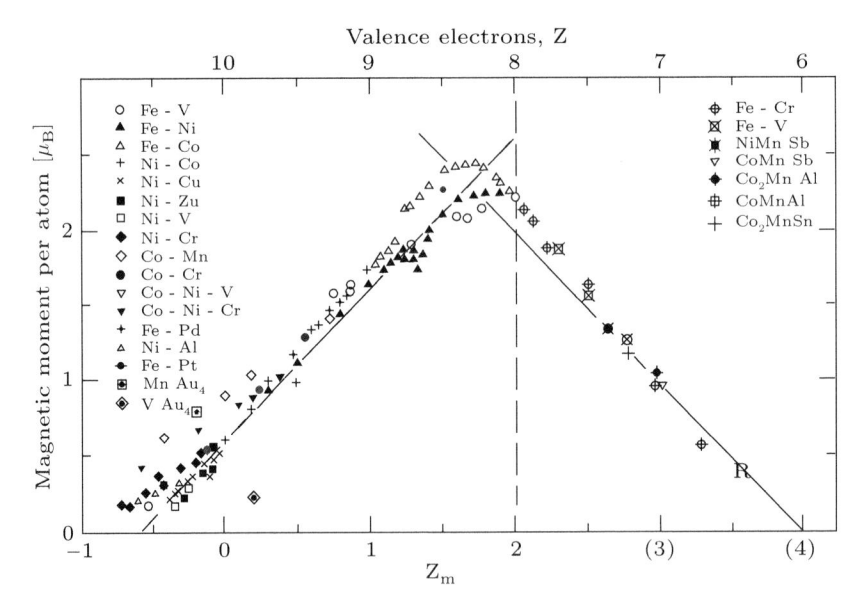

Figure 4.55 *Generalized Slater–Pauling curve. Z_m is the magnetic valence defined up to the value 2, see text. The left-hand side of the curve gives the average magnetic moment as a function of the average magnetic valence. The data have, with some exceptions, been taken from Williams et al. (1983b). The right-hand side gives the average magnetic moment as a function of the average valence, Z, see text. The data have been taken from Kübler (1984b) and from the original Slater–Pauling curve, Fig. 4.53.*

complication is, however, irrelevant to the magnetization, which depends on the fact that these bands lie below the Fermi energy and contribute five or zero electrons (per atom) to $N_{d\uparrow}$ while the gap stabilizes the number $N_{sp\uparrow}$.

We have chosen to illustrate the case when $N_{d\uparrow}$ = const. in the definition of magnetic valence and take as an example $Co_{1-x}Cr_x$ for $x = 0.2$. We calculate trivially that $Z_m = -0.4$, thus we expect the average magnetization to be 0.2 – 0.5 μ_B depending on the value of $N_{sp\uparrow}$. For the electronic structure calculation it is easiest to assume an ordered compound for which we take the Ni_4Mo crystal structure. It consists of a body-centered tetragonal unit cell with $c/a = \sqrt{2/5}$ that is constructed from six face-centered cubic unit cells. The body center is occupied with Cr, all other sites with Co (see e.g. Kübler, 1984). The calculation is carried to self-consistency varying the volume to approximately attain the total energy minimum which results in $a = 5.468$ Å. The ratios of atomic sphere radii are chosen according to Vegard's law resulting in nearly neutral spheres ($\sim \pm 0.1$ electronic charge). The magnetic moment calculated is for Co, $M_{Co} = 0.7\,\mu_B$ and for Cr, $M_{Cr} = -0.53\,\mu_B$, thus giving an average moment of $M = 0.45\,\mu_B$ per atom. The average moment is in good agreement with the estimate above. We graph the state density of $Cr_{0.2}Co_{0.8}$ in Fig. 4.56 showing also the partial Cr d density of states and illustrating the gap by showing the sp density of states in the inset. Although, clearly, the Cr d state density strongly peaks in both spin directions above the Fermi energy, E_F, its up-spin density below E_F is *not* zero due to hybridization with the Co d density. However, the down-spin state density of Cr below E_F is larger than that of the up-spin electrons, resulting in the negative magnetic moment (antiparallel to the Co moments) thus mimicking a situation which looks like $N_{d\uparrow}$ = const. This example shows how a complicated situation is described in simple terms.

Using different *ab initio* computational techniques Dederichs *et al.* (1991) extract essentially the same physical picture. In an interesting and exhaustive review they show the results of KKR Green's function calculations for dilute alloys and for concentrated disordered alloys using a self-consistent KKR-CPA method. The methodology used in this work is briefly reviewed in Sec. 5.5.2 of this treatise. It is rewarding to see that a complementary approach leads essentially to the same understanding.

We next discuss a number of peculiar cases. Among them are the systems $MnAu_4$ and VAu_4 which we placed on (or near) the generalized Slater–Pauling curve. Experimentally they are magnetic (Adachi *et al.*, 1980) but with the magnetic valences from Table 4.11 they are not. An electronic structure calculation for VAu_4 (Kübler, 1984a) revealed that the repulsive potential of V apparently is not strong enough in gold to push the V d ↑ bands above the Fermi level and, as a consequence, V acquires a calculated magnetic moment of 1.8 μ_B giving an average moment of 0.36 μ_B which should be compared with one of the experimental values of $\sim 0.2\,\mu_B$. As with the case of chemical valence, the magnetic valence of an atom may depend on the environment, so if we assume V in Au not to have the value quoted in Table 4.11, but rather with $2\,N_{d\uparrow} = 10$ the value $Z_m = 5$, we estimate for VAu_4, $Z_m = 0.2$ which is the coordinate used in Fig. 4.55. The case of $MnAu_4$ may be treated in a similar way by assigning the value $Z_m = 3$ to the magnetic valence of Mn in gold. Note, however, not all ferromagnetic systems find their place in Fig. 4.55; thus, for instance, a well-known case with interesting magneto-optical

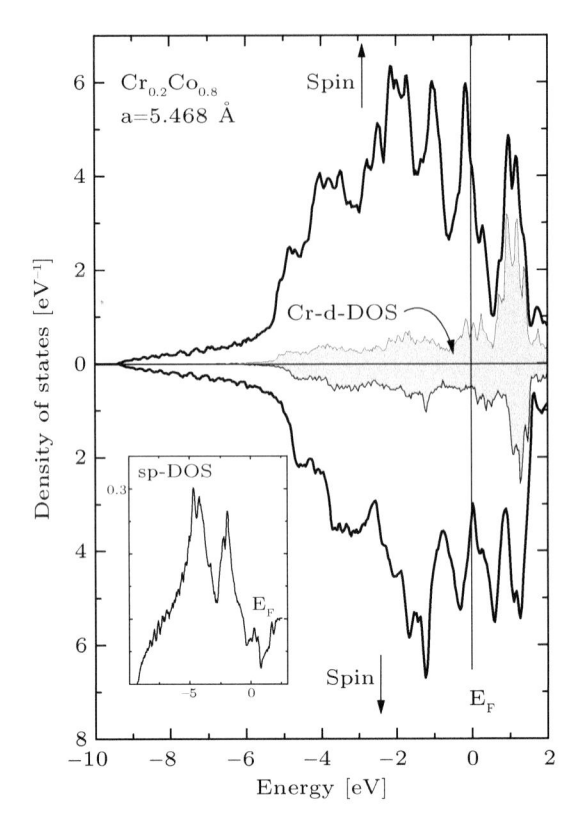

Figure 4.56 *Spin-up and spin-down state densities of $Co_{0.8}Cr_{0.2}$ and those of the d states of Cr (shaded) calculated in the Ni_4Mo crystal structure. Inset: density of states of sp electrons.*

properties is MnBi (Chen and Stutius, 1974; Coehoorn and de Groot, 1985; Köhler and Kübler, 1996) which possesses an average magnetization of $M = 1.92\,\mu_B$, but cannot plausibly be explained with the concept of magnetic valence.

The sensitivity of V to the environment is illustrated by another example. Some time ago, a prediction was made about VPd_3; Williams *et al.* (1981), on the basis of LDA-ASW band-structure calculations, suggested that it, too, should be ferromagnetic, but experimentally it was found to be nonmagnetic (Burmeister and Sellmyer, 1982). The discrepancy finds a simple explanation in that the calculations were carried out assuming the crystal structure to be Cu_3Au while the experimentally observed structure is $TiAl_3$ which differs from the former only by another next-nearest neighbor arrangement. For the latter structure, band calculations correctly give a nonmagnetic solution (Kübler, 1984b).

The most exhaustive and useful application of the generalized Slater–Pauling concept concerns alloy systems of Fe, Co, and Ni with *metalloids*, both in crystalline and amorphous form. Extensive documentation for $Ni_{1-x}B_x$, $Ni_{1-x}P_x$, $Co_{1-x}M_x$ with

$M = B$, Si, Sn, P, and Au, $Fe_{1-x}M_x$ with $M = $ Al, Au, B, Ga, Be, Si, C, Ge, Sn, and P can be found in the original publication by Malozemoff *et al.* (1984a) who also extract typical values for $2N_{sp\uparrow}$ for the Fe and Co systems.

4.4.1.2 Constant minority-electron count

We now turn to the right part of the generalized Slater–Pauling curve shown in Fig. 4.55 which is characterized by a constant value of $N_\downarrow$, Eqn (4.128) and corresponds to the left side of the conventional Slater–Pauling curve, Fig. 4.53. In this case the reason for the constant $N_\downarrow$ is a deep valley in the down-spin host state density that is quite common for the bcc coordinated systems. If we assume $N_\downarrow = 3$ in alloying we get the straight $-45°$ line in Fig. 4.55. We may illustrate the valley for the case of FeCr which we easily simulate using the CsCl crystal structure. Minimizing the total energy we obtain a lattice constant of $a = 2.810$ Å and an average moment of $M = 0.9\,\mu_B$ which is to be compared with the value of $M = 1\,\mu_B$ from Fig. 4.55 with $Z = 7$. An experimental value is close by. The density of states shown in Fig. 4.57 clearly reveals the strong valley in the down-spin spectrum and is especially pronounced in the host's partial Fe d ↓ DOS. The value of $N_\downarrow = 3$ comes about because of the spin-down states only those of t_{2g} character are occupied.

Besides some binary compounds, the generalized Slater–Pauling curve also contains ternaries; those shown are Heusler alloys. Heusler alloys (or Heusler compounds as they are frequently called) are magnetic intermetallic compounds, usually containing Mn. They either have the $L2_1$ or the $C1_b$ crystal structure (which are cubic) with the generic formula X_2MnY or XMnY, respectively, where $X = $ Co, Ni, Cu, Pd, Pt, etc. and $Y = $ Al, Sn, In, Sb. The crystal structure of the Heusler compounds in the $L2_1$ structure is shown in Fig. 4.58. The $C1_b$ structure is obtained from $L2_1$ by removing every second atom of the inscribed cube such that a tetrahedron is formed by the X-atoms. However, the magnetization of only a small subset of them is explained by the Slater–Pauling curve, but those that are appear quite interesting. Thus in the group X_2MnY the compounds Co_2MnAl and Co_2MnSn possessing an average magnetization of 1 μ_B and 1.25 μ_B, respectively, agree with the Slater–Pauling curve. Extensive calculations for a large number of Heusler compounds by Kübler *et al.* (1983a) revealed a strong valley, or nearly a perfect gap, in the down-spin state density of these two compounds. We will take up the discussion of these interesting cases on the basis of new publications in Section 4.4.2.2. Here we first continue with the compounds having the $C1_b$ structure, where the "near-gap" turns into a perfect gap in cases like NiMnSb and PtMnSb, as was first discovered by de Groot *et al.* (1983a). Since the down-spin electrons possess all the features of a semiconductor, the up-spin electrons, however, being perfectly metallic, de Groot *et al.* named these compounds *half-metallic ferromagnets*. Because of the gap the magnetization of the cell must be an integer and is precisely 4 μ_B which places NiMnSb, PdMnSb, and PtMnSb in Fig. 4.55 with an average moment of 1.33 μ_B at the spot marked NiMnSb on the right side of the generalized Slater–Pauling curve. The experimental values also given by de Groot *et al.* (1983a) are in good agreement with the theoretical predictions. We select one of these compounds and show in Fig. 4.59 the total spin-up and spin-down state densities of PtMnSb, illustrating the Mn d spectrum as shaded curves. The gap in

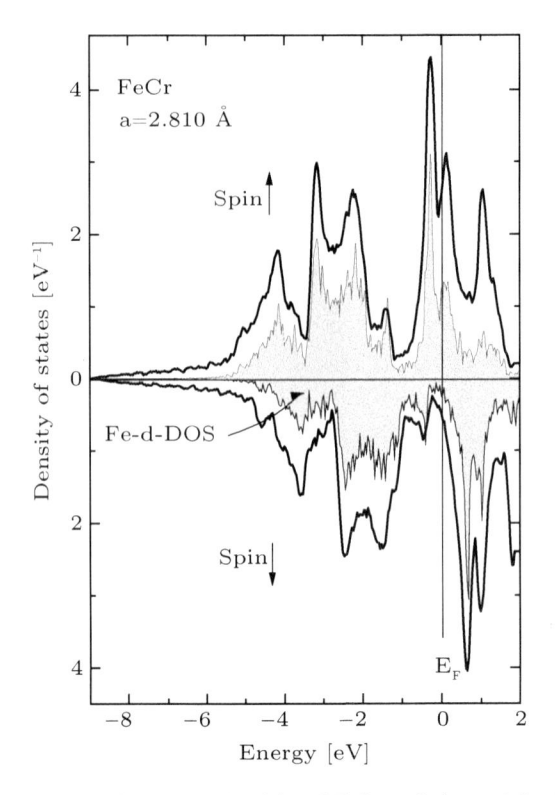

Figure 4.57 *Spin-up and spin-down state densities of FeCr and those of the d states of Fe (shaded) calculated in the CsCl crystal structure.*

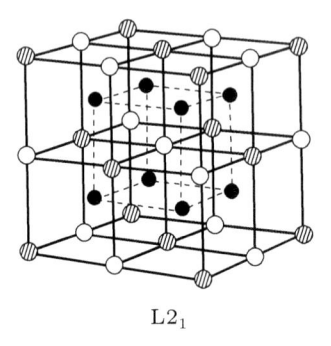

L2$_1$

Figure 4.58 *Crystal structure of a Heusler compound, X_2YZ, in the $L2_1$ structure. The Bravais lattice is fcc, the X-atoms occupy the black corners of the inscribed cube, Y go into the shaded corners of the large cube, and Z occupy the white center sites, both forming an NaCl - structure*

the down-spin electrons is clearly recognizable. We furthermore see that the Mn d spin-down electrons have very little weight below the Fermi energy. They are thus excluded from the Mn sphere and we infer that the result of this localized exclusion is an equally

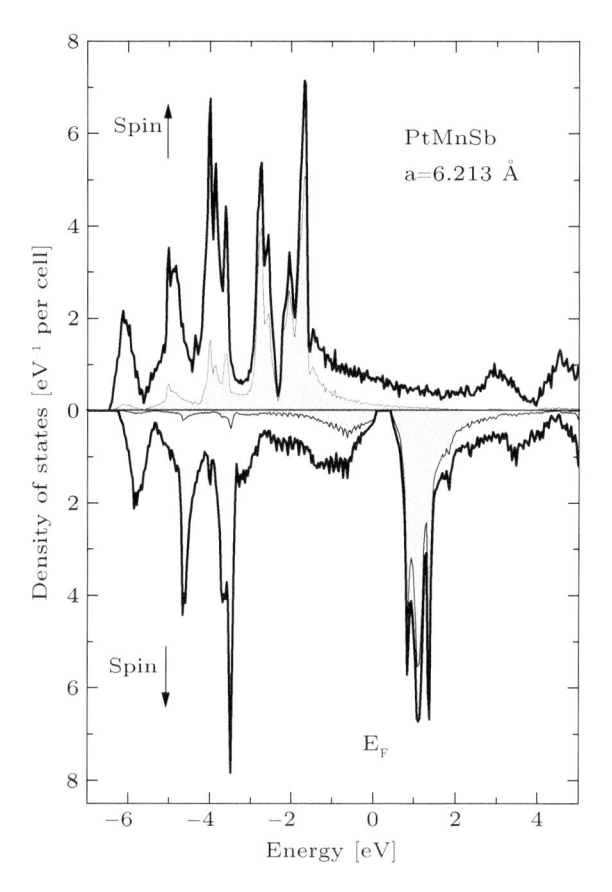

Figure 4.59 *Spin-up and spin-down state densities of PtMnSb and those of the d states of Mn (shaded) calculated in the $C1_b$ crystal structure.*

localized region of magnetization leading to a highly localized magnetic moment (Kübler et al., 1983).

Next we make an attempt to understand the mechanism responsible for the semiconducting gap in the down-spin spectrum. De Groot et al. (1983a) published the band structure of PtMnSb which shows that in the down-spin manifold nine electrons fill exactly one s band, the three p bands, and the five d bands resulting in $N_\downarrow = 3$ per atom and Eqn (4.128) gives with $Z = 7.333$ for the magnetization the value given above. The magnetic moment per cell, 4 μ_B, is almost perfectly localized on Mn. Next, to test the stability of the occupied down-spin bands, we may calculate the band structure for CoMnSb and find again a half-metallic ferromagnet for which the magnetization follows Eqn (4.128) again (Kübler, 1984b) giving a magnetic moment of 3 μ_B per cell. Except for a change in the bandwidth the occupied part of the down-spin band structure is essentially unchanged. Indeed, this can be continued (de Groot, 1991)

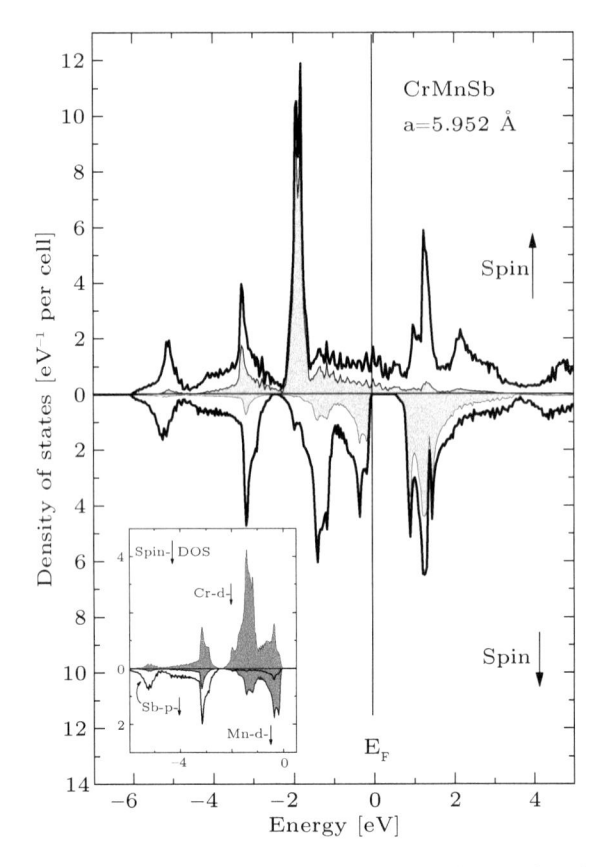

Figure 4.60 *Spin-up and spin-down state densities of CrMnSb and those of the d states of Mn (shaded) calculated in the $C1_b$ crystal structure. Inset: Hybridization of the spin-down bands of Cr d (shaded, upper panel) with Mn d (shaded, lower panel) and Sb p states (lower panel, line drawing).*

with FeMnSb, MnMnSb, and finally CrMnSb. Unfortunately, these systems cannot be experimentally produced in the $C1_b$ crystal structure that we assumed, but the result is astonishing: all these "computer" compounds are half-metallic ferromagnets, the moments per cell decreasing in integer steps until the value zero is reached for CrMnSb, the average moment following from Eqn (4.128). Obviously, CrMnSb must have exactly compensating magnetic moments and is, actually, a rather unusual antiferromagnet. We show its density of states in Fig. 4.60 for a calculated lattice constant. The size of the gap in the down-spin spectrum has increased compared with PtMnSb and the inset shows the strong hybridization of the Cr d states with those of the Sb p and Mn d electrons. Note how well the peaks coincide. The hybridization is so strong that we may speak of a Cr (in the other cases Mn, Fe, Co, Ni, etc.) *induced* Mn-Sb *covalent interaction*. We have chosen to show the results for CrMnSb because they are most pronounced in this case, but they are also clearly discernible in the other systems and were previously exhibited for CoMnSb (Kübler, 1984b).

It is of course a challenge to verify if the band structure predicted and described here is realized in nature. A number of experiments were conducted for this purpose, as for instance, those by Bona *et al.* (1985), Hanssen and Mijnarends (1986), and Kisker *et al.* (1987a,b). Although not yet totally conclusive—the problems seem to be connected with sample quality—little doubt remains about the correctness of the predictions. Initially the compound PtMnSb appeared to be of particular interest because of its exceptionally high magneto-optical Kerr effect (de Groot *et al.*, 1983b; Wijngaard *et al.*, 1989). This effect is still discussed controversially, yet it seems that the half-metallic band structure alone is not sufficient to describe it.

Finally we remark on the effects of temperature. The calculations described pertain to $T = 0$ since they are valid for the ground state. Increasing the temperature we excite besides phonons also magnons and at higher temperatures spin fluctuations, see Chap. 5. These excitations result in a finite density of states in the gap, very small for low temperatures, but growing when the temperature increases. The reason is, of course, the disordering of the magnetic structure. Rather simple calculations can simulate the expected effect by tilting the magnetic moments from their ideal ferromagnetic order using noncollinear moment arrangements. Obviously, the excitation of a magnon in a half-metallic ferromagnet is different from that in normal ferromagnets since the creation of states in the gap implies a strong electron magnon interaction. It is presently not clear theoretically what these excitations will do to transport properties of half-metallic ferromagnets; however, as expected, experimental studies show unique features (Moodera and Mootoo, 1994).

4.4.2 Selected case studies

4.4.2.1 CrO_2

The Heusler alloys described above are not the only known half-metallic systems. Thus, for instance Dijkstra *et al.* (1989) found that $KCrSe_2$ concerning its electronic structure belongs to the same class of compounds. However, it is the case of CrO_2 that attracted substantial theoretical (Schwarz, 1986; Kulatov and Mazin, 1990; Matar *et al.*, 1992; Nikolaev and Andreev, 1993; Brändle, 1993; van Leuken and de Groot, 1995; Uspenskii *et al.*, 1996; Lewis *et al.*, 1997; Korotin *et al.*, 1998; Mazin *et al.*, 1999a,b) and experimental attention (Kämper *et al.* 1987; Wiesendanger *et al.* 1990; Hwang and Cheong, 1997; Tsujioka *et al.* 1997; Ranno *et al.*, 1997; Suzuki and Tedrow, 1998, 1999; Coey *et al.*, 1998). The reason for this may have been the practical importance of CrO_2 as a magnetic recording material, but the fundamental interest in its half-metallic electronic structure and the implications of this cannot be stressed enough. Since the number of these compounds seems to be growing—we only mention as a further case $La_{1-x}Ca_xMnO_3$ under certain assumptions on x and the structure (Pickett and Singh, 1996)—we choose as a first example CrO_2 and highlight some of its electronic-structure properties.

Recently, the growing interest in half-metallic ferro- or ferrimagnetic systems has led to a flood of interesting publications of which we only choose one review article to comment upon here. This is the work by Katsnelson *et al.* (2008) who describe and classify a large set of these compounds: Heusler C_{1b} and $L2_1$ alloys, CrO_2, colossal-magnetoresistance materials, double perovskites, magnetite, anionogenic ferromagnets, pyrites, spinels, etc. Besides the electronic structure they formulate and discuss many-body effects that are not obvious from the band structure obtainable by the methods described in this monograph.

We resume the narrative now with CrO_2, which is a ferromagnetic metal with a Curie temperature of $T_c \simeq 390$ K (Kouvel and Rodbell, 1967). It crystallizes in the rutile structure which is tetragonal and contains two formula units of CrO_2 per cell. The Cr atoms form a body-centered tetragonal lattice and are surrounded by distorted oxygen octahedra. Its half-metallic band structure was first discovered by Schwarz (1986).

Now, in principle the determination of the electronic structure of transition metal oxides is beset with deep problems if one attempts to use the local density-functional approximation (LDA). The review article by Pickett (1989) is quite informative in this case. The problems originate from the electron–electron interaction (the electron correlation) which is apparently treated incorrectly in the LDA. This is not the place to go deeply into the field of electron correlations for which there are other useful treatises such as for instance that by Fulde (1991). Let us just mention that in the *insulating* transition metal oxides the treatment of *exchange* together with that of *screening* in density-functional theory is an as yet insufficiently well-understood problem. We have briefly remarked on this in Sec. 2.8.

The case of CrO_2 may be different. If this compound possesses metallic properties—as it experimentally does (Suzuki and Tedrow, 1998)—then, because of metallic screening, an LDA calculation has a fair chance of giving a good description of the electronic properties. A calculation by the ASW method is easily carried out (this method was also used by Schwarz (1986) and by Matar *et al.* (1992)) and gives the density of states shown in Fig. 4.61. We note in passing that the rutile crystal structure is fairly open so that in the atomic sphere approximation so-called empty spheres, S_E, are needed; see the remarks in Sec. 3.4.4. In choosing the coordinates of these S_E's it has been found advantageous to occupy the Wyckoff positions of highest symmetry that are not already taken up by atoms. This gives to a first approximation four S_E's in the position labeled c (see No. 136 "International Tables": Henry and Lonsdale, 1969) and to a second approximation another four S_E's in position g. The results, however, seem to be convergent with four S_E's. The main features of the electronic structure of CrO_2 are summarized as follows: In its formal valence state of 4+, Cr has two 3d electrons that line up to form a magnetic moment of 2 μ_B. Exactly this value—2 μ_B per CrO_2 or 4 μ_B per cell—is obtained due to a gap, E_G, at the Fermi energy, E_F, in the minority spin spectrum (spin ↓ in Fig. 4.61) of $E_G \sim 1.88$ eV, the valence band being ~ 0.38 eV above E_F. As can be seen from the shading, the metallic majority-spin states are almost entirely those of the Cr 3d electrons, the states straddling the Fermi energy being of t_{2g} parentage and those peaking at about 3 eV of e_g parentage. The Fermi energy barely misses a quasi-gap in the spin-↑ spectrum and the density of states at E_F is $\mathcal{N}(E_F) \simeq 2.0$ eV^{-1} per cell

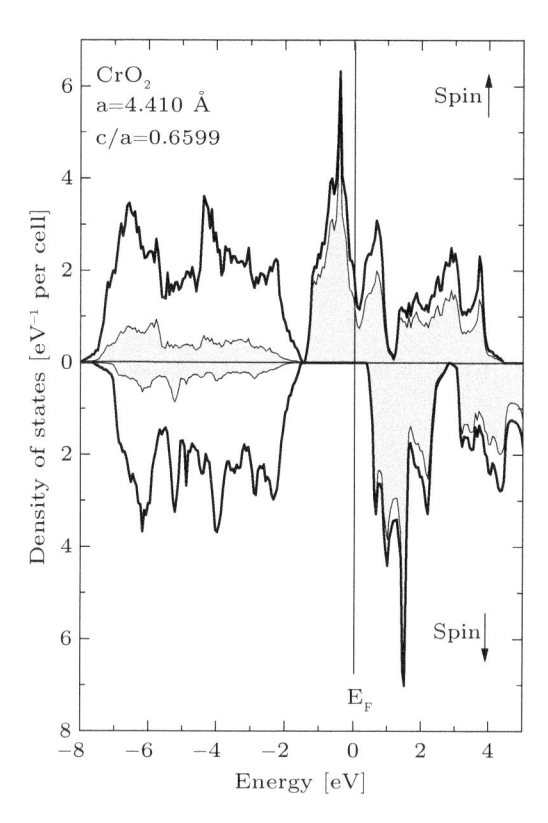

Figure 4.61 *Spin-up and spin-down state densities of CrO_2 per cell (solid line) and those of the d states of Cr (shaded) (two atoms). Experimental lattice constants used as indicated (from Landolt and Börnstein (1973)).*

corresponding to ~ 2.4 mJ/K^2 mol. The peak-to-peak distance of the sharp (t_{2g}) spin-↑ to spin-↓ features is ~ 1.9 eV representing the exchange splitting. The bands below -1.5 eV consist of O 2p electrons which hybridize and bond with Cr 3d electrons (shaded), but there are also hybrid (antibonding) 2p states above and near the Fermi energy that are recognizable by the differences between the shaded area and the solid lines. A calculation of the band structure constraining CrO_2 to be nonmagnetic results in a sharp peak at the Fermi energy with $\mathcal{N}(E_F) \simeq 2.4$ eV^{-1} per Cr. This is approximately the value that assumed nonmagnetic bcc Fe would have, which we know gives for the Stoner condition $I\mathcal{N}(E_F) > 1$ (see Sec. 4.2.2). Thus we can safely conclude that CrO_2 is an itinerant-electron ferromagnet, albeit with a magnetic moment well localized on the Cr site. A calculation of the total energy for assumed antiferromagnetic CrO_2 results in a value that is 0.16 eV per CrO_2 higher than that of the ferromagnetic with a magnetic moment of 1.71 μ_B per Cr, the oxygen polarization being exactly zero for symmetry reasons. If, finally, spin–orbit coupling is included in the calculations there is no change in Fig. 4.61 that is visible with the naked eye.

Before we attempt a comparison with experimental information, we select from other theoretical work a calculation that is methodologically different from the simple ASW technique and compare the results with those discussed above.

The work of Mazin *et al.* (1999a) on CrO_2 is important because they use the all-electron full potential LAPW method further testing the LSDA by also including the generalized gradient approximation (GGA) (see Sec. 2.8). Again the robustness of the half-metallic band structure is remarkable and the density of states they show is well represented by our Fig. 4.61. Noteworthy are a plot of the entire Fermi surface and results for the optical conductivity that agree in a broad sense with (presently not yet published) experimental data. The density of states at the Fermi energy these authors obtain is $\mathcal{N}(E_F) \simeq 1.9$ eV^{-1} per cell which is very close to our value.

Turning to more experimental work, Mazin *et al.* (1999a) discuss quantitatively the value of $\mathcal{N}(E_F)$ comparing with the specific heat measurements of Tsujioka *et al.* (1997) to obtain information about band mass renormalization effects. Tsujioka *et al.* analyze their data using for the specific heat the formula

$$c(T) = \gamma T + \beta T^3 + \alpha T^{3/2}. \qquad (4.136)$$

The $T^{3/2}$ term is Bloch's magnon contribution (see Sec. 5.2). They obtain $\gamma = 2.5 \pm 0.5$ mJ/K^{2}mol which is to be compared with the value of Mazin *et al.* of $\gamma = 2.24$ mJ/K^{2}mol or our $\gamma = 2.4$ mJ/K^{2}mol. The quotient $\gamma^{\text{theory}}/\gamma^{\text{exp}}$ is a measure of band mass renormalization effects which are large when $\gamma^{\text{theory}}/\gamma^{\text{exp}} \ll 1$. In fact, one would have expected a stronger renormalization from electron–electron and electron–magnon interactions, the value of $\gamma^{\text{theory}}/\gamma^{\text{exp}}$ perhaps being fortuitously close to 1, but the previously large renormalization from the calculation of Lewis *et al.* (1997) (see above) and another experimental estimate (for which it is hard to find the original reference) of $\gamma = 7$ mJ/K^{2}mol must be seriously questioned.

Further support for the half-metallic band structure of CrO_2 comes from a comparison of the calculated and measured diagonal part of the optical conductivity and reflectivity by Brändle *et al.* (1993) who find good agreement of measurements with early *ab initio* calculations by the same authors. The off-diagonal magneto-optical properties were calculated by Uspenskii *et al.* (1996) who obtained reasonable agreement with experiment. Finally, vacuum tunneling measurements by Wiesendanger *et al.* (1990) quite impressively also suggested nearly 100% spin polarization.

4.4.2.2 Heusler compounds

We now turn to recent work on a subset of the $L2_1$-Heusler compounds that had an astonishing comeback because of their high spin polarization and their possible applications in the field of spin electronics and magnetoresistive devices. We will say more about these applications in the later Sec. 4.5.3.3. Here we select nine Co_2-Heusler compounds listed in Table 4.12 and discuss the electronic structure of some of them in detail. A great number of other interesting Heusler compounds can be found in the literature, for instance in work by Galanakis *et al.* (2002) and Felser *et al.* (2007); the

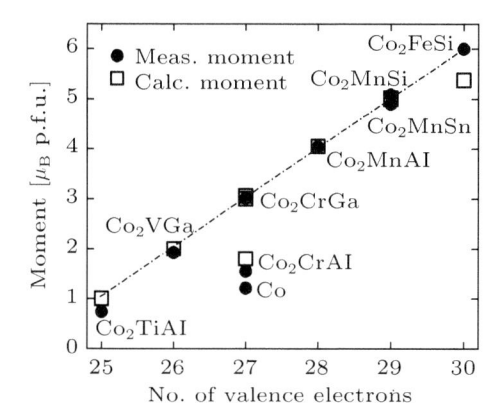

Figure 4.62 *Section of the Slater–Pauling curve describing a set of Co$_2$-Heusler compounds; after Kübler et al. (2007)*

latter focus their attention in particular on matters of applications. The review article by Katsnelson *et al.* (2008) mentioned before is also useful in this context.

We begin by drawing the attention to the magnetic moments per unit cell of our selection of Heusler compounds graphed as a function of the number of valence electrons per unit cell, N_V. This is shown in Fig. 4.62, which clearly shows a portion of the left-hand side of the conventional Slater–Pauling curve. The coordinates used here can easily be transformed to the ones used before.

Fig. 4.62 also reveals that the calculations are in good agreement with measured values of the magnetic moments. We furthermore can see from this figure and from Table 4.12 that most of these compounds possess integer or near-integer valued magnetic moments. Indeed, these are half-metallic ferromagnets (HMF).

It is of interest to understand why these compounds are HMF. Galanakis *et al.* (2002) gave a convincing explanation which we summarize as follows.

First, the hybridization of the *minority* states of the two Co atoms is considered, ignoring s and p electrons temporarily. The five Co d states are numbered $d1, d2, \ldots, d5$ and are shown in the upper part of the left-hand side of Fig. 4.63. Here $d1, d2, d3$ denote the t_{2g} and $d4, d5$ the e_g states, respectively. The hybridization in Co–Co results, in order of increasing energy, in two e_g and three t_{2g} states followed by the three "*ungerade*" t_{1u} and two e_u states. Denoting the two Co sites by a and b we describe these t_{2g} (t_{1u}) states by the orbital combination $d_{ia} + d_{ib}$ ($d_{ia} - d_{ib}$), with $i = 1, 2, 3$ and the e_g (e_u) by $d_{ia} + d_{ib}$ ($d_{ia} - d_{ib}$), with $i = 4, 5$.

Next, the minority Co–Co states are allowed to hybridize with Mn as shown in the lower portion on the left-hand side of Fig. 4.63. The double-degenerate e_g hybridize with the $d4$ and $d5$ states of Mn, which transform with the same representation. They create a low-lying, double-degenerate bonding e_g and a double-degenerate antibonding e_g of high energy. Similarly, the three Co–Co states of t_{2g} symmetry couple with the Mn $d1, d2, d3$ states to form bonding t_{2g} states of low energy and antibonding t_{2g} states

Table 4.12 *Collection of pertinent experimental and calculated data for nine representative Co_2-Heusler compounds, after Kübler et al. (2007). The quantity N_V is the number of valence electrons, the magnetic moments, M^{exp} and M^{calc} are given in μ_B per unit cell. The local moment for Co is denoted by $\mathcal{L}_{Co}$ and those of the other magnetic atoms in the cell by $\mathcal{L}_X$, all in units of μ_B. The Curie temperatures T_C^{SP} were calculated by the theory given in Sec. 5.4.7.*

Compound	N_V	a [Å]	M^{exp}	M^{calc}	$\mathcal{L}_{Co}$	$\mathcal{L}_X$	T_C^{SP}	T_C^{exp}
Co_2TiAl[a]	25	5.847	0.74	1.00	0.570	−.139	157	134
Co_2VGa[a]	26	5.779	1.92	2.00	0.914	0.172	368	352
Co_2VSn[a]	27	5.960	1.21	1.80	0.677	0.445	103	95
Co_2CrGa[b]	27	5.805	3.01	3.06	0.575	1.911	362	495
Co_2CrAl[a]	27	5.727	1.55	3.00	0.669	1.661	341	334
Co_2MnAl[a]	28	5.749	4.04	4.05	0.590	2.877	609	697
Co_2MnSi[a]	29	5.645	4.90	5.00	0.969	3.061	990	985
Co_2MnSn[a]	29	5.984	5.08	5.02	0.885	3.254	899	829
Co_2FeSi[c]	30	5.640	6.00	5.38	1.307	2.762	1185	1100

[a] Lattice constant and experimental Curie temperature from Buschow et al. (1981), van Engen et al. (1983), Webster (1971), Ziebeck and Neumann (2001)
[b] Lattice constants and Curie temperature from Umetsu et al. (2005)
[c] Lattice constant and Curie temperatures from Wurmehl et al. (2005a,b and 2006a).

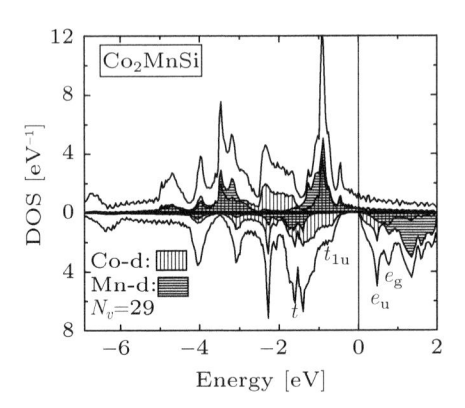

Figure 4.63 *Possible hybridization scheme of the minority d-states of Co_2MnZ on the left-hand side; the Co-Co hybridization is on the top left and below is that of Co_2-Mn, after Galanakis et al. (2002). On the right-hand side is the spin-resolved density of states of Co_2MnSi: upper panel is spin up, lower panel is spin down; after Kübler et al. (2007). Note on the left that the states just above and below the Fermi energy, labelled e_u and t_{1u}, do not hybridize, accordingly, on the right-hand side the minority states with the same labels above and below the Fermi energy (at energy origin) are purely Co d-states.*

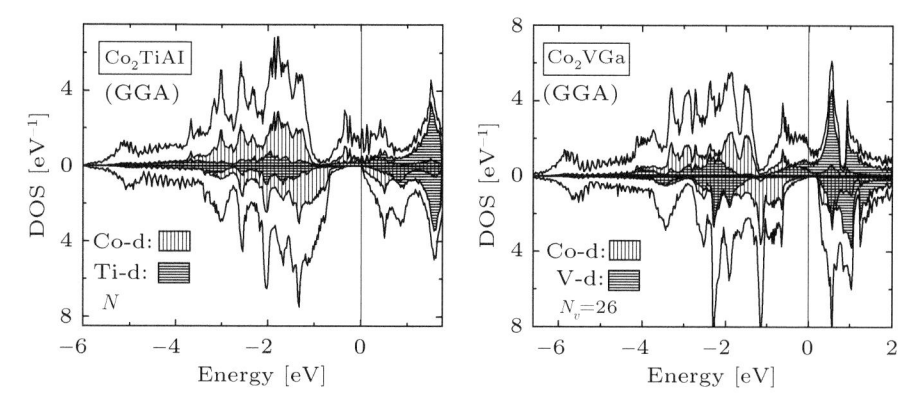

Figure 4.64 *Spin-resolved density of states of Co_2TiAl and Co_2VGa, upper panel is spin up, lower panel is spin down; after Kübler et al. (2007).*

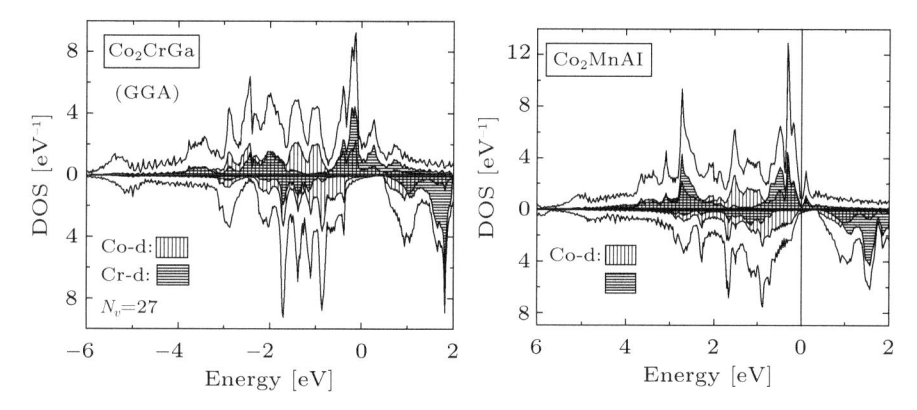

Figure 4.65 *Spin-resolved density of states of Co_2CrGa and Co_2MnAl, upper panel is spin up, lower panel is spin down; after Kübler et al. (2007).*

of high energy. Finally, the ungerade states of Co–Co, t_{1u} and e_u, cannot couple with any of the Mn-orbitals since there are none with an ungerade representation. There is no hybridization here. As indicated in the figure, the t_{1u} are occupied, while the e_u are above the Fermi energy. The densities of states shown on the right-hand side of Fig. 4.63 clearly reflect these states: one recognizes hybridization by common peaks in the states labeled Co and Mn and sees that common peaks are absent in the states straddling the gap. They are non-hybridizing. This is also so in Figs. 4.64 and 4.65, where we show examples for the valence electron numbers $N_V = 25, 26, 27$, and 28. The compounds containing Ti, V, and Cr behave just as those containing Mn: the states straddling the gap do not hybridize and are due to the ungerade Co–Co states. The role of the *sp* atoms is, of course, important as they determine the valence-electron concentration.

Returning to Fig. 4.62, we comment that Co_2FeSi possesses the largest magnetic moment observed so far in the Heusler compounds. The reason our calculations do neither reproduce the experimental value of 6 μ_B p.f.u. nor a HMF in this case is believed to be due to electron correlations that are incompletely described in the local density-functional approximation (LDA). Improving the LDA by using the generalized gradient approximation (GGA) does not change the situation (Kübler *et al.* 2007). However, if one boosts the electron correlation by applying Hubbard-parameters U to both Co and Fe (Wurmehl *et al.* 2005a) in an LDA+U calculation (see the comments at the end of Sec. 2.8.5) one succeeds in reproducing the experimental characteristics.

We will come back to the Heusler compounds in Sec. 4.5.3.3 where we discuss the tunnel magnetoresistance and in Sec. 5.2.2 where we deal with spin-wave spectra; in Sec. 5.4.8 we collect results of calculations of their Curie temperatures and discuss why they are so large for high valence-electron concentrations.

4.4.2.3 Double perovskites

Another important family of half-metallic ferrimagnets has been discovered in the ordered double perovskites (Kobayashi *et al.* 1998; Fang *et al.* 2001). They attracted enormous research interest because of their high spin polarization and possible applications in magnetoresistive devices (similar to the Heusler compounds). The crystal structure of these compounds can easily be derived from the $L2_1$ structure shown in Fig. 4.58. If we denote the double perovskite by the formula $A_2B'B''O_6$, where A = Sr,Ba,Ca, B' = Cr,Mn,Fe, and B'' = Mo,W,Re,Os, then the A-atoms occupy the black corners of the inscribed cube, B' and B'' occupy the other shaded and white fcc-sites. In addition, corner-sharing oxygen octahedra are introduced such that their centers coincide with the B'- and B''-atoms. Often the cubic structure is tetragonally distorted, sometimes even forming lower symmetric structures (Serrate *et al.* 2007). For simplicity we will limit our attention to the cubic structures.

Fig. 4.66 gives an overview of double perovskites arranged according to Curie temperatures and the number of valence electrons, N_V, which are supplied by the transition-metal elements B' and B''. The remaining electrons from the A and O atoms remain the same in each series, therefore, they were suppressed in the notation. Since the magnetic properties of $Sr_2FeB''O_6$, with B'' = Mo, W, and Re were described in great detail by Fang *et al.* (2001), we focus our attention on the series $Sr_2CrB''O_6$ with B'' = W, Re, Os, and Ir, presenting a selection of results obtained by Mandal *et al.* (2008), where more information can be found. It should be pointed out that the Ir-compound has not been prepared as yet and we will see that we cannot make a clear prediction about its likely properties.

Figure 4.67 shows a simplified level scheme which takes into account the crystal field splitting of an octahedral environment, viz. t_{2g} and e_g and assumes antiferromagnetic coupling of the two transition metals in a given double perovskite. The sequence of the densities of states graphed in Figs. 4.68 and 4.69 follows this simple scheme to a surprising extent. One clearly notices in Fig. 4.68 the increasing occupation of the minority t_{2g}- states in going from Sr_2CrWO_6 to Sr_2CrReO_6. The half-metallic gap occurs in the majority states. The O-2p-states are seen at low energies from about

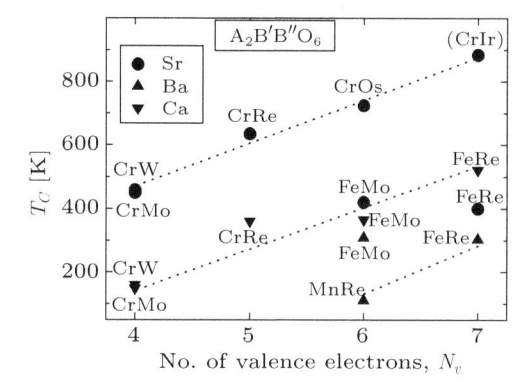

Figure 4.66 *Measured Curie temperatures versus the number of valence electrons for a selection of double perovskites. The names of these elements $B'B''$ in the compound $A_2B'B''O_6$ appear by the data points. The value in parenthesis is a projected one; after Mandal et al. (2008).*

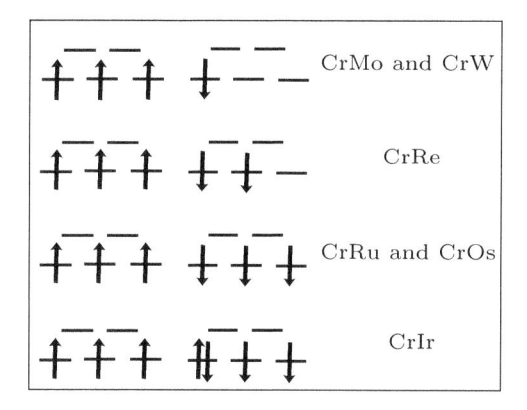

Figure 4.67 *Schematic level diagram for the series $Sr_2CrB''O_6$. Three lines symbolize the t_{2g} states and two the e_g states. The number of valence electrons is given by the sum of the arrows, which count the up-spin and down-spin occupations; after Mandal et al. (2008).*

−8 eV to −4 eV to hybridize weakly with the transition-metal states. As indicated in Fig. 4.67 the increasing occupation leads to a decreasing magnetic moment, from 2 μ_B in Sr_2CrWO_6 to 1 μ_B in Sr_2CrReO_6. The coupling of two transition metal atoms in a given compound is, indeed, antiferromagnetic; the calculated magnetic moments of the individual transition metals are given in Table 4.13.

Continuing with the compound Sr_2CrOsO_6 we note from Figs. 4.67 and 4.69 that the minority t_{2g} is now completely filled but still gapped, thus reducing the magnetic moment per unit cell to 0 μ_B: the moments exactly compensate and the system is insulating by means of a gap in the majority states. It was made by Krockenberger *et al.* (2007), who also computed the electronic structure, which is in agreement with Fig. 4.69 as far as the insulating and magnetic properties are concerned.

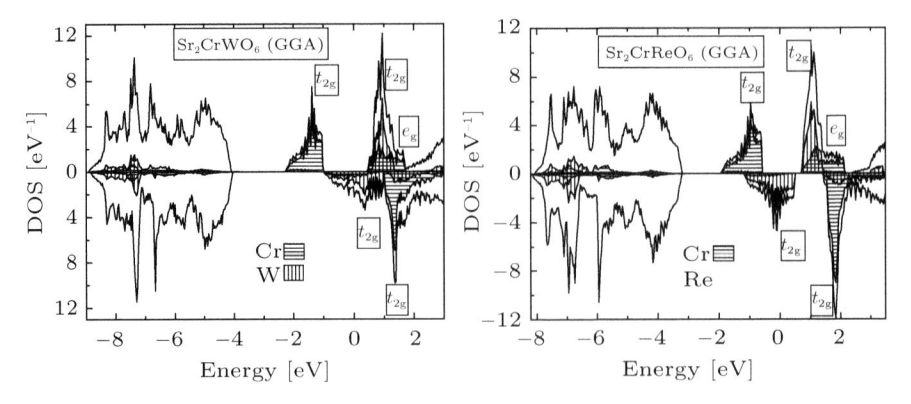

Figure 4.68 *Spin-resolved density of states of Sr_2CrWO_6 and Sr_2CrReO_6. Upper panels are spin up, lower panels spin down. The total number of valence electrons is $N_V = 4$ for the former and $N_V = 5$ for the latter, while the total magnetic moment decreases from 2 μ_b to 1 μ_B; after Mandal et al. (2008).*

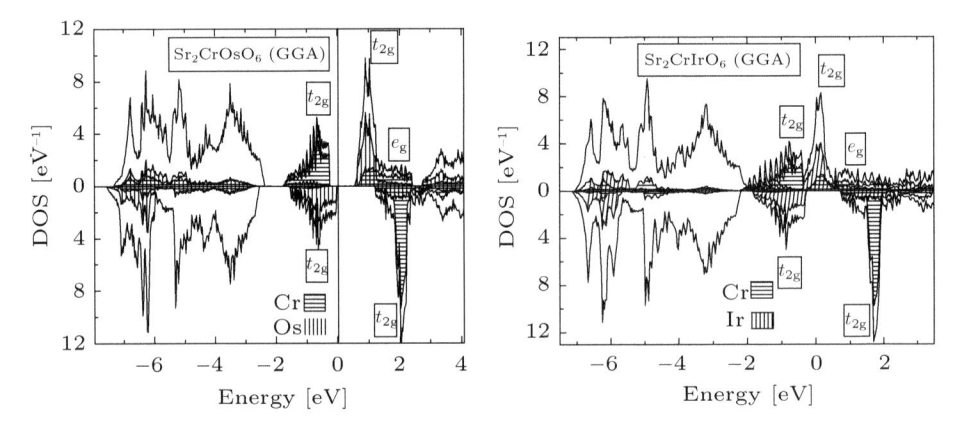

Figure 4.69 *Spin-resolved density of states of Sr_2CrOsO_6 and Sr_2CrIrO_6. The total number of valence electrons is $N_V = 6$ for the former and $N_V = 7$ for the latter, while the total magnetic moment increases from 0 μ_b to 1 μ_B; after Mandal et al. (2008).*

Although we obviously reached with Sr_2CrOsO_6 the end of a series, we might ask what another electron might do and therefore continue with the electronic structure of Sr_2CrIrO_6 (Mandal et al. 2008), for which the densities of states are shown on the right-hand side of Fig. 4.69: a half-metallic ferrimagnet (HMF) having a gap in the minority states (as usual) is obtained. One must emphasize that this system was assumed to be cubic with a plausible value for the lattice constant. A more reliable prediction needs a full optimization that was not attempted yet. A calculation of the Curie temperature by means of the theory given in Sec. 5.4.7 results in $T_C = 884$ K which is used as the projected value in Fig. 4.66.

Table 4.13 *Collection of pertinent experimental and calculated data for 4 double perovskites, $Sr_2B'B''O_6$ The quantity N_V is the total number of valence electrons supplied by B' and B''. The space group symmetry is given in the column, symm. The total calculated magnetic moment, M_{tot}^{calc} is given in μ_B per formula unit. The local moment of B' is denoted by $\mathcal{L}_{B'}$, that of B'' by $\mathcal{L}_{B''}$, and the induced moments of the 6 O atoms by $6\mathcal{L}_O$, all in units of μ_B. The calculated Curie temperatures, T_C^{calc}, (in K) were obtained by the theory given in Sec. 5.4.7.*

Compound	N_V	symm.	a[Å]	M_{tot}^{calc}	$\mathcal{L}_{B'}$	$\mathcal{L}_{B''}$	$6\mathcal{L}_O$	T_C^{calc}	$T_C^{\exp(a)}$
Sr_2CrWO_6	4	$Fm\bar{3}m^{(a)}$	$7.832^{(a)}$	2.0	2.557	−0.439	0.082	434	458
Sr_2CrReO_6	5	$I4/m^{(b)}$	$7.814^{(b,c)}$	1.0	2.423	−1.272	−0.151	742	620
Sr_2CrOsO_6	6	$Fm\bar{3}m^{(d)}$	$7.824^{(d)}$	0.0	2.443	−1.893	−0.550	881	725
Sr_2CrIrO_6	7	$Fm\bar{3}m^{(e)}$	$7.881^{(e)}$	1.0	2.296	−0.953	−0.343	884	

[a] From Serrate et al. (2007). [b] From Kato et al. (2004).
[c] Calculated from experimental atomic volume. [d] From Krockenberger et al. (2007).
[e] Assumed values.

However, this HMF-state is most likely metastable because a total-energy search returns a metallic antiferromagnet of the AFII structure, which corresponds in a k-vector notation to $\mathbf{K} = (\frac{1}{2}, \frac{1}{2}, \frac{1}{2})$. In this configuration the magnetic moments are aligned ferromagnetically in the (111) plane and alternate along the [111] direction. This result has to be questioned: correlations are likely to alter the electronic structure and result in a Mott–Hubbard insulator. LDA+U calculations were therefore employed but it was found that even in this case the antiferromagnet remains metallic. Comparable calculations by Fang et al. (2001) also show that the LDA+U method not always opens a gap at the Fermi energy.

Finally, the large Curie temperatures, especially for Sr_2CrReO_6 and Sr_2CrOsO_6, require remarkably large ferromagnetic exchange interactions between the Cr cations. This issue will be be discussed in Sec. 5.4.8. A very thorough review of nearly all properties of double perovskites is that by Serrate et al. (2007).

4.4.2.4 Invar

A small subset of itinerant magnets that always appear to involve Fe possesses the unique property of an anomalously small and almost temperature independent ("invariant") thermal expansion. This property is found in Fe-Ni alloys around the composition $Fe_{65}Ni_{35}$—recognizable by an anomaly in the Slater–Pauling curve, Fig. 4.55—and also in Fe_3Pt. It was discovered by Guillaume in 1897 and has intrigued solid state physicists ever since. The name Invar is quite common, but also Elinvar, which refers to similar invariant *elastic* properties, is of importance in this connection. Invar and Elinvar alloys are widely used in engineering applications, but their fundamental physical properties are still a challenge, which as Wassermann (1997) pointed out, have led to something like 20 theoretical models. A comprehensive review can be found in the book chapter by Wassermann (1990). Here we address the question: What distinguishes the small subset of itinerant magnets in which the phenomenon is observed? Without trying to be complete we give an answer by means of *ab initio* electronic structure calculations, however, concentrating on the salient facts only.

We begin by comparing the results of six different total energy calculations, first those of γ-Fe shown in Fig. 4.47, Sec. 4.3.3, then Fe_3Ni calculated in the Cu_3Au structure and finally FeNi calculated in the CuAu I structure. Note that these structures are fcc coordinated. This comparison is shown in Fig. 4.70, where the total energy differences that appear have been arranged with arbitrary origins to enhance the readability. A plot like this was first presented and discussed by Williams *et al.* (1982, 1983a), albeit calculated with less precision as it was meant to convey the basic idea. For FeNi we see that the magnetic state possessing a high moment (HM) is stable over a large volume range, the nonmagnetic state (NM) having higher total energy because of the *missing* magnetic pressure discussed in Sec. 4.3.1. Increasing the Fe concentration by 25% we come to Fe_3Ni and the corresponding total energy differences appearing in the center of Fig. 4.70. Here with decreasing volume the high-moment state abruptly ceases to exist giving way to a state of low moment (LM). The inset shows that the two states (HM and LM) are really different *and almost degenerate*. The moment discontinuity amounts to about 0.8 μ_B per atom, from ~ 1.0 μ_B per atom in the HM state to ~ 0.2 μ_B per atom in the LM state. A glance at the low portion of the figure, showing the case of γ-Fe, reveals that the HM-state has moved to a large volume, the total energy labeled NM/AF being that of the nonmagnetic state at equilibrium which changes to an antiferromagnetic state (AF) at larger volume, see Fig. 4.47. We also recall from Fig. 4.47 that for γ-Fe the states of lowest energy are *noncollinear* moment arrangements, the total energies of which lie close to the common tangent of the NM/AF-HM curves. We expect the same to be true for Fe_3Ni and indeed can verify the noncollinear states finding the total energy to follow again roughly the common tangent that we can imagine drawn in the inset in Fig. 4.70. Now it is quite plausible that for alloys Fe_xNi_{100-x} with increasing x the low-moment (LM) or non-magnetic (NM) states, relative to the HM states, move continuously down in energy from the values they have in FeNi to those in Fe_3Ni approaching the situation we see in γ-Fe. However, before the case of γ-Fe is reached the fcc coordinated crystals will become unstable and go over to bcc coordinated structures for which normally the label α is used.

In our view, the unique property of Invar systems is that, in a narrow concentration range, the high-moment (HM) and the low-moment (LM) states are both energetically stable and that Invar behavior reflects thermal excitations from a possibly noncollinear HM state to the lower-volume LM states. It is worthwhile to point out that our emphasis of transitions from a magnetic state to a smaller-volume less magnetic state is qualitatively similar to the interpretation of Invar by Weiss (1963) who, in terms of the localized magnetic moment picture, also postulated two states. Our calculations reconcile that interpretation with the itinerant-electron picture and put it on a solid theoretical foundation. What these thermal excitations might look like will be discussed in Chap. 5. Here we continue with those properties of Invar that can be inferred from electronic structure calculations for the ground state or for states "nearby."

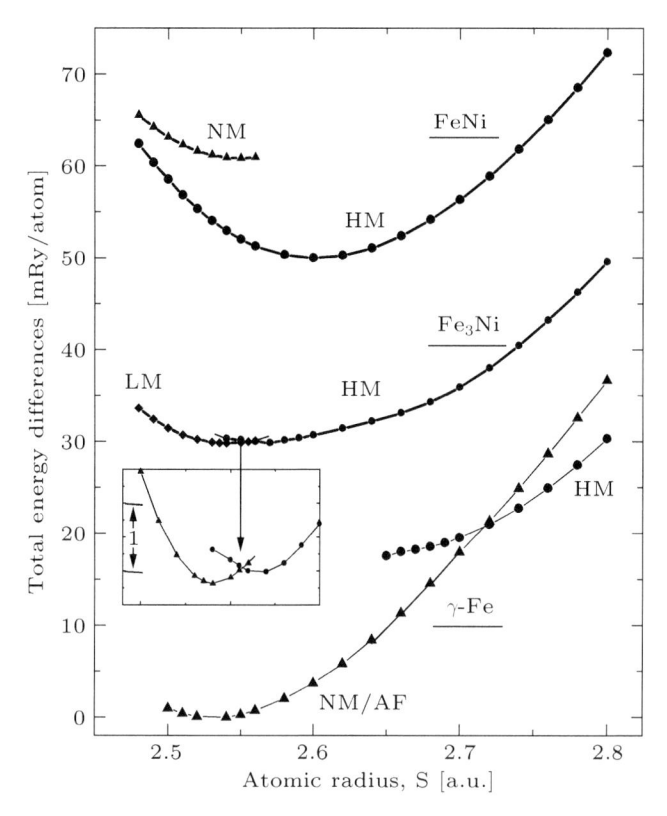

Figure 4.70 *Total energy differences as a function of the atomic sphere radius, S, for FeNi, Fe$_3$Ni, and γ-Fe, all fcc coordinated. Total energy origins are arbitrarily shifted to about 0 mRy, 30 mRy, and 50 mRy for the minimum values of γ-Fe, Fe$_3$Ni, and FeNi, respectively. NM: nonmagnetic, HM: high-moment state, LM: low-moment state, and AF: antiferromagnetic. Inset: blow-up of total energy minima for Fe$_3$Ni, scale shown is 1 mRy.*

The issue of the *stability* of two different magnetic states having different volumes can be discussed for Fe$_3$Ni by means of the density of states and by an analysis of the bonding properties of the states near the Fermi energy. We show in Fig. 4.71 the d state densities of Fe$_3$Ni for the two volumes corresponding to the total energy minima given in the inset of Fig. 4.70. It can be seen quite clearly that the Fermi energy is locked *exactly* into sharp valleys in the down-spin states for both the high-moment (HM) and low-moment (LM) states. We have seen this phenomenon frequently stabilizes the structure. Note that the state density features in the $\downarrow$ spin states are nearly identical for the two volumes and the Fermi energy chooses alternatively one of the valleys appearing in both the Fe and Ni systems. Entel *et al.* (1993) have analyzed the bonding properties in detail and singled out the t_{2g} states of Fe, which we therefore marked in Fig. 4.71. The alternative locking of the Fermi energy decreases the magnetization if the valley of higher energy is chosen in the $\downarrow$ electron system causing, obviously, the t_{2g} peak in the $\uparrow$ states to be unoccupied.

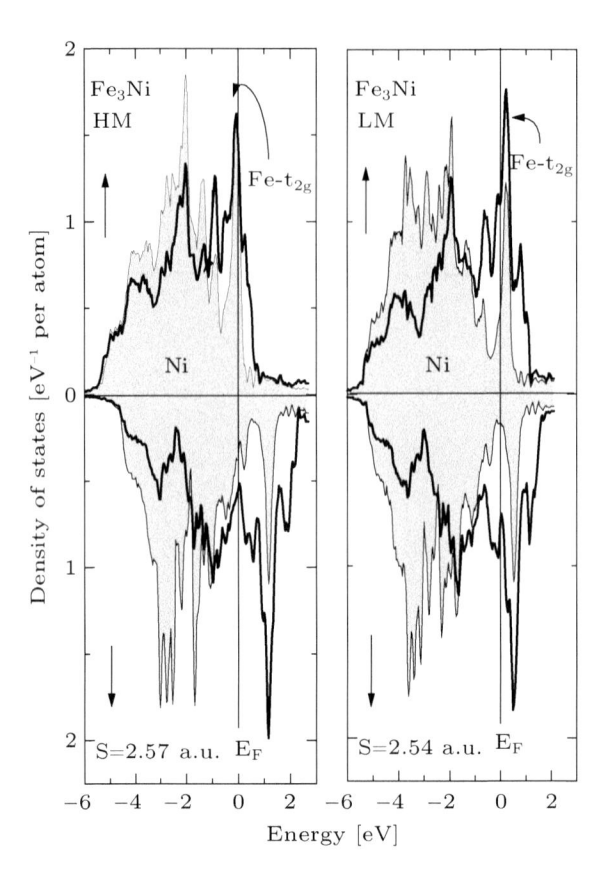

Figure 4.71 *Density of up (↑) and down (↓) spin d states per atom of Ni (shaded) and Fe (solid lines) in Fe_3Ni for the high-moment (HM) state with $S = 2.57$ a.u. (left) and for the low-moment (LM) state with $S = 2.54$ a.u. (right). These volumes correspond to the minima of the total energy shown in the inset of Fig. 4.70. The peaks originating from the Fe t_{2g} band states are marked appropriately.*

Since at the upper band edge the states are predominantly antibonding this occupation change will decrease the volume because nonbonding or even bonding orbitals will be occupied instead.

• Returning to Fig. 4.70 and paying attention now to the energetic order of the HM and LM states in the inset, we see that the ground state is actually the one with the low magnetic moment so that occupying the available HM states in Fe_3Ni would *increase* the volume. Furthermore the state of lowest energy will not be ferromagnetic because we found noncollinear order to possess still lower energy. In fact, the sequence of levels was initially found to be different. Entel *et al.* (1993) report the HM state to be about 1 mRy per atom *above* the LM state. Since their results were obtained by the ASW method, too, we must conclude that these calculations are very sensitive to computational details. One of the reasons for the small discrepancy might be the fact that in the earlier calculations no relativistic corrections were used in the Schrödinger equation, while the present results employed the scalar relativistic approximation. Of course, a change of the level sequence is expected at some critical concentration; however, its precise value cannot be inferred from the calculations discussed so far, except that our present results indicate a critical concentration of somewhat more than 25% Ni. For smaller concentrations of Ni the behavior typical for Invar should change to the opposite, and this is indeed observed experimentally. The alloys are then said to be *anti-Invar* (Acet *et al.*, 1994).

More recent calculations were motivated by the fact that the Fe-Ni alloys are not ordered and other concentrations than ours would be needed to establish the Invar mechanism beyond any doubt. Such calculations can be done using the CPA theory that is outlined in Sec. 5.5.2. and results exist, for instance by Johnson *et al.* (1990), Johnson and Shelton (1997), Abrikosov *et al.* (1995) etc. We choose to reproduce in Fig. 4.72 the CPA results by Schröter *et al.* (1995) which show the change of level

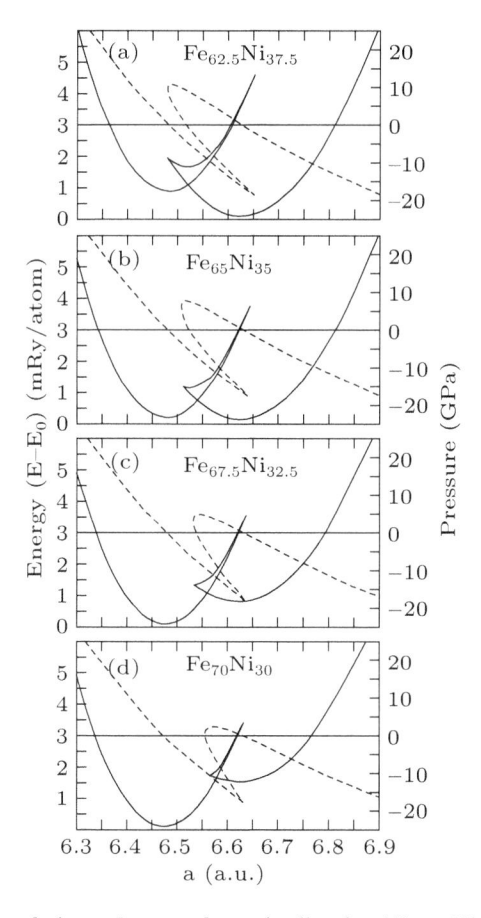

Figure 4.72 *Total energy relative to the ground state for disordered $Fe_{1-x}Ni_x$ from KKR-CPA calculations. Left minimum corresponds to the nonmagnetic, right minimum to the high-moment ferromagnetic state, apparent minimum in between is due to a saddle point. Dashed lines give the pressure. Invar region for $x \leqslant 0.35$. Reprinted with permission from Schröter et al. (1995), p. 194.*

sequence of HM and NM states with changing concentration x in $\text{Fe}_{1-x}\text{Ni}_x$. Notice that now the HM and NM levels become approximately equal for the alloy $\text{Fe}_{65}\text{Ni}_{35}$ so that for $x \leqslant 0.35$ the nonmagnetic state would be stable. The calculations by Johnson and Shelton (1997), however, we interpret to give a magnetic solution for the ground state, perhaps of the disordered local-moment type discussed for simple elemental metals in Sec. 5.5.2.

The calculated trends in the total energy are compared with the experimental phase diagram in Fig. 4.73. Part (a), from Wassermann (1990), shows as a function of the iron concentration the measured Curie, T_c, and Néel, T_N, temperatures as well as the magnetization, and identifies the Invar region. The shaded portion of the phase diagram is a complicated mixed magnetic phase with noncollinear order. The symbols γ and α denote concentration ranges where the alloys are fcc and bcc coordinated, respectively. Part (b) of Fig. 4.73, from Schröter et al. (1995) contains calculated total energy differences, $E_{NM} - E_{HM}$ being the KKR-CPA results of Schröter et al. (1995) and $E_{LM} - E_{HM}$ the LMTO-CPA results of Abrikosov et al. (1995). Dashed lines and crosses give the calculated magnetization. Although we will see in Chap. 5 that the total energy differences are not exactly a measure of the Curie temperatures, a comparison of the trends in (a) and (b) is still significant and quite impressive.

We close our discussion of Invar by pointing out that the compound Fe_3Pt has also attracted considerable attention. In contrast to the Fe-Ni system Fe_3Pt can be made as an ordered compound which allows the simpler electronic structure methods for ordered systems to be employed. The physical picture one finds is again that seen in Fe_3Ni. But noncollinear states are much more prominent in Fe_3Pt. For further theoretical details we refer to the work by Uhl et al. (1994), Uhl and Kübler (1997).

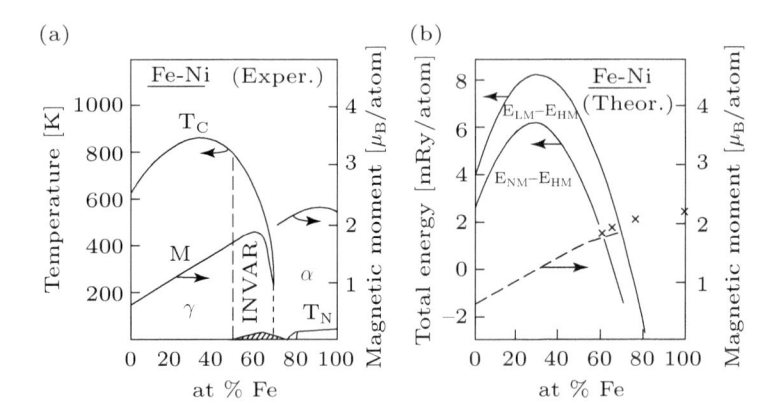

Figure 4.73 *(a) Experimental phase diagram after Wassermann (1990) showing the Curie, T_c, and Néel, T_N, temperatures as function of the Fe concentration and the magnetization. (b) Calculated magnetization (dashes and crosses) and total energy differences after Schröter et al. (1995) whose results are those marked $E_{NM} - E_{HM}$; values marked $E_{LM} - E_{HM}$ are from Abrikosov et al. (1995).*

4.5 Multilayers

The experimental advances made during the past decade in the preparation and characterization of thin films and surfaces made possible the fabrication of "designer solids" by controlling materials on the atomic scale, that is, layer by layer, row by row, and eventually atom by atom. A review article by Himpsel *et al.* (1998) covers this large field of magnetic nanostructures. Here we select a topic of atomically engineered magnetic systems which consist of ferromagnetic layers separated by thin films of nonmagnetic metals, the so-called spacer. They are made by various experimental methods: sputtering or molecular beam epitaxy which, however, are not at the center of interest here. The reader who needs information about the experimental techniques should consult the review by Himpsel *et al.* (1998) or, for instance, Vol. 121 of the *Journal of Magnetism and Magnetic Materials* which contains a great amount of information. Of interest here are two outstanding phenomena which we want to understand theoretically, the first being an oscillatory exchange coupling of the magnetic layers, mediated by the electrons in the spacer. The second is an exceptionally strong influence of the relative orientation of the magnetization on the electrical resistivity of the multilayers. This is called giant magnetoresistance (GMR). The GMR effect was independently discovered by Fert (Baibich *et al.*, 1988) and Grünberg (Binasch *et al.*, 1989), who were awarded the Physics Nobel price for their discovery in 2007.

4.5.1 Oscillatory exchange coupling

If two ferromagnetic layers like Fe or Co are separated by a few monolayers of a nonmagnetic metal consisting of e.g. Cu (called the spacer), then it is observed that the two ferromagnetic layers line up their magnetic moments either parallel or antiparallel depending on the thickness of the spacer, see Fig. 4.74 (a) and (b), which

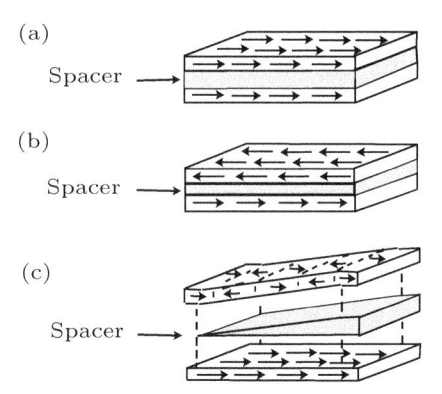

Figure 4.74 *(a) and (b): Schematic diagram of multilayers assuming different thickness of atomic dimensions, arrows giving directions of magnetization. (c) Schematic exploded view of a multilayer having a wedge-shaped spacer with varying thickness ranging typically from 0 – 10 nm.*

288 *Electronic Structure and Magnetism*

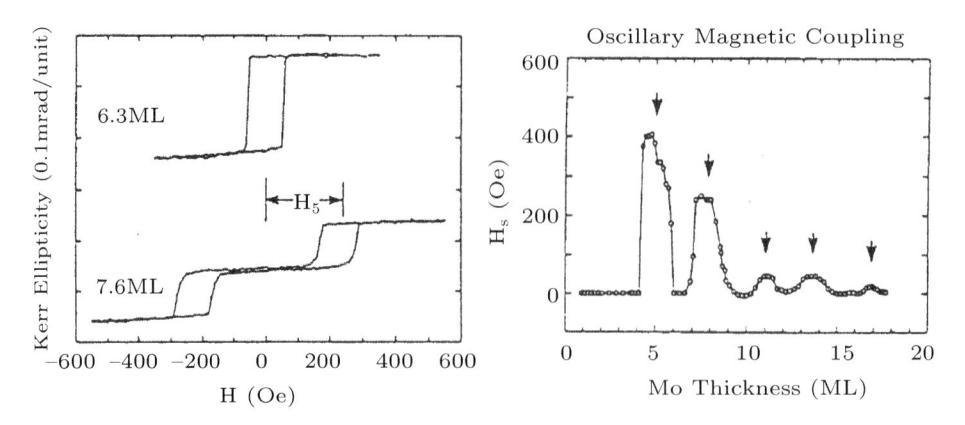

Figure 4.75 *Magnetic oscillations at Fe/Mo/Fe(100) trilayers determined by the magneto-optic Kerr effect. (a) Hysteresis loops characteristic of parallel and antiparallel coupling (top and bottom). H_s is the magnetic field to force antiparallel layers parallel. (b) Alternating antiparallel and parallel coupling (arrows and baseline, respectively). Reprinted with permission from Qiu et al. (1992).*

show schematically the arrangement of spacer and ferromagnet layers. This fascinating phenomenon was first discovered by Grünberg *et al.* (1986), followed by work of Parkin *et al.* (1990) and others. The switching of the direction of magnetization happens with atomic precision, as shown in Fig. 4.75, where it is seen that the addition of little more than an atomic layer to the spacer inverts the alignment of the Fe layers (Qiu *et al.*, 1992). Frequently, sandwich structures are fabricated in which the thickness of the spacer varies continuously from 0 to over 100 Å, as shown in a schematic exploded view in Fig. 4.74(c). Thus Unguris *et al.* (1991), by using scanning electron microscopy with polarization analysis, observed directly whether the top Fe layer is ferromagnetically coupled to the bottom layer and accurately determined the spacer thickness—in this experiment made of Cr—at which the coupling reverses. The effect has been observed for many combinations of ferromagnets and spacer material. Typical oscillation periods are about 10 Å, but shorter and longer periods have been observed, too, for instance short periods in Fe/Cu/Fe and Co/Cu/Co by Johnson *et al.* (1992), in Fe/Mo/Fe by Qiu *et al.* (1992), and a short as well as a long period in Fe/Cr/Fe by Unguris *et al.* (1991). More detailed information will be presented as we go farther afield.

4.5.1.1 RKKY exchange

The underlying physical picture for the coupling mechanism is the following: a ferromagnetic layer in contact with the spacer induces a (small) spin polarization of the conduction electrons in the spacer; this polarization is long ranged and thus interacts with the second ferromagnetic layer, thus giving rise to an effective exchange coupling. This concept, however, needs to be made quantitative.

In early attempts (Yafet, 1987; Fairbairn and Yip, 1990; Coehoorn, 1991; Herman and Schrieffer, 1992; Bruno and Chappert, 1992; Lee and Chang, 1994, 1995a,b) to understand the oscillatory exchange coupling an old but important mechanism was used that goes back to Ruderman, Kittel, Kasuya, and Yoshida (RKKY). Here one considers a number, n, of magnetic moments, $\mathbf{S}_i$ (i = 1, ..., n) immersed in a sea of conduction electrons, originally assumed to be free electrons. The interaction is written in the form

$$V(\mathbf{r},s) = A \sum_i \delta(\mathbf{r} - \mathbf{R}_i) \boldsymbol{\sigma} \cdot \mathbf{S}_i, \qquad (4.137)$$

where A is an adjustable intra-atomic exchange parameter, the delta function δ limits the interaction to be of contact form, and $\boldsymbol{\sigma}$ denotes the conduction electron Pauli spin operator. This interaction potential, although extremely well justified for its original purpose—nuclear spins coupled by conduction electrons (Ruderman and Kittel, 1954) or the localized magnetic moments of rare earths immersed in conduction electrons (Kasuya, 1956; Yosida, 1957)—is at best a phenomenological ansatz for the case at hand. It lacks *ab initio* character since the separation of *magnetic moment* on the one hand and *conduction electron* on the other is not really possible in an itinerant-electron system. However, it is an important model conceptually, of great value and worthwhile to consider here. We therefore derive what is called the RKKY-indirect exchange interaction starting with Eqn (4.137).

Let $\{\psi_k(\mathbf{r},s) = \psi_k(\mathbf{r})\chi(s)\}$ be the set of Bloch spinors describing the conduction electrons. Here χ is an eigenspinor to σ_z as used before. We use second-order perturbation theory to calculate the change of the one-particle energies, $\Delta\varepsilon_{ks}^{(2)}$, brought about by the potential $V(\mathbf{r},s)$, and subsequently—employing the force theorem—we determine the total energy by summing up the one-particle energies,

$$E_{\text{tot}} = \sum_s \sum_k f_k(\varepsilon_k + \Delta\varepsilon_{ks}^{(2)}). \qquad (4.138)$$

Here f_k denotes the Fermi–Dirac distribution function and we temporarily collect both the band index and the wave vector in the single letter k. The contribution of first-order perturbation theory drops out from the sum and the total energy is in second order

$$E_{\text{tot}} = E_{\text{tot}}^{(0)} + \sum_{ks} f_k \sum_{k'} \frac{\sum_{ij} |V_{k'k}|^2 \mathbf{S}_i \cdot \mathbf{S}_j}{\varepsilon_k - \varepsilon_{k'}}, \qquad (4.139)$$

where the factor $\mathbf{S}_i \cdot \mathbf{S}_j$ is obtained by manipulating the spin-part in Eqn (4.137) and

$$|V_{k'k}|^2 = A^2 \, \psi_k^*(\mathbf{R}_i) \, \psi_{k'}(\mathbf{R}_i) \, \psi_{k'}^*(\mathbf{R}_j) \, \psi_k(\mathbf{R}_j). \qquad (4.140)$$

The RKKY energy, E_{RKKY}, is then defined by the total energy difference and the RKKY exchange constants, $J(\mathbf{R}_{ij})$, by

$$E_{\text{tot}} - E_{\text{tot}}^{(0)} \equiv E_{\text{RKKY}} = -\sum_{ij} J(\mathbf{R}_{ij}) \, \mathbf{S}_i \cdot \mathbf{S}_j, \qquad (4.141)$$

where $\mathbf{R}_{ij} = \mathbf{R}_i - \mathbf{R}_j$. We read off from Eqn(4.139):

$$J(\mathbf{R}_{ij}) = 2A^2 \operatorname{Re} \sum_{n\mathbf{k}} \sum_{n'\mathbf{q}} \frac{f_{n\mathbf{k}}(1 - f_{n'\mathbf{k}+\mathbf{q}})}{\varepsilon_{n'\mathbf{k}+\mathbf{q}} - \varepsilon_{n\mathbf{k}}} \exp(i\mathbf{q} \cdot \mathbf{R}_{ij})$$

$$\cdot u^*_{n\mathbf{k}}(\mathbf{R}_i) u_{n'\mathbf{k}+\mathbf{q}}(\mathbf{R}_i) u^*_{n'\mathbf{k}+\mathbf{q}}(\mathbf{R}_j) u_{n\mathbf{k}}(\mathbf{R}_j),$$

(4.142)

where we have used the fact that the Bloch function can be written as

$$\psi_\mathbf{k}(\mathbf{r}) \equiv \psi_{n\mathbf{k}}(\mathbf{r}) = u_{n\mathbf{k}}(\mathbf{r}) \exp(i\mathbf{k} \cdot \mathbf{r}).$$

(4.143)

The function $u_{n\mathbf{k}}(\mathbf{r})$ is periodic with the crystal lattice and n denotes the band index.

There have been only a few attempts to evaluate Eqn (4.142) as it stands, the issue being the product of four wave functions. Herman and Schrieffer (1992), who also drop the assumption of localized magnetic moments, discuss possible effects arising from the wave functions, but they do not carry out any detailed calculations. We will come back to these effects. Other examples are the papers by Lee and Chang (1994 and 1995a,b) who, using approximations we will briefly discuss subsequently, also retain the wave function product. However, usually the next step is to use plane waves which amounts to replacing the product of four wave functions in Eqn (4.142) by $1/\Omega^2$, where Ω is the volume in which the plane waves are normalized. This step simplifies the formalism enormously and allows us to write

$$J(\mathbf{R}_{ij}) = \left(\frac{A}{\Omega}\right)^2 \operatorname{Re} \sum_\mathbf{q} \chi_0(\mathbf{q}) \exp(i\mathbf{q} \cdot \mathbf{R}_{ij}),$$

(4.144)

where

$$\chi_0(\mathbf{q}) = \sum_{nn'\mathbf{k}} \frac{f_{n\mathbf{k}} - f_{n'\mathbf{k}+\mathbf{q}}}{\varepsilon_{n'\mathbf{k}+\mathbf{q}} - \varepsilon_{n\mathbf{k}}}$$

(4.145)

is the unenhanced magnetic susceptibility (in units of μ_B^2) that was discussed in Sec. 4.2 and evaluated in Sec. 2.8 for free electrons, in which case it is basically the Lindhard function. It is this case that can be mastered with a moderate amount of mathematics, yielding insight into the possible origin of oscillatory exchange coupling.

4.5.1.2 Free electrons

The simplest possible case is to assume magnetic moments immersed in a gas of free electrons without restricting the geometry in any way. Then Eqn (4.145) becomes

$$\chi_0(\mathbf{q}) = \frac{\Omega}{(2\pi)^3} \int \frac{f_\mathbf{k} - f_{\mathbf{k}+\mathbf{q}}}{\varepsilon_{\mathbf{k}+\mathbf{q}} - \varepsilon_\mathbf{k}} d^3k.$$

(4.146)

Using the free-electron eigenvalues (for convenience with an effective mass m^*) $\varepsilon_\mathbf{k} = \hbar^2 \mathbf{k}^2/2m^*$ we easily derive (or follow Kittel, 1968)

$$\chi_0(\mathbf{q}) = \frac{3N}{4\varepsilon_F} \left\{ 1 + \frac{4k_F^2 - q^2}{4k_F\, q} \ln \left| \frac{2k_F + q}{2k_F - q} \right| \right\}, \tag{4.147}$$

where N is the number of electrons in the system, ε_F is the Fermi energy, $\varepsilon_F = \hbar^2 k_F^2/2m^*$, and k_F is the Fermi wave vector (connected with the electron density, n, by $k_F = (3\pi^2 n)^{1/3}$). Notice that $\chi_0(\mathbf{q})$ possesses a logarithmic singularity at twice the Fermi radius k_F. Next the Fourier transform indicated in Eqn (4.144) is to be carried out. After changing the sum to an integral as in the step from Eqn (4.145) to Eqn (4.146), one does the integration in the complex plane, all details being given by Kittel (1968). Then the RKKY exchange coupling between a magnetic moment at the origin and a magnetic moment on a spherical shell of radius R is obtained as

$$J(R) = \frac{16\, A^2\, m^*\, k_F^4}{(2\pi)^3\, \hbar^2} F(2k_F\, R), \tag{4.148}$$

where

$$F(x) = \frac{x \cos x - \sin x}{x^4}. \tag{4.149}$$

The result is clearly seen to be oscillatory going asymptotically as $J(R) \sim \cos(2k_F R)/R^3$. The period, λ, is related to the Fermi wavelength, $\lambda_F = 2\pi/k_F$, by $\lambda = \lambda_F/2$; it is thus connected with the logarithmic singularity in the susceptibility. In the next step we consider a three-dimensional system again, but assume the magnetic moments populate planes P_1 and P_2 being separated by a spacer similar to the situation depicted in Fig. 4.74(a) or (b), in which we may label the bottom layer with P_1 and the top layer with P_2. Quite generally the interlayer coupling is obtained by specifying the summation indices in Eqn (4.141) such that we sum up all pairs ij, i running in P_1 and j in P_2. One then defines the coupling energy per unit area as

$$E_C = I_C \cos \theta, \tag{4.150}$$

where θ is the angle between the direction of magnetization of the bottom and top layers and the interlayer coupling constant is given by

$$I_C = \frac{d}{V_0} S^2 \sum_{j \in P_2} J(\mathbf{R}_{0j}), \tag{4.151}$$

where 0 denotes the origin assumed in layer P_1, d is the spacing between two adjacent atomic layers, and V_0 is the volume of the atomic unit cell. The sign convention chosen is such that positive (negative) J gives ferromagnetic (antiferromagnetic) coupling, see Eqn (4.141).

In the simple model we deal with here we assume the spacer consists of free electrons. In addition we replace the ferromagnetic layers by a continuous uniform distributions of spins in P_1 and P_2 which allows us to exchange the summation in Eqn (4.151) by an integration,

$$\sum_{j \in P_2} \rightarrow \frac{d}{V_0} \int_{P_2} d^2 \mathbf{R}_{\|}, \qquad (4.152)$$

where $\mathbf{R}_{\|}$ is the in-plane projection of the vector $\mathbf{R}_{0j} = \mathbf{R}_0 - \mathbf{R}_j$. We then may proceed with Eqn (4.148) and denoting the distance between the planes P_1 and P_2 by z we integrate

$$\int_{P_2} F(2k_F \sqrt{z^2 + R_{\|}^2}) d^2 \mathbf{R}_{\|} = \frac{2\pi}{(2k_F)^2} \int_{2k_F z}^{\infty} F(x) x \, dx \qquad (4.153)$$

by parts to obtain for the coupling constant, I_C, defined by Eqn (4.151)

$$I_C = I_C(z) = 2 I_0 \, d^2 k_F^2 \left[\frac{\pi}{2} - \text{Si}(2k_F z) + \frac{\sin(2k_F z)}{(2k_F z)^2} - \frac{\cos(2k_F z)}{2k_F z} \right], \qquad (4.154)$$

where

$$\text{Si}(t) = \int_0^t \frac{\sin x}{x} dx$$

is the sine integral function and

$$I_0 = \left(\frac{A}{V_0}\right)^2 S^2 \frac{m^*}{4\pi^2 \hbar^2}. \qquad (4.155)$$

The asymptotic value of $[\pi/2 - \text{Si}(x)]$ is $[\sin x/x^2 + \cos x/x]$ whence

$$I_C = I_0 \frac{d^2}{z^2} \sin(2k_F z) \text{ as } z \longrightarrow \infty. \qquad (4.156)$$

We see there is a single oscillation period $\lambda = \lambda_F/2$ and the coupling decays as z^{-2}, z being the distance between the two magnetic planes P_1 and P_2. It should be noticed that the electron density of the spacer only appears in the oscillation period but not in the constant I_0.

4.5.1.3 Aliasing

At this stage we may attempt a comparison with experimental data. We therefore show in Fig. 4.76 interlayer exchange coupling obtained by Parkin and Mauri (1991) for multilayers where the spacer consists of Ru and the magnetic layers of $Ni_{80}Co_{20}$. The

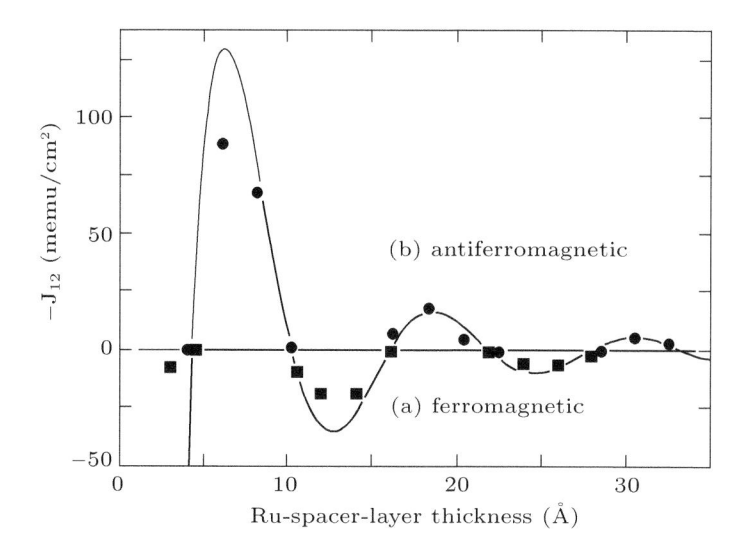

Figure 4.76 *Interlayer exchange coupling strength, J_{12}, (corresponding in our notation to I_C) for $Ni_{80}Co_{20}$ layers separated by Ru spacers, from Parkin and Mauri (1991) (with permission). The solid line is a fit as explained in the text.*

data are fitted with the function $J_{12} \sim \sin(\phi + 2\pi z/\Lambda)/z^p$ giving values $p = 1.8$ and $\Lambda = 11.5$ Å. While the value of p is in satisfactory agreement with Eqn (4.156) the value of Λ is much larger than the oscillation period, which one estimates from the lattice constant of Ru to be $\lambda = \pi/k_F = 2.45$ Å. However, this discrepancy finds a simple explanation in the discrete, atomic nature of the multilayer (Coehoorn, 1991).

The argument is that the spacer thickness z in Eqn (4.156) is not a continuous variable, but is a multiple of the spacer separation d, i.e. $z = Nd$, where N is an integer. We show in Fig. 4.77 a plot of $\sin(2\pi z/\lambda)/z^2$ for a ratio $\lambda d = 1.15$ with wavelength $\lambda = 2.5$ Å, and interplanar spacing, d, plotted as a function of the continuous variable z. Dots sample the function at the discrete locations of the atomic layers. By connecting the dots we see that the discrete sampling results in a much longer *effective* wavelength, Λ, than the value of λ we started with. This effect is known as *aliasing* and the theory of Fourier analysis allows us the write down a connection of the effective wavelength Λ with the interplanar spacing d and the original λ,

$$\Lambda = \frac{1}{|1/\lambda - n/d|}, \quad \text{with } n \geq 1 \text{ such that } \Lambda \geq 2d. \tag{4.157}$$

To give an idea about the justification of this result (Bruno and Chappert, 1992) let z be a continuous variable and let $f(z)$ possess the Fourier transform $F(q)$; if we denote by f_N the values obtained by sampling $f(z)$ at the equally spaced intervals $z = Nd$ (with integer N) then the Fourier transform, $\tilde{F}(q)$ of f_N, is

294 *Electronic Structure and Magnetism*

$$\tilde{F}(q) = \sum_{n=-\infty}^{\infty} F(q + 2\pi n/d), \tag{4.158}$$

which is not hard to verify by calculating

$$f_N = \int_{-\infty}^{\infty} \tilde{F}(q) \exp(iq\,Nd)\,dq$$

and observing that the resulting δ-functions sample $f(z)$ as desired. To describe the function $f_{N=z/d}$—which in Fig. 4.77, for example, looks like the function defined by

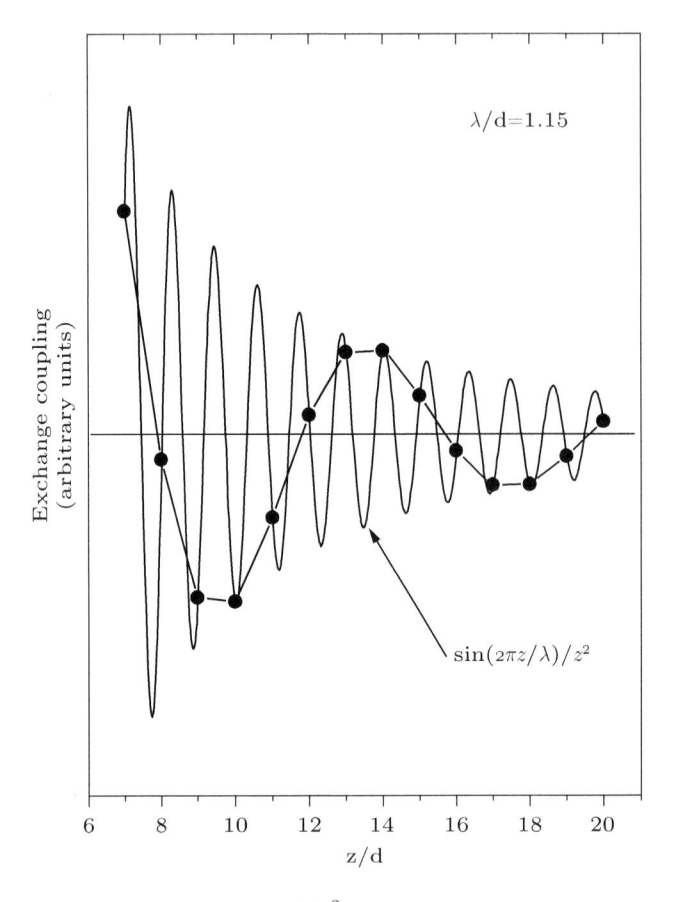

Figure 4.77 *Plot of the function* $\sin(2\pi z/\lambda)/z^2$ *for a wavelength,* λ, *and interplanar spacing,* d, *ratio of* $\lambda/d = 1.15$, *with* $\lambda = 2.5$ Å. *Dots sample the function at the discrete locations of the atomic layers. The aliasing effect is visible in the function defined by the connected dots, for which one reads off* $\Lambda/d = 7.8$ *in agreement with Eqn (4.158).*

the connected dots—as a function of the continuous variable z, $\tilde{F}(q)$ is back-Fourier transformed, but only values of q in the interval $-\pi/d \leq q \leq \pi/d$ are allowed, since periods smaller than $2d$ are physically meaningless; in other words, $2\pi/\Lambda$ must be in the first one-dimensional Brillouin zone that has the length d in the z-direction. Equation (4.157) summarizes these conditions.

Although by now a certain amount of understanding has been reached, we cannot be content with the theory presented so far. Ignoring for the moment doubts about the general applicability of RKKY-type theories (mentioned at the beginning of this section), we stress that our basic assumption about the electronic structure of the multilayer is vastly oversimplified. Noting again that the periodicity of the oscillatory exchange coupling in both Eqns (4.148) and (4.156) originates from the Fermi surface through the logarithmic singularity in the susceptibility function $\chi_0(\mathbf{q})$, we expect the correct topology of the Fermi surface—just as in the case of bulk chromium in Sec. 4.3.4—to lead to important effects not captured by the simplified theory we dealt with so far. The next step is therefore to introduce some improvements of the theory that begin to take into account the real band structure.

4.5.1.4 Fermi surface effects

Let us rewrite the coupling energy $E_C = I_C \cos\theta$ where I_C is given by Eqn (4.151) using for the exchange constants Eqn (4.142) but dropping the product of the wave functions again, i.e. we begin with

$$I_C = \frac{S^2 A^2 d}{V_0 (2\pi)^6} \operatorname{Re} \sum_{n,n'} \int d\mathbf{k} \int d\mathbf{q} \, \frac{f_{n\mathbf{k}} - f_{n'\mathbf{k+q}}}{\varepsilon_{n'\mathbf{k+q}} - \varepsilon_{n\mathbf{k}}} \sum_{\mathbf{R}_\parallel \in P_2} \exp(iq_z z + i\mathbf{q}_\parallel \mathbf{R}_\parallel). \quad (4.159)$$

Next we assume (i) the Fermi surface is formed by one band only, so we may drop the sum over the band indices, n and n', and (ii) the Brillouin zone is that appropriate for the bulk spacer. Assumption (i) is well justified for the noble metals, but assumption (ii) is valid only in the limit of very thick spacers. The Brillouin zone integrals are now carried out in the periodic zone scheme, for which the sum on $\mathbf{R}_\parallel$ gives zero unless $\mathbf{q}_\parallel = 0$. By rearranging the integrations it is not hard to see now that one can write the coupling constant I_C as

$$I_C = \mathcal{C} S^2 \int_{\text{2DBZ}} j_z(\mathbf{k}_\parallel) \, d\mathbf{k}_\parallel, \quad (4.160)$$

where $\mathcal{C}$ denotes a collection of constants, 2DBZ indicates that the integration is taken over the two-dimensional Brillouin zone, and $j_z(\mathbf{k}_\parallel)$ is given by

$$j_z(\mathbf{k}_\parallel) = \operatorname{Re} \int_{-\pi/d}^{\pi/d} \int_{-\pi/d}^{\pi/d} dq_z \, dk_z \, \frac{f_{\mathbf{k}_\parallel k_z} - f_{\mathbf{k}_\parallel q_z}}{\varepsilon_{\mathbf{k}_\parallel q_z} - \varepsilon_{\mathbf{k}_\parallel k_z}} \exp\left[i(q_z - k_z)z\right]. \quad (4.161)$$

The choice of a prismatic Brillouin zone, having as base the 2DBZ and height d, enables the integrations to be separated as shown in Eqns (4.160) and (4.161). Bruno and Chappert (1992) now proceed to evaluate $j_z(\mathbf{k}_\parallel)$ by using a complex contour integration and subsequently perform the integration in Eqn (4.160) employing the stationary-phase approximation (see also Chap. 1). The result of these somewhat tedious manipulations is

$$I_C = I_0 \frac{d^2}{z^2} \sum_\alpha \frac{m^*_\alpha}{m^*} \sin(q_z^\alpha z + \psi_\alpha), \qquad (4.162)$$

where q_z^α is a nesting vector which connects two extremum points of the Fermi surface along the z-direction and I_0 is given by Eqn (4.155). The quantity m^*_α is an effective mass and ψ_α a phase shift that are determined by the curvature and topology of the Fermi surface at the stationary points labeled by α (the reader interested in more details should consult the paper by Bruno and Chappert, 1992).

It is important to notice that the formula of Bruno and Chappert, Eqn (4.162), in agreement with experimental findings, describes situations in which more than one oscillation period is possible. In fact, the agreement is even quantitative. We show in Fig. 4.78 the well-known Fermi surface topology of copper which, with somewhat changed dimensions, also applies to Ag and Au (Ashcroft and Mermin, 1976). The number of oscillations we can read off from Fig. 4.78 are: one long period for a (111)-growth direction, two oscillation periods (one short and one long) for a (001)-growth direction, and four oscillation periods for (110)-growth directions. Table 4.14, collected by Bruno (1995), summarizes the periods obtained by the RKKY theory and those measured.

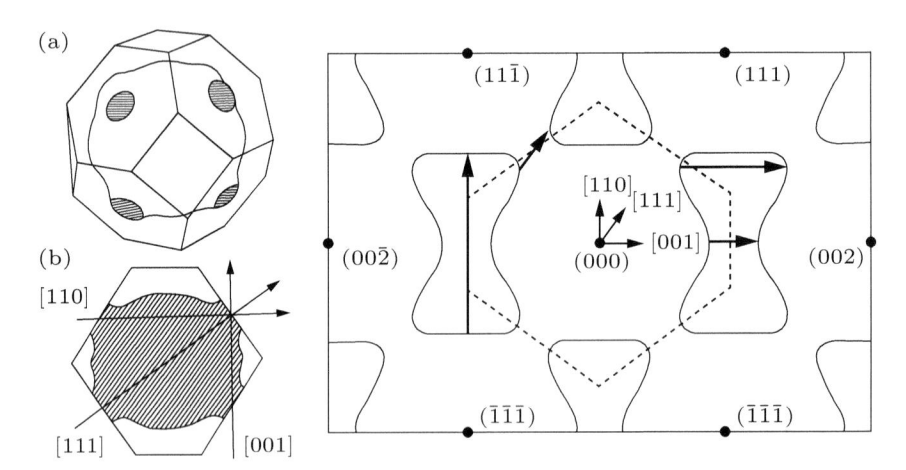

Figure 4.78 *Fermi surface of Cu: (a) Standard representation in the reduced zone. (b) Cross-sectional view in the reduced zone indicating the three important growth directions for multilayers. Right-hand side: Cross-sectional view in the periodic zone scheme; arrows give the nesting vectors that determine the period of the oscillations (after Herman and Schrieffer, 1992).*

Table 4.14 *List of theoretical oscillation periods, in units of monolayers (ML), and experimental oscillation periods (in ML). Fermi surface data from experimental dHvA measurements, except for bcc Cu, after Bruno (1999).*

Spacer	Theoretical periods	System	Experimental periods	Reference
Cu(111)	$\Lambda = 4.5$	Co/Cu/Co	$\Lambda = 6$	1
		Co/Cu/Co	$\Lambda = 5$	2
		Fe/Cu/Fe	$\Lambda = 6$	3
Cu(001)	$\Lambda_1 = 2.6$			
		Co/Cu/Co	$\Lambda = 6$	4
	$\Lambda_2 = 5.9$			
		Fe/Cu/Fe	$\Lambda = 7.5$	5
			$\Lambda_1 = 2.6$	
		Co/Cu/Co		6
			$\Lambda_2 = 8$	
Au(001)	$\Lambda_1 = 2.6$		$\Lambda_1 = 2$	
		Fe/Au/Fe		7
	$\Lambda_2 = 8.6$		$\Lambda_2 = 7.5$	
Ag (001)	$\Lambda_1 = 2.4$		$\Lambda_1 = 2.4$	
		Fe/Ag/Fe		8
	$\Lambda_2 = 5.6$		$\Lambda_2 = 5.6$	
bcc Cu(001)[a]	$\Lambda_1 = 2.2$			
		Fe/Cu/Fe	$\Lambda = 2$	9
	$\Lambda_2 = 2.6$			
Cu(110)	$\Lambda_1 = 2.1$			
	$\Lambda_2 = 2.5$			
	$\Lambda_3 = 3.3$			
	$\Lambda_4 = 9.6$			
		Co/Cu/Co	$\Lambda = 9.4$	9
		Co/Cu/Co	$\Lambda = 9.4$	10

[a] ASW calculations of the bulk Fermi surface of bcc Cu (Johnson et al., 1992).

References: (1) Mosca et al., 1991; (2) Parkin et al., 1991; (3) Petroff et al., 1991; (4) de Miguel et al., 1991; (5) Bennett et al., 1990; (6) Johnson et al., 1992; (7) Fuss et al., 1992; (8) Unguris et al., 1991; (9) Johnson et al., 1992; (10) Okuno and Inomata, 1993.

These results are indeed impressive; however, further improvements are difficult to achieve. For instance, one would like to have a theory that is valid also for cases when the spacer does not consist of a noble metal. Of course one can always *assume* that RKKY in the simple form presented above continues to hold even when, as in the early experiments by Unguris *et al.* (1991) and also by Wolf *et al.* (1993), the spacer consists of e.g. chromium. On the theoretical side the neglect of the wave functions was discussed qualitatively by Herman and Schrieffer (1992), but no detailed calculations were undertaken. This was, however, done by Lee and Chang (1994) who used a tight-binding description of the band structure of Cu and verified Eqn (4.162) for the case at hand, but, most importantly, they brought to attention how the superlattice which the multilayer constitutes breaks up the Fermi surface and leads to interesting effects that are better studied in a different context.

A theory that encompasses the results of the RKKY theory has been developed that does not suffer from the shortcomings of RKKY. It can be made to be *ab initio* and it provides an interpretative tool by combining *ab initio* elements with model aspects, thus providing insight into the physical mechanisms important for magnetic multilayers. We therefore turn to a discussion of this theory next.

4.5.2 Oscillatory exchange coupling in the quantum-well picture

To begin with, we digress briefly into *ab initio* total energy-based theories because they are simple to understand and they give insight into the essence of the problem. The idea is to calculate the total energy of the system for different directions of the magnetizations in neighboring magnetic layers, and to identify the energy difference with the interlayer exchange coupling. This is nothing but an improved way of estimating the right-hand side of Eqn (4.141) using for the left-hand side an—in principle—exact total energy difference. The geometries chosen are in almost all cases repeated layers of spacer and ferromagnet for which, as a function of the thickness of the spacer and/or that of the ferromagnet, the total energy differences are obtained. Calculations of this type were either performed using semi-empirical tight-binding models (e.g. Stoeffler and Gautier, 1991, 1993) and *ab initio* schemes like ASW and LMTO (e.g. Herman *et al.*, 1991; Herman *et al.*, 1993; Sticht *et al.*, 1993; van Schilfgaarde and Herman, 1993). The technical difficulty, especially in the latter approach, is that the energy differences for different magnetic configurations are very small (of the order of meVs), while the total energy is huge. Furthermore, when the spacer thickness is increased the size of the unit cell increases with the number of atomic layers, making the calculations progressively more difficult; in particular, it is the numerical convergence that becomes a serious problem. Thus one is limited to a relatively small layer thickness, a fact that renders the investigation of long-period oscillations quite hard. However, resorting to fitting procedures using the RKKY formula given in Eqn (4.162) did provide a solution to the problem, that, indeed, allowed the observed short and long periods in Fe/Cr/Fe multilayers (Unguris *et al.*, 1991; Wolf *et al.*, 1993) to be traced back to Fermi surface nesting of chromium that we discussed in Sec. 4.3.4 (see Fig. 4.38). In short, one problem we locate here is that of interpreting numerical data using models whose reliability must be postulated thus leaving open some quantity of doubt.

Another problem that constitutes a serious challenge is that the coupling strengths obtained from total energy calculations are typically one order of magnitude larger than those obtained experimentally. To overcome this difficulty one invoked effects of roughness which the spacer and ferromagnetic layers possess in the region of contact. This is most likely a correct assumption but is still being investigated with the theory that we will present now.

4.5.2.1 Confined states in multilayers

It is not our aim to describe the events leading up to the development of the theory in any historical order, nor do we want to raise priority issues here. Rather, we begin with a question that can be answered using a simple *ab initio* calculation. We assume we want to grow on fcc Cu a few layers of fcc Fe in the (001)-orientation, continuing this with fcc Cu then fcc Fe again, etc., and thus produce a multilayer system. The band structure of separate bulk fcc Cu and fcc Fe is shown in Fig. 4.79, not plotted in the usual Brillouin zone but rather in a way that is adapted to the multilayer geometry we want to study, see the details in the figure caption. Now we ask what the band structure of a particular multilayer looks like in comparison with that of the bulk constituents shown in Fig. 4.79. For this purpose a simple, periodic-multilayer geometry is constructed consisting of two layers of fcc Fe followed by four layers of fcc Cu stacked in the (0,0,1)-direction. The tetragonal unit cell then contains six atoms with $c/a = 4.24264$. This is still small enough to easily calculate (on a PC) self-consistently the total energy for a periodic arrangement that has as basic building block something that looks like Fig. 4.74(a) or Fig. 4.74(b). The total energy is slightly (about 4 meV per cell) in favor of the antiparallel orientation depicted in Fig. 4.74(b). However, the convergence was not established in detail since it is only the band structure we are interested in. This is shown in Fig. 4.80, where for the parallel moment arrangement we depict in (a) the spin-up states and in (b) the spin-down states; (c) shows the states for the antiparallel moment arrangement defined by the **q**-vector $\mathbf{q} = (0, 0, 0.5)\, 2\pi/c$. The labels are defined as in Fig. 4.79. Certainly, the band structure looks complicated on first sight, but its salient features can be understood as follows. The directions parallel to the multilayer are labeled by the (long) lines ΓX and $Z' Z$. Here, with only little effort, a correspondence of the bands with those shown along the same lines in Fig. 4.79 can be recognized (except perhaps for the energy range between -2 and -4 eV where the d bands get very crowded). However, the bands along the short lines $Z \Gamma$ and $X Z'$, which mark directions perpendicular to the multilayer, have almost all become totally flat. Having no dispersion they must describe localized states, states that are *confined* to the layer. Electrons in these states do not propagate across layers. However, not all states are confined. Thus one can see a band near the Fermi energy (chosen as the energy origin) along $Z \Gamma$ possessing dispersion not much different from the bands near 0 eV labeled Δ_1 in Fig. 4.79(a) and (b). It is precisely the fact that these Δ_1 states exists at nearly the same energy in Cu and spin-up Fe that explains the dispersion in Fig. 4.80(a): a spin-up electron of this energy can freely travel across the layers staying in the same state and hence is *not* confined. The corresponding

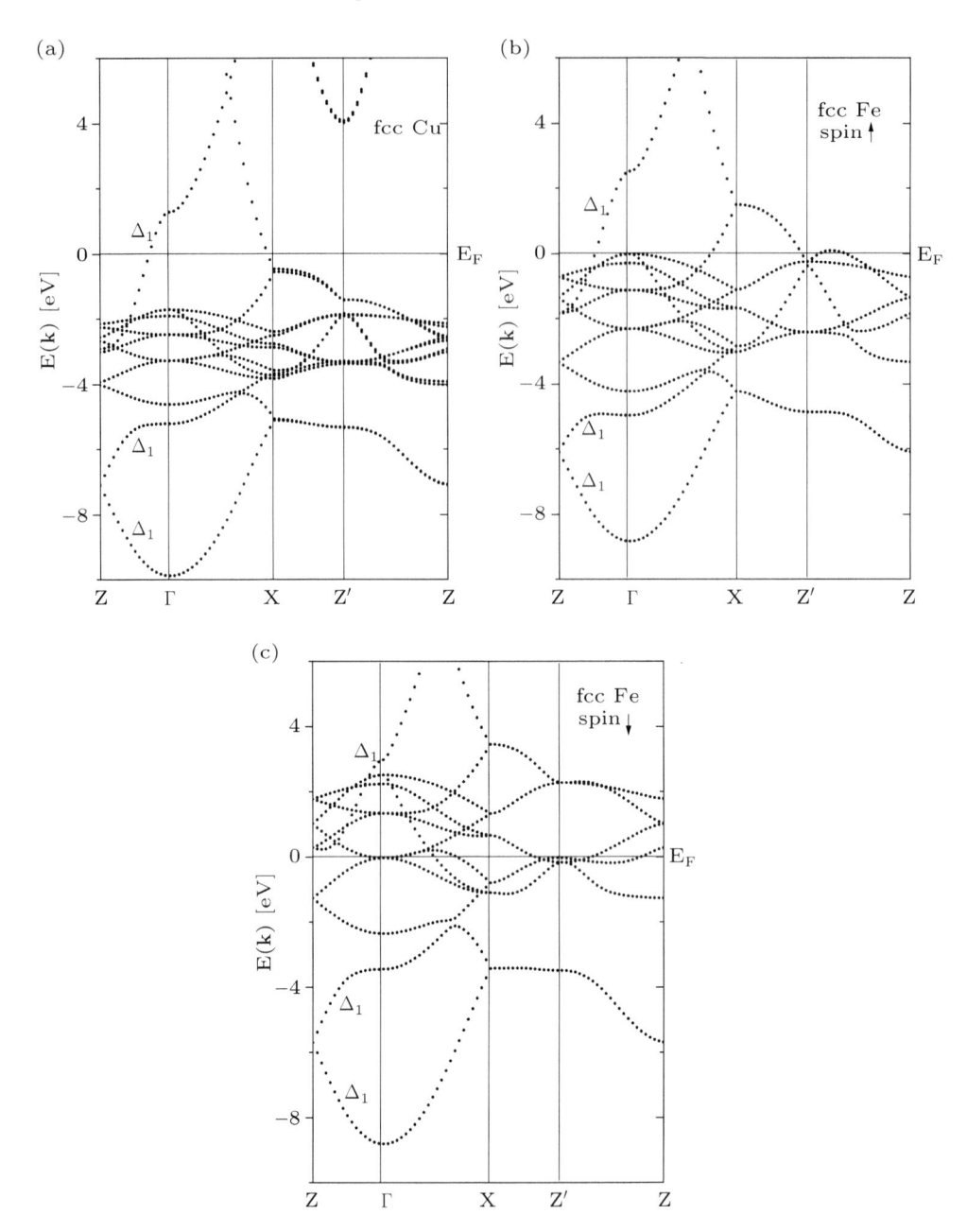

Figure 4.79 Band structure of (a) fcc Cu, (b) fcc Fe, spin up, and (c) fcc Fe, spin down. The unit cell used is the CuAuI cell consisting of a body-centered prism with a square base of dimension $\sqrt{2}\,a$ by $\sqrt{2}\,a$ and height a, where a is the fcc lattice constant. The labels are defined as $Z = (0,0,0.5)\,2\pi/c$, $\Gamma = (0,0,0)$, $X = (0.5,0,0)\,2\pi/a$, $Z' = (0,0.5,0.5\,a/c)\,2\pi/a$, where $c = a_{\text{fcc}}$ and $a = a_{\text{fcc}}/\sqrt{2}$. For a_{fcc} we have chosen the experimental lattice constant of Cu at which fcc Fe is ferromagnetic, see the HS-FM state in Fig. 4.47, Sec. 4.3.4.

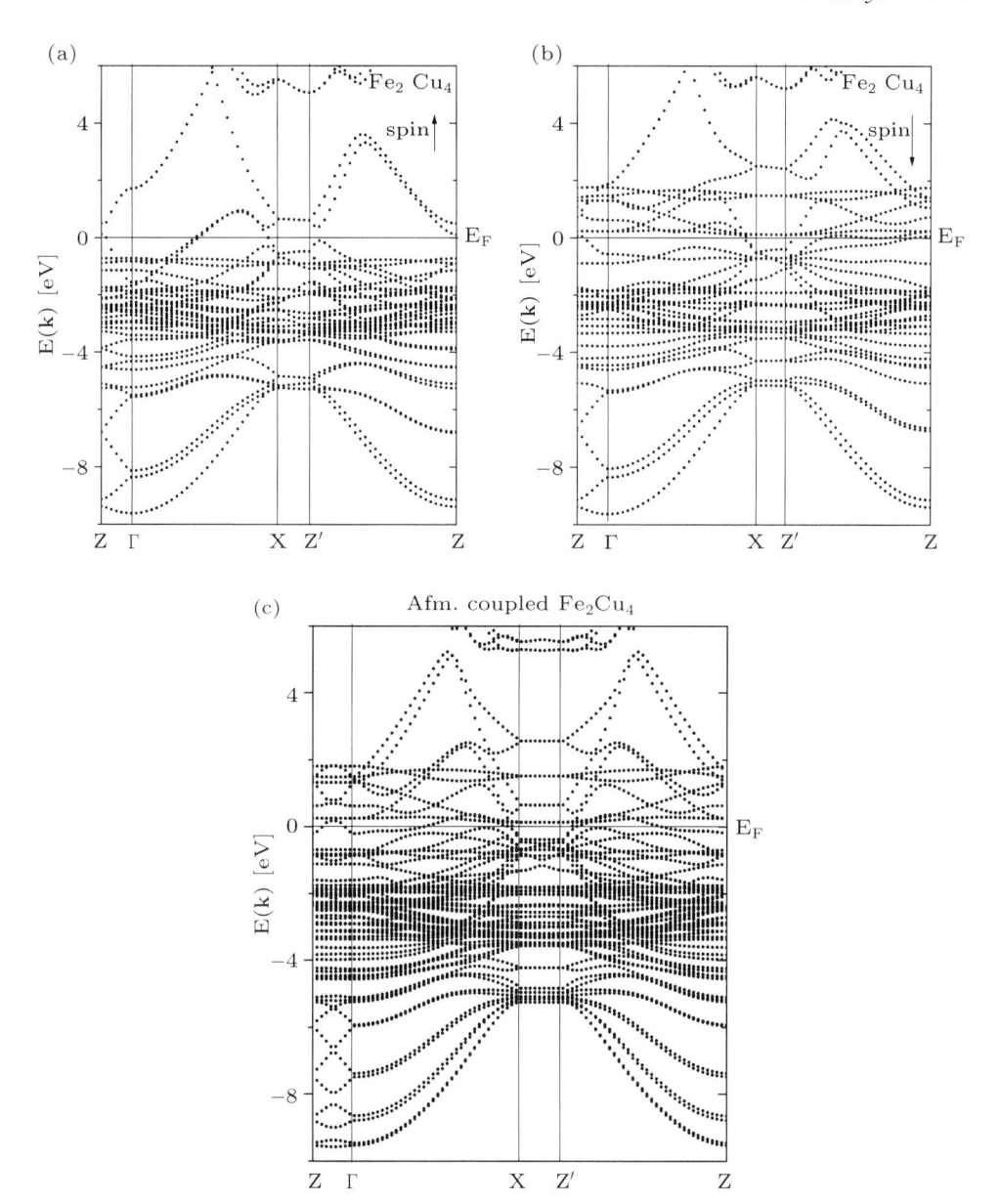

Figure 4.80 *Band structure of (a) spin-up bands of a Fe/Cu multilayer possessing two layers of Fe and four layers of Cu stacked along (001) in fcc coordination with $c/a = 4.24264$ and parallel moments; labels are given in the caption of Fig. 4.79; (b) same as (a) but spin down, and (c) for antiparallel moment arrangements of ferromagnetic, successive Fe layers. This state is defined by $\mathbf{q} = (0, 0, 0.5)\, 2\pi/c$.*

spin-down electron is partially reflected because the Δ_1 state in Fig. 4.79(c) above 0 eV does not cross the Fermi level. This explains the difference between the spin-up and spin-down states in the multilayer and shows that *the spacer carries information about the electron spin*. The energy bands shown in Fig. 4.80(c) for antiparallel spins in the ferromagnetic layers describe both spin-up and spin-down electrons. The states derived from Δ_1 remain dispersive, but the expected folding from the zone appropriate for (a) and (b) into that appropriate for (c) makes the bands appear narrower. Clearly, confined states remain confined. Another simple CoCu multilayer that conveys a similar picture is considered in the appendix.

As early as 1991 a model was formulated by Edwards *et al.* (1991) to understand the oscillatory exchange coupling by assuming the states for the spin-down electron to be completely confined in the spacer layer. Since the size of the spacer plays an important role in the confinement, one envisioned interference effects that are associated with size quantization of the energy. In 1993 these states were first seen in photoemission and inverse photoemission experiments by Ortega *et al.* (1993) (see also Himpsel *et al.*, 1998) who called the electron states in question *quantum-well states* and connected them with magnetic coupling through a spacer. A unified view was given by Bruno (1993a,b,c and 1995) whose theory we largely follow here. Very similar ideas were put forward by Lang *et al.* (1993), Nordström *et al.* (1994), Kudrnovsky *et al.* (1994), and Stiles (1993, 1996).

We begin with a simple model that allows us to describe changes of the density of states, $\Delta \mathcal{N}(\varepsilon)$, due to—partial or total—electron confinement. It is then shown how from a knowledge of $\Delta \mathcal{N}(\varepsilon)$ we can calculate the total energy difference which leads to oscillatory exchange coupling without the assumptions made in Sec. 4.5.1.1. This simple model will then be put on solid ground using a Green's function method.

4.5.2.2 A simple model

We consider a one-dimensional, highly simplified quantum well consisting of a spacer characterized by the potential $V = 0$, having width D, and bounded by the potentials $V_B \neq 0$ and $V_T \neq 0$ on the bottom and top, respectively, see Fig. 4.74(a) or (b). We assume an electron of wave vector $k^+ (>0)$ propagates from bottom to top and is partially reflected down by the potential V_T emerging with a complex amplitude $r_T = |r_T| \exp(i\phi_T)$. The reflected wave of wave vector $k^- = -k^+$ is in turn reflected by the potential V_B where it emerges with the amplitude $r_B = |r_B| \exp(i\phi_B)$. The modulus $|r_B|, |r_T|$ of the reflection coefficients give the magnitude of the reflected waves, while ϕ_B, ϕ_T express the phase shift due to the reflection. Interference between the waves due to multiple reflections causes a modification of the density of states in the spacer that is estimated as follows. The phase shift resulting from a complete round trip in the spacer is

$$\Delta\phi = qD + \phi_T + \phi_B \qquad (4.163)$$

where $q = k^+ - k^-$. The interference is constructive if $\Delta\phi = 2n\pi$ (with n an integer) resulting in the density of states increasing, whereas, if $\Delta\phi = (2n+1)\pi$, it is destructive and the density of states will decrease. It is plausible, therefore, to assume the density of states varies as a function of D as $\cos(qD + \phi_T + \phi_B)$. This is a clever ansatz due to

Bruno (1995) which allows us to sum up the effects of multiple reflections by making the further assumptions that the effect is proportional to $|r_T r_B|$, to the width D, and to the density of states per unit energy and unit width, $(2/\pi)(\mathrm{d}q/\mathrm{d}\varepsilon)$ (where the factor of 2 accounts for the spin degeneracy). Thus

$$\Delta \mathcal{N}(\varepsilon) = \frac{2D}{\pi} \frac{\mathrm{d}q}{\mathrm{d}\varepsilon} \sum_{N=1}^{\infty} |r_T r_B| \cos N(qD + \phi_T + \phi_B)$$

$$= \frac{2}{\pi} \operatorname{Im}\left[\mathrm{i} D \frac{\mathrm{d}q}{\mathrm{d}\varepsilon} \sum_{N=1}^{\infty} |r_T r_B|^N \exp(\mathrm{i} N q D) \right] \quad (4.164)$$

$$= \frac{2}{\pi} \operatorname{Im}\left[\mathrm{i} \frac{\mathrm{d}q}{\mathrm{d}\varepsilon} D \frac{r_T r_B \exp(\mathrm{i}qD)}{1 - r_T r_B \exp(\mathrm{i}qD)} \right]$$

where we have made some simple substitutions and summed the geometric series that originates from the multiple reflections. We will see next that we need the change of the density of states integrated up to some energy ε, i.e.

$$\Delta N(\varepsilon) = \int^{\varepsilon} \Delta \mathcal{N}(\varepsilon') \, \mathrm{d}\varepsilon' \quad (4.165)$$

which is easily seen from Eqn (4.164) to be given by

$$\Delta N(\varepsilon) = -\frac{2}{\pi} \operatorname{Im} \ln[1 - r_T r_B \exp(\mathrm{i}qD)]. \quad (4.166)$$

Clearly, the quantity $\Delta N(\varepsilon)$ is oscillatory as a function of D, the amplitude being given by the reflection coefficients r_T and r_B and the wavelength, Λ, by $q = k^+ - k^-$ through $\Lambda = 2\pi/q$.

Knowledge of $\Delta N(\varepsilon)$ allows us at once to write down the change in total energy; for this we take a formula from Chap. 5 which relates the free-particle thermodynamic grand potential to the Fermi–Dirac distribution function f and the integrated density of states as

$$\Omega_0 = -\int_{-\infty}^{\infty} f(\varepsilon) N(\varepsilon) \, \mathrm{d}\varepsilon. \quad (4.167)$$

We will later comment on the derivation of this equation which appears in Chap. 5 as Eqn (5.291). In the limit $T = 0$ the grand potential becomes the total energy, E for which we thus obtain $E = \int^{\varepsilon_F} N(\varepsilon) \, \mathrm{d}\varepsilon$, hence from Eqn (4.166)

$$\Delta E = \frac{2}{\pi} \operatorname{Im} \int^{\varepsilon_F} \ln\left[1 - r_T r_B \exp(\mathrm{i}qD)\right] \mathrm{d}\varepsilon. \quad (4.168)$$

This again we call the coupling energy.

It is not hard to generalize the result to a three-dimensional layered system. In this case translational invariance parallel to the planes gives the wave vector $\mathbf{k}_\parallel$ as a good quantum number. For a given $\mathbf{k}_\parallel$ we are facing the situation described above and merely have to sum over $\mathbf{k}_\parallel$. This gives for the integrated density of states per unit area

$$\Delta N(\varepsilon) = -\frac{1}{2\pi^3} \operatorname{Im} \int d^2\mathbf{k}_\parallel \ln\left[1 - r_T\, r_B \exp(iq_\perp D)\right] \qquad (4.169)$$

and for the coupling energy per unit area

$$\Delta E = \frac{1}{2\pi^3} \operatorname{Im} \int d^2\mathbf{k}_\parallel \int^{\varepsilon_F} \ln\left[1 - r_T\, r_B \exp(iq_\perp D)\right] d\varepsilon. \qquad (4.170)$$

Now we consider a paramagnetic layer sandwiched between two ferromagnetic barriers B and T, the reflection coefficients of which are now spin dependent. We consider only the two possibilities shown in Fig. 4.74(a) and (b) calling the configuration (a) *ferromagnetic* (F) and (b) *antiferromagnetic* (AF).

Introducing the spin-dependent reflection coefficients $r_T^\uparrow, r_T^\downarrow$ for the top barrier and $r_B^\uparrow, r_B^\downarrow$ for the bottom we immediately write down from Eqn (4.170) the total energy change per unit area for the ferromagnetic configuration as

$$\Delta E_\mathrm{F}$$
$$= \frac{1}{4\pi^3} \operatorname{Im} \int d^2\mathbf{k}_\parallel \int^{\varepsilon_F} d\varepsilon \ln\left\{\left[1 - r_T^\uparrow r_B^\uparrow \exp(iq_\perp D)\right]\left[1 - r_T^\downarrow r_B^\downarrow \exp(iq_\perp D)\right]\right\}, \qquad (4.171)$$

whereas for the antiferromagnetic configuration

$$\Delta E_\mathrm{AF}$$
$$= \frac{1}{4\pi^3} \operatorname{Im} \int d^2\mathbf{k}_\parallel \int^{\varepsilon_F} d\varepsilon \ln\left\{\left[1 - r_T^\uparrow r_B^\downarrow \exp(iq_\perp D)\right]\left[1 - r_T^\downarrow r_B^\uparrow \exp(iq_\perp D)\right]\right\}. \qquad (4.172)$$

Thus the interlayer exchange coupling, $I_c = \Delta E_\mathrm{F} - \Delta E_\mathrm{AF}$, is

$$I_c = \frac{1}{4\pi^3} \operatorname{Im} \int d^2\mathbf{k}_\parallel \int^{\varepsilon_F} d\varepsilon \ln \frac{\left[1 - r_T^\uparrow r_B^\uparrow \exp(iq_\perp D)\right]\left[1 - r_T^\downarrow r_B^\downarrow \exp(iq_\perp D)\right]}{\left[1 - r_T^\uparrow r_B^\downarrow \exp(iq_\perp D)\right]\left[1 - r_T^\downarrow r_B^\uparrow \exp(iq_\perp D)\right]}. \qquad (4.173)$$

For weak confinement, i.e. for small reflection coefficients, this is easily simplified to read

$$I_c = -\frac{1}{\pi^3} \operatorname{Im} \int d^2\mathbf{k}_\parallel \int^{\varepsilon_F} d\varepsilon\, \Delta r_T\, \Delta r_B \exp(iq_\perp D), \qquad (4.174)$$

where $\Delta r_T = (r_T^\uparrow - r_T^\downarrow)/2$ and $\Delta r_B = (r_B^\uparrow - r_B^\downarrow)/2$.

It is this last expression for the interlayer exchange coupling that can be worked out completely in the limit of *large spacer thickness*. In this case the exponential factor in Eqn (4.174) oscillates rapidly with energy ε and wave vector $\mathbf{k}_\parallel$ leading to cancellations of the contributions to the integrals except at the Fermi energy ε_F where the energy integral is terminated. We therefore expand the energy in a Taylor series; denoting Cartesian components with the index j we have $\varepsilon = \varepsilon_F + \sum_j (k_j - k_{jF}) \, d\varepsilon/dk_j + \cdots$, which gives in lowest order for $q_\perp = k_\perp^+ - k_\perp^-$:

$$q_\perp = q_{\perp F} + 2 \frac{\varepsilon - \varepsilon_F}{\hbar v_{\perp F}^{+-}}, \tag{4.175}$$

where we have used the standard expression for the group velocity, $v_{\perp F}^\pm = (1/\hbar)(\partial \varepsilon / \partial k_\perp^\pm)_{\varepsilon = \varepsilon_F}$ and

$$\frac{2}{v_{\perp F}^{+-}} = \frac{1}{v_{\perp F}^+} - \frac{1}{v_{\perp F}^-}. \tag{4.176}$$

In the limit of large D the energy integral is now easily obtained and the exchange coupling becomes

$$I_c = \frac{1}{2\pi^3} \operatorname{Im} \int d^2 \mathbf{k}_\parallel \frac{i\hbar v_{\perp F}^{+-}}{D} \Delta r_T \Delta r_B \exp(iq_{\perp F} D). \tag{4.177}$$

Note that the quantity $q_{\perp F}$ spans the Fermi surface and the velocity $v_{\perp F}^{+-}$ is, through Eqn (4.176), a combination of the group velocities at the extremities $k_\perp^+$ and $k_\perp^-$.

Finally, it is again the assumed large spacer thickness that allows the remaining integration over $\mathbf{k}_\parallel$ to be carried out using the stationary phase approximation (see also Chap. 1). The quantity $q_{\perp F}$ is stationary near critical vectors $\mathbf{k}_\parallel^C$ about which one expands $q_{\perp F}$ as

$$q_{\perp F} = q_{\perp F}^C - \frac{(k_x - k_x^C)^2}{\varkappa_x^C} - \frac{(k_y - k_y^C)^2}{\varkappa_y^C}. \tag{4.178}$$

By a proper choice of axes, terms mixed in x and y cancel out; $\varkappa_x^C$ and $\varkappa_y^C$ are combinations of radii of curvature of the Fermi surface at $(\mathbf{k}_\parallel^C, k_\perp^{+C})$ and $(\mathbf{k}_\parallel^C, k_\perp^{-C})$. One now obtains from Eqn (4.177)

$$I_c = \operatorname{Im} \sum_C \frac{\hbar v_\perp^C \varkappa_C}{4\pi^2 D^2} \Delta r_T^C \Delta r_B^C \exp(iq_\perp^C D), \tag{4.179}$$

where the quantities $q_\perp^C$, $v_\perp^C$, Δr_T^C, and Δr_B^C correspond to the critical vector $\mathbf{k}_\parallel^C$ and

$$\varkappa_C = \sqrt{\varkappa_x^C \varkappa_y^C}. \tag{4.180}$$

306 *Electronic Structure and Magnetism*

Equation (4.179) is the desired result for this simplified model. One sees that the interlayer exchange coupling, I_c, consists of a superposition of oscillatory terms for which we can read off the wavelengths as $\Lambda_C = 2\pi/q_\perp^C$. All that is needed is the bulk Fermi surface of the spacer material. Thus, by considerations completely different from those used for the RKKY theory in Sec. 4.5.1.1, we again obtain the data collected before in Table 4.12 which shows that the number of different wavelengths is at most four if the spacer is made of a noble metal.

We will now show how essentially the same results can be obtained by means of a rigorous theory, thus putting the above results on solid grounds. For the job at hand we need the method of Green's functions which we will again make use of in Chap. 5.

4.5.2.3 Green's functions

The method of Green's functions that was already briefly discussed in Sec. 3.3.2 supplies a very powerful and general tool to tackle the electronic problem under consideration. As we will see, it can be made as precise as one wishes, but can also be approximated to yield intuitive concepts that can be visualized as scattering processes leading, for instance, to reflection and transmission at interfaces. Denoting by ε the single-particle energy and by $\mathcal{H}$ the effective single-particle Hamiltonian, we define the Green's function, $G(\varepsilon)$, as

$$G(\varepsilon) = (\varepsilon - \mathcal{H})^{-1}. \tag{4.181}$$

Formally, if we know the set of eigenkets $\{|k\rangle\}$ of $\mathcal{H}$, then the closure relation,

$$\sum_k |k\rangle \langle k| = 1, \tag{4.182}$$

is used to show easily that

$$G(\varepsilon) = \sum_k \frac{|k\rangle \langle k|}{\varepsilon - \varepsilon_k}, \tag{4.183}$$

where ε_k are the eigenenergies. For translationally invariant systems we may associate the ket $|k\rangle$ with the Bloch ket $|n\,\mathbf{k}\rangle$, n being the band index and $\mathbf{k}$ the wave vector. If necessary we may add a spin index as well.

An important formula relates the density of states, $\mathcal{N}(\varepsilon)$, with the trace of the Green's function:

$$\mathcal{N}(\varepsilon) = -\frac{1}{\pi} \operatorname{Im} \operatorname{Tr} G(\varepsilon + \mathrm{i}0^+). \tag{4.184}$$

This is easily seen by using the formula

$$\frac{1}{x + \mathrm{i}0^+} = \frac{\mathrm{P}}{x} - \pi \mathrm{i}\delta(x) \tag{4.185}$$

and comparing with the definition of the density of states in Chap. 3. (This formula, Eqn (4.185), we remind the reader, is to be used under an integral that includes the origin, P denoting the principal value and δ the Dirac delta function.)

In beginning to make a connection with the problem at hand, we now assume that the Hamiltonian describes a host material which we want to disturb with a perturbation layer that may consist of an arbitrary stacking of different materials. For the following the only restriction we impose is an in-plane translational invariance so that the wave vector parallel to the layer, $\mathbf{k}_\parallel$, remains a good quantum number. Writing now for the Hamiltonian

$$\mathcal{H} = \mathcal{H}_0 + V_A, \tag{4.186}$$

where V_A describes the perturbation due to the layer we introduced, we next relate the Green's function of Eqn (4.181) with the Green's function, $G_0(\varepsilon)$, of the host material,

$$G_0(\varepsilon) = (\varepsilon - \mathcal{H}_0)^{-1}. \tag{4.187}$$

As in Sec. 3.3.2 this is done with the t-matrix, $T_A(\varepsilon)$, which allows us to express $G(\varepsilon)$ with $G_0(\varepsilon)$ as

$$G(\varepsilon) = G_0(\varepsilon) + G_0(\varepsilon)\, T_A(\varepsilon)\, G_0(\varepsilon) \tag{4.188}$$

and relate the t-matrix with the perturbation by

$$T_A(\varepsilon) = V_A + V_A\, G_0(\varepsilon)\, V_A + V_A\, G_0(\varepsilon)\, V_A\, G_0(\varepsilon)\, V_A + \cdots$$
$$= V_A\, [1 - G_0(\varepsilon)\, V_A]^{-1}. \tag{4.189}$$

Here we have summed a geometric series.

Having in mind a multilayer we next introduce a second perturbation layer, described by the potential V_B, parallel to the first one so that we can describe a spacer of N atomic layers sandwiched between two magnetic layers, A and B, of arbitrary thickness. We must, however, require that the perturbation potentials drop rapidly to zero in the spacer. The Green's function we are interested in now is

$$G(\varepsilon) = (\varepsilon - \mathcal{H}_0 + V_A + V_B)^{-1}. \tag{4.190}$$

By expanding this equation or by considering explicitly multiple scattering in the spacer back and forth between the bordering perturbations A and B, we write, temporarily dropping the energy variable ε in the notation,

$$G = G_0 + G_0 T_A G_0 + G_0 T_B G_0 + G_0 T_A G_0 T_B G_0 + G_0 T_B G_0 T_A G_0 + \cdots$$
$$+ G_0 T_A G_0 T_B G_0 T_A G_0 + G_0 T_B G_0 T_A G_0 T_B G + \cdots$$
$$= G_0 + G_0 T_A G_0 + G_0 T_B G_0 \qquad (4.191)$$
$$+ G_0 T_A (1 - G_0 T_B G_0 T_A)^{-1} G_0 T_B (1 + G_0 T_A) G_0$$
$$+ G_0 T_B (1 - G_0 T_A G_0 T_B)^{-1} G_0 T_A (1 + G_0 T_B) G_0.$$

Here the t-matrix T_A is given by Eqn (4.189) and similarly T_B. Equation (4.191) can conveniently be written as

$$G(\varepsilon) = G_0(\varepsilon) + G_A(\varepsilon) + G_B(\varepsilon) + G_{AB}(\varepsilon), \qquad (4.192)$$

where

$$G_A(\varepsilon) = G_0(\varepsilon) T_A(\varepsilon) G_0(\varepsilon), \qquad (4.193)$$

and similarly for $G_B(\varepsilon)$. These Green's functions express the effect of the A or B perturbations alone, while the Green's function $G_{AB}(\varepsilon)$ contains all the terms of Eqn (4.191) that involve *both* T_A and T_B; we therefore interpret $G_{AB}(\varepsilon)$ as an interference term that is responsible for the interaction between A and B. Hence, by means of Eqn (4.184), we see that the change of the density of states due to the interference term is

$$\Delta \mathcal{N}(\varepsilon) = -\frac{1}{\pi} \operatorname{Im} \operatorname{Tr} G_{AB}(\varepsilon + i 0^+). \qquad (4.194)$$

Considering the last two terms on the right-hand side of Eqn (4.191) that define G_{AB} we can show by a direct differentiation and using the cyclic property of the trace that

$$\operatorname{Tr} G_{AB}(\varepsilon) = \frac{d}{d\varepsilon} \operatorname{Tr} \ln \left[1 - G_0(\varepsilon) T_A(\varepsilon) G_0(\varepsilon) T_B(\varepsilon) \right] \qquad (4.195)$$

thus obtaining for the integrated density of states change defined by Eqn (4.165)

$$\Delta N(\varepsilon) = -\frac{1}{\pi} \operatorname{Im} \operatorname{Tr} \ln \left[1 - G_0(\varepsilon) T_A(\varepsilon) G_0(\varepsilon) T_B(\varepsilon) \right]_{\varepsilon \to \varepsilon + i 0^+}. \qquad (4.196)$$

We obtain the desired interlayer coupling as before from a suitable total energy difference, or at finite temperatures from a difference of grand potentials. Although Eqn (4.167), strictly speaking, is only valid for noninteracting particles we will again use this relation to express the grand potential change (per unit area) due to the interference terms as

$$\Delta \Omega =$$
$$\frac{1}{4\pi^3} \operatorname{Im} \int d^2 \mathbf{k}_\| \int_{-\infty}^{\infty} d\varepsilon \, f(\varepsilon) \operatorname{Tr} \ln \left[1 - G_0(\varepsilon) T_A(\varepsilon) G_0(\varepsilon) T_B(\varepsilon) \right]_{\varepsilon \to \varepsilon + i 0^+}. \qquad (4.197)$$

This can be justified by applying the *force theorem* to changes of the magnetic configuration, i.e. Eqn (4.197) gives a good approximation when it is used in forming the difference of two different magnetic configurations. Bruno (1999) has made quite some effort to discuss this point. Furthermore, as before, we integrate over the good quantum numbers $\mathbf{k}_\parallel$ and the trace is assumed to be carried out over the remaining degrees of freedom.

The final steps are somewhat tedious and require basis functions to obtain matrix elements for the argument of the logarithm in Eqn (4.197). Thus using the decomposition defined by Eqn (4.183) we begin with

$$\langle k\,\sigma | G_0(\varepsilon)\,T_A(\varepsilon)\,G_0(\varepsilon)\,T_B(\varepsilon) | k\,\sigma \rangle$$
$$= \sum_{\mathbf{k}'_\perp\,\sigma'\,n'} \frac{\langle \mathbf{k}_\parallel\,\mathbf{k}_\perp\,\sigma\,n | T_A(\varepsilon) | \mathbf{k}_\parallel\,\mathbf{k}'_\perp\,\sigma'\,n' \rangle \langle \mathbf{k}_\parallel\,\mathbf{k}'_\perp\,\sigma'\,n' | T_B(\varepsilon) | \mathbf{k}_\parallel\,\mathbf{k}_\perp\,\sigma\,n \rangle}{\left(\varepsilon + \mathrm{i}0 - \varepsilon_{\mathbf{k}_\parallel\,\mathbf{k}_\perp\,n}\right)\left(\varepsilon + \mathrm{i}0 - \varepsilon_{\mathbf{k}_\parallel\,\mathbf{k}'_\perp\,n'}\right)} \quad (4.198)$$

where we have temporarily reintroduced band indices, n and n', added spin indices, σ and σ', and explicitly split off the good quantum number $\mathbf{k}_\parallel$. Since $G_0(\varepsilon)$ is the unperturbed bulk Green's function we may next use Bloch functions in the form

$$\langle \mathbf{r} | \mathbf{k}_\parallel\,\mathbf{k}_\perp\,\sigma\,n \rangle = \sum_{L\,\mathbf{R}} a_{L\,n}(\mathbf{k}_\parallel\,\mathbf{k}_\perp)\,\varphi_L(\mathbf{r} - \mathbf{R})\,\exp\left[\mathrm{i}(\mathbf{k}_\parallel + \mathbf{k}_\perp) \cdot \mathbf{R}\right] |\sigma\rangle \quad (4.199)$$

where the functions $\varphi_L(\mathbf{r} - \mathbf{R})$ constitute a local basis like the LMTOs or ASWs, L as in Chap. 3 stands for (l, m), the angular momentum and magnetic quantum numbers, the coefficients $a_{L\,n}(\mathbf{k}_\parallel\,\mathbf{k}_\perp)$ are the basic expansion coefficients, and $\mathbf{R}$ is a translation vector. For simplicity in writing we next drop the band index n and the wave vector $\mathbf{k}_\parallel$ in the notation. Then the matrix elements in Eqn (4.198) are at once seen to be given by

$$\langle \mathbf{k}_\perp\,\sigma | T_A | \mathbf{k}'_\perp\,\sigma \rangle = \sum_{L_1\,L_2} a^*_{L_1}(\mathbf{k}_\perp)\,a_{L_2}(\mathbf{k}'_\perp)\,T_{A\,L_1 L_2\,\sigma}\,\exp\left[\mathrm{i}\,\mathbf{R}_A \cdot (\mathbf{k}'_\perp - \mathbf{k}_\perp)\right] \quad (4.200)$$

and

$$\langle \mathbf{k}'_\perp\,\sigma | T_B | \mathbf{k}_\perp\,\sigma \rangle = \sum_{L_3\,L_4} a^*_{L_3}(\mathbf{k}'_\perp)\,a_{L_4}(\mathbf{k}_\perp)\,T_{B\,L_3 L_4\,\sigma}\,\exp\left[\mathrm{i}\,\mathbf{R}_B \cdot (\mathbf{k}_\perp - \mathbf{k}'_\perp)\right] \quad (4.201)$$

where we made use of the assumption that the perturbation potentials V_A and V_B drop rapidly to zero in the spacer so that the matrix element

$$T_{A\,L_1 L_2\,\sigma} = \sum_{\mathbf{R}\,\mathbf{R}'} \int \mathrm{d}^3\mathbf{r}\,\langle \sigma | \varphi^*_{L_1}(\mathbf{r} - \mathbf{R})\,T_A\,\varphi_{L_2}(\mathbf{r} - \mathbf{R}') | \sigma \rangle \quad (4.202)$$

is only nonvanishing if $\mathbf{R} = \mathbf{R}' = \mathbf{R}_A$, and similarly for $T_{B\,L_3 L_4\,\sigma}$ which is only nonvanishing if $\mathbf{R} = \mathbf{R}' = \mathbf{R}_B$. Note that these matrix elements depend on the spin direction because the perturbations V_A and V_B are assumed to be ferromagnetic. Substituting Eqn (4.200) and Eqn (4.201) into Eqn (4.198) we encounter the sum

$$S = \sum_{k'_\perp} \frac{a_{L_2}(k'_\perp)\, a^*_{L_3}(k'_\perp)\, \exp(-i k'_\perp D)}{\varepsilon + i0 - \varepsilon_{k'_\perp}} \qquad (4.203)$$

where $D = R_B - R_A > 0$ is the thickness of the spacer. This sum is transformed to an integral in the standard way, i.e.

$$S = \frac{d N_\perp}{2\pi} \int_{-\pi/d}^{\pi/d} \frac{a_{L_2}(k'_\perp)\, a^*_{L_3}(k'_\perp)\, \exp(-i k'_\perp D)}{\varepsilon + i0 - \varepsilon_{k'_\perp}}\, d k'_\perp \qquad (4.204)$$

where $N_\perp$ is the number of layers in the spacer and d is the thickness of each layer. This integral can be evaluated in the complex $k'_\perp$ plane (Bruno, 1995). For this one expands $\varepsilon_{k'_\perp}$ near the pole as $\varepsilon_{k'_\perp} = \varepsilon \pm k_\perp^\pm (d\varepsilon_{k'_\perp}/dk'_\perp)$, then—because of the exponential in the integrand—an integration path is chosen in the lower half-plane along the real axis from $-\pi/d$ to π/d down along $\pi/d + \operatorname{Im} k'_\perp$ closing at $-\infty$, and back to the real axis along $-\pi/d + \operatorname{Im} k'_\perp$. By means of Cauchy's integral theorem Eqn (4.204) becomes

$$S = i d N_\perp \sum_{k_\perp^-} \frac{a_{L_2}(k_\perp^-)\, a^*_{L_3}(k_\perp^-)}{\hbar v_\perp} \exp(-i k_\perp^- D) \qquad (4.205)$$

where $\hbar v_\perp = d\varepsilon_{k_\perp}/dk_\perp$. Note that a band index is implied and the sum is now taken over the residues which in the simplest case may consist of one term only.

Having done this we inspect Eqn (4.198) and, extracting another sum over $k_\perp$ from the trace in Eqn (4.197), we find an integral analogous to Eqn (4.204) which we deal with in the same way except now the integration path is closed in the upper complex $k_\perp$ half-plane. We can then formulate the final result by defining "reflection matrices"

$$\tilde{R}_X^\pm = \begin{pmatrix} R_{X\uparrow}^\pm & 0 \\ 0 & R_{X\downarrow}^\pm \end{pmatrix} \qquad (4.206)$$

with

$$R_{X\sigma}^\pm = \frac{\pm i d N_\perp}{\hbar v_\perp} \left\langle k_\perp^\mp \sigma | T_X(\varepsilon_{k_\perp^\pm} + i0) | k_\perp^\pm \sigma \right\rangle \qquad (4.207)$$

where $X = A$ or $X = B$ and $k_\perp^\pm$ denotes the residues. Then, assuming the magnetization of plane B to form an angle θ with that of plane A we can express the grand potential difference as

4.5 Multilayers

$$\Delta\Omega(\theta) = \frac{1}{4\pi^3} \operatorname{Im} \int d^2 \mathbf{k}_\parallel \int_{-\infty}^{\infty} d\varepsilon\, f(\varepsilon)\, \operatorname{Tr}$$

$$\ln\left\{1 - \sum_{k_\perp^- k_\perp^+} \tilde{R}_A^{\mp} U(\theta)\, \tilde{R}_B^{\pm} U(\theta) \exp\left[i\,(k_\perp^+ - k_\perp^-)\,D\right]\right\}. \quad (4.208)$$

Here we have used the spin rotation matrix

$$U(\theta) = \begin{pmatrix} \cos\frac{\theta}{2} & \sin\frac{\theta}{2} \\ -\sin\frac{\theta}{2} & \cos\frac{\theta}{2} \end{pmatrix}. \quad (4.209)$$

Equation (4.208) is the central result of this subsection.

In principle one can now calculate numerically the interlayer exchange coupling by forming the difference

$$I_c = \Delta\Omega(\theta = 0) - \Delta\Omega(\theta = 180°) \quad (4.210)$$

using Eqn (4.208), but the remaining integrations are not trivial to carry out. Here, however, our aim is to justify the simple model of Sec. 4.5.2.2 and thus the new interpretation of the data collection in Table 4.12. We therefore deal with the limit of large spacer thickness and with reflection matrices, $\tilde{R}_X^\pm$, that are in some sense small. It is not hard to form the difference given in Eqn (4.210) using Eqn (4.208); expanding the logarithm and retaining only the leading terms we derive with only little effort for the interlayer exchange coupling the expression

$$I_c = -\frac{1}{\pi^3} \operatorname{Im} \int d^2 \mathbf{k}_\parallel \int_{-\infty}^{\infty} d\varepsilon\, f(\varepsilon) \sum_{k_\perp^- k_\perp^+} \Delta R_A^{\mp} \Delta R_B^{\pm} \exp\left[i\,(k_\perp^+ - k_\perp^-)\,D\right] \quad (4.211)$$

where $\Delta R_A^{\mp} = (R_{A\uparrow}^{\mp} - R_{A\downarrow}^{\mp})/2$ and similarly $\Delta R_B^{\pm} = (R_{B\uparrow}^{\pm} - R_{B\downarrow}^{\pm})/2$. If we can get by with only a single pair of vectors $\mathbf{k}_\perp^+$ and $\mathbf{k}_\perp^-$ then Eqn (4.211) assumes the same form at $T = 0$ as Eqn (4.174). Since the assumptions necessary to derive Eqn (4.179) were of a rather general nature we need not write out the remaining steps to arrive at the final formula

$$I_c = \operatorname{Im} \sum_C \frac{\hbar v_\perp^C \varkappa_C}{4\pi^2 D^2} \Delta R_A^{\mp C} R_B^{\pm C} \exp(i\, q_\perp^C D). \quad (4.212)$$

All quantities are defined as in Eqn (4.179) except that the differences of reflection coefficients introduced *ad hoc* there are now defined by matrix elements of the *t*-matrices T_A and T_B. Everything said after Eqn (4.180) is, of course, valid here as well.

4.5.3 Giant magnetoresistance

We have seen that the exchange coupling, depending on the thickness of the spacer, can favor either a parallel or an antiparallel arrangement of the ferromagnetic layers as sketched in Fig. 4.74. By means of an external magnetic field one can now change the relative orientation continuously from an antiparallel arrangement as in Fig. 4.74(b), to one that is parallel as in Fig. 4.74(a). If the electrical resistivity of such a stack is measured as a function of the relative orientation of the magnetization, or, what is the same, as a function of the magnetic field applied to change the orientation, it is in general found to drop from a given value for an antiparallel to a lower value for a parallel orientation. This effect, although of different magnitude, is observed regardless of whether the current is in the plane of the stack, the so-called CIP orientation, or perpendicular to the plane, the so-called CPP orientation, and it has been called *giant magnetoresistance* (GMR). A useful description of the experimental situation and an assessment of the theoretical background can be found in a review article by Levy (1994), see also Himpsel *et al.* (1998).

To describe the effect quantitatively one may define the ratio

$$\text{GMR} = \frac{\Delta R}{R} = \frac{\rho^{\text{AP}} - \rho^{\text{P}}}{\rho^{\text{P}}} = \frac{\sigma^{\text{P}}}{\sigma^{\text{AP}}} - 1 \qquad (4.213)$$

where ΔR is the change in the resistance when the magnetization changes from antiparallel to parallel orientations, ρ^{AP} and ρ^{P} are the corresponding resistivities and σ^{AP} as well as σ^{P} the corresponding conductivities. This definition of the quantity GMR is not unique, but it is used quite frequently. We show a schematic drawing of GMR as a function of the external magnetic field, B, in Fig. 4.81, indicating the orientation of the magnetization of two ferromagnetic planes below the B-axis. For typical cases the scale of the GMR-axis in Fig. 4.81 might be as large as 100% at low

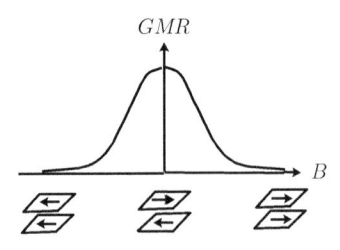

Figure 4.81 *Schematic drawing of the giant magnetoresistance, GMR, Eqn (4.213), as a function of the external magnetic field, B. Scale of GMR is typically 100% at low temperatures.*

temperatures, see for instance Parkin *et al.* (1991), but the effect becomes smaller at room temperature where it may be of the order of 10%, see Levy (1994). A theoretical explanation of this remarkable effect has been given by a number of authors, one group making use of a fully quantum mechanical approach (Bauer *et al.*, 1994; Scheb *et al.*, 1995; Butler *et al.*, 1995, 1996; Zhang and Levy, 1998) which is provided by the Kubo–Greenwood formalism (Kubo, 1957; Greenwood, 1958), the other (Zahn *et al.*, 1995, 1998; Mertig, 1999) using a semi-classical approach based on the Boltzmann equation, see for instance Ziman (1969). This collection of work on the GMR is not claimed to be complete!

Although a fully quantum mechanical approach (where a current–current correlation function must be evaluated) is in principle preferable, we will not follow this line of thought, but rather turn to the semi-classical theory; it is elementary, easier to describe, and quite transparent providing insight into the physics of the problem that is tightly connected with the electronic structure of the system. Furthermore, the results are not in contradiction with those of the quantum mechanical approach.

4.5.3.1 The Boltzmann equation

To be reasonably self-contained we begin with a brief derivation of the Boltzmann transport equation; full details can be found in the book by Ziman (1969).

Central to the Boltzmann equation is the distribution function which we denote by $f_\mathbf{k}(\mathbf{r},t)$; it allows us to formulate the relation between the current density, $\mathbf{J}(\mathbf{r},t)$, and the external electric field, $\mathbf{E}(\mathbf{r},t)$, as

$$\mathbf{J}(\mathbf{r},t) = \frac{e}{\Omega} \sum_\mathbf{k} \mathbf{v}_\mathbf{k} f_\mathbf{k}(\mathbf{r},t) \qquad (4.214)$$

where we assume we know the dependence of $f_\mathbf{k}(\mathbf{r},t)$ on $\mathbf{E}(\mathbf{r},t)$, e is the charge of the electron, Ω the volume (not to be confused with the usage of the same symbol in the previous subsection), and $\mathbf{v}_\mathbf{k}$ is the group velocity which is related with the band structure, $\varepsilon_\mathbf{k}$, by

$$\mathbf{v}_\mathbf{k} = \frac{1}{\hbar} \nabla_\mathbf{k} \varepsilon_\mathbf{k} \qquad (4.215)$$

where we have ignored the band index. In the absence of external fields we require that the distribution function $f_\mathbf{k}$ becomes the Fermi–Dirac function, i.e.

$$f_\mathbf{k}(\mathbf{r},t) \longrightarrow f(\varepsilon_\mathbf{k}) = \frac{1}{\exp \frac{\varepsilon_\mathbf{k} - \mu}{k_\mathrm{B} T} + 1} \qquad (4.216)$$

where μ is the chemical potential, k_B Boltzmann's constant, and T the temperature. Next it is assumed that the distribution function changes due to *diffusion*, external *fields*, and *scattering* processes. In equilibrium the total rate of change must be zero, so we write accordingly

$$\left(\frac{\partial f_{\mathbf{k}}}{\partial t}\right)_{\text{diff}} + \left(\frac{\partial f_{\mathbf{k}}}{\partial t}\right)_{\text{field}} + \left(\frac{\partial f_{\mathbf{k}}}{\partial t}\right)_{\text{scatt}} = 0. \qquad (4.217)$$

The sum of the first two terms above is traditionally evaluated using the Liouville equation, but a simple way to make the result plausible is to expand the function $f_{\mathbf{k}-\dot{\mathbf{k}}t}(\mathbf{r} - \mathbf{v}_{\mathbf{k}}t, 0)$ into a Taylor series which gives for Eqn (4.217)

$$-\dot{\mathbf{k}}\nabla_{\mathbf{k}} f_{\mathbf{k}} - \mathbf{v}_{\mathbf{k}} \nabla_{\mathbf{r}} f_{\mathbf{k}} + \left(\frac{\partial f_{\mathbf{k}}}{\partial t}\right)_{\text{scatt}} = 0. \qquad (4.218)$$

With Newton's law for the force, $\hbar \dot{\mathbf{k}}$, we obtain the Boltzmann equation as

$$\frac{e}{\hbar}(\mathbf{E} + \mathbf{v}_{\mathbf{k}} \times \mathbf{B}) \nabla_{\mathbf{k}} f_{\mathbf{k}} + \mathbf{v}_{\mathbf{k}} \nabla_{\mathbf{r}} f_{\mathbf{k}} = \left(\frac{\partial f_{\mathbf{k}}}{\partial t}\right)_{\text{scatt}} \qquad (4.219)$$

where **B** is the external magnetic field.

This equation is now linearized using as an ansatz

$$f_{\mathbf{k}}(\mathbf{r}) = f(\varepsilon_{\mathbf{k}}) - \phi(\mathbf{k}, \mathbf{r}) \left(\frac{\partial f(\varepsilon)}{\partial \varepsilon}\right)_{\varepsilon = \varepsilon_{\mathbf{k}}} \qquad (4.220)$$

where $f(\varepsilon)$ is the Fermi–Dirac function defined in Eqn (4.216).

We now drop those terms in the Boltzmann equation that are irrelevant in the present context. There is first the gradient term, $\nabla_{\mathbf{r}} f_{\mathbf{k}}$, which is important if a *thermal* gradient is to be treated (see Ziman, 1969); we assume this to be zero. Second, we drop the external magnetic field. This may be surprising at first sight. However, we are not interested in the usual magnetoresistance; in fact, the magnetic field used in the experiment is fairly small and only serves to change the relative orientation of magnetization of the ferromagnetic planes.

In first order we now only need

$$\nabla_{\mathbf{k}} f_{\mathbf{k}} = \left(\frac{\partial f(\varepsilon)}{\partial \varepsilon}\right)_{\varepsilon = \varepsilon_{\mathbf{k}}}, \nabla_{\mathbf{k}} \varepsilon_{\mathbf{k}} = \left(\frac{\partial f(\varepsilon)}{\partial \varepsilon}\right)_{\varepsilon = \varepsilon_{\mathbf{k}}} \mathbf{v}_{\mathbf{k}} \hbar \qquad (4.221)$$

and hence consider instead of Eqn (4.219) the simpler case

$$e\,\mathbf{E} \cdot \mathbf{v}_{\mathbf{k}} \left(\frac{\partial f(\varepsilon)}{\partial \varepsilon}\right)_{\varepsilon = \varepsilon_{\mathbf{k}}} = \left(\frac{\partial f_{\mathbf{k}}}{\partial t}\right)_{\text{scatt}}. \qquad (4.222)$$

In order to proceed we must obviously now deal with the scattering term on the right-hand side of this equation. To determine the rate of change of the distribution function, $(\partial f_{\mathbf{k}}/\partial t)_{\text{scatt}}$, we assume the probability is given by $P_{\mathbf{k}\mathbf{k}'}$ that a scattering event changes the state of an electron having the wave vector $\mathbf{k}$ to a state with vector $\mathbf{k}'$. We exclude spin-flip scattering and hence need not overburden the notation with spin indices which we can add later on. The rate of change of the distribution function due to these processes is $-P_{\mathbf{k}\mathbf{k}'} f_{\mathbf{k}} (1 - f_{\mathbf{k}'})$ where we enforce the Pauli principle by the factors involving the distribution function. But opposite processes of scattering into the (unoccupied) state at $\mathbf{k}$ from (occupied) states at $\mathbf{k}'$ give a rate of change of $+P_{\mathbf{k}'\mathbf{k}} f_{\mathbf{k}'} (1 - f_{\mathbf{k}})$. Because of microscopic reversibility, i.e. $P_{\mathbf{k}\mathbf{k}'} = P_{\mathbf{k}'\mathbf{k}}$, we can thus write

$$\left(\frac{\partial f_{\mathbf{k}}}{\partial t}\right)_{\text{scatt}} = \sum_{\mathbf{k}'} \left[f_{\mathbf{k}'} (1 - f_{\mathbf{k}}) - f_{\mathbf{k}} (1 - f_{\mathbf{k}'}) \right] P_{\mathbf{k}\mathbf{k}'} = \sum_{\mathbf{k}'} (f_{\mathbf{k}'} - f_{\mathbf{k}}) P_{\mathbf{k}\mathbf{k}'}. \quad (4.223)$$

Now admitting only *elastic* scattering and using the linearization, Eqn (4.220), we obtain from Eqn (4.222):

$$e\,\mathbf{E} \cdot \mathbf{v}_{\mathbf{k}} = \sum_{\mathbf{k}' \text{ on FS}} [\phi(\mathbf{k}) - \phi(\mathbf{k}')] P_{\mathbf{k}\mathbf{k}'}, \quad (4.224)$$

where we indicated that the sum is restricted to the Fermi surface.

Defining a *relaxation time*, $\tau_{\mathbf{k}}$, by means of

$$\tau_{\mathbf{k}}^{-1} = \sum_{\mathbf{k}' \text{ on FS}} P_{\mathbf{k}\mathbf{k}'} \quad (4.225)$$

we can write Eqn (4.224) in the form

$$e\,\mathbf{E} \cdot \mathbf{v}_{\mathbf{k}} = \phi(\mathbf{k}) \tau_{\mathbf{k}}^{-1} - \sum_{\mathbf{k}' \text{ on FS}} \phi(\mathbf{k}') P_{\mathbf{k}\mathbf{k}'}. \quad (4.226)$$

We derive a formal solution of this equation by defining a new vector $\mathbf{\Lambda}_{\mathbf{k}}$ by means of

$$\phi(\mathbf{k}) = e\,\mathbf{E} \cdot \mathbf{\Lambda}_{\mathbf{k}} \quad (4.227)$$

which we substitute into Eqn (4.226) to obtain

$$\mathbf{\Lambda}_{\mathbf{k}} = \tau_{\mathbf{k}} \left(\mathbf{v}_{\mathbf{k}} + \sum_{\mathbf{k}' \text{ on FS}} P_{\mathbf{k}\mathbf{k}'} \mathbf{\Lambda}_{\mathbf{k}'} \right). \quad (4.228)$$

Suppose we can solve this integral equation. Then Eqn (4.227) may be substituted into the linearized equation for the distribution function, Eqn (4.220), and the current is then written out with Eqn (4.214). Observing that the contribution to the current that involves the Fermi–Dirac function must vanish, we obtain

$$\mathbf{J} = \frac{e^2}{\Omega} \sum_{\mathbf{k}} (\mathbf{E} \cdot \mathbf{\Lambda}_{\mathbf{k}}) \mathbf{v}_{\mathbf{k}} \left(-\frac{\partial f(\varepsilon_{\mathbf{k}})}{\partial \varepsilon_{\mathbf{k}}} \right) \quad (4.229)$$

from which, comparing with $\mathbf{J} = \boldsymbol{\sigma} \mathbf{E}$, we finally obtain the *conductivity tensor* for zero temperature as

$$\boldsymbol{\sigma} = \frac{e^2}{\Omega} \sum_{\mathbf{k}} \boldsymbol{\Lambda}_{\mathbf{k}} \circ \mathbf{v}_{\mathbf{k}} \delta(\varepsilon_{\mathbf{k}} - \varepsilon_F) \qquad (4.230)$$

where we have repeatedly used the fact that for $T \to 0$ we get $-\partial f(\varepsilon_{\mathbf{k}})/\partial \varepsilon_{\mathbf{k}} = \delta(\varepsilon_{\mathbf{k}} - \varepsilon_F)$. We note that the conductivity describes a single spin channel. The transition probabilities $P_{\mathbf{k}\mathbf{k}'}$ may be obtained from scattering theory which gives through Fermi's golden rule (Sakurai, 1985),

$$P_{\mathbf{k}\mathbf{k}'} = \frac{2\pi}{\hbar} c |T_{\mathbf{k}\mathbf{k}'}|^2 \delta(\varepsilon_{\mathbf{k}} - \varepsilon_{\mathbf{k}'}) \qquad (4.231)$$

where $T_{\mathbf{k}\mathbf{k}'}$ is the single-atom scattering t-matrix used before and c is the concentration of scattering centers which we may assume to be dilute here. The solution of Eqn (4.228) is a difficult problem for which Sondheimer (1962) has worked out the conditions for exact solutions. They are not met for the general expression given by Eqn (4.231), but we will below approximate $P_{\mathbf{k}\mathbf{k}'}$ to be of the form $P_{\mathbf{k}\mathbf{k}'} = p(\mathbf{k}) p(\mathbf{k}')$. In this case it is easily shown by substitution that

$$\boldsymbol{\Lambda}_{\mathbf{k}} = \tau_{\mathbf{k}} \mathbf{v}_{\mathbf{k}}. \qquad (4.232)$$

In showing this one makes use of the solubility condition for Eqn (4.228) (Sondheimer, 1962):

$$\sum_{\mathbf{k}' \text{ on FS}} \mathbf{v}_{\mathbf{k}} = 0. \qquad (4.233)$$

Hence Eqn (4.230) becomes

$$\boldsymbol{\sigma} = \frac{e^2}{\Omega} \sum_{\mathbf{k}} \mathbf{v}_{\mathbf{k}} \circ \mathbf{v}_{\mathbf{k}} \tau_{\mathbf{k}} \delta(\varepsilon_{\mathbf{k}} - \varepsilon_F). \qquad (4.234)$$

4.5.3.2 *Approximate evaluation of the conductivity*

Following Zahn *et al.* (1995) we begin by approximating the relaxation time by an average over the Fermi surface. Introducing spin indices we define this averaged relaxation time by

$$\tau^\sigma = \frac{\sum_{\mathbf{k}} \tau_{\mathbf{k}}^\sigma \delta(\varepsilon_{\mathbf{k}\sigma} - \varepsilon_F)}{\sum_{\mathbf{k}} \delta(\varepsilon_{\mathbf{k}\sigma} - \varepsilon_F)}. \qquad (4.235)$$

Returning now to Eqn (4.213) we assume the resistivity of the parallel moment arrangement, σ^P, is given by a spin-up resistor (transport only by the majority electrons) in parallel with a spin-down resistor (transport only by the minority electrons). This goes back to ideas of Mott (1964) concerning the transport properties of ferromagnetic alloys. We thus write

$$\sigma^P = \sigma^\uparrow + \sigma^\downarrow \qquad (4.236)$$

with

$$\sigma^\sigma = \frac{e^2}{\Omega} \tau^\sigma \sum_{\mathbf{k}} \mathbf{v}_\mathbf{k}^\sigma \circ \mathbf{v}_\mathbf{k}^\sigma \, \delta(\varepsilon_{\mathbf{k}\sigma} - \varepsilon_F). \qquad (4.237)$$

In the antiparallel configuration the electronic states are degenerate and the conductivity becomes

$$\sigma^{AP} = 2 \frac{e^2}{\Omega} \tau^{AP} \sum_{\mathbf{k}} \mathbf{v}_\mathbf{k}^{AP} \circ \mathbf{v}_\mathbf{k}^{AP} \, \delta(\varepsilon_\mathbf{k} - \varepsilon_F). \qquad (4.238)$$

If we assume z to be the growth direction of the layered system, then CIP corresponds to the xx or yy components of the conductivity tensor and CPP to the zz component. In Eqns (4.237) and (4.238) the velocity components are then chosen accordingly and we arrive at the following expression for the quantity GMR defined in Eqn (4.213):

$$\mathrm{GMR}_j = \frac{\sum_\sigma \tau^\sigma \sum_\mathbf{k} (v_{\mathbf{k}j}^\sigma)^2 \, \delta(\varepsilon_{\mathbf{k}\sigma} - \varepsilon_F)}{2 \tau^{AP} \sum_\mathbf{k} (v_{\mathbf{k}j}^{AP})^2 \, \delta(\varepsilon_\mathbf{k} - \varepsilon_F)} - 1 \qquad (4.239)$$

where $j = $ CIP or CPP. We now explore some consequences of this equation.

Let us assume that the relaxation times $\tau^\uparrow$, $\tau^\downarrow$, and τ^{AP} are all equal. This means that we look for an *intrinsic* reason for the GMR that is due to the electronic structure of the system, not derivable from a spin-dependent property of the scattering centers. In this case the relaxation times cancel in Eqn (4.239) and the GMR is determined by the Fermi surface and Fermi velocities as a function of the magnetization configuration. Zahn *et al.* (1995) calculated the quantity GMR in this case *ab initio* using a (100)-oriented multilayer system of repeated Fe_3Cr_n with $n = 2, 4, \ldots, 12$; they obtained GMR $\simeq 200\%$ in CIP and GMR $\simeq 700\%$ in CPP configurations which is in agreement with the experimentally obtained maximum CIP-GMR (220%) for ultra-thin Fe layers (Schad *et al.*, 1994). Zahn *et al.* (1998) report similar results for Co/Cu (001) multilayers, giving values for the intrinsic GMR of GMR $\simeq 30\%$ in CIP and GMR $\simeq 125\%$ in CPP configurations.

We can understand these results by using the band structure for the Fe/Cu multilayer shown in Figs. 4.79 and 4.80 which is qualitatively similar to that of a Co/Cu system.

318 *Electronic Structure and Magnetism*

As was pointed out in Sec. 4.5.2.1, when the *majority* electrons in the Co-layer hop across the interface to Cu they can do so easily because states of the same energy are available and as a consequence the bands are dispersive; this is not so for the *minority* electrons which are therefore much more confined to a given layer than the majority electrons. We may conclude that it is predominantly the majority electrons that provide the high-conductivity channel in the CPP situation. In the antiparallel configuration the band structure is spin degenerate, but all bands possess many gaps and are flat, which will result in a much reduced conductivity.

In an attempt to view this somewhat more quantitatively we may further approximate the expression given in Eqn (4.239) by using the *local* density of states at the Fermi energy for each layer in the system, $\mathcal{N}_n^\sigma(\varepsilon_F)$, see Eqn (3.246) for $\varepsilon = \varepsilon_F$, where n labels the layer, and use Fermi surface averaged velocities, similarly defined for each layer, which gives approximately

$$\text{GMR} \cong \frac{\sum_n \mathcal{N}_n^\uparrow(\varepsilon_F) \left\langle v_{\mathbf{k}}^{\uparrow 2} \right\rangle_n + \sum_n \mathcal{N}_n^\downarrow(\varepsilon_F) \left\langle v_{\mathbf{k}}^{\downarrow 2} \right\rangle_n}{2 \sum_n \mathcal{N}_n^{\text{AP}}(\varepsilon_F) \left\langle v_{\mathbf{k}}^{\text{AP} 2} \right\rangle_n} - 1. \qquad (4.240)$$

Note that we look at the intrinsic case where the relaxation times are assumed to be equal and thus cancel out. It is a relatively straightforward matter to calculate the local state densities at ε_F for a moderately large multilayer. One finds at once that in Fe/Cu or Co/Cu the minority state densities are much larger than those of the majority electrons. The calculation by Zahn *et al.* (1998) for Co_9Cu_7 which we have used to draw Fig. 4.82, is quite instructive here. We see in Fig. 4.82 that, while the local state densities at ε_F for the majority (spin-up) electrons are rather uniform over the Co and Cu sites, the minority (spin-down) electrons are large at the Co sites but small for Cu. This supports the point of view that the minority electrons are effectively confined to the Co layer. This picture is completed by the calculations of Zahn *et al.* (1998) who show that the *wave functions* indicate unambiguously both the extended states in the majority channel and the confinement in the minority channel; the Cu electrons are weakly spin polarized and show the *same* confinement as the Co states. Another important detail emerges concerning states at the Co-Cu *interface*; as the pronounced maxima seen in Fig. 4.82 in the state densities at ε_F for the minority Co electrons indicate, there exist strongly confined quantum-well states located at the interface which are responsible for these maxima. These states, as we will see, are extremely important for the GMR since they undergo strong scattering from defects at the interface.

To complete the determination of the quantity GMR given in Eqn (4.240) we need values for the velocity averages. Zahn *et al.* (1998) analyze their wave functions and reach the following conclusions. In general, extended states are quite similar to free-electron states possessing averaged Fermi velocities that are roughly the same in the plane and perpendicular to the plane. This changes for the quantum-well and interface states; while their in-plane velocities are of the same order of magnitude as those of the extended states, the velocities perpendicular to the planes are drastically reduced. The quantum-well and interface states thus contribute to the CIP conductivity but give only a small

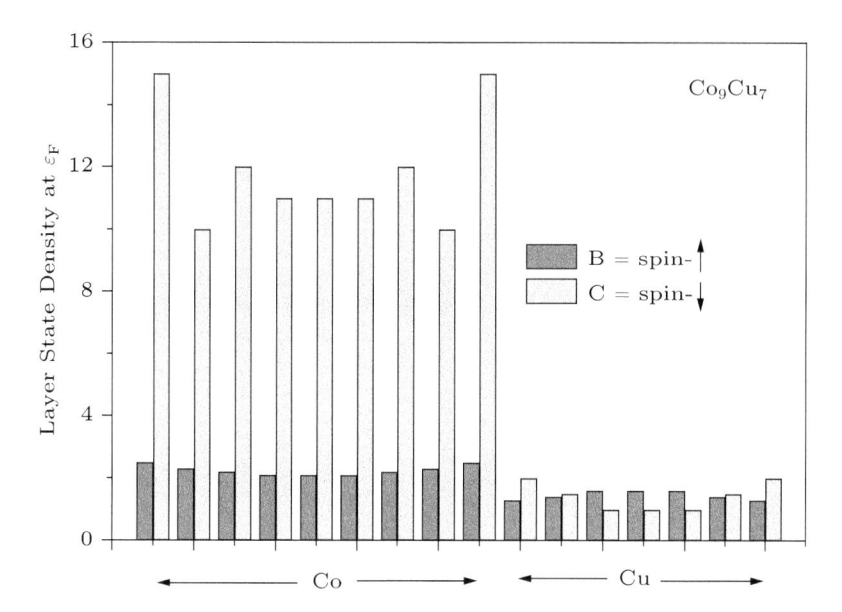

Figure 4.82 *Local density of states at the Fermi energy of a Co_9Cu_7 multilayer. Data extracted from the paper of Zahn et al. (1998). Dark gray shading: spin-up; light gray shading: spin-down.*

contribution to the CPP conductivity. Since extended states dominate in the majority channel for parallel magnetization, the CIP and CPP conductivities are nearly of the same order. For antiparallel magnetization the CIP conductivity is reduced somewhat, extended states being now somewhat more localized and quantum-well states having considerable in-plane velocities. But for antiparallel magnetization the CPP conductivity is reduced drastically since quantum-well and interface states prevent conduction.

We next investigate *extrinsic* effects using initially the expressions given by Eqn (4.237) and (4.238) but now scaling the spin-dependent scattering effects with an anisotropy ratio that is defined by

$$\beta = \frac{\tau^\uparrow}{\tau^\downarrow}. \quad (4.241)$$

For $\beta < 1$ the majority electrons experience stronger scattering than the minority electrons and vice versa for $\beta > 1$. This way the intrinsically high conductivity of the spin-up channel can change and become the low-conductivity channel. The anisotropy ratios were calculated by Mertig *et al.* (1993, 1994) for a variety of defects and were found to be in the range from 0.1 to 10.

Thus Zahn *et al.* (1995), by combining alternating Fe layers with Cr defects ($\beta = 0.11$) and with Cu defects ($\beta = 3.68$) in Fe_3Cr_n, showed that the GMR is drastically reduced and, in the case of a CIP geometry, can even change sign. Experimentally, the inverse effect was obtained for Fe/Cr/Fe/Cu/Fe/Cu multilayers by George *et al.* (1994) in CIP geometry.

We can go one more step toward an *ab initio* determination of the GMR effect, although the final goal still seems quite some distance away.

The relaxation time can in principle be determined by evaluating the transition probabilities given by Eqn (4.231). This is a local quantity since it is given by the scatterer at some place $\mathbf{r}_i$ in the spacer or the ferromagnet. To start with we may use the Born approximation (Sakurai, 1985) for the transition matrix and write

$$T^\sigma_{\mathbf{k}\mathbf{k}'}(\mathbf{r}_i) = \langle \psi^\sigma_\mathbf{k} | \Delta V^\sigma(\mathbf{r}_i) | \psi^\sigma_{\mathbf{k}'} \rangle \qquad (4.242)$$

where $\Delta V^\sigma(\mathbf{r}_i)$ is the perturbation of the potential at the impurity site and $\psi^\sigma_\mathbf{k}$ is the unperturbed Bloch function of the system. If we focus our attention on the influence of the superlattice wave function on the relaxation time, we do not need the details of the scattering potential but may characterize it by a scale factor t^σ, writing

$$\Delta V^\sigma(\mathbf{r}_i) = t^\sigma \delta(\mathbf{r} - \mathbf{r}_i) \qquad (4.243)$$

where δ is the Dirac δ-function. The spin anisotropy ratio is now

$$\beta = \left(\frac{t^\downarrow}{t^\uparrow}\right)^2 \qquad (4.244)$$

and working out the scattering probability, $P_{\mathbf{k}\mathbf{k}'}$, we see that it factorizes, so that Eqn (4.232) solves the integral equation with a relaxation time given by (4.225) and the spin-dependent conductivity becomes

$$\boldsymbol{\sigma}^\sigma = \frac{e^2}{\Omega} \sum_\mathbf{k} \mathbf{v}^\sigma_\mathbf{k} \circ \mathbf{v}^\sigma_\mathbf{k} \tau^\sigma_\mathbf{k} \delta(\varepsilon_{\mathbf{k}\sigma} - \varepsilon_F). \qquad (4.245)$$

The conductivity for the antiparallel moment arrangement is now obvious. The relaxation time in the Born approximation is easily worked out and seen to depend on the position of the scattering center, $\mathbf{r}_i$:

$$[\tau^\sigma_\mathbf{k}(\mathbf{r}_i)]^{-1} = 2\pi c \, |\psi^\sigma_\mathbf{k}(\mathbf{r}_i)|^2 \, \mathcal{N}^\sigma_i(\varepsilon_F) (t^\sigma)^2 + \tau^{-1} \qquad (4.246)$$

where $\mathcal{N}^\sigma_i(\varepsilon_F)$ is the local density of states at the Fermi energy and a spin-independent term has been added. In Fig. 4.82 we see the strong position dependence of $\mathcal{N}^\sigma_i(\varepsilon_F)$ which will now be reflected in $[\tau^\sigma_\mathbf{k}(\mathbf{r}_i)]^{-1}$. Choosing different values for β, Zahn *et al.* (1998) determined local GMR ratios, i.e. separately evaluating Eqn (4.213) layer by layer by making use of Eqn (4.245) and (4.246). The results are partly reproduced in Fig. 4.83. The local values for the GMR are evidence of the locally different properties of the multilayer: the CPP-GRM shown for $\beta = 1$, i.e. no spin anisotropy, weakly reflects the interface states. With spin-dependent scattering $\beta > 1$ the minority electrons are scattered more strongly than the majority electrons and the existing spin anisotropy of

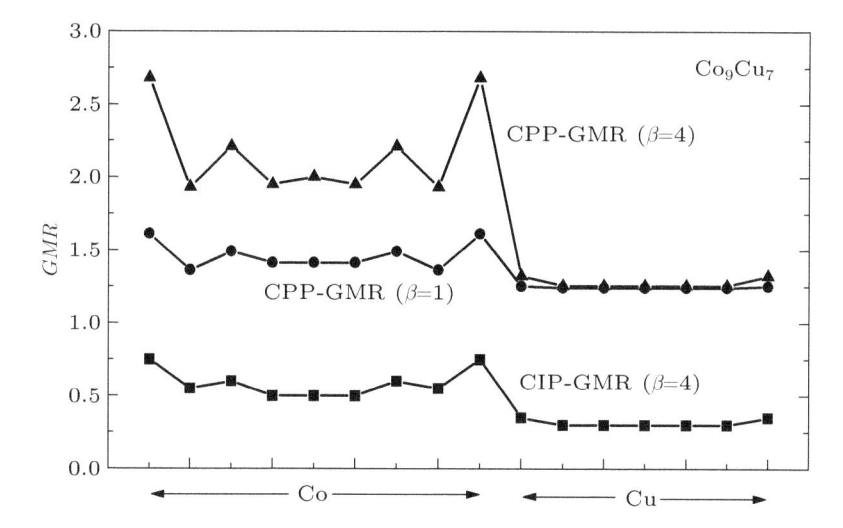

Figure 4.83 *Layer resolved GMR of Co_9Cu_7. Data from Zahn et al. (1998). Triangles: CPP-GMR, anisotropy ratio $\beta = 4$, dots: CPP-GMR, anisotropy ratio $\beta = 1$, squares: CIP-GMR, anisotropy ratio $\beta = 4$.*

the local state densities at ε_F is amplified and leads to an even stronger enhancement of the GMR amplitudes showing the pronounced effects of the quantum-well and interface states.

Butler *et al.* (1996), using the Kubo–Greenwood theory for CoCu multilayers, also show locally resolved conductivities for parallel and antiparallel moment arrangements. The physical picture that evolves is basically the same as that described by Zahn *et al.* Still, it would be highly desirable to arrive at a fully *ab initio* theory that could lead to an even more complete understanding of the physics that governs giant magnetoresistance. Numerous theoretical questions remain unanswered.

4.5.3.3 Tunnel magnetoresistance

Giant magnetoresistance effects in *magnetic tunnel junctions* (MTJ) have in the recent decade become an extremely active field of research and, in fact, it appears that MTJs are replacing magnetoresistance devices based on metallic magnetic multilayers. A MTJ consists of a thin insulating layer (the tunnel barrier) sandwiched between two ferromagnetic metal layers (the electrodes). The resistance of a MTJ depends on the relative alignment of the magnetic moments of the electrodes and is called the tunnel magnetoresistance (TMR).

The TMR was apparently first observed by Julliere (1975) but subsequently received only little attention. The discovery of the giant magnetoresistance (GMR) in metallic magnetic multilayers by Baibich *et al.* (1988) and Binasch *et al.* (1989) changed the situation when the GMR was beginning to be used industrially in the read heads of hard disk drives. The desire to increase the effect at room temperature led to much experimental and theoretical research on the GMR and on magnetic tunnel junctions. The work of Miyazaki and Tezuka (1995) and Moodera et al. (1995), who observed sizable magnetoresistance effects at room temperature in MTJs, kicked off new and ongoing research on tunnel magnetoresistance.

In what follows we made use of two clear review articles by Heiliger et al. (2006) and Yuasa (2008), which contain numerous references of experimental and theoretical work on the TMR and much information on basic and applied research issues that we cannot discuss here.

To begin we consider the simplest possible model to formulate a guide into the physics; this is the one-dimensional rectangular barrier problem well known from elementary quantum mechanics (see for instance Merzbacher, 1970 for full detail—or do it yourselves).

An electron is assumed to be incident from the left; its state is given on the left of the potential barrier by plane waves describing incident and reflected waves, damped exponentials inside the barrier, and a transmitted plane wave on the right. Because of piecewise constant potentials these are all simple exponentials with real or imaginary arguments, where all coefficients are fixed by matching value and slope separately on each side of the barrier. The transition coefficient is explicitly given by the energy of the incident electron, width, and height of the potential well.

Applied to a real, magnetic tunnel junction this model can be assumed to idealize a cross-section of the electrode-insulator-electrode system, the k-vector of the incident wave corresponding now to k_z, assuming the insulator layer normal to the z-axis. The width approximates the thickness of the insulator, whose material properties determine the height of the barrier. The plane waves are replaced by Bloch states, for which $\mathbf{k}_\parallel$ are still good quantum numbers. These states are spin polarized and are filled up to the Fermi energy, E_F, to describe the ferromagnetic electrodes. To calculate the transmission coefficient one should match values and slopes of the Bloch waves on the left and right of the insulator to solutions of the Schrödinger equation inside the insulator. An insulator, by definition, possesses an energy gap at E_F; one is thus led to determine eigenfunctions in the gap. This has already been pointed out by Heine (1965) in the context of metal–semiconductor junctions and surface states. Assuming this quite formidable problem has been solved and the transition probability, T, has been obtained, one calculates the tunnel conductance by using the Landauer formula (1987) (Imry and Landauer, 1999), which was derived from the Kubo formula by Fisher and Lee (1981) as well as Baranger and Stone (1989), to name only two of the many papers on this subject. One obtains for the conductivity G:

$$G = \frac{e^2}{h} T. \qquad (4.247)$$

Here e^2/h is the conductance quantum (per spin). This formula is valid for one transition channel. For describing a real MTJ one sums over all channels (Butler et al., 2001):

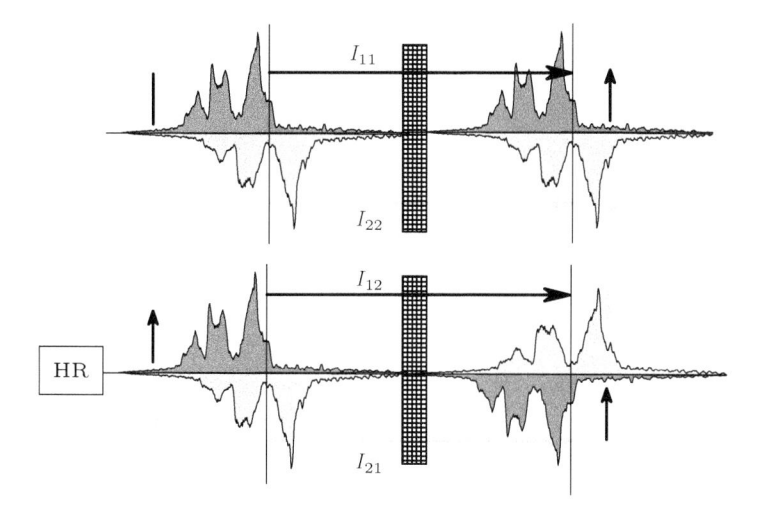

Figure 4.84 *Schematic view of a tunnel junction using densities of states of Fe. LR: moments are aligned parallel resulting in low resistance state. HR: moments are aligned antiparallel resulting in high resistance state. I_{ij} with $i = 1, 2 = \uparrow, \downarrow$ and $j = 1, 2 = \uparrow, \downarrow$ symbolize the possible tunnel transitions for electrons at the Fermi energy.*

$$G = \frac{e^2}{h} \sum_{Ik_\parallel} T_I(k_\parallel). \quad (4.248)$$

A schematic overview is given in Fig. 4.84 where also the index I is defined.

The first successful experiments by Miyazaki and Tezuka (1995) as well as Moodera *et al.* (1995) used tunnel barriers made of amorphous Al-O. There is no crystallographic symmetry in the barrier and at the interfaces to the electrodes. Therefore, Bloch states with various symmetries couple to the evanescent states (decaying tunnel states) in Al-O and thus have finite tunneling probabilities. These processes are regarded as being *incoherent* whereas processes that can be described by the steps outlined above are called *coherent*.

In 2001 the first-principles calculations by Butler *et al.* (2001) as well as Mathon and Umerski (2001) predicted that epitaxial MTJs with crystalline MgO tunnel barriers would have magnetoresistance ratios (MR) over 1000% thus drawing the attention to coherent magnetic tunneling.

The crucial step in the description of coherent tunneling is the identification of the quantum states in the electrodes and the barrier that are coupled at the interfaces and are dominant in the tunneling processes. A free-hand drawing of the situation appropriate for a Fe(001)|MgO(001)|Fe(001)-junction is shown in Fig. 4.85 obtained from the work of Butler *et al.* (2001). Although the Δ_1- and Δ_5- state appearing here are easily identified in the band structure of bcc iron, for instance by inspection of Fig. 4.29, their existence

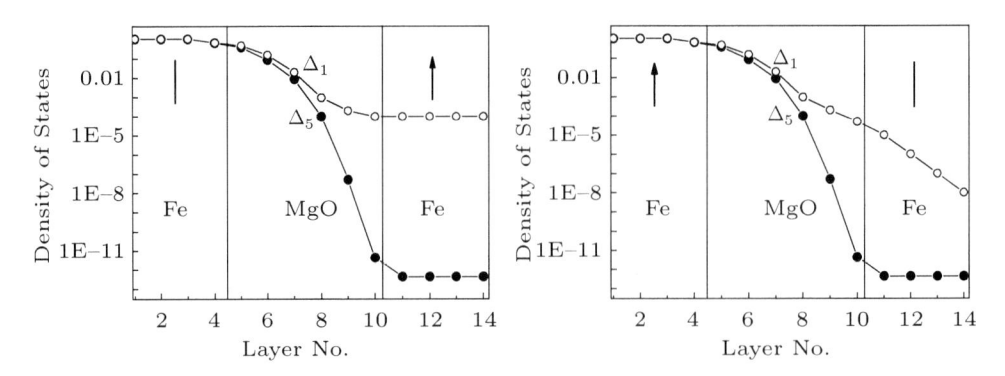

Figure 4.85 *Free-hand rendering of the tunneling density of states obtained by Butler et al. (2001). LR: I_{11}-transition for $k_\parallel = 0$ in Fe(100)|MgO|Fe(100). HR: I_{12}-transition for $k_\parallel = 0$ in Fe(100)|MgO|Fe(100).*

in the barrier is the result of calculations. The states inside the barrier, for energies inside the gap (!), can be viewed as Bloch functions with complex **k**-vectors. They are thus not propagating and often called evanescent. MacLaren et al. (1999) and Mavropoulos et al. (2000) describe these states for ZnSe-barriers in great detail.

It is seen in Fig. 4.85 LR that the transmission through the Δ_5-state is much lower than the transition through the Δ_1-state. When the magnetic moment of the right electrode (Fig. 4.85 HR) is reversed the Δ_5-transmission is approximately unchanged, but the Δ_1-transition leads to a decaying state in the right electrode because there is no Δ_1-state near the Fermi energy in the minority states of Fe, see Fig. 4.29.

For a complete description of the TMR for parallel magnetized electrodes, we have to add to the I_{11}-transitions, defined in Fig. 4.84 and sketched in Fig. 4.85 LR, the I_{22}-transitions, i.e. the minority → minority transitions, see Eqn (4.248). They are not shown here, but can easily be envisioned to go through a Δ_5-state having a low transmission probability, so that the Δ_1-transition in the majority channel is dominant in the sum, leading to low resistance (LR), compared with the high-resistance (HR) when the magnetic moment of the right electrode is reversed. In this case one has to complete the picture by adding the I_{21}-transitions defined in Fig. 4.84, which again tunnel through a Δ_5-state having low transmission.

After the theoretical predictions by Butler et al. (2001) and Mathon and Umerski (2001) there have been several experimental attempts to fabricate epitaxial Fe|MgO|Fe, but the room temperature MR ratio did not exceed about 70%. The experimental difficulties had to do with the quality of the Fe(001)/MgO(001) interface. Zhang et al. (2003), by first-principles calculations, identified the oxidized interfacial Fe monolayer as the limiting factor preventing an effective coupling of the Δ_1-states across the interface. Fabrication of high-quality epitaxial Fe(001)|MgO(001)|Fe(001) MTJs by using MBE growth under ultrahigh vacuum solved the problem (Yuasa et al., 2004; Yuasa, 2008). A representative result is shown in Fig. 4.86, which is a freehand drawing from the original. A MR ratio of up to 180% was obtained. Parkin et al. (2004), also using MgO-barriers,

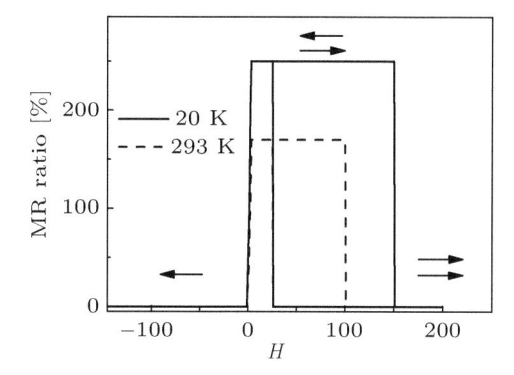

Figure 4.86 *Freehand rendering of magnetoresistance curves for an epitaxial Fe(001)|MgO(001)| Fe(001) MTJ for two temperatures, shown by Yuasa (2008). The arrows give the magnetization directions. The bias voltage is given to be 10 mV.*

but different electrodes and different methods, succeeded in measuring even higher MR ratios of 220% at room temperature. Finally, Yuasa *et al.* (2006) measured MR ratios at room temperature above 400% with epitaxial Co(001)|MgO(001)|Co(001) MTJs consisting of metastable bcc-Co electrodes. Their results look qualitatively like those shown in Fig. 4.86, except for a different MR ratio scale.

Turning now to the *half-metallic ferromagnets* it appears that Heusler compounds (see Sec. 4.4.2.2 and 5.4.8) are ideally suited because of their high Curie temperatures. In fact we expect to find a very large tunnel magnetoresistance because *ideally* only the transition I_{11} (Fig. 4.84) should dominate, I_{12} should be extremely small, and I_{22} as well as I_{21} should vanish entirely, provided both electrodes are made of the same HMF. Such a MTJ should be a perfect spin valve. However, these expectations apply only to zero temperature since at finite temperatures the spin-wave excitations that are both measured and calculated require states to be present (for $T > 0$) in the gap of the minority electrons, see Sec. 5.2.2 on spin waves in Heusler compounds.

Out of the wealth of experimental material we select two exceptional papers, which we believe point into the right direction.

Sakurabe *et al.* (2006) fabricated MTJs having Co_2MnSi electrodes and an Al-O barrier using a magnetron sputtering system. At low temperatures they found a very large MR ratio of 570% that, unfortunately, falls off to 67% at room temperature. This leads to the question why temperature plays such an unusually large role here. Since the Curie temperature of Co_2MnSi is 985 K it is unlikely that the temperature dependence of the bulk magnetization causes this effect. A glance at the densities of states of Co_2MnSi in Fig. 4.63 reveals, however, that the Fermi energy is extremely close to the upper band edge in the minority states, thus, as Sakurabe *et al.* (2006) pointed out, a minority–minority tunnel channel (I_{22}) is likely to open by a small bias voltage and by thermal fluctuations, which at room temperature are of the order of magnitude of the energy difference between the Fermi energy and the band edge. The density of states of Co_2MnAl shown in Fig. 4.65 suggest, therefore, that the Fermi energy might be moved to the center of the gap if electrodes are made of $Co_2MnAl_xSi_{1-x}$ with some suitable value of x. If the stringent experimental conditions allow such a tunnel junction to be made, the unusually large temperature effect in the MR ratio should become smaller.

So high-quality Heusler compounds having the Fermi energy in the middle of the gap *and* coherent tunneling through crystalline MgO should result in improved performance. The band structure of Co_2MnSi and Co_2FeSi (other compounds and alloys should be checked) possess in the majority states Δ_1- and Δ_5-bands crossing the Fermi energy. They originate from (*sp*) Γ_1- and Γ_{15}-states some 3 to 4 eV above the Fermi energy.

Tezuke *et al.* (2006) went in this direction and fabricated epitaxially grown magnetic tunnel junctions using as electrodes $Co_2FeAl_{0.5}Si_{0.5}$ Heusler alloys and as the tunnel barrier MgO. With an elaborate growth technique they made sure that the electrodes possess the desired $L2_1$ crystal structure. They succeeded in obtaining MR ratios of 175% at room temperature. Clearly, more interesting results are to be expected in this field.

We close this section with remarks on the magnetoresistance effects in the double perovskites. The minority-spin electrons in these compounds lie in t_{2g} hybrid bands (Sec. 4.4.2.3; Kobayashi *et al.*, 1998; Sarma *et al.*, 2000; Fang *et al.*, 2001; and Mandal *et al.*, 2008), which supply both the conduction electrons and are responsible for the ferromagnetic interactions whereas the majority spin electrons are gapped at the Fermi energy. Therefore strong magnetoresistance effects are expected in the ferromagnetic phase. Since Serrate *et al.* (2007) reviewed these effects thoroughly, we limit ourselves to pointing out some salient facts.

The fabrication of tunnel junctions with electrodes based on double perovskites has met with technical problems (Serrate *et al.*, 2007); therefore, most of the studies on spin-dependent transport were performed in polycrystalline ceramics. For instance, Kobayashi *et al.* (1998) reported magnetoresistance (MR) ratios of 42% at 4.2 K and 10% at room temperature for polycrystalline samples of Sr_2FeMoO_6. The evidence is strong that the MR in the double perovskites is caused by spin-dependent scattering at the grain boundaries and the conduction mechanism is electron tunneling across insulating grain boundaries of the granular material. The simplest picture to explain this effect is described by Serrate *et al.* (2007) as follows. "An ensemble of grains separated by insulating grain boundaries can be viewed as a network of electrodes embedded in an insulating matrix. At the coercive field the overall magnetization of the material is zero and the magnetizations of the grains point randomly...This constitutes a higher resistance state compared to the low resistance state achieved above the saturation field, when all the magnetizations of neighboring grains are parallel. Hence, under the application of an external magnetic field, the sample undergoes a resistive decrease as the magnetization approaches saturation." A detailed microscopic picture that is amendable to first-principles electronic structure calculations is at present missing.

4.6 Relativistic effects

A proper treatment of the effects of relativity begins with the four-component Dirac (1926) formalism. We have seen in Sec. 3.4.7.3 how this general theory can be reduced to the simpler two-component formalism the outstanding feature of which is *spin–orbit coupling*. For reasons of simplicity we will here discuss relativistic effects in the two-component formalism which, as we showed in Sec. 3.4.7.3, involves what is called the scalar relativistic Hamiltonian, $\mathcal{H}_{sc}$, Eqn (3.351), plus the spin–orbit Hamiltonian, $\mathcal{H}_{so}$, for which we refer to Eqn (3.352). In some magnetic systems containing heavy elements like uranium we need to add another *empirical* Hamiltonian, $\mathcal{H}_{orb}$, Eqn (3.355), which accounts for effects of the incompletely quenched orbital degrees of freedom to the magnetic moment. In this section we are not trying to review all effects ascribed to the spin–orbit Hamiltonian $\mathcal{H}_{so}$ (and perhaps $\mathcal{H}_{orb}$), but rather select aspects which possess interesting and significant symmetry properties that lead to noncollinear magnetic order. Here we largely follow the review of Sandratskii (1998).

Spin–orbit coupling, furthermore, supplies the physical origin for *magneto-crystalline anisotropy* and a number of spectroscopic effects. Magneto-crystalline anisotropy is the source of what is commonly associated with *permanent magnets*, namely the fact that the magnetization has a direction that is tied to the crystalline lattice. We will deal with this important phenomenon at some length.

Spectroscopic effects like the Kerr effect (the change of the direction of polarization of light when it is reflected from a ferromagnetic surface) are described by the frequency-dependent conductivity tensor that can be determined to a good approximation using density-functional theory. We deal with this topic now.

4.6.1 Magneto-optical properties

The theoretical basis we need here is extracted from the Kubo formula (Kubo, 1957). This important formula was derived by Kubo by quantizing the classical Boltzmann equation discussed in Sec. 4.5.3.1. It is interesting to follow the first steps of this derivation. Thus we first replace the variables **k** and **r** in the Boltzmann equation by the canonical coordinates p and x and write

$$\frac{\partial f}{\partial t} = -\sum \left(\frac{\partial f}{\partial x} \frac{\partial H}{\partial p} - \frac{\partial f}{\partial p} \frac{\partial H}{\partial x} \right), \qquad (4.249)$$

where H is the Hamiltonian and we use the classical Hamilton equations

$$\dot{x} = \frac{\partial H}{\partial p}, \dot{p} = -\frac{\partial H}{\partial x}.$$

The sum is implied to be extended over the components of x and p and one recognizes that the summand is the classical Poisson bracket $\{f, H\}$. The well-known quantization rule is to replace the Poisson bracket by the commutator while transforming the operators to the Heisenberg picture. Finally, replacing the classical distribution function f by the density matrix n we obtain the quantized version of Eqn (4.249)

$$\frac{\partial n}{\partial t} = \frac{1}{i\hbar}[n, H] \tag{4.250}$$

where the Heisenberg picture is implied, i.e. an operator A in the Schrödinger picture reads in the Heisenberg picture $U^\dagger(t)AU(t)$ with $U(t) = \exp(-iHt/\hbar)$.

In the linear response theory that yields the susceptibility as well as the conductivity one next linearizes the density matrix $n = n_0 + \Delta n$, introducing a time-dependent periodic perturbation by means of

$$H_F = H - AF(t). \tag{4.251}$$

The response of an observable B is given by $\langle \Delta B(t) \rangle = \text{Tr}(\Delta n(t)B)$. The susceptibility is now defined by $\langle \Delta B(t) \rangle = \chi_{BA}(\omega) F \exp(-i\omega t)$ and is derived to be

$$\chi_{BA}(\omega) = \frac{1}{i\hbar} \int_0^\infty dt\, e^{i\omega t} \langle [A(-t), B] \rangle \tag{4.252}$$

This is the famous *Kubo formula*.

To obtain the conductivity we specify the perturbation by using an electric field $\mathbf{E}(t)$

$$AF(t) = \sum_i e_i \mathbf{r}_i \mathbf{E}(t), \tag{4.253}$$

where e_i is the charge of the particle at the position $\mathbf{r}_i$. The conductivity is now a tensor because of the vector $\mathbf{E}$. Putting the things together and expanding all quantities in terms of the eigenvalues and eigenfunctions of the unperturbed Hamiltonian we arrive at a formula that is useful for the calculations we have in mind (when trying to derive this formula remember that the observables are defined in the Heisenberg picture):

$$\sigma_{\alpha\beta}(\omega) = \frac{ie^2}{m^2 V} \sum_{\vec{k} \in BZ} \sum_{\substack{n,m \\ n \neq m}} \frac{f(\epsilon_{n,\vec{k}}) - f(\epsilon_{m,\vec{k}})}{\epsilon_{m,\vec{k}} - \epsilon_{n,\vec{k}}} \tag{4.254}$$

$$\cdot \frac{<n,\vec{k}|p_\beta|m,\vec{k}><m,\vec{k}|p_\alpha|n,\vec{k}>}{\omega - \frac{\epsilon_{m,\vec{k}} - \epsilon_{n,\vec{k}}}{\hbar} + i\delta}.$$

Here e is the electron charge and m its mass, V is the volume of the Brillouin zone (BZ), and $f(\epsilon)$ is the Fermi–Dirac (FD) distribution function. The eigenvalues $\epsilon_{m,\vec{k}}$ are labelled by the band index m and the vector $\vec{k}$ inside the BZ. For simplicity in writing we imply a spin index. The matrix elements of the momentum operator are expressed with the Bloch eigenfunctions $\psi_{n\vec{k}}(\mathbf{r})$

$$<n,\vec{k}|p_\beta|m,\vec{k}> = \int \psi^*_{n\vec{k}}(\mathbf{r})[\mathbf{p} + (\frac{\hbar}{4mc^2})[\vec{\sigma} \times \nabla V(\mathbf{r})]]_\beta \psi_{m\vec{k}}(\mathbf{r})\, d\mathbf{r} \qquad (4.255)$$

where a correction due to spin–orbit coupling has been added to the momentum operator $\mathbf{p} = -i\hbar\nabla$. Spin–orbit coupling must be included in the band-structure calculations for the magneto-optical effects to be nonzero. The additional term in Eqn (4.255) is of less importance, but it leads to spin-flip transitions. The difference of the two FD-distribution functions in the numerator of the conductivity tensor gives nonvanishing contributions only for interband transitions. Intraband transitions are included by adding to $\sigma_{\alpha\beta}(\omega)$ the empirical Drude conductivity

$$\sigma_D(\omega) = \frac{\sigma_0}{1 - i\omega\tau_D} \qquad (4.256)$$

where the constants τ_D and σ_0 are to be obtained from experimental data; the latter, if desired, may be calculated with the plasma frequency.

Of the many applications of the conductivity we now select the Kerr effect (Kerr, 1878) that is found when polarized light is reflected from the surface of a magnetic material. For those interested in more, a wealth of information is found in the review article by Ebert (1996) and in a collection of papers edited by Ebert and Schütz (1996).

The rotation angle of the direction of polarization of the reflected light measured from the incoming polarization is given by the Kerr angle. For simplicity we assume that both the magnetization and the incoming wave vector are perpendicular to the surface. Then the complex Kerr angle is given by (Reim and Schoenes, 1990)

$$\Phi_K \equiv \phi_K + i\varepsilon_K = \frac{-\sigma_{xy}}{\sigma_{xx}\sqrt{1 + \frac{4\pi i}{\omega}\sigma_{xx}}}. \qquad (4.257)$$

where ϕ_K is the real Kerr rotation and ε_K is the so-called Kerr ellipticity; furthermore, the crystal is assumed to be cubic with the z-axis parallel to the magnetization.

Beginning with the magneto-optical properties of Fe and Ni we treat these elementary metals as a benchmark. Oppeneer et al. (1992a) describe the numerical difficulties stemming from the evaluation of the Brillouin zone summations and the evaluation of the momentum matrix elements. For the former problem there exist useful algorithms in the literature (Oppeneer and Lodder, 1987; Kaprzyk and Mijnarends, 1986). The evaluation of the matrix elements depends on the method by which the electronic band structure is evaluated. Oppeneer and collaborators used the ASW-method described in detail in Chap. 3. In this case the integral in Eqn (4.255) consists of contributions from each of the atomic spheres in the unit cell that are evaluated for both head and tails of the basis functions summed over the full set of quantum number l, m, σ; for details see the original paper by Oppeneer et al. (1992a), where the corrections stemming from the interstitial space are also treated.

The Kerr rotations for iron and nickel are shown in Fig. 4.87, where experimental data are compared with calculated results. Concentrating on Fe first we see that the

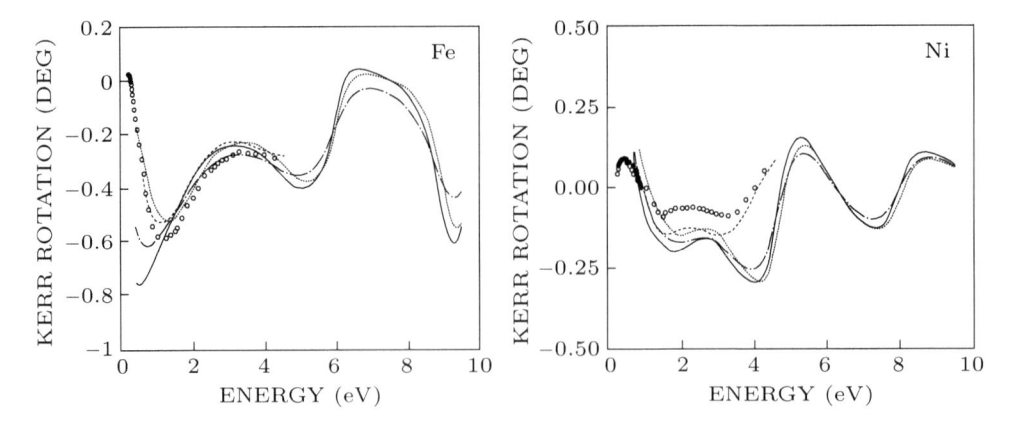

Figure 4.87 *The Kerr rotation of bcc Fe and fcc Ni. Circles and short dashes are experimental data (Krinchik and Artem'ev, 1968 and van Engen et al., 1983). Different values of the relaxation constant δ result in the calculated results, where the dotted curve includes the Drude term, after Oppeneer et al. (1992b)*

experimental data lie within a spread of the theoretical results, the spread being due to different values of the relaxation constants, δ. This is an empirical input chosen as $\delta \simeq 0.04$ Ry. Also note that the empirical Drude conductivity is needed to describe the low-energy part successfully. Except for these empirical inputs the calculations are *ab initio*. The good overall agreement speaks for the density-functional approximation that is at the basis of our theory and encourages us to apply the theory to interesting compounds (which we will do next). However, we take it with a grain of salt when we compare theory with experiment for nickel in Fig. 4.87. Here is a systematic disagreement that is not really understood. We have seen before in Fig. 4.34 that Ni is a problem; especially, an overestimated exchange splitting is observed. Oppeneer *et al.* (1992a,b) in a set of calculations varied the strength of the spin–orbit (SO) interaction, the magnetic moment, and the lattice constant to find the ensuing changes of the Kerr rotation. The SO interaction was found to be related linearly to the Kerr angle, so that a certain amount of scaling is possible, but a convincing picture is not in sight. Thus a certain measure of caution is asked for without becoming overly philosophical.

We now turn our attention to certain compounds that we discussed before in Sec. 4.4.1.2 where we found ferromagnets that gap in one spin direction, the so-called *half-metallic ferromagnets*. The density of states for the case of interest is that of PtMnSb shown in Fig. 4.59, which indicates that electric conduction happens only in one spin direction. NiMnSb is found to have the same property. The isoelectric PdMnSb is only partially gapped, that is the gap begins slightly above the Fermi energy. Antonov *et al.* (1997) compared the band structure of the three compounds. At the time of their discovery in 1983 the search for efficient materials for the fabrication of optical magnetic discs was enthusiastically pursued, so the discovery of de Groot *et al.* (1983a) of half-metallic ferromagnets with PtMnSb possessing a large Kerr rotation was noted with great

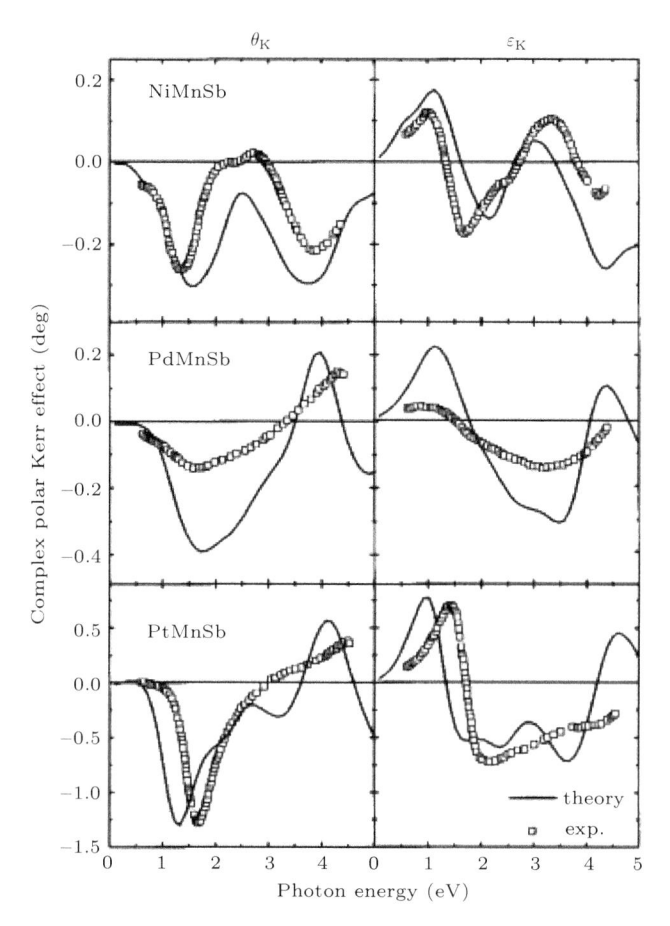

Figure 4.88 *The experimental and theoretical Kerr rotations and ellipticities of the three compounds NiMnSb, PdMnSb, and PtMnSb, after Antonov et al. (1997).*

interest. We compare in Fig. 4.88 the Kerr rotation and ellipticity of NiMnSb, PdMnSb, and PtMnSb showing both theory and experiment.

The Kerr rotation of PtMnSb is indeed large but the peak position of the calculated θ_K is slightly shifted to lower energies. θ_K for NiMnSb is much smaller and possesses two peaks, both overestimating the experimental data. In view of the half-metallic band structure the difference between NiMnSb and PtMnSb is astonishing. PdMnSb is again different, which is perhaps not surprising. The large Kerr rotation of PtMnSb was discussed by many authors, most recently perhaps by Katsnelson *et al.* (2008); a consensus, however, about any outstanding physical mechanisms has not been reached. Antonov *et al.* (1997) investigated the contributions to the Kerr rotation coming from the various parts of Eqn (4.257) by looking in detail at its numerator and denominator and found that it is the denominator which dominates at about 1 eV.

The search for materials suitable for magneto-optical applications was carried out in the 1980s by a number of groups. It was discovered that other compounds also possess a large Kerr rotation. As an example we examine MnBi, which is expected to show interesting effects because of the large spin–orbit coupling through the heavy element bismuth. MnBi has the hexagonal NiAs-structure (space-group 194). A recent paper on MnBi is that by McGuire *et al.* (2014), which reports that MnBi possesses this crystal structure at higher temperatures (room temperature, say) when the magnetic moment becomes parallel to the c-axis. At low temperatures it has a moment of nearly $3.9\mu_B$ oriented in the ab-plane. A spin reorientation transition appears at about 100 K and the magnetism abruptly disappears at 630 K. Köhler and Kübler (1996) calculated the Kerr rotation for MnBi assuming the room-temperature structure; their result is shown in Fig. 4.89. The Kerr rotation is indeed very large. In contrast to experiment the calculated data exhibit only one large peak, which explains a part of the experimental values, failing, however, entirely at higher energies. Oppeneer *et al.* (1996) also calculated the Kerr effect of MnBi with different methods and obtained the same results. From their discussion and the information that Di *et al.* (1992) carried out the measurements at 80 K it is fairly obvious that the calculations cannot explain the experimental data because of the spin reorientation transition. Furthermore, the stoichiometry of the samples was not the ideal one used in the calculations. More work will be done, especially on temperature effects. Simulations of the Kerr effect for cases such as MnBi for changing spin orientations seems to be a challenging problem.

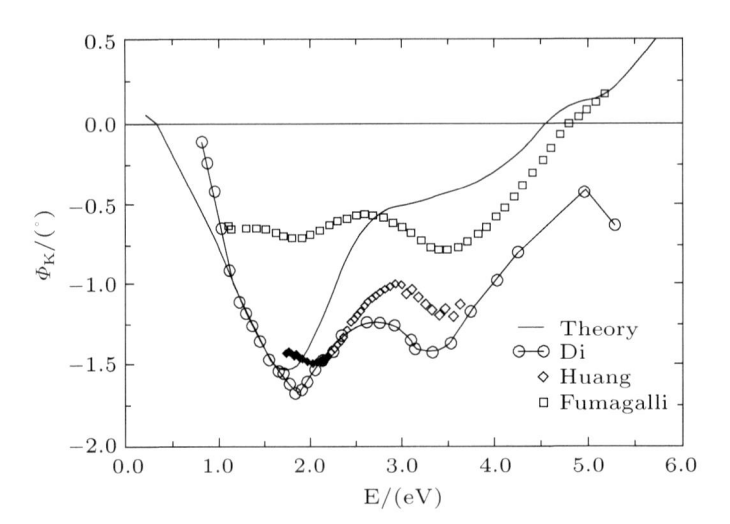

Figure 4.89 *The Kerr rotation of hexagonal MnBi. The theoretical result is compared with three sets of experimental data, those of Di et al. (1992), Huang et al. (1995), and Fumagalli et al. (1996), after Köhler and Kübler (1996).*

4.6.2 Symmetry properties of spin–orbit coupling

The Hamiltonian we now consider consists of two parts. First we have the scalar relativistic Kohn–Sham Hamiltonian of a magnet assumed to possess an arbitrary configuration of magnetic moments. For this we write

$$\hat{\mathcal{H}}_{sc}(\mathbf{a}_\nu, \mathbf{e}_\nu) = \sum_\nu U^+(\theta_\nu, \phi_\nu) \begin{pmatrix} \mathcal{H}_{sc}^{\nu\uparrow}(\mathbf{r}_\nu) & 0 \\ 0 & \mathcal{H}_{sc}^{\nu\downarrow}(\mathbf{r}_\nu) \end{pmatrix} U(\theta_\nu, \phi_\nu). \quad (4.258)$$

Here $U(\theta_\nu, \phi_\nu)$ is the standard spin-1/2 rotation matrix which describes the transformation between a global and a local coordinate system of the ν-th atom, the spin orientation of which is—with respect to the z-axis of the global system—given by the polar angles θ_ν and ϕ_ν, see Eqn (2.188). The quantities $\mathcal{H}_{sc}^{\nu\uparrow}(\mathbf{r}_\nu)$ and $\mathcal{H}_{sc}^{\nu\downarrow}(\mathbf{r}_\nu)$ are the standard atomic scalar relativistic Hamiltonians (spin-up, spin-down) in the local frame of reference for the atom at site ν, see Eqn (3.351). They contain the mass–velocity, the Darwin term, and the effective one-particle potential which, as usual, is given by functional derivatives of the total energy and is spin-diagonal in the local frame of reference connected with atom ν. (In writing down Eqn (4.258) and the following Eqn (4.259) we assume the atomic sphere approximation but neglect the unit step functions used elsewhere in this treatise.)

Note that the scalar relativistic Hamiltonian possesses the same symmetry properties as the nonrelativistic Hamiltonian which is obtained from the former by neglecting the mass–velocity and Darwin terms. Therefore the symmetry aspects to be discussed here will be the same for both cases.

The second part of the Hamiltonian describes spin–orbit coupling which we write in the form

$$\hat{\mathcal{H}}_{so} = \sum_\nu U^+(\theta_\nu, \phi_\nu) \left\{ \sum_\alpha M_{\alpha\nu} \sigma_\alpha \hat{l}_{\alpha\nu} \right\} U(\theta_\nu, \phi_\nu). \quad (4.259)$$

Here σ_α and $\hat{l}_{\alpha\nu}$ are the Cartesian components of the Pauli spin matrices and the angular momentum, respectively, in the local frame of reference connected with atom ν and the coefficients $M_{\alpha\nu}$ can be found in Sec. 3.4.7.3., by reading them off Eqn (3.352).

When dealing with compounds made of heavy elements we may have to add $\mathcal{H}_{orb}$, Eqn (3.355), to Eqn (4.259) by adding it inside $\{\sum_\alpha M_{\alpha\nu} \sigma_\alpha \hat{l}_{\alpha\nu}\}$.

Traditionally the symmetry properties of nonrelativistic Hamiltonians are described in terms of ordinary irreducible representations of the space group relevant to the crystal at hand, whereas for the relativistic problem the so-called double-valued irreducible representations are used. This difference in the type of representation is due to the different choice of the functions on which the symmetry operations are performed. In the relativistic case the transformed function is always a *spinor* function whereas it is a *scalar* wave function in the nonrelativistic case for which, correspondingly, only the transformation properties in real space need be considered.

It is the spinor form of the wave function that is of prime importance for the case of noncollinear magnetism, whence we must use the double-valued irreducible representation. However, there is a subtle difference in the symmetry properties of the problem depending on whether or not the spin–orbit coupling is retained in the Hamiltonian. To describe this properly one introduces a generalized set of operators which allows an independent transformation of the spin and space variables (Sandratskii, 1991). For the group of such operators we will use the term spin-space group (SSG).

We define the action of an operator of the SSG on a two-component spinor function as follows

$$\{\alpha_S|\alpha_R|\mathbf{t}\}\,\psi(\mathbf{r}) = U(\alpha_S)\,\psi(\alpha_R|\mathbf{t}^{-1}\mathbf{r}) \tag{4.260}$$

where ψ is a two-component spinor function, U, as before, is the spin-1/2 rotation matrix, α_S and α_R are, respectively, spin and space rotations, and $\mathbf{t}$ is a space translation. Operators of the usual space group are those with $\alpha_S = \alpha_R$.

One easily proves that a transformation of the scalar relativistic Hamiltonian (i.e. not including spin–orbit coupling) of a noncollinear magnet with the operations given in Eqn (4.260) leaves the form of the Hamiltonian invariant, i.e.

$$\hat{\mathcal{H}}_{sc}(\mathbf{a}'_\nu, \mathbf{e}'_\nu) = \{\alpha_\mathbf{S}|\alpha_\mathbf{R}|\mathbf{t}\}\,\hat{\mathcal{H}}_{sc}(\mathbf{a}_\nu, \mathbf{e}_\nu)\,\{\alpha_\mathbf{S}|\alpha_\mathbf{R}|\mathbf{t}\}^{-1} \tag{4.261}$$

where $\mathbf{a}_\nu$ are atomic positions and $\mathbf{e}_\nu$ directions of magnetic moments corresponding to the untransformed Hamiltonian. We say that two magnetic crystals with atomic positions connected by the relation $\mathbf{a}'_\nu = \alpha_\mathbf{R}\,\mathbf{a}_\nu + \mathbf{t}$ and directions of magnetic moment connected by $\mathbf{e}'_\nu = \alpha_\mathbf{S}\,\mathbf{e}_\nu$ are equivalent in the scalar relativistic case if Eqn (4.261) holds. Note that the operations $\alpha_\mathbf{S}$ and $\alpha_\mathbf{R}$ are independent.

However, a transformation of the spin–orbit coupling Hamiltonian with an SSG operator does not reproduce the form of the Hamiltonian. Only when $\alpha_\mathbf{S} = \alpha_\mathbf{R}$ do we restore the form-invariance of the Hamiltonian and find equivalent magnetic crystals:

$$\{\alpha_\mathbf{R}|\alpha_\mathbf{R}|\mathbf{t}\}\,\hat{\mathcal{H}}_{so}(\mathbf{a}_\nu, \mathbf{e}_\nu)\,\{\alpha_\mathbf{R}|\alpha_\mathbf{R}|\mathbf{t}\}^{-1} = \hat{\mathcal{H}}_{so}(\mathbf{a}'_\nu, \mathbf{e}'_\nu). \tag{4.262}$$

In most cases we are interested in the symmetry properties of one particular magnetic configuration and not in establishing the equivalence of different magnetic configurations. In this case the atomic positions and corresponding atomic moments must be the same before and after the transformation and Eqns (4.261) and (4.262) reduce to commuting Hamiltonian and symmetry operators.

In what follows we will show how symmetry arguments help to unravel the physical picture behind the computational results obtained for a selection of different examples that represent cases of key interest.

4.6.3 Noncollinear magnetic structures in uranium compounds

4.6.3.1 The case of U_3X_4

Since relativistic effects are expected to be large in compounds that contain uranium, we look here for interesting examples. It is in particular in U_3P_4 and U_3As_4 that complex noncollinear structures are observed (Burlet et al., 1981) and we start with a discussion of U_3P_4 subsequently turning to results of calculations for the magnetic properties of U_3As_4 and U_3Sb_4.

The compounds U_3X_4 crystallize in the Th_3P_4 structure which has a bcc lattice with a basis formed by two formula units. In multiples of the lattice parameter a the coordinates of the six U atoms are: $(\frac{3}{8} 0 \frac{1}{4})$; $(\frac{1}{4} \frac{3}{8} 0)$; $(0 \frac{1}{4} \frac{3}{8})$; $(\frac{1}{8} 0 \frac{3}{4})$; $(\frac{3}{4} \frac{1}{8} 0)$; $(0 \frac{3}{4} \frac{1}{8})$ and the eight P atoms are located at $(u\,u\,u)$; $(\frac{1}{2}+u\,\frac{1}{2}-u\,-u)$; $(\frac{1}{2}-u\,-u\,\frac{1}{2}+u)$; $(-u\,\frac{1}{2}+u\,\frac{1}{2}-u)$ and $(\frac{1}{4}+u\,\frac{1}{4}+u\,\frac{1}{4}+u)$; $(\frac{3}{4}+u\,\frac{1}{4}-u\,\frac{3}{4}-u)$; $(\frac{1}{4}-u\,\frac{3}{4}-u\,\frac{3}{4}+u)$; $(\frac{3}{4}-u\,\frac{3}{4}+u\,\frac{1}{4}-u)$, with $u = 1/12$. The unit cell looks rather complicated and need not be shown here in detail. The uranium atoms possess magnetic moments and U_3P_4 develops a large ferromagnetic component along the (111)-axis; however, the moments are not collinear but form angles with respect to the (111)-axis as shown in Fig. 4.90. When one attempts to calculate the magnetic structure aligning all magnetic moments along the (111)-axis then the scalar relativistic and the relativistic Hamiltonians (meaning in the latter case inclusion of $\hat{\mathcal{H}}_{so}$) lead to drastically different results. Free to rotate, the magnetic moments deviate from the initial parallel directions in the relativistic case but stay parallel in the scalar relativistic case. To appreciate the difference between these two cases one may formulate the following nearly self-evident statement (Sandratskii, 1998): *"The symmetry of the initial Kohn–Sham Hamiltonian must be preserved during calculations."*

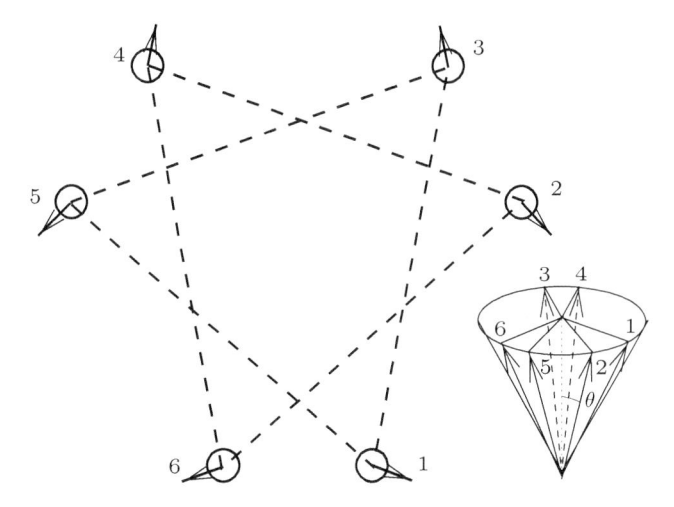

Figure 4.90 *Projection of the atomic positions and magnetic moments of the U atoms in U_3X_4 onto the (111)-plane (X = P and As). At the right the cone formed by the magnetic moments of the U atoms is shown, the cone's axis being parallel to the (111)-axis (with permission from Sandratskii, 1998).*

This means on the one hand that if the combined symmetry of the crystal and magnetic structure is so high that a deviation of magnetic moments from the initial directions would lead to perturbing the invariance of the Hamiltonian with respect to at least one symmetry operator then this deviation cannot take place. On the other hand, if a deviation of the magnetic moments from the initial directions is allowed by *all* symmetry operations present, then there are no symmetry reasons for staying with the initial magnetic configuration and the magnetic moments will start to rotate tending to assume the state of lowest total energy. This simulated annealing of the magnetic structure will continue until the system finds the ground-state magnetic structure that is from, a symmetry point of view, "accidental."

These general statements will help to explain the behavior of the magnetic moments in U_3P_4. We start with the scalar relativistic case. As was shown above the symmetry basis of a scalar relativistic problem is formed by the spin-space group, i.e. separate transformations of the spin and space variables are allowed. We can formulate the result in a very general way: starting the scalar relativistic calculation with a collinear configuration we will never obtain a deviation of magnetic moments from the initial direction. Indeed, independent of the crystal structure any spin rotation by an arbitrary angle ϕ about the direction of the magnetic moments $\{C_\phi|\varepsilon|0\}$ is a symmetry operation. This group of symmetry operations gives the symmetry basis for treating the spin projection of an electron state as a good quantum number (Sandratskii, 1991). Deviations of any magnetic moment from this direction would break the symmetry with respect to operations $\{C_\phi|\varepsilon|0\}$ and are, therefore, forbidden.

The situation changes drastically in the presence of spin–orbit coupling because of the reduction of the symmetry basis from the spin-space group to the usual space groups which transforms spin and space variables in an identical way i.e. $\alpha_S = \alpha_R$. This means that the question of stability of a magnetic structure cannot be answered without an analysis of the particular crystal structure. To illustrate the important symmetry properties it is sufficient to consider the simple picture shown in Fig. 4.90 where the projections onto the (111)-plane of the positions of the six U atoms are shown. Let us assume that initially all magnetic moments are parallel to the (111)-axis which is perpendicular to the plane of the paper. Then the following operations leave the magnetic and crystal structures invariant: the rotations by 120° and 240° about the (111)-axis and the reflections in the planes containing the (111)-axis accompanied by time reversal. Of importance is the observation that none of these operations leaves the position of any particular atom unchanged. Because of this, symmetry imposes no restrictions on the direction of the magnetic moment of a particular atom but only on the orientation of the atomic moments relative to each other and to the crystal lattice. The deviation of the moments from the (111)-axis resulting in a noncollinear magnetic configuration does not change the symmetry of the crystal. Therefore, the ferromagnetic ($\theta = 0$) structure is—from the point of view of symmetry—not isolated from structures possessing a nonzero θ. In Fig. 4.90 we show the projections of the magnetic moments on the (111)-plane obtained in the self-consistent calculation; these calculated deviations from the (111)-axis evidently do not destroy the symmetry of the atomic configuration. To emphasize that the magnetic moments possess components parallel to the (111)-axis, we also show

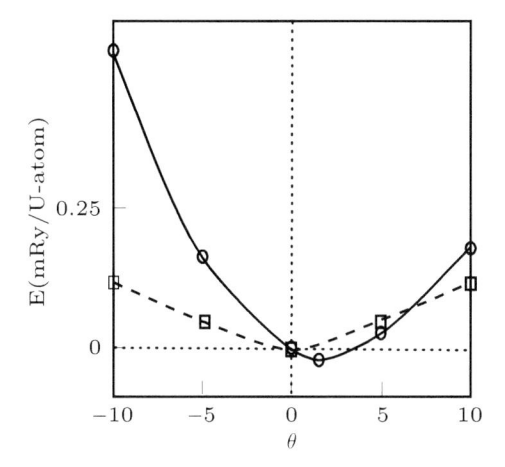

Figure 4.91 *Total energy of U_3P_4 as a function of angle θ. The solid line was calculated with and the dashed line without SOC (with permission from Sandratskii, 1998).*

the cone formed by the six U moments with the axis of the cone parallel to the (111)-direction.

More understanding of the system's behavior can be gained with the help of Fig. 4.91, which shows the result of total energies when the directions of the magnetic moments are constrained to some values of θ near equilibrium (see Fig. 4.90 for a definition of θ). Figure 4.91 shows that scalar relativistic calculations give a total energy curve symmetric with respect to a change of the sign of θ. This means an extremum of the total energy for the ferromagnetic configuration $\theta = 0$ is predetermined by symmetry. In the relativistic case, however, the total energy as a function of θ is not symmetric about $\theta = 0$. In fact, for reasons of symmetry no extremum of the total energy at $\theta = 0$ is expected. As a result, the position of the extremum of the total energy curve as a function of θ is "accidental" i.e. not determined by the symmetry of the problem.

Results of calculations for U_3As_4 (Sandratskii and Kübler, 1997b) are very similar to those for U_3P_4 and again, in agreement with experiment, supply a noncollinear cone structure having the axis of the cone parallel to the (111)-direction. However, in contrast to U_3P_4 and U_3As_4, the experimental properties of U_3Sb_4 (see e.g. Henkie et al., 1987 and Maslanka et al., 1989) are different in that the easy axis is no longer the (111)-axis but is observed to be (001) and the experimental findings are discussed in terms of a collinear magnetic structure with all magnetic moments parallel to the (001)-direction. Concerning the collinearity, however, calculations do not support this point of view. Starting out with all moments parallel to the (001)-direction one sees in the calculations that the magnetic moments of the first group of U atoms (atoms 1–4), if free to rotate, deviate from their initial direction and form a cone structure shown schematically in Fig. 4.92 where the projection of the U atoms onto the (001)-plane is presented. Atoms 5 and 6 keep their directions along the (001)-axis.

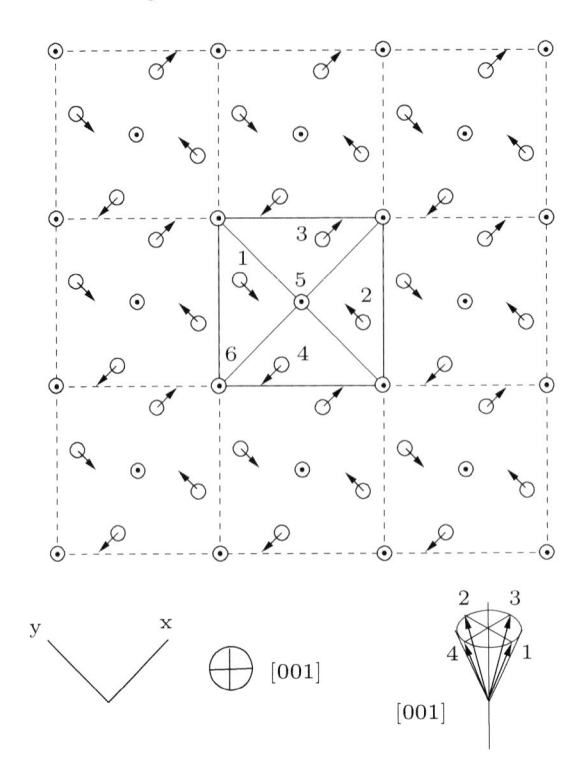

Figure 4.92 *Projection of the atomic positions and magnetic moments of the U atoms in U_3X_4 onto the (001)-plane (X = Sb). At the right the cone formed by the magnetic moments of the U atoms 1–4 is shown, the cone's axis being parallel to the (001)-axis (with permission from Sandratskii, 1998).*

Now our symmetry principle helps us to understand this observation. Indeed, as is seen in Fig. 4.92, the system has a fourfold symmetry axis which passes through the position of atom 5. (A detailed symmetry analysis shows that, to keep the crystal structure invariant, the 90° rotation about the z-axis needs to be accompanied by the inversion.) To preserve this symmetry the direction of the moment of atom 5 (and equivalent to it atom 6) must be kept along the (001)-axis. On the other hand, atoms 1–4 transform into each other under rotation about this axis. Therefore, only relative directions of the moments of these atoms and not the directions of the particular moments are restricted by this symmetry axis. In particular, the noncollinear structure shown in Fig. 4.92 has the entire symmetry of the initial structure.

Although the possibility of the magnetic structure of U_3Sb_4 being noncollinear had already been discussed by Oleksy (1984) on the basis of a symmetry analysis of a phenomenological spin Hamiltonian, in later treatments of the experimental results this structure was considered as one of the possible alternatives not realized in reality. Our calculation and our symmetry analysis show unequivocally that in the case of U_3Sb_4 we deal again with noncollinearity predetermined by symmetry, hence the magnetic structure of U_3Sb_4 is unlikely to be collinear. Note that the result of newer neutron diffraction measurements on U_3Sb_4 by Gukasov *et al.* (1996) give an estimate of the deviation angle somewhat bigger than the error bar of the experiment, which can be treated as a qualitative confirmation of our conclusion.

4.6.3.2 The case of UPdSn

The compound UPdSn was extensively studied by Robinson *et al.* (1991, 1992, 1993, 1994), de Boer *et al.* (1992), Havela *et al.* (1992), and Troc *et al.* (1995), experimentally revealing a number of interesting physical properties. The main attention focused on two magnetic phase transitions which were found to be accompanied by lattice distortions.

The paramagnetic state has the hexagonal GaGeLi crystal lattice. The uranium atoms lie on a simple hexagonal lattice with two uranium layers per crystallographic unit cell. Below 45 K UPdSn becomes magnetic with a noncollinear antiferromagnetic structure. In this phase (which we will refer to as structure I) all magnetic moments of the uranium atoms lie parallel to a plane and compensate one another completely. Simultaneously orthorhombic lattice distortions are detected. At 20 K a second phase transition is observed. Here the magnetic structure (structure II) is still noncollinear and compensated; however, the magnetic moments deviate from the plane developing components perpendicular to it. Thus, contrary to the U_3X_4 compounds, both noncollinear magnetic structures of UPdSn possess no ferromagnetic component.

A detailed account of the crystal structure and calculations for UPdSn can be found in a paper by Sandratskii and Kübler (1997b). They start electronic structure calculations with the undistorted hexagonal lattice having initially all uranium magnetic moments directed along the z-axis and forming a collinear antiferromagnetic structure (see Fig. 4.93(a)) with an orthorhombic magnetic unit cell. Subsequently, allowing the moments to rotate, they are found to deviate immediately from the z-axis keeping their equivalence and the compensated character of the magnetic structure, however. The resulting self-consistent directions of the magnetic moments are shown in Fig. 4.93(a); they form a magnetic structure which is very similar to the experimental structure I. Again, the symmetry principle formulated above helps to expose the physical reasons for the instability of the initial collinear structure. The reasoning is as follows.

Beginning with a symmetry analysis of the initial collinear antiferromagnetic structure, we consider two types of symmetry transformations: unitary transformations which do not contain time reversal and anti-unitary transformations which are products of a unitary transformation with the time-reversal operation. The system at hand possesses four unitary symmetry transformations and four anti-unitary transformations collected in Table 4.15. Note that the vector $\tau_2 = \frac{1}{2}(a, b, 0)$ is a lattice translation of the hexagonal chemical lattice which in the magnetic case must be accompanied by time-reversal to be a symmetry operator (operator 5 in Table 4.15).

The following symmetry property of the initial collinear structure is important in our case: all nontrivial unitary transformations, 2–4, and three of the four nonunitary transformations, 5, 6, and 8; they do not leave the atomic positions unchanged but transfer every atom into the position of another atom. Simultaneously, the moment of the atom is transformed to assume a direction corresponding to the new atomic position.

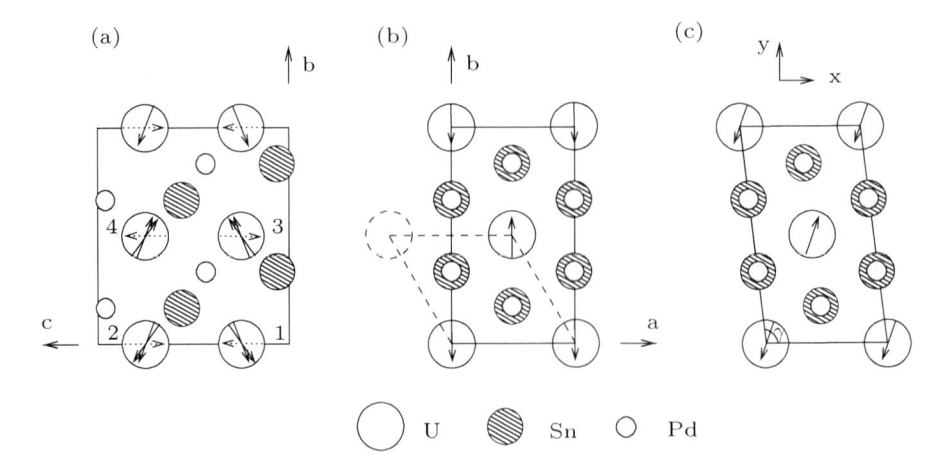

Figure 4.93 *Projections of the crystal and magnetic structure of UPdSn onto the xy- and yz-planes. (a) Projection of the orthorhombic unit cell onto the yz-plane. Dotted arrows show the initial magnetic structure used to start the calculation; thick arrows show the resulting self-consistent directions of the magnetic moments. Thin arrows show the experimental magnetic structure (Robinson et al., 1991). (b) Projection of the orthorhombic unit cell onto the xy-plane. Both experimental and theoretical projections of the magnetic moments are parallel to the y-axis. (c) Projection of the monoclinic unit cell onto the xy-plane. Arrows show schematically the deviations of the magnetic moments from the yz-plane (after Sandratskii and Kübler, 1997a).*

Therefore these operations do not impose any restrictions on the directions of particular atomic moments but only on their relative directions. These restricting relations are collected in the last column of Table 4.15.

The only nontrivial symmetry operation that keeps atoms in their initial positions is the anti-unitary transformation 7. This symmetry operation requires that $m_x^i = 0$ for each atom i. Because of this, the restrictions on the relative directions of the atomic moments resulting from the other symmetry operations take the following form: $m_y^1 = m_y^2 = -m_y^3 = -m_y^4$ and $m_z^1 = -m_z^2 = m_z^3 = -m_z^4$. This means that the initial collinear structure has the same symmetry as the noncollinear structures that satisfy these relations. The noncollinear magnetic structure detected experimentally belongs to this type. Thus, according to the symmetry principle, it is highly improbable that the total energy will attain a minimum for the collinear structure. Rather, the magnetic moments will deviate from collinearity, but the details of this deviation can only be ascertained by self-consistent calculations giving the minimum of the total energy.

Next we study the influence of lattice distortions on the magnetic structure. We start with the orthorhombic distortions which were observed to accompany the magnetic structure I. Following the experiment of Robinson *et al.* (1993) we introduce a small variation of the lattice parameters a and b such that the relation $b = \sqrt{3}\,a$ valid for the ideal hexagonal lattice is no longer satisfied. This distortion does not affect the symmetry of the system because the magnetic structure has already lowered the symmetry of the crystal

4.6 *Relativistic effects* 341

Table 4.15 *Symmetry properties of orthorhombic UPdSn.*

	Operation	Transposition of U atoms	Restriction on U moments
1	$\{\varepsilon\|0\}$	no	no
2	$\{C_{2z}\|\boldsymbol{\tau}_1\}$	$1 \leftrightarrow 4 ; 2 \leftrightarrow 3$	$\begin{pmatrix} m_x \\ m_y \\ m_z \end{pmatrix}_i = \begin{pmatrix} -m_x \\ -m_y \\ m_z \end{pmatrix}_j ; i \leftrightarrow j$
3	$\{\sigma_x\|\boldsymbol{\tau}_2\}$	$1 \leftrightarrow 3 ; 2 \leftrightarrow 4$	$\begin{pmatrix} m_x \\ m_y \\ m_z \end{pmatrix}_i = \begin{pmatrix} m_x \\ -m_y \\ -m_z \end{pmatrix}_j ; i \leftrightarrow j$
4	$\{\sigma_y\|\boldsymbol{\tau}_3\}$	$1 \leftrightarrow 4 ; 2 \leftrightarrow 3$	$\begin{pmatrix} m_x \\ m_y \\ m_z \end{pmatrix}_i = \begin{pmatrix} -m_x \\ m_y \\ -m_z \end{pmatrix}_j ; i \leftrightarrow j$
5	$\{\varepsilon\|\boldsymbol{\tau}_2\} R$	$1 \leftrightarrow 3 ; 2 \leftrightarrow 4$	$\mathbf{m}_i = -\mathbf{m}_j ; \quad i \leftrightarrow j$
6	$\{C_{2z}\|\boldsymbol{\tau}_3\} R$	$1 \leftrightarrow 2 ; 3 \leftrightarrow 4$	$\begin{pmatrix} m_x \\ m_y \\ m_z \end{pmatrix}_i = \begin{pmatrix} m_x \\ m_y \\ -m_z \end{pmatrix}_j ; i \leftrightarrow j$
7	$\{\sigma_x\|0\} R$	no	$m_x = 0 ; \quad$ all i
8	$\{\sigma_y\|\boldsymbol{\tau}_1\} R$	$1 \leftrightarrow 4 ; 2 \leftrightarrow 3$	$\begin{pmatrix} m_x \\ m_y \\ m_z \end{pmatrix}_i = \begin{pmatrix} m_x \\ -m_y \\ m_z \end{pmatrix}_j ; i \leftrightarrow j$

$$\varepsilon = \begin{pmatrix} 1 & & \\ & 1 & \\ & & 1 \end{pmatrix} ; C_{2z} = \begin{pmatrix} -1 & & \\ & -1 & \\ & & 1 \end{pmatrix} ; \sigma_x = \begin{pmatrix} -1 & & \\ & 1 & \\ & & 1 \end{pmatrix} ; \sigma_y = \begin{pmatrix} 1 & & \\ & -1 & \\ & & 1 \end{pmatrix}$$

$\boldsymbol{\tau}_1 = \frac{1}{2}(a,b,c); \quad \boldsymbol{\tau}_2 = \frac{1}{2}(a,b,0); \quad \boldsymbol{\tau}_3 = \frac{1}{2}(0,0,c)$ R time-reversal operation.

from hexagonal to orthorhombic. As a result no qualitative changes of the magnetic structure were observed due to the orthorhombic lattice distortion and quantitative changes appear to be very small, too. The magnetic structure is very close to the structure shown in Fig. 4.93(a) and is still in good agreement with the experimental structure I. A basically different response is obtained when the crystal undergoes a monoclinic distortion. In agreement with experiment the b-side of the basal rectangle (Fig. 4.93) may be rotated by 0.4° about the c-axis. Then starting the calculations with the magnetic structure I, one finds after only the first iteration that all uranium moments depart from the yz-plane but stay mutually equivalent and compensate the magnetic structure.

Again, a symmetry analysis helps to understand this process. The monoclinic distortion decreases the symmetry of the system such that from the eight operations of the orthorhombic structure only four are left over. These are the operations numbered 1, 2, 5, and 6 in Table 4.13. Operation 5 demands equivalence of atom 1 to atom 3 and atom 2 to atom 4. Simultaneously, the moments of the equivalent atoms must be antiparallel: $\mathbf{m}_1 = -\mathbf{m}_3$ and $\mathbf{m}_2 = -\mathbf{m}_4$. Operation 2 is responsible for the equivalence of atoms 1 and 4 and the following relation between the components of the magnetic moments: $m_x^1 = -m_x^4$, $m_y^1 = -m_y^4$, $m_z^1 = m_z^4$. Further symmetry operations do not lead to additional restrictions. Thus we see the important difference between the orthorhombic and the monoclinic structures of UPdSn: in the monoclinic structure there is no symmetry operation demanding the x-component of the magnetic moments to be zero. This means that a deviation of the magnetic moments from the yz-plane does not change the symmetry of the system and therefore will occur according to our symmetry principle. Thus the result of the calculation for the monoclinically distorted lattice and the corresponding symmetry analysis are in agreement with the experimental data.

The deviations of the magnetic moments from parallel directions in the case of U_3X_4 and UPdSn caused by the combined symmetry properties of the crystal *and* the magnetic structures is reminiscent of the effect of weak ferromagnetism in Fe_2O_3, connected with the names Dzyaloshinsky and Moriya. We therefore show next that the symmetry analysis given together with the method of calculation also explains this interesting case.

4.6.4 Weak ferromagnetism

The phenomenon of weak ferromagnetism has been known for something like 40 years (see e.g. Dzyaloshinsky, 1958 and Moriya, 1960). It is characterized by a small net magnetic moment resulting from a collection of atomic magnetic moments that nearly cancel each other. Weak ferromagnetism was first observed in hematite, α-Fe_2O_3. Dzyaloshinsky (1958) showed that it was an intrinsic effect due to particular symmetry properties of the crystal structure and the magnetic moment arrangements. He suggested the following model Hamiltonian

$$\mathcal{H} = I_{ij}\mathbf{S}_i\mathbf{S}_j + \overset{*}{\mathbf{d}}_{ij} \cdot [\mathbf{S}_i \times \mathbf{S}_j] + \mathbf{S}_i \cdot \mathbf{K}_i \cdot \mathbf{S}_i \qquad (4.263)$$

establishing a basis for further work on weak ferromagnetism. In Eqn (4.263) the indices i and j (implying appropriate summation) number the atoms in the lattice, I_{ij} and

$\mathbf{d}_{ij}$ are the symmetric and antisymmetric exchange constants, respectively, and the tensor $\mathbf{K}_i$ contains information about the single-ion magneto-crystalline anisotropy. The first term of the Hamiltonian Eqn (4.263), the symmetric exchange, is supposed to lead to a compensated magnetic configuration. The next two terms, the antisymmetric exchange and the magneto-crystalline anisotropy terms, respectively, can lead to a small ferromagnetic moment in an otherwise antiferromagnetic crystal.

Moriya (1960) showed that Dzyaloshinsky's explanation can be given a microscopic footing by means of Anderson's perturbation approach to magnetic superexchange. The density-functional theory (DFT) techniques explained in this treatise provide all components necessary for a first-principles study of weak ferromagnetism. Considering spin–orbit coupling and allowing for arbitrary noncollinear configurations of the magnetic moments it is possible to explain the phenomenon without any phenomenological parameters or a perturbation treatment; a canted magnetic structure appears within the usual DFT self-consistency cycle as that magnetic configuration which supplies the minimum of the total energy.

Moriya also showed that depending on the type of the crystal structure either of two mechanisms, antisymmetric exchange (second term on the right-hand side of Eqn (4.263)) or magneto-crystalline anisotropy (third term on the right-hand side of Eqn (4.263)), can be the origin for the canting of magnetic moments. Thus, in the case of α-Fe_2O_3 it is the antisymmetric exchange that plays the dominant role, whereas in the case of NiF_2 antisymmetric Heisenberg exchange is ruled out in favor of the magneto-crystalline anisotropy which is here responsible for the appearance of the ferromagnetic component.

Our next topic is thus a first-principles study of both types of weak ferromagnetism. The compound α-Fe_2O_3 is chosen as a classical example for weak ferromagnetism caused by the antisymmetric exchange interaction. The triangular antiferromagnet Mn_3Sn is considered as an example where the antisymmetric exchange contributions from different atoms cancel perfectly and, therefore, cannot be the reason for the weak ferromagnetism observed here. In this case it is the magneto-crystalline anisotropy term that is responsible for the effect.

4.6.4.1 Hematite (α-Fe_2O_3)

The crystal structure of Fe_2O_3 is shown in Fig. 4.94. This is a rhombohedral lattice with a basis of two formula units per unit cell. The following 12 point-group operations enter the space group characterizing the symmetry of the atomic positions: the identity operation, rotations by 120° and 240° about the z-axis, rotations by 180° about three axes in the xy-plane ($y = 0$ and $y = \pm\sqrt{3}\,x$) and these six operations multiplied by the inversion.

We start the discussion with the results of a scalar relativistic calculation (Sandratskii and Kübler, 1996a,b) for four collinear magnetic structures in which the moments of Fe are aligned along the symmetry axis in Fig. 4.94, being arranged in the following sequences: $(++++)$, $(+-+-)$, $(+--+)$, and $(++--)$. Here $+$ and $-$ designate up and down directions with respect to some axis which may, for instance, be the z-axis in Fig. 4.94. In agreement with experiment it was found that the structure

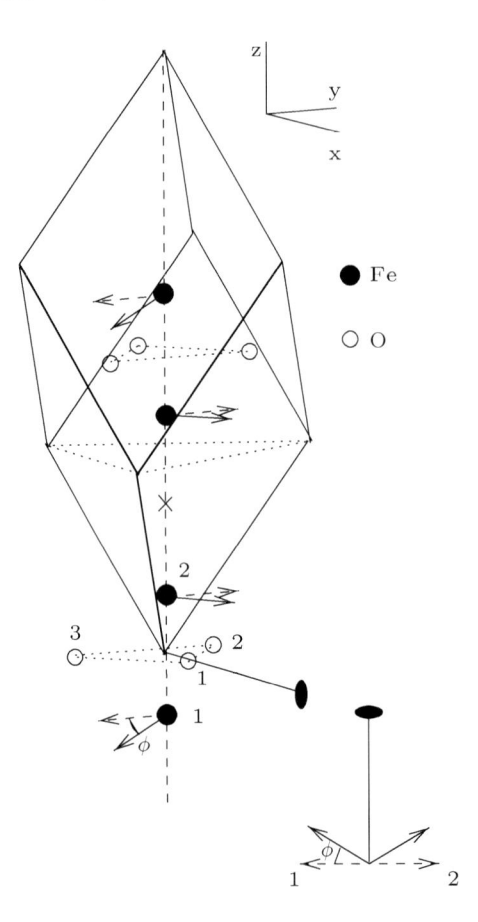

Figure 4.94 *The unit cell of Fe_2O_3. The cross on the diagonal of the rhombohedron shows the point of inversion. The solid line passing through the first oxygen atom indicates a twofold symmetry axis. The collinear (dashed arrow) and canted (solid arrow) directions of the Fe atoms are shown. The canting of the Fe moments in the xy-plane is demonstrated differently in the lower right corner of the figure (after Sandratskii and Kübler, 1996a).*

$(+--+)$ possesses the lowest total energy. Therefore only calculations for the $(+--+)$ configuration will be discussed further. As was shown above, any collinear structure is stable in a scalar relativistic calculation. So, no deviations of the magnetic moments are obtained in this case.

If we next switch on spin–orbit coupling and choose the z-axis as the direction of the magnetic moments, the following operations remain in the symmetry group of the Kohn–Sham Hamiltonian: the identity, rotations about the z-axis, and these operations multiplied by the inversion. The symmetry operations which are of special importance for us are the rotations about the z-axis: none of them change the position of any of the four Fe atoms lying on the axis of rotation. The directions of the magnetic moments are parallel to this axis and, therefore, are not changed either. It is clear that any deviation of the magnetic moments from the z-axis will destroy the invariance of the crystal with respect to this operation. As the symmetry of the Hamiltonian cannot become lower during iterations this change is forbidden by symmetry. This result agrees with the experimental observation of a collinear antiferromagnetic structure with moments oriented parallel to the z-axis below 260 K, the so-called Morin transition (Morin, 1950).

The situation changes drastically when the moments are parallel to the y-axis arranged again in the sequence $(+--+)$. Now the following symmetry operations are left in the group of the Kohn–Sham Hamiltonian: the identity, the 180° rotation about the x-axis, and these operations multiplied by inversion. Inversion transforms the atoms of the upper Fe_2O_3 molecule into the atoms of the lower molecule, see Fig. 4.94. Since the magnetic moments are axial vectors, they do not change under this transformation. Hence, the magnetic moments of two Fe_2O_3 molecules must be parallel, so we may look at the magnetic moments of one molecule only, for instance, of that being below the inversion center in Fig. 4.94. The only condition imposed on the moments of the Fe atoms by symmetry is the transformation of the moment of atom 1 into that of atom 2 by the rotation through 180° about the x-axis (see Fig. 4.94). However to fulfill this condition it is not necessary for the atomic moments to be parallel to the y-axis nor to remain collinear. Indeed, calculations show that a collinear starting configuration does not lead to zero off-diagonal elements of the density matrices, $\tilde{n}_\nu$. Correspondingly, in this simulated annealing process, the magnetic moments move and deviate from their collinear starting directions toward the direction of the x-axis (see Fig. 4.94) until an "accidental" magnetic structure is obtained that possesses a minimum in the total energy. The total energy might again look like Fig. 4.91, replacing the angle θ by ϕ, but in reality the total energy changes are so small that they cannot yet be resolved numerically. The principle, however, is clarified by the symmetry arguments just outlined.

4.6.4.2 The case of Mn_3Sn

The important part of the crystal structure of Mn_3Sn is shown as a projection in Fig. 4.95. Mn_3Sn belongs to that class of compounds in which the experimentally observed weak ferromagnetism cannot be attributed to the antisymmetric exchange interaction because of complete cancellation of the contributions of different atoms (Tomiyoshi and Yamaguchi, 1982). In an early paper on Mn_3Sn by Sticht *et al.* (1989), a number of different magnetic configurations were investigated that are allowed by Landau's theory of phase transitions, but spin–orbit coupling was neglected. These calculations showed, however, that compared with other magnetic arrangements, the triangular configurations possess a distinctly lower total energy. This simplifies our task and allows us to restrict the relativistic calculations to triangular structures only, of which four configurations are depicted in Fig. 4.95. Calculations without spin–orbit coupling showed that all four configurations are equivalent, i.e. they are degenerate having equal total energies and local magnetic moments. Another property of all four configurations is their great stability during iterations.

Again, as before, the situation changes drastically when spin–orbit coupling (and in this case orbital polarization as well, see Sandratskii and Kübler (1996b) for more detail)

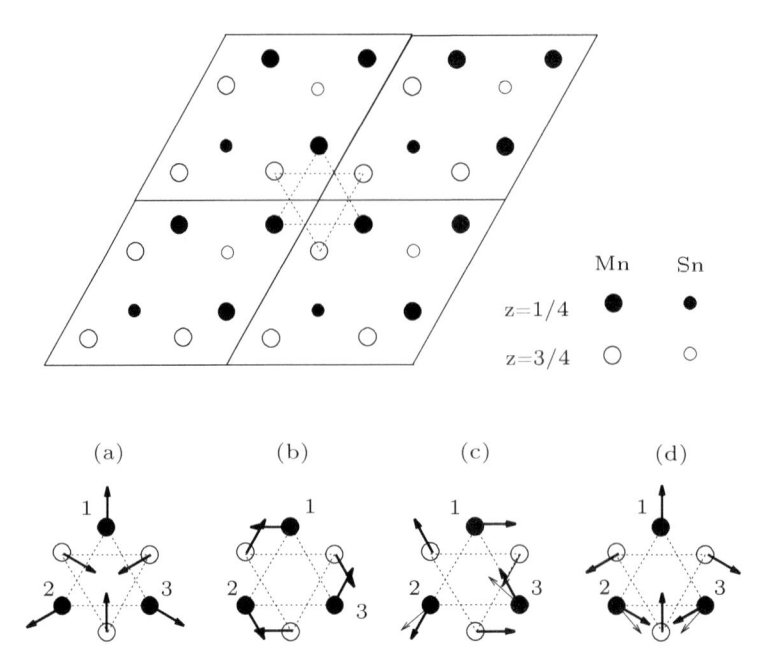

Figure 4.95 *Hexagonal crystal structure of Mn_3Sn shown as a projection and triangular magnetic structures. Rotations of the magnetic moments leading to weak ferromagnetism in structures (c) and (d) are shown only for atoms in the $z = 0.25$ plane (thin arrows). Moments of the atoms in the $z = 0.75$ plane are parallel to the moments of the corresponding atoms of the $z = 0.25$ plane (after Sandratskii and Kübler, 1996b).*

is taken into account. First, the degeneracy is lifted, i.e. all four magnetic configurations become inequivalent and, second, for two of them ((c) and (d) in Fig. 4.95) the magnetic moments deviate slightly from the initial directions as indicated in Fig. 4.95.

The stability of the antiferromagnetic triangular structures and their equivalence obtained in the scalar relativistic calculations is again explained easily using the spin-space groups and the principle of preserving the symmetry of the Kohn–Sham Hamiltonian during the iteration process. Indeed, structure (a) transforms into structure (b), and (c) into (d) by a pure spin rotation through 90° about the hexagonal axis and (a) transforms into (d) by a 180° spin rotation about the direction of the spin of atom 1. Thus, in the scalar relativistic approximation, all four structures are equivalent.

Concerning the property of the stability of a given magnetic structure, it is easily seen that for all four structures depicted in Fig. 4.95 a deviation of any magnetic moment from the directions shown destroys at least one symmetry operation. In particular, for all atoms of structures (a) and (b) and for atom 1 of structures (c) and (d) these operations are the combined spin-space rotations through 180° about the spins of these atoms. For atoms 2 and 3 of structures (c) and (d) the symmetry operation is more involved and consists of a 180° spin rotation about the direction of the spin of the atom accompanied by a 180°

space rotation about another axis, i.e. the spin and space parts of the transformation are different.

In the relativistic case (i.e. including spin–orbit coupling and orbital polarization), however, the operations of the spin-space group do not preserve the form invariance of the Kohn–Sham Hamiltonian, the operations that remain being those of the usual space groups with $\alpha_S = \alpha_R$. But no space-group operation transforms any of the four magnetic structures into another one, whence all four magnetic structures now become inequivalent. To decide about the stability of a given magnetic configuration in this case, we again check whether one of the symmetry operations of the configuration will be destroyed by a deviation of the moment. Now we see that the two types of structures (a), (b) ("direct structures") and (c), (d) ("inverse structures") are distinguished by the "handedness" of the spin rotations: in the case of (a), (b) a $+120°$ rotation turns the spin of atom 1 into that of atom 2 etc., whereas it is a $-120°$ rotation in the case of (c), (d). Indeed, transformations guaranteeing the stability of the directions of all moments of the direct structures in the scalar relativistic case have coinciding spin and space transformations and continue to be symmetry operations also in the relativistic case. The same holds true for atom 1 of the inverse structures. However, the operations responsible for the stability of the other two moments in the scalar-relativistic case do not apply here and the moments can move without decreasing the symmetry of the Hamiltonian although the deviations of the moments of two atoms are not independent. This completes the discussion of the symmetry arguments which were at the center of interest here. More details concerning the numerical results can be found in the paper by Sandratskii and Kübler (1996b).

4.6.5 Magneto-crystalline anisotropy

So far in this section we have been concerned with the orientation of the magnetic moments *relative* to each other. We now want to discuss the energy involved in rotating the magnetization from a direction of low energy—the so-called *easy* axis—to a direction of high energy—the so-called *hard* axis. In the case of transition metal ferromagnets, to which we now return, the energy difference involved is typically of the order of 10^{-6} to 10^{-3} eV/atom; this *anisotropy energy* is thus a very small contribution to the total magnetic energy.

It was Harvey Brooks who showed in a pioneering paper some 80 years ago (Brooks, 1940) that the anisotropy energy is due to *spin–orbit coupling* which breaks the rotational invariance with respect to the spin quantization axis. Another mechanism is the magnetic dipole interaction between spins. This interaction contains a term depending on the orientation of the spins with respect to the line joining them and thus is structure dependent since it depends on the actual position of the spins in the lattice (see for instance Bruno (1993c)). In cubic crystals, however, the magnetic dipole interactions cancel out to a first approximation because of the symmetrical arrangement of nearest neighbors and in second-order perturbation theory they give too small a magnitude. They cannot, however be ignored in systems with reduced symmetry as in surfaces, thin films, and multilayers. We focus our attention here on the bulk ferromagnets Fe, Co, and

Ni and, therefore, consider only the effect of spin–orbit coupling that is described by the Hamiltonian, $\mathcal{H}_{so}$, see Eqn (4.259), in which the spin-rotation matrices can now be used to change the direction of magnetization with respect to a quantization axis that may be attached to a crystallographic axis.

Beginning with the salient experimental facts we first write out the total energy as it is usually expressed:

$$E(\mathbf{e}) = K_0 + f(\alpha_1 \alpha_2 \alpha_3) \tag{4.264}$$

where $\alpha_1, \alpha_2, \alpha_3$ are the direction cosines of the magnetization that points along the vector $\mathbf{e}$ and f is some function to be specified. It was already Brooks who pointed out that the form of this function may be deduced from the symmetry class of the crystal without any assumptions as to the mechanism responsible for the anisotropy. For cubic crystals Eqn (4.264) is written as

$$E(\mathbf{e}) = K_0 + K_1(\alpha_1^2\alpha_2^2 + \alpha_2^2\alpha_3^2 + \alpha_3^2\alpha_1^2) + K_2\alpha_1^2\alpha_2^2\alpha_3^2$$
$$+ K_3(\alpha_1^2\alpha_2^2 + \alpha_2^2\alpha_3^2 + \alpha_3^2\alpha_1^2)^2 + \cdots \tag{4.265}$$

with the coordinate axes taken along the cubic axes. For systems with hexagonal closed packed (hcp) symmetry it is more customary to use spherical polar coordinates, θ and ϕ, with respect to the c and a-axis, respectively, and write

$$E(\mathbf{e}) = K_0 + K_1 \sin^2\theta + K_2 \sin^4\theta + (K_3 + K_3' \cos 6\phi)\sin^6\theta + \cdots. \tag{4.266}$$

The experimental values of the anisotropy constants for Fe, Co, and Ni are given in Table 4.16.

The easy axis of Fe is seen to be the (100)-direction and that of fcc Co and Ni the (111)-direction, while that of hcp Co is along the c-axis. It should be noted that, due to the lower symmetry, the anisotropy energy of hcp Co is one order of magnitude larger than that of Fe, fcc Co, and Ni.

Table 4.16 *Anisotropy constants of Fe, Co, and Ni at $T = 4.2$ K in $\mu eV/atom$.*

	Fe (bcc)	Co (hcp)	Co (fcc)	Ni (fcc)
K_1	4.02 [a]	30.6 [d]; 53.3 [b]	−5.2 [e]	−8.63 [a]
K_2	0.0144 [a]	7.31 [b]	1.8 [e]	3.95 [a]
K_3	6.6×10^{-3} [a]	–	–	0.238 [a]
K_3'	–	0.84 [c]	–	–

[a] Escudier (1975); [b] Rebouillat (1972); [c] Paige et al. (1984); [d] Weller et al. (1994); [e] Suzuki et al. (1994) (extrapolated to $T = 0$).

The determination of the anisotropy energy by *ab initio* density-functional methods is an extremely difficult task. The usual procedure would be to determine the total energy choosing the magnetization along two nonequivalent directions (at least), for example the (001) and (111)-directions in a cubic crystal. But since the total energy for an atom is of the order of 4×10^4 eV/atom, while the anisotropy energy is of order 10^{-6} eV/atom, the required numerical accuracy is excessively high. Thus a number of attempts were made using the *force theorem*, i.e. with a self-consistent reference energy one determines the change in energy by a non-self-consistent calculation, evaluating the difference of the band energies only. One would think that the small changes in the double-counting terms cancel out of the calculation. That this is not so is borne out by very detailed calculations for bulk Fe, Co, and Ni carried out by Daalderop *et al.* (1990). Indeed, for Ni and hcp Co these calculations result in the wrong easy axis, while for Fe the correct easy axis is obtained but the energy difference is too small by a factor of 3.

Although Wang *et al.* (1993) pointed out the possible reasons for the failure of calculations of this type and proposed a scheme that was expected to remedy the situation, the bulk systems apparently were not recalculated using their methodology. However, Wang *et al.* successfully applied their method to monolayers, where the anisotropy energy is considerably larger.

It seems, however, that the problem can be solved by *brute force* after all, i.e. obtaining the necessary total energy differences from a fully self-consistent treatment that includes the spin–orbit interaction. Trygg *et al.* (1995), who also included the orbital polarization term, $\mathcal{H}_{orb}$, Eqn (3.355), and Halilov *et al.* (1998a) showed that these calculations are indeed possible in spite of an enormous number of k-points needed to do the Brillouin zone integrations. The total energy differences, ΔE, are obtained choosing the magnetization along the (001) and (111)-directions in which case $\Delta E = E(001) - E(111)$ in the cubic crystal. For hcp Co Trygg *et al.* select the [0001] and the [10$\bar{1}$0] directions, i.e. $\Delta E = E(0001) - E(10\bar{1}0)$. The details of the calculations differ, in one case (Trygg *et al.*) a full potential method is used and the number of k-points in the full Brillouin zone is 15 000 for Fe, fcc Co, and hcp Co, and about 25 000 for Ni; in the other case (Halilov *et al.*) the LMTO method is shown to be accurate enough with something like 372 248 k-points in the full zone. The results of these calculations are collected in Table 4.17 where they are also compared with experimental energy differences extracted

Table 4.17 *Magneto-crystalline anisotropy: energy differences, ΔE, in μeV/atom, obtained from Table 4.14, labeled exp., and total energy differences from ab initio calculations, labeled calc.*

	Fe (bcc)	Co (hcp)	Co (fcc)	Ni (fcc)
exp.	−1.34	−65	1.66	2.7
calc. [a]	−2.6	–	2.4	1.0
calc. [b]	−1.8	−110	2.2	−0.5

[a] Halilov *et al.* (1998a); [b] Trygg *et al.* (1995)

4.7 Berry phase effects in solids

from the data given in Table 4.16. It should be noticed that Trygg *et al.* obtain the wrong easy axis for Ni; however, the results of Halilov *et al.* indicate that this is just a matter of an insufficient number of k-points: brute force wins! It seems that the essential physics has been captured by these calculations and the remaining discrepancy is not alarming.

In the past 20 to 25 years an important and interesting branch of physics grew out of the concept of the *Berry phase* (Berry, 1984). It was discovered that it is at the foundation of a modern theory of polarization (Resta, 1994) and in magnetism it provides a new basis for an explanation of the anomalous Hall effect (Xiao *et al.*, 2010 and Nagaosa *et al.*, 2010). Furthermore, it was found that it reaches deeply into questions connected with topology in electronic structure and it opens new avenues for thermoelectric and thermomagnetic effects.

We begin by sketching the derivation of the Berry curvature following closely Berry's paper. Consider the time-dependent Schrödinger equation that is assumed to depend on a parameter $\mathbf{R}$ which moves along a path in some given space

$$i\hbar \frac{\partial}{\partial t}\psi(t) = H(\mathbf{R})\psi(t). \quad (4.267)$$

The motion is supposed to be such that no transitions happen in the system, a so-called adiabatic motion. Such a path is drawn on a torus in Fig. 4.96. Next we write for each value of $\mathbf{R}$ the time-independent Schrödinger equation,

$$H(\mathbf{R})|n(\mathbf{R})\rangle = E_n(\mathbf{R})|n(\mathbf{R})\rangle \quad (4.268)$$

and try for the time-dependent wave function

$$\psi(t) = \exp[\frac{-i}{\hbar}\int_0^t E_n(\mathbf{R}(t'))dt'] \cdot \exp[i\gamma_n(t)]|n(\mathbf{R})\rangle \quad (4.269)$$

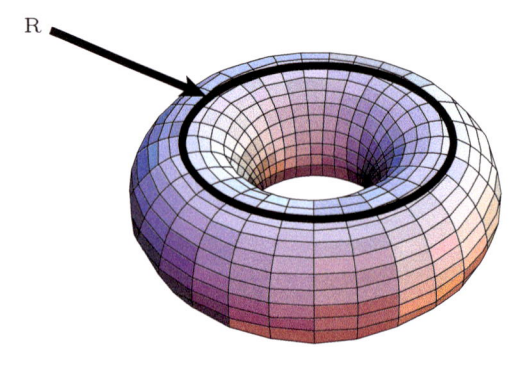

Figure 4.96 *A path* $\mathbf{R}$ *along which the system is supposed to move around the torus.*

Substituting into eqn (4.267) we obtain easily

$$\frac{d}{dt}\gamma(t) = i\langle n(\mathbf{R})|\nabla_{\mathbf{R}} n(\mathbf{R})\rangle \frac{d}{dt}\mathbf{R}(t).$$

Integrating along a path C, where C is any closed path as sketched in Fig. 4.96, and going around completely we finally have

$$\gamma(C) = i\oint \langle n(\mathbf{R})|\nabla_{\mathbf{R}} n(\mathbf{R})\rangle d\mathbf{R}. \qquad (4.270)$$

This is called the Berry phase; it vanishes when the state vectors are real.

We next specify the space for the vector $\mathbf{R}$ to be the *reciprocal space* used in energy band theory (Zak, 1989). The eigenfunctions for an electron in a periodic solid are Bloch functions $\psi_{n,\mathbf{k}}(\mathbf{r})$ which satisfy eqn (3.12). We see at once that the function

$$u_{n,\mathbf{k}}(\mathbf{r}) = e^{-i\mathbf{k}\cdot\mathbf{r}} \psi_{n,\mathbf{k}}(\mathbf{r}) \qquad (4.271)$$

is crystal-periodic, for which one easily derives the Schrödinger equation

$$\left[\frac{(\mathbf{p}+\hbar\mathbf{k})^2}{2m} + V_{KS}\right] u_{n,\mathbf{k}}(\mathbf{r}) = \epsilon_{n,\mathbf{k}}\, u_{n,\mathbf{k}}(\mathbf{r}). \qquad (4.272)$$

This is the so-called $\mathbf{k}\cdot\mathbf{p}$ equation used in many applications. The effective Hamiltonian is $\mathbf{k}$ dependent and V_{KS} is the Kohn–Sham Potential. The Berry phase for a band with index n is now defined by

$$\gamma_n = i\int_{BZ} \langle u_{n,\mathbf{k}}|\nabla_{\mathbf{k}} u_{n,\mathbf{k}}\rangle d\mathbf{k}, \qquad (4.273)$$

where BZ means a path in the Brillouin zone, from one side to the opposite, for instance. Strictly speaking, γ_n should not be called Berry phase since the path is not necessarily closed. Therefore, it is sometimes called a *Zak phase* after Zak (1989), who first introduced Berry's concept into band theory.

The following two definitions complete the foundation; first there is the vector

$$\mathbf{A}(\mathbf{k}) = i\sum_{n\in occ} \langle u_{n,\mathbf{k}}|\nabla_{\mathbf{k}} u_{n,\mathbf{k}}\rangle, \qquad (4.274)$$

which is called the *Berry connection*, and second the curl of $\mathbf{A}$

$$\mathbf{\Omega}(\mathbf{k}) = \nabla_{\mathbf{k}} \times \mathbf{A}(\mathbf{k}) \qquad (4.275)$$

which is called the *Berry curvature*. Sometimes it is useful to drop the sum over the occupied states and use the Berry curvature for the band n

$$\boldsymbol{\Omega}_n(\mathbf{k}) = i\, \nabla_\mathbf{k} \times \langle u_{n,\mathbf{k}} | \nabla_\mathbf{k} u_{n,\mathbf{k}} \rangle = i \langle \nabla_\mathbf{k} u_{n,\mathbf{k}} | \times | \nabla_\mathbf{k} u_{n,\mathbf{k}} \rangle. \tag{4.276}$$

Eqn (4.275) looks very much like the equation that connects the magnetic field $\mathbf{B}$ with with the vector potential, therefore some see the Berry curvature as a fictitious magnetic field. The Berry connection $\mathbf{A}$ is central for the modern theory of polarization, whereas the Berry curvature plays an important role in metallic magnetism. Before we get into these details we sketch the derivation of a formula for the Berry curvature that is useful for numerical applications. First expand

$$\langle \nabla_\mathbf{k} u_{n,\mathbf{k}} | \times | \nabla_\mathbf{k} u_{n,\mathbf{k}} \rangle = \sum_{m \neq n} \langle \nabla_\mathbf{k} u_{n,\mathbf{k}} | u_{m,\mathbf{k}} \rangle \times \langle u_{m,\mathbf{k}} | \nabla_\mathbf{k} u_{n,\mathbf{k}} \rangle \tag{4.277}$$

and derive by differentiating the time-independent Schrödinger equation

$$\langle u_{m,\mathbf{k}} | \nabla_\mathbf{k} u_{n,\mathbf{k}} \rangle = \frac{\langle u_{m,\mathbf{k}} | \nabla_\mathbf{k} H | u_{n,\mathbf{k}} \rangle}{(\epsilon_m - \epsilon_n)} \quad m \neq n.$$

Substituting into eqn (4.7) we obtain

$$\boldsymbol{\Omega}_n(\mathbf{k}) = \Im \sum_{m \neq n} \frac{\langle u_{n,\mathbf{k}} | \hat{\mathbf{v}} | u_{m,\mathbf{k}} \rangle \times \langle u_{m,\mathbf{k}} | \hat{\mathbf{v}} | u_{n,\mathbf{k}} \rangle}{(\epsilon_m - \epsilon_n)^2}, \tag{4.278}$$

where the velocity operator is $\hat{\mathbf{v}} = \nabla_\mathbf{k} H / \hbar$. This formula for the Berry curvature is used by most numerical applications.

4.7.1 The anomalous Hall effect

The Hall effect was discovered by Hall (1879) and describes a current-carrying conductor in a magnetic field that produces a voltage perpendicular to the direction of the current and the magnetic field because of the Lorenz force. Figure 4.97 shows a sketch of the experimental setup. Later Hall (1881) discovered that his effect was larger by an order of magnitude if the conductor was a ferromagnet; the setup is still as before but a magnetization vector is to be imagined parallel to the field $\mathbf{B}$ in the figure. One frequently describes the Hall resistivity with the simple relation

$$\rho_{xy} = R_0 B_z + R_S M_z \tag{4.279}$$

where a Cartesian coordinate system is implied with the z-axis parallel to $\mathbf{B}$ or $\mathbf{M}$ and R_0 and R_S are coefficients that describe the magnitude of the normal or anomalous Hall effect, respectively.

An early theory of the anomalous Hall effect (AHE) is that by Karplus and Luttinger (1954), which in a somewhat involved way explains the effect. It is the Berry curvature that describes the physics clearly. The simplest way to deal with the transport is by using semi-classical Boltzmann theory and considering the motion of electrons in the solid by

4.7 Berry phase effects in solids

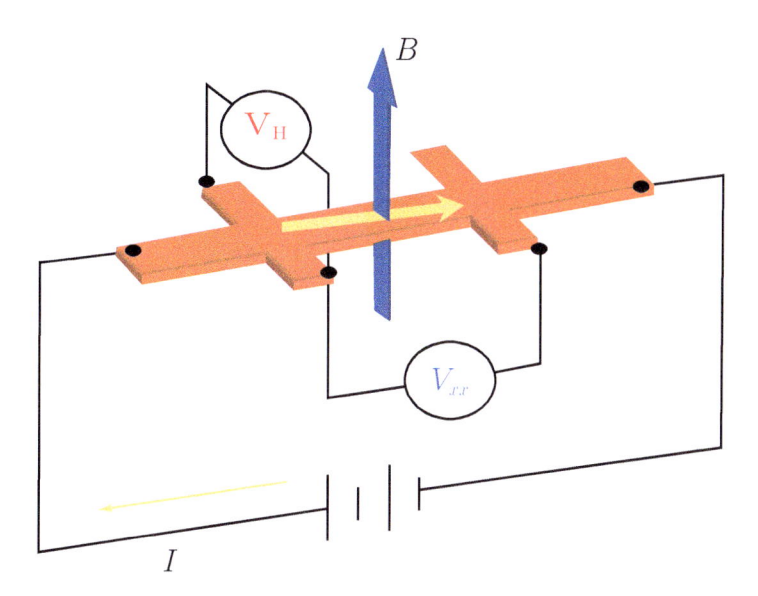

Figure 4.97 *Schematic view of a Hall effect setup. For the anomalous Hall effect replace the vector* **B** *by the magnetization* **M**.

wave packets as is detailed carefully by Xiao *et al.* (2010). We omit the details of the derivation and state the results for the equations of motion of the wave packet's center of mass as

$$\dot{\mathbf{r}} = \mathbf{v} - \dot{\mathbf{p}} \times \mathbf{\Omega}/\hbar \qquad (4.280)$$
$$\dot{\mathbf{p}} = e\mathbf{E} + e\dot{\mathbf{r}} \times \mathbf{B}$$

where $\mathbf{v} = \frac{\partial \epsilon(\mathbf{k})}{\hbar \partial \mathbf{k}}$ is the usual band velocity and $\mathbf{\Omega}$ is the Berry curvature. **E** and **B** are the external electric and magnetic fields. We notice that there is an additional term to the band velocity, which restores the symmetry of these two equations with respect to **r** and **p**. So again, the Berry curvature looks like a magnetic field. We will later see, however, that $\mathbf{\Omega}$ can harbour monopoles, unlike **B**. The physics due to the additional velocity becomes quite transparent when we set the Lorenz term to zero and write for the first part of eqn (4.280)

$$\mathbf{v}_n(\mathbf{k}) = \frac{\partial \epsilon_n(\mathbf{k})}{\hbar \partial \mathbf{k}} - \frac{e}{\hbar} \mathbf{E} \times \mathbf{\Omega}_n(\mathbf{k}). \qquad (4.281)$$

The additional term is called the anomalous velocity and is transverse to the electric field; it gives rise to the anomalous Hall effect. In fact, writing the distribution function as the unperturbed part $f_n(\mathbf{k})$ plus the shift $\delta f_n(\mathbf{k})$ proportional to the electric field and the relaxation time we express the current in terms of the velocity as

$$\mathbf{J} = -\frac{e^2}{\hbar}\mathbf{E} \times \sum_{n,\mathbf{k}} \mathbf{\Omega}_n(\mathbf{k}) f_n(\mathbf{k}) + \frac{e}{\hbar}\sum_{n,\mathbf{k}} \frac{\partial \epsilon_n(\mathbf{k})}{\partial \mathbf{k}} \delta f_n(\mathbf{k}) \tag{4.282}$$

where the sum on $\mathbf{k}$ is over the Brillouin zone. From the first term we extract the *anomalous Hall conductivity*

$$\sigma_{xy} = -\frac{e^2}{\hbar} \int \frac{d^3k}{(2\pi)^3} \sum_{n \in occ} \Omega_n^z(\mathbf{k}) f_n(\mathbf{k}), \tag{4.283}$$

for the current in the x-direction and the voltage in the y-direction, the Berry curvature being in the z-direction. The second term, apart from the longitudinal conductivity, also contributes to the Hall conductivity through special scattering events discussed below. This conductivity formula is dissipationless, which is quite remarkable.

Now turning to experimental data for the anomalous Hall effect (AHE), we expect other terms to play a role through the second term of Eqn (4.282). These scattering events must be transverse to the current; there are two of them, explained in detail by Nagaosa *et al.* (2010) and called side-jump and skew-scattering events. Although the situation is quite controversial one finds that spin–orbit coupling (SOC) is the active ingredient in both processes; in the case of skew scattering right- and left-handed scattering events with respect to the magnetization have different transition probabilities and thus cause a transverse asymmetry. The side-jump processes occur in transverse scattering of wave packets; if they have a wave vector along $\mathbf{k}$ they can suffer a displacement $\perp \mathbf{k}$ upon scattering. Both effects are usually called extrinsic, while the AHE is called intrinsic. The electrical conductivity is usually obtained through the resistivity. Thus one obtains the Hall conductivity for a cubic crystal as

$$\sigma_{xy} = \frac{\rho_{xy}}{\rho_{xx}^2 + \rho_{xy}^2} \approx \frac{\rho_{xy}}{\rho_{xx}^2}, \tag{4.284}$$

the simplification being possible since in most cases $\rho_{xx} \gg \rho_{xy}$. Other symmetries than cubic are treated analogously. We see that even though the conductivity σ_{xy} can be calculated *ab initio* through eqns (4.283) and (4.278) one needs the resistivity ρ_{xx}, usually, but not necessarily, obtained from experiment, to get a theoretical value for $\rho_{xy} \approx \sigma_{xy}\rho_{xx}^2$. Alternatively experiment yields both values on the right hand side of eqn (4.284). Table 4.18 contains a collection of measured and calculated values of the anomalous Hall conductivity for the elementary magnets Fe, Co, and Ni. The calculations were entirely *ab initio*. The large experimental value for Fe cited in reference (a) was obtained from an iron whisker, the smaller one is cited by Gosálbez-Martinez *et al.* (2015) in a very advanced treatment of bcc Fe, which we come back to later. Co is interesting because of the orientation dependence of both the measured and calculated values. One is tempted to assign the underestimate of the latter to extrinsic effects, but generally the belief is that the intrinsic effects dominate the AHE. However, the work by Weischenberg *et al.* (2011) sheds a different light on the situation. Using a Gaussian disorder model

Table 4.18 *Calculated and measured conductivities in $(\Omega cm)^{-1}$ of the anomalous Hall effect. Experimental data are found in references (a) to (c).*

	exp.	calc.
bcc Fe	700 – 1030	750[a]
hcp Co $\mathbf{M} \parallel \mathbf{c}$	813	481[b]
hcp Co $\mathbf{M} \perp \mathbf{c}$	150	116[b]
fcc Ni	-1100	-1066[c]

(a) Yao *et al.*(2004)
(b) Roman *et al.* (2009)
(c) Fuh and Guo (2011)

they obtained an expression for the *side-jump conductivity* that is entirely given by the band structure and requires no adjustable parameters. Added to the intrinsic AHC the remaining discrepancies disappear in Table 4.18. In particular, the larger value for bcc Fe is verified. The story of Ni, however, seems to be more complicated and is not settled. Thus it appears that one cannot simply ignore the side-jump conductivity. A great deal of effort has been spent on theory and measurements for alloys and compounds, which we briefly discuss next, but throughout ignoring effects of the side-jump conductivity.

Before we do that it seems of interest to explain a simple but less precise method to estimate the AHE. It is a finite-difference approach which can be extracted from the work of Resta (1994). First we construct the function

$$U_{\mathbf{j}}(\mathbf{k}) = \det[\langle u_{n,\mathbf{k}} | u_{m,\mathbf{k}+\mathbf{j}} \rangle] \tag{4.285}$$

using the crystal-periodic eigenfunctions of the Kohn–Sham Hamiltonian, where $n, m \in$ occupied states and $\mathbf{k}$ as well as $\mathbf{k+j}$ are inside the Brillouin zone. From this so-called link-variable (Fukui and Hatsugai, 2007) one can show that the j-th component of the Berry connection is given by

$$A_j(\mathbf{k}) = \Im \ln U_j(\mathbf{k}). \tag{4.286}$$

The z-component of the curl of $\mathbf{A}(\mathbf{k})$ can now be expressed as

$$\Omega^z(\mathbf{k}) = \Im \ln \frac{U_y(\mathbf{k}+\mathbf{k}_x)U_x(\mathbf{k})}{U_y(\mathbf{k})U_x(\mathbf{k}+\mathbf{k}_y)}, \tag{4.287}$$

except for a scaling factor that is due to Δk. The anomalous Hall conductivity is then obtained from

$$\sigma_{xy} = -\frac{e^2}{\hbar} \int_{BZ} \frac{d^3k}{(2\pi)^3} \Omega^z(\mathbf{k}) f(\mathbf{k}). \tag{4.288}$$

Table 4.19 *Collection of experimental and calculated data relevant for the Hall conductivity.* $M^{\rm exp}$ *is the the experimental and* $M^{\rm calc}$ *the calculated magnetic moment in* μ_B *per f.u.,* $\sigma_{xy}^{\rm exp}$ *is the measured Hall conductivity in* $(\Omega{\rm cm})^{-1}$*, while* $\sigma_{xy}^{\rm calc}$ *are calculated values.*

Compound	$M^{\rm exp}$	$M^{\rm calc}$	$\sigma_{xy}^{\rm exp}$	$\sigma_{xy}^{\rm calc}$
Co$_2$MnAl	4.04	4.045	2000[a]	1800[b]
Co$_2$MnGa	4.0	4.02	1600[c]	1460[d]
Rh$_2$MnAl	–	4.066	–	1500[b]
Co$_2$VGa	1.92	1.95	137[c]	66[b] 140[c]
Mn$_2$CoAl	2.0	1.97	22[e]	3[e]
Co$_3$Sn$_2$S$_2$	0.87	0.99	1130[f]	1100[f]

(a) Vidal *et al.* (2011)
(b) Kübler and Felser (2012)
(c) Manna *et al.* (2018)
(d) Eqn (4.288)
(e) Ouardi *et al.* (2013)
(f) Liu *et al.* (2018)

Some tuning of the finite differences supplies an optimal value of Δk together with the number of k-points in the BZ.

Five systems chosen in Table 4.19 are cubic Heusler compounds, for which the structure was explained in Fig. 4.58, except for Mn$_2$CoAl which has the inverted Heusler structure, where the atoms Mn-Mn-Co-Al occupy the Wyckoff positions a-c-b-d (c and d label the corners of the inscribed cube in the figure). Notice the huge spread of values for the anomalous Hall conductivity. The reason for the huge conductivities of the first three compounds and the last one will be the topic of a later Sec. 4.9.1 where also the small values of the remaining two find an explanation. The compound Co$_3$Sn$_2$S$_2$ is a *Shandite*, its space group is R-3m, no.166 in the International Tables of space groups. It consists of Co-Kagome layers stacked along the hexagonal axis. The data used in Table 4.19 were obtained from single crystals. More recent data are available from thin films which, although of interest, are ignored here in the interest of brevity.

4.7.2 The anomalous Hall effect in antiferromagnets

The anomalous Hall effect in ferromagnets is large and finite because magnetism breaks time-reversal symmetry. Thus we do not expect an AHE to exist in antiferromagnets. This can be different if their magnetic order is noncollinear. In fact, Chen *et al.* (2014) showed an AHE exists for the cubic non-collinear Mn$_3$Ir and Kübler and Felser (2014) for hexagonal Mn$_3$Sn that is described in Fig. 4.95.

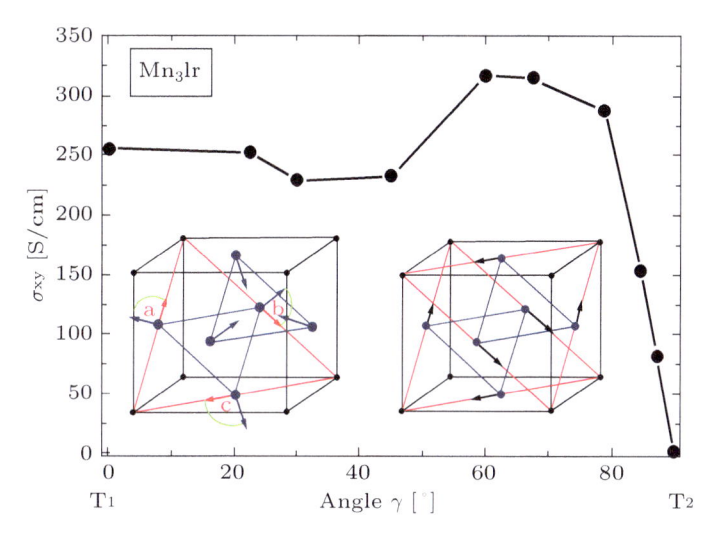

Figure 4.98 *The AHE in antiferromagnetic Mn_3Ir. The two magnetic configurations T_1 and T_2 were first described and measured by Krén et al. (1968). The conductivity σ_{xy} was calculated by Eqn (4.288) changing the configuration by means of the angle $\gamma = a = b = c$, left insert.*

Beginning with Mn_3Ir we show in Fig. 4.98 the calculated anomalous Hall conductivity (AHC) for two antiferromagnetic configurations shown in the insert. The conductivity σ_{xy} is calculated with Eqn (4.288) and is seen to change from a large value for the configuration T_1 on the left to zero for T_2 on the right when tuning the configuration continuously by changing the angle γ. Concurrently a small ferromagnetic component with a moment of $0.038\mu_B$ appears for T_1 decreasing to zero at T_2. We see that time reversal is broken. The moment is in the (111)-direction, so the values in Fig. 4.98 represent only one component of the conductivity, which is also in the (111)-direction. Its magnitude is therefore about 430 S/cm. This is in fair agreement with the findings of Chen *et al.* (2014) who further elaborate the magnetic structure by showing that it consists of Kagome lattices stacked along the (111)-direction.

Going further we note that Krén *et al.* (1968) described the compounds Mn_3Rh and Mn_3Pt as cubic noncollinear antiferromagnets with the same magnetic structure as Mn_3Ir. In fact, the AHC is calculated for Mn_3Rh to be about 350 S/cm and for Mn_3Pt 76 S/cm (Kübler, 2014, unpublished). Figure 4.99 demonstrates the role of symmetry; the Berry curvature is compared for the two states T_1 and T_2 and clearly shows that the Berry curvature does not cancel out in the state T_1. Unfortunately, no experimental work exists at the time of this writing that proves the anomalous Hall effect is reality for the Mn_3Ir-type compounds.

Another related class of antiferromagnetic compounds that order noncollinearly were first described by Fruchart and Bertaut (1978) and were called *metallic perovskites* or in more modern work *antiperovskites* by Gurung *et al.* (2019). The crystal structure is simple cubic just like Mn_3Ir but contains a nitrogen or carbon atom at the center of

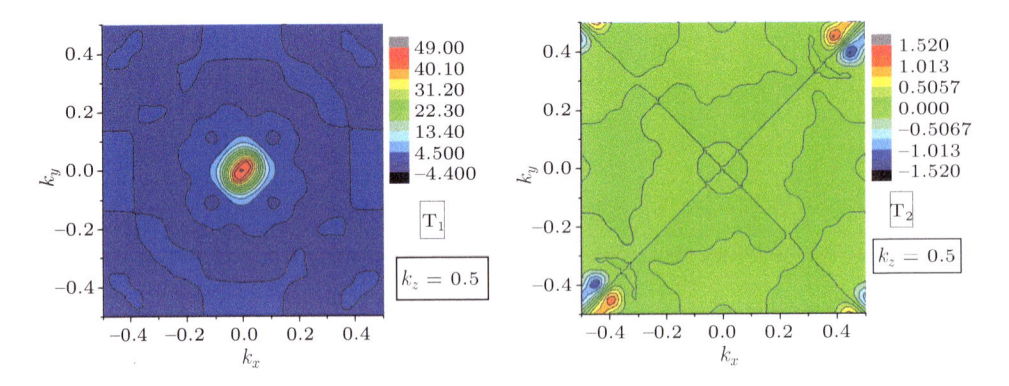

Figure 4.99 *Plots of the Berry curvature in units of e^2/h in the top plane of the Brillouin zone for Mn_3Rh.*

the cube—examples are Mn_3GaN, Mn_3SnN, Mn_3NiN, etc. Of the two noncollinear antiferromagnetic configurations (Γ_{4g} and Γ_{5g} corresponding to T_1 and T_2, respectively) only one, Γ_{4g}, allows a nonvanishing anomalous Hall conductivity; Gurung et al. (2019) calculate $\sigma_{xy} = 40$ S/cm for Mn_3GaN and $\sigma_{xy} = 133$ S/cm for Mn_3SnN, for example. These authors present a modern and interesting discussion of this class of materials. Also noteworthy is the recent and far-reaching theoretical work of Zhou et al. (2019). Experimental work on thin films of Mn_3NiN by Boldrin et al. (2019) demonstrates that the AHE indeed exists in the metallic perovskites; the value for the anomalous Hall conductivity is measured to be 22 S/cm in the magnetic state Γ_{4g}. The effect is much more pronounced in two hexagonal noncollinear antiferromagnets Mn_3Sn and Mn_3Ge, which we discuss next.

The symmetry properties of the triangular antiferromagnetic structure of Mn_3Sn was explained in great detail in Sec. 4.6.4.2. It is important to remember that there are two types of magnetic structures, the "direct structures" Fig. 4.95 (a,b) and the "inverse structures" (c,d). The latter are accompanied by orbital polarization that leads to small ferromagnetic moments breaking time-reversal symmetry. Although the total-energy calculations of Zhang et al. (2013) favor the cubic versions of Mn_3Sn or Mn_3Ge hexagonal, single crystals can and were grown slightly off-stoichiometric (Tomiyoshi and Yamaguchi, 1982 and Brown et al., 1990, 1992). These differences from the ideal structures were ignored in the calculations by Kübler and Felser (2014). Assuming $P6_3/mmc$ symmetry (Space Group No 194) for Mn_3Sn and Mn_3Ge they predicted an anomalous Hall effect in the "inverse structures." Because of an unusual choice of axes in their early work comparison with later experiments is confusing and somewhat misleading. Corrected now, Fig. 4.100 from Kübler and Felser (2017) shows the two inverse structures either of which is realized by the experimental work of Nakatsuji et al. (2015) for Mn_3Sn, Kiyohara et al. (2016) for Mn_3Ge, and Nayak et al. (2016) also for Mn_3Ge.

Table 4.20 gives an overview of relevant data. The Néel temperatures of the two compounds are 365 K for Mn_3Ge and 420 K for Mn_3Sn. The measurements for Mn_3Ge

(a)   (b)

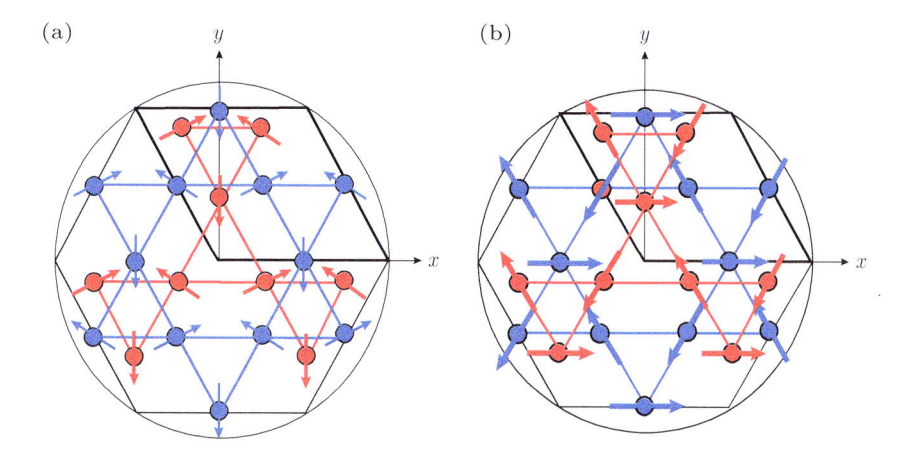

Figure 4.100 *The two inverse structures for hexagonal* Mn_3Sn, *also called chiral antiferromagnets. The same applies to* Mn_3Ge. *There are two formula units per unit cell, one layer of Mn is at* $z = 0$ *(blue colour) together with Sn atoms at the center and the corners of the hexagon. Another layer of Mn is at* $0.5c/a$ *with the other Sn atom (not drawn) at* $(-\frac{1}{2}, \frac{1}{2\sqrt{3}}, \frac{c}{2a})$ *etc. Compared with Fig. 4.95 the coordinate system is shifted.*

Table 4.20 *Calculated and measured conductivities in* $(\Omega cm)^{-1}$ *of the anomalous Hall effect for* Mn_3Ge *and* Mn_3Sn.

	σ_{xz}	σ_{zy}
$Mn_3Ge^{(a)}$	380	310
$Mn_3Ge^{(b)}$	500	110
$Mn_3Ge^{(c)}$	330	-
$Mn_3Ge^{(d)}$	560	10
$Mn_3Sn^{(e)}$	75	100
$Mn_3Sn^{(c)}$	133	-
$Mn_3Sn^{(d)}$	-	310

(a) Kiyohara *et al.* (2016)
(b) Nayak *et al.* (2016)
(c) calc. by Zhang *et al.* (2017)
(d) calc. by our Eqn (4.288)
(e) Nakatsuji *et al.* (2015)

were made at 2 K and higher, but for Mn_3Sn the lowest temperature was 100 K because at very low temperatures the detailed spin structure is unknown (Nakatalsuji *et al.*, 2015). The simulations were done at 0 K and give a finite AHC only for the inverse (chiral) structures. The conductivity is a pseudovector, so σ_{xy} points out of the planar layers

360 *Electronic Structure and Magnetism*

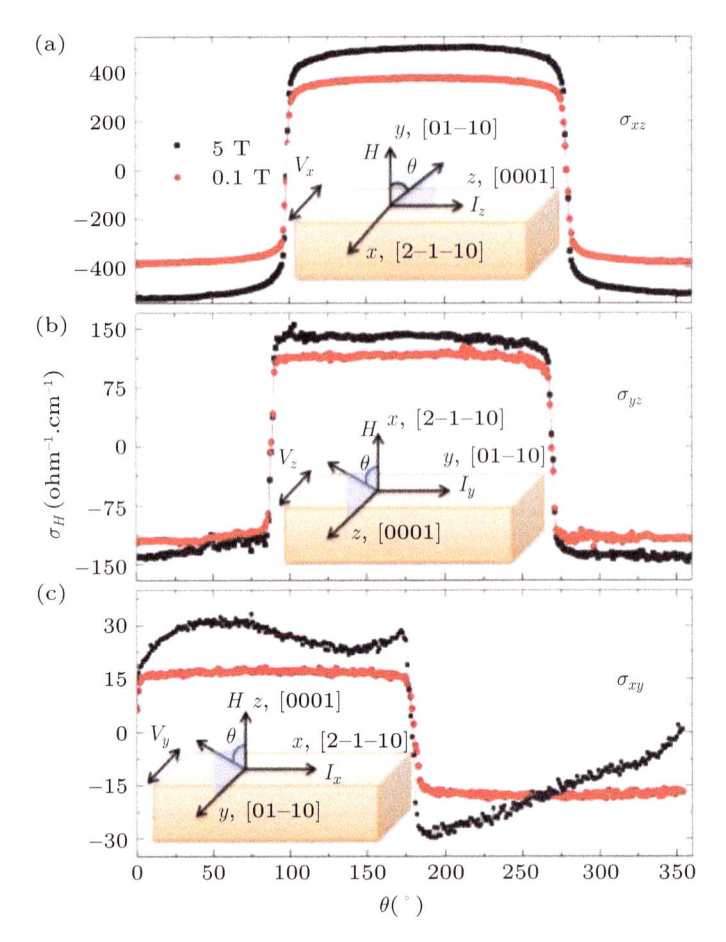

Figure 4.101 *Angular dependence of the AHE of Mn_3Ge, reproduced from Nayak et al. (2016) with permission from AAAS. The direction of the magnetic field is indicated in the insets in each of the panels A, B, and C; see text for details.*

sketched in Fig. 4.100; it is calculated to be zero. This agrees with the measurements of Nakatsuji et al. (2015) as well as Kiyohara et al. (2016). The conductivity σ_{xz} is in the y-direction, represented by Fig. 4.100(a); σ_{zy} is in the x- direction and is represented by Fig. 4.100(b).

Nayak et al. (2016) give an illustrative demonstration of the directional dependence of the AHE of Mn_3Ge in a figure, which we reproduce in Fig. 4.101. The explanation of the physics involved largely follows Nayak et al. (2016). To begin with it should be said that theoretically no external magnetic field should be necessary to observe the AHE. However, the rather delicate spin arrangements seen here lead to the formation of antiferromagnetic domains, in each of which the Hall voltage will be different, thus cancelling out in the sample and leading to a vanishing AHC. It appears that a weak in-plane ferromagnetic moment helps a small magnetic field in one direction within the manganese plane to align the domains thus establishing a large AHE. In particular, looking at Fig. 4.100(a) imagine an applied field in the y-direction. Upon reversal of its direction it will reverse all magnetic moments and thus the sign of the AHC. The same is true for a field in the x-direction and reversing the sign of the AHC looking at Fig. 4.100(b). Turning now to Fig. 4.101 we see that it is possible to change the sign of the AHC continuously within one of the categories σ_{xz} or σ_{yz} or even σ_{xy}, the latter of which theory predicts to vanish. In part A of the figure the field is in the y-direction ($\theta = 0$) and the current out of plane in the z-direction; the voltage, V_x, is measured in the xz-plane. When θ increases there remains a decreasing component of the field in the y-direction, which changes sign at $\theta = 90°$, whereupon the AHC changes sign because the chirality changes sign. At $\theta = 270°$ the situation reverses back. Apparently the spin structure sketched in Fig. 4.100(a) locks in and stays nearly constant in the changing field. The situation is analogous in part B of the figure, where the spin structure shown in Fig. 4.100(b) locks in, except that the signal is smaller. The calculations suggest a much smaller AHC. Part C of the figure applies to a field rotating in the zy-plane; here a component of the field remains parallel to the y-axis except for $\theta = 0$ and $180°$. To see the signal change at these values of θ is astonishing since for the planar spin structures the conductivity σ_{xy} is expected to vanish. The finite AHC seems to imply that a slight canting of the spins occurs out of the plane. This case deserves further theoretical attention.

4.8 Weyl fermions

In this section we deal with deeper properties of energy bands that were discovered in this 21st century but partly described much earlier. So did Herring (1937, 1940), who defined and discussed by means of symmetry arguments *accidental degeneracy* that occurs when certain energy bands cross. Much later Nielsen and Ninomiya (1981, 1983) introduced the Weyl Hamiltonian to the crossing of energy bands in solids with far-reaching consequences. It was in this century that experimental results moved the field to fruition and led to a satisfactory theoretical picture.

We begin with Weyl (1929), who postulated the Hamiltonian

$$\mathcal{H} = \pm c\,\vec{\sigma}\cdot\mathbf{p}\,,$$

where c is the speed of light, $\vec{\sigma}$ is the Pauli spin vector, and $\mathbf{p}$ is the momentum operator. This Hamiltonian, in contrast to that of Dirac, sets up a two-component theory for massless particles, perhaps neutrinos of some kind. But they were never detected in particle physics. There are, however, applications of Weyl's Hamiltonian in other fields, see for instance Volovik (2003).

For the low-energy applications in solid state physics one slightly modifies the Weyl Hamiltonian (Wan *et al.*, 2011 and Turner and Vishwanath, 2013) and writes

$$\mathcal{H} = \sum_{i=1}^{3}(\mathbf{v}_i \cdot \mathbf{p})\sigma_i \,, \qquad (4.289)$$

where $\sigma_i, i=1,2,3$ are the three Pauli matrices. Three linearly independent velocities $\mathbf{v}_i$ are introduced that replace the $\pm$ in the original and define the helicity by means of $\kappa = \mathrm{sign}[\mathbf{v}_1 \cdot (\mathbf{v}_2 \times \mathbf{v}_3)]$. $\kappa = +1$ defines right-handed helicity, $\kappa = -1$ left-handed helicity. It appears that the term chirality is also used instead of helicity. The Hamiltonian of Eqn (4.289) is easily diagonalized giving two energy eigenvalues

$$E_\pm = \pm \sqrt{\sum_{i=1}^{3}(\mathbf{v}_i \cdot \mathbf{k})^2} \qquad (4.290)$$

As function of the momentum $\mathbf{k}$ the energy $E_\pm$ is plotted in Fig. 4.102 for both $\kappa = 1$ and $\kappa = -1$. The connection with the crossing bands is a free-hand drawing; the reason for drawing two "Weyl cones" is that they always come in pairs of opposite helicity (Nielsen and Ninomiyam 1983 and Turner and Vishwanath, 2013), a fact that will become clear soon. Of importance is the connection of the Weyl node with the Berry curvature. Berry (1984) showed that the singularity at the crossing point leads to a magnetic monopole, which in our case exists in reciprocal space (k-space or momentum space). Furthermore, the energy band involved acquires a topological property that is given as an integer number, C, called the Chern number. We may see this by placing a spinor $|\zeta_-\rangle$ at the $\kappa = -1$ and another $|\zeta_+\rangle$ at the $\kappa = +1$ Weyl nodes, where

$$|\zeta_-\rangle = \begin{pmatrix} \cos\frac{\theta}{2}\, e^{-i\phi/2} \\ \sin\frac{\theta}{2}\, e^{i\phi/2} \end{pmatrix}, \ |\zeta_+\rangle = \begin{pmatrix} -\sin\frac{\theta}{2}\, e^{-i\phi/2} \\ \cos\frac{\theta}{2}\, e^{i\phi/2} \end{pmatrix}, \qquad (4.291)$$

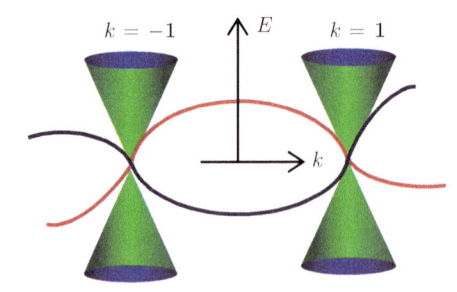

Figure 4.102 *Sketch of a Weyl scenario. Embedded in the Brillouin zone two crossings are assumed of nondegenerate energy bands. The bands are fitted to energy cones (Weyl cones, Eqn (4.290)) that are drawn in an energy, E, versus wave-vector, k, diagram. The colour code does not capture their different helicities, $\kappa = \pm 1$.*

and calculate for each case the Berry connection, that is $A_\theta^+ = i\langle\zeta_+|\partial_\theta|\zeta_+\rangle = 0$ and $A_\phi^+ = i\langle\zeta_+|\partial_\phi|\zeta_+\rangle = \frac{1}{2}\cos\theta$ as well as $A_\theta^- = i\langle\zeta_-|\partial_\theta|\zeta_-\rangle = 0$ and $A_\phi^- = i\langle\zeta_-|\partial_\phi|\zeta_-\rangle = -\frac{1}{2}\cos\theta$. From the Berry connection we obtain the Berry curvature

$$\Omega_{\theta\phi}^\pm = \partial_\theta A_\phi^\pm - \partial_\phi A_\theta^\pm = \pm\frac{1}{2}\sin\theta. \quad (4.292)$$

The Chern number is defined as the value of the integral divided by 2π of the Berry curvature over a sphere enclosing each Weyl point

$$C_\pm = \frac{\pm 1}{2\pi}\int_0^{2\pi}d\phi\int_0^\pi d\theta \frac{1}{2}\sin\theta = \pm 1.$$

Thus the Chern number of the Weyl point is ± 1. For a realistic band structure the value of $|C|$ can be an integer larger than 1; it signals a topological non-trivial case provided $|C| > 0$. We can now argue why the Weyl nodes always come in pairs. Using Gauss' law for a surface that encloses the entire Brillouin zone one obtains the sum of all Weyl points, which has to be zero because of neutrality of the sample. Thus the minimum number of Weyl points is two if there are any at all, and they must have opposite chirality as indicated in Fig. 4.102.

The definition of the **Chern number** is of crucial importance and is put on solid ground now. The Chern number for a band n is defined by

$$C_n(S) = \frac{1}{2\pi}\oint_S dS\, \mathbf{n}\cdot\boldsymbol{\Omega}_n(\mathbf{k}), \quad (4.293)$$

where S is a closed surface and $\mathbf{n}$ is the surface normal. The band is supposed to be nondegenerate over the surface S. Haldane (2004) gives an equivalent definition by means of

$$C_n \mathbf{G}_{a,n} = -\frac{1}{2\pi}\int_{BZ}\Omega_n^a(\mathbf{k})d^3k, \quad (4.294)$$

where $\mathbf{G}_{a,n}$ is the a-component of a reciprocal lattice vector, while Ω_n^a is the a-component of the Berry curvature of band n (which is a pseudovector, we remember). The integration is over the Brillouin zone.

To illustrate Haldane's formula we assume a value of $C_n = 1$ and take an fcc lattice assuming the Berry curvature is in the z-direction for the AHC to be σ_{xy} for a magnetic metal with a filled (or nearly filled) band. Then writing out Eqn (4.283) for one band only and setting $f_n = 1$ we obtain

$$\sigma_{xy} = -\frac{e^2}{\hbar}\frac{1}{(2\pi)^2}\int_{BZ}\frac{d^3k}{2\pi}\Omega_n^z(\mathbf{k}) = \frac{e^2}{h}C_n\frac{2\vec{z}}{a}, \quad (4.295)$$

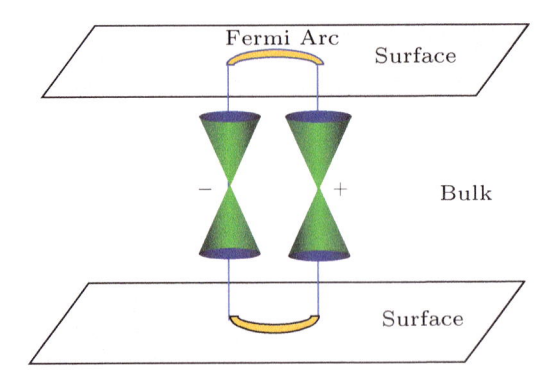

Figure 4.103 *Weyl nodes communicate with the surface. Projection of the Weyl nodes to the surface forms a Fermi arc. Berry flux flows from + to −.*

where a is the lattice constant and $\vec{z}$ is a unit vector in the z direction. Assuming finally $C_n = 1$ and $a = 0.575$ nm (a value typical for a Heusler compound) we calculate $\sigma_{xy} = 1345$ S/cm.[1] This is seen to be of the order of the AHC of the Heusler compounds listed in Table 4.4. We will later exhibit their Weyl nodes.

Returning to the main topic we now discuss the **surface states** of the model. Figure 4.103 is very schematic but is supposed to give a clue. Projections of the Weyl points to the surface Brillouin zone are connected by what is called a Fermi arc. These surface states occur when bulk bands are not too close in energy to avoid hybridization. The situation is best explained by means of an example. We show in Fig. 4.104 the computed surface states of $Co_3Sn_2S_2$ together with its crystal structure. The later is hexagonal with the space group R-3m (No. 166) which can be transformed to rhombohedral, for which the surface calculations were done by Xu *et al.* (2018). The surface states and the Weyl nodes can be seen most clearly in part (b) of the figure. The Fermi arcs in the other sections are found by first getting oriented with the help of part (d); the arcs can then be seen in parts (a) and (c).

There is another important phenomenon that must be described and this is the **chiral anomaly** (Nielsen and Ninomiya, 1983). We first note the formula

$$\frac{\partial}{\partial t} n_{+,-}(\mathbf{r}) = \pm \frac{e^2}{h^2} \mathbf{E} \cdot \mathbf{B}. \qquad (4.296)$$

It gives the rate of change of the number of particles $n_{+,-}$ at a given Weyl point of positive/negative chirality as a function of the product $\mathbf{E} \cdot \mathbf{B}$. We see the number of particles is not conserved depending on the orientation of the electro-magnetic field. This is known as the *Adler–Bell–Jackiw* anomaly in field theory. Turner and Vishwanath (2013) give a compelling derivation for condensed matter that we partially follow now.

---

[1] The Hall conductance is $e^2/h = 3.874046 \cdot 10^{-5}$ S.

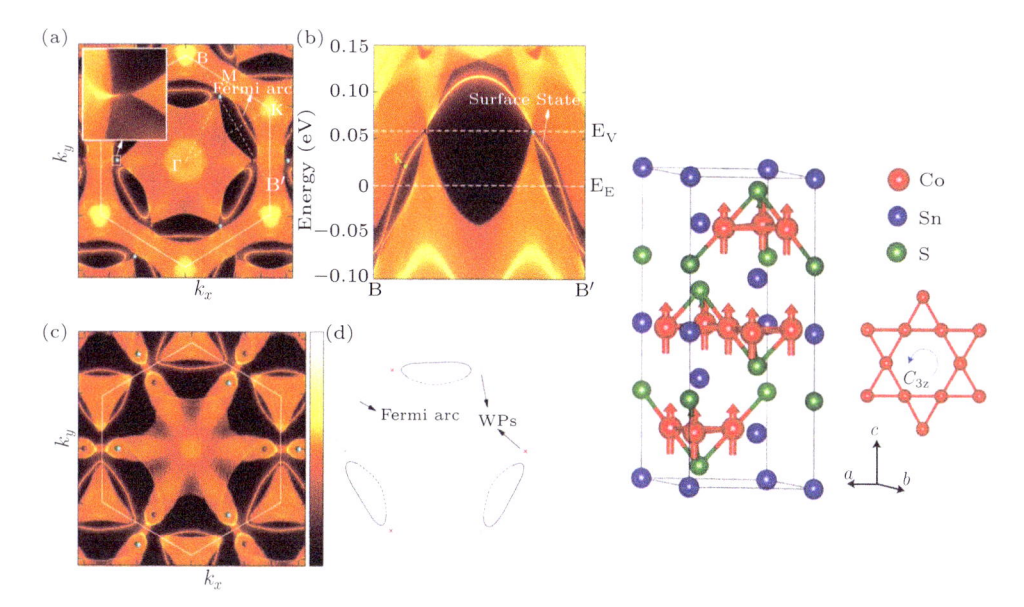

Figure 4.104 *Calculated Fermi arcs and surface states for $Co_3Sn_2S_2$. After Xu et al. (2018). The hexagonal crystal structure is depicted on the right. Parts (a) to (c) were obtained with the rhombohedral variant. (a) Sn-terminated surface FS with energy fixed at the Weyl point; (b) energy for the Sn-terminated surface along a path crossing the Weyl points connected by a Fermi arc; (c) Sn-terminated surface FS at the charge-neutrality point.*

Consider a Weyl semimetal of cross-sectional area A and length L_z in an electric and magnetic field applied along the z-direction. The magnetic field gives Landau levels as seen in Fig. 4.105(a); only the $N=0$ level is of importance. They provide one-dimensional chiral modes with the linear dispersion $\epsilon_{r,l} = \pm v p_z$ (see the figure). The electric field accelerates the electrons and the Fermi momenta increase accordingly $\dot{p}_{F,r,l} = eE_z$, creating a larger Fermi sea at the right and a smaller one at the left Weyl point, so that the electrons move from the left Weyl point to the right. The rate of increase is

$$\frac{dN_+}{dt} = eE_z \frac{AB}{\Phi_0} \frac{L_z}{h}, \qquad (4.297)$$

where L_z/h is the density of states in a chiral mode and AB/Φ_0 is the density of modes; $\Phi_0 = h/e$ is the flux quantum. The number density is obtained from Eqn (4.297) by dividing out the volume AL_z, resulting in Eqn (4.296). In Fig. 4.105(b) the chiral anomaly is shown using measurements for $Co_3Sn_2S_2$. The magnetoresistance is seen to decrease as the angle between **B** and **E** (given as current) is decreased to zero; for **B** parallel to **E** the magnetoresistance is negative, i.e. anomalous. The same physics is shown in part (c) giving the conductance instead of the resistance.

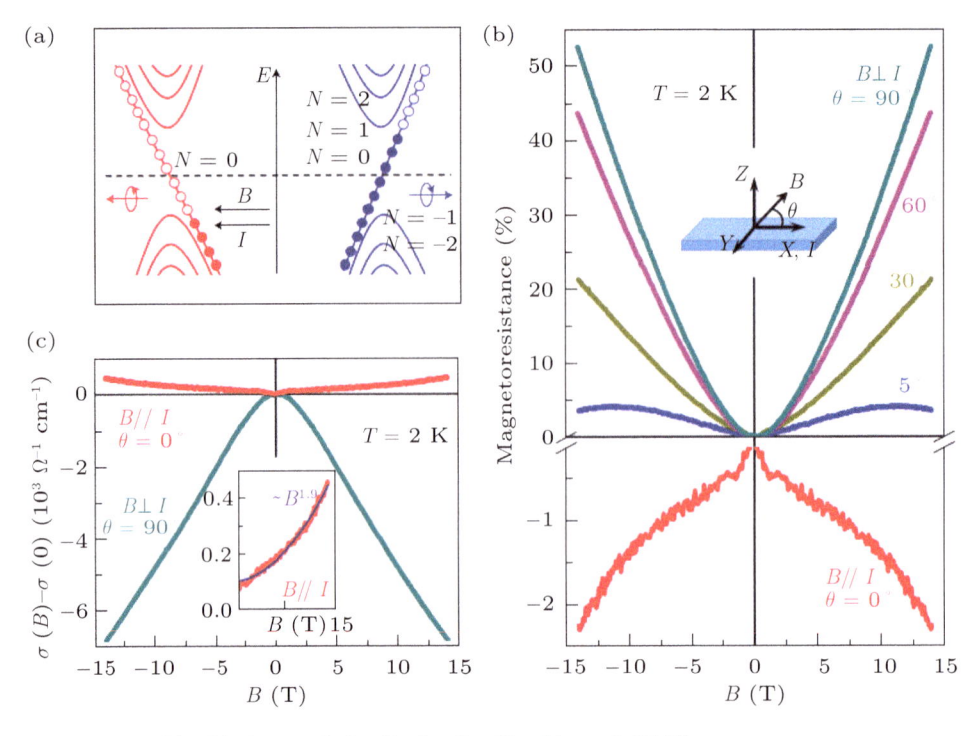

Figure 4.105 *The chiral anomaly for $Co_3Sn_2S_2$. After Liu et al. (2018)*

We summarize: Band crossings lead to a new type of fermion that is describable with the Weyl Hamiltonian. Its electronic states possess positive or negative helicity (chirality) which leads to right- or left-handed particles. The topology of the electronic states is embodied in the Chern number which is a nonzero integer. Outstanding properties show up in surface states that result in Fermi arcs which we expect to be seen in photoemission experiments. The electrical conductivity in a magnetic field shows a chiral anomaly and the anomalous Hall conductivity tends to be large.

4.9 Real-case Weyl fermions

We now collect examples for Weyl fermions as they occur in nature, i.e. as they are made, measured, and understood in the laboratory. Guided by the size of the anomalous Hall effect we begin with bcc iron, see Table 4.18. In Fig. 4.106 we give a combination of graphs: (a) is the Berry curvature in the Γ-H-plane, taken from Yao *et al.* (2004), (b) from the same publication shows points of large Berry curvature. Gosálbez-Martínez, Souza, and Vanderbilt (2015) in a very thorough study of bcc Fe found a large number of band crossings, but they resulted in Fermi sheets with zero Chern numbers except for one pair of Weyl points having Chern numbers of ± 1, which lead to Fermi sheets with

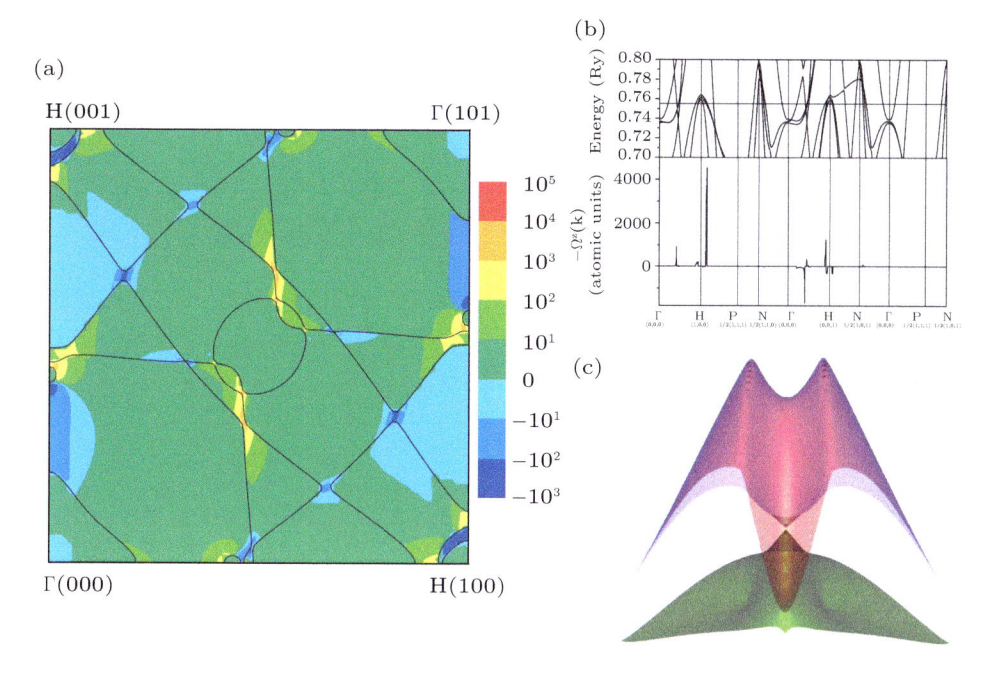

Figure 4.106 *Berry curvature and Weyl point (WP) for bcc Fe. (a) and (b) are from Yao et al. (2004). (a) The Berry curvature, $-\Omega^z(\mathbf{k})$, in the (010)-plane in atomic units. (b) Band structure and Berry curve along symmetry lines. (c) WP calculated with the ASW method, situated below the Fermi energy. The point shown is at $\mathbf{k} = (0.585, 0.0, 0.585)\, 2\pi/a$, $\Delta E = -0.68$ eV and has $C = 1$. It is viewed from the partner WP at $(0.415, 0.0, 0.415)$ at the same energy having $C = -1$.*

$C \pm 1$ located along Δ close to the Fermi surface at the left edge in (a). They conclude that bcc Fe is a topologically nontrivial metal.

We repeated parts of their work with the ASW method and found a pair of Weyl points at different positions, near the Fermi surface depicted in the center of (a). The Weyl point inside the first Brillouin zone at $\mathbf{k}=(0.415, 0.0, 0.415)\, 2\pi/a$ has $C = -1$, the corresponding one at the inverted position has $C = 1$. These values for C were obtained using Eqn (4.293) with a small sphere for S. The location of our Weyl points does not agree with that found by Gosálbez-Martinez et al. (2015). Different computational methods lead to small changes of the energy bands, which may become more pronounced because of the approximation necessary to account for spin–orbit coupling (SOC). Without SOC, i.e. a relativistic effect, the Berry curvature vanishes. We also conclude that bcc Fe is a topologically nontrivial metal, which because of the non-zero Chern numbers possesses a large anomalous Hall conductivity.

4.9.1 Topological effects in magnetic compounds

Weyl fermions were first discovered in a solid state system by Xiangang Wan et al. (2011) for pyrochlore iridates. This is a seminal publication containing a complete description of

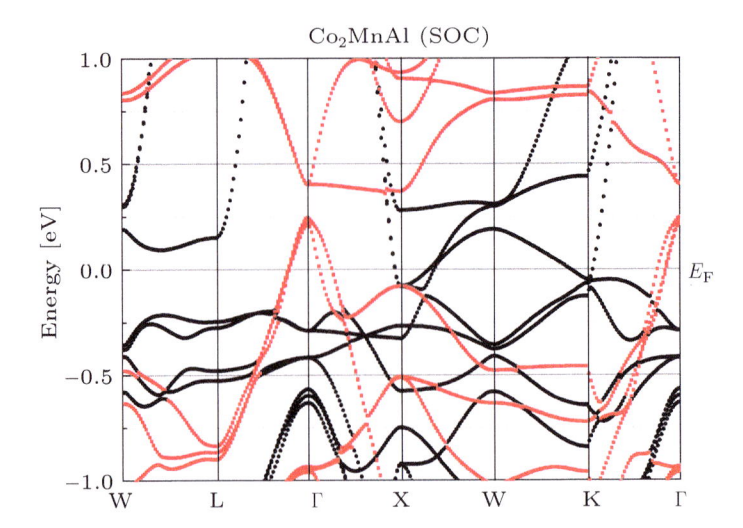

Figure 4.107 *The band structure of Co_2MnAl. The ASW method was used. Minority-spin states are in red, majority-spin states are in black.*

the new physics, but pyrochlore iridates are quite complicated hosting many Weyl points. So it is not surprising that simpler compounds were searched for with a small number of Weyl points. Thus Zhijun Wang *et al.* (2016) predicted that the Heusler compound Co_2ZrSn should have a minimum number of Weyl points and supplied a complete set of calculations. Rather than delving into their calculations we turn to the compounds listed in Table 4.19 that are experimentally well characterized (accepting that they contain one more magnetic atom). The case of $Co_3Sn_2S_2$ was discussed in Sec. 4.8, except for details concerning the bulk Weyl points. These are nicely pictured in the publication by Liu *et al.* (2018) to which we refer for further information.

So we begin with Co_2MnAl and show its band structure in Fig. 4.107, following largely the publication of Kübler and Felser (2016). It is important to do the calculation including spin–orbit coupling (SOC), for otherwise the Berry curvature vanishes at each point in momentum space. The presence of SOC breaks the symmetry, so spin-up and spin-down are no longer good quantum numbers. However, one can determine for each point **k** the magnetic moment and define majority- and minority-spin states. These are distinguished in the figure.

The Fermi surface of Co_2MnAl consist of minority-states (red) centered at the Γ-point and small pockets of majority-states (black) near X and K. It is in these states where we find Weyl points. In fact, plotting band no. 28 and 29 keeping $k_y = 0.5$ we obtain the right-hand portion of Fig. 4.108. The two bands (surfaces) are centered at L. Because of the finite mesh points there appear gaps, two of which are protected by symmetry and thus form Weyl points. This is demonstrated by zooming in at one of the points. The result is shown on the left of the figure, where the location of the Fermi energy should be noted. It is 30 meV below the Weyl node. A calculation of the Chern number by means of

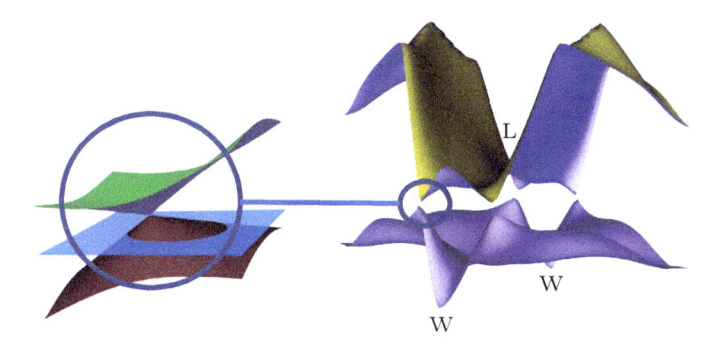

Figure 4.108 *The band-structure of Co_2MnAl in a three-dimensional rendering. On the right, centered at the point L, band no. 28 and 29 are graphed as seen from the $k_y = 0.5$ plane. On the left is shown the zoomed Weyl point including the Fermi energy. The symmetry point $W = (0, 0.5, 1)$ is located at the left, the one on the right is $W = (0.5, 1, 0)$.*

Eqn (4.293) gives $C = -2$ for a small sphere around the Weyl point indicated at the left in Fig. 4.108. It is located slightly below symmetry point $W = (0, 0.5, 1)$; the other one visible in Fig. 4.108 is slightly above the symmetry point $W = (1, 0.5, 0)$ and has $C = 2$. For symmetry reasons there are 16 Weyl points (see also Co_2MnGa to be discussed next), inverting the k-point ($\mathbf{k}_W \to -\mathbf{k}_W$) changes the sign of the Chern number. The calculations described here employed a magnetization out-of-plane, (001). When the magnetization changes to (010), for instance, the symmetry changes and so do the Weyl points. Li *et al.* (2020) in a systematic study showed that the Hall conductivity this way becomes tunable over a wide range; as a weak magnetic field rotates the magnetization, transitions are induced between Weyl points and nodal rings thus changing the Hall conductivity. Since this happens at room temperature the effect may be of interest for applications. Unfortunately, there are presently no measurements available concerning surface states and the chiral anomaly.

We now continue the discussion and turn to Co_2MnGa showing its spin-filtered band structure in Fig. 4.109. There are five publications which describe the Weyl points for this interesting compound. Without claiming to be complete, these are Chang *et al.* (2017), Manna *et al.* (2018), Belopolski *et al.* (2019), Sakai *et al.* (2018), and Guin *et al.* (2019a). The latter two mainly describe the Nernst effect, which, as we will discuss later on, depends strongly on the AHE. We proceed exactly as for Co_2MnAl and construct Fig. 4.110, which shows the two highest occupied bands. In part (a) one sees two Weyl points, one of which is enlarged in part (b). The location of the zoomed point is $(0, 0.5, 0.8)$ with the Chern number $C = -2$, the other one is at $(1, 0.5, 0.20)$ with $C = 2$. The former appears by symmetry at 8 points in the Brillouin zone, the latter at 16 points shared by the first and neighboring Brillouin zones. Thus there are 16 Weyl points all together. The magnetization in this case is along (001). If the direction is changed, the number of Weyl points decreases due to lower symmetry. Thus for instance, the number of Weyl points becomes 8 if the magnetization is along (100). The two Heusler

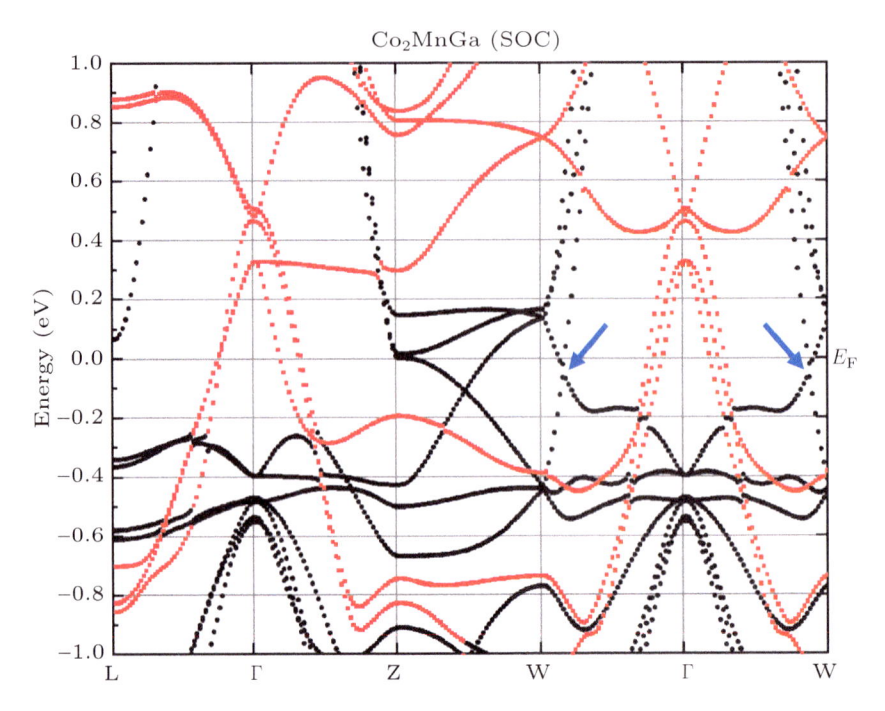

Figure 4.109 *The band structure of Co_2MnGa. The ASW method was used. Majority-spin states are in black, minority-spin states are in red. The arrows near the W points mark the relevant band crossings.*

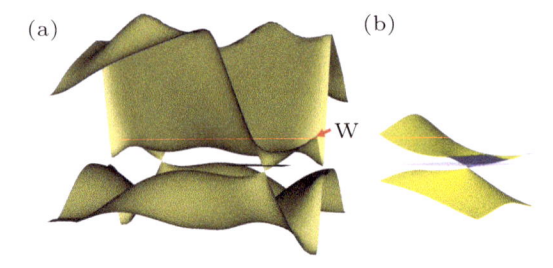

Figure 4.110 *Parts of the band structure of Co_2MnGa in three-dimensional rendering. Part (a) is oriented as in Fig. 4.108, the symmetry point shown is given by $W = (0, 0.5, 1)$. (b) shows the zoomed Weyl point, which is 100 meV below the Fermi energy.*

compounds Co_2MnAl and Co_2MnGa are quite similar except that the Fermi energy is below the Weyl points in the former and above in the latter, see Figs. 4.108 and 4.110.

Recent angular resolved photoemission spectra (ARPES) by Belopolski *et al.* (2019) revealed interesting "drumhead" surface states for Co_2MnGa and node lines. Just like the Weyl points the node lines depend on the direction of the magnetization. We show in Fig. 4.111 twice a portion of the three-dimensional band structure, which demonstrates

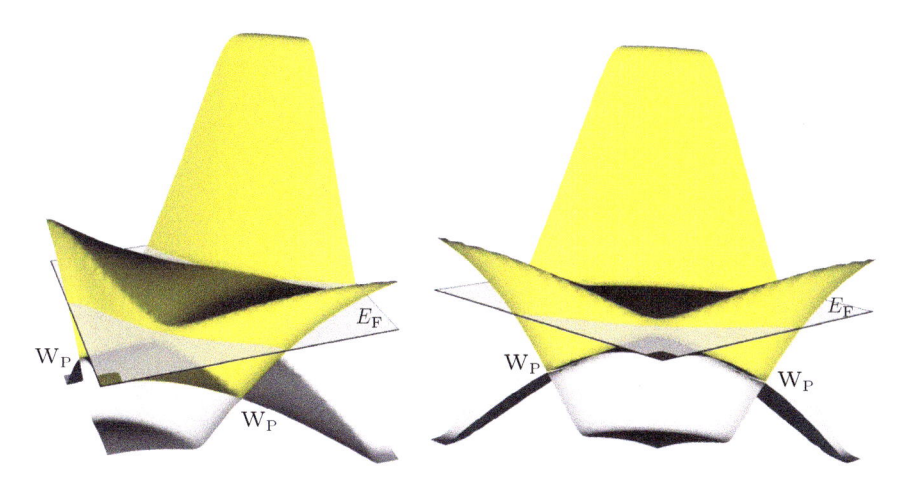

Figure 4.111 *Node line and Weyl points, W_P, of Co_2MnGa. Two views of the same bands; the magnetization is along (010) and $k_y = 0$. Fermi energy is marked E_F.*

a node line connecting two Weyl points. If the direction of the magnetization is changed from the value assumed to, for instance, (001) then the Weyl point on the left in the figure becomes gapped and the node line opens up. As a consequence the Weyl point on the right becomes isolated and appears as in Fig. 4.110. To demonstrate the node line for a magnetization direction (001) one would have to look at the $k_z = 0$ plane.

The chiral anomaly has not been clearly seen for Co_2MnGa. The difficulty seems to be that the "normal" positive magnetoresistance cannot be detected, instead it turns out to be negative for the magnetic field **B** perpendicular to the current (Manna *et al.*, 2018).

4.9.2 Topology of antiferromagnetic compounds

We return to Sec. 4.7.2 and particularly to Table 4.20. We remember the anomalous Hall conductivities were large for Mn_3Ge and intermediate for Mn_3Sn. For completeness we will investigate both cases for the existence of Weyl nodes. Beginning with Mn_3Sn we show its band structure in Fig. 4.112 also giving the Brillouin zone and the magnetic structure employed. We follow Kübler and Felser (2017) and, especially for the surface states, the work of Yang *et al.* (2017). Since Mn_3Sn and Mn_3Ge are metals (in contrast to semimetals), it is of interest to view the Weyl points together with the Fermi surface. This interesting connection is motivated by the observation that the AHC seems to depend on all states below the Fermi surface, which—as Haldane (2004) observed—is at odds with Landau's Fermi-liquid theory. It asserts that charge transport in metals involves only quasi-particles with energy within $k_B T$ of the Fermi surface. Haldane shows that the non-quantized part of the Hall conductivity indeed obeys Landau's theory.

Haldane's theory is easy to state as long as it deals with the simple case of two dimensions, which is sufficient to see the principle. He uses the basic relation $\mathbf{\Omega} = \nabla \times \mathbf{A}$ in the integral for the conductivity and integration by parts to show that the

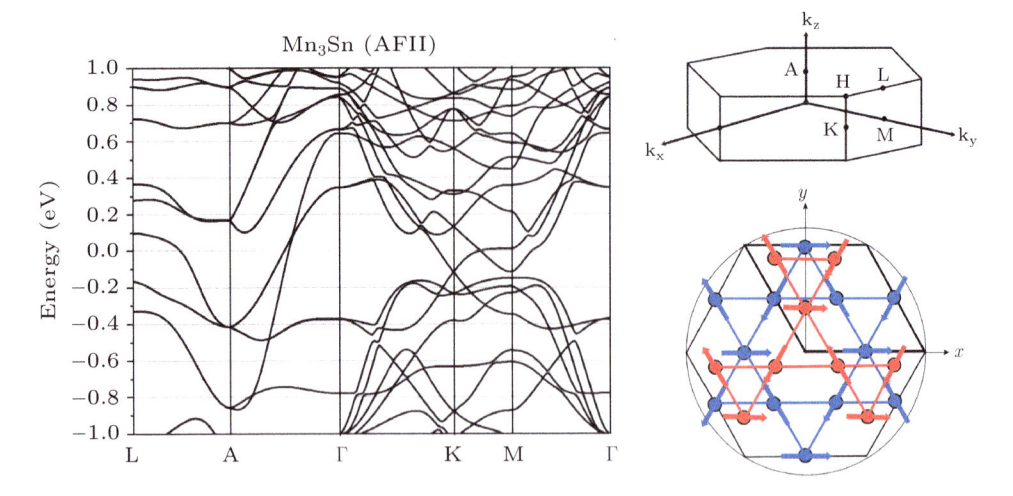

Figure 4.112 *Band structure of Mn_3Sn, the Brillouin zone, and the structure assumed. See also Fig. 4.100(b).*

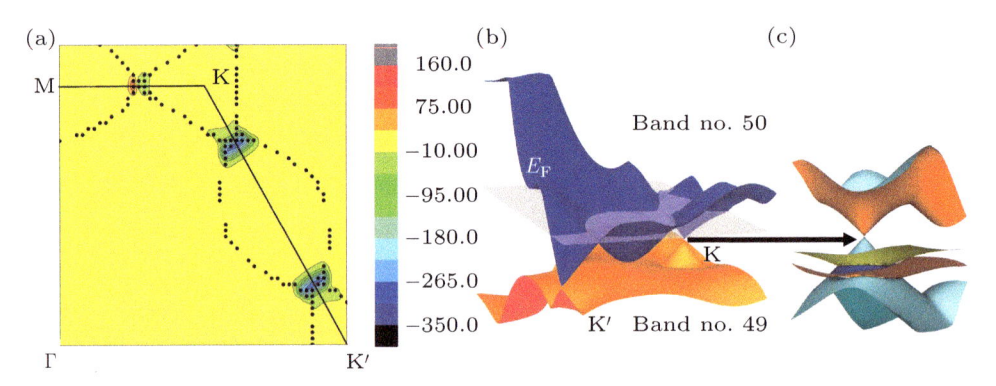

Figure 4.113 *Mn_3Sn: (a) Fermi surface contour and Berry curvature $\Omega^x(k_x,k_y,k_z=0)/2\pi$. (b) Three-dimensional band structure and the Fermi energy, E_F. (c) Zoomed Weyl point at the symmetry point $K=(\frac{1}{3},\frac{1}{\sqrt{3}},0)$; another Weyl point can be seen at $K'=(\frac{2}{3},0,0)$.*

two-dimensional Hall conductivity depends on the curvature of the Fermi contours. This is nicely brought out by the calculated results which are shown in Fig. 4.113. One sees in part (a) that the spots due to the Berry curvature do not appear at the location of the Weyl points, K and K', but at positions of the Fermi contour of strong curvature. Actually it is not clear if these points should be called Weyl points since they occur at symmetry points where the representation is two-dimensional. So this is not an accidental crossing of bands. However Fig. 4.113(c) shows that two rather flat bands cut the cone in two concentric discs of radii $k_1 \simeq 0.09$ and $k_2 \simeq 0.11$. Spin–orbit coupling largely lifts the degeneracy leaving a number of crossings that are identified as $W_1^{\pm}$ and

4.9 Real-case Weyl fermions   373

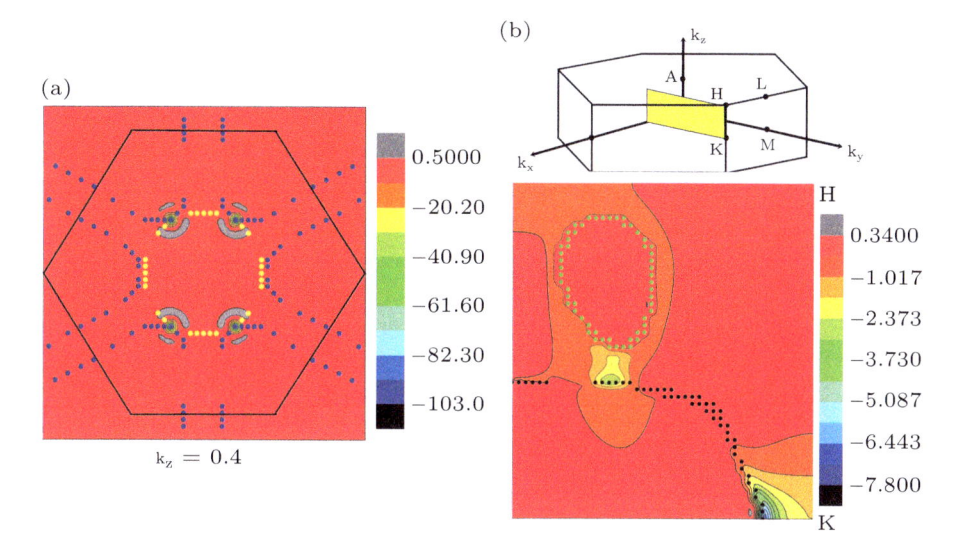

Figure 4.114 *Mn$_3$Sn: Bulk Fermi contours and Berry curvature signals. (a) The $k_z = 0.4$ plane where a second Fermi contour appears (yellow dots). (b) The second Fermi contour is nearly circular (green dots), the Brillouin zone above shows the plane (yellow) for this plot.*

$W_2^\pm$ by Yang *et al.* (2017). They used a different magnetic order (Fig. 4.100(a) instead of (b)) so the positions do not quite agree. But we agree that the Weyl points lie in the $k_z = 0$ plane.

To demonstrate that the Berry curvature appears at the Fermi surface way above the $k_z = 0$ plane we show Fig. 4.114. As one moves away from the $k_z = 0$ plane a higher state is being occupied forming a second Fermi surface, as is seen in part (a) of the figure. At the touching points of the two surfaces (yellow and black dots) Berry curvature appears. Part (b) of the figure shows the same fact from a different perspective.

Before we discuss the surface states we take a brief look at Mn$_3$Ge and show its magnetic order and the band structure in Fig. 4.115. It is very similar to Mn$_3$Sn but there are differences; remember that the anomalous Hall conductivity of Mn$_3$Ge is much larger. In Fig. 4.116(a) one sees the Fermi contours together with the Berry curvature in the $k_z = 0$ plane that is seen to occur not at K or K' but at locations of large curvature of the Fermi contours of which there are two (yellow and red dots). Part (b) of the figure shows the energy states at the symmetry points K and K' and what we loosely call a Weyl point at K in part (c) of the figure. The states that cut the cone are not as flat as in the case of Mn$_3$Sn. They generate a number of crossings that we identify as Weyl points. But there are more in the interior of the Brillouin zone.

Yang *et al.* (2017) used the VASP-code to obtain the bulk band structure and then a properly adjusted tight-binding method. The surface states were calculated on a semi-infinite surface, in other words, the surface band structure corresponds to the bottom surface of a half-infinite crystal. They obtained the momentum-resolved local density of states on the surface layer by using a Green's function method. The salient points

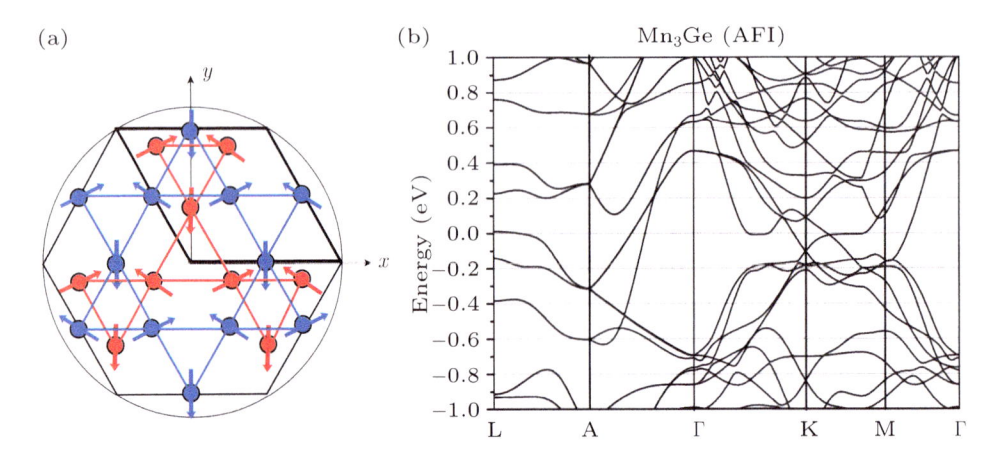

Figure 4.115 Band structure of Mn_3Ge, and the structure assumed. See also Fig. 4.100(a).

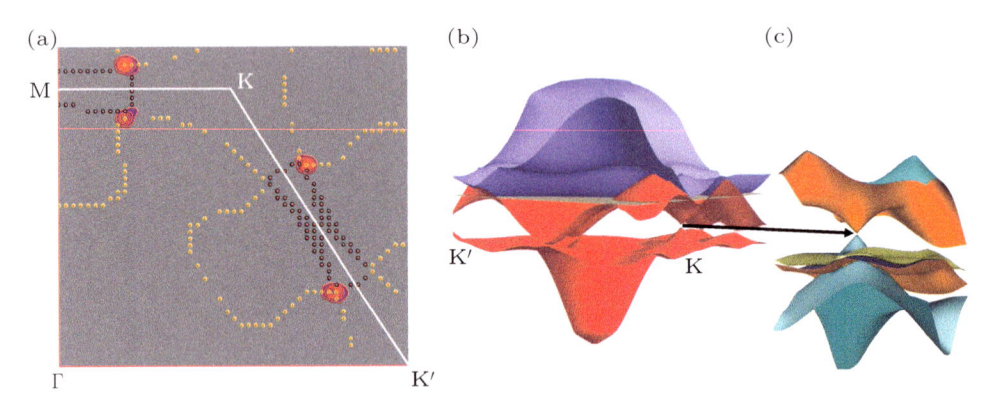

Figure 4.116 Mn_3Ge: (a) Fermi surface contour and Berry curvature $\Omega^y(k_x, k_y, k_z = 0)/2\pi$. (b) Three-dimensional band structure and the Fermi energy, E_F. (c) Zoomed Weyl point at the symmetry point $K = (\frac{1}{3}, \frac{1}{\sqrt{3}}, 0)$; another Weyl point can be seen at $K' = (\frac{2}{3}, 0, 0)$.

of their calculations are seen in Fig. 4.117 where we show the results for Mn_3Sn and Mn_3Ge side by side. The number of Weyl points for the latter is found to be much larger than the former. It is sufficient to concentrate on one pair of Weyl points which is done in the figure for the pair W_1^+ and W_1^-. Their separation is of the order of the disk described above. The Fermi arcs are nicely seen. Unfortunate there are no experimental photoemission results for the surface states. The ARPES data given by Kuroda et al. (2017) are inconclusive; however, their conductivity measurements clearly demonstrate that Mn_3Sn in the chiral state is topologically nontrivial. We remember that the chiral anomaly is a prominent effect in the ferromagnet $Co_3Sn_2S_2$. Fig. 4.105 displays the data together with an explanation. For the antiferromagnet Mn_3Sn the measurements are shown in Fig. 4.118, for Mn_3Ge they do not exist at present.

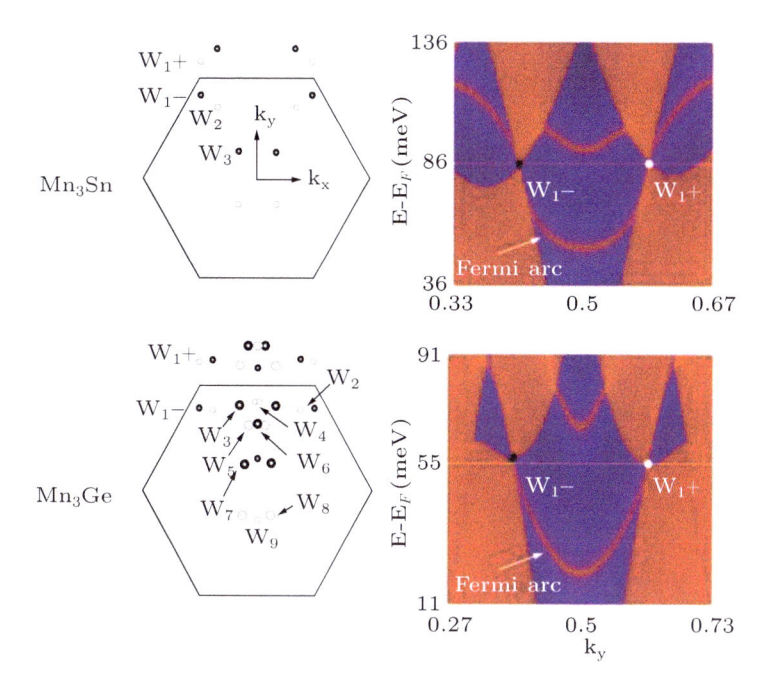

Figure 4.117 *Surface states of Mn_3Sn and Mn_3Ge after Yang et al. (2017) showing from left to right the location of the Weyl points and for $W_1^\pm$ the Fermi arcs.*

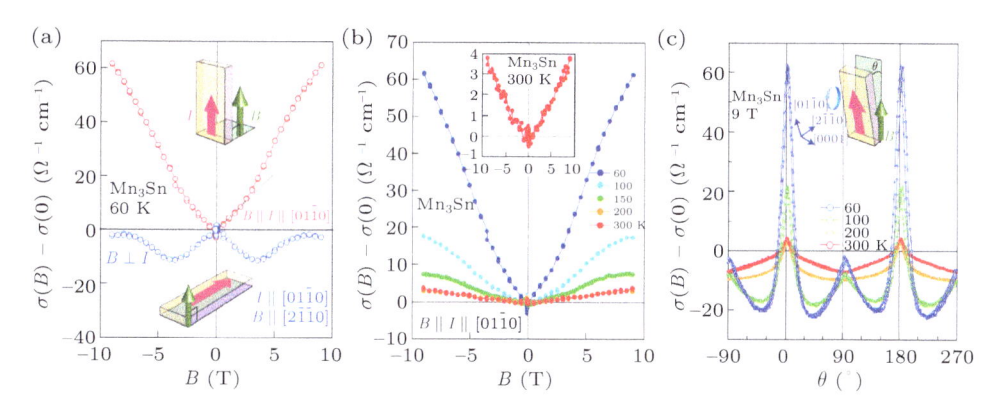

Figure 4.118 *Mn_3Sn: The chiral anomaly, from Kuroda et al. (2017). (a) Longitudinal conductivity for two orientations of the magnetic field given in the insets. (b) Longitudinal conductivity as a function of the magnetic field at different temperatures. (c) Longitudinal conductivity at 9 T for varying orientations of the field.*

There are further *experimentally verified* antiferromagnets that possess a noncollinear order; one notable case is Mn_5Si_3. It was discovered by Sürgers et al. (2014, 2016) and has an AHC of $\sigma_{xy} = 102$ S/cm. Its topology is not known.

4.9.3 The Nernst effect

We come back to the anomalous Hall effect that—we remind the reader—is due to the Berry curvature, $\Omega^z(\mathbf{k}) = \sum_n \Omega_n^z(\mathbf{k})$, where n labels the bands. For convenience we state again that the anomalous Hall conductivity is given by

$$\sigma_{xy} = -\frac{e^2}{\hbar} \int_{\text{BZ}} \frac{d^3 k}{(2\pi)^3} \Omega^z(\mathbf{k}) f(\mathbf{k}), \tag{4.298}$$

where $f(\mathbf{k})$ is the Fermi distribution function and we assume for simplicity that the field is in the z-direction.

In the presence of a temperature gradient an intrinsic Hall current results as illustrated in Fig. 4.119; it is called the anomalous Nernst effect (ANE). It is but one of a large number of thermoelectric and thermomagnetic effects that have a famous history. The interested reader may consult an early paper by Callen (1948) and the book by Ziman (1967). Furthermore, Sommerfeld and Frank (1931) is a weathered but useful reference and Behnia and Aubin (2016) address modern aspects, as do Fu *et al.* (2020) in very recent work.

The anomalous Nernst conductivity (ANC) is shown by Xiao *et al.* (2006) to be given by

$$\alpha_{xy} = \frac{1}{T}\frac{e}{\hbar} \sum_n \int_{\text{BZ}} \frac{d^3 k}{(2\pi)^3} \Omega_n^z(\mathbf{k}) \left[(\epsilon_n(\mathbf{k}) - \mu) f(\mathbf{k}) + k_B T \log(1 + e^{-\beta(\epsilon_n(\mathbf{k}) - \mu)}) \right], \tag{4.299}$$

where $\beta = 1/k_B T$ and μ is the chemical potential. Integration by parts and use of Eqn 4.298 converts this equation to the useful form

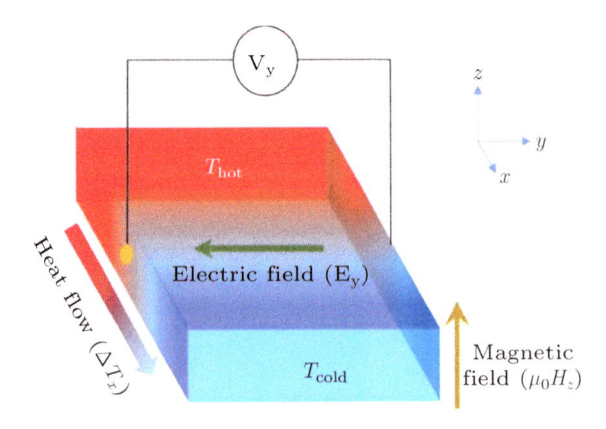

Figure 4.119 *Schematic diagram describing the Nernst effect after Guin et al. (2019b); the magnetic field may also be read as magnetization.*

4.9 Real-case Weyl fermions

$$\alpha_{xy} = -\frac{1}{e}\int d\epsilon \frac{\partial f}{\partial \mu}\sigma_{xy}(\epsilon)\frac{\epsilon-\mu}{T}. \quad (4.300)$$

For low temperatures a Sommerfeld expansion may be used to obtain

$$\alpha_{xy} = \frac{\pi^2}{3}\frac{k_B}{e}k_B T\left[\frac{d}{d\epsilon}\sigma_{xy}(\epsilon)\right]_{\epsilon=\epsilon_F}. \quad (4.301)$$

Such a relation between the electric and the thermoelectric conductivity is known as a Mott relation (Mott and Jones, 1936). Notice that Eqns 4.300 and 4.301 may be read without the connection to the Berry curvature, i.e. the transverse conductivity σ_{xy} may be of a quite general origin.

To access experimental data we define the Seebeck coefficient $S_{xx} = \frac{V_x}{\nabla T}$, where V_x is the longitudinal voltage, and the Nernst thermopower $S_{yx} = \frac{L_x V_y}{L_y \nabla T}$ where L_x is the distance between the two temperature leads and L_y is the distance between the two voltage wires (see Fig. 4.119 for the orientation). The Nernst coefficient α_{yx} is then entirely expressed in terms of experimental data by

$$\alpha_{yx} = \frac{S_{yx}\rho_{xx} - S_{xx}\rho_{yx}}{\rho_{xx}^2 + \rho_{yx}^2}. \quad (4.302)$$

Guin *et al.* (2019a) published an impressive collection of data, which we reproduce in Fig. 4.120. We select the compound $Co_3Sn_2S_2$ as our first example and show in Fig. 4.121(a) the measured hysteresis of the AHC, σ_{xy} for five temperatures. In (b) the

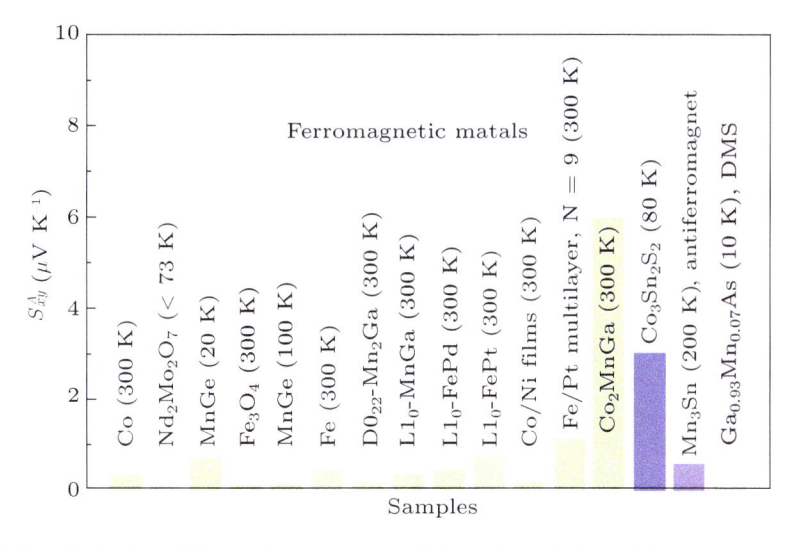

Figure 4.120 *Collection of Nernst thermopower coefficients, from Guin et al. (2019a).*

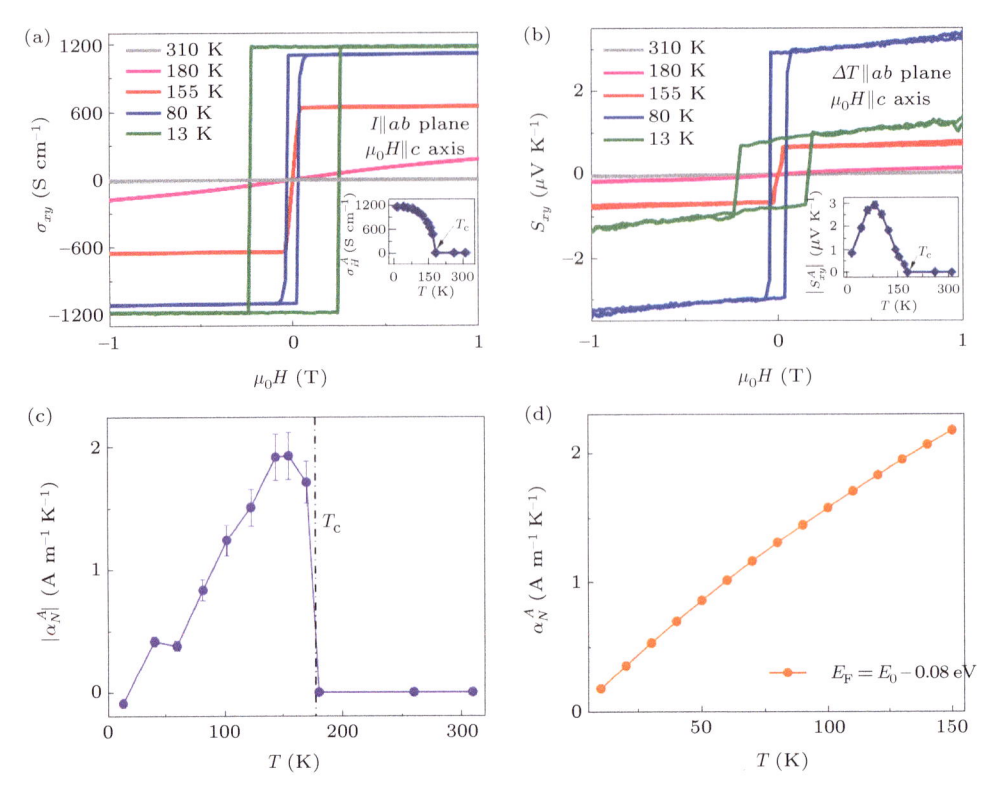

Figure 4.121 $Co_3Sn_2S_2$: Hysteresis loops of (a) the AHC and (b) the Nernst thermopower coefficient measured at different temperatures. (c) The measured and (d) calculated anomalous Nernst conductivity as a function of the temperature. Plots collected from the paper of Guin et al. (2019a).

measured hysteresis of the Nernst thermopower, S_{xy}, is plotted at the same temperatures, the insert giving S_{xy}, as a function of the temperature. Extracting the anomalous Nernst conductivity (ANC), α_{xy}, we obtain part (c) and a calculation using Eqn 4.299 yields part (d). The results of the calculation are in good agreement with the measurements. But notice that the Fermi energy has been decreased by 80 meV. The reason for this rather minute change is the sensitive numerical dependence of α_{xy} on the value of the Fermi energy.

We now continue the discussion dealing with Heusler compounds for which we show a selection in Table 4.21. These are only few cases out of a truly gigantic set of Heusler data calculated by Noky et al. (2020). Our selection suffices to show the principle. There are two columns each for the anomalous Hall conductivity (AHC) and the anomalous Nernst conductivity (ANC) where one set of values is obtained with the *ab initio* value of the Fermi energy and the other one is maximized by means of changing the Fermi energy. So examining for instance Co_2MnGa in Fig. 4.119 we see that the Nernst thermopower is very large, being about $S^A_{xy} \simeq 6$ μVK^{-1} at 300 K, which corresponds

Table 4.21 *Collection of experimental data of the AHC in S/cm and calculated values by Noky et al. (2020) giving a range for the AHC and the ANC in A m^{-1} K^{-1}.*

Compound	AHC^{exp}	AHC^{calc}	AHC$^{\text{calc}}_{max}$	ANC^{calc}	ANC$^{\text{calc}}_{max}$
Co$_2$MnAl	2000[a]	1631	1739	1.93	4.13
Co$_2$MnGa	1600[b]	1310	1472	0.05	4.79
Rh$_2$MnAl	–	1723	2064	2.26	6.06
Co$_2$VGa	137[b]	131	528	0.16	1.12
Mn$_2$CoAl	22[c]	6	116	0.22	0.44

[a] Vidal et al. (2011)
[b] Manna et al. (2018)
[c] Ouardi et al. (2013)

to $\alpha_{yx} \simeq 7$ A m^{-1}K^{-1}, which is of the order of magnitude of the maximal value in Table 4.21. Sakai et al. (2018) also investigated Co$_2$MnGa finding very similar results, the value for α_{xy}, however, being almost exactly given by the ANC$_{max}$ in Table 4.21.

The calculations of Noky et al. and our Table show that generally the ANC is large when the AHC is large, although this does not strictly follow from the Mott relation, which involves a derivative.

Returning to the collection of experimental results, Fig. 4.120, we continue with the Nernst effect observed in the noncollinear antiferromagnet Mn$_3$Sn, listed in the figure. The anomalous Hall effect for this compound and for Mn$_3$Ge was discussed in Sec. 4.7.2 and Fig. 4.98 sketches the two relevant magnetic configurations. The label S_{xy} used in Fig. 4.120 has to be replaced in this case by either S_{zx} or S_{yz}. Li et al. (2017) published an impressive set of measurements on Mn$_3$Sn that are reproduced in Fig. 4.122. It shows nicely how the anomalous Nernst effect (b and c) is connected with the anomalous Hall effect (a and b). The relation can be understood by means of the Mott relation, Eqn 4.301. The Righi–Leduc effect shown in (e) and (f) of the figure is the thermal Hall effect given by another Mott relation,

$$\kappa_{yz} = \frac{\pi^2}{3} \frac{k_B^2 T}{e^2} \sigma_{yz}. \quad (4.303)$$

The measurements shown were taken at room temperature. At lower temperatures the signals are larger, Li et al. (2017), see also Ikhlas et al. (2017), who, in particular, emphasized the dependence of the thermoelectric effects on the stoichiometry of the samples. Turning now to Mn$_3$Ge we discuss the work by Xu et al. (2020) and that by Wuttke et al. (2019) who convey different albeit important messages.

Xu et al. (2020) begin their discussion with a figure much like Fig. 4.122 except for the scales. The more interesting point is the temperature dependence of the coefficients which is shown in Fig. 4.123. The quotient displayed in part C of the figure should be equal to $L_0 = \frac{\pi^2}{3}(\frac{k_B}{e})^2$ from Eqn (4.303). Up to about 100 K this ratio is constant. We

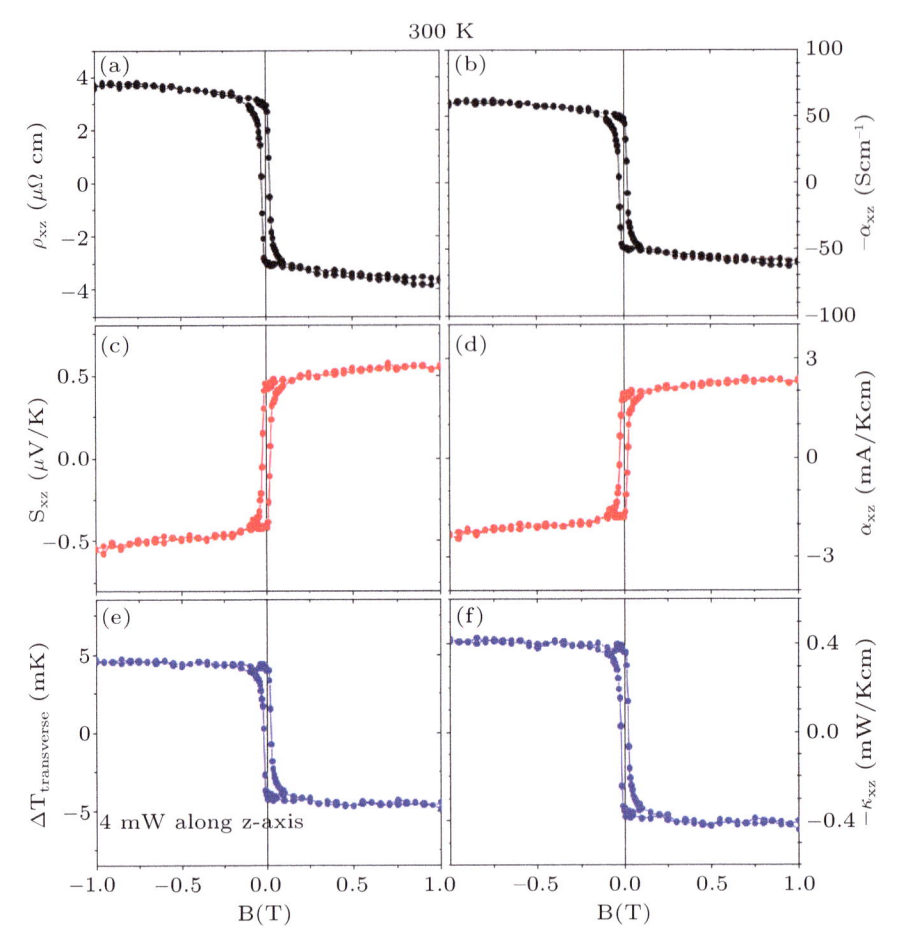

Figure 4.122 Mn_3Sn from Li et al. (2017): Measured anomalous Hall effect (a) and (b), anomalous Nernst effect (c) and (d), as well as Righi–Leduc effect (e) and (f) at room temperature.

say that up to 100 K the Wiedemann–Franz law is obeyed; it obviously is not for higher temperatures. The authors make a serious attempt to understand the deviation. In fact, the Mott relation is only valid at low temperatures, even though it seems to do a good job for Mn_3Sn. One finds the necessary formulation derived by Qin et al. (2011) and in a useful overview by Zhang (2016):

$$\kappa_{yz} = \frac{1}{\hbar T} \sum_n \int_{BZ} \frac{d^3k}{(2\pi)^3} \int_{\varepsilon_{n\mathbf{k}}}^{\infty} d\varepsilon \, \mathbf{\Omega}_n^x(\mathbf{k})(\varepsilon - \mu)^2 \frac{\partial f}{\partial \varepsilon}. \qquad (4.304)$$

Using this equation, Xu et al. (2020) elaborate the deviation from the Wiedemann–Franz law.

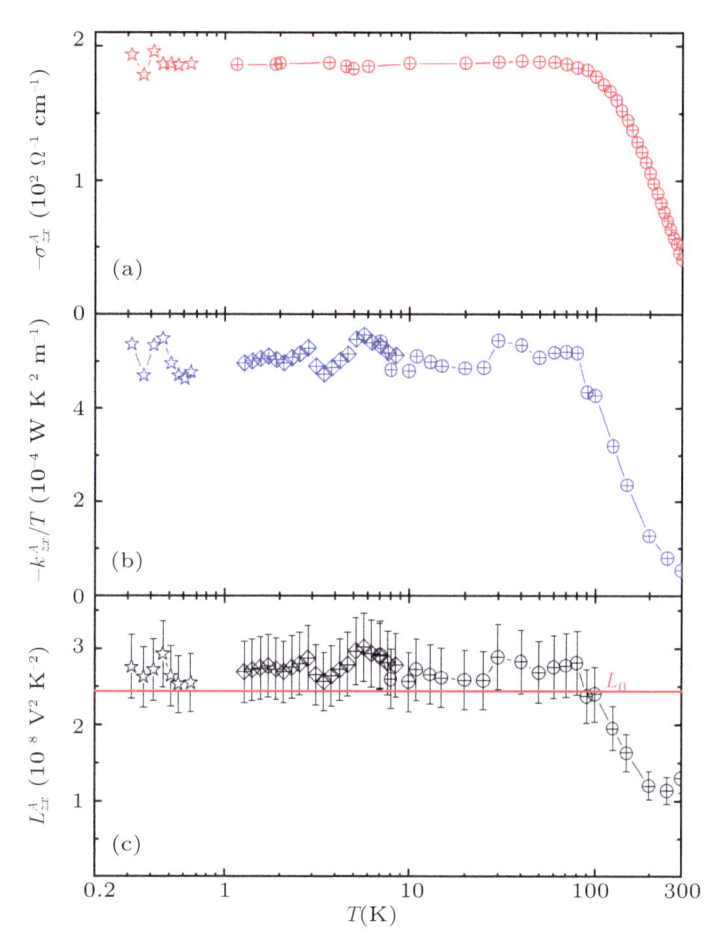

Figure 4.123 *Mn$_3$Ge: Reproduced from Xu et al. (2020). (A) the anomalous Hall conductivity, (B) the Righi–Leduc conductivity, and (C) the quotient* $L^A_{zx} = \frac{\kappa^A_{zx}}{T\sigma^A_{zx}}$.

The work of Wuttke *et al.* (2019) concentrates entirely on understanding the measured anomalous Nernst conductivity of Mn$_3$Ge in terms of the Berry curvature and the Weyl nodes, basically by evaluating Eqn (4.299) using a realistic model for the electronic structure. We will not go into further details because the work is better appreciated by viewing the original reference. We finish this section by drawing attention to the work of Weischenberg *et al.* (2013) (especially the supplementary information), who determined the role of side-jump contributions to the anomalous Nernst conductivity. This problem was briefly discussed before in connection with Table 4.18 and the anomalous Hall effect. It appears that these extra contributions cannot be ignored for Fe, Co, and Ni, but are less important for compounds, such as FePt. Much more work is needed here for deeper insight.

4.9.4 Remark about topology

In Sec. 4.9.1 we described *topological properties* without defining this term properly. To remedy this we now take a brief excursion into *topology*, a field of mathematics concerned with properties of objects that remain invariant under continuous deformations. What kind of objects are relevant for the quantum-mechanical state of matter?

At the time of this writing, the year 2020, it is exactly 40 years after the discovery of the Quantum Hall Effect (QHE) by von Klitzing *et al.* (1980). This discovery was the beginning of a new field of condensed matter physics that is still thriving, see also the rieview by von Klitzing *et al.* (2020). The Hall conductivity of a two-dimensional insulator in a strong magnetic field was found to be quantized in units of e^2/h, e being the charge of the electron and h Planck's constant. Neither the quality of the surface of the insulator nor the chemical composition seemed to be of importance for the great precision with which the unit of conductance, e^2/h, was obtained. Thouless *et al.* (1982), by applying the Kubo formula to a model consisting of a two-dimensional periodic

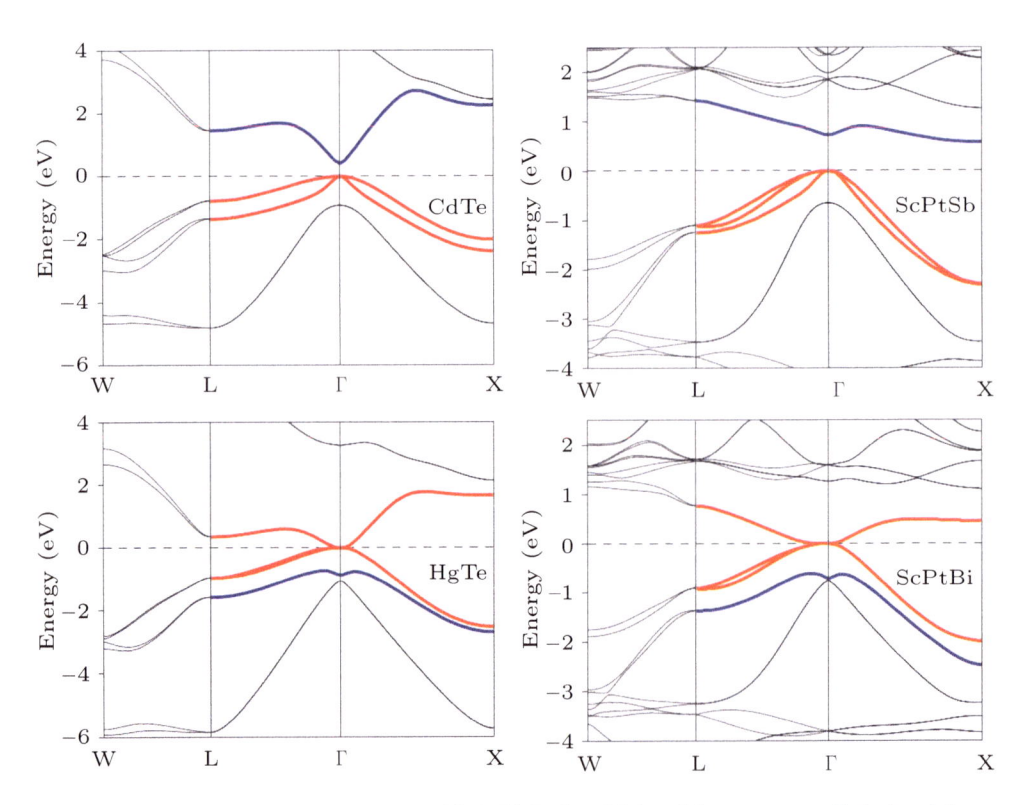

Figure 4.124 *Band inversion in HgTe and ScPtBi. In the vicinity of the zone center Γ the band drawn in red has Γ_8, the one in blue Γ_6 symmetry. Comparing CdTe with HgTe (or ScPtSb with ScPtBi) we see that bands Γ_6 and Γ_8 are inverted. Dashed line marks the Fermi energy. CdTe and ScPtSb are normal (trivial) insulators, HgTe and ScPtBi are topological insulators. From Chadov et al. (2010).*

potential in a magnetic field, calculated the Hall conductivity and found it to be integer multiples of e^2/h. From the present perspective they determined the Berry curvature where the integer multiples are Chern numbers. Simon (1983), from a mathematical perspective, gives an interesting explanation. The quantum states are numbered by Landau levels, which were given in great detail in Sec. 1.2.2, Eqn (1.41). The nontrivial topology comes about through the magnetic field and is expressed by nonzero Chern numbers.

Haldane (1988) constructed a two-dimensional model that possessed topologically nontrivial states without a magnetic field. The Hamiltonian he proposed breaks time-reversal symmetry, i.e. describes a magnetic insulator where the electrons occupy Bloch functions. Topological insulators are obtained when bands invert. This inversion has been studied thoroughly, albeit only little in connection with magnetic systems. For didactic purposes we show an example for nonmagnetic systems, the cubic compounds CdTe and HgTe, as well as the half-Heusler compounds ScPtSb and ScPtBi. The band structure of these compounds is displayed in Fig. 4.124. The change of the position of the coloured bands, the inversion, is brought about by the stronger spin–orbit coupling due to the heavier elements Hg compared with Cd or Bi compared with Sb. So this is clearly a relativistic effect. To obtain a real insulator in the topological cases one needs to break the band touching at Γ, which may be done by a tetragonal distortion. To prove the nontrivial topology one needs to do a careful analysis. When inversion symmetry is missing as in the example given in Fig. 4.124 one may follow Fu and Kane (2006) and determine Z_2, a topological invariant that is analogous to the Chern invariant discussed before. We will not go into further detail now but simply state that a topological insulator

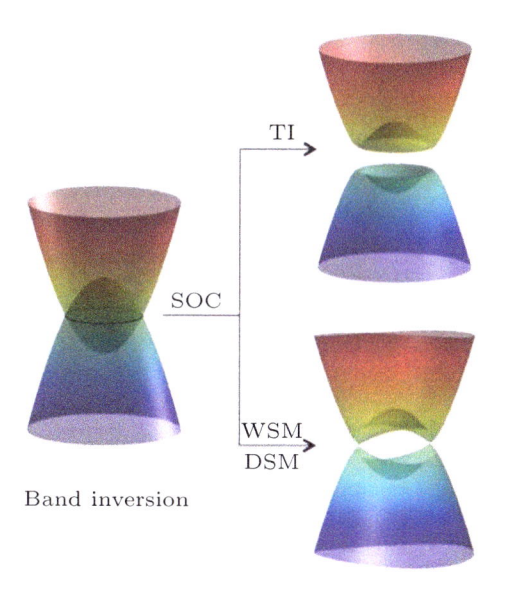

Figure 4.125 *From topological insulator to Weyl semimetal. After Yan and Felser (2017).*

has an inverted bulk band structure but the surface states are gapped so that the surface is conducting.

A prediction for a two-dimensional magnetic topological insulator was made by Yu *et al.* (2010). They used *ab initio* calculations on semiconductor layers of Bi_2Te_3, Bi_2Se_3, and Sb_2Te_3 to determine the inverted band structure and the surface states. The doping with Cr or Fe to break the time-reversal symmetry leads to split bands through the exchange field. The proposal led Chang *et al.* (2013) to fabricate samples of $Cr_{0.15}(Bi_{0.1}Sb_{0.9})_{1.85}Te_3$ which indeed showed the quantized anomalous Hall effect (QAHE) without applying an external magnetic field.

The three-dimensional QAHE is one of the future projects; first results exist and are encouraging. Fig. 4.125 finally demonstrates the close connection of the topological insulator with Weyl or Dirac semimetals.

5
Magnetism at Finite Temperatures

5.1 Density-functional theory at $T > 0$

Our task now is to describe the thermal properties of magnets. We therefore return to the beginning of Chap. 4 where some of the goals were stated and continue with the discussion of Stoner theory. At first sight it appears quite natural to extend the approach there. Following Stoner we thus might initially use the simple model density of states from free electrons and go to finite temperatures employing the well-known Fermi–Dirac statistics. In particular, we could generalize the expression for the magnetic moment, Eqn (1.119), using a molecular field as in Sec. 4.1. We might then extend the picture with realistic density of states functions and calculated exchange constants of the type introduced in Chap. 4. Indeed, these steps can and will be justified by means of Mermin's (1965) generalization of density-functional theory to finite temperatures. We will give an outline of this theory here even though we presently do not know a physically meaningful approximation to the finite-temperature exchange-correlation potential. The discussion given in this introductory section, therefore, serves to define our limits of understanding but does not, unfortunately, lead to a useful algorithm, although we will need the results later on in Sec. 5.5.2.

As in Sec. 2.5 a many-electron system in an external potential is considered. We remind the reader of Sec. 2.7 where the connection of the external potential with the density matrix was shown to be

$$V[\tilde{n}] = \sum_{\alpha\beta} \int v_{\alpha\beta}^{\text{ext}}(\mathbf{r})\, n_{\beta\alpha}(\mathbf{r}) \mathrm{d}\mathbf{r}, \tag{5.1}$$

where $\alpha, \beta = 1, 2$ are spin indices and $n_{\beta\alpha}(\mathbf{r})$ are the elements of the density matrix, $\tilde{n}$, which defines the particle density through the trace

$$n(\mathbf{r}) = \mathrm{Tr}\,\tilde{n}(\mathbf{r}) \tag{5.2}$$

and the vector of the magnetization by

$$\mathbf{m}(\mathbf{r}) = \mathrm{Tr}\,\boldsymbol{\sigma}\tilde{n}(\mathbf{r}), \tag{5.3}$$

where $\boldsymbol{\sigma}$ is given by the Pauli spin matrices.

Mermin (1965) laid the formal foundations for the proof that in the grand canonical ensemble at a given temperature T and chemical potential μ the equilibrium density matrix $\tilde{n}(\mathbf{r})$, i.e. the equilibrium particle density $n(\mathbf{r})$ and the equilibrium magnetization $\mathbf{m}(\mathbf{r})$, are determined by the external potential and magnetic field that make up $v_{\alpha\beta}^{\mathrm{ext}}(\mathbf{r})$. The correct $n(\mathbf{r})$ and $\mathbf{m}(\mathbf{r})$ minimize the Gibbs grand potential Ω:

$$\Omega[\tilde{n}] = \sum_{\alpha\beta} \int v_{\alpha\beta}^{\mathrm{ext}}(\mathbf{r})\,\tilde{n}_{\beta\alpha}(\mathbf{r})\mathrm{d}\mathbf{r} + \iint \mathrm{d}\mathbf{r}\mathrm{d}\mathbf{r}' \frac{n(\mathbf{r})n(\mathbf{r}')}{|\mathbf{r}-\mathbf{r}'|} - \mu \int \mathrm{d}\mathbf{r}\, n(\mathbf{r}) + G[\tilde{n}], \tag{5.4}$$

where G is a unique functional of charge and magnetization at a given temperature T and chemical potential μ. Here we omit the proof and refer the reader interested in these details to Mermin's (1965) original paper or to reviews like that by Rajagopal (1980) or Kohn and Vashishta (1983). We mention, however, that the functional property embodied in Eqn (5.4) is easily shown to originate from the basic inequality

$$\Omega[\rho] > \Omega[\rho_0] \tag{5.5}$$

which is valid for all grand canonical density matrices, ρ (not to be confused with the density matrices $\tilde{n}(\mathbf{r})$), positive definite with unit trace, and the correct one, ρ_0. The classical version of this inequality was proved by Gibbs and a short direct proof of the quantum inequality can be found in Mermin's paper (1965).

The quantity $G[\tilde{n}]$ in Eqn (5.4) is, just as the corresponding quantity in Chap. 2, written as the sum of three terms, i.e.

$$G[\tilde{n}] = T_0[\tilde{n}] - TS_0[\tilde{n}] + \Omega_{xc}[\tilde{n}] \tag{5.6}$$

with T_0, S_0 being respectively the kinetic energy and entropy of a system of noninteracting electrons with density matrix $\tilde{n}$ at a temperature T. The quantity Ω_{xc} is the exchange and correlation contribution to the Gibbs grand potential.

Proceeding as in Chap. 2 we now construct the minimum of the grand potential using a system of noninteracting electrons moving in an effective potential. We thus assume we can determine single-particle functions $\{\psi_{i\alpha}(\mathbf{r})\}$ that permit us to write the elements of the density matrix as

$$n_{\beta\alpha}(\mathbf{r}) = \sum_{i=1}^{\infty} \psi_{i\beta}(\mathbf{r})\,\psi_{i\alpha}^{*}(\mathbf{r})\,f(\varepsilon_i), \tag{5.7}$$

where α and β, due to the electron spin, take on the values 1 and 2 and $f(\varepsilon) = [1 + \exp \beta(\varepsilon - \mu)]^{-1}$ is the Fermi–Dirac distribution function ($\beta = 1/k_\text{B}T$). We obtain the single-particle spinor functions by solving the Schrödinger equation

$$\sum_\beta [-\delta_{\alpha\beta} \nabla^2 + v'_{\alpha\beta}(\mathbf{r}) - \varepsilon_i \, \delta_{\alpha\beta}] \, \psi_{i\beta}(\mathbf{r}) = 0 \tag{5.8}$$

and attempt to determine the potential, v', by minimizing the grand potential. To do this we rewrite Eqn (5.4) as

$$\Omega[\tilde{n}] = \sum_{\alpha\beta} \int v^\text{ext}_{\alpha\beta}(\mathbf{r}) \, n_{\beta\alpha}(\mathbf{r}) \, \mathrm{d}\mathbf{r} + \iint \mathrm{d}\mathbf{r}\,\mathrm{d}\mathbf{r}' \, \frac{n(\mathbf{r})\,n(\mathbf{r}')}{|\mathbf{r} - \mathbf{r}'|} + \Omega_{xc}[\tilde{n}] \\ - \sum_{\alpha\beta} \int v'_{\alpha\beta}(\mathbf{r}) \, n_{\beta\alpha}(\mathbf{r}) \, \mathrm{d}\mathbf{r} + \Omega_0[\tilde{n}], \tag{5.9}$$

where $\Omega_0[\tilde{n}]$ is the grand potential of noninteracting electrons, which we write as

$$\Omega_0[\tilde{n}] = T_0[\tilde{n}] + V'[\tilde{n}] - T S_0[\tilde{n}], \tag{5.10}$$

$V'[\tilde{n}]$ originating from the trial potential v' and all other symbols being defined above. Standard thermodynamics yields

$$\Omega_0[\tilde{n}] = -\beta^{-1} \sum_{i=1}^{\infty} \ln[1 + \exp \beta(\mu - \varepsilon_i)], \tag{5.11}$$

where, as usual $\beta = 1/k_\text{B}T$. We now vary Ω in Eqn (5.9) and, observing that a variation of the energy eigenvalues cancels a variation of the trial potential, v', we obtain the effective potential as

$$v'_{\alpha\beta}(\mathbf{r}) \equiv v^\text{eff}_{\alpha\beta}(\mathbf{r}) = v^\text{ext}_{\alpha\beta}(\mathbf{r}) + 2\,\delta_{\alpha\beta} \int \frac{n(\mathbf{r}')}{|\mathbf{r} - \mathbf{r}'|}\,\mathrm{d}\mathbf{r}' + v^{xc}_{\alpha\beta}(\mathbf{r}), \tag{5.12}$$

where

$$v^{xc}_{\alpha\beta}(\mathbf{r}) = \frac{\delta}{\delta n_{\beta\alpha}(\mathbf{r})} \Omega_{xc}[\tilde{n}], \tag{5.13}$$

and we have absorbed into Ω_{xc} contributions from the kinetic energy and the entropy that are not contained in Eqn (5.10) (see also Chap. 2). With the result for the effective potential we may finally rewrite the grand potential as

$$\Omega[\tilde{n}] = -\beta^{-1} \sum_{i=1}^{\infty} \ln[1 + \exp \beta(\mu - \varepsilon_i)] - \iint d\mathbf{r}\, d\mathbf{r}' \, \frac{n(\mathbf{r})\, n(\mathbf{r}')}{|\mathbf{r} - \mathbf{r}'|}$$
$$- \sum_{\alpha\beta} \int d\mathbf{r} v_{\alpha\beta}^{xc}(\mathbf{r})\, n_{\beta\alpha}(\mathbf{r}) + \Omega_{xc}[\tilde{n}]$$
(5.14)

which should be compared with Eqn (2.156).

Formally the problem now appears to be solved: the potential in the Schrödinger equation, Eqn (5.8), is given by the Eqns (5.12) and (5.13) in which the density matrix (and hence the density itself) is given by Eqn (5.7). For a ferromagnet at zero temperature we know that the basic Schrödinger equation, in the local-density approximation, is diagonal in spin space and hence easily solved. But at finite temperatures the formal expression for the exchange-correlation potential, $v_{\alpha\beta}^{xc}(\mathbf{r})$, Eqn (5.13), gives no clue concerning its general properties. If we assume it remains diagonal in spin space and use the zero-temperature exchange-correlation potential we may again solve the Schrödinger equation, temperature in this case entering only through the Fermi–Dirac distribution in Eqn (5.7). This is not a hard calculation and one obtains for the temperature, T_c^{Stoner}, where the magnetization vanishes, the values given in Table 5.1. Comparing with the experimental Curie temperatures, T_c^{exp}, we see that the results are entirely unsatisfactory.

The obvious disagreement apparent in Table 5.1 is part of what is normally called the failure of Stoner theory. We have chosen here to discuss the problem using a naive version of Mermin's finite-temperature density-functional theory and see that our basic assumption of a diagonal effective potential is most likely the weak part of this treatment. Restricting the effective potential to diagonal form implies that the magnetization decreases as the temperature increases because of excitations that are of the order of the exchange splitting, i.e. of rather high energies, since the exchange splitting is large on the scale of the Curie temperatures, see Chap. 4. We are thus led to look for low-energy excitations to explain the magnetic phase transition.

Awareness of low-energy excitations arose in the 1970s predominantly through the pioneering work of Moriya (1985), Hubbard (1979), Hasegawa (1979a,b), Korenman et al. (1977), Gyorffy et al. (1985), Edwards (1982), and others. The broad consensus reached was that *orientational fluctuations of the local magnetization* represent the essential

Table 5.1 Comparison of the calculated Stoner temperatures, T_c^{Stoner}, with the measured Curie temperatures, T_c^{exp}, for the elemental ferromagnets.

	T_c^{Stoner} (K)	T_c^{exp} (K)
Fe	5300	1039
Co	4000	1390
Ni	2900	630

ingredients to a thermodynamic theory. This cannot be incorporated in a straightforward way into the Stoner–Mermin theory, but many physically sound approximations had been suggested. Still, the detailed approaches seemingly differed considerably, some like Gyorffy *et al.* (1985) emphasizing *disordered local moments*, while others like Korenman *et al.* (1977) a *fluctuating local band picture*.

We will not follow the historical development, but, trying to remain in the larger context of density-functional theory, we begin by deriving the necessary tools to calculate the spin-wave spectrum of an itinerant-electron ferromagnet. Having thus obtained the low-energy excitations of the system, we can subsequently discuss approximations to describe its thermal properties. It should be noticed at the outset that we do *not* initially postulate a model Hamiltonian, like an Ising or Heisenberg model. A full account of the latter approach can e.g. be found in the book by Lovesey (1984, Vol. 2).

5.2 Adiabatic spin dynamics

We return to the basic equations that allow us to calculate the spin-density matrix and the total energy of a noncollinear magnet, i.e. the Schrödinger equation, Eqn (2.203) and the total energy, Eqn (2.206). In the form appropriate for computations there are Eqns (3.307)–(3.309). We first rewrite the Schrödinger equation in a way that is easier to manipulate. This is

$$\sum_{\gamma=1}^{2}[\delta_{\alpha\gamma}(-\nabla^2+v)+(\boldsymbol{\sigma}\cdot\mathbf{B})_{\alpha\gamma}]\psi_{i\gamma}=\varepsilon_i\,\psi_{i\alpha}. \quad (5.15)$$

The only change is the use of an effective potential that we denote here by $\mathbf{B}$. From the product $\boldsymbol{\sigma}\cdot\mathbf{B}$ with the vector $\boldsymbol{\sigma}$ formed by the Pauli spin matrices, we see that it acts like a magnetic field. It is easy to show from the previously used effective potential in Eqn (2.203) that $\mathbf{B}$ is given by

$$\mathbf{B}=(\sin\theta\cos\varphi\,\mathbf{e}_x+\sin\theta\sin\varphi\,\mathbf{e}_y+\cos\theta\,\mathbf{e}_z)B, \quad (5.16)$$

where θ and φ are the polar angles used before, depending in principle on each point $\mathbf{r}$ in the crystal, and $B=B(\mathbf{r})$ replaces the previously used Δv, i.e. it is given by the functional derivative of the exchange-correlation potential,

$$B=\frac{1}{2}\left(\frac{\delta E_{xc}}{\delta n_1}-\frac{\delta E_{xc}}{\delta n_2}\right), \quad (5.17)$$

where n_1 and n_2 as before denote the eigenvalues of the spin-density matrix, $\tilde{n}$, Eqn (5.7) (which we evaluate at zero temperature). The quantities $\mathbf{e}_x$, $\mathbf{e}_y$, and $\mathbf{e}_z$ are unit vectors fixed in space and $v=v(\mathbf{r})$ in Eqn (5.15) is the spin-independent part of the effective potential.

We are now ready to state the essential approximation. If we set out to describe the motion of the magnetization as a function of time at each point **r** in the crystal, we must determine the equation of motion for **m**(**r**) given by Eqn (5.3). We thus look for the equation of motion of its constituents and are led to the time-dependent Schrödinger equation for the wave functions $\psi_{i\alpha}$, which in principle should give the time dependence of the spin-density matrix, $\tilde{n}$, and thus that of **m**(**r**). However, we are not in a position to carry out a full time-dependent density-functional theory and, perhaps, do not even have to do so. Rather, knowing that the bandwidths of the d electrons in the crystal are of the order of electron volts, while the low-lying excitations we look for are spin waves which have energies of the order of milli-electron volts, we may assume that the motion of the electrons is much faster than that of the magnetic moments. This assumption implies that we imagine different timescales govern the physics of the electrons and the magnetic moments. We faced a situation like this before when the motion of the electrons was separated from that of the nuclei by means of the Born–Oppenheimer or adiabatic approximation (see Chap. 2). It has become common to use the term *adiabatic* here. Nevertheless, it is important to emphasize that a systematic adiabatic approach has not been developed for the spin problem, simply because — in contrast to the nuclear motion—there is no large mass governing the timescale. The approach to take is to carry out the necessary calculations and compare the results with experimental data which will decide whether or not the approximations are tolerable.

In the adiabatic approximation (see for instance Schiff, 1955) one assumes the time-independent Schrödinger equation holds at any instant of time for a potential that parametrically depends on time. The energy eigenvalues are then supposed to depend on time in such a way that no levels cross and transitions to other levels do not occur.

In our case the time-dependent parameters in the Schrödinger equation are the angles θ and ϕ in Eqn (5.16). The essential assumption now is that the time-dependent Schrödinger equation holds in the context of density-functional theory in the adiabatic approximation, that is

$$i\frac{\partial \psi_{i\alpha}}{\partial t} = \sum_{\gamma=1}^{2}[\delta_{\alpha\gamma}(-\nabla^2 + v) + (\boldsymbol{\sigma}\cdot\mathbf{B})_{\alpha\gamma}]\psi_{i\gamma} \quad \text{for} \quad \alpha = 1, 2 \qquad (5.18)$$

describes the time evolution of the state given by the spinor components ψ_{i1} and ψ_{i2} having energy $\varepsilon_i = \varepsilon_i(\theta, \phi)$. It has been shown by Runge and Gross (1984) and later in a thorough review article by Gross *et al.* (1996) that density-functional theory of time-dependent phenomena can be justified extremely well, so that Eqn (5.18) is indeed valid in our context.

We can now work out the equation of motion for the magnetization by evaluating

$$\frac{\partial}{\partial t}\mathbf{m}(\mathbf{r}) = \mathrm{Tr}\,\boldsymbol{\sigma}\frac{\partial}{\partial t}\tilde{n} = \sum_{i=1}^{\infty} f(\varepsilon_i)\sum_{\alpha\beta}\boldsymbol{\sigma}_{\alpha\beta}\frac{\partial}{\partial t}[\psi_{i\beta}(\mathbf{r})\,\psi_{i\alpha}^*(\mathbf{r})], \qquad (5.19)$$

where $f(\varepsilon_i)$ is the Fermi–Dirac distribution function (for which we may assume $T = 0$) and $\boldsymbol{\sigma}_{\alpha\beta}$ are the elements of the Pauli spin matrices. If we ignore the possible occurrence of a Berry phase (Berry, 1984) in products of the type $[\partial \psi_{i\beta}(\mathbf{r})/\partial t]\, \psi_{i\alpha}^*(\mathbf{r})$ when $\alpha \neq \beta$, we obtain with Eqn (5.18) after some manipulations

$$\frac{\partial}{\partial t}\mathbf{m}(\mathbf{r}) = -2\,\mathbf{m}(\mathbf{r}) \times \mathbf{B}(\mathbf{r}) + \mathrm{Tr}\,(\boldsymbol{\sigma}\,\tilde{g}), \tag{5.20}$$

where the elements of $\tilde{g}$ are

$$g_{\alpha\beta} = -\mathrm{i}\sum_{i=1}^{\infty} f(\varepsilon_i)\,\boldsymbol{\nabla}\left(\psi_{i\alpha}\boldsymbol{\nabla}\psi_{i\beta}^* - \mathrm{c.c.}\right). \tag{5.21}$$

They are seen to represent the divergence of the current which is a consequence of the quantum mechanical continuity equation, Eqn (2.85), and thus describe, through n_{conv} in Sec. 2.4 (the convective part of the density), the rate of change of an *orbital* contribution to the magnetization. Since we are only concerned with the spin part of the magnetization we ignore $\mathrm{Tr}\,(\boldsymbol{\sigma}\,\tilde{g})$ and continue with

$$\frac{\partial}{\partial t}\mathbf{m}(\mathbf{r}) = -2\,\mathbf{m}(\mathbf{r}) \times \mathbf{B}(\mathbf{r}). \tag{5.22}$$

The formalism has been quite general so far, but now we begin to specialize it to the case where atoms occupy a lattice with a basis. We thus pick a point $\mathbf{r}$ in the atom having basis vector $\boldsymbol{\tau}_\nu$ in the unit cell given by the vector $\mathbf{R}_n$, i.e. we define $\mathbf{r}_{n\nu} = \mathbf{r} - \mathbf{R}_n - \boldsymbol{\tau}_\nu$. Next we try a spin-spiral form for the magnetization and write

$$\mathbf{m}(\mathbf{r}_{n\nu}) = m(\mathbf{r}_\nu)\,\{\sin\theta_\nu \cos[\mathbf{q}\cdot(\mathbf{R}_n + \boldsymbol{\tau}_\nu) + \varphi_\nu]\,\mathbf{e}_x \\ + \sin\theta_\nu \sin[\mathbf{q}\cdot(\mathbf{R}_n + \boldsymbol{\tau}_\nu) + \varphi_\nu]\,\mathbf{e}_y + \cos\theta_\nu\,\mathbf{e}_z\}. \tag{5.23}$$

This case was discussed in Sec. 3.4.7 where we developed a generalization of Bloch's theorem for a lattice possessing a magnetic moment given by

$$\mathbf{M}_{n\nu} = \int_{\Omega_\nu} \mathbf{m}(\mathbf{r}_{n\nu})\,\mathrm{d}\mathbf{r}, \tag{5.24}$$

i.e.

$$\mathbf{M}_{n\nu} = M_\nu\{\sin\theta_\nu \cos[\mathbf{q}\cdot(\mathbf{R}_n + \boldsymbol{\tau}_\nu) + \varphi_\nu]\,\mathbf{e}_x \\ + \sin\theta_\nu \sin[\mathbf{q}\cdot(\mathbf{R}_n + \boldsymbol{\tau}_\nu) + \varphi_\nu]\,\mathbf{e}_y + \cos\theta_\nu\,\mathbf{e}_z\}. \tag{5.25}$$

We remind the reader that M_ν is the magnitude of the magnetic moment at site τ_ν, $\mathbf{R}_n$ is a translation vector, $\mathbf{q}$ is the wave vector characterizing the spiral, and $\mathbf{q} \cdot \mathbf{R}_n + \varphi_\nu$ as well as θ_ν are polar angles as depicted in Fig. 3.21. Using the atomic sphere approximation (ASA) we approximate the volume Ω_ν by the appropriate atomic sphere (compare with Eqn (3.334)). The time-dependent parameters are the polar angles θ_ν, φ_ν and the magnitude of the magnetic moment, M_ν.

Assuming a ferromagnet we now perturb the equilibrium by forcing the magnetic moments into a spiral. This deviation from the ground state costs energy which we express as

$$\Delta E = \sum_{\ell \nu j \mu} J_{\ell \nu j \mu} \, \mathbf{M}_{\ell \nu} \cdot \mathbf{M}_{j \mu}, \tag{5.26}$$

where ℓ and j label translations (cells) and ν, μ basis atoms. The quantities $J_{\ell \nu j \mu}$ are called exchange constants as in the Heisenberg model. Much space will be devoted later on in Sec. 5.4.7.1 to their numerical determination. The perturbation gives rise to a magnetic field. In the atomic sphere labelled ν in cell ℓ it follows from $\mathbf{B}_{\ell \nu} = -\delta \Delta E / \delta \mathbf{M}_{\ell \nu}$, which gives

$$\mathbf{B}_{\ell \nu} = -2 \sum_{j \mu} J_{\ell \nu j \mu} \, \mathbf{M}_{j \mu} . \tag{5.27}$$

The equation of motion, Eqn (5.22), thus becomes in the ASA

$$\frac{\partial}{\partial t} \mathbf{M}_{\ell \nu} = 4 \sum_{j \mu} J_{\ell \nu j \mu} \, \mathbf{M}_{\ell \nu} \times \mathbf{M}_{j \mu} . \tag{5.28}$$

To solve the equation of motion it is convenient to define

$$M_{\ell \nu \pm} = M_{\ell \nu \, x} \pm M_{\ell \nu \, y} \tag{5.29}$$

and read off from Eqn (5.25)

$$M_{\ell \nu \pm} = M_\nu \sin \theta_\nu \exp[\pm \mathrm{i} \mathbf{q} \cdot (\mathbf{R}_\ell + \boldsymbol{\tau}_\nu) \pm \varphi_\nu]. \tag{5.30}$$

Using these definitions we obtain for the equation of motion

$$\frac{\partial}{\partial t} M_{\ell \nu \pm} = \pm 4\mathrm{i} \sum_{j \mu} J_{\ell \nu j \mu} [M_{\ell \nu \, z} M_{j \mu \pm} - M_{j \mu \, z} M_{\ell \nu \pm}] \tag{5.31}$$

and similarly for the z-component

$$\frac{\partial}{\partial t} M_{\ell \nu \, z} = 2\mathrm{i} \sum_{j \mu} J_{\ell \nu j \mu} [M_{\ell \nu +} M_{j \mu -} - M_{j \mu +} M_{\ell \nu -}]. \tag{5.32}$$

On both sides of these equations one now inserts the spiral, Eqn (5.30), and similarly the z-component. First one notices that

$$\frac{d}{dt}M_\mu = 0 \text{ for all } \nu, \qquad (5.33)$$

which follows directly from the equation of motion, Eqn (5.28). Next the equation for the time derivative of θ_ν is eliminated from the two equations that result from Eqns (5.31) and (5.32). With the definition

$$j_{\nu\mu}(\mathbf{q}) = \sum_j J_{\ell\nu j\mu} \cos[\mathbf{q}\cdot(\mathbf{R}_\ell - \mathbf{R}_j)] \qquad (5.34)$$

one then obtains

$$\sin\theta_\nu \frac{d}{dt}\varphi_\nu = 4\sum_\mu \{j_{\nu\mu}(\mathbf{q})\cos[\mathbf{q}\cdot(\boldsymbol{\tau}_\mu - \boldsymbol{\tau}_\nu) + \varphi_\mu - \varphi_\nu]M_\mu \cos\theta_\nu \sin\theta_\mu$$
$$- j_{\nu\mu}(0)M_\mu \cos\theta_\mu \sin\theta_\nu\}. \qquad (5.35)$$

This relation is now linearized by letting $\theta_\nu \to 0$. The eliminated equation for the time derivative of θ_ν shows that in this limit $d\theta_\nu/dt \to 0$. Writing $\varphi_\nu = \omega t$, where we introduced the frequency ω, we finally obtain

$$\omega\theta_\nu = \sum_\mu 4\{j_{\nu\mu}(\mathbf{q})\cos[\mathbf{q}\cdot(\boldsymbol{\tau}_\mu - \boldsymbol{\tau}_\nu)]M_\mu\theta_\mu - j_{\nu\mu}(0)M_\mu\theta_\nu\}. \qquad (5.36)$$

We apply these results first to the elementary ferromagnetic metals. The general case will be taken up again in Sec. 5.2.2.

5.2.1 Magnon spectra of bcc Fe, fcc Co, and fcc Ni

For the elementary ferromagnets we drop all basis vector indices and see that Eqn (5.36) simply becomes

$$\omega = 4Mj(\mathbf{q}), \qquad (5.37)$$

where M is the magnetic moment (at zero temperature). Since the principal quantity is the energy change, ΔE, when the ground state is perturbed by a spiral, Eqn (5.26), we rewrite ΔE using a spiral characterized by the wave vector $\mathbf{q}$. This is easily seen to be

$$\Delta E = M^2 \sin^2\theta j(\mathbf{q}) \doteq \Delta E(\mathbf{q}, \theta), \qquad (5.38)$$

where θ is the tilt angle of the spiral. Thus the dispersion relation for a spin wave, $\omega = \omega(\mathbf{q})$, becomes

$$\omega(\mathbf{q}) = \lim_{\theta \to 0} \frac{4}{M} \frac{\Delta E(\mathbf{q}, \theta)}{\sin^2 \theta}. \tag{5.39}$$

We remind the reader that M is the magnetic moment in units of μ_B.

The same expression has been derived for the magnon spectrum by Niu and Kleinman (1998) and, under very general assumptions, by Niu et al. (1999). Their derivation is extremely interesting and is completely different from ours and does not presuppose the atomic sphere approximation. It does, however, assume adiabaticity and is obtained by calculating the Berry curvature. For details of this fascinating derivation we refer the reader to the original literature.

The derivation by Halilov et al. (1998b), though, rests on assuming the Heisenberg model and, just as ours, apparently requires the atomic sphere approximation. Their formula agrees with Eqn (5.39) and so does that of Rosengaard and Johansson (1997) who used similar assumptions as Halilov et al. Another derivation of the same result is due to Antropov et al. (1995 and 1996) who also used an equation of motion method, as we did, but determined the total energy part differently. Another approach has been taken by Savrasov (1998) who obtained the dynamical spin susceptibility in self-consistent density-functional theory. He extracts the spin-wave dispersion from the imaginary part of the susceptibility and finds good agreement with experiment. Computationally our formula and that of Halilov et al. (1998b) are the same. Thus numerical results based on Eqn (5.39) (with the use of the ASW method) give within small numerical differences the same spectra as those obtained by Halilov et al. (1998b) who used the LMTO method. Since the latter results are quite complete and include a comparison with experimental data, we have chosen to show these in Figs. 5.1–5.3.

A note to the computations seems in order. One finds numerically that the total energy difference for small polar angles, θ, goes to zero as $\sin^2 \theta$, thus the calculation is simply

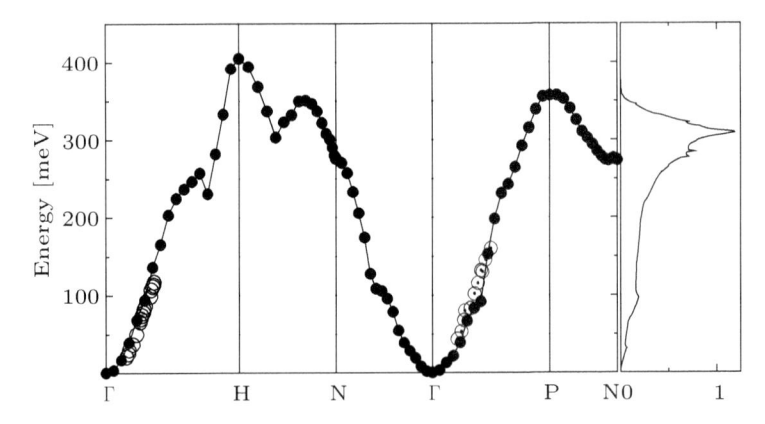

Figure 5.1 *Adiabatic magnon dispersion relations along high-symmetry lines and magnon densities of states (in states per (meV-cell) $\cdot 10^{-2}$) of bcc Fe. Closed circles are the calculated results of Halilov et al. (1998b), who kindly supplied this figure. Experimental data are marked by open symbols; pure Fe at 10 K: dotted circles (Loong et al., 1984), Fe with 12% Si at room temperature: blank circles (Lynn, 1975).*

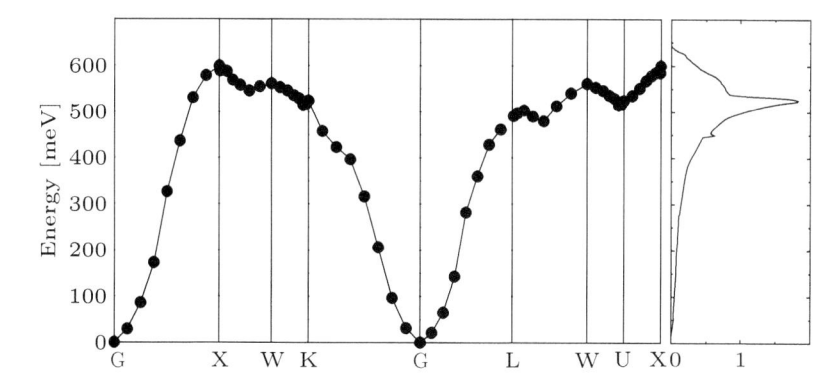

Figure 5.2 *Adiabatic magnon dispersion relations along high-symmetry lines and magnon densities of states (in states per (meV-cell) $\cdot 10^{-2}$) of fcc Co calculated by Halilov et al. (1998b), who kindly supplied this figure.*

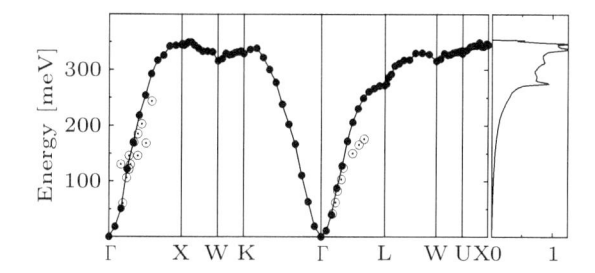

Figure 5.3 *Adiabatic magnon dispersion relations on high-symmetry lines and magnon densities of states (in states per (meV-cell) $\cdot 10^{-2}$) of fcc Ni. Closed circles are the calculated results of Halilov et al. (1998b), who kindly supplied this figure. Experimental data are marked as dotted circles and are those of Mook and Paul (1985).*

carried out at a finite angle, e.g. $\theta = 20°$ is a "save" choice in Eqn (5.39). Furthermore, it is observed that for small θ the magnetic moment is to a good approximation constant.

In the case of Fe and Ni, where parts of the spectra are known experimentally, the agreement is excellent in the long wavelength region and even good for shorter wavelengths. In the case of Ni, however, the Stoner spin-flip excitation continuum sets in at about 150 meV. This leads to a change of the spectrum that our adiabatic approach is unable to resolve. To account for this Cooke et al. (1985) determined the transverse magnetic susceptibility using a many-body treatment that is beyond the scope of our treatise here. Good agreement in this frequency range is also obtained by Savrasov (1998). We comment, however, that these authors used the term "optical magnon" for the frequency branch they obtained. This is somewhat misleading as we will see in the case of an hcp lattice which has a true optical branch. Note further that in all three metals the maximum magnon frequency is roughly five times larger than the Curie temperature. Thus short wavelength (high q) magnons will only be sparsely occupied even close to

Table 5.2 *Spin-wave stiffness constants, D, in meVÅ²*

	Calc.[1]	Calc.[2]	Exp.
bcc Fe	247	355	314[a]
fcc Co	502	535	510[b]
fcc Ni	739	715	550[c]

[1] Calculation by Rosengaard and Johansson (1997);
[2] Present ASW calculation;
[a] Stringfellow (1968); [b] Wohlfarth (1980); [c] Mitchell and Paul (1985).

the Curie temperature. For small values of q we can fit the calculated magnon energies with the form

$$\omega(\mathbf{q}) = D |\mathbf{q}|^2 (1 - \beta |\mathbf{q}|^2) \tag{5.40}$$

and thus obtain the spin-wave stiffness constants, D, which are collected in Table 5.2. Here we list, besides the experimental values, also those values obtained by Rosengaard and Johansson (1997). The agreement for Fe and Co is good, but Ni is off, presumably because of the proximity of the Stoner continuum mentioned above.

5.2.2 Magnon spectra for non-primitive lattices and compounds

For non-primitive lattices we have to account for the basis and thus return to the general result given in Eqn (5.36). If we define a quantity $\tilde{j}_{\nu\mu}$ by means of

$$\tilde{j}_{\nu\mu}(\mathbf{q}) = j_{\nu\mu}(\mathbf{q}) \cos[\mathbf{q} \cdot (\boldsymbol{\tau}_\mu - \boldsymbol{\tau}_\nu)] - \delta_{\mu\nu} \sum_\lambda j_{\mu\lambda}(0) M_\lambda / M_\mu \tag{5.41}$$

then Eqn (5.36) can be written as

$$\omega \theta_\nu = 4 \sum_\mu \tilde{j}_{\nu\mu} M_\mu \theta_\mu . \tag{5.42}$$

By multiplying both sides with $\sqrt{M_\nu}$ we see that the spin-wave spectrum as a function of the wave vector $\mathbf{q}$, $\omega(\mathbf{q})$, is obtained by diagonalizing a matrix $\mathcal{A}$ whose elements are given by

$$(A)_{\nu\mu} = 4\sqrt{M_\nu}\, \tilde{j}_{\nu\mu}(\mathbf{q})\, \sqrt{M_\mu} . \tag{5.43}$$

The same result was obtained by Halilov *et al.* (1998b).

5.2 Adiabatic spin dynamics

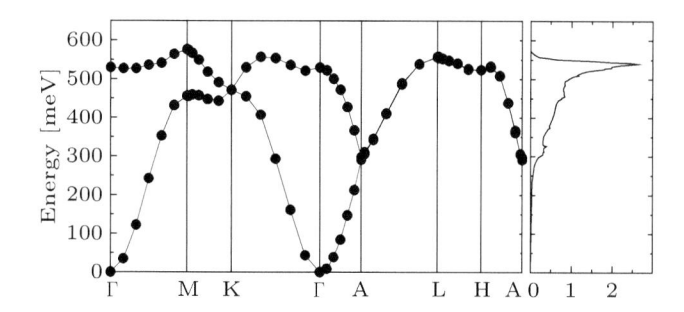

Figure 5.4 *Adiabatic magnon dispersion relations along high-symmetry lines and magnon densities of states (in states per (meV-cell)·10^{-2}) of hcp Co calculated by Halilov et al. (1998b), who kindly supplied this figure.*

We first apply this formula to *hcp Cobalt*. The crystal structure of this hexagonal closed packed form of Co is described by placing one atom at the origin and the other at $\tau = a(\sqrt{3}/3, 0, c/2a)$, where a and c are the lattice constants. The matrix to be diagonalized is 2×2 and the two exchange functions, $j_{11}(\mathbf{q})$ and $j_{12}(\mathbf{q})$, are obtained as described in Sec. 5.4.2.4. The results are shown in Fig. 5.4, which has been supplied by Halilov et al. (1998b). A second branch is clearly seen, which in analogy with phonon spectra is called the *optical branch*.

We next turn to the spin waves of a compound, beginning with a system that has only one magnetic atom in the unit cell. Its spectrum is therefore obtained by means of Eqn (5.39). Since there exists rather complete experimental information about the magnon spectra of Heusler alloys by Noda and Ishikawa (1976a,b) as well as Tajima et al. (1977), it is of interest to find out how well we are able to reproduce these data with our computational techniques.

We therefore choose as an example the ferromagnetic compound Cu_2MnAl whose electronic structure was obtained by Kübler et al. (1983) and discussed in Chap. 4. In attempts to calculate the magnon spectra by means of the force theorem as explained above one discovers quickly that small residual magnetic moments found on the Cu and Al sites change with the choice of the wave vector $\mathbf{q}$ so that the assumption of a constant magnetic moment is not well justified. Therefore it seems that one must resort to a full self-consistent total energy calculation in which the magnetic moment can be *constrained* to be constant.

A simple version of a *constrained total energy search* was discussed in Sec. 4.3; here we proceed as follows. Assume we desire the magnetic moment to have the value given by Eqn (5.28). We then minimize the functional

$$\tilde{E}[\tilde{n}] = E[\tilde{n}] - \sum_{n\nu} \boldsymbol{\lambda}_{n\nu} \cdot \left[\int_{\Omega_\nu} \mathbf{m}(\mathbf{r}_{n\nu})\, d\mathbf{r} - \mathbf{M}_{n\nu} \right], \tag{5.44}$$

where $E[\tilde{n}]$ is given by Eqn (2.206) and the vector quantities $\boldsymbol{\lambda}_{n\nu}$ are Lagrange multipliers. The Euler–Lagrange equation of this functional gives an effective single-particle Hamiltonian of the form

$$\mathcal{H} = -\nabla^2 \mathbf{I} + \sum_{n\nu} \left[v(\mathbf{r}_{n\nu}) + \boldsymbol{\sigma} \cdot \mathbf{B}(\mathbf{r}_{n\nu}) - \boldsymbol{\sigma} \cdot \boldsymbol{\lambda}_{n\nu} \right] \Theta_\nu \left(|\mathbf{r}_{n\nu}| \right), \tag{5.45}$$

where v, $\boldsymbol{\sigma}$, and $\mathbf{B}$ are defined following Eqn (5.15) and the unit step function Θ_ν has been defined above in connection with Eqn (5.32). When the fields $\mathbf{B}(\mathbf{r}_{n\nu})$ and $\boldsymbol{\lambda}_{n\nu}$ are parallel (as assumed in Sec. 4.3) the ASW or LMTO schemes can be applied directly after shifting the *local* potentials by $\pm \lambda_{n\nu}$ taking care that the desired values of the magnetic moments are obtained. Further technical details are explained in the paper by Uhl *et al.* (1994).

Using Eqn (5.39), but calculating the total energy difference, $\Delta E(\mathbf{q}, \theta)$, from such a constrained total energy search with the moments fixed at their ground-state values, we obtain the magnon spectrum shown in Fig. 5.5. The agreement is only moderate with rather large deviations from the experimental values along the line Σ. To a certain extent this is due to convergence difficulties (having possibly not used a sufficient number of k-points, for instance), but an interesting physical origin for the facts apparent in the figure cannot be excluded.

Other Heusler compounds with more than one magnetic atom exist for which the spin-wave spectra begin to be measured. The electronic structure of some Heusler compounds was obtained in Sec. 4.4 and in Sec. 5.4.7.1 as well as Sec. 5.4.8 details will be given for the determination of the exchange functions $j_{\nu\mu}(\mathbf{q})$. Since there are

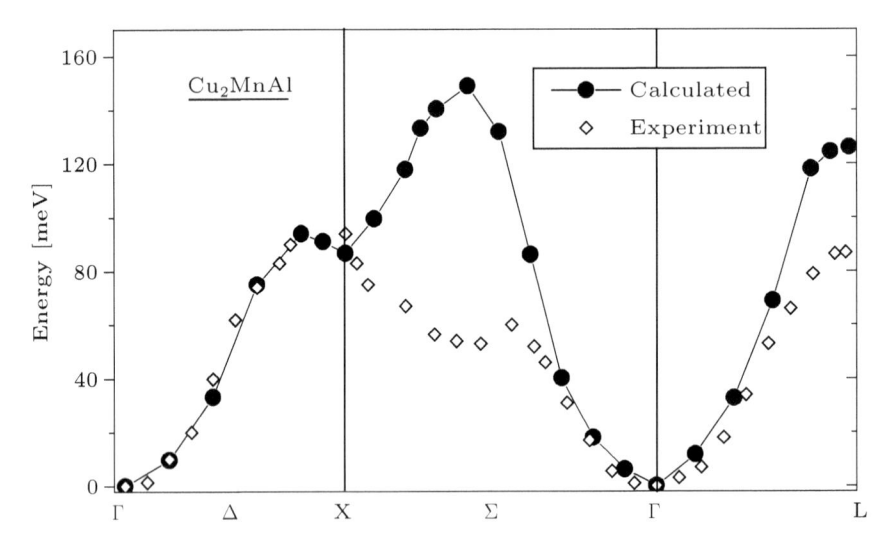

Figure 5.5 *Adiabatic magnon dispersion relations along high-symmetry lines of Cu_2MnAl. Experimental data points shown were obtained by Tajima et al. (1977) at 4.2 K.*

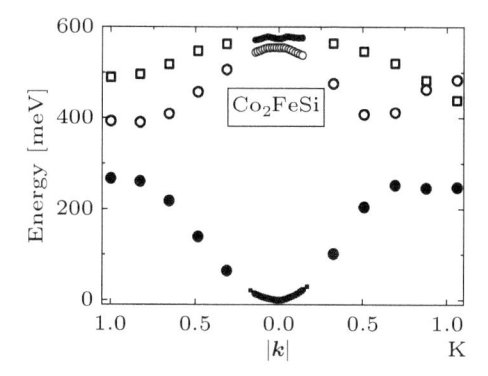

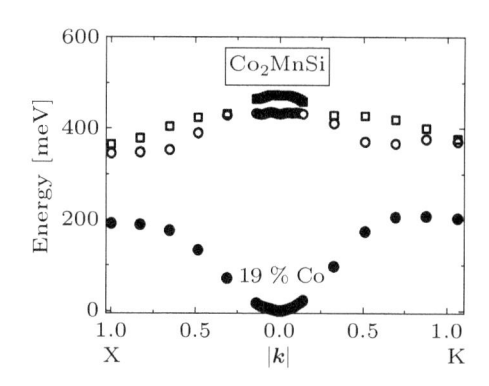

Figure 5.6 *Adiabatic magnon dispersion relations along high-symmetry lines of (a) Co_2FeSi and (b) Co_2MnSi. The highest branch is in each case entirely due to the Co moments; the Co weight in the acoustic branch is also given. The calculated spin-wave stiffness constant is $D = 715\,\mathrm{meV\mathring{A}^2}$ for (a) and $D = 770\,\mathrm{meV\mathring{A}^2}$ for (b).*

three magnetic atoms in Heusler compounds like Co_2FeSi the Eqn (5.43) requires diagonalizing a 3×3 matrix, which results in three branches. We choose to show the results of calculations for Co_2FeSi and Co_2MnSi and emphasize that Co_2FeSi, at least in the calculations, is not a half-metallic ferromagnet, whereas Co_2MnSi is. In Fig. 5.6 we show the calculated spectra along two symmetry lines. Co_2FeSi has the highest Curie temperature of all presently known Co_2-Heusler compounds (see Fig. 5.17). For Co_2FeSi the calculated spin-wave stiffness constant of $D = 715\,\mathrm{meV\mathring{A}^2}$ seems to be in good agreement with recent measurements giving $D = 705\,\mathrm{meV\mathring{A}^2}$ at room temperature (Hillebrands *et al.*, 2009). The calculated value for Co_2MnSi of $D = 770\,\mathrm{meV\mathring{A}^2}$ is somewhat too high compared with the measured room temperature value of $D = 580\,\mathrm{meV\mathring{A}^2}$. Neutron diffraction work on $La_{0.85}Sr_{0.15}MnO_3$ (Vasiliu-Doloc *et al.*, 1998), however, shows that the spin-wave stiffness constant depends on the temperature, comparable to the magnetization. But since the Curie temperatures of Co_2FeSi (1100 K) and Co_2MnSi (985 K) are high we do not expect the room temperature values to be much different from the $T = 0$ in these cases. Thus, more work is needed to understand $D(T)$.

5.2.3 Thermal properties of itinerant-electron ferromagnets at low temperatures

It is not a trivial matter to pass from the equation of motion, Eqn (5.22), to a description of thermal equilibrium defined by a suitable Gibbs distribution. Antropov *et al.* (1996) discuss this problem in some detail giving a number of useful references. We will, however, not dwell on this problem any further, but simply postulate that thermal equilibrium is properly described by quantizing the magnon spectrum and using the Planck distribution function for calculating the mean occupation number, $\langle n_\mathbf{q} \rangle$, for a magnon of energy $\omega_\mathbf{q}$, i.e.

$$\langle n_{\mathbf{q}} \rangle = \frac{1}{\exp(\omega_{\mathbf{q}}/k_{\mathrm{B}}T) - 1}. \tag{5.46}$$

Assuming that at low temperatures the magnons do not interact we add up the single-magnon energies to obtain the total energy as

$$E = \sum_{\mathbf{q}} \omega_{\mathbf{q}} \langle n_{\mathbf{q}} \rangle. \tag{5.47}$$

The total energy change given by Eqn (5.35) may then be equated with $\omega_{\mathbf{q}} \langle n_{\mathbf{q}} \rangle$ and using Eqn (5.39) for small but finite θ we obtain

$$\langle \theta_{\mathbf{q}}^2 \rangle = \frac{4}{M} \langle n_{\mathbf{q}} \rangle, \tag{5.48}$$

where for small θ the mean value of $\sin^2 \theta$ is written as $\langle \theta_{\mathbf{q}}^2 \rangle$. This equation now enables us to evaluate the decrease of the magnetization at finite temperatures, $\Delta M(T)$ defined as

$$\Delta M(T) = M - M \sum_{\mathbf{q}} \langle \cos \theta_{\mathbf{q}} \rangle \tag{5.49}$$

to be

$$\Delta M(T) \simeq M \sum_{\mathbf{q}} \langle \theta_{\mathbf{q}}^2 \rangle / 2 = 2 \sum_{\mathbf{q}} \langle n_{\mathbf{q}} \rangle. \tag{5.50}$$

The same result has been obtained by Halilov et al. (1998b) who, however, used a different method. To get the leading temperature dependence of the magnetization decrease one may proceed as in standard spin-wave theory (see e.g. Wagner, 1972 or Yosida, 1996): Using Eqn (5.46) and simplifying the magnon energies by keeping only the leading term in Eqn (5.40) one evaluates the integral

$$\int_0^\infty \frac{x^2 \, dx}{\exp x^2 - 1} = \frac{\sqrt{\pi}}{4} \varsigma(3/2), \tag{5.51}$$

where $\varsigma(3/2) \simeq 2.612$ is the Riemann zeta function, and obtains straightforwardly for the magnetization decrease (in μ_{B}) per unit volume at very low temperatures

$$\Delta M(T) \simeq 0.1173 \left(\frac{k_{\mathrm{B}} T}{D} \right)^{3/2}. \tag{5.52}$$

A similar calculation shows that the corresponding specific heat is proportional to $T^{3/2}$ also. Because of $\mathbf{q}^4$ and $\mathbf{q}^6$ terms in the expansion of the magnon energies there are also contributions proportional to $T^{5/2}$ and $T^{7/2}$. Our results do not differ from those of ferromagnets described by the Heisenberg model, for which this law was first obtained by Bloch, being therefore often referred to as Bloch's $T^{3/2}$ law. It is well satisfied experimentally particularly in metals with a high Curie temperature.

At elevated temperatures the occupation numbers $\langle n_{\mathbf{q}} \rangle$ and the amplitudes $\langle \sin \Theta_{\mathbf{q}} \rangle$ become large; the modes begin to interact and the magnon picture is to be replaced by a picture of strong long wavelength transverse spin fluctuations. We continue, however, to assume that the adiabatic approximation remains valid.

5.3 Mean-field theories

The determination of the low-temperature properties above rests on the assumption of noninteracting elementary excitations; thus Eqns (5.46) and (5.47) are consistent and are obtained by summing the simple partition function (e.g. Callen, 1985),

$$Z = \mathrm{Tr} \exp \left(-\sum_{q} \beta \omega_{\mathbf{q}} n_{\mathbf{q}} \right), \tag{5.53}$$

where, as before, Tr denotes the trace, the sum is restricted to the first Brillouin zone, $\beta = 1/(k_B T)$, and $n_{\mathbf{q}}$ are Bose–Einstein occupation numbers. (Extending the integral to infinity, as was done in Eqn (5.51), is an approximation permissible only at very low temperatures.) When the modes interact one cannot proceed so simply. Instead the strategy now is to identify a soluble model that is in some sense similar to the model of interest and then apply a controlled correction to calculate the effect of the difference of the two models. Such an approach is the Bogoliubov–Peierls variational principle, for which we now give a simple derivation (Callen, 1985).

We denote the Hamiltonian of the system by $\mathcal{H}$ and that of the soluble model by $\mathcal{H}_0$, the difference being $\mathcal{H}_1$ so that $\mathcal{H} = \mathcal{H}_0 + \mathcal{H}_1$. One then defines

$$\mathcal{H}(\lambda) = \mathcal{H}_0 + \lambda \mathcal{H}_1 \tag{5.54}$$

and sees that varying λ from 0 to 1 turns the model Hamiltonian continuously into that of the real system. The free energy corresponding to $\mathcal{H}(\lambda)$ is $F(\lambda)$ given by

$$F(\lambda) = -\beta^{-1} \ln \mathrm{Tr} \exp \left[-\beta \mathcal{H}(\lambda) \right]. \tag{5.55}$$

We next assume that the Hamiltonians $\mathcal{H}_0$ and $\mathcal{H}_1$ commute. For the noncommutative case a derivation can be found in the book by Feynman (1972, p. 67). The result is independent of this assumption. Looking for the dependence of the free energy on λ we differentiate $F(\lambda)$ and obtain for the first derivative

$$\frac{\mathrm{d}F(\lambda)}{\mathrm{d}\lambda} = \frac{\mathrm{Tr}\, \mathcal{H}_1 \exp\left[-\beta(\mathcal{H}_0 + \lambda \mathcal{H}_1)\right]}{\mathrm{Tr} \exp\left[-\beta(\mathcal{H}_0 + \lambda \mathcal{H}_1)\right]} \doteq \langle \mathcal{H}_1 \rangle \tag{5.56}$$

and for the second derivative

$$\frac{d^2 F(\lambda)}{d\lambda^2} = -\beta \left[\langle \mathcal{H}_1^2 \rangle - \langle \mathcal{H}_1 \rangle^2 \right] = -\beta \langle \mathcal{H}_1 - \langle \mathcal{H}_1 \rangle \rangle^2, \tag{5.57}$$

where the averages are taken with respect to the canonical weighting factor $\exp[-\beta \mathcal{H}(\lambda)]$. An immediate consequence of Eqn (5.57) is that

$$\frac{d^2 F(\lambda)}{d\lambda^2} \leq 0 \tag{5.58}$$

for all λ. Thus a plot of $F(\lambda)$ as a function of λ is everywhere concave and it follows that $F(\lambda)$ lies below the straight line tangent to $F(\lambda)$ at $\lambda = 0$, whence

$$F(\lambda) \leq F(0) + \lambda \left(\frac{dF(\lambda)}{d\lambda} \right)_{\lambda=0} \tag{5.59}$$

and taking $\lambda = 1$, we finally get

$$F \leq F_0 + \langle \mathcal{H} - \mathcal{H}_0 \rangle_0. \tag{5.60}$$

The quantity $\langle \mathcal{H} - \mathcal{H}_0 \rangle_0 = \langle \mathcal{H}_1 \rangle_0$ is defined as in Eqn (5.56), but with $\lambda = 0$. It is the average value of $\mathcal{H}_1$ in the soluble model system. Equation (5.60) is the Bogoliubov–Peierls inequality which can be used as a variational principle if the right-hand side is made as small as possible by means of variational parameters that may be used in the definition of the soluble model Hamiltonian $\mathcal{H}_0$.

5.3.1 A useful example

To illustrate the use of the variational principle, we choose for $\mathcal{H}_0$ the simple, classical trial Hamiltonian

$$\mathcal{H}_0 = -\sum_j \mathbf{h}_j \cdot \hat{\mathbf{e}}_j. \tag{5.61}$$

Here $\hat{\mathbf{e}}_j$ is a unit vector that points in the direction of the magnetic moment (of assumed unit length) at site j. The field $\mathbf{h}_j$ is a variational parameter to be determined by minimizing the free energy F_1 which we define to be given by the right-hand side of Eqn (5.60), i.e. we minimize

$$F_1 = F_0 + \langle \mathcal{H} - \mathcal{H}_0 \rangle_0. \tag{5.62}$$

Next we determine F_1. For this purpose we evaluate the classical partition function associated with $\mathcal{H}_0$ in a local frame of reference where the z-axis is oriented along $\mathbf{h}_j$, obtaining

$$Z_0 = \prod_j \int_0^{2\pi} \mathrm{d}\varphi_j \int_0^\pi \sin\theta_j \, \mathrm{d}\theta_j \exp(\beta h_j \cos\theta_j) = \prod_j \frac{4\pi}{\beta h_j} \sinh \beta h_j, \tag{5.63}$$

from which the free energy F_0 is seen to be

$$F_0 = -\beta^{-1} \sum_j \ln\left(\frac{4\pi}{\beta h_j} \sinh \beta h_j\right) \tag{5.64}$$

and the average $\langle \mathcal{H}_0 \rangle_0$ is calculated as

$$\langle \mathcal{H}_0 \rangle_0 = -\frac{\partial}{\partial \beta} \ln Z_0 = -\sum_j h_j m_j, \tag{5.65}$$

where m_j is found to be given by

$$m_j = L(\beta h_j), \tag{5.66}$$

that is, the average magnetic moment m_j is equal to $L(x)$, the classical *Langevin* function which is defined as

$$L(x) = \coth x - 1/x. \tag{5.67}$$

The free energy, F_1, given by

$$F_1 = -\beta^{-1} \sum_j \ln\left(\frac{4\pi}{\beta h_j} \sinh \beta h_j\right) + \sum_j h_j m_j + \langle \mathcal{H} \rangle_0 \tag{5.68}$$

must now be minimized with respect to h_j. Remembering that the last term on the right-hand side depends on the field h_j through the average $\langle \ \rangle_0$, we evaluate $\partial F_1/\partial h_i = 0$ and obtain

$$h_i = -\frac{\partial \langle \mathcal{H} \rangle_0}{\partial m_i}, \tag{5.69}$$

provided $\partial m_i/\partial h_i \neq 0$. The field thus obtained is called the *molecular* or *mean* field and to obtain the desired temperature dependence of the magnetization we must solve Eqn (5.66) and (5.69) self-consistently for which—obviously—the Hamiltonian $\mathcal{H}$ needs to be known. But the Hamiltonian $\mathcal{H}$ is an immensely complicated functional of the magnetization and is not known, even in the adiabatic approximation; thus further approximations are needed and it is this very point where numerous theories begin, to some of which we will return briefly later on.

To complete our illustration we choose for the Hamiltonian $\mathcal{H}$ the venerable, classical Heisenberg model . We do not claim that this really describes the physics of itinerant-electron magnets, but, surprisingly, we obtain rather decent estimates of the Curie temperatures for Fe, Co, and Ni, provided we calculate the exchange constants by means of the same total energy differences that determine the magnon spectra. Let us therefore assume that $\mathcal{H}$ is given by

$$\mathcal{H} = {\sum_{i,j}}' M^2 J_{ij} \hat{\mathbf{e}}_i \cdot \hat{\mathbf{e}}_j, \tag{5.70}$$

where the prime denotes omission of the term $i = j$ and assume the magnitude of the magnetic moment to be a constant of value M. The average $\langle \mathcal{H} \rangle_0$ is then easily calculated,

$$\langle \mathcal{H} \rangle_0 = {\sum_{i,j}}' M^2 J_{ij} m_i m_j \tag{5.71}$$

from which the derivative in Eqn (5.69) follows as

$$h_j = -2M^2 {\sum_{i}}' J_{ji} m_i = -2M^2 \left(\sum_i J_{ji} m_i - J_{jj} m_j \right). \tag{5.72}$$

To estimate the J_{ij} we examine the total energy of a spin spiral of the form in Eqn (5.25), simplifying this expression for the case of magnetic moments on a Bravais lattice appropriate for Fe, fcc Co, and Ni. We then express the total energy per atom as

$$E(\mathbf{q},\theta) = \frac{1}{N} \sum_{i,j} J(\mathbf{R}_i - \mathbf{R}_j) \mathbf{M}(\mathbf{R}_i) \cdot \mathbf{M}(\mathbf{R}_j), \tag{5.73}$$

in this way defining an exchange function $J(\mathbf{R}_i - \mathbf{R}_j)$ which will be quite long range. Substituting spin spirals for $\mathbf{M}(\mathbf{R}_i)$ and $\mathbf{M}(\mathbf{R}_j)$ and forming the total energy difference $\Delta E(\mathbf{q},\theta) \doteq E(\mathbf{q},\theta) - E(0,\theta)$, we obtain using some trigonometric identities

$$\Delta E(\mathbf{q},\theta) = \sum_i M^2 J(\mathbf{R}_i) \sin^2 \theta \left[\cos(\mathbf{q} \cdot \mathbf{R}_i) - 1 \right]. \tag{5.74}$$

We are thus led to define the cosine lattice transform

$$j(\mathbf{q}) = \sum_i J(\mathbf{R}_i) \cos(\mathbf{q} \cdot \mathbf{R}_i), \tag{5.75}$$

whence

$$\Delta E(\mathbf{q},\theta) = M^2 \sin^2 \theta \left[j(\mathbf{q}) - j(0) \right]. \tag{5.76}$$

From Eqn (5.75) we obtain the exchange function $J(\mathbf{R}_i)$ as

$$J(\mathbf{R}_i) = \sum_{\mathbf{q}} j(\mathbf{q}) \exp(i\mathbf{q} \cdot \mathbf{R}_i) \tag{5.77}$$

for the simple Bravais lattices to which we want to apply the formalism. Returning finally to Eqn (5.72), equating J_{ij} with $J(\mathbf{R}_i - \mathbf{R}_j)$ and assuming the mean field and the moments to be uniform, we obtain

$$h = 2M^2 m \sum_{\mathbf{q}} j(\mathbf{q}). \tag{5.78}$$

We now assume a small external magnetic field, h^{ext}. Then the magnetization is from Eqn (5.66)

$$m = L(\beta h + \beta h^{\text{ext}}). \tag{5.79}$$

For values where m, h, and h^{ext} are small, that is at and above the Curie temperature, we may expand the Langevin function as

$$\lim_{x \to 0} L(x) = \frac{x}{3}, \tag{5.80}$$

thus

$$m = \frac{1}{3} \beta \left(h + h^{\text{ext}} \right). \tag{5.81}$$

The susceptibility, $\chi(T)$, at and above the Curie temperature is now easily calculated from $\chi = \partial m / \partial h^{\text{ext}}$ which, using Eqn (5.78) and Eqn (5.81), gives

$$\chi(T) = \frac{1}{3k_{\text{B}}T - 2M^2 \sum_{\mathbf{q}} j(\mathbf{q})}. \tag{5.82}$$

Hence the Curie temperature, T_c, is

$$k_{\text{B}} T_c = \frac{2}{3} \sum_{\mathbf{q}} M^2 j(\mathbf{q}). \tag{5.83}$$

Just like the spin-wave spectra, the total energy differences can easily be computed, either using the force theorem where no magnetic moment change is possible, or by a self-consistent calculation, in which case the magnetic moment must be constrained to its equilibrium value. The sum on $\mathbf{q}$ is then taken over the first Brillouin zone and is best obtained using special k-points and discrete sampling. We should actually call the temperature obtained from Eqn (5.83) the Langevin temperature which we therefore denote by T_{Lang} and collect in Table 5.3 where, for ease in comparison, we also repeat the experimental Curie temperatures from Table 4.1. Obviously, this is a quick and easy way to estimate Curie temperatures.

We would like to comment that, although T_{Lang} and T_c are in fairly good agreement, at least concerning the trend, the magnetization profile does not agree with the experimental

Table 5.3 *Integrated total energy differences for Fe, Co, and Ni and the "Langevin" temperature from Eqn (5.83), T_{Lang}, which is to be compared with the experimental Curie temperatures, T_c (from Table 4.1).*

	bcc Fe	fcc Co	fcc Ni
$\sum_{\mathbf{q}} M^2 j(\mathbf{q})$ [mRy]	12.5	14.8	6.1
T_{Lang} [K]	1316	1558	642
T_c [K]	1044	1388	627

data. Referring to Fig. 4.1 we point out that the Langevin function will place the magnetization as a function of temperature below the 3/2 Brillouin function, since the former is the limiting function for the magnitude of the magnetic moment tending to infinity.

5.4 Spin fluctuations

5.4.1 A simple formulation

In this subsection we shall develop another mean-field theory that should describe itinerant electrons more realistically. We initially focus our attention on elemental systems like Fe, Co, and Ni and therefore begin by looking at the total energy gain shown in Fig. 4.23. There we stated in Eqn (4.98) that the total energy change, ΔE, can be approximately written as

$$\Delta E = \alpha_2 M^2 + \alpha_4 M^4 + \cdots, \qquad (5.84)$$

when we allow an initially nonmagnetic system to become ferromagnetic. Although we will find that this expansion is overly naive and that the determination of the coefficients α_2, and α_4 is much more involved than appears at this point, we still use this approach for didactic reasons and can later on, if necessary, redefine the coefficients. If we terminate the expansion with the fourth-order term as indicated, then the quantities α_2 and α_4 can be read off from Eqn (4.98) to be

$$\alpha_2 = -\frac{2 E_s}{M_s^2} \qquad (5.85)$$

and

$$\alpha_4 = \frac{E_s}{M_s^4}, \qquad (5.86)$$

where E_s is the saturation energy, i.e. the value of the total energy difference, ΔE, at the minimum, and M_s is the saturation magnetization at $T = 0$. These values change, of course, when the expansion of the calculated total energy as a function of the magnetization is carried to higher orders.

We now postulate that the total energy expansion, Eqn (5.84), remains valid when the magnetization M is replaced by the corresponding vector quantity

$$\mathbf{M}(\mathbf{R}) = M\,\mathbf{e}_z + \mathbf{m}(\mathbf{R}) = M\,\mathbf{e}_z + \sum_{j,\mathbf{k}} m_{j\mathbf{k}} \exp(i\mathbf{k}\cdot\mathbf{R})\,\mathbf{e}_j. \tag{5.87}$$

Here we imply that M is the macroscopic magnetization along some direction, say the z-direction, and $\mathbf{m}$ is a local deviation of the magnetization which is expanded in a Fourier series. Since $\mathbf{m}(\mathbf{R})$ is real we require the Fourier coefficients to obey $m_{j-\mathbf{k}} = m_{j\mathbf{k}}^*$. Physically they describe spin fluctuations. The quantities $\mathbf{e}_j$ ($j = 1, 2, 3$) are Cartesian unit vectors.

The Hamiltonian for the soluble model, $\mathcal{H}_0$, is now chosen as

$$\mathcal{H}_0 = \sum_{j,\mathbf{k}} a_{j\mathbf{k}} |m_{j\mathbf{k}}|^2, \tag{5.88}$$

where $a_{j\mathbf{k}}$ are variational parameters. This choice of $\mathcal{H}_0$ leads to Gaussian statistics which is known to allow all thermal averages and the partition function to be carried out analytically, as we will see.

We should next find a suitable approximation for the real Hamiltonian, $\mathcal{H}$. To this end we require that the total energy difference, Eqn (5.84), can be extracted from $\mathcal{H}$, which we write as the sum of two terms,

$$\mathcal{H} = \mathcal{H}_1 + \mathcal{H}_2. \tag{5.89}$$

The first term on the right-hand side is assumed to consist of single-site terms and, using Eqn (5.84), we write

$$\mathcal{H}_1 = \frac{1}{N} \sum_i \sum_{n=1}^{n_m} \alpha_{2n}\,\mathbf{M}(\mathbf{R}_i)^{2n}, \tag{5.90}$$

where $n_m = 2$, at least, and N is the number of atoms in the crystal. As before we must also include two-site interaction terms, as was done in Eqn (5.70). Generalizing this equation such that in addition to directional changes the magnitude of the magnetization is allowed to change as well, we write instead

$$\mathcal{H}_2 = \frac{1}{N} \sum_{il} J(\mathbf{R}_i - \mathbf{R}_l)\,\mathbf{M}(\mathbf{R}_i)\cdot\mathbf{M}(\mathbf{R}_l). \tag{5.91}$$

In keeping with the usual terminology we may call the $J(\mathbf{R} - \mathbf{R}')$ exchange constants and just as the coefficients α_{2n} of Eqn (5.90) we obtain the functions $J(\mathbf{R} - \mathbf{R}')$ from constrained total energy calculations using the same steps as in going from Eqn (5.73) to Eqn (5.74), i.e. we substitute a spin spiral characterized by the spiral wave vector $\mathbf{k}$ of the type given in Eqn (5.25) (suitably simplifying to describe an elementary Bravais lattice) and obtain the total energy corresponding to Eqn (5.91):

$$E_2(M, \mathbf{q}, \theta) = \sum_i J(\mathbf{R}_i) \, M^2 [\sin^2 \theta \cos(\mathbf{q} \cdot \mathbf{R}_i) + \cos^2 \theta]. \tag{5.92}$$

As before we define the transformed exchange constants as

$$j(\mathbf{q}) = \sum_i J(\mathbf{R}_i) \cos(\mathbf{q} \cdot \mathbf{R}_i), \tag{5.93}$$

thus obtaining

$$E_2(M, \mathbf{q}, \theta) = M^2 \left[j(\mathbf{q}) \sin^2 \theta + j(0) \cos^2 \theta \right]. \tag{5.94}$$

The total energy difference corresponding to both $\mathcal{H}_1$ and $\mathcal{H}_2$ is therefore

$$\Delta E(M, \mathbf{q}, \theta) = \sum_{n=1}^{n_m} \alpha_{2n} \, M^{2n} + j(\mathbf{q}) \, M^2 \sin^2 \theta, \tag{5.95}$$

the terms $j(0)$ being effectively zero since they are already contained in the first term of the right-hand side. In practice it is sufficient to use $\theta = 90°$ because the total energy is to a good approximation proportional to $\sin^2 \theta$ so that this factor cancels out. The total energy differences are then calculated for a selected set of the spiral wave vectors $\mathbf{q}$, carrying the expansions up to order $n_m = 2$. To convey an idea about the basic numerical input to what follows, we show some total energy differences for $\mathbf{q} = 0$, $\mathbf{q} = \frac{2\pi}{a}(0.125, 0.125, 0.125)$, $\mathbf{q} = \frac{2\pi}{a}(0.25, 0.25, 0.25)$, and $\mathbf{q} = \frac{2\pi}{a}(0, 0, 1)$ as a function of the magnetization for Ni in Fig. 5.7. It should be noted that the equilibrium magnetization, that is the value of M at the minimum, goes to zero long before $\mathbf{q}$ reaches the Brillouin zone boundary. Also the similarity with a typical Landau free energy plot should be noted (Callen, 1985).

We are now in the position to deal with the thermodynamics at hand and return to the Bogoliubov variational free energy defined in Eqn (5.62). We first need to evaluate the classical partition function, Z_0, associated with $\mathcal{H}_0$:

$$Z_0 = \prod_{j\mathbf{k}}' \int \mathrm{d}x_{j\mathbf{k}} \int \mathrm{d}y_{j\mathbf{k}} \, \exp(-\beta \mathcal{H}_0), \tag{5.96}$$

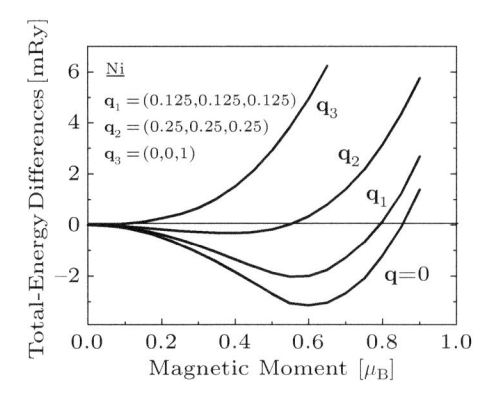

Figure 5.7 *Total energy differences for Ni as functions of the magnetization, M, for several values of the spiral vector* **q** *(given in units of $2\pi/a$). Polar angle $\theta = 90°$.*

where the prime restricts the product to consist of **k**-vector pairs (**k**, −**k**) and the independent variables $x_{j\mathbf{k}}$, and $y_{j\mathbf{k}}$ are defined by $m_{j\mathbf{k}} = m^*_{j-\mathbf{k}} = x_{j\mathbf{k}} + i y_{j\mathbf{k}}$, $j = 1$, 2, 3 numbering Cartesian coordinates.

It is the Gauss integral

$$\int_0^\infty x^{2n} \exp(-ax^2)\, dx = \frac{(2n-1)!!}{2^{n+1} a^n} \sqrt{\frac{\pi}{a}} \tag{5.97}$$

that leads at once to the simple result for the partition function

$$Z_0 = \prod_{j\mathbf{k}} \sqrt{\frac{\pi}{2\beta\, a_{j\mathbf{k}}}} \tag{5.98}$$

and hence to the zero-order free energy

$$F_0 = -\frac{1}{2\beta} \sum_{j\mathbf{k}} \ln \frac{\pi}{2\beta\, a_{j\mathbf{k}}}. \tag{5.99}$$

Furthermore, to evaluate the thermal average of the real Hamiltonian, averages of the type

$$\langle |m_{j\mathbf{k}}|^2 \rangle_0 = \frac{1}{2\beta\, a_{j\mathbf{k}}}, \tag{5.100}$$

and

$$\langle |m_{j\mathbf{k}}|^4 \rangle_0 = 3 \langle |m_{j\mathbf{k}}|^2 \rangle_0^2 \tag{5.101}$$

are needed which are easily verified using a standard formula for the averages and Eqns (5.88), (5.97), and (5.67). The notation $\langle \cdots \rangle_0$ indicates averages on the basis of the model Hamiltonian $\mathcal{H}_0$. With Eqn (5.100) we obtain at once

$$\langle \mathcal{H}_0 \rangle_0 = \frac{1}{2\beta} \sum_{j\mathbf{k}} 1. \tag{5.102}$$

The sums that now occur are restricted to the first Brillouin zone throughout, since the magnetic moments reside on a lattice as a consequence of which the total energy difference, Eqn (5.92), is periodic as a function of $\mathbf{q}$ with the periodicity volume being the first Brillouin zone. The sum in Eqn (5.102) is therefore the total number of modes.

In order to write out $\langle \mathcal{H} \rangle_0$ as concisely as possible it is worthwhile to define abbreviations of frequently occurring quantities. There is first the sum $\sum_\mathbf{k} \langle |m_{j\mathbf{k}}|^2 \rangle_0$ of the Cartesian component j which can be in the direction of the macroscopic magnetization or perpendicular to it. The former we call the longitudinal (l), the latter the transverse (t) fluctuations, i.e. we define

$$t^2 = \sum_\mathbf{k} \langle |m_{t\mathbf{k}}|^2 \rangle_0 \tag{5.103}$$

and

$$l^2 = \sum_\mathbf{k} \langle |m_{l\mathbf{k}}|^2 \rangle_0. \tag{5.104}$$

Next we define the quantities

$$\xi_T = \sum_\mathbf{k} j(\mathbf{k}) \langle |m_{t\mathbf{k}}|^2 \rangle_0 \quad \text{and} \quad \xi_L = \sum_\mathbf{k} j(\mathbf{k}) \langle |m_{l\mathbf{k}}|^2 \rangle_0 \tag{5.105}$$

that connect the exchange constants with the transverse and longitudinal fluctuations. Then a lengthy but straightforward calculation leads to

$$\langle \mathcal{H}_1 \rangle_0 = \sum_{n=1}^{n_m} \sum_{\substack{\mu\tau\lambda \\ \mu+\tau+\lambda=n}} \alpha_{2n} f_{\mu\tau\lambda} M^{2\mu} t^{2\tau} l^{2\lambda}, \tag{5.106}$$

and

$$\langle \mathcal{H}_2 \rangle_0 = \sum_{\substack{\mu\tau\lambda \\ \mu+\tau+\lambda=0}} f_{\mu\tau\lambda} M^{2\mu} t^{2\tau} l^{2\lambda} \left[(2\tau+2)\xi_T + (2\mu+2\lambda+1)\xi_L \right]. \tag{5.107}$$

5.4 Spin fluctuations

The coefficients $f_{\mu\tau\lambda}$ can be calculated with

$$f_{\mu\tau\lambda} = \frac{(\mu+\tau+\lambda)!}{\mu!\,\lambda!}\frac{2^\tau(2\mu+2\lambda-1)!!}{(2\mu-1)!!}. \tag{5.108}$$

The Bogoliubov variational free energy, F_1, can at this stage be expressed in terms of $\langle|m_{j\mathbf{k}}|^2\rangle$ which replaces the variational parameter $a_{j\mathbf{k}}$; this is so because of Eqn (5.100). We therefore collect

$$F_1 = -\frac{k_\mathrm{B}T}{2}\sum_{j\mathbf{k}}\left[1+\ln\left(\pi\left\langle|m_{j\mathbf{k}}|^2\right\rangle_0\right)\right] + \langle\mathcal{H}_1\rangle_0 + \langle\mathcal{H}_2\rangle_0. \tag{5.109}$$

Next there are the variational equations that need be determined. One of those is obtained from $\partial F_1/\partial M = 0$, the others arise from

$$\frac{\partial F_1}{\partial\langle|m_{j\mathbf{q}}|^2\rangle_0} = 0 \tag{5.110}$$

for $j=t$ and $j=l$. In the actual calculations we remember that, because of Eqns (5.103)–(5.105),

$$\frac{\partial l^2}{\partial\langle|m_{j\mathbf{q}}|^2\rangle_0} = \delta_{jl}\,,\quad \frac{\partial t^2}{\partial\langle|m_{j\mathbf{q}}|^2\rangle_0} = \delta_{jt},\quad\text{and}\quad \frac{\partial\xi_{nj}}{\partial\langle|m_{j\mathbf{q}}|^2\rangle_0} = j_n(\mathbf{q}), \tag{5.111}$$

where the δ_{il} are the Kronecker δ-functions.

To gain more insight into the theory and practice applications, it is of advantage to write out in detail the formulas that need to be evaluated numerically. We begin with the evaluation of $\partial F_1/\partial M = 0$ which gives

$$\frac{\partial F_1}{\partial M} = 2M\alpha_2 + 2M\left(2M^2+4t^2+6l^2\right)\alpha_4 = 0. \tag{5.112}$$

If $M\neq 0$, i.e. in the ordered state, we thus obtain

$$\frac{M^2}{M_s^2} = 1 - \frac{2t^2+3l^2}{M_s^2}, \tag{5.113}$$

where we have used the relations (5.85) and (5.86) to express the coefficients α_2 and α_4 in terms of the saturation magnetization, M_s, and the magnetic energy gain, E_s.

In the paramagnetic state, where $M=0$, the fluctuations l^2 and t^2 are degenerate. Thus we define the paramagnetic fluctuation, p^2, writing in this case $l^2 = t^2 = p^2$, and obtain from Eqn (5.113)

$$\frac{p^2}{M_s^2} = \frac{1}{5}. \tag{5.114}$$

We see that the relative paramagnetic fluctuation has the value 1/5 at the Curie point. This result was apparently first obtained by Moriya (1985).

The inverse of the uniform susceptibility is calculated from the free energy and is

$$\chi_0^{-1} = \frac{\partial^2 F_1}{\partial M^2} = 2\alpha_2 + 4\alpha_4(3M^2 + 2t^2 + 3l^2). \tag{5.115}$$

In the ferromagnetic state we eliminate the fluctuations with Eqn (5.113) and obtain

$$\chi_0^{-1} = 8\alpha_4 M^2 = 8\frac{E_s}{M_s^2}\frac{M^2}{M_s^2}, \tag{5.116}$$

so that at $T = 0$

$$\chi_{00}^{-1} = 8\frac{E_s}{M_s^2}. \tag{5.117}$$

In the paramagnetic state we obtain from Eqn (5.115)

$$\chi_0^{-1} = 2\alpha_2 + 20 p^2 \alpha_4. \tag{5.118}$$

We will exhibit its temperature dependence later, together with the low-temperature results.

We now turn to evaluating the optimization condition embodied in Eqn (5.110) and begin with the derivative of F_1 with respect to the longitudinal modes of the fluctuations $\langle |m_{l\,\mathbf{k}}|^2 \rangle$ which allows us—after some algebra—to express the longitudinal fluctuations, $l^2 = \sum_{\mathbf{k}} \langle |m_{l\,\mathbf{k}}|^2 \rangle$, as

$$l^2 = k_\text{B} T \sum_{\mathbf{k}} \chi_L(\mathbf{k}), \tag{5.119}$$

where we defined the nonuniform and temperature-dependent susceptibility through the relation

$$\chi_L^{-1}(\mathbf{k}) = 8\alpha_4 M^2 + 2j(\mathbf{k}). \tag{5.120}$$

Note that the usage of the term susceptibility is not justified at this stage but will be so later on. However, we see that the first term on the right-hand side is the inverse uniform susceptibility given in Eqn (5.116). The remaining terms will be discussed later.

The derivative of the free energy with respect to one of the transverse modes, say $\langle |m_{x\,\mathbf{k}}|^2 \rangle$, allows us to express the transverse fluctuations, $t^2 = \sum_{\mathbf{k}} \langle |m_{t\,\mathbf{k}}|^2 \rangle$, in the form

$$t^2 = k_\text{B} T \sum_{\mathbf{k}} \chi_T(\mathbf{k}), \tag{5.121}$$

where

$$\chi_T^{-1}(\mathbf{k}) = 8\,\alpha_4(t^2 - l^2) + 2j(\mathbf{k}). \tag{5.122}$$

In the derivation of both Eqn (5.120) and Eqn (5.122) we used the relation for the magnetization M, Eqn (5.112), to simplify the results. Obviously, the three relations (5.113), (5.119), and (5.121) constitute a set of self-consistency equations that must be solved simultaneously for the magnetization M and the fluctuations l^2 and t^2. This can only be done numerically.

In the paramagnetic case we set $M = 0$ in the free energy and differentiate with respect to the paramagnetic modes which we denote by $\langle |m_{p\mathbf{k}}|^2 \rangle$ obtaining for the paramagnetic fluctuations, $p^2 = \sum_{\mathbf{k}} \langle |m_{p\mathbf{k}}|^2 \rangle$, the relation

$$p^2 = k_B T \sum_{\mathbf{k}} \chi_P(\mathbf{k}), \tag{5.123}$$

where

$$\chi_P^{-1}(\mathbf{k}) = 2\,\alpha_2 + 20\,\alpha_4\,p^2 + 2j(\mathbf{k}). \tag{5.124}$$

Note that with the result for the inverse of the uniform susceptibility, χ_0^{-1}, Eqn (5.118), we can express the last equation as

$$\chi_P^{-1}(\mathbf{k}) = \chi_0^{-1} + 2j(\mathbf{k}), \tag{5.125}$$

thus indicating again a connection with the nonuniform susceptibility that will be explored later on in Sec. 5.4.5. The relations for the paramagnetic case constitute a separate set of self-consistency equations that must be treated numerically.

We finally write down a simple formula for the Curie temperature which is easily obtained by using Eqns (5.114) and (5.124) and is given by

$$k_B T_c = \frac{2M_s^2}{5} \left[\sum_{\mathbf{k}} \frac{1}{j(\mathbf{k})} \right]^{-1}. \tag{5.126}$$

We will see soon that this approximation for the Curie temperature is valid only when the phase transition is of second order.

A note concerning the notation is in order here. The sums like $\sum_{\mathbf{k}}$ are understood to be divided by the number of particles in the system. In practical calculations they are executed by weighted values of $\mathbf{k}$ such that $\sum_{\mathbf{k}} 1 = 1$. This simplifies the notation considerably.

Before turning to exploratory applications we essentially repeat the foregoing derivation for the case of *non-primitive lattices* like hcp Cobalt where we have to consider two atoms per unit cell, each possessing individual degrees of freedom. It is a consequence of this that the spin-wave spectrum of hcp Co consists of two branches which, in analogy with lattice vibrations, we called acoustic and optic, see Sec. 5.2.

The modifications of our theory set in as far back as Eqn (5.84) together with Eqns (5.85) and (5.86). This is so because the individual degrees of freedom of the magnetic atoms require a more general formulation of the fluctuations, which we now express as follows

$$\mathbf{M}_\tau(\mathbf{R}) = M_\tau\,\mathbf{e}_z + \sum_{n\,j\,\mathbf{k}} m_{n\,j\,\mathbf{k}}\, C_{n\,\tau} \exp(i\,\mathbf{k}\cdot\mathbf{R})\,\mathbf{e}_j. \quad (5.127)$$

Here $\mathbf{R}$ is a vector of the Bravais lattice and τ denotes a basis vector. Unfortunately, as in the theory of lattice vibrations, we cannot avoid the excessive accumulation of indices by which we denote, besides the basis, the normal mode, n, the Cartesian vector component, j, and the wave vector, $\mathbf{k}$. The coefficients $C_{n\,\tau}$ will be obtained later on as eigenvectors from a secular equation, thus defining the normal modes.

The new degree of freedom embodied in the basis-vector index τ requires that in an initial attempt we generalize Eqn (5.84) to read

$$\Delta E = \sum_\tau \left(A_\tau\, M_\tau^2 + B_\tau\, M_\tau^4 \right), \quad (5.128)$$

where the coefficients A_τ and B_τ replace α_2 and α_4; they are in principle obtained by total energy calculations. However, as was said in connection with the coefficients α_2 and α_4 in Sec. 5.4.1, their determination is nontrivial. In fact, they are relatively easy to get only for a case with two equivalent magnetic atoms like hcp Co. Furthermore, the simple formulas like Eqns (5.85) and (5.86) are in general no longer useful.

The model Hamiltonian, $\mathcal{H}_0$, is chosen as before but includes an additional sum over the normal modes

$$\mathcal{H}_0 = \sum_{n\,j\,\mathbf{k}} a_{n\,j\,\mathbf{k}} |m_{n\,j\,\mathbf{k}}|^2. \quad (5.129)$$

The real Hamiltonian which we wrote as a sum of two terms, $\mathcal{H} = \mathcal{H}_1 + \mathcal{H}_2$, is approximated as before by single-site terms that are, using Eqn (5.128), written as

$$\mathcal{H}_1 = \frac{1}{N} \sum_{\mathbf{R}\,\tau} \left[A_\tau\, \mathbf{M}_\tau(\mathbf{R})^2 + B_\tau\, \mathbf{M}_\tau(\mathbf{R})^4 \right], \quad (5.130)$$

and two-site interaction terms

$$\mathcal{H}_2 = \frac{1}{N} \sum_{\mathbf{R}\,\mathbf{R}'} \sum_{\tau\,\tau'} J_{\tau\,\tau'}(\mathbf{R}-\mathbf{R}')\, \mathbf{M}_\tau(\mathbf{R}) \cdot \mathbf{M}_{\tau'}(\mathbf{R}'). \quad (5.131)$$

As before the exchange constants $J_{\tau\tau'}(\mathbf{R}-\mathbf{R}')$ are obtained from the total energies of noncollinear spiral moment arrangements which are described by

$$\mathbf{M}_\tau(\mathbf{R}) = M_\tau[\sin\theta_\tau \cos(\mathbf{k}\cdot\mathbf{R}+\varphi_\tau)\mathbf{e}_x \\ + \sin\theta_\tau \sin(\mathbf{k}\cdot\mathbf{R}+\varphi_\tau)\mathbf{e}_y + \cos\theta_\tau \mathbf{e}_z]. \tag{5.132}$$

Substituting this expression into Eqn (5.131) we obtain for the total energy corresponding to $\mathcal{H}_2$:

$$E_2(\mathbf{k},\{M_\tau,\theta_\tau,\varphi_\tau\}) = \sum_{\tau\tau'} M_\tau M_{\tau'} [j_{\tau\tau'}(\mathbf{k})\sin\theta_\tau \sin\theta_{\tau'} \cos(\varphi_\tau - \varphi_{\tau'}) \\ + j_{\tau\tau'}(0)\cos\theta_\tau \cos\theta_{\tau'}], \tag{5.133}$$

which should be compared with Eqn (5.94), and $j_{\tau\tau'}(\mathbf{k})$ is defined by

$$j_{\tau\tau'}(\mathbf{k}) = \sum_{\mathbf{R}} J_{\tau\tau'}(\mathbf{R})\exp(i\mathbf{k}\cdot\mathbf{R}). \tag{5.134}$$

Note the angle φ_τ consists of $\mathbf{k}\cdot\tau$ plus some arbitrary phase. We now see that Eqn (5.133) for $\mathbf{k}=0$ and $\theta_\tau=0$ (all τ) gives another collinear contribution which we add to Eqn (5.128) to obtain finally the generalization of Eqn (5.84) that we must employ:

$$\Delta E = \sum_\tau \left(A_\tau M_\tau^2 + B_\tau M_\tau^4\right) + \sum_{\tau\tau'} j_{\tau\tau'}(0) M_\tau M_{\tau'}. \tag{5.135}$$

This means in general we constrain the magnetic moments separately and extract besides A_τ and B_τ also the coefficients $j_{\tau\tau'}(0)$ by, for instance, a least-squares fit of the total energy surface defined by $\Delta E(\{M_\tau\})$. The exchange constants $j_{\tau\tau'}(\mathbf{k})$ then follow from total energy differences of noncollinear configurations which, from Eqn (5.133), are of the form

$$E_2(\mathbf{k},\{M_\tau,\theta_\tau,\varphi_\tau\}) - E_2(0,\{M_\tau,\theta_\tau,\varphi_\tau\}) \\ = \sum_{\tau\tau'} M_\tau M_{\tau'}[j_{\tau\tau'}(\mathbf{k}) - j_{\tau\tau'}(0)]\sin\theta_\tau \sin\theta_{\tau'}\cos(\varphi_\tau - \varphi_{\tau'}), \tag{5.136}$$

with possible simplifications depending on properties of the case under investigation (see below) suffices. To avoid confusion with the finite-temperature treatment to follow, one should add an extra index to distinguish the zero-temperature results here, for instance $M_\tau \to M_{\tau\,T=0}$, but we hope not doing so will cause no major difficulties.

Assuming now that all coefficients have been determined we return to the Bogoliubov variational free energy. The partition function Z_0 associated with $\mathcal{H}_0$ is, generalizing Eqn (5.96) slightly by accounting for the normal modes labeled by n,

416 *Magnetism at Finite Temperatures*

$$Z_0 = \prod{}'_{njk} \int dx_{njk} \int dy_{njk} \, \exp(-\beta \mathcal{H}_0), \tag{5.137}$$

where again the prime restricts the product to consist of **k**-vector pairs (**k**, − **k**) and the independent variables x_{njk} and y_{njk} are defined by $m_{njk} = m^*_{nj-\mathbf{k}} = x_{njk} + i y_{njk}$, $j = 1, 2, 3$ numbering Cartesian coordinates. With these simple changes we see that we can now follow the same steps as before. We thus write down the Bogoliubov free energy immediately:

$$F_1 = -\frac{k_B T}{2} \sum_{njk} \left[1 + \ln\left(\pi \langle |m_{njk}|^2 \rangle_0\right)\right] + \langle \mathcal{H}_1 \rangle_0 + \langle \mathcal{H}_2 \rangle_0. \tag{5.138}$$

The averages $\langle \mathcal{H}_1 \rangle_0$ and $\langle \mathcal{H}_2 \rangle_0$ require some extra work but because of the lower-order expansion used here they appear simpler than Eqn (5.106) and (5.107) and are found to be given by

$$\begin{aligned}\langle \mathcal{H}_1 \rangle_0 + \langle \mathcal{H}_2 \rangle_0 &= \sum_\tau A_\tau (M_\tau^2 + 2 t_\tau^2 + l_\tau^2) \\ &+ \sum_\tau B_\tau \left[M_\tau^4 + 2 M_\tau^2 (2 t_\tau^2 + 3 l_\tau^2) + 8 t_\tau^4 + 4 t_\tau^2 l_\tau^2 + 3 l_\tau^4\right] \\ &+ \sum_{\tau\tau'} [j_{\tau\tau'}(0) M_\tau M_{\tau'} + 2 \xi_{\tau\tau'T} + \xi_{\tau\tau'L}],\end{aligned} \tag{5.139}$$

where we again distinguish transverse (t_τ^2) and longitudinal (l_τ^2) squared spin fluctuations of the atom labeled by τ. They are defined with respect to the z-axis of the crystal, l being parallel to z and t in the xy-plane, by

$$t_\tau^2 = \sum_{n\mathbf{k}} |C_{n\tau}|^2 \langle |m_{nt\mathbf{k}}|^2 \rangle_0 \quad \text{and} \quad l_\tau^2 = \sum_{n\mathbf{k}} |C_{n\tau}|^2 \langle |m_{nl\mathbf{k}}|^2 \rangle_0. \tag{5.140}$$

Furthermore, the ξ-functions are given by

$$\xi_{\tau\tau'j} = \sum_{n\mathbf{k}} j_{\tau\tau'}(\mathbf{k}) C^*_{n\tau'} C_{n\tau} \langle |m_{nj\mathbf{k}}|^2 \rangle_0 \quad \text{for } j = T \text{ and } L. \tag{5.141}$$

The final steps consist in formulating and solving the self-consistency equations that follow from the variation. The magnetic moments, M_τ, as functions of the temperature are obtained from $\partial F_1/\partial M_\tau = 0$ for all vectors τ in the basis. The variation with respect to the coefficients a_{njk} introduced through the model Hamiltonian, Eqn (5.129), is as before replaced by $\partial F_1/\partial \langle |m_{njk}|^2 \rangle_0 = 0$ because of the identity

$$\langle |m_{njk}|^2 \rangle_0 = 1/(2\beta a_{njk}), \tag{5.142}$$

which as in the case of Eqn (5.100) is easily verified.

The partial derivative of F_1 with respect to M_τ gives the coupled equations:

$$[A_\tau + 2B_\tau(M_\tau^2 + 2t_\tau^2 + 3l_\tau^2)] M_\tau + \sum_{\tau'} j_{\tau\tau'}(0) M_{\tau'} = 0 \tag{5.143}$$

while the partial derivatives with respect to the longitudinal and transverse fluctuations, $\langle |m_{nl\mathbf{k}}|^2\rangle_0$, and $\langle |m_{nt\mathbf{k}}|^2\rangle_0$, are easily obtained from Eqns (5.138)–(5.142). The result is conveniently written in the form

$$a_{nl\mathbf{k}} = \sum_{\tau\tau'} C_{n\tau'}^* H_{\tau'\tau}^{(l)} C_{n\tau} \quad \text{and} \quad a_{nt\mathbf{k}} = \sum_{\tau\tau'} C_{n\tau'}^* H_{\tau'\tau}^{(t)} C_{n\tau} \tag{5.144}$$

where we used for the left-hand side Eqn (5.142) and defined the matrices

$$H_{\tau'\tau}^{(l)} = [A_\tau + B_\tau(6 M_\tau^2 + 4t_\tau^2 + 6l_\tau^2)] \delta_{\tau\tau'} + j_{\tau\tau'}(\mathbf{k}) \tag{5.145}$$

and

$$H_{\tau'\tau}^{(t)} = [A_\tau + B_\tau(2 M_\tau^2 + 8t_\tau^2 + 2l_\tau^2)] \delta_{\tau\tau'} + j_{\tau\tau'}(\mathbf{k}). \tag{5.146}$$

The quadratic form one recognizes in Eqn (5.144) suggests trying a solution by the Rayleigh–Ritz variational procedure. The resulting eigenvalue equations are then

$$\sum_{\tau'}[H_{\tau\tau'}^{(j)} - a_{nj\mathbf{k}}\delta_{\tau\tau'}] C_{n\tau'} = 0 \quad \text{for} \quad j = l, t, \tag{5.147}$$

where the eigenvalues $a_{nj\mathbf{k}}$ are numbered by the mode index n, which is seen to be equal to the number of basis vectors, and the eigenvectors are given by $C_{n\tau}$ which obey $\sum_\tau C_{n\tau}^* C_{n'\tau} = \delta_{nn'}$.

It is not hard to derive a formula for the Curie temperature by the same reasoning that led to Eqn (5.126). Denoting the eigenvalues of the exchange matrix $j_{\tau\tau'}(\mathbf{k})$ by $j_n(\mathbf{k})$ and the saturation (or local) moments by $\mathcal{L}_\tau$ we find

$$k_B T_c = \frac{2}{5} \sum_\tau \mathcal{L}_\tau^2 \left[\sum_{\mathbf{k}n} \frac{1}{j_n(\mathbf{k})}\right]^{-1}. \tag{5.148}$$

Here we assumed that the relation given by Eqn (5.128) possesses local minima when the magnetization assumes the values $\mathcal{L}_\tau$. It should be noticed that the sum over the inverse exchange functions is summed over the mode indices n and the sum over the local moments over the basis vector index τ. As before, this relation is valid if the phase transition is of second order.

The basic equations are now complete, but the actual procedure to obtain solutions is somewhat more involved than before, in fact may be quite complicated for the general case. However, we will restrict ourselves to two basis vectors, i.e. investigate the hcp lattice. The procedure is as follows: in a first step, using LDA total energy calculations, one sets up the data base for the coefficients $\{A_\tau, B_\tau\}$ and for the exchange constants $j_{\tau\tau'}(\mathbf{k})$ permitting a sufficiently large number of $\mathbf{k}$-vectors. In the second step one solves Eqn (5.147). This supplies with Eqn (5.142) the temperature-dependent fluctuation components $\langle |m_{nj\mathbf{k}}|^2 \rangle_0$, which define with Eqn (5.140) the integrated fluctuations t_τ^2 and l_τ^2. These in turn suffice to complete the self-consistency problem for obtaining the temperature-dependent magnetization M_τ by means of Eqn (5.143) and supply the necessary input for Eqn (5.145) and Eqn (5.146) to continue iteratively.

5.4.2 Exploratory results for the elementary ferromagnets

Although the self-consistency equations for the macroscopic magnetization, Eqn (5.113), and the fluctuations, Eqns (5.119)–(5.122) are fairly complex and must be solved simultaneously, they present no major problem if they are treated iteratively. To obtain the exchange constants, $j(\mathbf{k})$, the force theorem is used in the generalized gradient approximation (GGA) to density-functional theory and the angle θ is chosen to be $\theta = 90°$. A recent discussion by Ruban et al. (2004) gives support for this particular choice of θ.

5.4.2.1 Nickel

So beginning with the case of nickel we collect in Fig. 5.8 and in Table 5.4 the pertinent results; the reduced macroscopic magnetization, M, as a function of the temperature is seen to decrease while the reduced transverse, t^2, and longitudinal, l^2, fluctuations increase until solutions to the self-consistency equations cease to exist slightly short of

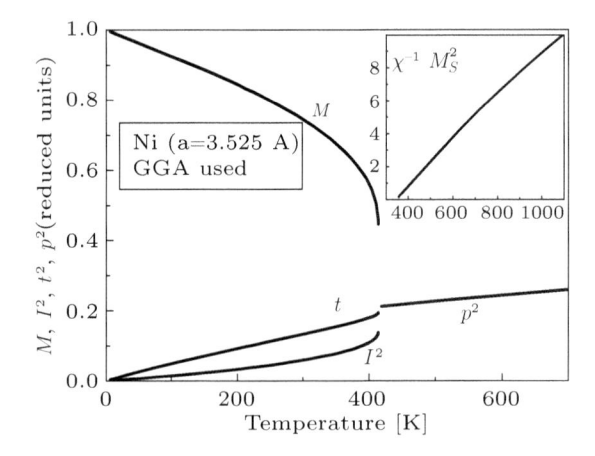

Figure 5.8 *Calculated magnetization data for fcc Ni as functions of temperature using the general gradient approximation (GGA). M is the reduced magnetization, l^2, t^2, and p^2 are the reduced fluctuations defined by Eqns (5.119), (5.121), and (5.123). Inset: inverse paramagnetic susceptibility multiplied by the saturation magnetization squared, M_s^2, in mRy.*

Table 5.4 *Calculated magnetic properties of the elementary ferromagnets, compared with experimental data from Table 4.1. T_c Curie temperature, M_s magnetic moment at $T = 0$, q_c "number of magnetic carriers," Eqns (4.18) and (4.20), E_s energy gain in magnetic state.*

		bcc Fe	fcc Co	bcc Co	hcp Co	fcc Ni
T_c [K]	calc.[a]	1180	1200	1205	1076	414
	exp.	1044	1388			627
M_s [μ_B]	calc.	2.20	1.65	1.69	1.60	0.63
	exp.	2.216		1.7[b]	1.715	0.616
q_c	calc.	1.40	1.11	1.07		0.26
	exp.	2.29	2.29			0.9
E_s [mRy]		26.0	13.9	17.1	10.8	4.4

[a] For Ni and fcc Co the GGA was used; [b] Bland *et al.* (1991).

414 K. The solution for the reduced paramagnetic fluctuations, p^2, from Eqns (5.123) and (5.124) is also shown starting at $T_c = 414$ K which we take to be the calculated Curie temperature. For earlier calculations (Uhl and Kübler, 1996) the local density-functional approximation (LSDA) was used, which resulted in a slightly lower Curie temperature of 370 K.

The general physical picture conveyed by Fig. 5.8 shows qualitatively the right features, i.e. longitudinal and transverse fluctuations of the magnetization lead to a phase transition. Furthermore, the inverse high temperature susceptibility given in the inset shows Curie–Weiss behavior. However, a number of details tell us that at this stage the theory is not really adequate.

First of all, the calculated Curie temperature is smaller than the measured value of $T_c = 627$ K (see Table 5.4). Furthermore, the calculated magnetization falls off too fast both at low temperatures and before reaching 414 K. The phase transition is indeed calculated to be of first order. Finally the Curie–Weiss law obtained has an incorrect slope. To assess the discrepancy we calculate from the slope of $\chi^{-1}M_s^2$ the number of magnetic carriers, q_c, defined by means of the Curie constant in Eqns (4.18) and (4.20) and compare it with the experimental value in Table 5.4. Indeed, this number is too small. Furthermore, if one uses Eqn (5.126) to calculate the Curie temperature one obtains $T_C = 341$ K. The reason for this discrepancy is the calculated nature of the phase transition, which is first order and not second order as assumed in the derivation of Eqn (5.126). Finally the intercept of the inverse Curie–Weiss law also indicates a calculated Curie temperature lower than 414 K.

One might ask how thermal expansion and the finite temperature of the electron system influence these results. These questions can be answered as follows: the Curie temperature of 414 K is the highest so far obtained by this method and it is calculated using the experimental lattice constant. This is justified by thermal expansion, indeed, the value of the Curie temperature at the theoretical lattice constant is still lower by about 20 K. One can also estimate the effect of the finite electron temperature by carrying out the calculations using the Fermi–Dirac distribution function appropriate for 400 K (for a justification of this step see Sec. 5.5.2): the effect is again a decrease of the Curie temperature by about 10 K (Weis, unpublished, 1998). Physically this is due to Stoner excitations. We will come back to the case of Ni later on in Sec. 5.4.6.1 with attempts to improve the theory and with a critical assessment of our present understanding.

5.4.2.2 fcc Cobalt

The calculated magnetic properties for fcc cobalt, the high-temperature form of Co, are collected in Table 5.4. As in the case of Ni we used the GGA, which results in a somewhat higher Curie temperature as compared with the LSDA, where $T_C = 950$ K was obtained. Since the reduced macroscopic magnetization as well as the transverse, longitudinal, and paramagnetic fluctuations as a function of the temperature look qualitatively very similar to Fig. 5.8 we do not graph these quantities here. Solutions to the self-consistency equations cease to exist near 1200 K. This is again smaller than the measured value of $T_c = 1388$ K. The calculated phase transition is, as in the case of Ni, of first order, which is not in agreement with experiment. The calculated high-temperature inverse susceptibility is nearly linear in the temperature, but the slope of χ^{-1} is again too high, so that the calculated number of magnetic carriers obtained is too low, see Table 5.4. Finally, if one uses Eqn (5.126) to calculate the Curie temperature one obtains $T_C = 900$ K. Thus, our theory does not capture correctly the magnetic properties of fcc Co where the challenge is an explanation of the extremely high experimental Curie temperature of 1388 K.

In spite of these obvious shortcomings the overall behavior is encouraging so that we attempt a prediction for a modification of Co, its bcc form, that can be grown by molecular beam epitaxy on GaAs substrates, as was first shown by Prinz (1985). The resulting samples of bcc Co are thin films that can be made thick enough to resemble bulk in the film center where Bland *et al.* (1991) estimated the magnetic moment of this form of Co to be approximately 1.7 μ_B. We collect the calculated magnetic properties of bulk bcc Co in Table 5.4, obtained in the LSDA using for a calculated equilibrium lattice constant $a = 2.783$ Å which is somewhat smaller than the lattice constant, $a = 2.827$ Å, estimated by Prinz (1985). The calculated Curie temperature is $T_c = 1205$ K, but the phase transition is again of first order. At the larger experimental atomic volume this estimate of the Curie temperature increases to $T_c = 1270$ K. Although we need not accept the predicted nature of the phase transition, we can say that the predicted Curie temperature will be at least of the order of magnitude of the experimental Curie temperature of fcc Co, perhaps even higher.

5.4.2.3 Iron

For bcc iron we use the LSDA and collect the calculated magnetization data in Table 5.4. Although the phase transition is again incorrectly found to be of first order, the calculated Curie temperature is of the right order of magnitude. Estimating the effect of the finite electron temperature by carrying out the calculations with the Fermi–Dirac distribution function appropriate for 1000 K, we find a further decrease of the calculated Curie temperature by about 150 K (Weis, unpublished, 1998). The calculated inverse paramagnetic susceptibility, even though linear, suffers from the same shortcomings as those calculated for Ni and Co, and leads to an underestimate of the number of magnetic carriers, see Table 5.4. We will return to the case of Fe with an alternative approach in Sec. 5.4.7.

5.4.2.4 hcp Cobalt

Since in this case the two basis atoms are identical (and stay so at finite temperatures), the magnetic moments are $M_1 = M_2 = M$ and Eqn (5.135) is simply $\Delta E = 2AM^2 + 2BM^4 + 2[j_{11}(0) + j_{12}(0)]M^2$, where $j_{11}(\mathbf{k})$ and $j_{12}(\mathbf{k})$ are the only distinct exchange constants. We see that we can choose $j_{11}(0) + j_{12}(0) = 0$ and describe the collinear total energy change entirely with A and B, for which, in fact, we may again use Eqns (5.85) and (5.86). The noncollinear total energy contribution, Eqn (5.133), gives

$$E_2(\mathbf{k}) = 2M^2 j_{11}(\mathbf{k}) + 2M^2 j_{12}(\mathbf{k}) \cos(\mathbf{k}\cdot\boldsymbol{\tau}), \tag{5.149}$$

where $\boldsymbol{\tau}$ is the basis vector characterizing the hcp lattice and we have made a choice for the angles φ_τ. Since, by virtue of Eqn (5.134), the exchange constants, $j_{11}(\mathbf{k})$ and $j_{12}(\mathbf{k})$, are periodic with the periodicity of the reciprocal lattice, but $\cos(\mathbf{k}\cdot\boldsymbol{\tau})$ is not, we may determine $j_{11}(\mathbf{k})$ and $j_{12}(\mathbf{k})$ from Eqn (5.149) with a judicious choice of reciprocal lattice vectors. Thus, denoting the reciprocal unit vectors by $\mathbf{A}, \mathbf{B}, \mathbf{C}$ we select, for instance, the reciprocal lattice vectors $\mathbf{K}_i$ ($i=1,\ldots,6$) from the set $\{0, \mathbf{A}, \mathbf{B}, \mathbf{C}, \mathbf{A}+\mathbf{C}, \mathbf{B}+\mathbf{C}\}$ and verify $\sum_i \cos(\mathbf{k}\cdot\boldsymbol{\tau}+\mathbf{K}_i\cdot\boldsymbol{\tau}) = 0$, but $\sum_i \cos^2(\mathbf{k}\cdot\boldsymbol{\tau}+\mathbf{K}_i\cdot\boldsymbol{\tau}) = 3$. Using these relations we immediately obtain the desired exchange constants from Eqn (5.149). This completes the data base for hcp Co. Turning now to the secular equation, Eqn (5.147), we write down the solution by inspection:

$$n=1 \; : \; C_{11} = 1/\sqrt{2} \quad C_{12} = 1/\sqrt{2} \quad \text{for} \; j=l,t \tag{5.150}$$

and

$$n=2 \; : \; C_{21} = 1/\sqrt{2} \quad C_{22} = -1/\sqrt{2} \quad \text{for} \; j=l,t. \tag{5.151}$$

Correspondingly for $n=1$ and $j=l$ (for instance) using Eqn (5.145)

$$a_{1l\mathbf{k}} = 1/(2\beta \langle |m_{1l\mathbf{k}}|^2 \rangle_0) = A + B[6M^2 + 4t^2 + 6l^2] + j_{11}(\mathbf{k}) + j_{12}(\mathbf{k}) \tag{5.152}$$

and for $n=2$ and $j=l$

$$a_{2l\mathbf{k}} = 1/(2\beta \langle |m_{2l\mathbf{k}}|^2 \rangle_0) = A + B[6M^2 + 4t^2 + 6l^2] + j_{11}(\mathbf{k}) - j_{12}(\mathbf{k}). \tag{5.153}$$

Obviously, the case $n=1$ corresponds to the acoustic branch, whereas $n=2$ corresponds to the optic branch. The transverse case, $j=t$, follows from Eqn (5.146) accordingly for both the acoustic and optic branches. After determining the fluctuations

t^2 and l^2 by means of Eqn (5.140) the self-consistency cycle is closed and the free energy is finally determined.

Results for hcp Co appear in Table 5.4. The phase transition is not much different from that of the elemental ferromagnets so we omit the details. However, having obtained the free energy for both fcc and hcp Co, we can compare the two cases provided we use the same approximation for the fcc and hcp faces, i.e. in both cases the LSDA. In Fig. 5.9 we show the calculated free energies for ferromagnetic and paramagnetic hcp and fcc Co choosing as energy origin the value of the free energy of the constrained fcc phase with $M = M_s$ for all temperatures. We notice that the free energies cross so that the stable low-temperature hcp phase becomes unstable at a calculated temperature of $T_0 = 590$ K above which the fcc phase is the stable one (Sandratskii et al., 1998). The experimental transition temperature is $T_0 = 703$ K (Stearns, 1986).

We argued in Sec. 4.3.1 that the hcp structure of Co is stabilized magnetically, and indeed, the calculated ground-state energy of hcp Co is lower by 1.7 mRy per atom (see also Fig. 5.9) than that of ferromagnetic fcc Co. We stress that spin fluctuations are the only ingredients to the results shown in Fig. 5.9. Thus, although we cannot eliminate other mechanisms for the phase transition, like e.g. phonons, which may indeed change the transition temperature, we conclude that the increasing spin fluctuations and

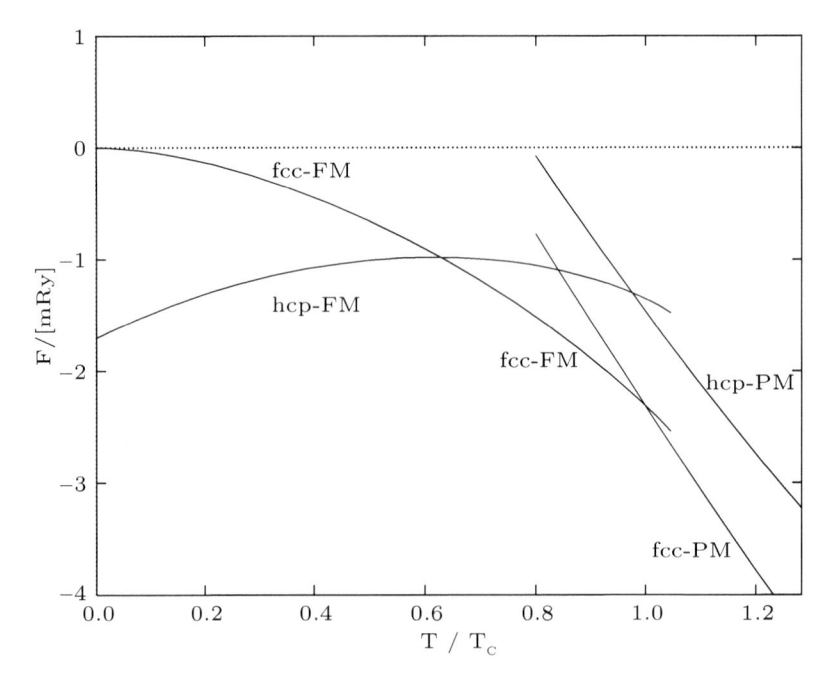

Figure 5.9 *Calculated free energies for ferromagnetic (FM) and paramagnetic (PM) hcp and fcc Co, after Sandratskii et al. (1998). The energy origin is the value of the free energy of the constrained fcc phase with $M = M_s$ for all temperatures.*

the decreasing magnetization at higher temperature at about 590 K restore the normal tendency of Co to be face-centered cubic.

5.4.3 Simple itinerant antiferromagnets

We now return to the spin-fluctuation formalism and write out the small changes needed for calculating thermal properties of simple antiferromagnets. By simple we mean itinerant magnets (preferably metallic) having one magnetic atom per chemical unit cell; examples to be treated are γ-Mn and FeRh.

In contrast to a ferromagnet, where we denote the ground-state magnetization by the vector

$$\mathbf{M} = M\,(0,0,1), \tag{5.154}$$

the antiferromagnet with a sublattice magnetization of $\mathbf{M}$ is described by

$$\mathbf{M} = M\,[\cos(\mathbf{Q}\cdot\mathbf{R}), \sin(\mathbf{Q}\cdot\mathbf{R}), 0], \tag{5.155}$$

where $\mathbf{R}$ is a lattice vector and, in the case of a collinearly ordered antiferromagnet, $2\mathbf{Q}$ equals a reciprocal lattice vector so that in this case $\cos(\mathbf{Q}\cdot\mathbf{R}) = \pm 1$ and $\sin(\mathbf{Q}\cdot\mathbf{R}) = 0$. If we continue to use the z-axis as a reference direction with respect to which we label longitudinal (l) and transverse (t) fluctuations then the coefficients in Eqn (5.106) change since both transverse fluctuations are in the plane of the sublattice magnetization, while in the ferromagnet it is the single longitudinal fluctuation that is parallel to the macroscopic magnetization. The necessary changes can be calculated to consist of changing the coefficients $f_{\mu\tau\lambda}$ in Eqn (5.106) to $f^A_{\mu\tau\lambda}$ to be obtained from the formula (Uhl and Kübler, 1997)

$$f^A_{\mu\tau\lambda} = \frac{(\mu+\tau+\lambda)!}{\mu!\,\tau!\,\lambda!}\,\frac{2^\tau (\mu+\tau)!\,(2\lambda-1)!!}{\mu!}. \tag{5.156}$$

Furthermore, Eqn (5.107) is to be replaced by

$$\langle \mathcal{H}_2 \rangle_0 = \sum_{\substack{\mu\tau\lambda \\ \mu+\tau+\lambda=0}} f^A_{\mu\tau\lambda}\, M^{2\mu}\, t^{2\tau}\, l^{2\lambda}\, [(2\mu+2\tau+2)\,\xi_T + (2\lambda+1)\,\xi_L]. \tag{5.157}$$

It is now straightforward to write out the changes in the subsequent formulas. It is then seen that the self-consistency equations remain of the form of Eqns (5.119), (5.121), and (5.123), and the sublattice magnetization comes from $\partial F_1/\partial M = 0$. Using the same abbreviations as before we find the details to be given by:

$$\frac{M^2}{M_s^2} = 1 - \frac{4t^2 + l^2}{M_s^2} \tag{5.158}$$

for the sublattice magnetization,

$$\chi_L^{-1}(\mathbf{k}) = 8\alpha_4(l^2 - t^2) + 2j(\mathbf{k}), \tag{5.159}$$

for the longitudinal susceptibility, and

$$\chi_T^{-1}(\mathbf{k}) = 4\alpha_4 M^2 + 2j(\mathbf{k}) \tag{5.160}$$

for the transverse susceptibility. The paramagnetic susceptibility (obviously) remains unchanged

5.4.3.1 fcc Mn

We illustrate the results by giving two examples, the first being fcc Mn whose ground-state properties were discussed in detail in Sec. 4.3.4, see for instance Fig. 4.44. Its experimental Néel temperature is $T_N \simeq 540$ K below which a tetragonal distortion sets in with a c/a ratio that reaches $c/a \simeq 0.945$ at $T = 0$. This lattice distortion is due to the tetragonally reduced symmetry of the antiferromagnetic spin configuration that is described by the vector $\mathbf{q} = (0, 0, 2\pi/c)$. Uhl and Kübler (1997) determined the c/a ratio self-consistently. Here, however, we examine the thermal properties without considering the distortion and summarize our results in Fig. 5.10. For these calculations the experimental estimate of the lattice constant of $a = 3.723$ Å (corresponding to an atomic sphere radius of 2.75 a.u.) was used which leads to a sublattice magnetization at $T = 0$ of $M_s = 2.4\,\mu_B$ and a gain in the total energy (compared with the nonmagnetic case) of $E_s = 10.6$ mRy. The calculated Néel temperature is in perfect agreement with the experimental estimate, but we believe this is fortuitous. Inspection of the figure reveals that the phase transition is of first order. From previous experience with the

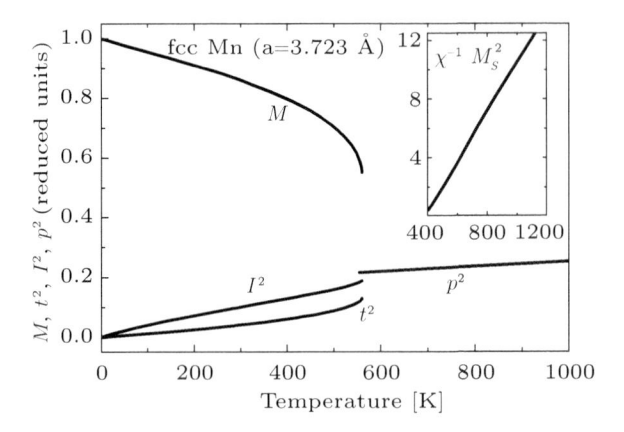

Figure 5.10 *Calculated magnetization data for antiferromagnetic fcc Mn as functions of temperature. M is the reduced sublattice magnetization, l^2, t^2, and p^2 are the reduced fluctuations. Inset: inverse paramagnetic susceptibility multiplied by the saturation-sublattice magnetization squared, M_s^2, in mRy.*

ferromagnets we must conclude that this is unphysical. We note in Fig. 5.10 that, in contrast to ferromagnets, the longitudinal fluctuations are larger than the transverse ones. This has an important consequence for the entropy of the system.

In our mean-field approach the entropy, S, is easily obtained, in fact calculating $S = -\partial F_1/\partial T$ we see immediately that the mean-field value of the entropy is given correctly in zero order, i.e.

$$S = S_0 = T^{-1}(\langle \mathcal{H}_0 \rangle_0 - F_0) \tag{5.161}$$

because of $F = U - TS$, where U is the internal energy. Thus, at a given temperature the entropy difference between the ferromagnetic (F) and the antiferromagnetic (AF) state is

$$\Delta S_{F-AF} = \frac{k_B}{2} \sum_{\mathbf{k}} 2 \ln \frac{\langle |m_{t\mathbf{k}}|^2 \rangle_0^{(F)}}{\langle |m_{t\mathbf{k}}|^2 \rangle_0^{(AF)}} + \frac{k_B}{2} \sum_{\mathbf{k}} \ln \frac{\langle |m_{l\mathbf{k}}|^2 \rangle_0^{(F)}}{\langle |m_{l\mathbf{k}}|^2 \rangle_0^{(AF)}} \tag{5.162}$$

from Eqns (5.99), (5.100), and (5.102). Since the transverse fluctuations are larger in the ferromagnet than in the antiferromagnet the entropy of the former will in general be larger (the first term on the right-hand side of Eqn (5.162) wins) and if all parameters are identical the free energy of the ferromagnet will be below that of the antiferromagnet, both coinciding at $T = 0$. Thus if we are given a system for which at $T = 0$ the calculated total energy of the antiferromagnet is only slightly lower than that of the ferromagnet, the free-energy curves can in principle cross at some finite temperature giving rise to a first-order phase transition from an antiferromagnetic to a ferromagnetic state. Obviously, the ferromagnet is stabilized by the entropy and we speak of an *entropy-stabilized state*. Note from Fig. 4.44 that this is unlikely to happen for fcc Mn. In the case of γ-Fe the high-spin ferromagnet crosses the antiferromagnetic state at a large atomic volume, see Fig. 4.47. So if this volume range becomes accessible experimentally one could expect to find an entropy-stabilized high-spin–high-temperature ferromagnet here. There is, however, a well-studied intermetallic compound, FeRh, which possesses a phase transition that seems to be of just the right kind. We therefore examine this case more closely.

5.4.3.2 FeRh

The intermetallic compound FeRh has received a great deal of experimental and theoretical attention. It possesses the CsCl structure and is antiferromagnetic of type AM2, characterized by $\mathbf{q} = (1,1,1)\pi/a$, at low temperatures. It is reported by Richardson *et al.* (1973) to undergo a first-order phase transition to a ferromagnetic state at $T_f = 328$ K accompanied by a small volume change and a strong entropy increase (Tu *et al.*, 1969; Richardson and Melville, 1972). The Curie temperature is found at $T_c = 670$ K (McKinnon *et al.*, 1970).

A good survey about the theoretical situation concerning FeRh is that by Hasegawa (1987), the electronic structure being described by him and by Koenig (1982), and total energy calculations were carried out by Moruzzi and Marcus (1992a,b) and Yuasa *et al.*

Magnetism at Finite Temperatures

(1995), while the present finite-temperature theory was first discussed by Uhl (1995) and Sandratskii *et al.* (1998).

In the CsCl structure where one atom occupies the center of a cube and the other the corners, there are two prominent antiferromagnetic moment arrangements labeled AM1 and AM2, the former being characterized by the vector $\mathbf{q} = (0,0,1)\,\pi/a$ and the latter by $\mathbf{q} = (1,1,1)\,\pi/a$. Placing the nonmagnetic Rh atom at the center of the cube we visualize the AM1 structure as consisting of ferromagnetic planes stacked along the z-axis with alternating magnetization, while the AM2 structure consists of ferromagnetic planes stacked along the body diagonal with alternating magnetization. A sketch of the two magnetic structures together with a simplified level scheme giving the total energy differences per atom at $T = 0$ between the antiferromagnetic states AM1, AM2, and the ferromagnet, F, is shown in Fig. 5.11. The calculated equilibrium atomic sphere

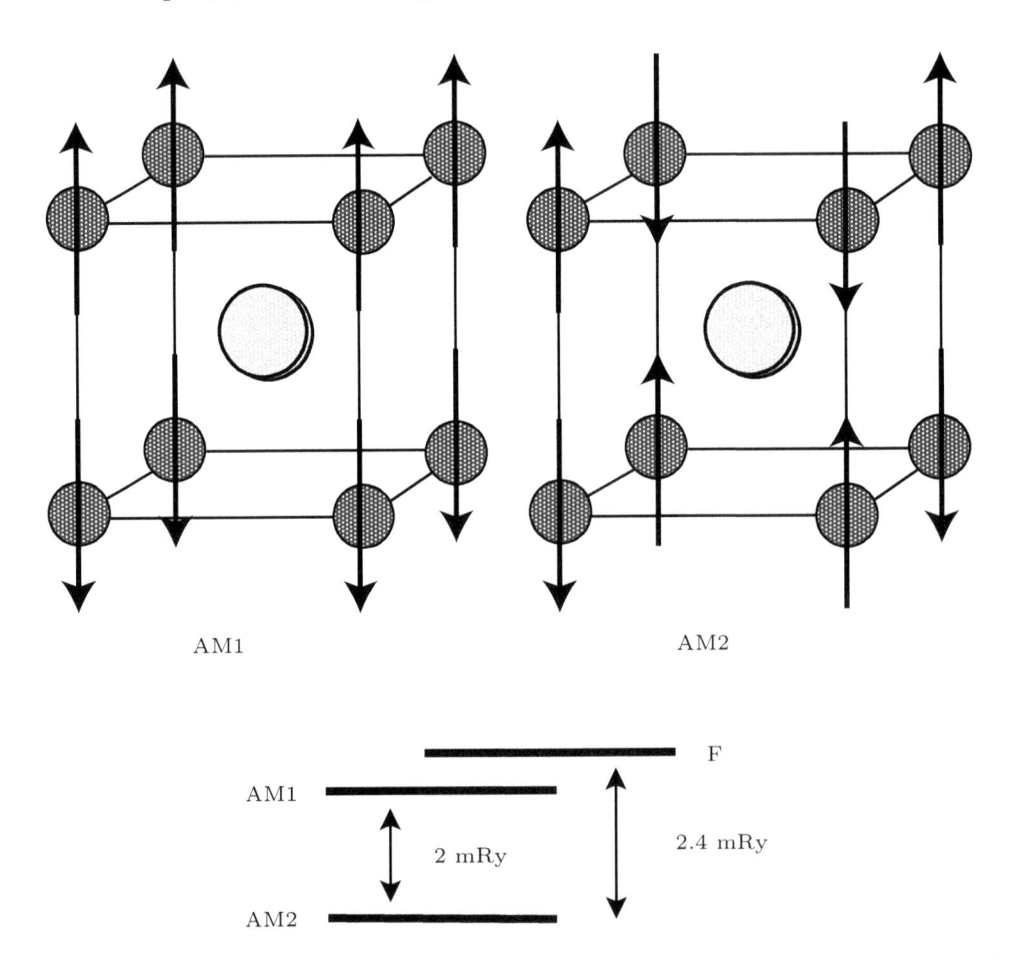

Figure 5.11 *The antiferromagnetic structures AM1 and AM2 of the FeRh crystal and simplified level scheme of the total energy differences (per atom) at $T = 0$ between the antiferromagnetic states, AM1, AM2, and the ferromagnet, F.*

radius for the AM2 state is found to be $S = 2.765$ a.u. with magnetic moments of Fe, $M_{Fe} = 2.99\,\mu_B$, and Rh, $M_{Rh} = 0$, while the ferromagnet possesses a slightly larger atomic volume with $S = 2.775$ a.u. and magnetic moments of $M_{Fe} = 3.10\,\mu_B$, $M_{Rh} = 1.06\,\mu_B$. Symmetry dictates that the magnetic moments of Rh vanish in the AM1 and AM2 states. The assumed nonmagnetic state is found at $E_s = 29.1$ mRy per atom above the AM2 state. To determine the magnetic properties at finite temperatures we first calculate the total energy per atom as a function of the magnetic moment per atom, $\bar{M} = (M_{Fe} + M_{Rh})/2$ and obtain the coefficients α_2, α_4, and α_6 for the AM1, AM2, and F states. Subsequently the exchange constants $j(\mathbf{q})$ are determined for a number of $\mathbf{q}$-vectors using an intermediate volume given by $S = 2.770$ a.u. For the total energy differences of the noncollinear configurations which we need in this step, the reference is, of course, the state having the lowest total energy (AM2). The free energy as a function of the temperature can then be determined and the result is shown in Fig. 5.12. Graphed here is the free energy counted from its values in the ferromagnetic state. We see that the AM2 state is stable up to a temperature of $T_f = 435$ K, where a first-order phase transition stabilizes the ferromagnetic state. This temperature should be compared with the experimental value of $T_f = 328$ K. The calculated Curie temperature is seen to occur

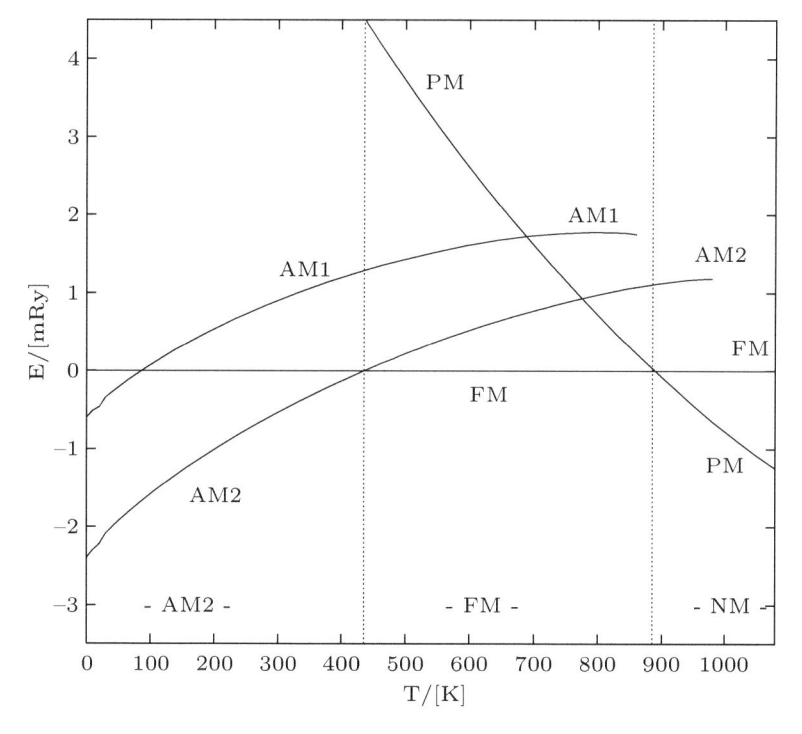

Figure 5.12 *Calculated free energies for the ferromagnetic (FM), the antiferromagnetic (AM1 and AM2), and the paramagnetic (PM) phases of FeRh. The energy origin is chosen to be the free energy of the ferromagnetic state, so that free-energy differences appear here (after Sandratskii et al., 1998).*

at $T_c = 885$ K where the paramagnetic free energy (PM) crosses the FM line. This value, just as T_f, is higher than the experimental Curie temperature of $T_c = 670$ K and the transition to the paramagnetic state is erroneously of first order, but the ferromagnet is clearly entropy stabilized and the general physical picture seems to be captured well by the present *ab initio* theory.

5.4.4 Previous, semi-empirical spin-fluctuation theories

Without claiming to be exhaustive we now add a brief description of previous, related theories that dealt with the physics of spin fluctuations, but were not *ab initio*. For a rather complete survey see the book by Moriya (1985), which we will discuss separately later on.

We begin with the work of Murata and Doniach (1972) who described spin fluctuations as a scalar quantity, $m(x)$, a function of the continuous space variable, x. The single-site Hamiltonian is expanded in powers of $m(x)^2$ and $m(x)^4$ as in our case, but inhomogeneities in the magnetization are assumed to give rise to a gradient term $[\nabla m(x)]^2$. The theory resembles a Ginzburg–Landau approach, but in contrast to this, the expansion coefficients here are not assumed to be temperature dependent. The thermodynamic properties of the model are obtained by means of the Bogoliubov variational method using a Gaussian trial Hamiltonian very similar to ours. Because of the gradient expansion the theory needs a cut-off parameter in the sum over the fluctuations. After a judicious choice of parameters, especially the cut-off, it explains qualitatively the specific heat of the weak ferromagnet Sc_3In. It should be noted that the Fourier-transformed gradient term resembles the spin-wave stiffness expression given by Eqn (5.40). Therefore, the approach of Murata and Doniach is in this respect comparable to the Debye theory for the specific heat of solids where the elementary excitations for long wavelengths are assumed to hold also for short wavelengths and a cut-off takes care of the proper number of normal modes. Lonzarich and Taillefer (1985) in a seminal paper extended the model of Murata and Doniach by introducing a magnetization *vector* otherwise making similar assumptions, especially about the gradient term which is now written as a sum over the vector components, $\sum_\nu |\nabla m_\nu(\mathbf{r})|^2$. The thermodynamic properties are again obtained by the variational method and there remains a cut-off parameter as before. The theory is applied to the weak ferromagnets Ni_3Al and $MnSi$ with good success because of carefully chosen parameters. We will return to this point later on in the section after next. An important new step, however, is their introducing a generalized, dynamical susceptibility and connecting it with the fluctuations by the *fluctuation-dissipation theorem* (Callen and Welton, 1951). We will deal with this aspect of the problem next and will put our susceptibility formulas into this general context as well. It is also noteworthy that Lonzarich and Taillefer (1985) derive a simple and interesting formula for the Curie temperature which is brought into contact with the Stoner temperature, see for instance Table 5.1, and a suitably defined spin-fluctuation temperature. Mohn and Wohlfarth (1987) propose a similar formula, but being derived differently it is claimed to be applicable to all itinerant magnets. This claim cannot however be corroborated.

Beginning with the work of Wagner (1989), who introduced volume fluctuations into the theory, a series of papers appeared in which parameters needed for modeling spin and volume fluctuations were obtained by means of total energy LDA band-structure calculations; all, however, using the gradient approximation embodied in the expression $\sum_\nu |\nabla m_\nu(\mathbf{r})|^2$ with a cut-off parameter (Schröter et al., 1990, 1992; Mohn et al., 1989, 1991; Kirchner et al., 1992; Weber et al., 1994). The main thrust of most of these papers was the Invar problem, see Sec. 4.4.2. Schröter et al. (1995), for instance, in an exceptionally complete description of the application of spin-fluctuation theory to Invar, give a set of Ginzburg–Landau coefficients that allow a determination of the fluctuations and of the magnetization profile $M(T)$. Although there are problems with the order of the phase transition it appears that the reduced magnetization is in good agreement with experimental data by Yamada et al. (1982).

5.4.5 Connection with the fluctuation-dissipation theorem

We now turn to a general discussion of fluctuations and place them into a larger context. This is of interest by itself and can lead to a crucial test to decide if a theory is formulated consistently. Indeed, we will see that the spin-fluctuation theory presented above passes this test. Furthermore, we may gain new insight that could point to future improvements of finite-temperature theories.

Originally derived by Callen and Welton (1951) for the relation between the generalized resistance and the fluctuations in linear dissipative systems and viewed as an extension of the Nyquist relation for the voltage fluctuations in electrical impedances, the fluctuation-dissipation theorem (as it is called today) is quite generally derived in linear response theory as a formal relation between fluctuations and the imaginary part of the generalized susceptibility.

It is proved in many textbooks, as for instance Becker and Sauter (1968), Jones and March (1973), White (1983), etc., so we just state the basic relation in the form of interest for us:

$$\langle |\mu_{j\mathbf{k}}|^2 \rangle = \frac{2\hbar}{\pi} \int_0^\infty d\omega \left[\frac{1}{2} + n(\omega) \right] \operatorname{Im} \chi_j(\mathbf{k}, \omega), \qquad (5.163)$$

where $n(\omega)$ is the Bose distribution function, $n(\omega) = [\exp(\hbar\omega\beta) - 1]^{-1}$, the factor $\frac{1}{2}$ originates from the zero-point vibrations, and $\chi_j(\mathbf{k},\omega)$ is the wave vector and frequency-dependent susceptibility in the j-direction; furthermore $\mu_{j\mathbf{k}}$ is the integrated Fourier transform of the exact fluctuations, by which we describe the space and time dependence of the local magnetization as $\mathbf{M} + \boldsymbol{\mu}(\mathbf{r},t)$. We do not want to overload the notation with too many indices, so we principally think of elementary magnets again. Since we are interested in the slow fluctuations we remove the fast fluctuations by a suitably chosen filter function which defines a convolution (Lonzarich and Taillefer, 1985) that leads to

$$\langle |m_{j\mathbf{k}}|^2 \rangle = \frac{2\hbar}{\pi} \int_0^\infty d\omega\, n(\omega) \operatorname{Im} \chi_j(\mathbf{k}, \omega). \qquad (5.164)$$

This involves the slow fluctuations $\langle |m_{j\mathbf{k}}|^2 \rangle$ for which the wave vector $\mathbf{k}$ is restricted to the first Brillouin zone; the filtering process has removed the fluctuations possessing wavelengths smaller than the lattice spacing, furthermore the zero-point fluctuations have been removed as well. It is not entirely clear if the relation (5.164) is still exact.

We first use Eqn (5.164) to show that the quantities we called susceptibilities in Sec. 5.4.1 following Eqn (5.120) really are what we called them. This not only satisfies our curiosity but points toward possible improvements and different theories.

For low frequencies ω and high temperatures T (small $\beta = 1/k_B T$) we can write $n(\omega) \approx k_B T/\hbar\omega$, then using the Kramers–Kronig relation that expresses the real part of the susceptibility in terms of the imaginary part,

$$Re\,\chi_j(\mathbf{k},\omega) = \frac{2}{\pi}\int_0^\infty d\omega' \frac{\omega'}{\omega'^2 - \omega^2} Im\,\chi_j(\mathbf{k},\omega') \qquad (5.165)$$

and assuming the high frequencies in the integral appearing in Eqn (5.164) are negligible because the susceptibility effectively vanishes for large ω, we obtain

$$\langle |m_{j\mathbf{k}}|^2 \rangle = k_B T\, Re\,\chi_j(\mathbf{k},0). \qquad (5.166)$$

Thus, indeed the susceptibilities $\chi_L(\mathbf{k})$, $\chi_T(\mathbf{k})$, and $\chi_P(\mathbf{k})$ in Sec. 5.4.1 are the static nonuniform susceptibilities as claimed and the overall formulation of the theory is consistent. Notice the disappearance of Planck's constant $\hbar$ in the result signals the classical limit.

In the next step we could, starting with Eqn (5.164), attempt to formulate a *dynamical* theory by constructing a reliable approximation for the frequency-dependent imaginary part of the susceptibility using as much of the band structure of the magnet as is known. This and much more was done by Moriya together with his students and is convincingly summarized in his book on spin fluctuations (Moriya, 1985). We certainly cannot do justice to his work, but want to point out a few key factors. The susceptibility, in a first attempt, is obtained in the *random-phase approximation* (RPA). Although to some extent comparable with the contents of our Secs. 4.2.1 and 4.2.2, exchange and correlation in the RPA are not treated using density-functional theory. It is subsequently found necessary to *renormalize* the theory by taking the free energy into account in a similar level of approximation as ours. But again apparently no attempt is made to use total energy results. The theory he arrives at is initially valid for weakly itinerant magnets like MnSi, TiBe$_2$, ZrZn$_2$, etc., but in a final step appears as a *unified theory* that is claimed to cover all itinerant magnets, but is no longer dynamical. The price paid for this generality is that the theory is not *ab initio*, but it is certainly very rich.

We are now ready to resume the discussion that we began in the previous subsection and return to the work of Lonzarich and Taillefer (1985). Their theory is less ambitious than Moriya's but also contains the interesting step toward a dynamic extension that is worth a short detour.

To do this we recall the expressions for the susceptibility in Sec. 5.4.1, i.e. Eqns (5.120), (5.122), and (5.125). Replacing the term $j(\mathbf{k})$ temporarily by the spin-wave stiffness term we can in each case write

$$\chi_j^{-1}(\mathbf{k}) = \chi_j^{-1}(0) + c_j\,\mathbf{k}^2 \qquad (5.167)$$

for $j = L, T, P$, the longitudinal, transverse, and paramagnetic susceptibilities; c_j is the appropriate spin-wave stiffness constant and $\chi_j^{-1}(0)$ depends in Eqn (5.120) on the magnetization and in Eqn (5.122) on the fluctuations. The model for the generalized susceptibility is now extended by adding a frequency-dependent term that is basically obtained from the frequency-dependent Lindhard function and rests on highly idealized parabolic bands (Lonzarich and Taillefer, 1985, see especially the appendix of this paper). Then Eqn (5.167) is written as

$$\chi_j^{-1}(\mathbf{k}) = \chi_j^{-1}(0) + c_j\mathbf{k}^2 - \mathrm{i}\,\frac{\omega}{\gamma_j\,k}. \qquad (5.168)$$

This form of the susceptibility is interesting because the integral in Eqn (5.164) can be done analytically. The result will be given in the next subsection. A consequence of this is that the fluctuations can be recalculated, but the calculated thermal properties are no longer *ab initio* because, in the work of Lonzarich and Taillefer a cut-off is needed in the integrals and at least the parameters γ_j need to be estimated. A discussion of whether or not the above approximation for the dynamic susceptibility might be useful for Fe has been given by Luchini *et al.* (1991). Unfortunately, it seems that this extension of the theory will not remove the unphysical first-order phase transitions.

It should have emerged by now that a new look at spin fluctuations should be attempted with *ab initio* calculations of the temperature-dependent dynamic susceptibility. In fact, this is feasible since time-dependent density-functional theory has been established (Runge and Gross, 1984; Gross *et al.*, 1996) so that the formulation of the susceptibility in Sec. 4.2.1 can be generalized. Still, the temperature dependence of exchange and correlation in density-functional theory is another problem that has to be solved with a self-consistent theory. In fact, this question brings us back to the beginning of this chapter. It might be that in some cases the temperature-dependent density-functional theory given in Sec. 5.1 suffices. An example for such a case seems to be chromium and some chromium alloys, for which Staunton *et al.* (1999) determined *ab initio* the dynamic paramagnetic spin susceptibility at finite temperatures, finding good agreement with inelastic neutron scattering data.

We next sketch the formulation of the dynamic approximation in a rather simplified form and apply it to the case of Ni where we gain some more insight into its magnetic phase transition. We, furthermore, will discuss the dynamic approximation for the weakly ferromagnetic compound Ni_3Al, whereby we can make a connection with the work of Lonzarich and Taillefer.

5.4.6 The dynamic approximation

We begin by rewriting the fluctuations, given in Eqns (5.103) and (5.104), together with (5.119), (5.121), and (5.123) as the single equation

$$\sum_{\mathbf{k}} \langle |m_{j\mathbf{k}}|^2 \rangle = k_B T \sum_{\mathbf{k}} \chi_j(\mathbf{k})^{-1} \qquad (5.169)$$

where we may distinguish the various cases by an appropriate choice of the index j.

The essential step is now the replacement of this equation by Eqn (5.164) making use of Eqn (5.167) and summing over $\mathbf{k}$. The resulting theory is commonly called the *dynamic approximation*. The importance of this step has been stressed repeatedly by Moriya (1985). In what follows we will replace the spin-wave term $c_j \mathbf{k}$ by our exchange function $2j(\mathbf{k})$ and restrict all sums over $\mathbf{k}$ to the first Brillouin zone, thus avoiding any cut-off parameters.

We now write the free energy in the form

$$F(M,T) = \alpha_2 M^2 + \alpha_4 M^4 + F_1(M,T). \qquad (5.170)$$

and employ a formula for the free energy part $F_1(M,T)$ that is due to Dzyaloshinskii and Kondratenko (1976). This is

$$F_1(M,T) = F_0(T) + \frac{1}{2} \sum_{j,\mathbf{k}} \int_{-\infty}^{\infty} \frac{d\omega}{2\pi} \mathrm{Im}\left\{ \ln \chi_j^{-1}(\mathbf{k},\omega) \right\} \coth\left(\frac{\omega}{2k_B T} \right) \qquad (5.171)$$

Here $F_0(T)$ does not depend on the magnetization and is, therefore, at this point of no concern. The inverse dynamic susceptibilities are approximated by

$$\chi_j^{-1}(\mathbf{k},\omega) = \chi_j^{-1}(\mathbf{k}) - \frac{i\omega}{\Gamma k} \qquad (5.172)$$

where the real part, $\chi_j^{-1}(\mathbf{k})$, will be seen to be given by the results of the static approximation, for $j = l$ and $j = t$ and $j = p$.

The complex part of the inverse dynamic susceptibility is a widely used approximation for the weakly ferromagnet metals, where long-wavelength fluctuations of small frequencies predominate. It rests on a broad experimental and theoretical basis (Moriya, 1985; Lonzarich and Taillefer, 1985). The latter has most concisely been summarized by Lonzarich and Taillefer who give the small ω and small q expansion of the dynamic Lindhard susceptibility as Eqn (5.172), valid for a single tight-binding band and omitting matrix elements. In this approximation one can express the dissipation constant Γ in terms of band-structure quantities as

$$\Gamma = (4/\pi V)\mu_B^2 N_F E_F / k_F \qquad (5.173)$$

where V is the volume of the primitive unit cell, N_F is the density of states at the effective Fermi energy E_F, k_F is the Fermi wave vector, and the magnetization is assumed to be small.

By taking next the second derivative of the free energy, Eqn (5.170) and (5.171) with respect to M one verifies that the uniform, static, longitudinal susceptibility is given by

Eqn (5.120), provided the fluctuations are connected with the susceptibility through the fluctuation-dissipation theorem, Eqn (5.164). The transverse susceptibility, however, is slightly different from Eqn (5.122). It is believed, though, that this equation is still a good approximation since numerically the two susceptibilities come out very nearly the same.

The remaining calculation is the evaluation of the fluctuation Eqn (5.164). With the above approximation for the dynamic susceptibility the frequency integration can be carried out analytically, as was done by Lonzarich and Taillefer (1985), who refer to a paper by Ramakrishnan (1974) obtaining $\langle |m_{j\mathbf{k}}|^2 \rangle = k_B T \chi_j(\mathbf{k}) g(z)/V$ where $g(z) = 2z(\ln z - 1/2z - \psi(z))$, $\psi(z)$ is Euler's psi function, and $z = \Gamma|\mathbf{k}|\chi_j^{-1}(\mathbf{k})/2\pi k_B T$. To a very good approximation one can write $g(z) \simeq 1/(1 + 5.63602z)$ which finally gives

$$\sum_{\mathbf{k}} \langle |m_{j\mathbf{k}}|^2 \rangle = k_B T \sum_{\mathbf{k}} \chi_j(\mathbf{k}) - \xi\Gamma \sum_{\mathbf{k}} |\mathbf{k}| \left[1 + \frac{\xi\Gamma|\mathbf{k}|}{k_B T}\chi_j^{-1}(\mathbf{k}) \right]^{-1} \quad (5.174)$$

The constant appearing is $\xi = 0.897$. Note that for $\Gamma = 0$ the results of the static approximation are obtained. With this equation for the fluctuations the numerical treatment of the self-consistency equations is just as straight-forward as before.

One could now argue that all important quantities are determined by the total energy and the band structure. However, the assumptions underlying the relation for Γ, Eqn (5.173), are heavily idealized, so that this relation can only be expected to give an order of magnitude estimate of the dissipation constant Γ, as we will see, indeed. In principle, the dynamic susceptibility can be easily formulated, but in the appropriate integral equation a kernel remains unknown. A desirable approach would be that of Savrasov (1998), who calculated the dynamical susceptibility *ab initio* but did not direct his attention to the problems addressed here. Similarly, the *ab initio* theory of Staunton *et al.* (2000) should be applied to ferromagnets.

Furthermore, even more serious, as it will turn out, is a conceptual error made in the determination of the Landau coefficients α_2 and α_4 from the total energy. This point will be exposed in detail in the following examples.

The determination of the exchange function $j(\mathbf{k})$, however, will be seen to be quite reliable thus turning out to be one of the strong points of the theory.

5.4.6.1 More on nickel

Although Ni is not understood to be a weakly ferromagnetic metal we start with this case since the calculations are quickly performed bringing out the weak points easily.

Returning, therefore, to Sec. 5.4.2.1 we first focus our attention on the order of the phase transition. We notice that the simple formula for the second-order phase transition, Eqn (5.126), only depends on the exchange average and the saturation ($T = 0$) value of the magnetization squared. So if, in a numerical experiment, one reduces the values of the Landau coefficients α_2 and α_4 by reducing the "saturation energy," E_s, in Eqns (5.85) and (5.86) then the minimum of the function defined by

$$f(M) = \alpha_2 M^2 + \alpha_4 M^4 \tag{5.175}$$

remains at a constant value M_s and the phase transition trivially changes from the calculated first order seen in Fig. 5.8 to a second-order transition moving the Curie temperature down to the lower value of 341 K, as obtained by Eqn (5.126), and decreasing the slope of the inverse susceptibility.

To continue it is convenient to turn to the *dynamic approximation* choosing a dissipation constant Γ such that the experimental Curie temperature is obtained assuring that the transition is of second order by using a value of E_s smaller than that given in Table 5.4. Thus Fig. 5.13(b) illustrates that the inverse susceptibility shows Curie–Weiss behavior; however, its value agrees with the experimental data (also given in the figure) only for a vastly reduced E_s.[1] Can one justify this reduction?

We recall that, by construction, the total energy gain shown in Fig. 4.23 and given by Eqn 5.84 is defined with respect to the nonmagnetic state. This state is the off-set for the quantity E_s. For Ni—or any other ferromagnet—the nonmagnetic state, however, is a computational *fiction*. Another reference state should be used, one that is magnetic with no long-range order. Moriya (1985, Ch.7) identified the reference state qualitatively by means of the Anderson condition which marks the appearance of randomly oriented local moments in metals. At this point there is no formal answer within the theory presented, but the appearance (or disappearance) of local moments can clearly be seen in the enhancement of the *non-uniform* susceptibility, $\chi(\mathbf{k})$. The susceptibility enhancement

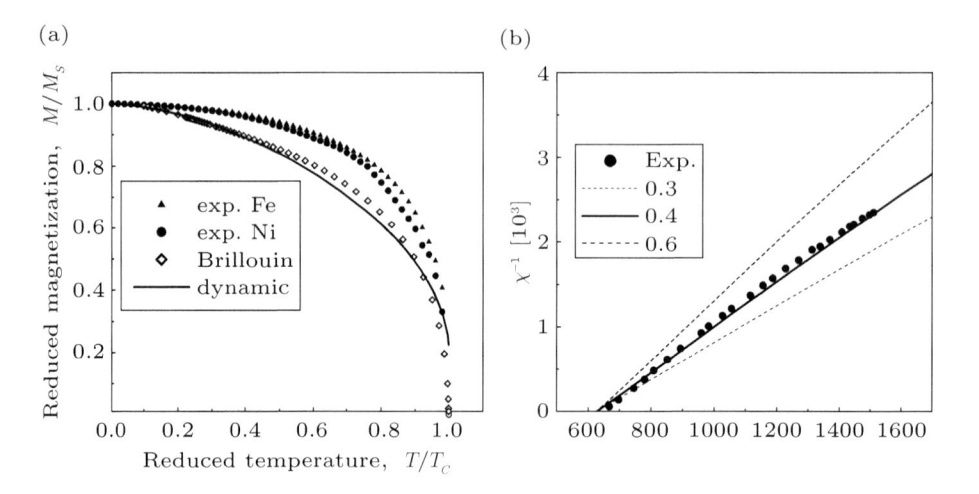

Figure 5.13 *(a) Reduced magnetization of Ni in the dynamic approximation is compared with experimental values and the Brillouin function. (b) Inverse susceptibility for Ni above the Curie point. Experimental data (circles) from Shimizu (1981); data calculated in the dynamic approximation are shown for different values of the parameter E_S in mRy.*

---

[1] The units used in Fig. 5.13(b) are obtained by means of 1 mRy/$\mu_B^2 = 4.208726 \times 10^2$ mol/emu.

was discussed in detail in Sec. 4.2.3.2. There we demonstrated that it peaks strongly at values of **k** where the magnetic moment becomes small, see Fig. 4.22 for Ni. The approximate relation Eqn (4.89) shows that the total energy is very flat as a function of M in those regions of the Brillouin zone where the susceptibility becomes large.

Proceeding empirically we may assume that it is this value of the total energy which determines the Landau coefficients and consequently E_s. We are thus led to using $\alpha_4 = -(2\chi_{max})^{-1}$, which results in $E_s \approx 0.4$ mRy. Fig. 5.13(b) shows that this leads to good agreement with the experimental Curie–Weiss susceptibility. Although this agreement could be fortuitous, we take it as indication for a useful procedure to apply to all weakly ferromagnet materials, which are described in the sequel of this chapter in Sec. 5.4.6.2.

In Fig. 5.13(a) we show the magnetization as a function of the temperature for Ni in reduced units. For comparison the experimental data for Fe and Ni are also shown together with the well-known Brillouin function valid in the mean-field approximation. One can see that especially at low temperatures the calculated curve does not agree with the experimental values. This is due to an improper treatment of the magnon excitations.

We complete the dynamic approximation for Ni by giving the value of the dissipation constant that leads to the experimental Curie temperature shown in Fig. 5.13(b); this is $\Gamma = 0.641$ μeVÅ^2. Although, differing from Eqn (5.172), the experimental wave-vector dependence proportional to k^{-2} was used in the calculations, we cannot expect agreement with the measured value of Γ, which is three orders of magnitude larger (Steinsvoll et al., 1984). This should not be surprising in view of the simplifications made in the modelling of the imaginary part of the susceptibility. Unfortunately our theory for Ni is no longer *ab initio*.

We add a brief remark on very recent *ab initio* work on Ni even though it is only in the later Sec. 5.5 that we can deal with the theoretical tools needed for a deeper understanding.

Ruban et al. (2007), using the static approximation, write the total energy, for which we used the Landau expansion, in terms of two quantities $J^{(0)}(\overline{m})$ and $J^{(1)}(\overline{m}, m_i)$. The first term is the energy of the reference state, for which a homogeneous disordered local moment (DLM) state is used (see Sec. 5.5.2). This state can be viewed as an alloy where each atom carries the moment $\overline{m}$ that is randomly oriented in space. In view of the work of Staunton and Gyorffy (1992), which is to be discussed later on in Sec. 5.6, it is not surprising to see that the function $J^{(0)}(\overline{m})$ does not possess a minimum other than $\overline{m} = 0$: Ni cannot sustain a DLM state. But longitudinal spin fluctuations on different atoms will lead to inhomogeneous states, which are described by the function $J^{(1)}(\overline{m}, m_i)$. It has been determined as the energy of a single impurity with magnetic moment m_i embedded in the homogeneous DLM effective medium characterized by the magnetic moment $\overline{m}$. The functions $J^{(1)}(\overline{m}, m_i)$ do not possess minima other than $m_i = 0$ either. So if these functions are the proper choice (and they may well be) then a *Landau expansion does not exist for Ni*. This is different for the case of Fe, as we will see. By means of carefully determined pair exchange interactions, which are similar to the Heisenberg exchange, Ruban et al. (2007) complete the Hamiltonian and determine the phase transition using the Monte Carlo method. The Curie temperature is obtained to be 615 K but the inverse susceptibility above T_C is too high. It still seems to be too early to draw far-reaching conclusions.

5.4.6.2 The weakly ferromagnetic compound Ni$_3$Al

A thoroughly studied itinerant-electron weak ferromagnet is Ni$_3$Al. In the Rhodes–Wohlfarth plot, Fig. 4.4, it should appear at the upper far left. Early experimental work is that by De Boer et al. (1969), followed by Buis et al. (1981) and others. Bernhoeft et al. (1983) did neutron scattering studies. In the theoretical work of Lonzarich and Taillefer (1985) Ni$_3$Al was used as the prime example.

Ni$_3$Al is simple cubic and has the Cu$_3$Au structure, which possesses three equivalent magnetic atoms. Thus three normal modes are taken into account in Eqns (5.127) and (5.147), of which two are found to be degenerate. The inverse longitudinal susceptibilities are twice

$$\chi_{\ell 1}^{-1}(\mathbf{k}) = 2\alpha_2 + 4\alpha_4(3M^2 + 2m_t^2 + 3m_\ell^2) + 2j_{11}(\mathbf{k}) - 2j_{12}(\mathbf{k}) \tag{5.176}$$

and once

$$\chi_{\ell 2}^{-1}(\mathbf{k}) = 2\alpha_2 + 4\alpha_4(3M^2 + 2m_t^2 + 3m_\ell^2) + 2j_{11}(\mathbf{k}) + 4j_{12}(\mathbf{k}). \tag{5.177}$$

Correspondingly, for the fluctuations perpendicular to the direction of the magnetization, twice

$$\chi_{t1}^{-1}(\mathbf{k}) = 2\alpha_2 + 4\alpha_4(M^2 + 4m_t^2 + m_\ell^2) + 2j_{11}(\mathbf{k}) - 2j_{12}(\mathbf{k}) \tag{5.178}$$

and once

$$\chi_{t2}^{-1}(\mathbf{k}) = 2\alpha_2 + 4\alpha_4(M^2 + 4m_t^2 + m_\ell^2) + 2j_{11}(\mathbf{k}) + 4j_{12}(\mathbf{k}). \tag{5.179}$$

The self-consistency step is defined by the equation

$$m_j^2 = k_B T \sum_\mathbf{k} \frac{1}{3}[2\chi_{j1}(\mathbf{k}) + \chi_{j2}(\mathbf{k})] \tag{5.180}$$

for $j = \ell$ and $j = t$. The Curie temperature in the static approximation is then given by

$$k_B T_c = \frac{6}{5} M_s^2 \cdot \left(\sum_\mathbf{k} \frac{2}{j_{11}(\mathbf{k}) - j_{12}(\mathbf{k})} + \sum_\mathbf{k} \frac{1}{j_{11}(\mathbf{k}) + 2j_{12}(\mathbf{k})} \right)^{-1}, \tag{5.181}$$

provided the transition is of second order.

The exchange functions are obtained from the total energy of a spin-spiral given in the case of Ni$_3$Al by

$$\Delta E = M_s^2 [3j_{11}(\mathbf{k}) + 2j_{12}(\mathbf{k}) \sum_{i=1}^{3} \cos(\mathbf{k} \cdot \boldsymbol{\tau}_i)]. \tag{5.182}$$

The quantity M_s is the Ni-moment in the ground state and $\boldsymbol{\tau}_i$ are the basis vectors. Since the exchange functions are periodic with the periodicity of the reciprocal lattice, they are obtained from Eqn (5.182) with a choice of three reciprocal lattice vectors, $\mathbf{K}_1 = (1,0,0)$, $\mathbf{K}_2 = (0,1,0)$, and $\mathbf{K}_3 = (0,0,1)$ in units of $2\pi/a$, where a is the lattice constant. Thus replacing $\mathbf{k}$ by $\mathbf{k} + \mathbf{K}_j$ $j = 1,2,3$ we get equations to solve for $j_{11}(\mathbf{k})$, finally eliminating the cosine by choosing all phases to be zero to determine $j_{12}(\mathbf{k})$.

The static approximation gives a second-order phase transition if the coefficient α_2 is obtained from the maximum of the non-uniform susceptibility as described for the case of Ni before. Its $\mathbf{k}$-dependence is very similar to that shown for Ni in Fig. 4.22; its maximum value of $\chi_{Max} \simeq 0.88 \cdot 10^{-3}$ emu/mol, however, occurs at $\mathbf{k} \simeq (0.1,0,0) 2\pi/a$.

The total-energy differences needed to evaluate the exchange functions are obtained by both the force theorem and constrained-moment calculations. Together with the lattice constant, the magnetic moment, and the Curie–Weiss law ratio, the static approximation Curie temperature is given in Table 5.5, where we also compare with experimental data. As expected, the constrained-moment calculations result in a larger estimate for T_c than the force theorem, which leads to $T_c = 12.7$ K only.

To obtain the experimental Curie temperature in the *dynamic approximation* the dissipation constant is chosen as $\Gamma = 0.15$ μeVÅ. The calculated temperature dependence of the magnetization is shown in Fig. 5.14(a) and is seen to agree nearly perfectly with the experimental data measured by De Boer et al. (1969). The value of Γ is, however, considerably smaller than the neutron-scattering value of $\Gamma = 3.3$ μeVÅ measured by Bernhoeft et al. (1983). Finally, the calculated inverse Curie–Weiss susceptibility shown in Fig. 5.14(b) (dashed line with $\alpha = -2507)^2$ gives an effective magnetic moment at $T > T_c$ larger than in the ground state, as it should, but the calculated ratio of q_c/q_s given in Table 5.5 is too small. Possible reasons for this discrepancy may be our empirical renormalization scheme and (or) Stoner excitations, which could decrease the coefficient α_2 considerably. To illustrate the dependence on α_2 we include in Fig. 5.14(b) (solid line) the results for a value of $\alpha_2 = -580$ which was found by Lonzarich and Taillefer (1985) to describe the experimental data. This value gives for the ratio q_c/q_s the rather acceptable result $q_c/q_s = 5$.

Having thus brought the Ginzburg–Landau approach of Lonzarich and Taillefer (1985) in close contact with our theory, we see that the cut-off parameter in the former is successfully replaced by zone integrations in the latter. The experimental data are explained well in both theories. We ask, however, how the Landau coefficients can be obtained reliably without the use of experimental data. Our use of the maximum of the nonuniform susceptibility is not implausible, but cannot be justified from first principles. Comparing differently disordered or partially ordered states, as was done by Ruban

---

[2] The conversion of the Gaussian units is achieved by using the relations 1 $\mu_B = 9.2741 \times 10^3$ GÅ^3 and 1mRy $= 2.17991 \times 10^{10}$ G^{2}Å^3

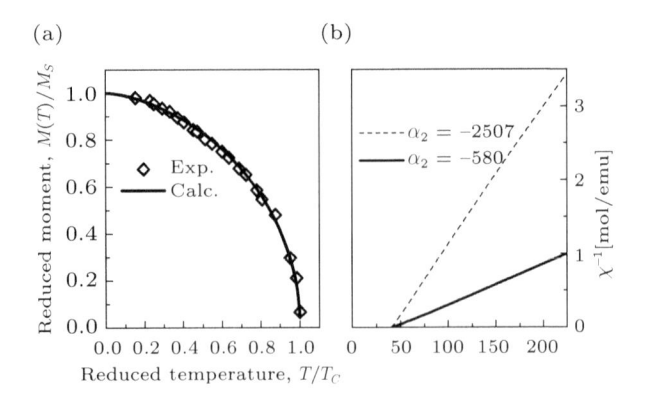

Figure 5.14 *(a) Reduced magnetic moment, $M(T)/M_s$, as a function of the reduced temperature, T/T_c of Ni_3Al in the dynamic approximation. The experimental data (circles) were taken from the paper by Lonzarich (1986) who used data by De Boer et al. (1969). (b) Calculated Curie–Weiss law using $\alpha_2 = -2507$ determined by means of the susceptibility maximum, $\chi_{Max} = 0.88 \cdot 10^{-3}$ emu/mol, and $\alpha_2 = -580$ from Lonzarich and Taillefer (1985).*

Table 5.5 *Calculated and experimental values for Ni_3Al: lattice constant, a, calculated spin-fluctuation, T_c in the static approximation, magnetic moment, M_s, and Curie–Weiss law ratio q_c/q_s.*

	a [Å]	T_c [K]	M_s [μ_B]	q_c/q_s
Calc.	3.5283	18.4	0.12	1.8
Exp.	3.568	41.0	0.075	7.5

et al. (2007) for Ni and Fe, might be a possibility but has—to our knowledge—not been attempted yet for Ni_3Al. Thus the physics of weakly ferromagnetic metals needs further attention.

A similar treatment of the unusual ferromagnet $ZrZn_2$ is described in a paper by Kübler (2004).

5.4.7 The spherical approximation

It should be clear by now that another approach is desirable; this should not have at the center a Landau expansion for the energies that govern the fluctuations and it should apply to the entire range of cases from very weak ferromagnets to the local moment limit.

Such a theory was proposed by Moriya and Takahashi (1978) (Moriya, 1985) who used the Stratonovich–Hubbard functional integral method. One can construct a simple form for the functional that does not contain the original model parameters. Instead one can formulate the appropriate functional in terms of on-site and off-site total energies obtainable in the LSDA or improved approximations. In a notation different from that of Moriya and Takahashi the formalism is quite general and may be described as follows. Here we freely make use of a paper by Kübler (2006).

A functional $\Psi = \Psi(M_\tau, \mathcal{L}_\tau, \mathbf{M}_{n\mathbf{k}})$ is constructed that depends on the magnetization of atom τ, M_τ, the size of the local moment of atom τ, $\mathcal{L}_\tau$, and the fluctuation vector $\mathbf{M}_{n\mathbf{k}} = (m_{xn\mathbf{k}}, m_{yn\mathbf{k}}, m_{zn\mathbf{k}})$ of the "normal mode" labelled n. The functional integral to be evaluated is then

$$\exp(-F/k_\mathrm{B}T) \propto \prod_\tau \int \mathrm{d}\mathcal{L}_\tau^2 \int \prod_{n\mathbf{k}} \mathrm{d}\mathbf{M}_{n\mathbf{k}} \exp[-\Psi(M_\tau, \mathcal{L}_\tau, \mathbf{M}_{n\mathbf{k}})/k_\mathrm{B}T] \quad (5.183)$$

which supplies the free energy, F. In writing down an expression for the functional Ψ one uses an important approximation in which Lagrange multipliers, $\lambda_{\alpha\tau}$, for each Cartesian component, $(\alpha = x, y, z)$, constrain the size of the magnetic moments to be near a most probable size. This is called the *spherical approximation*. This approximation appears in the literature first in 1949 (Montroll, 1949), then Berlin and Kac (1952). Lax (1955) used it for the Heisenberg model.

Thus one writes

$$\Psi = \sum_{\tau\tau'}\sum_{\mathbf{k}n} j_{\tau\tau'}(\mathbf{k}) C^*_{n\tau} C_{n\tau'} |\mathbf{M}_{n\mathbf{k}}|^2 + \sum_\tau E_\tau(M_\tau, \mathcal{L}_\tau^2)$$
$$- \sum_{\alpha\tau} \lambda_{\alpha\tau}(\mathcal{L}_{\alpha\tau}^2 + \delta_{\alpha z} M_\tau^2 - \sum_{n\mathbf{k}} |C_{n\tau}|^2 |m_{\alpha n\mathbf{k}}|^2). \quad (5.184)$$

Here $j_{\tau\tau'}(\mathbf{k})$ is the exchange energy as a function of the wave vector $\mathbf{k}$ that accounts for the off-site interactions whereas $E_\tau(M_\tau, \mathcal{L}_\tau^2)$ is the on-site energy of a given configuration of local moments and magnetization that, among other things, controls the longitudinal fluctuations.

The integral over $\mathbf{M}_{n\mathbf{k}}$ can now be carried out, however, the $\mathbf{k} = 0$ component $\mathbf{M}_{\tau\mathbf{q}=\mathbf{0}} = (0, 0, m_{\tau z 0}) = (0, 0, M_\tau)$ is singled out and identified as the macroscopic magnetization of atom τ. The result is

$$\exp(-F/k_\mathrm{B}T) \propto \int \mathrm{d}\mathcal{L}_\tau^2 \int \mathrm{d}M_\tau$$
$$\exp\left\{ -\frac{\sum_\tau [E_\tau(M_\tau, \mathcal{L}_\tau^2) - \sum_\alpha \lambda_{\alpha\tau} \mathcal{L}_{\alpha\tau}^2]}{k_\mathrm{B}T} - \frac{1}{2}\sum_{\alpha n\mathbf{k}} \ln \frac{\lambda_{\alpha n} + j_n(\mathbf{k})}{\pi k_\mathrm{B}T}\right\}. \quad (5.185)$$

Here the exchange function is defined by

$$j_n(\mathbf{k}) = \sum_{\tau\tau'} j_{\tau\tau'}(\mathbf{k}) C^*_{n\tau} C_{n\tau'}. \quad (5.186)$$

It allows the determination of the coefficients $C_{n\tau}$ by diagonalizing $j_{\tau\tau'}(\mathbf{k})$. Furthermore

$$\lambda_{\alpha n} = \sum_\tau \lambda_{\alpha\tau} |C_{n\tau}|^2 \quad (5.187)$$

and

$$M_\tau^2 = \sum_{n\alpha} |C_{n\tau}|^2 |m_{\alpha n \mathbf{k}=0}|^2 . \tag{5.188}$$

Next Eqn (5.185) is differentiated with respect to $\lambda_{\alpha\tau}$ and the result is equated to zero in order to determine the condition for the constraint that is implied by the spherical approximation. With $\mathcal{L}_\tau^2 = \sum_{\alpha=1}^{3} \mathcal{L}_{\alpha\tau}^2$ and $\sum_\tau |C_{n\tau}|^2 = 1$, one obtains for the local moment

$$\sum_\tau \mathcal{L}_\tau^2 = k_B T \sum_\alpha \sum_{\mathbf{k}n} \chi_{\alpha n}(\mathbf{k}), \tag{5.189}$$

where

$$\chi_{\alpha n}(\mathbf{k}) = 1/[2\lambda_{\alpha n} + 2j_n(\mathbf{k})]. \tag{5.190}$$

It is emphasized that for simplicity in writing thermally averaged quantities and the original variables are not distinguished in the notation used here. Next a saddle-point approximation is employed to evaluate the integral over $\mathcal{L}_\tau$. The saddle-point condition is obtained by taking the derivative with respect to $\mathcal{L}_{\alpha\tau}^2$. This gives

$$\lambda_{\alpha\tau} = \frac{\partial E_\tau(M_\tau, \mathcal{L}_\tau^2)}{\partial \mathcal{L}_{\alpha\tau}^2} + k_B T \sum_{\mathbf{k}n} \chi_{\alpha n}(\mathbf{k}) \frac{\partial j_n(\mathbf{k})}{\partial \mathcal{L}_{\alpha\tau}^2}. \tag{5.191}$$

The quantity $\chi_{\alpha n}(\mathbf{k})$ in Eqn (5.189) is identified as the susceptibility using the fluctuation-dissipation theorem in the static approximation.

Not much interesting work has so far been done on the low-temperature properties of the above equations. It is, however, quite easy to make progress by considering temperatures above the ordering temperature where $M_\tau = 0$ for all basis atoms τ. Then, because of isotropy, Eqn (5.189) becomes

$$\sum_\tau \mathcal{L}_\tau^2 = 3k_B T \sum_{\mathbf{k}n} \chi_n(\mathbf{k}), \tag{5.192}$$

where, from Eqn (5.190), the susceptibility is

$$\chi_n(\mathbf{k}) = 1/[2\lambda_n + 2j_n(\mathbf{k})]. \tag{5.193}$$

Using $\lambda_n = \sum_\tau \lambda_\tau |C_{n\tau}|^2$ the quantity λ_n is to be determined with

$$\lambda_\tau = 3\frac{\partial E_\tau(0, \mathcal{L}_\tau^2)}{\partial \mathcal{L}_\tau^2} + 3k_B T \sum_{\mathbf{k}n} \chi_n(\mathbf{k}) \frac{\partial j_n(\mathbf{k})}{\partial \mathcal{L}_\tau^2}. \tag{5.194}$$

At the Curie (Néel) temperature the first term on the right-hand side vanishes. Then ignoring the second term one sees that $\lambda_n = 0$ for all n. Thus the Curie temperature in the spherical approximation is obtained as

$$k_\mathrm{B} T_c = \frac{2}{3} \sum_\tau \mathcal{L}_\tau^2 \left[\sum_{\mathbf{k}n} \frac{1}{j_n(\mathbf{k})} \right]^{-1}. \tag{5.195}$$

At this stage, however, the size of the local moment, $\mathcal{L}_\tau$, is not known unless one succeeds in solving Eqns (5.192) to (5.194) self-consistently. A number of comments are in order.

For the case of a primitive lattice one might assume $\mathcal{L}^2 = S(S+1)$; Eqn (5.195) with this choice is then known as the RPA-formula which can be derived quantum mechanically for a spin-$S = 1/2$ system, for which $\mathcal{L}^2 = S(S+1)$ (Tahir-Keli and ter Haar, 1962; Tahir-Keli and Jarrett, 1964; Tyablikov, 1967). For itinerant-electron systems, however, a relation in terms of S is not defined. In spite of this, in many applications of the RPA-formula one simply takes $\mathcal{L}^2 = M_s^2$, where M_s is the saturation magnetization. For the case of itinerant electrons in nonprimitive lattices Eqn (5.195) is new. It is the important result of this section. Note the Eqn (5.148) for weakly ferromagnetic systems, which is smaller than Eqn (5.195) by a factor 3/5. For the Heisenberg model Rusz et al. (2005) have recently derived a different formula for the Curie temperature for nonprimitive lattices using the RPA with Tyablikov decoupling.

More insight is obtained if Eqns (5.192) to (5.194) are solved. Assuming the exchange functions $j_n(\mathbf{k})$ and the on-site energy $E_\tau(0, \mathcal{L}_\tau^2)$ are known then $\mathcal{L}_\tau^2$ and the uniform susceptibility $\chi_{0n}^{-1} = 2\lambda_n$ can in principle be obtained in terms of the temperature.

To demonstrate that for the case of a nearly localized ferromagnet the formalism can be exploited to calculate the local moment *together* with the Curie–Weiss law for the susceptibility, one may assume that for this case an expansion of the energy function $E(0, \mathcal{L}^2)$ about the value of $\mathcal{L}^2/M_s^2 = 1$ is meaningful. The derivative of the exchange function with respect to $\mathcal{L}^2$ in Eqn (5.194) can then be estimated from $\partial j(\mathbf{k})/\partial M_s^2$, i.e. replacing $\mathcal{L}$ by M_s. The latter derivative has been obtained numerically by finite differences using constrained total-energy calculations to values of the magnetic moment, $M_s \pm \Delta M$ for bcc Fe.

Employing the expansion

$$E(0, \mathcal{L}^2) = \frac{1}{4} \chi_{eff}^{-1} \cdot (\mathcal{L}^2/M_0^2 - 1), \tag{5.196}$$

the result shown in Fig. 5.15 for the inverse susceptibility for bcc Fe is obtained. The value used here for the expansion coefficient is $\chi_{eff} \simeq 0.67 \cdot 10^{-4}$ emu/mol. The Curie temperature is obtained as $T_c = 1018$ K which is to be compared with the experimental value of 1044 K. The magnetic moment at the calculated Curie temperature is calculated to be $\mathcal{L} \simeq 1.07 M_s$. It should be stressed that the only input parameter is χ_{eff} which together with the *ab initio* exchange function $j(\mathbf{k})$ explains both the slope and the Curie temperature. It is interesting to observe that χ_{eff} is of the order of magnitude of a Brillouin-zone average of the non-uniform susceptibility $\chi(\mathbf{k})$ discussed in Sec. 4.2.3.

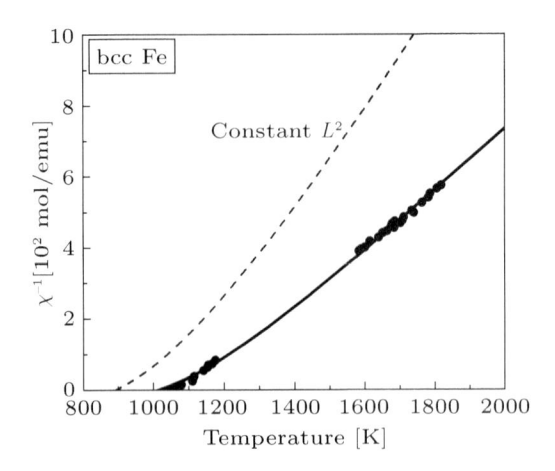

Figure 5.15 *Calculated inverse susceptibility of iron. The dashed curve is the inverse susceptibility in the spherical approximation with constant $\mathcal{L}^2$, the solid curve is obtained by solving Eqns (5.192) to (5.194) self-consistently for the case of a primitive lattice. Dots are the experimental data from Shimizu (1981).*

This section is closed with a remark about the mean-field approximation (MFA) for the multi-sublattice case which is of interest here. A classical review article is that by Anderson and Hasegawa (1955) and of the many modern applications of the MFA which use the LSDA only the recent paper by Sasioglu *et al* (2005) is mentioned here.

The MFA for the multi-sublattice case is obtained by including the $T=0$ magnetic moments in the definition of the exchange constants, i.e. $J_{\tau\tau'}(\mathbf{k}) \doteq M_\tau j_{\tau\tau'}(\mathbf{k}) M_{\tau'}$. The summed exchange $\sum_\mathbf{k} J_{\tau\tau'}(\mathbf{k})$ represents the mean field at site τ due to site τ' and is then diagonalized, the largest eigenvalue, $J_{\max}$, determining the Curie temperature in the MFA:

$$k_B T_c^{\mathrm{MF}} = (2/3) J_{\max}. \tag{5.197}$$

5.4.7.1 Exchange in detail

For practical applications of the theory to multi-sublattice cases the exchange functions $j_{\tau\tau'}(\mathbf{k})$ need be determined from spiral calculations and the total energy. We have dealt with some examples before, like hcp Co and Ni_3Al, but now want to go through a number of cases more systematically (Kübler, 2006).

We always start with the off-site total energy, $E_{\mathrm{OS}}(\mathbf{k})$, which is written by means of spin-spirals in the following way

$$E_{\mathrm{OS}}(\mathbf{k}) = \sum_{\tau\tau'} M_\tau M_{\tau'} [j_{\tau\tau'}(\mathbf{k}) \sin\theta_\tau \sin\theta_{\tau'} \cos(\varphi_\tau - \varphi_{\tau'})$$
$$+ j_{\tau\tau'}(0) \cos\theta_\tau \cos\theta_{\tau'}], \tag{5.198}$$

where M_τ, θ_τ, and φ_τ are the polar coordinates for the magnetic moment vector of atom τ. For brevity the energy is called spiral energy in the following.

Two magnetic atoms per cell

For the case of two magnetic atoms in the unit cell (two sublattices), like FeNi, CoNi, and NiMnSb, one determines the three functions $j_{11}(\mathbf{k})$, $j_{22}(\mathbf{k})$, and $j_{12}(\mathbf{k})$ by calculating the spiral energies four times for a set of k-points that should be dense enough to calculate the sums, like that occurring in Eqn (5.195), reliably. First one chooses $\theta_1 = \theta_2 = \theta$ and all azimuthal angles equal to zero, calling the resulting energy $E_0(\mathbf{k})$. Next the azimuthal angle is chosen to be $\varphi = \mathbf{k} \cdot \boldsymbol{\tau}$ and the result is denoted by $E_1(\mathbf{k})$. A third k-scan results in the energy $E_2(\mathbf{k})$ where the sign of the term $\cos(\mathbf{k} \cdot \boldsymbol{\tau})$ is changed with a choice of a reciprocal lattice vector obeying $\mathbf{K} \cdot \boldsymbol{\tau} = \pi$. Here one uses the fact that the exchange functions are periodic in reciprocal space. A last scan is carried out with $\theta_1 = 0$ and $\theta_2 = \theta$ denoting the energy by $E_3(\mathbf{k})$. The desired exchange functions can now be determined by subtracting the energy origin

$$E_{\text{OS}}(\mathbf{k}=0) = M_1^2 j_{11}(0) + M_2^2 j_{22}(0) + 2M_1 M_2 j_{12}(0). \tag{5.199}$$

The result is then written out in terms of the spiral energy differences $\Delta_i(\mathbf{k})$, $i = 0, 1, ...3$. Defining

$$S_a(\mathbf{k}) = \frac{1}{2} \sum_{i=1}^{2} \Delta_i(\mathbf{k}) \tag{5.200}$$

we obtain

$$j_{12}(\mathbf{k}) = [\Delta_0(\mathbf{k}) - S_a(\mathbf{k})] / 2M_1 M_2 \sin^2 \theta, \tag{5.201}$$

$$j_{11}(\mathbf{k}) = j_{11}(0) + [S_a(\mathbf{k}) - \Delta_3(\mathbf{k}) + 2M_1 M_2 j_{12}(0) F(\theta)] / M_1^2 \sin^2 \theta \tag{5.202}$$

and

$$j_{22}(\mathbf{k}) = j_{22}(0) + \left[4M_1 M_2 j_{12}(0) \sin^2 \frac{\theta}{2} + \Delta_3(\mathbf{k})\right] / M_2^2 \sin^2 \theta \tag{5.203}$$

where

$$F(\theta) = \sin^2 \theta - 2 \sin^2 \frac{\theta}{2}. \tag{5.204}$$

The exchange eigenvalues are given by the roots of a quadratic equation, i.e.

$$j_{n=1,2} = (j_{11} + j_{22})/2 \pm \sqrt{j_{12}^2 + [(j_{11} - j_{22})/2]^2}, \tag{5.205}$$

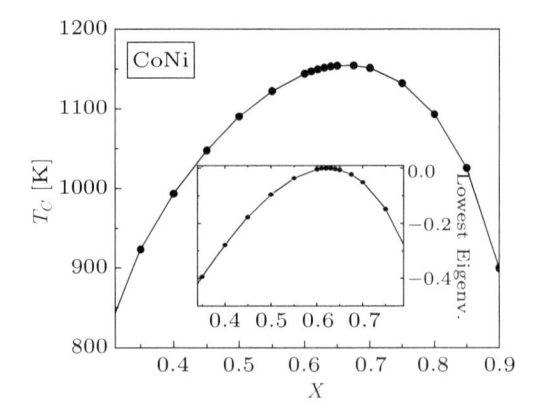

Figure 5.16 *The Curie temperature in the spherical approximation of CoNi as a function of the interpolation parameter x. The inset shows the lowest eigenvalue at $\mathbf{k} = 0$ as a function of x.*

where, for simplicity in writing, the dependence on $\mathbf{k}$ is implied. Equations (5.202) and (5.203) contain the as yet undetermined coefficients $j_{11}(0)$ and $j_{22}(0)$. Their sum is fixed by

$$M_1^2 j_{11}(0) + M_2^2 j_{22}(0) = \frac{1}{2} \sum_{i=1}^{2} E_i(0). \tag{5.206}$$

Thus one can determine the separate values by interpolating requiring that the lowest eigenvalue at $\mathbf{k}$ vanishes. The inset of Fig. 5.16 gives the lowest eigenvalue at $\mathbf{k} = 0$ as a function of the interpolation variable x. The result of this procedure is seen to be unique and the spherical approximation for the Curie temperature of CoNi is obtained as $T_c = 1149$ K, which should be compared with the experimental value of $T_c = 1140$ K.

Three magnetic atoms, two being equivalent

An example for this case is a Heusler compound, for instance Co_2MnSi, where the equivalent moments are those of Co and the crystal structure is cubic (L2$_1$). Denoting the compound by Co_2XY we need to determine the four exchange functions $j_{11}(\mathbf{k})$ (for the exchange interaction between the X atoms), $j_{22}(\mathbf{k})$ (for Co), $j_{12}(\mathbf{k})$ (for the exchange interaction between Co and X atoms), and $j_{23}(\mathbf{k})$ (for the exchange interaction between the Co atoms having different basis vectors). For this we do eight spiral energy scans using the following variables. One scan for $\theta > 0$ and $\varphi = 0$ for all atoms, labeled $E_0(\mathbf{k})$ and another one for $\theta = 0$ for atom X and the same $\theta > 0$ as well as $\varphi = 0$ for all atoms, labeled $E_{00}(\mathbf{k})$. Four more scans, labeled $E_i(\mathbf{k}), i = 1$ to 4, are done with $\theta > 0$ and $\varphi_\tau - \varphi_{\tau'} = \mathbf{k}' \cdot (\boldsymbol{\tau} - \boldsymbol{\tau}')$ choosing $\mathbf{k}' = \mathbf{k} + \mathbf{K}_i$ with $\mathbf{K}_1 = 0$ and $\mathbf{K}_2$ to $\mathbf{K}_4$ being the reciprocal lattice vectors $(0, 2, 0)$, $(1, 1, -1)$, and $(1, 1, 1)$ in units of $2\pi/a$. Finally E_5 and E_6 are obtained with $\theta = 0$ for atom X and $\theta > 0$ for the other atoms using $\varphi_\tau - \varphi_{\tau'} = \mathbf{k}' \cdot (\boldsymbol{\tau} - \boldsymbol{\tau}')$ with $\mathbf{K}_5 = 0$ and $\mathbf{K}_6 = (1, 1, 1)$.

If we denote the spiral energy differences by $\Delta_i(\mathbf{k})$ and define

$$S_a(\mathbf{k}) = \frac{1}{4}\sum_{i=1}^{4}\Delta_i(\mathbf{k}) \qquad (5.207)$$

and

$$S_b(\mathbf{k}) = \frac{1}{2}\sum_{i=5}^{6}\Delta_i(\mathbf{k}) \qquad (5.208)$$

together with

$$f_1 = 4M_1 M_2 (1 - \cos\theta) \qquad (5.209)$$

and

$$f_2 = f_1 \cos\theta \qquad (5.210)$$

then one derives

$$j_{23}(\mathbf{k}) = [\Delta_{00}(\mathbf{k}) - S_b(\mathbf{k})]/2M_2^2 \sin^2\theta \qquad (5.211)$$

and

$$j_{12}(\mathbf{k}) = [\Delta_0(\mathbf{k}) + S_b(\mathbf{k}) - S_a(\mathbf{k}) - \Delta_{00}(\mathbf{k})]/4M_1 M_2 \sin^2\theta \qquad (5.212)$$

as well as

$$j_{11}(\mathbf{k}) = j_{11}(0) + [S_a(\mathbf{k}) - S_b(\mathbf{k}) + j_{12}(0)f_2]/M_1^2 \sin^2\theta \qquad (5.213)$$

and

$$j_{22}(\mathbf{k}) = j_{22}(0) + j_{23}(0) + [S_b(\mathbf{k}) + j_{12}(0)f_1]/M_2^2 \sin^2\theta. \qquad (5.214)$$

A sum rule for the constants $j_{11}(0)$ and $j_{22}(0)$ allows one to adjust these such that the lowest branch of eigenvalues is zero for $\mathbf{k} = 0$ as in the case for two sublattices. The three branches are given for each wave vector $\mathbf{k}$ by the eigenvalues $j_{n=1}(\mathbf{k}) = j_{22}(\mathbf{k}) - j_{23}(\mathbf{k})$ and

$$j_{n=2,3} = (j_{11} + j_{22} + j_{23})/2 \pm \sqrt{2j_{12}^2 + (j_{11} - j_{22} - j_{23})^2/4} \qquad (5.215)$$

where, for simplicity in writing, the $\mathbf{k}$-dependence of the exchange functions is implied. The generalization to symmetries other than $L2_1$ is obvious.

Four magnetic atoms, three being equivalent

An example for this case is the Cu_3Au structure, as for instance Ni_3Fe, where the equivalent moments are those of Ni. The single magnetic moment is denoted with the label 1. There are now ten spiral energy scans needed. The derivation of the expressions for the exchange functions in terms of energy differences follows the same scheme as in the previous cases, albeit with the appropriate reciprocal-lattice vectors of the Cu_3Au structure and an obvious choice for the two additional calculations. One defines

$$S_a(\mathbf{k}) = \frac{1}{4} \sum_{i=1}^{4} \Delta_i(\mathbf{k}) \tag{5.216}$$

and

$$S_b(\mathbf{k}) = \frac{1}{4} \sum_{i=5}^{8} \Delta_i(\mathbf{k}) \tag{5.217}$$

and derives for the simpler case where all angles θ are chosen to be $90°$:

$$j_{11}(\mathbf{k}) = j_{11}(0) + [S_a(\mathbf{k}) - S_b(\mathbf{k})]/M_1^2, \tag{5.218}$$

$$j_{22}(\mathbf{k}) = j_{22}(0) + [S_b(\mathbf{k}) + 6M_1 M_2 j_{12}(0) + 6M_2 j_{23}(0)]/3M_2^2, \tag{5.219}$$

$$j_{12}(\mathbf{k}) = [\Delta_0(\mathbf{k}) + S_b(\mathbf{k}) - \Delta_9(\mathbf{k}) - S_a(\mathbf{k})]/4M_1 M_2, \tag{5.220}$$

and

$$j_{23}(\mathbf{q}) = [\Delta_9(\mathbf{k}) - S_b(\mathbf{k})]/6M_2^2. \tag{5.221}$$

Finally, the exchange eigenvalues are given by twice $j_{n=1,2} = j_{22} - j_{23}$ and

$$j_{n=3,4} = (j_{11} + j_{22} + 2j_{23})/2 \pm \sqrt{3j_{12}^2 + [(j_{11} - j_{22} - 2j_{23})/2]^2}, \tag{5.222}$$

where, for simplicity in writing, the k-dependence of the exchange functions is implied again. The open exchange constants are determined as before.

5.4.8 Collection of results

For the results that follow the spiral energies have been determined by using the force theorem and a spiral tilt angle of $\theta = 20°$ except for the Cu_3Au structures where $\theta = 90°$ was chosen. The reason for the particular choice of the magnetic compounds listed in

Table 5.6 *Collection of pertinent experimental and calculated data for 8 selected magnetic compounds. T_c^{SP} and T_c^{MF} denote calculated Curie temperatures in the spherical and mean-field approximation, respectively.*

Compound	lattice	a[Å]	c/a	$M_1[\mu_B]$	$M_2[\mu_B]$	T_c^{SP}[K]	T_c^{MF}[K]	T_c^{exp}[K]
FeNi[a]	CuAu	2.481	$\sqrt{2}$	2.551	0.600	968	1130	790
CoNi[a]	CuAu	2.459	$\sqrt{2}$	1.643	0.673	1149	1538	1140
FeNi$_3$[a]	AuCu$_3$	3.489		2.822	0.588	986	1290	870
CoNi$_3$[a]	AuCu$_3$	3.473		1.640	0.629	733	925	920
NiMnSb[b]	C1$_b$	5.920		3.697	0.303	968[c]	1281	730
Mn$_2$VAl[d]	L2$_1$	5.875		−0.769	1.374	580	663	760
Mn$_3$Ga[e]	L2$_1$	5.823		−2.744	1.363	314	482	
Mn$_3$Ga[e]	DO22	3.772	1.898	−2.829	2.273	762	1176	730

[a] Lattice constants and experimental Curie temperatures from Bonnenberg *et al.* (1986)
[b] Lattice constants and experimental Curie temperatures from de Groot *et al.* (1983b)
[c] $T_c^{SP} = 1091$ K if moment of Ni is neglected
[d] Lattice constants and experimental Curie temperatures from Weht and Pickett (1999)
[e] Wurmehl et al. (2006b), Balke *et al.* (2007)

Table 5.6 is the need to establish a certain level of confidence for the numerical procedure employed here. Thus FeNi and CoNi serve as examples for the two-sublattice case. The Heusler compound NiMnSb can also be treated as a two-sublattice case if one is interested in the role of the small Ni-moment. The two examples for the Cu$_3$Au structure are FeNi$_3$ and CoNi$_3$. The remaining cases possess three sublattices with two being equivalent.

It is seen in Table 5.6 that the Curie temperature of FeNi is overestimated. This is also so for FeNi$_3$ whereas CoNi is well described by the spherical approximation. The Curie temperature of CoNi$_3$ is underestimated, the experimental value being near the mean-field approximation. Thus, in these binary compounds, the presence of Fe leads to overestimated Curie temperatures.

The system NiMnSb is the much-studied half-metallic C1$_b$-Heusler compound (de Groot *et al.*, 1983a). A recent estimate by Sasioglu *et al.* (2005) is with 900 K close to our value in the spherical approximation. Ignoring the magnetic moment of Ni a rather high value for T_c^{SP} is obtained which does not agree with the results of Sasioglu *et al.* (2005).

The next cases in Table 5.6 are Heusler compounds like the prototypical Pd$_2$MnSn. Mn$_2$VAl is a half-metallic ferrimagnet for which Weht and Pickett (1999) calculated the electronic structure. The calculations underlying the present estimate of the Curie temperature were obtained in the LSDA, i.e. in contrast to the work of Weht and Pickett who used the generalized gradient approximation (GGA). Our calculations resulted in a smaller energy gap; this probably is the reason why even the the mean-field approximation underestimates the experimental Curie temperature. Generally it supplies an upper bound. The net magnetic moment obtained in the LSDA here is 1.98 μ_B to be compared with the 2 μ_B by Weht and Pickett. Our results support the statement of Weht and Pickett (1999) that the GGA makes a qualitative difference in the predicted behavior of Mn_2VAl.

In the case of Mn_3Ga, Mn occupies two different lattice sites: MnI on the Mn-site and MnII on the Pd-sites in the prototype Pd_2MnSn. It was predicted to be ferrimagnetic and half-metallic with a nearly zero net moment (Wurmehl et al., 2005a,b). Our LSDA calculations for Mn_3Ga give, indeed, a spin-polarization of 98% at the Fermi energy and an estimated Curie temperature of about room temperature. Recent measurements show, however, that Mn_3Ga prefers the tetragonal DO22-structure (Wurmehl et al. 2006b; Balke et al. 2007), for which indeed the LSDA calculations show a relative stability by 0.1 eV per formula unit. In the DO22-structure the spin-polarization is still of the order of 66% and the estimated Curie temperature is quite high with $T_c^{SP} = 762$ K. This value is correlated with a large increase of the magnetic moment of MnII labelled M_2 in Table 5.6. It should be mentioned that the Curie temperature was originally a prediction; the measurements were made after the appearance of the calculations by Kübler (2006).

We set out to enquire into the usefulness of Eqn (5.195) to supply reliable estimates of the Curie temperature of magnetic compounds. Although as it stands it—strictly speaking—contains the local moments squared, $\mathcal{L}_\tau^2$, as unknown quantities. For our estimates they are replaced by the $T = 0$ saturation moments. This is expected to work reasonably well for magnets in the local-moment limit. Elementary Fe, Co, and Ni to varying degrees are certainly not in this limit, but their binary compounds seem—at least in some sense—near this limit. The Heusler compounds discussed in Sec. 4.4.2.2 are expected to be in the local-moment limit. This statement is supported by Fig. 5.17, where we compare the measured with the calculated Curie temperatures for a large set of Co_2-Heusler compounds. We see very good agreement over a very large range of temperatures with rather few exceptions.

This leads to to the question as to whether the calculations described so far can be useful in distinguishing different exchange mechanisms. These, it should be emphasized, appear in the calculations in mixed form and are not easily untangled. This is in contrast to the model-Hamiltonian approach where different perturbation treatments give different exchange interactions, depending on the relative magnitude of the relevant exchange parameters. Still, it helps our understanding if such an analysis is attempted on the basis of calculated trends. These might reveal RKKY-type interactions (see Sec. 4.5.1.1) or concepts like *antiferromagnetic superexchange* (Anderson, 1950, 1963) and *ferromagnetic double exchange* through charge carriers (Zener, 1951; Anderson and Hasegawa, 1955; de Gennes, 1960; Anderson, 1963).

It is most likely Zener's theory that provides the key concept. In the wording of de Gennes (1960) it is summarized by three points: "(1) intra-atomic exchange is strong so that the only important configurations are those where the spin of each carrier is parallel to the local ionic spin." This is Hund's rule. "(2) The carriers do not change their spin orientation when moving; accordingly they can hop from one ion to the next only if the two ionic spins are not antiparallel; (3) when hopping is allowed the ground state energy is lowered (because the carriers are then able to participate in the binding). This results

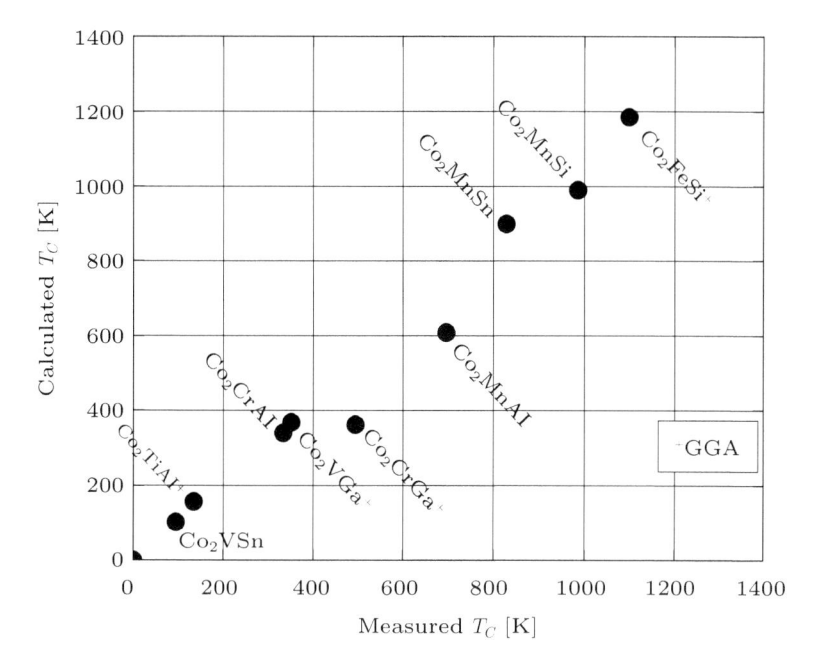

Figure 5.17 *Calculated versus measured Curie temperatures of a collection of Co_2-Heusler compounds. For the calculations the spherical approximation was used. From Kübler et al. (2007).*

in a lower energy for ferromagnetic configurations." This exchange mechanism through carriers is called double exchange, sometimes also kinetic exchange. It obviously applies to metals and, in particular, to half-metallic ferromagnets (HMF).

In contrast to this is the concept of superexchange (Anderson, 1950, 1963; Kramers, 1934) that in its pure form applies to insulators and explains antiferromagnetism. It operates through an intermediary nonmagnetic ion, often, but not necessarily, oxygen. In the oxygen ion the spins are antiparallel so that the electrons of one of the magnetic ions can move over to the other magnetic ion via the oxygen bridge only if their spins are antiparallel. If the spins were parallel the Pauli principle would partly forbid such a transition.

Turning now to Fig. 5.18(a) we see that the Curie temperatures increase with increasing valence electron concentration, N_V. This is what one expects of the double exchange mechanism (DE). But this view is too simple. Indeed, the exceptions to be seen for $N_V = 27$ in both Fig. 5.18(a) and Fig. 4.62 signal a break in the mechanism.

We may use Eqn (5.195) and look separately at the factors $\sum_\tau \mathcal{L}_\tau^2$ and $[\sum_{\mathbf{k}n} 1/j_n(\mathbf{k})]^{-1}$. These are plotted as "prefactor" and "exchange average," respectively, in the insets of Fig. 5.18(b). Moving up from with $N_V = 25$ one notices a sharp drop at $N_V = 27$. The low Curie temperature, T_C, of Co_2VSn could be attributed to the fact that this compound is not a HMF (see Table 4.12). But Co_2CrAl is calculated to be a HMF as Fig. 5.19 demonstrates. Its T_C is calculated (341 K) to be in very good agreement with the

450 *Magnetism at Finite Temperatures*

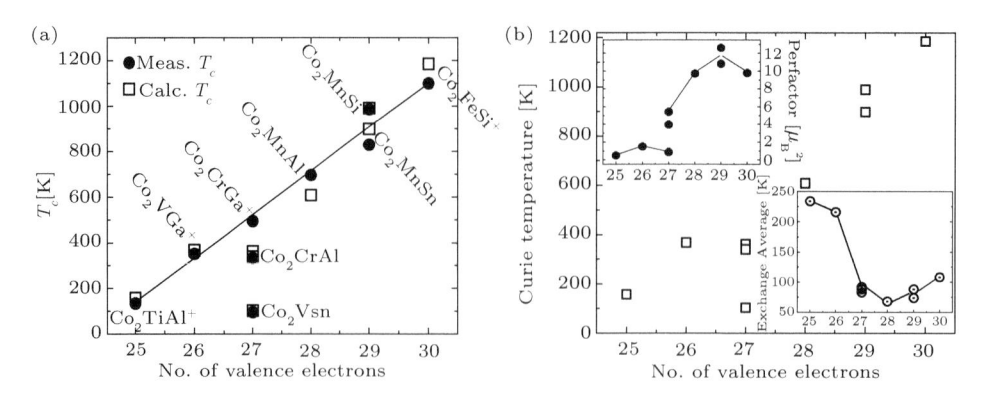

Figure 5.18 *(a) Calculated (squares) and measured (circles) Curie temperatures versus the number of valence electrons. The Heusler compounds marked with an asterisk were computed using the GGA. (b) Calculated Curie temperatures (squares) versus the number of valence electrons. The insets give the computed values of the two parts of Eqn (5.195), the prefactor and the exchange average; from Kübler et al. (2007).*

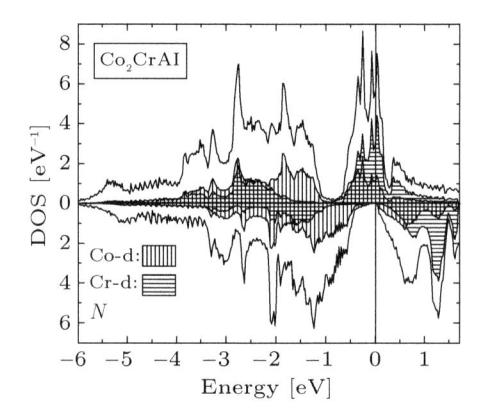

Figure 5.19 *Spin-resolved density of states of Co_2CrAl: upper panel is spin up, lower panel is spin down.*

measured (334 K) value, which is low. Figure 5.19 shows in the majority states huge state densities due to both Cr and Co at or near the Fermi energy, E_F. In Co_2CrGa (Fig. 4.64) these densities are lower. Unfortunately our calculation (in contrast to experiment) does not lead to a higher T_C here. Still, we connect the relatively low T_C with the high majority state density at or near E_F, the exchange mechanism still being most likely DE.

From $N_V = 28$ upward both the calculated and measured Curie temperatures follow a nearly linear upward trend that one expects from double exchange. Of course, other possibilities exist. Sasioglu *et al.* (2008) in a comparative study of Cu_2MnZ and Pd_2MnZ, where Z = In, Sn, Sb, and Te, argue for an RKKY mechanism. It is not clear whether or not their reasoning can be applied to the Co_2-Heusler compounds, where the half-metallic property seems crucial for the high values of T_C.

We finally turn to the double perovskites, for which the electronic structure was discussed in Sec. 4.4.2.3. Here we raise the question why the Curie temperatures graphed in Fig. 4.66 and partially collected in Table 4.13 are so large.

First we note that in all the compounds $Sr_2CrB''O_6$, where $B'' = $ W, Re, Os, and Ir, the Cr-moments are antiparallel to the the B''- moments. The same is true for all other cases given in Fig. 4.66 (Kobayashi et al., 1998; Sarma et al., 2000; Fang et al., 2001; Mandal et al., 2008). This is generally ascribed to a strong superexchange mechanism (SE). The ferromagnetic coupling between the 3d-elements is then believed to be due to the DE mechanism. Sarma et al. (2000) and Fang et al. (2001) provide some more details that, however, do not change the picture greatly. Note that the results calculated in the spherical approximation and contained in Table 4.13—and more results by Mandal et al. (2008)—are in good to fair agreement with the experimental data. The half-metallic property certainly plays a large role here, comparable to the Heusler compounds.

The surprising result is the ferrimagnetic insulator Sr_2CrOsO_6. It was synthesized by Krockenberger et al. (2007) with a Curie temperature of 725 K. The calculations on the basis of the electronic structure shown in Fig. 4.69 gave 881 K. With no conduction electrons available we cannot invoke the double exchange mechanism. Thus, at present, the ferrimagnetic with a large Curie temperature remains unexplained.

5.5 Magnetic skyrmions

The topic we are to address now had a beginning in 1989 when Bogdanov and Yablonskii (1989) considered thermodynamic stable "vortices" in magnetic crystals. Ishikawa et al. (1976) much earlier discovered experimentally fascinating magnetic structures in MnSi, which these authors described as *helical*. Bak and Jensen (1980) as well as Schaub and Mukamel (1985) presented a theory for incommensurate structures and Binz et al. (2006) poposed a theory for the helical spin state of MnSi. The term *skyrmion* was used by Rössler et al. (2006) to describe a particle-like structure in magnetic metals such as MnSi. The name derives from the work of Skyrme (1961) who postulated topologically protected particles in continuous fields. Pokrovsky (1979) apparently was the first to apply the skyrmion model to solid-state matter.

A magnetic skyrmion by present understanding is a local whirl or vortex of the spin configuration in a magnetic material. As we will see, a skyrmion can be defined by a topological number, the so-called skyrmion number, which is a measure of the winding of the magnetic moments. It is topologically protected by the skyrmion number; it behaves as a particle that can be moved, created, and annihilated. It has attracted a tremendous amount of research, largely because of its possible technical applications. Recent reviews are by Fert et al. (2017), Everschor-Sitte et al. (2018), and Göbel et al. (2020). A concise mathematical treatment is by Melcher (2014).

We show in Fig. 5.20 the phase diagram of MnSi that was obtained by Mühlbauer et al. (2009) using neutron diffraction measurements. It is especially the A-phase that hosts the interesting spin structure, two-dimensional magnetic skyrmions. Before going to a

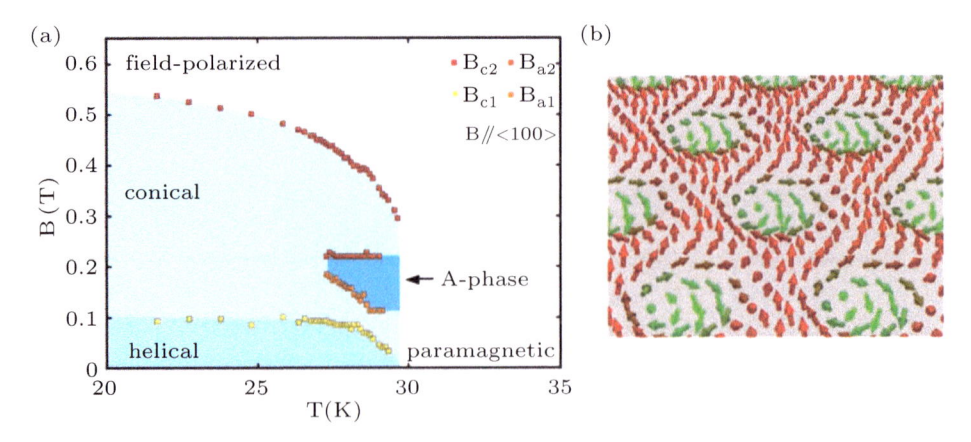

Figure 5.20 *A: The phase diagram of MnSi as a function of the temperature and the magnetic field. B: A schematic view of the triangular crystal of skyrmions that develops in the A-phase bounded by the the values of the magnetic field indicated in the figure. Taken from Mühlbauer et al. (2009).*

detailed mathematical description we list in Table 5.7, collected by Nagaosa and Tokura (2013), a number of compounds that host magnetic skyrmions. All compounds in this list possess the B20 crystal structure, which have the symmetry P2$_1$3 and the space group no. 198. Magnetic skyrmions are also found in other symmetry groups, but these are of the first generation (others will be mentioned later on in Sec. 5.5.3). In Fig. 5.21 we depict the crystal structure of B20 MnSi showing the entiomorphous pair L-R on the left and R-L on the right, L-R meaning that the Mn atoms are positioned along a left-handed and Si along a right-handed axis and opposite for R-L. Concerning chiral crystal structures we refer the interested reader to a paper by Flack (2003) to find a brief treatment of chiral structures, an explanation of the terminology, and important references. We note that all compounds in Table 5.7 are metallic except for Cu_2OSeO_3 which is an insulating ferrimagnet. The Néel temperatures of the compounds span a large range and so do the helical periods. MnGe is an exception with $\lambda = 3$ nm which is small enough to encourage an *ab initio* calculation of the type described at the end of Chap. 3 with a result possibly similar to Fig. 3.21. To do this, however, we need to repeat some basic facts first and go back to Chap. 4. In connection with weak ferromagnetism we learned that relativistic effects and the breaking of inversion symmetry leads to the famous Dzyaloshinsky (1958)–Moriya (1960) (DM) interaction, which is given in Eqn 4.263 and was discussed there.

If for the moment we consider a pair of spins, S, which suspend an angle θ, we can approximate their energy by

$$E(\theta) = JS^2 \cos\theta + DS^2 \sin\theta, \qquad (5.223)$$

where J is the Heisenberg exchange constant and D the DM interaction. The minimum of E is easily seen to be at $\theta_0 = \arctan(D/J)$, so that the helical period $\lambda = 2\pi a/\theta_0$ becomes $\lambda = 2\pi a/\arctan(D/J)$, where a is the lattice constant. If we solve this simple relation for D/J plugging in the measured λ and the lattice constant a, we obtain an idea

Table 5.7 *Helimagnets: List of transition temperatures, T_N, and helical periods, λ.*

Materials		T_N(K)	λ(nm)	References
MnSi	Bulk	30	18	(a)
	Epitaxial thin film	45	8.5	(b)
$Mn_{1-x}Fe_xSi$	$x = 0.06$	16.5	12.5	(c)
	$x = 0.8$	10.6	11	(c)
	$x = 0.10$	6.8	10	(c)
$Fe_{1-x}Co_xSi$	$x = 0.10$	11	43	(d)(e)
	$x = 0.5$	36	90	(d)(e)
	$x = 0.6$	24	174	(d)(e)
	$x = 0.7$	7	230	(d)(e)
MnGe	$T = 20K$	170	3	(f)
	$T = 100K$	–	3.4	(f)
	$T = 150K$	–	5.5	(f)
$Mn_{1-x}Fe_xGe$	$x = 0.35$	150	4.7	(g)
	$x = 0.5$	185	14.5	(g)
	$x = 0.7$	210	77	(g)
	$x = 0.84$	220	220	(g)
FeGe	Bulk	278	70	(h)
Cu_2OSeO_3	Bulk	59	62	(i)

(a) Ishikawa *et al.* (1976)
(b) Li *et al.* (2013)
(c) Grigoriev *et al.* (2009)
(d) Beille *et al.* (1983)
(e) Onose *et al.* (2005)
(f) Kanazawa *et al.* (2011)
(g) Shibata *et al.* (2013)
(h) Lebech *et al.* (1989)
(i) Adams *et al.* (2012)

about the size of the ratio D/J, which for MnGe and $\lambda = 3$ nm comes out to be of order 1. For MnSi, however, it is an order of magnitude smaller.

An early theory by Nakanishi *et al.* (1980) demonstrated that the DM interaction in a ferromagnet having the B20 structure indeed leads to a helical ground state. In order to explain the helical state *ab initio* we may consider a calculation of a general

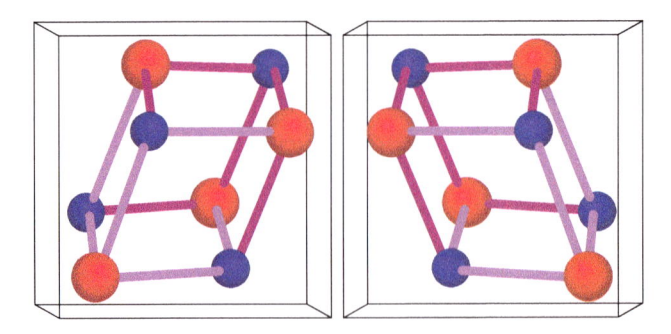

Figure 5.21 *The two chiral crystal structures of a B20 crystal, space group no. 198, $P2_13$ symmetry. For MnSi red marks the Mn atom, blue Si. Notice the missing center of inversion.*

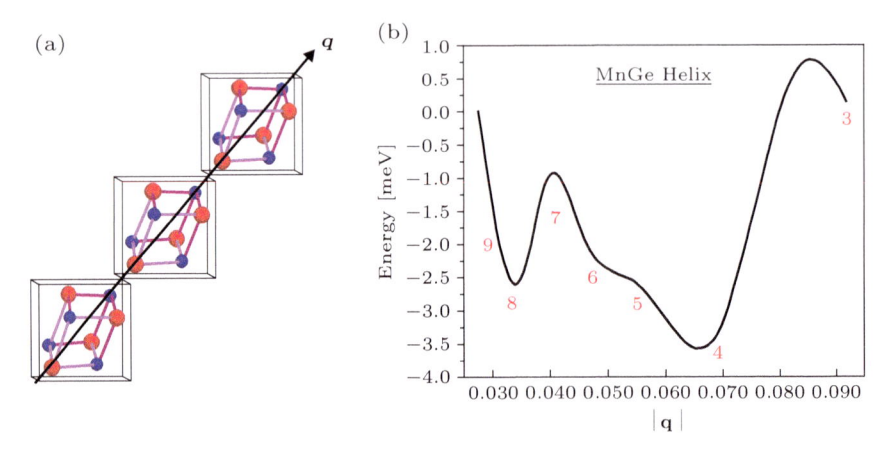

Figure 5.22 *Total-energy change of a helimagnet. (a) Three unit cells stacked along the body diagonal. (b) Total-energy change as a function of $\mathbf{q}$ for various numbers (red) of unit cells stacked as in (a). Note that for the actual calculation only one cell is needed.*

spin-spiral as discussed at the end of Chap. 3. However, we encounter a problem: the spin–orbit coupling, which is so important here, breaks the symmetry of the special unitary group SU(2) so that the generalized Bloch theorem for a spiral structure can no longer be applied. A supercell calculation is presently a natural choice and was undertaken by Bornemann et al. (2019) to describe a spin-spiral for MnGe. Their calculation, however, did not give the desired result. We will come back to calculations of this type. Sandratskii (2017) suggested an approximation to be made on the spin–orbit interaction such that it no longer breaks the SU(2) symmetry. This is achieved by dropping the terms $\sigma_x \hat{L}_x$ and $\sigma_y \hat{L}_y$ in Eqn (3.352). Only a numerical test will show if this approximation will lead to an answer that reflect the salient physics. We show in Fig. 5.22 the result of such a calculation. Here the vector $\mathbf{q}$ was chosen along the body diagonal of the unit cell with $\theta = 90°$, i.e. the magnetic moment, calculated to be $2\mu_B$

per Mn, is perpendicular to the body diagonal. This particular choice is advantageous for symmetry reasons. As shown in the figure we imagine that we choose the vector **q** by stacking three unit cells along the body diagonal as sketched in (a), then successively increasing this number. The total-energy change with respect to the ferromagnetic state is shown in part (b) of the figure. One sees that a total-energy minimum occurs between 4 and 5 unit cells corresponding to a helical period of $\lambda = 3.1$ nm and $\lambda = 3.9$ nm. This is of the correct order of magnitude, but may be fortuitous. Clearly, this atomistic approach will not succeed in explaining a swirling structure as shown in Fig. 5.20, therefore one replaces the atomistic model by a micromagnetic one, i.e. one expresses the Hamiltonian in the continuum approximation. Bornemann et al. (2019) sketched a derivation of this model, which in the more standard notation of Nagaosa and Tokura (2013) leads to the Hamiltonian

$$H = \int d^3 r \left[\frac{J}{2} (\nabla \mathbf{n}) + D \mathbf{n} \cdot (\nabla \times \mathbf{n}) - \mathbf{B} \cdot \mathbf{n} \right], \quad (5.224)$$

where $\mathbf{n} = \mathbf{n}(\mathbf{r})$ is the spin density at the position $\mathbf{r}$ and $\mathbf{B}$ is the magnetic field. Actually the continuum approximation is justified when D/J is small as in MnSi. So there may be a problem for MnGe, which we ignore for the present.

5.5.1 Formal properties of magnetic skyrmions

We next come to the topological classification of magnetic skyrmions. Its swirling structure is characterized by the skyrmion number, also called winding number (Braun, 2012; Nagaosa and Tokura, 2013), defined by

$$N_{Sk} = \frac{1}{4\pi} \int \int d^2 r \, \mathbf{n} \cdot \left(\frac{\partial \mathbf{n}}{\partial x} \times \frac{\partial \mathbf{n}}{\partial y} \right). \quad (5.225)$$

By assuming the spin density is normalized to 1, following Nagaosa and Tokura (2013), we express the shape of a skyrmion in the following form

$$\mathbf{n}(\mathbf{r}) = [\cos \Phi(\phi) \sin \Theta(r), \sin \Phi(\phi) \sin \Theta(r), \cos \Theta(r)]. \quad (5.226)$$

Substituting this expression into Eqn (5.225) we obtain using the in-plane polar coordinates $\mathbf{r} = (r \cos \phi, r \sin \phi)$:

$$N_{Sk} = \frac{1}{4\pi} \int_0^\infty dr \int_0^{2\pi} d\phi \frac{d\Theta(r)}{dr} \sin \Theta \frac{d\Phi(\phi)}{d\phi} = \frac{1}{2} [\cos \Theta]_\infty^0 \frac{1}{2\pi} [\Phi(\phi)]_0^{2\pi}. \quad (5.227)$$

This result allows us to classify the skyrmion structure by assuming e.g. the spin points up at the center $r = 0$ and down at $r = \infty$ then $\cos \Theta |_\infty^0 = 2$. The other factor describes the vorticity, $m = [\Phi(\phi)]_0^{2\pi}/2\pi$, and the helicity, γ, by defining

$$\Phi(\phi) = m\phi + \gamma. \quad (5.228)$$

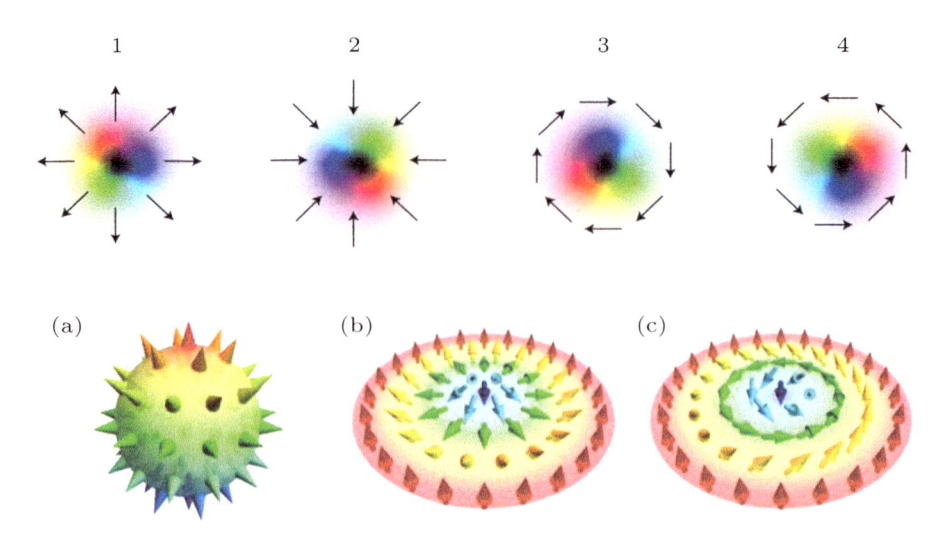

Figure 5.23 *Elements of the skyrmion spin structure. The sketches numbered 1 to 4 correspond to $m = 1$ and $\gamma = 0, \pi, -\pi/2, \pi/2$ taken from Nagaosa and Tokura (2013). (a) Hedgehog, obtained from (b) $m = 1$ and $\gamma = 0$, (c) $m = 1$ and $\gamma = \pi/2$. Taken from Everschor-Sitte et al. (2018).*

The vorticity therefore determines the skyrmion number as $N_{Sk} = m$. We show in the upper section of Fig. 5.23 the "Φ-contributions" to the skyrmion structure for $m = 1$ and $\gamma = 0, \pi, -\pi/2, \pi/2$ corresponding to the numbers 1 to 4. The lower part of the figure shows the combined effect of $\cos\Theta(r)$ and Φ; (a) is called a hedgehog and is obtained from (b) by wrapping the two-dimensional disk onto the unit sphere, (b), using $m = 1$ and $\gamma = 0$. This structure is called a Néel skyrmion and is observed on interfaces (Heinze et al., 2011). Part (c) is obtained by using $m = 1$ and $\gamma = \pi/2$, it is called a Bloch skyrmion and is observed in MnSi (Mühlbauer et al., 2009). We stress that these depict two-dimensional structures; in the third space dimension they are tubes. Structures obtained with $m = -1$ are called antiskyrmions. Note that the structures depicted in Fig. 5.20B form a lattice of Bloch skyrmions. In the review articles by Everschor-Sitte et al. (2018) and Göbel et al. (2020) the interested reader may find a wealth of further structures.

The next topic is concerned with the electrodynamics arising from skyrmions and their motion. An early paper dealing with this problem is by Volovik (1987), who derived quite generally for spin-polarized electrons what is now called the emerging electromagnetic field (EEMF)

$$b_\alpha = \frac{1}{2}\varepsilon^{\alpha\beta\gamma}\mathbf{n}\cdot(\partial_\beta \mathbf{n} \times \partial_\gamma \mathbf{n})\,, \tag{5.229}$$

where $\partial_\mu = \partial/\partial x_\mu$ and $\varepsilon^{\alpha\beta\gamma}$ is the totally antisymmetric tensor in three dimensions. Correspondingly there is an emergent electric field given by

$$e_\alpha = \mathbf{n} \cdot (\partial_\alpha \mathbf{n} \times \partial_t \mathbf{n}) \ . \tag{5.230}$$

Notice that the skyrmion number N_{Sk} given by Eqn (5.225) is the integral over the unit sphere of the emergent magnetic field in the z-direction.

The EEMF is of great importance for the anomalous Hall effect, which in this case is designated as *topological* Hall effect (Neubauer *et al.*, 2009; Schulz *et al.*, 2012; Nagaosa and Tokura, 2013). The motion of skyrmions leads to a time-dependent change of the emergent magnetic field and thus to an induced emergent electric field (Yokouchi *et al.*, 2020).

5.5.2 The phase transition

We now turn to an early mean-field theory for the phase transition. Following Rössler *et al.* (2006), the magnetic free-energy density is written in the form

$$f = Am^2(\nabla \mathbf{n})^2 + \lambda A(\nabla m)^2 + f_D(\mathbf{m}) + f_o(m) \ , \tag{5.231}$$

where the first and second term describe the magnetic stiffness. As in Eqn (5.224) $\mathbf{n} = \mathbf{n}(\mathbf{r})$ is the unit vector along the magnetization $\mathbf{m} = m\mathbf{n}$ and $A = J/2$. The coefficient λA with $\lambda < 1$ is introduced to describe the softening of the amplitude of the magnetization. (For MnSi $\lambda = 0.4$ is chosen.) The term $f_D(\mathbf{m}) = D\mathbf{M} \cdot (\nabla \times \mathbf{m})$ describes the chiral interaction with D being the DM constant. The last term, $f_o(m)$, represents the Landau expansion of the free energy in terms of even powers of m, which is written as $f_o = \alpha(T - T_C)m^2 + \backslash betam^4 + ...$ and leads in the absence of the chiral interaction to the usual mean-field value of the Curie temperature. To simplify the interaction we first set $m = 1$ and look at the part of the free-energy density that appears in Eqn (5.224), i.e. the first and third term of Eqn (5.231). Substituting Eqn (5.226) into Eqn (5.224) we obtain after some manipulations

$$H = 2\pi \int r dr \{ A[(\frac{d\Theta}{dr})^2 + \frac{\sin^2\Theta}{r^2}] + D[\frac{d\Theta}{dr} + \frac{\sin\Theta\cos\Theta}{r}] - B_z \cos\Theta \} \ . \tag{5.232}$$

This function is now minimized with standard methods of the calculus of variation resulting in the Euler equation

$$A[\frac{d^2\Theta}{dr^2} + \frac{1}{r}\frac{d\Theta}{dr} - \frac{\sin\Theta\cos\Theta}{r^2}] - \frac{D\sin^2\Theta}{r} - B_z \sin\Theta = 0 \ . \tag{5.233}$$

This second-order differential equation is solved numerically; assuming the boundary conditions $\Theta(r=0) = \pi$ and $\Theta(r \to \infty) = 0$, one obtains a smooth function for $\Theta(r)$, whose shape depends on the coefficients A, D, and the magnetic field, B_z, see for instance Wilson *et al.* (2014). This describes a single skyrmion in a frozen magnet and looks like Fig. 5.23(b) or (c) since the vorticity is undetermined.

For a more realistic description one needs to include the varying size of the magnetic moment $\mathbf{m}(\mathbf{r}) = m(r)\mathbf{n}(\mathbf{r})$. The analysis is slightly more involved but straightforward. One obtains now two Euler equations, one for $\Theta(r)$ and another one for $m(r)$ (Rössler et al., 2011):

$$A[\frac{d^2\Theta}{dr^2} + \frac{1}{r}\frac{d\Theta}{dr} - \frac{\sin\Theta\cos\Theta}{r^2}] - \frac{D\sin^2\Theta}{r} - B_z\sin\Theta + A\frac{2}{m}\frac{dm}{dr}[\frac{d\Theta}{dr} - 1] = 0 \quad (5.234)$$

and

$$A[\frac{\lambda}{m}(\frac{d^2m}{dr^2} + \frac{dm}{rdr}) + (\frac{d\Theta}{dr})^2 + \frac{\sin^2\Theta}{r^2}] + D[\frac{d\Theta}{dr} + \frac{\sin\Theta\cos\Theta}{r}] - \frac{B_z\cos\Theta}{m} + \frac{\partial f_0}{m\partial m} = 0 \quad (5.235)$$

The solution of these equations for $\Theta(r)$ and $m(r)$ are discussed by Rössler et al. (2011) and depend as before on the boundary conditions, which were chosen to be $\Theta(0) = \pi$ as well as $m(R) = m_o$, the magnetization at saturation. We omit all details and show instead in Fig. 5.24(a) the shape of the function Θ and the variation of the magnetic moment m as function of r. Fig. 5.24(b) demonstrates that the stabilization of the skyrmion depends sensitively on the variation of the size of the magnetic moment. What is missing in the mean-field approximation so far are magnetic fluctuations. It was shown by Mühlbauer et al. (2009) that these are essential in stabilizing the lattice of skyrmions.

For the formulation of a theory of fluctuations we follow Chaikin and Lubensky (1995) and use the functional integral approach to write the partion function as

$$\mathcal{Z} = \int \mathcal{D}\mathbf{r}\exp\{-\beta[\mathcal{H}_{mf}(\mathbf{m}(\mathbf{r})) - \int d^3\mathbf{B}\cdot\mathbf{m}(\mathbf{r}) + \mathcal{H}_{fl}]\}, \quad (5.236)$$

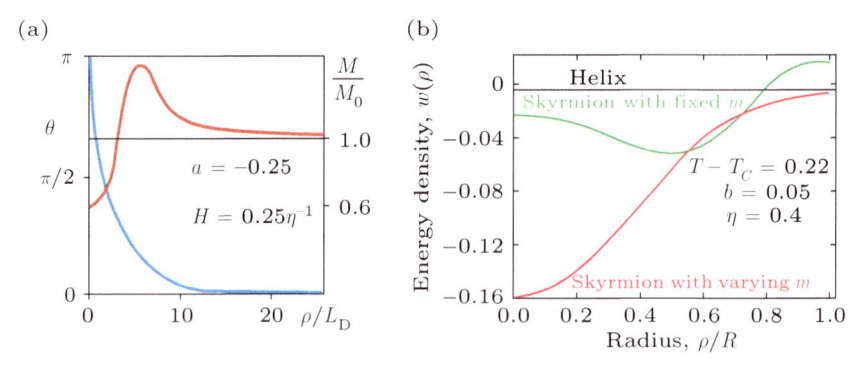

Figure 5.24 (a) The solution for $\Theta(r)$ and $m(r)$ of the Euler equations, reprinted from Rössler et al. (2011) and (b) the change of the energy density as a function of r, after Rössler et al. (2006).

where $\beta = 1/k_B T$ and $\mathbf{r}$ ideally takes all possible paths, $\mathbf{m}(\mathbf{r})$ being the corresponding spin density. The mean-field Hamiltonian, $\mathcal{H}_{mf}$, is by now familiar and is

$$\mathcal{H}_{mf} = \int d^3 r \{r_o \mathbf{m}(\mathbf{r})^2 + u\mathbf{m}(\mathbf{r})^4 + J[\nabla \mathbf{m}(\mathbf{r})]^2 + 2D\mathbf{m}(\mathbf{r}) \cdot [\nabla \times \mathbf{m}(\mathbf{r})]\} . \quad (5.237)$$

The fluctuations are calculated by expanding to second order in $\delta\mathbf{m}(\mathbf{r})$ around the saddle points of the mean-field free energy,

$$\mathcal{H}_{fl} = \frac{1}{2} \sum_{i,j} \int d^3 r \delta m_i(\mathbf{r}) G_{ij} \delta m_j(\mathbf{r}) , \quad (5.238)$$

where i, j label the components of $\delta\mathbf{m}(\mathbf{r})$ and

$$G_{ij} = \frac{\partial^2 F_{mf}}{\partial M_i \partial M_j}\Big|_\mathbf{M} \quad (5.239)$$

where $\mathbf{M}$ denotes the mean-field value of the magnetization and F_{mf} is the mean-field free energy. The fluctuation contribution to the latter is next obtained by Gaussian integration, which, we remind the reader, uses the formula

$$\int (\prod_{k=1}^{n}) e^{-\frac{1}{2}x_k G_{kk'} x_{k'}} = \prod_{p=1}^{n} (\frac{2\pi}{G_p})^{1/2} = (2\pi)^{n/2} (\det G)^{-1/2} , \quad (5.240)$$

where p denotes eigenvalues. Finally, taking the logarithm of the partition function $\mathcal{Z}$, dropping an unimportant constant and simplifying, we obtain for the free energy

$$F(\mathbf{M}, T) = F_{mf}(\mathbf{M}, T) + \frac{1}{2} \log \det G(\mathbf{M}, T) , \quad (5.241)$$

with $G(\mathbf{M}, T)$ given by Eqn (5.239).

The work necessary to evaluate these formulas is enormous; it is described in great detail in the Supplement to Mühlbauer et al. (2009). Disregarding the previous ansatz for the spin density, Eqn (5.226), these authors are inspired by their neutron-diffraction results, which show Bragg spots of six-fold symmetry in the A-phase of the MnSi crystal. They therefore expand the spin density, $\mathbf{M}(\mathbf{r})$, in terms of the hexagonal basis vectors as sketched in Fig. 5.25A. In detail they formulate

$$\mathbf{M}(\mathbf{r}) = \mathbf{M}_f + \sum_{i=1}^{3} \mathbf{M}_{\mathbf{Q}_i}(\mathbf{r} + \Delta\mathbf{r}) , \quad (5.242)$$

with

$$\mathbf{M}_{\mathbf{Q}_i} = A[\mathbf{n}_{i1} \cos(\mathbf{Q}_i \cdot \mathbf{r}) + \mathbf{n}_{i2} \sin(\mathbf{Q}_i \cdot \mathbf{r})] . \quad (5.243)$$

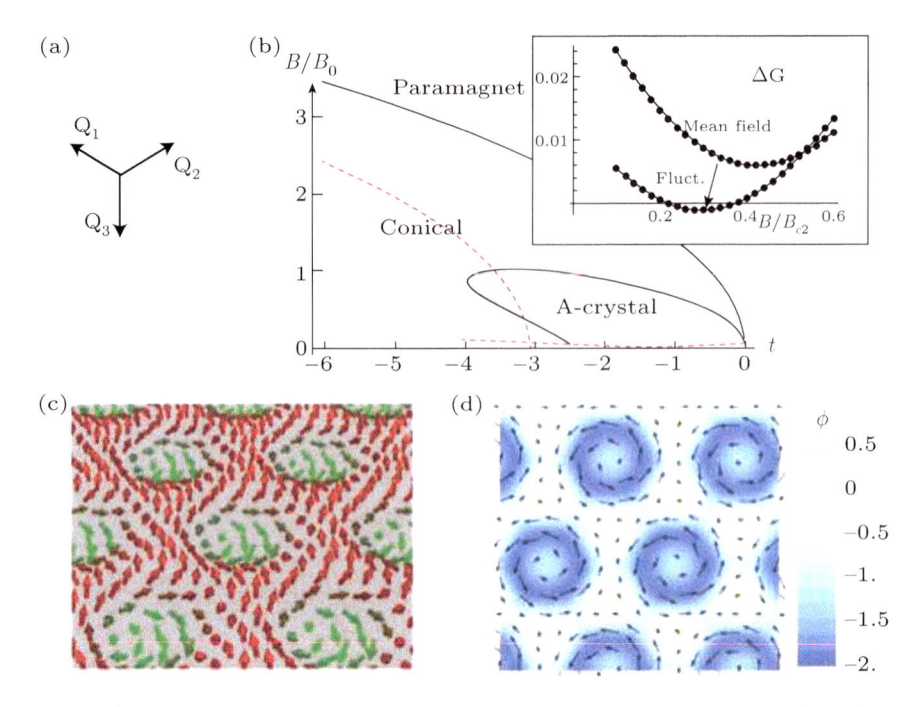

Figure 5.25 *The skyrmion lattice of MnSi, reprinted from Mühlbauer et al. (2009). (A) the hexagonal basis vectors of the spin order in the A-phase. (B) Theoretical phase diagram as a function of the reduced magnetic field and t, which is proportional to $T - T_c$. In the shaded region the A-phase is stable. The inset gives the energy difference (ΔG is ΔF in our notation) between the A-phase and the conical phase as a function of the field at the temperature $t = -3.5$. (C) Real space picture of the spin arrangement in the A-phase, x-y plane. (D) Calculated skyrmion density per unit area for the A-phase.*

This represents a chiral helix with amplitude A, wave vector $\mathbf{Q}_i$, and two orthogonal unit vectors $\mathbf{n}_{i1}$ and $\mathbf{n}_{i2}$, which are also orthogonal to $\mathbf{Q}_i$. We omit further detail and draw attention to Fig. 5.25B, which clearly demonstrates that the fluctuations stabilize the skyrmion lattice in the A-phase; this is the outstanding feature of the phase diagram shown in Fig. 5.20A.

5.5.3 New developments

We come back to the case of MnGe that was briefly discussed at the beginning of this section. The magnetic structure of MnGe below $T_N \simeq 170$K consists of three-dimensional skyrmions, in contrast to the two-dimensional ones of MnSi. The skyrmion structure is shown in Fig. 5.26 together with its phase diagram.

A combination of measurements led to this picture; Kanazawa et al. (2012) did small-angle neutron diffraction measurements and Tanigaki et al. (2015) used Lorentz transition electron measurements for real-space properties. The larger context is elaborated in a review article and its supplement by Kanazawa et al. (2016).

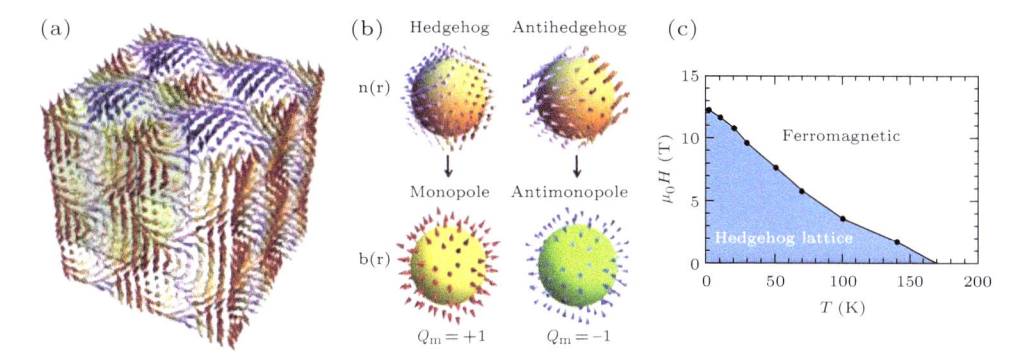

Figure 5.26 *Lattice of skyrmions of MnGe. (A) Spin orientaton of a $2 \times 2 \times 2$ magnetic unit cell. (B) Hedgehog and antihedgehog spin arrangements $\mathbf{n}(\mathbf{r})$ realized in the spin structure of MnGe. $\mathbf{b}(\mathbf{r})$ is the corresponding emergent magnetic field acting on the conduction electrons. Taken from Kanazawa et al. (2016). (C) The phase diagramm of MnGe, taken from Kanazawa et al. (2017).*

Here (see also Park and Han, 2011) the magnetic structure is modeled by three orthogonal helical structures of the same length q:

$$\begin{aligned}\mathbf{M}(\mathbf{r}) &= M_0(\cos qz, \sin qz, 0) + M_0(\sin qy, 0, \cos qy) + M_0(0, \cos qx, \sin qx) \\ &= M_0(\sin qy + \cos qz, \sin qz + \cos qx, \sin qx + \cos qy),\end{aligned} \quad (5.244)$$

where M_0 is the amplitude of the helical moment. In contrast to the two-dimensional skyrmion crystal, there occur points in the three-dimensional magnetization where $\mathbf{M}(\mathbf{r}) = 0$. Set for instance $(qx, qy, qz) = (\pi/4, 3\pi/4, -3\pi/4)$. This has an interesting consequence: each hedgehog or antihedgehog is accompanied by a monopole or correspondingly by an antimonopole, for which the magnetic charge $Q_m = \frac{1}{4\pi} \int_\mathbf{S} \mathbf{b} \cdot d\mathbf{S} = \pm 1$, see Fig. 5.26B. The emergent magnetic field, $\mathbf{b}$, is calculated by means of Eqn (5.229) using $\mathbf{n}(\mathbf{r}) = \mathbf{M}(\mathbf{r})/|\mathbf{M}(\mathbf{r})|$ and the sphere $\mathbf{S}$ encloses the singularity $\mathbf{M}(\mathbf{r}) = 0$. The details of the derivation are found in the supplement to Kanazawa et al. (2016).

It is worth noting that the topology is defined for a continuous function $\mathbf{n}(\mathbf{r})$ and is not rigorous for $\mathbf{n}_i$ on the atomic lattice. Even though the periodic modulation is relatively small, of the size of 3 nm to 6 nm, it should become clear by comparing Fig. 5.22 with Fig. 5.26A that 1/8th of the latter cube contains about 64 atomic unit cells. It is therefore not surprising that any calculations on the atomic scale, such as those of Bornemann et al. (2019) and Choi et al. (2019), are very difficult and perhaps not clarifying.

In view of the monopoles the magnetization will be unusual and interesting with anomalies in physical properties, as the anomalous Hall effect etc. The magnetic configuration under a magnetic field in the z-direction can approximately be descibed by the formula

$$\mathbf{M}_{m_z}(\mathbf{r}) = \mathbf{M}(\mathbf{r}) + (0, 0, m_z), \quad (5.245)$$

where m_z describes the uniform magnetization induced by the external magnetic field. The latter does not destroy the topological singularity; up to some critical field it shifts the singularities and thus the monopoles/antimonopoles. At finite temperatures we expect that the spin structure and the corresponding monopole positions thermally fluctuate this way resulting in fluctuating emergent magnetic fields.

Having dealt with the theory at some length, we will not describe the various anomalous experimental properties in any detail but refer to the literature. Thus the Nernst and Hall effects are covered thoroughly by Shiomi *et al.* (2013), whereas the topological Hall effect, the large magnetoresistance, and elastic anomalies are exposed by Kanazawa *et al.* (2016). We close this section by leaving the B20 compounds and briefly look at other crystal structures that may—or do in fact—host magnetic skyrmions. There is a class of Heusler compounds that are not centro-symmetric, an example being Mn_2RhSn. It has the $I\bar{4}m2$ structure with two non-equivalent Wyckoff positions occupied by Mn atoms: Mn_I at $2b(0,\frac{1}{2},0)$ and M_{II} at $2d(0,\frac{1}{2},\frac{3}{4})$. Sn occupies the $2a(0,0,0)$ and Rh the $2c(0,\frac{1}{2},\frac{1}{4})$ position. Meshcheriakova *et al.* (2014) found experimentally and theoretically the spin arrangement in the ground state to be noncollinear, the moment of Mn_{II} deviating from the *c*-axis by about 55°, whereas that of Mn_I stays parallel to the *c*-axis. The compound undergoes a spin-reorientation transition, induced by a temperature change and suppressed by an external magnetic field. Because of the presence of the Dzyaloshinsky—Moriya exchange interaction one expects to see skyrmions in Mn_2RhSn, but none were found. Another interesting compound is $Mn_{1.4}PtSn$. Vir *et al.* (2019) describe the crystal structure as non-centrosymmetric spacegroup no. 122, $I\bar{4}2d$, having D_{2d} symmetry.

The single crystal of $Mn_{1.4}PtSn$ was analyzed by neutron diffraction and measurements of the anomalous Hall effect succeeded in separating the topological Hall effect from the anomalous (**k**-space) Hall effect, because the crystal undergoes a spin-reorientation transition at $T \approx 170$ K, below which the magnetic order is noncollinear becoming ferromagnetic collinear above this temperature, as is shown in Fig. 5.27.

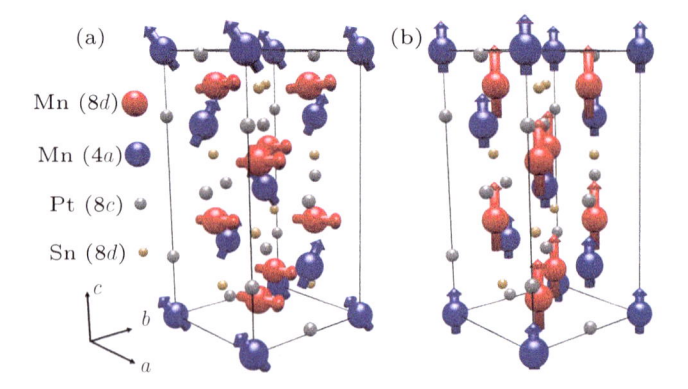

Figure 5.27 *The magnetic structure of $Mn_{1.4}PtSn$ based on neutron-structure refinement for (a) 2 K < T < T_{SR} and (b) T_{SR} < T < T_C, where $T_{SR} = 170$ K and $T_C = 392$ K, reprinted from the supplement to Vir et al. (2019). Note, the red arrows in part (b) should point down, so that the order is ferrimagnetic above the spin-rearrangement transition.*

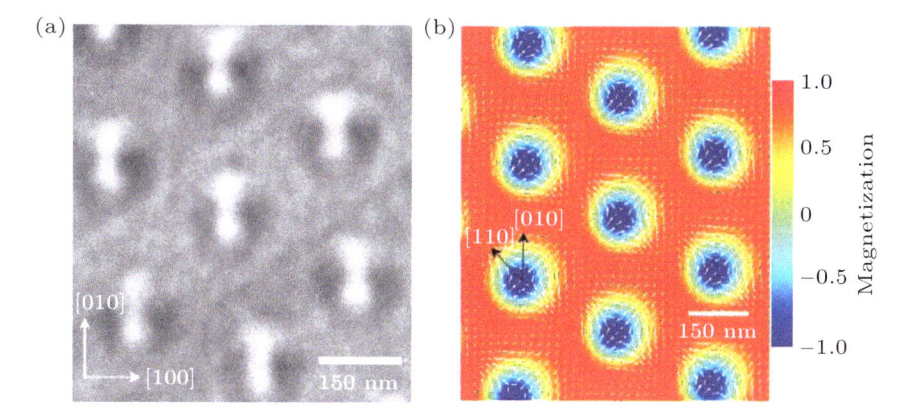

Figure 5.28 *The lattice of antiskyrmions in $Mn_{1.4}Pt_{0.9}Pd_{0.1}Sn$. (A) Under-focused LTEM image, (B) theoretical simulation. The color represents the magnetization component normal to the sample plane. Reprinted from Nayak et al. (2017).*

Ab initio density-functional calculations for $Mn_{1.5}PtSn$ support the low-temperature state. Although antiskyrmions were found in thin plates of $Mn_{1.4}PtSn$, Sukhanov et al. (2020) discovered a fractal magnetic domain pattern in pure bulk $Mn_{1.4}PtSn$. A discussion of this finding would go too far, so we refer the reader to the cited paper for further information and instead continue with a slight variant of $Mn_{1.4}PtSn$ where antiskyrmions were indeed discovered by Nayak et al. (2017); this is the inverse Heusler compound $Mn_{1.4}Pt_{0.9}Pd_{0.1}Sn$. More recent details on this topic can be found in a paper by Saha et al. (2019).

The magnetic properties of the compound $Mn_{1.4}Pt_{0.9}Pd_{0.1}Sn$ are very similar to those of $Mn_{1.4}PtSn$, especially the type of noncollinear order with the characteristic spin-reorientation transition at a slightly lower temperature than in $Mn_{1.4}PtSn$. Nayak et al. (2017) used Lorentz transmission electron microscopy (LTEM) to see two-dimensional antiskyrmions of size of about 150 nm above the spin-reorientation transition. Micromagnetic simulations were carried out using OOMMF-software (Donahue and Porter, 1999; Rohart and Thiaville, 2013). This method is based on a Fourier approach. A result of this simulation is shown in Fig. 5.28. A great number of systems other than Heusler compounds are listed in the review article by Everschor-Sitte et al. (2018). Of interest here are metallic multilayers. However, these are beyond the scope of this section.

5.6 High-temperature approaches

5.6.1 Short-range order

It was especially the magnetism of Fe—particularly its nature above the Curie temperature —that led to a long series of papers by Heine and his group. The following list is not claimed to be complete but fairly representative: Heine and Samson (1980), You and Heine (1982), Holden and You (1982), Heine and Samson (1983), Small and Heine (1984), Haines *et al.* (1985b, 1986), Heine and Joynt (1988), Luchini and Heine (1989), Heine *et al.* (1990), Luchini *et al.* (1991), and Chana *et al.* (1991). Although following a variety of different objectives, these papers have as a common basis the role played by the band structure in determining the physical properties of itinerant magnets. In one way or another these papers have been shaping the approach used in this treatise.

We want to select two key points of particular interest: validity and restrictions of the Heisenberg model in describing the thermodynamic properties of itinerant-electron magnets and the degree of short-range order above the Curie temperature. The controversy about these points has not yet been settled entirely.

We begin by stressing (You and Heine, 1982) that the calculations of Korenman *et al.* (1977), Capellmann (1979), Hubbard (1979), and Hasegawa (1979a,b) referred to at the end of Sec. 5.1 have been more restricted in their practical execution than their theoretical conception. Assumptions have had to be made either in representing the detailed electronic structure or in modeling the distribution of magnetic moment directions near T_c, or both. In this way configurations could be averaged and thermodynamic quantities calculated. It is here where the work of Heine's group developed its importance because they had a fast computational scheme for calculating electronic structure in substantial clusters of 500–1000 atoms with arbitrary kinds of order or disorder. Technically, spin-1/2 rotation matrices were used in the—by now familiar—way to define any desired magnetic moment arrangement including spin spirals, so that total energies could be obtained. The earlier work, however, although self-consistent, could by present standards not be called *ab initio* in the density-functional sense. Still, it soon emerged (Small and Heine, 1984) that only for *small* deviations of the magnetic order from the totally ferromagnetic state can one extract from total-energy differences (or the related couple method) the total energy, E, in the form of the Heisenberg model

$$E = -\sum_{ij} J_{ij} \hat{\mathbf{e}}_i \cdot \hat{\mathbf{e}}_j, \qquad (5.246)$$

where i, j label the atoms and $\hat{\mathbf{e}}_i$ is the direction of the magnetic moment, the notation of which has been absorbed into the exchange constants J_{ij}. In other words, the Heisenberg model can only be justified as the Taylor series expansion of the total energy to second order about the ferromagnetic ground state, i.e.

$$J_{ij}^{\mu\nu} = -\frac{1}{2} \frac{\partial^2 E(\{\hat{\mathbf{e}}_l\})}{\partial e_i^\mu \, \partial e_j^\nu} \quad \text{where } \mu, \nu = x, y, z. \qquad (5.247)$$

$E(\{\hat{\mathbf{e}}_l\})$ denotes the total energy calculated for a number of configurations $\{\hat{\mathbf{e}}_l\}$. The J_{ij}, incidentally, were found to be of fairly long range, up to fifth nearest neighbors, say. This is not in contradiction with the results of Sec. 5.4. More importantly, Luchini and Heine (1989) found that the exchange interactions as defined in Eqn (5.247) are strongly affected through a multiatom effect by the amount of order in the shells surrounding the atom whose exchange constants are to be calculated. These results were later corroborated by Heine *et al.* (1990) using fully self-consistent density-functional calculations with spin-spiral configurations. They showed that the nearest-neighbor exchange constant J_1 is weaker for Fe, Co, and Ni in the ferromagnetic state while it increases as one approaches a more highly disordered state. For such a disordered state the term *disordered local moment* state has become common and we will come back to it in the next subsection. These spiral calculations showed that the effect is substantial, up to a factor of two in J_1. Chana *et al.* (1991), furthermore, show that the smaller next-nearest-neighbor exchange constant, J_2, *decreases* as the disorder in the surrounding shells increases. This behavior of J_1 and J_2 is, of course, not captured by the Heisenberg model and is speculated to play an important role in the high-temperature properties, especially in the type of short-range order above the Curie temperature. This issue, although still not settled, may be summarized as follows.

There is considerable *experimental* evidence for the existence of substantial short-range order in Fe and Ni above their Curie temperatures, T_c. We do not mean the critical fluctuations very close to T_c but temperatures in the range $T \sim 1.2\,T_c$ for Fe and up to $\sim 2\,T_c$ in Ni. The controversy was started by the inelastic neutron-scattering data by Mook *et al.* (1973) on Ni and Lynn (1975) on Fe. They observed the persistence of spin waves in this temperature range. As usual, there were serious attempts to interpret the data differently, especially by Shirane (1984) and Shirane *et al.* (1986), but other evidence from angle-resolved photoemission studies by Haines *et al.* (1985a,b) and Kisker *et al.* (1985) strongly suggest the existence of short-range order of the above type.

On the *theoretical* side, short-range order for lengths less than about 25 Å was proposed a long time ago by Korenman *et al.* (1977) and Capellmann (1979). However, as Heine and Joynt (1988) point out, the short-range order observed is not on as massive a scale as that of Korenman and Capellmann, but is much more subtle. It is argued that there is no theoretical reason for a system to order on an atomic scale; with a small shift in the energy–entropy balance, the system can order on a *coarser* length scale without reaching the dimensions postulated by Korenman and Capellmann. It is in the view of Heine and collaborators precisely the dependence of the exchange constants J_1 and J_2 on disorder that might explain the short-range order of the size anticipated.

Qualitatively, for coarse magnetic disorder to happen, magnetic configurations which deviate only moderately from the ferromagnetic state need to have relatively low energy, whereas highly disordered configurations need to be more "expensive," i.e. should have a relatively high energy. To make this argument semi-quantitative we should look at the energy–entropy balance. For this purpose we take an expression for the entropy from the paper by Heine and Joynt (1988) who assumed that the system above the Curie temperature can be described as being filled with Bloch walls of thickness π/q where q is the magnitude of the wave vector of a macroscopically populated spin-wave mode that we approximate with a spiral configuration of the same q. The number of such configurations that can be formed is the same as that of a superparamagnet which consists of domains of size π/q on a side. Each domain can point in $(2S_B + 1)$ directions, where (for Fe) $S_B = \frac{1}{2} \cdot 2.2$ times the number of atoms in the domain. The entropy per atom of such a model is then

$$\frac{S}{N} = k_\mathrm{B} \left(\frac{2\theta}{\pi}\right)^3 \ln\left[2.2\left(\frac{\pi}{2\theta}\right)^3 + 1\right]. \qquad (5.248)$$

Here θ is the average nearest-neighbor angle. This expression for the entropy should be regarded as a reasonable interpolation formula between the limits of perfect order and perfect disorder, the latter being represented by $\theta = \pi/2$, for which Eqn (5.248) gives the well-known result $S/N = k_\mathrm{B}\ln(2S+1)$, where S is the atomic spin. Unfortunately, we cannot compare these assertions with the entropy calculated in Sec. 5.4, where it is obtained as a function of the temperature.

We can now discuss the free energy by combining $S(\theta)$ with spiral energies, which we associate here with the internal energy $U(\theta)$. Heine and Joynt (1988) show an exaggerated situation, but we want to see if a simple (and quick), *ab initio* total energy calculation for spin spirals approximates their results. This is not obvious on first sight, since simple means "packing" Fe "densely" with spin spirals as we do not want to bother with supercells. Indeed, the following simple analysis fails for Ni. Thus we show in Fig. 5.29 the free energy $U(\theta)$ for bcc Fe only, represented by the total energy (minus that of the ferromagnet) of a spin-spiral configuration with q in the z-direction, $q = 0$ corresponding to $\theta = 0$ and $q = (1,0,0)2\pi/a$ to $\theta = \pi$. The free energy, $F(\theta)$, and $TS(\theta)$ from Eqn (5.248) are plotted for $T = 1200$ K.

The result is only marginally convincing, but we can, with some confidence, repeat the points of Heine and Joynt (1988). They remark that as the temperature increases from below, the phase transition will set in when the free energy becomes zero. This may first occur at an average nearest-neighbor angle θ_c where walls of thickness $\pi/2\theta_c$ are formed spontaneously and randomly creating disorder at all q less than the corresponding q_c. At the corresponding value of θ_c the curves U and TS touch as can be seen in Fig. 5.29 giving a shallow minimum in the free energy at $\theta_c \simeq 46°$ for $T \simeq 1200$ K (Heine and Joynt obtain $\theta_c = 60°$). They note that if $U(\theta)$ had the shape of the Heisenberg form in Fig. 5.29, then the entropy curve would first hit the free energy at $\theta = \pi/2$ so that one gets disorder at all q, i.e. disorder at the atomic scale. However, the value of T for this to happen would be very slightly higher and the disordered local moment state would seem to set in at this somewhat more elevated temperature. We believe the different possibilities can no longer simply be distinguished using these rather rough arguments. Still, the evidence is in favor of short-range order which is also suggested by the results for the nonuniform susceptibility in Sec. 4.2.3. Furthermore, Heine and Joynt (1988) compare their theory with experimental results for the specific heat, the uniform static susceptibility, photoemission, and neutron scattering, finding agreement where a meaningful comparison can be made.

In spite of these statements it is still of great interest to review the theory that uses as a basic postulate a *disordered local moment state*. This theory is extremely well formulated; it puts into proper perspective some basic notions like the adiabatic principle and other important concepts that are also valid for the treatment of spin fluctuations given in Sec. 5.4. This theory, furthermore, is likely to be developed further; its execution is based on the powerful Green's function formalism which we introduced in Sec. 3.3.2 and began

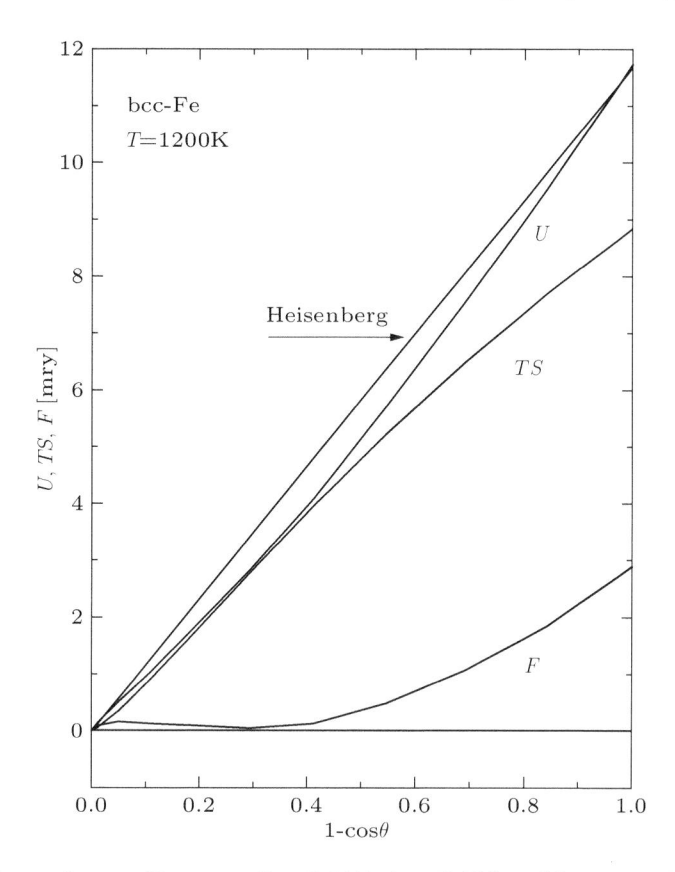

Figure 5.29 *Internal energy, U, entropy, Eqn (5.248), times T, TS, and free energy, F, as functions of the nearest-neighbor angle, θ. The line labeled "Heisenberg" is the Heisenberg form for the internal energy $\propto 1 - \cos\theta$.*

to use in Sec. 4.5. We are therefore prepared to embark on this mathematically elegant and useful methodology, but will limit ourselves to the salient facts only.

5.6.2 The disordered local moment state

The formalism to be described now will eventually make use of the variational principle that we discussed in Sec. 5.3. In fact our final aim is to calculate certain derivatives of essentially the averaged Hamiltonian $\langle \mathcal{H} \rangle_0$ after having chosen a model Hamiltonian $\mathcal{H}_0$. As we will see this will enable us to calculate the susceptibility above the Curie temperature. But there are two important differences to the method outlined above. *First*, the LDA Kohn–Sham equations must be solved for random magnetic moment configurations. This requires a substantial technical effort which we outline next. *Second*, the variational principle is recast in a form that uses the grand potential Ω instead of the Hamiltonian $\mathcal{H}$. We will briefly outline the basic concepts and close this chapter by showing how the susceptibility is formulated and assess the quality of the numerical results obtained so far. Our presentation relies heavily on a paper by Gyorffy et al. (1985).

5.6.2.1 The coherent potential approximation (CPA)

We begin by referring back to the Korringa–Kohn–Rostoker (KKR) theory discussed in Sec. 3.3.2, in particular to Eqn (3.97), and, changing the notation to agree with common usage (see e.g. Oguchi et al., 1983 or Gyorffy et al., 1985), we define what is usually called the scattering path operator τ (or the inverse of the KKR matrix) by

$$\tau = (t^{-1} - \bar{\mathbf{G}})^{-1} \tag{5.249}$$

or

$$\tau \cdot (t^{-1} - \bar{\mathbf{G}}) = 1 \tag{5.250}$$

implying a suitable matrix representation (the operator τ should not be confused with the basis vector which, unfortunately, we denoted by the same symbol in Sec. 5.4). In contrast to Sec. 3.3.2 we now need a vector notation because of the spin degrees of freedom. To see this we recall from Sec. 2.7, Eqn (2.199), that we can write the effective Kohn–Sham potential at the site n as

$$\mathbf{v}_n(\mathbf{r}) = \frac{1}{2}\left[v_+(\mathbf{r}) + v_-(\mathbf{r})\right]\mathbf{1} + \frac{1}{2}\left[v_+(\mathbf{r}) - v_-(\mathbf{r})\right](\hat{\mathbf{e}}_n \cdot \boldsymbol{\sigma}), \tag{5.251}$$

where as before $v_\pm(\mathbf{r})$ are the usual spin-up/-down (muffin-tin or atomic sphere) potentials, $\mathbf{1}$ is the 2×2 unit matrix, $\hat{\mathbf{e}}_n$ is a unit vector in the direction of the local spin-quantization axis, and $\boldsymbol{\sigma}$ is the vector spanned by the Pauli matrices. The single-site t-matrix occurring in Eqn (5.249) can be expressed in the same way as the effective potential, i.e. we can write

$$\mathbf{t}_n = \frac{1}{2}\left[t_+ + t_-\right] \times \mathbf{1} + \frac{1}{2}\left[t_+ - t_-\right] \times (\hat{\mathbf{e}}_n \cdot \boldsymbol{\sigma}). \tag{5.252}$$

For the matrix elements of $t_\pm$ for spherically symmetric scatterers see Sec. 3.3.2; they are diagonal in the angular momentum quantum numbers being related with the phase shifts of the l-th partial wave and now also carry a spin index, $\pm$. The quantity $\bar{\mathbf{G}}$ is given by $\bar{\mathbf{G}} = \bar{G}\,\mathbf{1}$ where $\bar{G}$ is the usual KKR structure Green's function, denoted in Sec. 3.3.2 by B and given by Eqn (3.123) (except for a constant). We see that if the directions $\hat{\mathbf{e}}_n$ vary from site to site being frozen in some random fashion, then the problem is equivalent to solving the time-independent Schrödinger equation for a random potential. In full generality this is an impossibly difficult task, but reliable approximations exist that have been developed to solve the *alloy* problem when two species of atoms randomly occupy lattice sites. This is the coherent potential approximation (CPA) which we need to become familiar with now (Soven, 1967; Velicky et al., 1968; Ehrenreich and Schwartz, 1976).

Suppose we are given a situation where the concentration of A-atoms is x, then that of the B-atoms is $(1-x)$ and we define the average of the inverse KKR matrix as

$$\langle \tau \rangle = x\,\tau^A + (1-x)\,\tau^B \doteq \tau_C, \tag{5.253}$$

where we defined the scattering path operator of a coherent medium, τ_C. Physically we replaced the random crystal potential by an effective medium that is described with a single inverse KKR matrix τ_C. Starting from here we must next derive the corresponding single-site t-matrix, $\mathbf{t}_C$, for we want to write like Eqn (5.249):

$$\tau_C = (\mathbf{t}_C^{-1} - \bar{\mathbf{G}})^{-1}, \tag{5.254}$$

which describes scattering in a coherent potential. There are a number of different ways to proceed. A good discussion can be found in the paper by Gyorffy *et al.* (1985), who also point out that the CPA gives a high-quality description of a disordered alloy. We, however, do not want to go into such great detail and therefore compress the discussion by using as a starting point an impurity problem (Temmerman, 1982). This turns out to be still somewhat lengthy, but it yields a transparent derivation. Thus we consider N_0 impurities characterized by the KKR matrix τ^I in the unperturbed medium (or host) characterized by τ. We can write the equivalent of Eqn (5.249) as

$$\tau^I = \begin{pmatrix} \psi^I & -G \\ -\tilde{G} & \vartheta \end{pmatrix}^{-1}, \tag{5.255}$$

where ψ^I is the $N_0 \times N_0$ impurity block given by the elements

$$(\psi^I)_{LL'}^{nn'} = (t^I)_{nL}^{-1}\,\delta_{LL'}\,\delta_{nn'} - \bar{G}_{LL'}(\mathbf{R}_n - \mathbf{R}_{n'}), \tag{5.256}$$

and n, n' refer to the impurity sites whose single-site t-matrix is denoted by t^I. The quantities G and $\tilde{G}$ denote the off-diagonal block consisting only of the structure Green's function and ϑ is the unperturbed-site block. For the unperturbed system the scattering path operator is written similarly as

$$\tau = \begin{pmatrix} \psi & -G \\ -\tilde{G} & \vartheta \end{pmatrix}^{-1} \tag{5.257}$$

with

$$\psi_{LL'}^{nn'} = t_{nL}^{-1}\,\delta_{LL'}\,\delta_{nn'} - \bar{G}_{LL'}(\mathbf{R}_n - \mathbf{R}_{n'}). \tag{5.258}$$

Taking the difference,

$$\Delta = \tau^{-1} - (\tau^I)^{-1} = \begin{pmatrix} \psi - \psi^I & 0 \\ 0 & 0 \end{pmatrix}, \tag{5.259}$$

we read off from the preceding equations

$$\Delta_{LL'}^{nn'} = \begin{cases} [t_{nL}^{-1} - (t^I)_{nL}^{-1}]\delta_{LL'}\delta_{nn'} & \text{for } n = 1, \ldots, N_0 \\ 0 & \text{elsewhere} \end{cases}. \tag{5.260}$$

It is now easy to express τ^I in terms of τ and Δ:

$$\tau^I = \tau + \tau\Delta\tau^I \tag{5.261}$$

i.e.

$$\tau^I = \tau(1 - \Delta\tau)^{-1}. \tag{5.262}$$

Next it is of advantage to define a matrix K by

$$K = (1 - \Delta\tau)^{-1}\Delta \tag{5.263}$$

because it allows us to express τ^I in the form

$$\tau^I = \tau + \tau K\tau. \tag{5.264}$$

The K-matrix is not only useful for the impurity problem, but it also allows us to reformulate in a compact way the CPA condition, Eqn (5.253). For this let the impurity atoms be the atoms labeled A and B which we put in the coherent medium characterized by the scattering path operator τ_C. Then we can define the K-matrices

$$K_A = (1 - \Delta_A\tau_C)^{-1}\Delta_A \tag{5.265}$$

and

$$K_B = (1 - \Delta_B\tau_C)^{-1}\Delta_B. \tag{5.266}$$

Forming

$$\langle K \rangle = x K_A + (1-x) K_B = 0 \tag{5.267}$$

we easily derive with Eqn (5.253):

$$\langle K \rangle = 0. \tag{5.268}$$

This is the CPA condition that implicitly determines $\mathbf{t}_C$. Using the above formulas we verify for the single-site t-matrices:

$$\mathbf{t}_C^{-1} = x\,\mathbf{t}_A^{-1} + (1-x)\,\mathbf{t}_B^{-1} + (\mathbf{t}_C^{-1} - \mathbf{t}_A^{-1})\,\tau^C(\mathbf{t}_C^{-1} - \mathbf{t}_B^{-1}). \tag{5.269}$$

Technically this equation is solved by iteration. It is what we set out to derive in order to solve Eqn (5.254).

Next we need to determine the density of states and the electron density which are necessary for making the calculation self-consistent and later for determining the generalized grand potential. To do this a formula originally due to Lloyd and Smith (1972) is used.

5.6.2.2 Lloyd's formula

From Sec. 4.5 we therefore recall Eqn (4.184) for the density of states, $\mathcal{N}(\varepsilon)$, which we repeat here for convenience,

$$\mathcal{N}(\varepsilon) = -\frac{1}{\pi} \operatorname{Im} \operatorname{Tr} G(\varepsilon + \mathrm{i} 0^+). \tag{5.270}$$

This rather formal relation can be used to express the difference between the density of states of the system including the scatterers, $\mathcal{N}(\varepsilon)$, and without the scatterers, $\mathcal{N}_0(\varepsilon)$. The formal derivation of this difference was discussed by Faulkner (1977) who stated the mathematical background in detail. Here it suffices to remind the reader that the Green's function (or "resolvent operator") in Eqn (5.270) is defined by

$$G(z) = (z - \mathcal{H})^{-1} \tag{5.271}$$

if $\mathcal{H}$ includes the scatterers and is

$$G_0(z) = (z - \mathcal{H}_0)^{-1} \tag{5.272}$$

if $\mathcal{H}_0$ does not include the scatterers. It is now possible but nontrivial to evaluate the expression $\operatorname{Im} \operatorname{Tr} [G(\varepsilon + \mathrm{i} 0^+) - G_0(\varepsilon + \mathrm{i} 0^+)]$ rigorously to obtain (Lloyd and Smith, 1972; Faulkner, 1977; Gyorffy et al., 1984) for the integrated density of states, $N(\varepsilon) = \int^\varepsilon \mathcal{N}(\varepsilon') \, \mathrm{d}\varepsilon'$

$$N(\varepsilon) = N_0(\varepsilon) - \frac{1}{\pi} \operatorname{Im} \operatorname{Tr} \ln \det(\mathbf{t}^{-1} - \bar{\mathbf{G}}). \tag{5.273}$$

Here we used the relation $\det M = \exp \operatorname{Tr} \ln M$ valid for the determinant of an operator M, $\mathbf{t}^{-1}$ is the inverse of the "on the energy shell" t-matrix and $\bar{\mathbf{G}} = \bar{G}\,\mathbf{1}$ where $\bar{G}$ are the usual, real-space, KKR structure constants, independent of the potential; the Tr-operation is taken over the spin degrees of freedom, and the determinant is to be taken with respect to the site indices and the angular momentum labels (both of which we suppressed in the notation) (Gyorffy et al., 1985). The quantity $N_0(\varepsilon)$ is the integrated free-particle density of states. For the alloy problem the wording should be changed, replacing "spin degrees of freedom" by the appropriate alloy degrees of freedom which is a sum over A and B. Equation (5.273) is often called Lloyd's formula although it is not the original form given by Lloyd and Smith (1972).

5.6.2.3 The CPA integrated density of states

We next apply Lloyd's formula to our problem by adding and subtracting the inverse of the coherent t-matrix, $\mathbf{t}_C^{-1}$, to $\mathbf{t}^{-1} - \bar{\mathbf{G}}$ in Eqn (5.273), then factoring out $\mathbf{t}_C^{-1} - \bar{\mathbf{G}} = \tau_C^{-1}$ to obtain

$$N(\varepsilon) = N_0(\varepsilon) - \frac{1}{\pi} \operatorname{Im} \operatorname{Tr} \left\{ \ln \det(\mathbf{t}_C^{-1} - \bar{\mathbf{G}}) + \ln \det[\mathbf{1} - (\mathbf{t}_C^{-1} - \mathbf{t}^{-1})\tau_C] \right\}. \quad (5.274)$$

What is left to do is to make the CPA explicit. To do this we carry out the trace over the spin degrees of freedom. Following Gyorffy et al. (1985) we therefore define instead of the alloy concentration, x and $1-x$, the spin-up and spin-down probabilities as

$$P_\uparrow \doteq P_1 = \frac{1}{2}(1+m) \quad \text{and} \quad P_\downarrow \doteq P_2 = \frac{1}{2}(1-m), \quad (5.275)$$

where we have assumed a uniform magnetization m, and finally obtain for the density of states

$$N(\varepsilon) = N_0(\varepsilon) - \frac{1}{\pi} \operatorname{Im} \operatorname{Tr} \ln \det(\mathbf{t}_C^{-1} - \bar{\mathbf{G}}) - \frac{1}{\pi} \operatorname{Im} \sum_{\sigma=1}^{2} P_\sigma \ln \det(1 - \Delta_\sigma \tau_C^\sigma). \quad (5.276)$$

We have abbreviated $\Delta_\sigma = (t_C^{-1} - t_\sigma^{-1})$ for $\sigma = 1 = \uparrow$ and $\sigma = 2 = \downarrow$, t_σ being the single-site t-matrix in the spin-σ effective potential. This integrated density of states formula is the central result that enables us not only to determine the Fermi energy, ε_F, by equating $N(\varepsilon_F)$ with the total number of electrons per unit cell, but is also needed to construct the generalized grand potential. Before we consider this last important step we mention that the above theory for the electronic structure is not quite complete because in order to make the calculations self-consistent we need the averaged spin-up and spin-down electron densities. In a KKR-CPA calculation it is obtained from the Green's function again which can be shown to give a compact representation that can be written as (Staunton et al., 1986)

$$n_\sigma(\mathbf{r}) = -\frac{1}{\pi} \operatorname{Im} \int^{\varepsilon_F} d\varepsilon \sum_L Z_{L\sigma}(\mathbf{r})^2 \langle \tau_{LL}^\sigma \rangle \quad (5.277)$$

where, as in Eqn (5.276), σ is the spin index, $Z_{L\sigma}(\mathbf{r})$ is the (real) single-site wave function (the radial solution of Sec. 3.3.2 in the spin-σ effective potential times the real spherical harmonics) and

$$\langle \tau_{LL}^\sigma \rangle = \left([1 + \Delta_\sigma \tau_C]^{-1} \tau_C \right)_{LL} \quad (5.278)$$

where Δ_σ was defined after Eqn (5.276). We merely list this formula for the sake of completeness. The reader interested in the derivation should consult the paper by

Faulkner and Stocks (1980) as well as Stocks and Winter (1984). The material presented there can with a little effort be connected with the basic theory given in Sec. 3.3.2.

5.6.2.4 The spin susceptibility in the paramagnetic state

We now assemble the remaining parts of the theory that complete the disordered local moment picture for itinerant-electron ferromagnets. For this we go back to the very beginning of this chapter, Sec. 5.1, and call to mind Eqn (5.14) for the Gibbs grand potential $\Omega[\tilde{n}]$ that was minimized by solving the temperature-dependent Kohn–Sham equations, Eqn (5.8), for some unknown set of magnetic configurations giving via Eqn (5.7) the elements of the density matrix, $\tilde{n}$, for each atomic sphere. These configurations, however, will change in time and it is here that the *adiabatic approximation* is used in the way described in Sec. 5.2. In the present context we may state it as follows.

Two different timescales are established, one being short measuring an average hopping time of the electrons, the other being sufficiently long to form a nonzero magnetization average that allows us to talk about magnetic moments that change their orientation slowly. A set of orientations is specified by a set of unit vectors $\{\hat{e}_i\}$ the index i labeling the site and

$$\hat{e}_i = \frac{\int_{V_i} \mathbf{m}(\mathbf{r}) \, d\mathbf{r}}{\left|\int_{V_i} \mathbf{m}(\mathbf{r}) \, d\mathbf{r}\right|} \tag{5.279}$$

where the magnetization $\mathbf{m}(\mathbf{r})$ is given by Eqn (5.3) and V_i denotes the atomic sphere at site i. The situation has been described thoroughly by Staunton *et al.* (1986) who point out that the many-electron system, although ergodic, does not cover its phase space uniformly in time. "It can be visualized as being confined for long times near 'points' which are characterized by particular arrangements of the orientations of such finite 'local' moments at every site and then moving rapidly to another similar point. One can therefore speak of 'temporarily broken ergodicity' (Palmer, 1982)." These states of temporarily broken ergodicity are labeled by the unit vectors $\{\hat{e}_i\}$ given by Eqn (5.279) and pointing along the local moments' orientations. The connection with Eqn (5.251) is now clear: the unit vectors $\{\hat{e}_i\}$ define the local spin quantization axes, i.e. the local frames of reference in which the density matrix $\tilde{n}$ is diagonal.

A useful scheme is now established by first assuming the *short-time description* of the system labeled by $\{\hat{e}_i\}$ is given by the finite-temperature density-functional grand potential defined in Sec. 5.1. The minimized grand potential is now denoted by $\bar{\Omega}\{\hat{e}_i\}$. Second, a prescription for the *long-time evolution* of the system in the reduced phase space for the $\{\hat{e}_i\}$ is developed using the Bogoliubov–Peierls variational mean-field theory outlined in Sec. 5.3.

Both parts of the theory are basically complete. We know how to solve the problem of obtaining $\bar{\Omega}\{\hat{e}_i\}$ for a large set of configurations, as for instance spin spirals of the type used in Sec. 5.4 or random configurations as described in this section. In fact, we could have introduced the theory outlined in Sec. 5.4. with the same terms. Of course, we must make approximations for the exchange-correlation contribution $\Omega_{\text{ex}}[\tilde{n}]$ for which the best available zero temperature, local spin-density-functional expression is presently used. This will be so until useful finite-temperature forms become available.

For the second part of the scheme the averages over the ensemble of orientational configurations are obtained by using probabilities defined by

$$P\{\hat{\mathbf{e}}_i\} = \frac{\exp\left(-\beta \bar{\Omega}\{\hat{\mathbf{e}}_i\}\right)}{\prod_j \int d\hat{\mathbf{e}}_j \exp\left(-\beta \bar{\Omega}\{\hat{\mathbf{e}}_j\}\right)}. \tag{5.280}$$

The free energy, F, is consequently determined from

$$F = -\beta^{-1} \ln \prod_j \int d\hat{\mathbf{e}}_j \exp\left(-\beta \bar{\Omega}\{\hat{\mathbf{e}}_j\}\right). \tag{5.281}$$

Clearly, the grand potential $\bar{\Omega}\{\hat{\mathbf{e}}_i\}$ acts like a classical spin Hamiltonian. We therefore derive the Bogoliubov–Peierls variational principle exactly as in Sec. 5.3 but replacing the Hamiltonian occurring there by $\bar{\Omega}\{\hat{\mathbf{e}}_i\}$. With this replacement the theory includes Stoner excitations through the finite temperature in the single-particle contribution Ω_0 (see the details below). The inequality now is obviously

$$F \leq F_0 + \langle \bar{\Omega} \rangle_0 - \langle \Omega_M \rangle_0 \tag{5.282}$$

where as before a "model" grand potential, Ω_M must be chosen which defines the corresponding free energy F_0 and the averages $\langle \ \rangle_0$ through Eqn (5.281) and (5.280) in which case $\bar{\Omega}$ is to be replaced by Ω_M. Indeed, except for the most recent development by Staunton and Gyorffy (1992), Gyorffy's group in the papers cited above used for the model Ω_M what they called the vector model. It is given by our Eqn (5.61), i.e.

$$\Omega_M = -\sum_i (\mathbf{h}_i + \mathbf{h}_i^{\text{ext}}) \cdot \hat{\mathbf{e}}_i \tag{5.283}$$

where we merely added an external field $\mathbf{h}_i^{\text{ext}}$. We remind the reader that the field $\mathbf{h}_i$ is the variational parameter to optimize Eqn (5.282). We can now simply take over the results from Sec. 5.3, except that we have to complete the notation there by adding appropriate vector indices. The formula we need is the result of the optimization, Eqn (5.69), which we now write in full vector notation as

$$\mathbf{h}_i(\{\mathbf{m}_i\}) = -\nabla_{\mathbf{m}_i} \langle \bar{\Omega} \rangle_0. \tag{5.284}$$

This is the molecular field. The magnetization is related to the molecular field by Eqn (5.66) which we write as

$$\mathbf{m}_i = L(\beta|\mathbf{h}_i + \mathbf{h}_i^{\text{ext}}|) \frac{\mathbf{h}_i + \mathbf{h}_i^{\text{ext}}}{|\mathbf{h}_i + \mathbf{h}_i^{\text{ext}}|} \tag{5.285}$$

where L is the Langevin function defined in Eqn (5.67).

We are now in the position to approach the central result. This is the dimensionless susceptibility tensor $\bar{\chi}_{i\alpha,j\gamma}$ which is obtained by taking the derivatives of the various components (labeled by Greek indices) of $\mathbf{m}_i$ with respect to the components of the external field $\mathbf{h}_j^{\text{ext}}$. Near and above the Curie temperature the magnetization and the molecular field are small so that we can expand the Langevin function, Eqn (5.80), obtaining

$$\mathbf{m}_i = \frac{1}{3}\beta(\mathbf{h}_i + \mathbf{h}_i^{\text{ext}}). \tag{5.286}$$

The susceptibility, $\bar{\chi}_{i\alpha,j\gamma} \doteq \partial m_{i\alpha}/\partial h_{j\gamma}^{\text{ext}}$, is now easy to derive. Using the relation of the molecular field with the magnetization given by Eqn (5.284) we verify that

$$\bar{\chi}_{i\alpha,j\gamma} = \frac{1}{3}\beta\,\delta_{ij}\,\delta_{\alpha\gamma} + \frac{1}{3}\beta \sum_{k\lambda} S^{(2)}_{i\alpha,k\lambda}\,\bar{\chi}_{k\lambda,j\gamma}, \tag{5.287}$$

where, using the notation of Gyorffy et al. (1985), we defined

$$S^{(2)}_{i\alpha,k\lambda} = \frac{\partial h_{i\alpha}}{\partial m_{k\lambda}} = -\left(\frac{\partial^2 \langle\bar{\Omega}\rangle_0}{\partial m_{i\alpha}\,\partial m_{k\lambda}}\right)_{\{\mathbf{m}_j^0\}}. \tag{5.288}$$

evaluated at a set of vectors $\{\mathbf{m}_j^0\}$ that satisfy Eqn (5.285) or Eqn (5.286). Above the Curie temperature the system is homogeneous and hence $S^{(2)}_{i\alpha,k\lambda}$ is diagonal in the component indices α, λ and only depends on the coordinate difference $\mathbf{R}_i - \mathbf{R}_k$. Consequently, Eqn (5.287) is solved by taking its lattice Fourier transform which gives

$$\chi(\mathbf{q}, T) = \frac{\mu_B^2}{3k_B T - S^{(2)}(\mathbf{q}, T)}, \tag{5.289}$$

where we have included a factor μ_B^2 to fix the units. Equation (5.289) is the central result: To calculate the susceptibility, the Fourier-transformed Eqn (5.288) must be determined. In the absence of an external field and above the Curie temperature one may then set the magnetization equal to zero.

It remains to calculate $S^{(2)}(\mathbf{q}, T)$ from $\langle\bar{\Omega}\rangle_0$. The principles of this step are utterly clear although the technical details necessary to complete the theory and bring them to numerical fruition are nontrivial. We concentrate on the basics and, referring to the grand potential given in Eqn (5.14) we rewrite the first term on the right-hand side, the single-particle contribution, Ω_0, given by Eqn (5.11), as

$$\Omega_0[\tilde{n}] = -\beta^{-1} \int_{-\infty}^{\infty} \ln\left[1 + \exp\beta(\mu - \varepsilon)\right] \mathcal{N}(\varepsilon)\,d\varepsilon. \tag{5.290}$$

This can be expressed using instead of the density of states, $\mathcal{N}(\varepsilon)$, the integrated density of states, $N(\varepsilon) = \int^\varepsilon \mathcal{N}(\varepsilon')\,d\varepsilon'$, or $dN(\varepsilon)/d\varepsilon = \mathcal{N}(\varepsilon)$. Thus by integrating by parts we easily verify

$$\Omega_0[\tilde{n}] = -\int_{-\infty}^{\infty} f(\varepsilon) N(\varepsilon) \, d\varepsilon \qquad (5.291)$$

where $f(\varepsilon)$ is the Fermi–Dirac distribution function. Clearly, substituting for the integrated density of states, $N(\varepsilon)$, the result given in Eqn (5.276), we obtain that part of the averaged free-particle grand potential that depends on the magnetization:

$$\begin{aligned}\langle \bar{\Omega}_0 \rangle_0 &= \frac{1}{\pi} \int_{-\infty}^{\infty} f(\varepsilon) \, \text{Im Tr} \ln \det(\mathbf{t}_C^{-1} - \bar{\mathbf{G}}) \, d\varepsilon \\ &+ \frac{1}{\pi} \int_{-\infty}^{\infty} f(\varepsilon) \sum_{\sigma=1}^{2} P_\sigma \, \text{Im} \ln \det \left(1 - \Delta_\sigma \, \tau_C^\sigma \right) d\varepsilon.\end{aligned} \qquad (5.292)$$

To complete the expression for the grand potential we should work out the remaining terms on the right-hand side of Eqn (5.14), which requires knowledge of the density supplied in principle by Eqn (5.277). Gyorffy et al. (1985) make a strenuous effort to derive these terms, only to conclude that they can be neglected. Since these terms are the finite-temperature version of the double-counting terms it is presumably the force theorem again that justifies their being negligible. Thus the susceptibility above the Curie temperature follows to a good approximation from

$$S_{ik}^{(2)} = -\left(\frac{\partial^2 \langle \bar{\Omega}_0 \rangle_0}{\partial m_i \, \partial m_k}\right)_{\{m_j\} \to 0}. \qquad (5.293)$$

The amount of technical detail involved in carrying out the remaining differentiation and the Fourier transform is still substantial. However, since at this point the principal features of the theory seem clear, we omit further derivations and refer the interested reader to the original literature, i.e. papers by Gyorffy et al. (1985) and Staunton et al. (1986). We come to an end by discussing some results obtained in the disordered local moment picture.

5.6.2.5 Results

Staunton et al. (1986) report that for bcc Fe local moments of magnitude 1.85 μ_B are self-consistently maintained in the disordered local moment state. These authors incorporated a temperature of $T = 1200$ K in the Fermi–Dirac distribution function and calculated the susceptibility $\chi(\mathbf{q}, T)$. The uniform ($q = 0$) susceptibility as a function of temperature can then be well fitted to Curie–Weiss behavior. This gives a Curie temperature of $T_c = 1280$ K and a Curie constant from which q_c, the number of magnetic carriers, is obtained as $q_c = 1.20$, or, the effective magnetic moment in the paramagnetic state as $\mu^{\text{eff}} = 1.97 \, \mu_B$ (experimental value: $\mu^{\text{eff}} = 3.13 \, \mu_B$). These values should be compared with those collected in Table 5.4. The situation is encouraging although the Curie temperature is somewhat overestimated and the value of q_c is too small. Interestingly, the trends are the same as in the spin-fluctuation theory presented in Sec. 5.4.1: overestimating T_c slightly but underestimating q_c substantially. The calculations reported for Ni by Staunton *et al.* (1986), unfortunately, failed if carried out as described above. The details depend somewhat on the chosen lattice constant and the precision of the calculation, but the results are characterized by a very small local moment that vanishes at finite temperatures. At some level of approximation it is found, for instance, that a local moment of magnitude 0.22 μ_B, only, is self-consistently maintained in the disordered local moment state at $T = 0$. Applying then a temperature of $T = 700$ K (or even smaller: 500 K is mentioned, too) in the Fermi–Dirac function, i.e. including Stoner excitations, it is found that no local moment is established and, as a consequence, the susceptibility shows an unphysical instability. This outcome of the calculations is not too surprising since we found in Sec. 4.2.3 that the moment of Ni vanishes in a spin spiral with $q \gtrsim$ half the Brillouin zone radius which corresponds to an average angle $< 90°$ between adjacent Ni moments. Calculations for Co have not been reported to our knowledge.

Staunton and Gyorffy (1992) remedied the situation by going beyond the mean-field approximation that is prescribed by the choice of the model grand potential Ω_M given in Eqn (5.283). This step is not done by inventing another model, but by breaking the impact of the model on the form of the average $\langle \bar{\Omega} \rangle_0$. What is meant by this statement can be explained using the fluctuation-dissipation theorem. We recall from Sec. 5.4.5 that the fluctuation is connected with the real part of the nonuniform susceptibility, $\chi(\mathbf{q})$, through Eqn (5.166), i.e. $\langle |m_\mathbf{q}|^2 \rangle = k_B T \chi(\mathbf{q})$. This implies in real space that the average of the magnetization at sites i and j, $\langle m_i m_j \rangle$, is not given by $\langle m_i m_j \rangle_0$ as required by mean-field theory which results in an *uncorrelated* average: $\langle m_i m_j \rangle_0 = \langle m_i \rangle_0 \langle m_j \rangle_0$. It is the treatment of this average that is dealt with differently in a way that apparently goes back to Onsager (Brout and Thomas, 1967). Referring for the hard details to the paper by Staunton and Gyorffy (1992) we here try to picture the idea only.

5.6.2.6 Onsager cavity-field approximation

We begin by going back to the expression given in Eqn (5.286) which in the mean-field approximation relates the magnetization at the site i, with the molecular field and the external field to linear order as $\mathbf{m}_i = \frac{1}{3}\beta(\mathbf{h}_i + \mathbf{h}_i^{\text{ext}})$. Instead of writing out the molecular field, $\mathbf{h}_i$, appropriate for the itinerant electrons, it is of advantage to first explain the approximation by using the Heisenberg model, Eqn (5.246), for which the molecular field is (see also Sec. 5.3) $\mathbf{h}_i = 2\sum_j J_{ij} \mathbf{m}_j$. The key point now is to exclude the effects of the magnetization on the site i and write

$$\mathbf{m}_i = \frac{1}{3} \beta \, \mathbf{h}_i^C \tag{5.294}$$

where $\mathbf{h}_i^C$ is called the cavity field defined by

$$\mathbf{h}_i^C = 2 \sum_j J_{ij}(\mathbf{m}_j - \delta\mathbf{m}_j^{(i)}) + \mathbf{h}_j^{\text{ext}}, \tag{5.295}$$

where $\delta \mathbf{m}_j^{(i)}$ is the magnetization induced at site j by the field at site i, thus

$$\delta \mathbf{m}_j^{(i)} = \chi_{ji}\, \mathbf{h}_i^C. \tag{5.296}$$

Moreover,

$$\mathbf{m}_j = \chi_{jj}\, \mathbf{h}_j^C \tag{5.297}$$

whence

$$\delta \mathbf{m}_j^{(i)} = \chi_{ji}\, \chi_{ii}^{-1}\, \mathbf{m}_i. \tag{5.298}$$

Substituting this result into Eqn (5.294) and taking the derivative of both sides of Eqn (5.295) with respect to the external field we obtain an equation for the susceptibility tensor χ_{ij} from which, after taking the lattice Fourier transform, one obtains at and above the Curie temperature (Staunton and Gyorffy, 1992)

$$\chi(\mathbf{q}) = \frac{1}{3k_{\mathrm B}T - J(\mathbf{q}) + \lambda}, \tag{5.299}$$

where

$$\lambda = \chi_{ii}^{-1} \int d\mathbf{q}\, J(\mathbf{q})\, \chi(\mathbf{q}). \tag{5.300}$$

This set of equations is known to represent an improvement on the mean-field approximation for the Heisenberg Hamiltonian. The analogous generalization of the mean-field theory for the disordered local moment picture is more involved, unfortunately. However, the main points can be summarized as follows.

We begin again with the mean-field equation (5.286) which we rewrite using the CPA grand potential $\langle \bar{\Omega} \rangle_0$ as

$$m_{i\alpha} = \frac{\beta}{3}\left[-\sum_{j\beta} \left(\frac{\partial^2 \langle \bar{\Omega} \rangle_0}{\partial m_{i\alpha}\, \partial m_{j\beta}} \right)_{\{\mathbf{m}_j^{(0)}\}} m_{j\beta} + h_{i\alpha}^{\mathrm{ext}} \right], \tag{5.301}$$

where as before the Greek indices label Cartesian coordinates and the derivatives are evaluated at a set of vectors $\{\mathbf{m}_j^0\}$ that satisfy Eqn (5.285) or Eqn (5.286). The cavity-field approximation is now made by replacing $m_{j\beta}$ on the right-hand side by $m_{j\beta} - \delta m_{j\beta}^{(i)}$ where as before $\delta m_{j\beta}^{(i)}$ is obtained from Eqn (5.298). However, the theory is more complicated than that because Staunton and Gyorffy write the local magnetization as the sum of two parts, one describing how the magnitude of the local moments responds to the external field, the other how they tend to align in the field. They denote the first part by $\boldsymbol{\mu}_i = (4\pi)^{-1} \int d\hat{\mathbf{e}}_i \, \delta\mu_i(\hat{\mathbf{e}}_i) \, \hat{\mathbf{e}}_i$ and the second by $\mathbf{m}_i = \langle \hat{\mathbf{e}}_i \rangle$. The susceptibility, correspondingly, now consists of two parts as well, $\chi_{ij} = \chi_{ij}^\mu + \chi_{ij}^m$, and the Onsager cavity fields to be subtracted are $\delta\boldsymbol{\mu}_j^{(i)}$ and $\bar{\mu} \, \delta\mathbf{m}_j^{(i)}$, where $\bar{\mu}$ is the average local moment self-consistently maintained in the disordered local moment state. It is clear that the equation corresponding to (5.301) will contain derivatives with respect to twice the μ's, twice the m's and mixed μ and m. We omit the rather lengthy formulas for the general case, but pick out the result for Ni for which the average local moment vanishes, $\bar{\mu} = 0$. This leads to a considerable simplification, so that the susceptibility is determined solely by χ^μ. The Fourier-transformed result is

$$\chi(\mathbf{q}) = \chi^\mu(\mathbf{q}) = \frac{\chi^0(\mathbf{q})}{1 - \gamma(\mathbf{q}) + \Lambda}. \tag{5.302}$$

Here

$$\Lambda = (\chi_{ii}^0)^{-1} \int d\mathbf{q} \, \gamma(\mathbf{q}) \, \chi(\mathbf{q}) \tag{5.303}$$

and $\gamma(\mathbf{q})$ is the Fourier transform of the rate of change of the magnitude of the local moment at site i: μ_i when the moment changes at the site j: $\gamma_{ij} = \partial\mu_i / \partial\mu_j$. The quantity χ_{ij}^0 is the Pauli susceptibility, $\chi^0(\mathbf{q})$ being its Fourier transform. All quantities are available from the self-consistent KKR-CPA calculation.

The *results* given for bcc Fe constitute a slight improvement over those obtained in the mean-field approximation. The Curie temperature obtained with the Onsager cavity-field approximation is listed in Table 5.8 in the column labeled Ref. 1. For Ni, where Staunton and Gyorffy (1992) find $\bar{\mu} = 0$ the susceptibility given by Eqn (5.302) shows approximate Curie–Weiss behavior and gives an effective paramagnetic moment of 1.21 μ_B which is to be compared with the experimental value of 1.6 μ_B, or the calculated number of magnetic carriers $q_c = 0.6$, which is in better agreement with the experimental value of $q_c = 0.9$ than the calculated value listed in Table 5.4 (Sec. 5.4.2.1). The calculated Curie temperature is listed in Table 5.8. Surprisingly, it is very similar

Table 5.8 *Collection of calculated and experimental (from Table 4.1) Curie temperatures in K for the elemental ferromagnets.*

	Ref. 1	Ref. 2	Ref. 3	Table 5.3	Table 5.4	exp.
bcc Fe	1015	1037	1060	1316	1095	1044
fcc Co		1250	1080	1558	1012	1388
fcc Ni	450	430	510	642	412	627

Ref. 1: Staunton and Gyorffy (1992)
Ref. 2: Halilov *et al.* (1998a)
Ref. 3: Rosengaard and Johansson (1997) (Monte Carlo calc.)

to the value calculated in Sec. 5.4.1 using the spin-fluctuation theory and repeated in Table 5.8 for convenience.

For ease of comparison we collect in Table 5.8 besides experimental values the Curie temperatures obtained by us in Sec. 5.4.1 (labeled Table 5.4) and by other groups. The results of Halilov *et al.* (1998a) labeled as Ref. 2 are rough estimates very similar to our results from Sec. 5.3 labeled Table 5.3. The results by Rosengaard and Johansson in the column labeled Ref. 3 are based on total energy data obtained from density-functional calculations using spin spirals similar to Sec. 5.4.1. However, the thermodynamics was carried out with a Monte Carlo technique applied at low temperatures. The results are interesting because they are so similar compared with the other calculations. The magnetization profile they obtained is better than that of Sec. 5.4.1, but still does not match the experimental profiles.

References

Abrikosov, A.A. (1988). *Fundamentals of the theory of metals*. North-Holland, Amsterdam.
Abrikosov, I.A., Eriksson, O., Söderlind, P., Skriver, H.L., and Johansson, B. (1995). *Phys. Rev. B* **51**, 1058.
Acet, M., Zähres, H., Wassermann, E.F., and Pepperhoff, W. (1994). *Phys. Rev. B* **49**, 6012.
Adachi, K., Matsui, M., and Fukuda, Y. (1980). *J. Phys. Soc. Japan* **48**, 62.
Adams, T., Chacon, A., Wagner, M., Bauer, A., Brandl, G., Pedersen, B., Berger, H., Lemmens, P., and Pfleiderer, C. (2012). *Phys. Rev. Lett.* **108**, 237204.
Aharoni, A. (2000). *Introduction to the Theory of Ferromagnetism*. Oxford University Press, Oxford.
Andersen, O.K. (1975). *Phys. Rev. B* **12**, 3060.
Andersen, O.K. (1984). In *Electronic structure of complex systems* (ed. P. Phariseau and T.M. Temmerman). Plenum, New York, p. 1.
Anderson, P.W. (1950). *Phys. Rev.* **79**, 350.
Anderson, P.W. and Hasegawa, H. (1955). *Phys. Rev.* **100**, 675.
Anderson, P.W. (1963). In *Magnetism I* (ed. G. Rado and H. Suhl). Academic Press, New York, Chap. 2.
Anisimov, V.I., Antropov, V.P., Liechtenstein, A.I., Gubanov, V.A., and Postnikov, A.V. (1988). *Phys. Rev. B* **37**, 5598.
Anisimov, V.I., Zaanen, J., and Andersen, O.K. (1991). *Phys. Rev. B* **44**, 943.
Antonov, V.N., Oppeneer, P.M., Yaresko, A.N., Perlov, A.Ya., and Kraft, T. (1997). *Phys. Rev. B* **56**, 13012.
Antropov, V.P., Katsnelson, M.I., van Schilfgaarde, M., and Harmon, B.N. (1995). *Phys. Rev. Lett.* **75**, 729.
Antropov, V.P., Katsnelson, M.I., van Schilfgaarde, M., Harmon, B.N., and Kuznezov, D. (1996). *Phys. Rev. B* **54**, 1019.
Asada, T. (1995). *J. Magn. Magn. Mat.* **140–144**, 47.
Asada, T. and Terakura, K. (1992). *Phys. Rev. B* **46**, 13599.
Asano, S. and Yamashita, J. (1971). *J. Phys. Soc. Japan* **31**, 1000.
Asano, S. and Yamashita, J. (1973). *Prog. Theor. Phys.* **49**, 373.
Ashcroft, N.W. and Mermin, N.D. (1976). *Solid state physics*. Holt, Rinehart and Winston, Philadelphia.
Austin, B.J., Heine, V., and Sham L. J. (1962). *Phys. Rev.* **127**, 276.
Bachelet, G.B., Hamann, D.R., and Schlüter, M. (1982). *Phys. Rev. B* **26**, 4199.
Baibich, M.N., Broto, J.M., Fert, A., Nguyen Van Dau, F., Petroff, F., Etienne, P., Creuzet, G., Friedrich A., and Chazelas, J. (1988). *Phys. Rev. Lett.* **61**, 2472.
Bak, P. and Jensen, M.H. (1980). *J. Phys. C* **13**, L881.
Balke, B., Fecher, G.H., Winterlik, J., and Felser, C. (2007). *Appl. Phys. Letters* **90**, 152504.
Baranger, H.U. and Stone, A.D. (1989). *Phys. Rev. B* **40**, 8169.
Bardeen, J. (1936). *Phys. Rev.* **50**, 1098.
Bauer, G.E., Brataas, A., Scheb, K.M., and Kelly, P.J. (1994). *J. Appl. Phys.* **75**, 6704.
Baym, G. (1973). *Lectures on quantum mechanics*. Benjamin/Cummings Publishing Company, Menlo Park.
Becke, A. (1992). *J. Chem. Phys.* **96**, 2155.
Becker, R. and Sauter, F. (1968). *Theorie der Elektrizität*, Vol. 3. B.G. Teubner, Stuttgart.

Behnia, K. and Aubin, H. (2016). *Rep. Prog. Phys.* **79**, 046502.
Beille, J., Voiron, J., and Roth, M. (1983). *Solid State Comm.* **47**, 399.
Belopolski, I., Manna, K., Sanchez, D.S., Chang, G., Ernst, B., Yiu, J., Zhang, S.S., Cochran, T., Shumiya, N., Zheng, H., Singh, B., Bian, G., Multer, D., Litskevich, M., Zhou, X., Huang, S.-M., Wang B., Chang, T.-R., Xu, S.-Y., Bansil, A., Felser, C., Liu, H., and Hasan, M.Z. (2019). *Science* **365**, 1278.
Bennett, W.R., Schwarzacher, W., and Egelhoff, W.F. (1990). *Phys. Rev. Lett.* **65**, 3169.
Berger, L. (1965). *Phys. Rev.* **137**, A220.
Berlin, T.H. and Kac, M. (1952). *Phys. Rev.* **86**, 821.
Bernhoeft, N.R., Lonzarich, G.G., Mitchell, P.W., and Paul, D.McK, (1983). *Phys. Rev. B* **28**, 422.
Berry, M.V. (1984). *Proc. Roy. Soc. London A* **392**, 45.
Binasch, G., Grünberg, P., Saurenbach, F., and Zinn, W. (1989). *Phys. Rev. B* **39**, 4828.
Binz, B., Vishwanath, A., and Aji, V. (2006). *Phys. Rev. Lett.* **96**, 207202.
Bjorken, J.D. and Drell, S.D. (1964). *Relativistic quantum mechanics*. McGraw-Hill, New York.
Bland, J.A.C., Bateson, R.D., Riedi, P.C., Graham, R.G., Lauter, H.J., Penfold, J., and Shackleton, C. (1991). *J. Appl. Phys.* **69**, 4989.
Bloch, F. (1929). *Z. Physik* **52**, 555.
Blundell, S. (2001). *Magnetism in Condensed Matter*. Oxford University Press, Oxford.
Bogdanov, A.N. and Yablonskii, D.A. (1989). *Sov. Phys. JETP*, **68**, 101.
Boldrin, D., Samathrakis, I., Zemen, J., Mihai, A., Zou, B., Johnson, F., Esser, B.D., McComb, D.W., Petrov, P.K., Zhang, H., and Cohen, L.F. (2019). *Phys. Rev. Materials* **3**, 094409.
Bona, G.L., Meier, F., Taborelli, M., Bucher, E., and Schmidt, P.H. (1985). *Solid State Commun.* **56**, 391.
Bonnenberg, D., Hempel, K.A., and Wijn, H.P.J. (1986). *Alloys between Fe, Co or Ni*. In *Landolt-Börnstein, New Series III/19a, Chap. 1.2.1* (ed. K.-H. Hellwege). Springer-Verlag, Berlin.
Bornemann, M., Grytsiuk, S., Baumeister, P.F., dos Santos Dias, M., Zeller, R., Lounis, S., and Blügel, S. (2019). *J. Phys. Condens. Matter* **31**, 485801.
Bouckaert, L.P., Smoluchowski, R., and Wigner, E.P. (1936). *Phys. Rev.* **50**, 58.
Bozorth, R.M. (1951). *Ferromagnetism*. D. Van Nostrand Company, New York.
Bradley, C.J. and Cracknell, P. (1972). *The mathematical theory of symmetry in solids*. Clarendon Press, Oxford.
Brändle, H., Weller, D., Scott, J.C., Sticht, J., Oppeneer, P.M., and Güntherodt, G. (1993). *Int. J. Mod. hys.* B **7**, 345.
Braun, H.-B. (2012). *Advances in Physics* **61**, 1.
Brinkman, W. and Elliott, R.J. (1966). *Proc. R. Soc. A* **294**, 343.
Brooks, H. (1940). *Phys. Rev.* **58**, 909.
Brout, R. and Thomas, H. (1967). *Physics* (Long Island City, N.Y.) **3**, 317.
Brown, P.J., Nunez, V., Tasset F., Forsyth, J.B., and Radhakrishna, P.(1990). *J. Phys. Condens. Matt.* **2**, 9409.
Brown, P.J., Ouladdiaf, B., Ballou, R., Deportes, J., and Makrosyan, A.S. (1992). *J. Phys. Condens. Matter* **4**, 1103.
Bruno, P. and Chappert, C. (1992). *Phys. Rev. B* **46**, 261.
Bruno, P. (1993a). *Europhys. Lett.* **23**, 615.
Bruno, P. (1993b). *J. Magn. Magn. Mat.* **121**, 248.
Bruno, P. (1993c). In *Magnetismus von Festkörpern und Grenzflächen* (ed. R. Hölzle). Forschungszentrum Jülich.
Bruno, P. (1995). *Phys. Rev. B* **52**, 411.

Bruno, P. (1999). In *Magnetische Schichtsysteme in Forschung und Anwendung* (ed. R. Hölzle). Forschungszentrum Jülich.
Buis, N., Franse, J.J.M. and Brommer, E. (1981). *Physica B+C*, **106**, 1.
Burdick, G.A. (1963). *Phys. Rev.* **129**, 138.
Burke, K. (1997). In *Electronic density functional theory: recent progress and new directions* (ed. J. F. Dobson, G. Vignale, and M.P. Das). Plenum Press, New York.
Burlet, P., Rossat-Mignod, J., Troc, R., and Henkie, Z. (1981). *Solid State Commun.* **39**, 745.
Burmeister, W.L. and Sellmyer, D.J. (1982). *J. Appl. Phys.* **53**, 2024.
Buschow, K.H.J. and van Engen, P.G. (1981) *J. Magn. Magn. Mater.* **25**, 90.
Butler, W.H., Zhang, X.-G., Nicholson, D.M.C., and MacLaren, J.M. (1995). *Phys. Rev. B* **52**, 13399.
Butler, W.H., Zhang, X.-G., Nicholson, D.M.C., Schulthess, T.C., and MacLaren, J.M. (1996). *Phys. Rev. Lett.* **76**, 3216.
Butler, W.H., Zhang, X.-G., Schulthess, T.C., and MacLaren, J.M. (2001). *Phys. Rev. B* **63**, 054416.
Cade, N.A. (1980). *J. Phys. F: Metal Phys.* **10**, L187.
Cade, N.A. (1981). *J. Phys. F: Metal Phys.* **11**, 2399.
Callaway, J. (1964). *Energy band theory*. Academic Press, New York.
Callaway, J. (1974). *Quantum theory of the solid state*. Academic Press, New York.
Callaway, J. (1981). *Institute of Physics Conf. Ser.* **55**. Institute of Physics, Bristol.
Callaway, J. and March, N.H. (1984). In *Solid state physics*, vol. 38 (ed. F. Seitz, D. Turnbull, and H. Ehrenreich). Academic Press, Orlando, p. 136.
Callaway, J. and Wang, C.S. (1977). *Phys. Rev. B* **16**, 2095.
Callen, H.B. (1948). *Phys. Rev.* **73**, 1349.
Callen, H.B. (1985). *Thermodynamics and an introduction to thermostatistics*, 2nd edition. John Wiley & Sons, New York.
Callen, H.B. and Welton, T.A. (1951). *Phys. Rev.* **83**, 34.
Capellmann, H. (1979). *Solid State Commun.* **30**, 7 and *Z. Phys. B* **35**, 269.
Capellmann, H. (1987). *Metallic magnetism*. Springer-Verlag, Berlin.
Ceperley, D.M. and Alder, B.J. (1980). *Phys. Rev. Lett.* **45**, 566.
Chadov, S., Qi, X., Kübler, J., Fecher, G.-H., Felser, C., and Zhang, S.-C. (2010). *Nature Materials* **9**, 541.
Chaikin, P.M. and Lubensky, T.C. (1995). *Principles of Condensed Matter Physics*. Cambridge University Press, Cambridge.
Chana, K.S., Samson, J.H., Luchini, M.U., and Heine, V. (1991). *J. Phys. Condens. Matter* **3**, 6455.
Chang, C.-Z., Zhang, J., Feng, X., Shen, J., Zhang, Z., Guo, M., Li, K., Ou, Y., Wei P., Wang, L.-L., Ji, Z.-Q., Feng, Y., Ji, S., Chen, X., Jia, J., Dai, X., Fang, Z., Zhang S.-C., He, K., Wang, Y., Lu, L., Ma, X.-C., and Xue, Q.-K. (2013). *Science* **340**, 167.
Chang, G., Xu, S.-Y., Zhou, X., Huang, S.-M., Singh, B., Wang, B., Belopolski, I., Yin, J., Zhang, S., Bansil, A., Lin, H., and Hasan, M.Z. (2017). *Phys. Rev. Lett.* **119**, 156401.
Chen, T. and Stutius, W.E. (1974). *IEEE Trans. Magn.* **10**, 581.
Chen, H., Niu, Q., and MacDonald, A. H. (2014). *Phys. Rev. Lett.* **112**, 017205.
Chodorow, M.I. (1939). *Phys. Rev.* **55**, 675.
Choi, H., Tai, Y.-Y., and Zhu, J.-X. (2019). *Phys. Rev. B* **99**, 134437.
Coehoorn, R. (1991). *Phys. Rev. B* **44**, 9331.
Coehoorn, R. and de Groot, R.A. (1985). *J. Phys. F: Metal Phys.* **15**, 2135.
Coey, J.M.D., Berkowitz, A.E., Balcells, L.I., Putris, F.F., and Barry, A. (1998). *Phys. Rev. Lett.* **80**, 3815.

Cohen, M.L. and Heine, V. (1970). In *Solid state physics*, vol. 24 (ed. H. Ehrenreich, F. Seitz, and D. Turnbull). Academic Press, New York, pp. 37–248.
Coldwell-Horsfall, R.A. and Maradudin, A.A. (1960). *J. Math. Phys.* **1**, 395.
Coleridge, P.T., Molenaar, J., and Lodder, A. (1982). *J. Phys. C: Sol. State Phys.* **15**, 6943.
Cooke, J.F., Blackman, A.K., and Morgan, T. (1985). *Phys. Rev. Lett.* **54**, 718.
Cortona, P., Doniach, S., and Sommers, C. (1985). *Phys. Rev. A* **31**, 2842.
Crangle, J. and Goodman, G.M. (1971). *Proc. Roy. Soc. A* **321**, 477.
Daalderop, G.H.O., Kelly, P.J., and Schuurmans, M.F.H. (1990). *Phys. Rev. B* **41**, 11919.
Davenport, J.W. (1984). *Phys. Rev. B* **29**, 2896.
de Boer, F.R., Schinkel, C.J., Biesterbos, J., and Probst, S. (1969). *J. Appl. Phys.* **40**, 1049.
de Boer, F.R., Brück, E., Nakotte, H., Andreev, A.V., Sechovsky, V., Havela, L., Nozar, P., Denissen, C.J.M., Buschow, K.H.J., Vaziri, B., Meissner, M., Malette, H., and Rogl, P. (1992). *Physica B* **176**, 275.
Dederichs, P.H., Blügel, S., Zeller, R., and Akai, H. (1984). *Phys. Rev. Lett.* **53**, 2512.
Dederichs, P.H., Zeller, R., Akai, H., and Ebert, H. (1991). *J. Magn. Magn. Mat.* **100**, 241.
de Gennes, P.-G. (1960). *Phys. Rev.* **118**, 141.
de Groot, R.A. (1991). *Physica B* **172**, 45.
de Groot, R.A., Mueller, F.M., van Engen, P.G., and Buschow, K.H.J. (1983a). *Phys. Rev. Lett.* **50**, 2024.
de Groot, R.A., Mueller, F.M., van Engen, P.G., and Buschow, K.H.J. (1983b). *J. Appl. Phys.* **55**, 2151.
de Miguel, J.J., Cebollada, A., Gallego, J.M., Miranda, R., Schneider, C.M., Schuster, P., and Kirschner, J. (1991). *J. Magn. Magn. Mat.* **93**, 1.
Derlet, P.M. and Dudarev, S.L. (2007). *Progr. Mat. Science* **52**, 299.
Di, G.Q., Iwata, S., Tsunashima, S., and Uchiyama, S. (1992). *IEEE Trans. Magn.* **7** (10), 792.
Dietz, E., Gerhardt, U., and Maetz, C.J. (1978). *Phys. Rev. Lett.* **40**, 892.
Dijkstra, J., van Bruggen, C.F., Haas, C., and de Groot, R.A. (1989). *Phys. Rev. B* **40**, 7973.
Dingle, R.B. (1952). *Proc. Roy. Soc. A* **211**, 517.
Dirac, P.A.M. (1926). *Proc. Roy. Soc. London A* **112**, 661.
Doniach, S. and Sommers, C. (1982). In *Proceedings of the international conference on valence fluctuations of solids* (ed. L.M. Falicov, W. Hanke, and M.B. Maple). North-Holland, Amsterdam.
Dreizler, R.M. and Gross, E.K.U. (1990). *Density functional theory*. Springer-Verlag, Berlin.
Duschanek, H., Mohn, P., and Schwarz, K. (1989). *Physica B* **161**, 139.
Dzyaloshinskii,, J.E., and Kondratenko, P.S. (1976). *Sov. Phys. -JETP* **43**, 1036.
Dzyaloshinsky, I. (1958). *J. Phys. Chem. Solids* **4**, 241.
Eastman, D.E., Janak, J.F., Williams, A.R., Coleman, R.V., and Wendin, G. (1979). *J. Appl. Phys.* **50**, 7423.
Ebert, H. (1996). *Rep. Prog. Phys.* **59**, 1665.
Ebert, H., and Schütz, G. (eds.) (1996). *Spin-orbit influenced spectroscopies of magnetic solids*, (Lecture Notes in Physics 466). Springer-Verlag, Heidelberg.
Edwards, D.M. (1982). *J. Phys. F: Metal Phys.* **12**, 1789.
Edwards, D.M., Mathon, J., Muniz, R.B., and Phan, M.S. (1991). *Phys. Rev. Lett.* **67**, 493.
Ehrenreich, H. and Schwartz, L.M. (1976). In *Solid State Physics*, vol. 31 (ed. H. Ehrenreich, F. Seitz, and D. Turnbull). Academic Press, New York, p. 149.
Ellerbrock, R.D., Fuest, A., Schatz, A., Keune, W., and Brand, R.A. (1995). *Phys. Rev. Lett.* **74**, 3053.
Entel, P., Hoffmann, E., Mohn, P., Schwarz, K., and Moruzzi, V.L. (1993). *Phys. Rev. B* **47**, 8706.
Eriksson, O., Johansson, B., and Brooks, M.S.S. (1989). *J. Phys. Condens. Matter* **1**, 4005.

Eschrig, H. (1989). *Physica C* **159**, 545.
Eschrig, H. (1996). *The fundamentals of density functional theory*. Teubner, Stuttgart.
Escudier, P. (1975). *Ann. Phys.* (Paris) **9**, 125.
Everschor-Sitte, K., Masell, J., Reeve, R.M., and Kläui, M. (2018). *J. Appl. Phys.* **124**, 240901.
Eyert, V. (2012). *The augmented spherical wave method: a comprehensive treatment.*, 2nd edition Springer-Verlag Berlin Heidelberg.
Fairbairn, W.M. and Yip, S.Y. (1990). *J. Phys. Condens. Matter* **2**, 4197.
Fang, Z., Terakura, K., and Kanamori, J. (2001). *Phys. Rev. B* **63**, 180407(R).
Faulkner, J.S. (1977). *J. Phys. C: Sol. State Phys.* **10**, 4661.
Faulkner, J.S. and Stocks, G.M. (1980). *Phys. Rev. B* **21**, 3222.
Fawcett, E. (1988). *Rev. Mod. Phys.* **60**, 209.
Fazekas, P. (1999). *Electron correlation and magnetism*. World Scientific, Singapore.
Fernando, G.W., Davenport, J.W., Watson, R.E., and Weinert, M. (1989a). *Phys. Rev. B* **40**, 2757.
Fernando, G.W., Quian, G.-X., Weinert, M., Wang, M., and Davenport, J.W. (1989b). *Phys. Rev. B* **40**, 7985.
Feder, R., Rosicky, F., and Ackermann, B. (1983). *Z. Phys. B* **52**, 31.
Felser, C., Fecher G.H., and Balke B. (2007). *Angew. Chem. Int. Ed.* **46**, 668.
Fert, A., Reyren, N., and Cros, V. (2017). *Nature Reviews Materials* **2**, 1.
Feynman, R.P. (1972). *Statistical mechanics – a set of lectures*. Benjamin, Reading, Mass.
Fisher, D.S. and Lee, P.A. (1981). *Phys. Rev. B* **23**, 6851.
Flack, H.D. (2003). *Helvetica Chimica Acta* **86**, 905.
Freeman, A.J. and Watson, R.E. (1961). *Acta Crystallogr.* **14**, 231.
Friedel, J. (1958). *Nuovo Cimento* **10**, Suppl. 2, 287.
Fruchart, D. and Bertaut E.F. (1978). *J. Phys. Soc. Japan* **44**, 781.
Fry, J.L., Brener, N.E., Laurent, D.G., and Callaway, J. (1981). *J. Appl. Phys.* **52**, 2101.
Fu, L. and Kane, C.L. (2006). *Phys. Rev. B* **74**, 195312.
Fu, C., Sun, Y., and Felser, C. (2020). *APL Materials* **8**, 040913.
Fumagalli, P. (1996). *Habilitationsschrift* Rheinisch Westfälische Universität Aachen.
Fukui, T. and Hatsugai, Y. (2007). *J. Phys. Soc. Japan* **76**, 053702.
Fulde, P. (1991). *Electron correlations in molecules and solids*, Springer Series in Solid State Sciences, vol. 100. Springer-Verlag, Heidelberg.
Fulde, P., Keller, J., and Zwicknagl, G. (1988). In *Solid state physics*, vol. 41 (ed. H. Ehrenreich, F. Seitz, and D. Turnbull). Academic Press, New York, p. 1.
Fuss, A., Demokritov, S., Grünberg, P., and Zinn, W. (1992). *J. Magn. Magn. Mat.* **103**, L221.
Fuh, H-R. and Guo G-Y. (2011). *Phys. Rev. B* **84**, 144427.
Galanakis, I., Dederichs, P.H., and Papanikolaou, N. (2002). *Phys. Rev. B* **66**, 174429.
Gell-Mann, M. and Brueckner, K.A. (1957). *Phys. Rev.* **106**, 364.
George, J.M., Pereira, L.G., Barthélémy, A., Petroff, F., Steren, L., Duvail, J.L., and Fert, A. (1994). *Phys. Rev. Lett.* **72**, 408.
Göbel, B., Mertig, I., and Tretiakov, O.A. (2020). *Physics Reports* https://doi.org/10.1016/j.physrep.2020.10.001.
Gordon, W. (1928). *Z. Phys.* **50**, 630.
Gosálbez-Martínez, D., Souza, I., and Vanderbilt, D. (2015). *Phys. Rev. B* **92**, 085138.
Greenwood, A.D. (1958). *Proc. Phys. Soc. London* **71**, 585.
Grigoriev, S.K., Dyadkin, V.A., Moskvin, E.V., Lamago, D., Wolf, Th., Eckerlebe, H., and Maleyev, S.K. (2009). *Phys. Rev. B* **79**, 144417.
Gross, E.K.U., Dobson, J.F., and Petersilka, M. (1996). In *Density functional theory* (ed. R.F. Nalewajski). Springer-Verlag, Berlin, p. 1.

Grünberg, P., Schreiber, R., Pang, T., Brodsky, M.B., and Sowers, H. (1986). *Phys. Rev. Lett.* **57**, 2442.
Gschneidner, K.A. (1964). In *Solid state physics*, vol. 16 (ed. H. Ehrenreich, F. Seitz, and D. Turnbull). Academic Press, New York, p. 275.
Gubanov, V.A., Liechtenstein, A.I., and Postnikov, A.V. (1992). *Magnetism and the electronic structure of crystals*. Springer-Verlag, Berlin.
Guillaume, C.E. (1897). *Compt. Rend. Acad. Sci.* **125**, 235.
Guin, S.N., Vir, P., Zhang Y., Kumar, N., Watzman, S.J., Fu, C., Liu, E., Manna, K., Schnelle, W., Gooth J., Shekhar, C., Sun, Y., and Felser, C. (2019a). *Adv. Mater.* **2019 a**, 1806622.
Guin, S.N., Manna, K., Noky, J., Watzman S.J., Fu, C., Kumar, N., Schnelle, W., Shekhar, C., Sun, Y., Gooth, J., and Felser, C. (2019b). *NPG Asia Materials* **11**, 16.
Gukasov, A., Wisniewski, P., and Henkie, Z. (1996). *J. Phys. Condens. Matter* **8**, 10589.
Gunnarsson, O. (1976). *J. Phys. F: Metal Phys.* **6**, 587.
Gunnarsson, O. and Lundqvist, B.I. (1976). *Phys. Rev. B* **13**, 4274.
Gurung, G., Shao, D-F., Paudel, T.R., and Tsymbal, E.Y. (2019). *Phys. Rev. Materials* **3**, 044409.
Gyorffy, B.L., Kollar, J., Pindor, A.J., Stocks, G.M., Staunton, J., and Winter, H. (1984). In *Electronic structure of complex systems* (ed. P. Phariseau and T.M. Temmerman). NATO ASI Series, Vol. B **113**, 593.
Gyorffy, B.L., Pindor, A.J., Staunton, J., Stocks, G.M., and Winter, H. (1985). *J. Phys. F: Metal Phys.* **15**, 1337.
Hafner, J. (2008). *J. Comput. Chem.* **29**, 2044.
Hafner, J. and Hobbs, D. (2003). *Phys. Rev. B* **68**, 014408.
Hafner, R., Spisak, D., Lorenz, R., and Hafner, J. (2002). *Phys. Rev. B* **65**, 184432.
Haines, E.M., Clauberg, R., and Feder, R. (1985a). *Phys. Rev.Lett.* **54**, 932.
Haines, E.M., Heine, V., and Ziegler, A. (1985b). *J. Phys. F: Metal Phys.* **15**, 661.
Haines, E.M., Heine, V., and Ziegler, A. (1986). *J. Phys. F: Metal Phys.* **16**, 1343.
Halilov, S.V., Perlov, A.Ya., Oppeneer, P.M., Yaresko, A.N., and Antonov, V.N. (1998a). *Phys. Rev. B* **57**, 9557.
Halilov, S.V., Eschrig, H., Perlov, A.Y., and Oppeneer, P.M. (1998b). *Phys. Rev. B* **58**, 293.
Hall, E. (1879). *Am. J. Math.* **2**, 287.
Hall, E. (1881). *Philos. Mag.* **12**, 157.
Haldane, F.D.M. (1988). *Phys. Rev. Lett.* **61**, 2015.
Haldane, F.D.M. (2004). *Phys. Rev. Lett.* **93**, 206602.
Ham, F.S. and Segall, B. (1961). *Phys. Rev.* **124**, 1786.
Hamada, N. (1981). *J. Phys. Soc. Japan* **50**, 77.
Hamann, D.R. (1979). *Phys. Rev. Lett.* **42**, 662.
Hamann, D.R. (1989). *Phys. Rev. B* **40**, 2980.
Hanssen, K.E.H.M. and Mijnarends, P.E. (1986). *Phys. Rev. B* **34**, 5009.
Harrison, W.A. (1966). *Pseudopotentials in the Theory of Metals*. Benjamin, New York.
Harrison, W.A. (1980). *Electronic structure and the properties of solids: the physics of the chemical bond*. Freeman, San Francisco.
Hartree, D.R. and Hartree W. (1948). *Proc. Roy. Soc. (London)* A**193**, 299.
Hasegawa, H. (1979a). *J. Phys. Soc. Japan* **46**, 1504.
Hasegawa, H. (1979b). *Solid State Commun.* **31**, 597.
Hasegawa, H. (1987). *J. Magn. Magn. Mat.* **66**, 175.
Hathaway, K.B., Jansen, H.J.F., and Freeman, A.J. (1985). *Phys. Rev. B* **31**, 7603.

Havela, L., Almeida, T., Naegele, J.R., Sechovsky, V., and Brück, E. (1992). *J. Alloy Compounds* **181**, 205.
Havens, G.G. (1933). *Phys. Rev.* **43**, 992.
Hedin, L. and Lundqvist, B.I. (1971). *J. Phys. C: Sol. State Phys.* **4**, 2064.
Heiliger, C., Zahn, P., and Mertig, I. (2006). *Materials Today* **9**, 48.
Heine, V. (1965). *Phys. Rev.* **138**, A1689.
Heine, V. (1967). *Phys. Rev.* **153**, 673.
Heine, V. (1970). In *Solid state physics*, vol. 24 (ed. H. Ehrenreich, F. Seitz, and D. Turnbull). Academic Press, New York, p. 1.
Heine, V. and Joynt, R. (1988). *Europhys. Lett.* **5**, 81.
Heine, V., Liechtenstein, A.I., and Mryasow, O.N. (1990). *Europhys. Lett.* **12**, 545.
Heine, V. and Samson, J.H. (1980). *J. Phys. F: Metal Phys.* **10**, 2609.
Heine, V. and Samson, J.H. (1983). *J. Phys. F: Metal Phys.* **13**, 2155.
Heine, V. and Weaire, D. (1970). In *Solid state physics*, vol. 24 (ed. H. Ehrenreich, F. Seitz, and D. Turnbull). Academic Press, New York, p. 249.
Heinze, S., von Bergmann, K., Menzel, M., Brede, J., Kubetzka, A., Wiesendanger, R., Bihlmayer, G., and Blügel, S. (2011). *Nature Physics* **7**, 713.
Heisenberg, W. (1926). *Z. Phys.* **38**, 441.
Held, K., Keller, G., Eyert, V., Vollhardt, D., and Anisimov V.I. (2001). *Phys. Rev. Lett.* **86**, 5345.
Henkie, Z., Maslanka, R., Oleksy, Cz., Przystawa, J., de Boer, F.R., and Franse, J.J.M. (1987). *J. Magn. Magn. Mat.* **68**, 54.
Henry, N.F.M. and Lonsdale, K. (1969). *International tables for x-ray crystallography*. Kynoch Press, Birmingham.
Henry, W.E. (1952). *Phys. Rev.* **88**, 559.
Herman, F. and Schrieffer, R. (1992). *Phys. Rev. B* **46**, 5806.
Herman, F., Sticht, J., and van Schilfgaarde, M. (1991). *J. Appl. Phys.* **69**, 4783.
Herman, F., Sticht, J., and van Schilfgaarde, M. (1993). *Int. J. Mod. Physics B* **7**, 425.
Herring, C. (1937). *Phys. Rev.* **52**, 365.
Herring, C. (1940). *Phys. Rev.* **57**, 1169.
Herring, C. (1966). In *Magnetism IV*, Chap. V and XIII (ed. G. Rado and H. Suhl). Academic Press, New York.
Himpsel, F.J. (1991). *Phys. Rev. B* **44**, 5966.
Himpsel, F.J. and Eastman, D.E. (1980). *Phys. Rev. B* **21**, 3207.
Himpsel, F.J., Knapp, J.A., and Eastman, D.E. (1979). *Phys. Rev. B* **19**, 2919.
Himpsel, F.J., Ortega, J.E., Mankey, G.J., and Willis, R.F. (1998). *Adv. Phys.* **47**, 511.
Hirai, K. (1997). *J. Phys. Soc. Japan* **66**, 560.
Hobbs, D., Hafner, J., and Spisak D. (2003). *Phys. Rev. B* **68**, 014407.
Hohenberg, P. and Kohn, W. (1964). *Phys. Rev. B* **136**, 864.
Holden, A.J. and You, M.V. (1982). *J. Phys. F: Metal Phys.* **12**, 195.
Honda, N., Tanji, Y., and Nakagawa, Y. (1976). *J. Phys. Soc. Japan* **41**, 1931.
Huang, D., Zhang, X.W., Luo, C.P., Yang,, H.S., and Wang, Y.J. (1995). *J. Appl. Phys.*, **75**, 6351.
Hubbard, J. (1979). *Phys. Rev. B* **20**, 4584.
Hwang, H.Y. and Cheong, S.W. (1997). *Science* **278**, 1607.
Ikhlas, M., Tomita, T., Suzuki, M.-T., Nishio-Hamane, D., Arita, R., Otani, Y., and Nakatsuji, S. (2017). *Nature Physics* **13**, 1085.
Imry, I. and Landauer, R. (1999). *Rev. Mod. Phys.* **71**, S306.
Ishikawa, Y., Tajima, K., Bloch, D., and Roth., M. (1976). *Solid State Commun.* **19**, 525.

Janak, J.F. (1977) *Phys. Rev. B* **16**, 255.
Janak, J.F. (1978). *Phys. Rev. B* **18**, 7165.
Jarlborg, T. (1986). *Solid State Commun.* **57**, 683.
Jarlborg, T. and Peter, M. (1984). *J. Magn. Magn. Mat.* **42**, 89.
Johansson, L.I., Petersson, L.-G., Bergren, K.-F., and Allen, J.W. (1980). *Phys. Rev. B* **22**, 3294.
Johnson, D.D., Pinski, F.J., Staunton, J.B., Gyorffy, B.L., and Stocks, G.M. (1990). *Physical metallurgy of controlled expansion invar-type alloys* (ed. K.C. Russel and D.F. Smith). The Minerals, Metals and Materials Society.
Johnson, D.D. and Shelton, W.A. (1997). *The invar effect*: a centennial symposium (ed. J. Wittenauer). The Minerals, Metals and Materials Society.
Johnson, M.T., Purcell, S.T., McGee, N.W.E., Coehoorn, R., aan de Stegge, J., and Hoving, W. (1992). *Phys. Rev. Lett.* **68**, 2688.
Johnson, W.B., Anderson, J.R., and Papaconstantopoulos, D.A. (1984). *Phys. Rev. B* **29**, 5337.
Jones, W. and March, N.H. (1973). *Theoretical solid state physics*. John Wiley & Sons, London.
Julliere, M. (1975). *Phys. Lett. A* **54**, 225.
Kämper, K.P., Schmitt, W., Güntherodt, G., Gambino, R.J., and Ruf, R. (1987). *Phys. Rev. Lett.*, **59**, 2788.
Kanazawa, N., Onose, Y., Arima, T., Okuyama, D., Ohoyama, K., Wakimoto, S., Kakurai, K., Ishiwata, S., and Tokura, Y. (2011). *Phys. Rev. Lett.*, **106**, 156603.
Kanazawa, N., Kim, J.-H., Inosov, D.S., White, J.S., Egetenmeyer N., Gavilano, J., Ishiwata, S., Onose, Y., Arima, T., Keimer B., and Tokura Y. (2012). *Phys. Rev. B*, **86**, 134425.
Kanazawa, N., Nii, Y., Zhang, X.-X., Mishchenko A.S., De Filippis, G., Kagawa, F., Iwasa, Y., Nagaosa, N., and Tokura Y. (2016). *Nature Communications* **7**, 11622.
Kanazawa, N., Seki, S., and Tokura, Y. (2017). *Adv. Mater.* **2017**, 1603227.
Kaprzyk, S. and Mijnarends, P.E. (1986). *J. Phys. C* **19**, 1283.
Karplus, R. and Luttinger, J. M. (1954). *Phys. Rev.* **95**, 1154.
Kasper, J.S. and Roberts, B.W. (1956). *Phys. Rev.* **101**, 537.
Kasuya, T. (1956). *Progr. Theor. Phys.* (Kyoto) **16**, 45.
Kato, H., Okuda, T., Okimoto, Y., Tomioka, Y., Oikawa, K., Kamiyama, T., and Tokura Y. (2004). *Phys. Rev. B* **69**, 184412.
Katsnelson, M.I. and Lichtenstein, A.I. (2000). *Phys. Rev. B* **61**, 8906.
Katsnelson, M.I., Irkhin, V. Yu., Chioncel, L., Lichtenstein, A.I., and de Groot, R.A. (2008). *Rev. Mod. Phys.* **80**, 315.
Kerr, J. (1878) *Phil. Mag.* **5**, 161.
Kaufman, L., Clougherty, E.V., and Weiss, R.J. (1963). *Acta Metall.* **11**, 323.
Kirchner, V., Weber, W., and Voitländer, J. (1992). *J. Phys. Condens. Matter* **4**, 8097.
Kisker, E. (1983). *J. Phys. Chem.* **87**, 3598.
Kisker, E., Carbone, C., Flipse, C.F., and Wassermann, E.F. (1987a). *J. Magn. Magn. Mat.* **70**, 21.
Kisker, E., Clauberg, R., and Gudat, W. (1985). *Z. Phys. B* **61**, 453.
Kisker, E., Wassermann, E.F., and Carbone, C. (1987b). *Phys. Rev. Lett.* **58**, 1784.
Kittel, C. (1966). *Introduction to solid state physics*, 3rd edition. John Wiley & Sons, New York.
Kittel, C. (1967). *Quantum theory of solids*. John Wiley & Sons, New York.
Kittel, C. (1968). In *Solid state physics*, vol. 22 (ed. F. Seitz, D. Turnbull, and H. Ehrenreich). Academic Press, New York, p. 1.
Kittel, C. (1986). *Introduction to solid state physics*. Wiley & Sons, New York.
Kiyohara, N., Tomita, T., and Nakatsuji, S. (2016). *Phys. Rev. Applied* **5**, 064009.

Knïpfle, K., Sandratskii, L.M., and Kübler, J. (2000). *Phys. Rev. B* **62**, 5564.
Kobayashi, K.-I., Kimura, T., Sawada, H., Terakura, K., and Tokura, Y. (1998). *Nature* **395**, 677.
Köhler, J. and Kübler, J. (1996). *J. Phys. Condens. Matter* **8**, 8681.
Kohn, W. and Rostoker, N. (1954). *Phys. Rev.* **94**, 1111.
Kohn, W. and Sham, L.J. (1965). *Phys. Rev. A* **140**, 1133.
Kohn, W. and Vashishta, P. (1983). In *Theory of the Inhomogeneous Electron Gas* (ed. S. Lundqvist and N.H. March). Plenum Press, New York.
Koelling, D.D. and Harmon, B.N (1977). *J. Phys. C: Sol. State Phys.* **10**, 3107.
Koenig, C. (1982). *J. Phys. F: Metal Phys.* **12**, 1123.
Korenman, V., Murray, J.L., and Prange, R.E. (1977). *Phys. Rev. B* **16**, 4032.
Körling, M. and Ergon, J. (1996). *Phys. Rev. B* **54**, R8293.
Korotin, M.A., Anisimov, V.I., Khomskii, D.I., and Sawatzky, G.A. (1998). *Phys. Rev. Lett.* **80**, 4305.
Korringa, J. (1947). *Physica* **13**, 392.
Koster, G.F. (1957). In *Solid state physics*, vol. 5 (ed. F. Seitz and D. Turnbull). Academic Press, New York, p. 173.
Kouvel, J.S. and Rodbell, D.S. (1967). *J. Appl. Phys.* **38**, 979.
Krakauer, H., Posternak, M., and Freeman, A.J. (1979). *Phys. Rev. B* **19**, 1706.
Kramers, H.A. (1934). *Physica*, **1**, 182.
Krasko, G.L. (1987). *Phys. Rev. B* **36**, 8565.
Krasko, G.L. (1989). *Solid State Commun.* **70**, 1099.
Krén, E., Kádár, G., Pál, L., Sólyom, J., Szabó, P., and Tarnoczi, T. (1968). *Phys. Rev.*, **171**, 574.
Krén, E. and Kádár, G. (1970). *Solid State Commun.* **8**, 1653.
Kresse, G and Furthmüller, J. (1996). *Comp. Mat. Science* **6**, 15.
Krinchik, G.S. and Artem'ev, V.A. (1968). *Sov. Phys.-JETP*, **26**, 1080.
Krockenberger, Y., Mogare, K., Reehuis, M., Tovar, M., Jansen, M., Vaitheeswaran, G., Kanchana, V., Bultmark, F., Delin, A., Wilhelm, F., Rogalev, A., Winkler, A., and Alff, L. (2007). *Phys. Rev. B* **75**, 020404(R).
Kubo, R. (1957). *J. Phys. Soc. Japan* **12**, 570.
Kübler, J. (1981). *Phys. Lett. A* **81**, 81.
Kübler, J. (1984a). *J. Magn. Magn. Mat.* **45**, 415.
Kübler, J. (1984b). *Physica* **127B**, 257.
Kübler, J. (2004). In *Handbook of Magnetism and Advanced Magnetic Materials* (ed. S. Parkin and H. Kronmüller) Volume 1: *Fundamentals and Theory* John Wiley & Sons, London.
Kübler, J. (2006). *J. Phys. Condens. Matter* **18**, 9795.
Kübler, J. and Eyert, V. (1992). *Electronic structure calculations*. In *Electronic and magnetic properties of metals and ceramics* (ed. K.H.J. Buschow). VCH Verlagsgesellschaft, Weinheim.
Kübler, J., Höck, K.-H., Sticht, J., and Williams, A.R. (1988a). *J. Phys. F: Metal Phys.* **18**, 469.
Kübler, J., Höck, K.-H., Sticht, J., and Williams, A.R. (1988b). *J. Appl. Phys.* **63**, 3482.
Kübler, J., Fecher, G.H., and Felser, C. (2007). *Phys. Rev. B* **76**, 024414.
Kübler, J. and Felser, C. (2012). *Phys. Rev. B* **85**, 012405 .
Kübler, J. and Felser, C. (2014). *Europhys. Lett.* **108**, 67001.
Kübler, J. and Felser, C. (2016). *Europhys. Lett.* **114**, 47005.
Kübler, J. and Felser, C. (2017). *Europhys. Lett.* **120**, 47002.
Kübler, J., Williams, A.R., and Sommers, C.B. (1983). *Phys. Rev. B* **28**, 1745.
Kudrnovsky, J., Drchal, V., Turek, I., and Weinberger, P. (1994). *Phys. Rev. B* **50**, 16105.
Kulatov, K. and Mazin, I.I. (1990). *J. Phys. Condens. Matter* **2**, 343.
Kuroda, K., Tomita, T., Suzuki, M.-T., Bareille, C., Nugroho, A.A., Goswami, P., Ochi, M., Ikhlas, M., Nakayama, M., Akebi, S., Noguchi, R., Ishii, R., Inami, N., Ono, K., Kumigashira,

H., Varykhalov A., Muro, T., Koretsune, T., Arita, R., Shin, S., Kondo T., and Nakatsuji, S. (2017). *Nature Materials* **16**, 1090.

Landauer, R. (1987). *Z. Phys. B – Condensed Matter* **68**, 217.

Landolt, H. and Börnstein, R. (1962). *Numerical data and functional relationships in science and technology*, vol. II, Part 9, 6th edition (ed. K.-H. Hellwege and A.M. Hellwege). Springer-Verlag, Berlin, pp. I–7.

Landolt, H. and Börnstein, R. (1973). *Numerical data and functional relationships in science and technology*, New Series III/7b1 (ed. K.-H. Hellwege). Springer-Verlag, Berlin.

Landolt, H. and Börnstein, R. (1986). *Numerical data and functional relationships in science and technology*, New Series III/19a, Chap. 1.1.1 (ed. K.-H. Hellwege). Springer-Verlag, Berlin.

Lang, P., Nordström, L., Zeller, R., and Dederichs, P.H. (1993). *Phys. Rev. Lett.* **71**, 1927.

Lawson, A.C., Larson, A.C., Aronson, M.C., Johnson, S., Fisk, Z., Canfield, P.C., Thompson, J.D., and von Dreele, R.B. (1994). *J. Appl. Phys.* **76**, 7049.

Lax, M. (1955). *Phys. Rev.* **97**, 629.

Lebech, B., Bernhard, J., and Freltoft, T. (1989). *J. Phys. Condens. Matter*, **1**, 6105.

Lee, B. and Chang, Y.-C. (1994). *Phys. Rev. B* **49**, 8868.

Lee, B. and Chang, Y.-C. (1995a). *Phys. Rev. B* **51**, 316.

Lee, L. and Chang, Y.-C. (1995b). *Phys. Rev. B* **52**, 3499.

Lejaeghere, K., Bihlmayer, G., Björkmann, T., Blaha, P., Blügel, S., and numerous other authors. (2016). *Science* **351**, 1415.

Levy, M. (1982). *Phys. Rev. A* **26**, 1200.

Levy, M. and Perdew, J.P. (1985). *Phys. Rev. A* **32**, 2010.

Levy, P.M. (1994). In *Solid state physics*, vol. 47 (ed. H. Ehrenreich, F. Seitz, and D. Turnbull). Academic Press, New York, p. 367.

Lewis, S.P., Allen, P.B., and Sazaki, T. (1997). *Phys. Rev. B* **55**, 10253.

Li, D., Freitag, M., Pearson, J., Qiu, Z.Q., and Bader, S.D. (1994). *Phys. Rev. Lett.* **72**, 3112.

Li, Y., Kanazawa, N., Yu, X. Z., Tsukazaki, Y., Kawasaki, M., Ichikawa, M., Jin, X.F., Kagawa, F., and Tokuro, Y. (2013). *Phys. Rev. Lett.* **110**, 117202.

Li, X., Xu, L., Ding, L., Wang, J., Shen, M., Lu, X., Zhu, Z., and Behnia, K. (2017). *Phys. Rev. Lett.* **119**, 056601.

Li, P., Koo, J., Ning, W., Li, J., Miao, L., Min, L., Zhu, Y., Wang, Y., Alem, N., Liu, C.-X., Mao, Z., and Yan, B. (2020) *Nature Communications* **11**, 3476.

Lieb, E.H. and Oxford, S. (1981). *Int. J. Quantum Chem.* **19**, 427.

Liebmann, R. (1986). *Statistical mechanics of periodic frustrated Ising systems*. Springer-Verlag, Berlin.

Liebsch, A. (1979). *Phys. Rev. Lett.* **43**, 1431.

Lifshitz, I.M. and Kosevich, A.M. (1955). *Zh. Eksp. & Teor. Fiz.* **29**, 730.

Lindhard, J. (1954). *J. Kgl. Danske Videnskap. Selskab., Mat.-fys. Medd.* **28**, 8.

Liu, E., Sun, Y., Kumar, N., Muechler, L., Sun, A., Jiao, L., Yang, S-Y., Liu D., Liang, A., Xu, Q., Kroder, J., Süss, V., Borrmann, H., Shekhar, C., Wang, Z., Xi, C., Wang, W., Schnelle, W., Wirth, S., Chen, Y, Goennenwein, S.T.B., and Felser, C. (2018). *Nature Physics* **14**, 1125.

Lloyd, P. and Smith, P.V. (1972). *Adv. Phys.* **21**, 69.

Lomer, W.M. (1962). *Proc. Phys. Soc. London* **86**, 489.

Lonzarich, G.G. and Taillefer, L. (1985). *J. Phys. C: Sol. State Phys.* **18**, 4339.

Lonzarich, G.G. (1986). *J. Magn. Magn. Mat.* **54**, 612.

Loong, C.-K., Carpenter, J.M., Lynn, J.W., Robinson, R.A., and Mook, H.A. (1984). *J. Appl. Phys.* **55**, 1895.

Loucks, T.L. (1967). *Augmented plane wave method*. Benjamin, New York.
Lovesey, S.W. (1984). *Theory of neutron scattering from condensed matter*, vol.1: *Nuclear scattering*; vol.2: *Polarization effects and magnetic scattering*. Clarendon Press.
Luchini, M.U. and Heine, V. (1989). *J. Phys. Condens. Matter* **1**, 8961.
Luchini, M.U., Heine, V., and McMullan, G.J. (1991). *J. Phys. Condens. Matter* **3**, 8647.
Ludwig, W. and Falter, C. (1988). *Symmetries in physics*. Springer-Verlag, Berlin.
Lynn, J.W. (1975). Phys. Rev. B **11**, 2624.
Ma, S.-K. and Brueckner, K.A. (1968). *Phys. Rev.* **165**, 18 (1968).
MacDonald, A.H. and Vosko, L. (1979). *J. Phys. C: Sol. State Phys.* **12**, 2977.
MacDonald, A.H., Pickett, W.E., and Koelling, D.D. (1980). *J. Phys. C: Sol. State Phys.* **13**, 2675.
MacLaren, J.M., Zhang, X.-G., Butler, W.H., and Wang, X. (1999). *Phys. Rev. B* **59**, 5470.
Maetz, C.J., Gerhardt, U., Dietz, E., Ziegler, A., and Jelitto, R.J. (1982). *Phys. Rev. Lett.* **48**, 1686.
Malozemoff, A.P., Williams, A.R., and Moruzzi, V.L. (1984a). *Phys. Rev. B* **29**, 1620.
Malozemoff, A.P., Williams, A.R., Moruzzi, V.L., and Terakura, K. (1984b). *Phys. Rev. B* **30**, 6565.
Malozemoff, A.P., Williams, A.R., Terakura, K., Moruzzi, V.L., and Fukamichi, K. (1983). *J. Magn. Magn. Mat.* **35**, 192.
Mandal, T.K., Felser, C., Greenblatt, M., and Kübler, J. (2008). *Phys. Rev. B* **78**, 134431.
Manna, K., Muechler, L., Kao, T-H., Stinshoff, R., Zhang, Y., Gooth, J., Kumar, N., Kreiner, G., Koepernik, K., Car, R., Kübler, J., Fecher, G. H., Shekhar, C., Sun Y., and Felser, C. (2018). *Phys. Rev. X* **8**, 041045.
March, N.H. (1983). In *Theory of the Inhomogeneous Electron Gas* (ed. S. Lundqvist and N.H. March). Plenum Press, New York.
Marcus, P.M. and Moruzzi, V.L. (1988a). *J. Appl. Phys.* **63**, 4045.
Marcus, P.M. and Moruzzi, V.L. (1988b). *Phys. Rev. B* **38**, 6949.
Marshall, W. (1967). *Theory of magnetism in transition metals*. Academic Press, New York.
Mårtensson, H. and Nilsson, P.O. (1984). *Phys. Rev. B* **30**, 3047.
Maslanka, R., Henkie, Z., Franse, J.J.M., Verhoff, R., Oleksy, Cz., and Przystawa, J. (1989). *Physica B* **159**, 181.
Matar, S., Demazeau, G., Sticht, J., Eyert, V., and Kübler, J. (1992). *J. Phys. I* France **2**, 315.
Mattheiss, L.F., Wood, J.H., and Switendick, A.C. (1968). In Methods in computational physics, vol. 8 (ed. B. Alder, S. Fernbach, and M. Rotenberg). Academic Press, New York, p. 64.
Mattis, D.C. (1965). *Theory of magnetism*. Harper and Row, New York.
Mattis, D.C. (1981). *Theory of magnetism I*. Springer-Verlag, Heidelberg.
Mattis, D.C. (1985). *Theory of magnetism II*. Springer-Verlag, Heidelberg.
Mavropoulos, P., Papanikolaou, N., and Dederichs, P.H. (2000). *Phys. Rev. Lett.* **85**, 1088.
Mazin, I.I., Singh, D.J., and Ambrosch-Draxl, C. (1999a). *Phys. Rev. B* **59**, 411.
Mazin, I.I., Singh, D.J., and Ambrosch-Draxl, C. (1999b). *J. Appl. Phys.* **85**, 6220.
McKinnon, J.B., Melville, D., and Lee, E.W. (1970). *J. Phys. C* **1**, 46.
Melcher, C. (2014). *Proc. R. Soc. A* **470**, 20140394.
Mermin, N.D. (1965). *Phys. Rev.* **137**, A1441.
Mertig, I. (1999). In *Magnetische Schichtsysteme in Forschung und Anwendung* (ed. R. Hölzle). Forschungszentrum Jülich.
Mertig, I., Zeller, R., and Dederichs, P.H. (1993). *Phys. Rev. B* **47**, 16178.
Mertig, I., Zeller, R., and Dederichs, P.H. (1994) *Phys. Rev. B* **49**, 11767.
Merzbacher, E. (1970). *Quantum mechanics*. John Wiley & Sons, New York.
Meshcheriakova, O., Chadov, S., Nayak, A.K., Rössler, U.K., Kübler, J., Adré, G., Tsirlin, A.A., Kiss, J., Hausdorf, S., Kalache, A., Schnelle, W., Nicklas, M., and Felser, C. (2014). *Phys. Rev. Lett.*, **113**, 087203.

Messiah, C. (1976). *Quantum mechanics*, vol. 1 (9th print). North-Holland, Amsterdam.
Messiah, C. (1978). *Quantum mechanics*, vol. 2 (9th print). North-Holland, Amsterdam.
Methfessel, M. and Kübler, J. (1982). *J. Phys. F: Metal Phys.* **12**, 141.
Miedema, A.R. and Niessen, A.K. (1983). *Comp. Coupling Phase Diagrams and Thermochem.* (CALPHAD) **7**, 27.
Mitchell, P. and Paul, D. McK. (1985). *Phys. Rev. B* **32**, 3272.
Miyazaki, T. and Tezuka, N. (1995). *J. Magn. Magn. Mater.* **139**, L231.
Mohn, P., Schwarz, K., and Wagner, D. (1989). *Physica B* **161**, 153.
Mohn, P., Schwarz, K., and Wagner, D. (1991). *Phys. Rev. B* **43**, 3318.
Mohn, P., Schwarz, K., Uhl, M., and Kübler, J. (1997). *Solid State Commun.* **102**, 729.
Mohn, P. and Wohlfarth, E.P. (1987). *J. Phys. F: Metal Phys.* **17**, 2421.
Mohn, P. (2003). *Magnetism in the Solid State: an Introduction.* Springer-Verlag, Berlin.
Montroll, E.W. (1949). *Nuovo cimento* **6**, Suppl. 2. 265.
Moodera, J.S. and Mootoo, D.M. (1994). *J. Appl. Phys.* **76**, 6101.
Moodera, J.S., Kinder L.R., Wong T.M., and Meservey, R. (1995). *Phys. Rev. Lett.* **74**, 3273.
Mook, H.A., Lynn, J.W., and Nicklow, R.M. (1973). *Phys. Rev. Lett.* **30**, 556.
Mook, H.A. and Paul, D. McK. (1985). *Phys. Rev. Lett.* **54**, 227.
Mori, M. and Tsunoda, Y. (1993). *J. Phys. Condens. Matter* **5**, L77.
Morin, F. (1950). *Phys. Rev.* **78**, 819.
Moriya, T. (1960). *Phys. Rev.* **120**, 91.
Moriya, T. (1964). *Sol. State Comm.* **2**, 239.
Moriya, T. (1985). *Spin fluctuations in itinerant electron magnetism.* Springer-Verlag, Berlin.
Moroni, S., Ceperley, D.M., and Senatore, G. (1995). *Phys. Rev. Lett.* **75**, 689.
Moruzzi, V.L. (1986). *Phys. Rev. Lett.* **57**, 2211.
Moruzzi, V.L. and Marcus, P.M. (1990a). *Phys. Rev. B* **42**, 8361.
Moruzzi, V.L. and Marcus, P.M. (1990b). *Phys. Rev. B* **42**, 10322.
Moruzzi, V.L., Janak, J.F., and Williams, A.R. (1978). *Calculated electronic properties of metals.* Pergamon Press, New York.
Moruzzi, V.L. and Marcus, P.M. (1992a). *Phys. Rev. B* **46**, 2864.
Moruzzi, V.L. and Marcus, P.M. (1992b). *Solid State Commun.* **83**, 735.
Moruzzi, V.L. and Marcus, P.M. (1993). In *Handbook of magnetic materials* (ed. K.H.J. Buschow). North-Holland, Amsterdam, p. 97.
Moruzzi, V.L., Marcus, P.M., and Kübler, J. (1989). *Phys. Rev. B* **39**, 6957.
Moruzzi, V.L., Marcus, P.M., Schwarz, K., and Mohn, P. (1986). *Phys. Rev. B* **34**, 1784.
Mosca, D.H., Petroff, F., Fert, A., Schroeder, P.A., Pratt, W.P., Laloe, R., and Lequien, S. (1991). *J. Magn. Magn. Mat.* **94**, L1.
Mott, N.F. (1935). *Proc. Phys. Soc. London* **47**, 571.
Mott, N.F. (1964). *Adv. Phys.* **13**, 325.
Mott, N.F. and Jones, H. (1936). *The theory of the properties of metals and alloys.* Clarendon Press, Oxford.
Mryasov, O.N., Liechtenstein, A.I., Sandratskii, L.M., and Gubanov, V.A. (1991). *J. Phys. Condens. Matter* **3**, 7683.
Mühlbauer, S., Binz, B., Jonietz, F., Pfleiderer, C., Rosch, A., Neubauer, A., Georgii, R., and Böni, P. (2009). *Science* **323**, 915.
Müller, S., Bayer, P., Reischl, C., Heinz, K., Feldmann, B., Zillgen, H., and Wuttig, M. (1995). *Phys. Rev. Lett.* **74**, 765.
Murata, K.K. and Doniach, S. (1972). *Phys. Rev. Lett.* **29**, 285.

Nagaosa, N., Sinova, J., Onoda, S., MacDonald, A.H., and Ong, N.P. (2010). *Rev. Mod. Phys.* **82**, 1539.
Nagaosa, N. and Tokura, Y. (2013). *Nature Nanotechnology* **8**, 899.
Nakamura, H., Yoshimoto, K., Shiga, M., Nishi, M., and Kakurai, K. (1997). *J. Phys. Condens. Matter* **9**, 4701.
Nakanishi, O., Yanase, A., Hasegawa, A., and Kataoka, M. (1980) *Solid State Commun.* **35**, 995.
Nakatsuji, S., Kiyohara, N., and Higo, T. (2015). *Nature* **527**, 212.
Nayak, A.K., Fischer, J.E., Sun, Y., Yan, B., Karel, J., Komarek, A.C., Shekhar, C., Kumar, N., Schnelle, W., Kübler, J., Felser, C., and Parkin, S.S.P. (2016). *Sci. Adv.* **2**, e1501870.
Nayak, A.K., Kumar, V., Ma, T., Werner P., Pippel, E., Sahoo, R., Damay, F., Rössler U.K., Felser, C., and Parkin, S.S.P. (2017). *Nature*, **548**, 561.
Neubauer, A., Pfleiderer, C., Binz, B., Rosch, A., Ritz, R., Niklowitz, P.G., and Böni, P. (2009). *Phys. Rev. Lett* **102**, 186602.
Nielsen, H.B. and Ninomiya, M. (1981). *Nucl. Phys. B* **193**, 173.
Nielsen, H.B. and Ninomiya, M. (1983). *Phys. Lett. B* **130**, 389.
Nikolaev, A.V. and Andreev, B.V. (1993). *Phys. Solid State* **35**, 603.
Niu, Q. and Kleinman, L. (1998). *Phys. Rev. Lett.* **80**, 2205.
Niu, Q., Wang, X., Kleinman, L., Liu, W.-M., Nicholson, D.M.C., and Stocks, G.M. (1999). *Phys. Rev. Lett.* **83**, 207.
Noda, Y. and Ishikawa, Y. (1976a). *J. Phys. Soc. Japan* **40**, 690.
Noda, Y. and Ishikawa, Y. (1976b). *J. Phys. Soc. Japan* **40**, 699.
Noky, J., Zhang, Y., Gooth, J., Felser, C., and Sun, Y. (2020). *Comp. Materials* **6**, 77.
Nolting, W. (1986). *Quantentheorie des Magnetismus*. B.G. Teubner, Stuttgart.
Nordström, L., Lang, P., Zeller, R., and Dederichs, P.H. (1994). *Phys. Rev. B* **50**, 13058.
Obermaier, G.M. and Schellnhuber, H.J. (1981). *Phys. Rev. B* **23**, 5185.
Oguchi, T., Terakura, K., and Hamada, N. (1983). *J. Phys. F: Metal Phys.* **13**, 145.
Okuno, S.N. and Inomata, K. (1993). *Phys. Rev. Lett.* **70**, 1711.
Oleksy, Cz. (1984). *Acta Phys. Polonica A* **66**, 665.
Onsager, L. (1952). *Philos. Mag.* **43**, 1006.
Onsager, L., Mittag, L., and Stephen, M.J. (1966). *Ann. Phys.* (Leipzig) **18**, 71.
Onose, Y., Takeshita, N., Terakura, C., Takagi, H., and Tokura, Y. (2005). *Phys. Rev. B*, **72**, 224431
Oppeneer, P.M., Antonov, V.N., Kraft, T., Eschrig, H., Yaresko, A.N., and Perlov, A.Y. (1996). *J. Appl. Phys.* **80**, 1099.
Oppeneer, P.M. and Lodder, A. (1987). *J. Phys. F*, **17**, 1880.
Oppeneer, P.M., Maurer T., Sticht, J., and Kübler J. (1992a). *Phys. Rev. B*, **45**, 10924.
Oppeneer, P.M., Sticht J., Maurer T., and Kübler J. (1992b). *Z. Phys. B: Condensed Matter* **88**, 309.
Ortega, J.E., Himpsel, F.J., Mankey, G.J., and Willis, R.F. (1993). *Phys. Rev. B* **47**, 1540.
Ouardi, S., Fecher, G.H., Felser, C., and Kübler, J. (2013). *Phys. Rev. Lett.* **110**, 100401.
Paige, D.M., Szpunar, B., and Tanner, B.K. (1984). *J. Magn. Magn. Mat.* **44**, 239.
Palmer, R.G. (1982). *Adv. Phys.* **31**, 669.
Papaconstantopoulos, D.A. (1986). *Handbook of the band structure of elemental solids*. Plenum Press, New York.
Park, J.-H. and Han, J. H. (2011) *Phys. Rev. B* **83**, 184406.
Parkin, S.S.P., Bhadra, R., and Roche, K.P. (1991). *Phys. Rev. Lett.* **66**, 2152.
Parkin, S.S.P. and Mauri, D. (1991). *Phys. Rev. B* **44**, 7131.
Parkin, S.S.P., Kaiser, C., Panchula, A., Rice, P.M., Hughes, B., Samant, M., and Yang, S.-H. (2004) *Nature Mater.* **3**, 862.
Parkin, S.S.P., More, N., and Roche, K.P. (1990). *Phys. Rev. Lett.* **64**, 2304.

Pauling, L. (1938). *Phys. Rev.* **54**, 899.
Pekeris, C.L. (1959). *Phys. Rev.* **115**, 1216.
Perdew, J.P. (1986). *Phys. Rev. B* **33**, 8822.
Perdew, J.P. (1991). In *Electronic structure of solids '91* (ed. P. Ziesche and H. Eschrig). Akademie Verlag, Berlin, p. 11.
Perdew, J.P. and Burke, K. (1996). *Int. J. Quantum Chem.* **57**, 309.
Perdew, J.P., Burke, K., and Ernzerhof, M. (1996a). *Phys. Rev. Lett.* **77**, 3865.
Perdew, J.P., Burke, K., and Wang, Y. (1996b). *Phys. Rev. B* **54**, 16533.
Perdew, J.P., Ernzerhof, M., Zupan, A., and Burke, K. (1998). *J. Chem. Phys.* **108**, 1522.
Perdew, J.P. and Kurth, S. (1998). In *Density functionals: theory and applications* (ed. D.P. Joubert). Springer-Verlag, Berlin.
Perdew, J.P., Ruzsinszky, A., Csonka, G.I., Vydrov, O.A., Scuseria, G.E., Constantin, L.A., Zhou, X., and Burke, K. (2008). *Phys. Rev. Lett.* **100**, 136406.
Perdew, J.P. and Wang, Y. (1986). *Phys. Rev. B* **33**, 8800.
Perdew, J.P. and Wang, Y. (1992). *Phys. Rev. B* **45**, 13244.
Perdew, J.P. and Zunger, A. (1981). *Phys. Rev. B* **23**, 5048.
Petroff, F., Barthelemy, A., Mosca, D.H., Lottis, D.K., Fert, A., Schroeder, P.A., Pratt, W.P., Laloe, R., and Lequien, S. (1991). *Phys. Rev. B* **44**, 5355.
Pettifor, D.G. (1977a). *J. Phys. F: Metal Phys.* **7**, 613.
Pettifor, D.G. (1977b). *J. Phys. F: Metal Phys.* **7**, 1009.
Phariseau, P. and Temmerman, W.M. (eds.) (1984). *The electronic structure.* NATO Advanced Science Institute Series; B; 113. Plenum Press, New York.
Pickett, W.E. (1989). *Rev. Mod. Phys.* **61**, 433.
Pickett, W.E. and Singh, D.J. (1996). *Phys. Rev. B* **53**, 1146.
Pines, D. (1964). *Elementary excitations in solids.* W.A. Benjamin, New York.
Podgorny, M. (1989). *Physica B* **161**, 105 and 109.
Pokrovsky, V.L. (1979). *Advances in Physics* **28**, 595.
Poulson, U.K., Kollar, J., and Andersen, O.K. (1976). *J. Phys. F: Metal Phys.* **6**, L241.
Prinz, G.A. (1985). *Phys. Rev. Lett.* **54**, 1051.
Qin, T., Niu, Q., and Shi, J. (2011). *Phys. Rev. Lett.* **107**, 236601.
Qiu, Z.Q., Pearson, J., Berger, A., and Bader, S.D. (1992). *Phys. Rev. Lett.* **68**, 1398.
Rajagopal, A.K. (1980). *Adv. Chem. Phys.* **41**, 59.
Rajagopal, A.K. and Callaway, J. (1973). *Phys. Rev. B* **7**, 1912.
Ramana, M.V. and Rajagopal, A.K. (1983). *Adv. Chem. Phys.* **54**, 231.
Ranno, L., Barry, A., and Coey, J.M.D. (1997). *J. Appl. Phys.* **81**, 5774.
Rebouillat, J.P. (1972). *IEEE Trans. Magn.* **8**, 630.
Reim, W. and Schoenes, J. (1990). In *Ferromagnetic materials* (eds. K.H.J. Buschow, E.P. Wohlfarth) **5**, 133. North-Holland, Amsterdam.
Resta, R. (1994). *Rev. Mod. Phys.* **66**, 899.
Rhodes, P. and Wohlfarth, E.P. (1963). *Proc. Roy. Soc.* **A273**, 247.
Richardson, M.J. and Melville, D. (1972). *J. Phys. F: Metal Phys.* **2**, 337.
Richardson, M.J., Melville, D., and Ricodeau, J.A. (1973). *Phys. Lett.* **46A**, 153.
Robinson, R.A., Lawson, A.C., Buschow, K.H.J., de Boer, F.R., Sechowsky, V., and von Dreele, R.B. (1991). *J. Magn. Magn. Mat.* **98**, 147.
Robinson, R.A., Lawson, A.C., Goldstone, J.A., and Buschow, K.H.J. (1993). *J. Magn. Magn. Mat.* **128**, 143.
Robinson, R.A., Lawson, A.C., Lynn, J.W., and Buschow, K.H.J. (1992). *Phys. Rev. B* **45**, 2939.

Robinson, R.A., Lynn, J.W., Lawson, A.C., and Nakotte, H. (1994). *J. Appl. Phys.* **75**, 6589.
Roman, E., Makrousov, Y, and Souza, I. (2009). *Phys. Rev. Lett.* **103**, 097203.
Rosengaard, N.M. and Johansson, B. (1997). *Phys. Rev. B* **55**, 14975.
Rössler, U.K., Bogdanov, A.N. and Pfleiderer, C. (2006). *Nature* **442**, 797.
Rössler, U.K., Leonov, A.A., and Bogdanov, A.N. (2011). *J. Phys.: Conference Series* **303**, 012105.
Ruban, A.V., Khmelevskyi, S., Mohn, P., and Johansson, B. (2007). *Phys. Rev. B* **75**, 054402.
Ruban, A.V., Shallcross, S., Simak, S.I., and Skriver, H.L. (2004). *Phys. Rev. B* **70**, 125115.
Ruderman, M.A. and Kittel, C. (1954). *Phys. Rev.* **96**, 99.
Runge, E. and Gross, E.K.U. (1984). *Phys. Rev. Lett.* **52**, 997.
Rusz, J., Turek, I., and Divis, M. (2005). *Phys. Rev. B* **71**, 174408.
Saha, R., Srivastava, A.K., Ma, T., Jena, J., Werner, P., Kumar, V., Felser, C., and Parkin S.S.P. (2019). *Nature Commun.*, **10**, 1038.
Sakai, A., Mizuta, Y.P., Nugroho, A.A., Sihombing, R., Koretsune, T., Suzuki, M.-T., Takemori, N., Ishii, R., Nishio-Hamane, D., Arita, R., Goswami, P., and Nakatsuji, S. (2018). *Nature Physics* **14**, 1119.
Sakurabe, Y., Hattori, M., Oogane, M., Ando, Y., Kato, H., Sakuma, A., Miyazaki, T., and Kubota, H. (2006). *Appl. Phys. Lett.* **88**, 192508.
Sakurai, J.J. (1967). *Advanced quantum mechanics*. Addison-Wesley, Redwood City.
Sakurai, J.J. (1985). *Modern quantum mechanics*. Benjamin/Cummings, Menlo Park.
Sandratskii, L.M. (1986a). *Phys. Stat. Sol. (b)* **135**, 167.
Sandratskii, L.M. (1986b). *J. Phys. F: Metal Phys.* **16**, L43.
Sandratskii, L.M. (1991). *J. Phys. Condens. Matter* **3**, 8565, 8587.
Sandratskii, L.M. (1998). *Adv. Phys.* **47**, 91.
Sandratskii, L.M. and Kübler, J. (1992). *J. Phys. Condens. Matter* **4**, 6927.
Sandratskii, L.M. and Kübler, J. (1996a). *Europhys. Lett.* **33**, 447.
Sandratskii, L.M. and Kübler, J. (1996b). *Phys. Rev. Lett.* **76**, 4963.
Sandratskii, L.M. and Kübler, J. (1997a). *Phys. Rev. B* **55**, 11395.
Sandratskii, L.M. and Kübler, J. (1997b). *J. Phys. Condens. Matter* **9**, 4897.
Sandratskii, L.M., Uhl, M., and Kübler, J. (1998). *Itinerant electron magnetism: fluctuation effects* (ed. D. Wagner, W. Brauneck, and A. Solontsov). Kluwer Academic Publishers, Dordrecht.
Sandratskii, L.M.(2017). *Phys. Rev. B* **96**, 024450.
Sarma, D.D., Mahadevan, P., Saha-Dasgupta, T., Ray, S., and Kumar, A. (2000). *Phys. Rev. Lett.* **85**, 2549.
Sasioglu, E., Sandratskii, L.M., Bruno, P., and Galanakis, I. (2005). *Phys. Rev. B* **72**, 184415.
Sasioglu, E., Sandratskii, L.M., and Bruno, P. (2008). *Phys. Rev. B* **77**, 064417.
Savrasov, S.Y. (1998). *Phys. Rev. Lett.* **81**, 2570.
Schad, R., Potter, C.D., Beliën, P., Verbanck, G., Moshchalkov, V.V., and Bruynseraede, Y. (1994). *Appl. Rev. Lett.* **64**, 3500.
Schaub, B. and Mukamel, D. (1985). *Phys. Rev. B* **32**, 6385.
Scheb, K.M., Kelly, P.J., and Bauer, G.E.W. (1995). *Phys. Rev. Lett.* **74**, 586.
Schellnhuber, H.-J., Obermaier, G.M., and Rauh, A. (1981). *Phys. Rev. B* **23**, 5191.
Schiff, L.I. (1955). *Quantum mechanics*. McGraw-Hill, New York.
Schröter, M., Ebert, H., Akai, H., Entel, P., Hoffmann, E., and Reddy, G.G. (1995). *Phys. Rev. B* **52**, 188.
Schröter, M., Entel, P., and Mishra, S.G. (1990). *J. Magn. Magn. Mat.* **87**, 163.
Schröter, M., Schmitz, B., and Entel, P. (1992). *J. Magn. Magn. Mat.* **104**, 747.

Schulz, T., Ritz, R., Bauer, A., Halder, M., Wagner, M., Franz, C., Pfleiderer, C., Everschor, K., Garst, M., and Rosch A. (2012). *Nature Physics* **8**, 301.
Schwarz, K. (1986). *J. Phys. F: Metal Phys.* **16**, L211.
Schwarz, K., Blaha, P., and Madsen, G.K.H. (2002). *Computer Physics Commun.* **147**, 71.
Schwarz, K., Mohn, P., Blaha, P., and Kübler, J. (1984). *J. Phys. F: Metal Phys.* **14**, 2659.
Segall, B. and Ham, F.S. (1968). In *Methods in computational physics*, vol. 8 (ed. B. Alder, S. Fernbach, and M. Rotenberg). Academic Press, New York, p. 251.
Serrate, D., De Teresa, J.M., and Ibarra, M.R. (2007). *J. Phys.: Condens. Matter* **19**, 023201.
Shibata, K., Yu, X.Z., Hara, T., Morikawa, D., Kanazawa, N., Kimoto, K., Ishiwata, S., Matsui, Y., and Tokura Y. (2013) *Nature Nanotechnology*, **8**, 723.
Shimizu, M. (1981). *Rep. Prog. Phys.* **44**, 329.
Shiomi, Y., Kanazawa, N., Shibata, K., Onose, Y., and Tokura, Y. (2013). *Phys. Rev. B*, **88**, 064409.
Shirane, G. (1984). *J. Magn. Magn. Mater.* **45**, 33.
Shirane, G., Böni, P., and Wicksted, J.P. (1986). *Phys. Rev. B* **33**, 1881.
Simon, B. (1983). *Phys. Rev. Lett.* **51**, 2167.
Singh, A.K., Manuel, A.A., and Walker, E. (1988). *Europhys. Lett.* **6**, 67.
Singwi, K.S., Sjölander, A., Tosi, M.P., and Land, R.H. (1970). *Phys. Rev. B* **1**, 1044.
Skomski, R. (2008). *Simple Models of Magnetism*. Oxford University Press, Oxford.
Skriver, H.L. (1984). *The LMTO method: muffin-tin orbitals and electronic structure*. Springer-Verlag, Berlin.
Skriver, H.L. (1985). *Phys. Rev. B* **31**, 1909.
Skyrme, T.H. (1961). *Proc. R. Soc. London Ser.A* **260**, 127.
Slater, J.C. (1936). *Phys. Rev.* **49**, 537.
Slater, J.C. (1937). *Phys. Rev.* **51**, 846.
Slater, J.C. (1965). *Quantum theory of molecules and solids*, vol. 2. McGraw-Hill, New York.
Slater, J.C. (1972). *Symmetry and energy bands in crystals*. Dover, New York.
Slater, J.C. (1974). *Quantum theory of molecules and solids*, vol. 4. McGraw-Hill, New York.
Slater, J.C. and Koster, G.F. (1954). *Phys. Rev.* **94**, 1498.
Small, L.M. and Heine, V. (1984). *J. Phys. F: Metal Phys.* **14**, 3041.
Söderlind, P., Ahuja, R., Eriksson, O., Wills, J.M., and Johansson, B. (1994). *Phys. Rev. B* **50**, 5918.
Sommerfeld, A. and Frank, N.H. (1931). *Rev. Mod. Phys.* **3**, 1.
Sondheimer, E.H. (1962). *Proc. Roy. Soc. A* **268**, 100.
Soven, P. (1967). *Phys. Rev.* **156**, 809.
Springford, M. (ed.) (1980). *Electrons at the Fermi surface*. Cambridge University Press, Cambridge.
Städele, M., Majewski, J.A., Vogl, P., and Görling, A. (1997). *Phys. Rev. Lett.* **79**, 2089.
Staunton, J.B. (1994). *Rep. Prog. Phys.* **57**, 1289.
Staunton, J.B. and Gyorffy, B.L. (1992). *Phys. Rev. Lett.* **69**, 371.
Staunton, J.B., Gyorffy, B.L., Stocks, G.M., and Wadsworth, J. (1986). *J. Phys. F: Metal Phys.* **16**, 1761.
Staunton, J.B., Poulter, J., Ginatempo, B., Bruno, E., and Johnson, D.D. (1999). *Phys. Rev. Lett.* **82**, 3340.
Stearns, M.B. (1986). *Fe, Co, Ni*. In *Landolt-Börnstein, New Series III/19a*, Chap. 1.1.2 (ed. K.-H. Hellwege). Springer-Verlag, Berlin.
Stenzel, E. and Winter, H. (1986). *J. Phys. F: Metal Phys.* **16**, 1789.
Stewart, A.L. (1963). *Adv. Phys.* **12**, 299.
Sticht, J., Herman, M.F., and Kübler, J. (1993). *Int. J. Mod. Physics B* **7**, 456.
Sticht, J., Höck, K.-H., and Kübler, J. (1989). *J. Phys. Condens. Matter* **1**, 8155.

Stiles, M.D. (1993). *Phys. Rev. B* **48**, 7238.
Stiles, M.D. (1996). *Phys. Rev. B* **54**, 14679.
Stixrude, L., Cohen, R.E., and Singh, D.J. (1994). *Phys. Rev. B* **50**, 6442.
Stocks, G.H. and Winter, H. (1984). In *Electronic structure of complex systems* (ed. P. Phariseau and T.M. Temmerman). Plenum, New York, p. 463.
Stoeffler, D. and Gautier, F. (1991). *Phys. Rev. B* **44**, 10389.
Stoeffler, D. and Gautier, F. (1993). *J. Magn. Magn. Mat.* **121**, 259.
Stoner, E.C. (1938). *Proc. Roy. Soc. London A* **165**, 372.
Stoner, E.C. (1939). *Proc. Roy. Soc. London A* **169**, 339.
Strange, P. (1999). *Relativistic quantum mechanics.* Cambridge University Press, Cambridge.
Strange, P., Staunton, J., and Gyorffy, B.L. (1984). *J. Phys. C: Sol. State Phys.* **17**, 3355.
Stringfellow, M.W. (1968). *J. Phys. C: Solid State Phys.* **1**, 950.
Sukhanov, A.S., Zuniga Cespedes B.E., Vir, P., Cameron, A.S., Heinemann, A., Martin, N., Chaboussant, G., Kumar, V., Milde, P., Eng, L.M., Felser, C., and Inosov, D.C. (2020). *Phys. Rev. B*, **102**, 174447.
Sürgers, C., Fischer, G., Winkel, P., and v. Löhneisen, H. (2014). *Nature Communications* **5**, 3400.
Sürgers, C., Kittler, W., Wolf, T., and v. Löhneisen, H. (2016). *AIP Advances* **6**, 055604
Suzuki, K. and Tedrow, P.M. (1998). *Phys. Rev. B* **58**, 11597.
Suzuki, K. and Tedrow, P.W. (1999). *Appl. Phys. Lett.* **74**, 428.
Suzuki, T., Weller, D., Chang, C.-A., Savoy, R., Huang, T., Gurney, B.A., and Speriosu, V. (1994). *Appl. Phys. Lett.* **64**, 2736.
Tahir-Keli, R.A. and ter Haar, D. (1962). *Phys. Rev.* **127**, 88 and 95.
Tahir-Keli, R.A. and Jarrett, H.S. (1964). *Phys. Rev.* **135**, A1096.
Tajima, K., Ishikawa, Y., Webster, P.J., Stringfellow, M.W., Tocchetti, D., and Zeabeck, K.R.A. (1977). *J. Phys. Soc. Japan* **43**, 483.
Takeda, T. (1978). *Z. Phys. B* **32**, 43.
Temmerman, W.M. (1982). *J. Phys. F: Metal Phys.* **12**, L25.
Temmerman, W.M., Svane, A., Szotek, Z., and Winter, H. (1998). In *Electronic density functional theory: Recent progress and new directions* (ed. J.F. Dobson, G. Vignale, and M.P. Das). Plenum Press, New York, p. 327.
Tanigaki, T., Shibata, K., Kanazawa, N., Yu, X., Onase, Y., Park, H. S., Shindo, D., and Tokura, Y. (2015). *Nano letters* **15**, 5438
Terakura, K. (1976). *J. Phys. F: Metal Phys.* **76**, 1385.
Terakura, K. (1977). *Physica* **91B**, 162.
Terakura, K. and Kanamori, J. (1971). *Prog. Theor. Phys.* **46**, 1007.
Thiry, P., Chandesris, D., Lecante, J., Guillot, C., Pinchaux, R., and Pétroff Y. (1979). *Phys. Rev. Lett.* **43**, 82.
Thouless, D.J., Kohmoto, M., Nightingale, M.P., and den Nijs, M. (1982). *Phys. Rev. Lett.* **49**, 405.
Tomiyoshi, S. and Yamaguchi, Y. (1982). *J. Phys. Soc. Japan* **51**, 2478.
Topp, W.C. and Hopfield, J.J. (1973). *Phys. Rev. B* **7**, 1295.
Toulouse, G. (1977). *Comm. Phys.* **2**, 115.
Troc, R., Tran, V.H., Kolenda, M., Kruk, R., Latka, K., Szytula, A., Rossat-Mignod, J., Bonnet, M., and Büchner, B. (1995). *J. Magn. Magn. Mat.* **151**, 102.
Trygg, J., Johansson, B., Eriksson, O., and Wills, J.M. (1995). *Phys. Rev. Lett.* **75**, 2871.
Tsujioka, T., Mizokawa, T., Okamoto, J., Fujimori, A., Nohara, M., Takagi, H., Yamaura, K., and Takano, M. (1997). *Phys. Rev. B* **56**, R15509.
Tsunoda, Y. (1989). *J. Phys. Condens. Matter* **1**, 10427.
Tsunoda, Y., Nishioka, Y., and Nicklow, R.M. (1993). *J. Magn. Magn. Mat.* **128**, 133.

Tu, P., Heeger, A.J., Kouvel, J.S., and Comly, J.B. (1969). *J. Appl. Phys.* **40**, 1368.
Turner, A.M., Donohu, A.W., and Erskine, J.L. (1984). *Phys. Rev. B* **29**, 2986.
Turner, A.M. and Vishwanath, A. (2013). arXiv:1301.0330v1.
Tyablikov, S.V. (1967). Methods in the quantum theory of magnetism. Springer, Boston, MA.
Uhl, M. (1995). Dissertation, unpublished, Darmstadt.
Uhl, M. (1998). unpublished material.
Uhl, M. and Kübler, J. (1996). *Phys. Rev. Lett.* **77**, 334.
Uhl, M. and Kübler, J. (1997). *J. Phys. Condens. Matter* **9**, 7885.
Uhl, M., Sandratskii, L.M., and Kübler, J. (1992). *J. Magn. Magn. Mat.* **103**, 314.
Uhl, M., Sandratskii, L.M., and Kübler, J. (1994). *Phys. Rev. B* **50**, 291.
Umetsu, R.Y., Kobayashi, K., Fujita A., Oikawa, K., Kainuma, R., Ishida, K., Endo, N., Fukamichi, K., and Sakuma, A. (2005). *Phys. Rev. B* **72**, 214412.
Unguris, J., Celotta, R.J., and Pierce, D.T. (1991). *Phys. Rev. Lett.* **67**, 140.
Uspenskii, Y.A., Kulatov, E.T., and Halilov, S.V. (1996). *Phys. Rev. B* **54**, 474.
van Engen, P.G., Buschow K.H.J., and Erman, M. (1983). *J. Magn. Magn. Mater.* **30**, 374.
van Leuken, H. and de Groot, R.A. (1995). *Phys. Rev. B* **51**, 7176.
van Schilfgaarde, M. and Herman, F. (1993). *Phys. Rev. Lett.* **71**, 1923.
Van Vleck, J.H. (1932). *The theory of electric and magnetic susceptibilities*. Oxford University Press, Oxford.
Vasiliu-Doloc, L., Lynn, J.W., Moudden, A.H., de Leon-Guevaran, A.M., and Revcolevschi, A. (1998). *Phys. Rev. B* **58**, 14913.
Velicky, B., Kirkpatrick, S., and Ehrenreich, H. (1968). *Phys. Rev.* **175**, 747.
Vidal, E.V., Stryganyuk, G., Schneider, H., Felser, C., and Jakob, G. (2011). *Appl. Phys. Lett.* **99**, 132509.
Vignale, G. and Rasolt, M. (1987). *Phys. Rev. Lett.* **59**, 2360.
Vignale, G. and Rasolt, M. (1988). *Phys. Rev. B* **37**, 10685.
Vir, P., Gayles, J., Sukhanov, A.S., Kumar, N., Damay, F., Sun, Y., Kübler, J., Shekhar, C., and Felser, C. (2019). *Phys. Rev. B* **99**, 140406.
Volovik, G.E. (1987). *J. Phys. C* **20**, L83.
Volovik, G.E. (2003). *The Universe in a Helium Droplet*. Clarendon Press. Oxford.
von Barth, U. (1984). In *Electronic structure of complex systems* (ed. P. Pharisea and T.M. Temmerman). Plenum, New York, p. 67.
von Barth, U. and Hedin, L. (1972). *J. Phys. C: Sol. State Phys.* **5**, 1629.
von Klitzing, K., Dorda, G., and Pepper, M. (1980). *Phys. Rev. Lett.* **6**, 494.
von Klitzing, K., Chakraborty, T., Kim, P., Madhavan, V., Dai, X., McIver, J., Tokura, Y., Savary, L., Smirnova D., Rey, A.M., Felser, C., Gooth, J., and Qi, X. (2020). *Nature Reviews Physics*, **2**, 397.
Vonsovski, S.V. (1974). *Magnetism*. Wiley & Sons, New York.
Vosko, S.H., MacDonald, A.H., and Liu, K.L. (1978). *Transition Metals* (Inst. Phys. Conf. Ser. **39**). Institute of Physics, Bristol, p. 33.
Vosko, S.H., Wilk, L., and Nusair, M. (1980). *Can. J. Phys.* **58**, 1200.
Wagner, D. (1972). *Introduction to the theory of magnetism*. Pergamon Press, New York.
Wagner, D. (1989). *J. Phys. Condens. Matter* **1**, 4635.
Wan, X., Turner, A.M., Vishwanath A., and Savrasov, Y. (2011). *Phys. Rev. B* **83**, 205101.
Wang, C.S., Klein, B.M., and Krakauer, H. (1985). *Phys. Rev. Lett.* **54**, 1852.
Wang, C.S., Wu, R., and Freeman, A.J. (1993). *Phys. Rev. Lett.* **70**, 869.

Wang, Z., Vergniory, M.G., Kushwaha, S., Hirschberger, M., Chulkov, E.V., Ernst, A., Ong, N.P., Cava, R.J., and Bernevig, B.A. (2016). *Phys. Rev. Lett.* **117**, 236401.
Wassermann, E.F. (1990). *INVAR: moment-volume instabilities in transition metals and alloys*. In *Ferromagnetic materials*, vol. 5 (ed. K.H.J. Buschow and E.P. Wohlfarth). North-Holland, Amsterdam.
Wassermann, E.F. (1997). *The Invar effect: a centennial symposium* (ed. J. Wittenauer). The Minerals, Metals and Materials Society.
Weber, W., Kirchner, B., and Voitländer, J. (1994). *Phys. Rev. B* **50**, 1090.
Webster, P.J. (1971). *J. Phys. Chem. Solids* **32**, 1221.
Weht, R. and Pickett, W.E. (1999). *Phys. Rev. B* **60**, 13006.
Weinberger, P. (1990). *Electron scattering theory for ordered and disordered matter*. Clarendon Press, Oxford.
Weis, J. (1998). Diplomarbeit, Darmstadt (unpublished).
Weischenberg, J., Freimuth, F., Blügel, S., and Makrousov, Y. (2013). *Phys. Rev. B* **87**, 060406.
Weischenberg, J., Freimuth, F., Sinova, J., Blügel, S., and Makrousov, Y. (2011). *Phys. Rev. Lett.* **107**, 106601.
Weiss, R.J. (1963). *Proc. Phys. Soc.* **82**, 281.
Weller, D., Harp, G.R., Farrow, R.F.C., Cebollada, A., and Sticht, J. (1994). *Phys. Rev. Lett.* **72**, 2097.
Weyl, H. (1929). *Phys. Z.* **56**, 330.
White, R.M. (1983). *Quantum theory of magnetism*. Springer-Verlag, Berlin.
Wiesendanger, R., Güntherodt, H.-J., Güntherodt, G., Gambino, R.J., and Ruf, R. (1990). *Phys. Rev. Lett.* **65**, 247.
Wigner, E.P. and Seitz, F. (1955). In *Solid State Physics*, vol. 1 (ed. F. Seitz and D. Turnbull). Academic Press, Orlando, p. 97.
Wijngaard, J.H., Haas, C., and de Groot, R.A. (1989). *Phys. Rev. B* **40**, 9318.
Williams, A.R., Kübler, J., and Gelatt Jr., C.D. (1979). *Phys. Rev. B* **19**, 6094.
Williams, A.R., Zeller, R., Moruzzi, V.L., Gelatt Jr., C.D., and Kübler, J. (1981). *J. Appl. Phys.* **52**, 2067.
Williams, A.R., Moruzzi, V.L., Gelatt Jr., C.D., Kübler, J., and Schwarz, K. (1982). *J. Appl. Phys.* **53**, 2019.
Williams, A.R., Moruzzi, V.L., Gelatt Jr., C.D., and Kübler, J. (1983a). *J. Magn. Magn. Mat.* **31–34**, 88.
Williams, A.R., Moruzzi, V.L., Malozemoff, A.P., and Terakura, K. (1983b). *IEEE Transactions on Magnetism* **MAG-19**, 1983.
Williams, A.R., Malozemoff, A.P., Moruzzi, V.L., and Matsui, M. (1984). *J. Appl. Phys.* **55**, 2353.
Williams, A.R. and von Barth, U. (1983). In *Theory of the Inhomogeneous Electron Gas* (ed. S. Lundqvist and N.H. March). Plenum Press, New York.
Wohlfarth, E.P. (1980). *Iron, Cobalt and Nickel*. In *Ferromagnetic Materials*, vol. 1 (ed. E.P. Wohlfarth). North-Holland, Amsterdam.
Wolf, J.A., Leng, O., Schreiber, R., Grünberg, P.A., and Zinn, W. (1993). *J. Magn. Magn. Mat.* **121**, 253.
Wurmehl, S., Fecher, G.H., Kandpal, H.C., Ksenofontov V., Felser, C., Lin, H.-J., and Morais, J. (2005a). *Phys. Rev. B* **72**, 184434.
Wurmehl, S., Fecher, G.H., Kandpal, H.C., Ksenofontov V., Felser, C., and Lin, H.-J. (2006a). *Appl. Phys. Lett.* **88**, 032503.

Wurmehl, S., Fecher, G.H., Ksenofontov V., Casper, F., Stumm, U., Felser, C., Lin, H.-J., and Hwu Y. (2005b). *J. Appl. Phys.* **99**, 08J103.

Wurmehl, S., Kandpal, H.C., Fecher, G.H., and Felser, C.(2006b). *J. Phys.: Condens. Matter* **18**, 6171.

Wuttke, C., Caglieris, F., Sykora, S., Scaravaggi, F., Wolter, A.U.B., Manna, K., Süss V., Shekhar, C., Felser, C., Büchner, B., and Hess, C. (2019). *Phys. Rev. B* **100**, 085111.

Xiao, D., Yao, Y., Fang, Z., and Niu, Q. (2006). *Phys. Rev. Lett.* **97**, 026603.

Xiao, D., Chang, C., and Niu, Q. (2010). *Rev. Mod. Phys.* **82**, 1959.

Xu, Q., Liu, E., Shi, W., Muechler, L., Gayles, J., Felser, C., and Sun, Y. (2018). *Phys. Rev. B* **97**, 235416.

Xu, L., Li, X., Lu, X., Collignon, C., Fu, H., Koo, J., Fauqué, B., Yan, B., Zhu, Z., and Behnia, K. (2020). *Sci. Adv.* **6**, eaaz3522.

Yafet, Y. (1987). *Phys. Rev. B* **36**, 3948.

Yamada, H., Yasui, M., and Shimizu, M. (1980). *Transition metals* (Inst. Phys. Conf. Ser. 55). Institute of Physics, Bristol, p. 177.

Yamada, O., Ono, F., Nakai, H., Maruyama, H., Arae, F., and Ohta, K. (1982). *Solid State Commun.* **42**, 473.

Yamada, T., Kunitomi, N., Nakai, Y., Cox, D.E., and Shirane, G. (1970). *J. Phys. Soc. Japan* **28**, 615.

Yamaoka, T., Mekata, M., and Takaki, H. (1974). *J. Phys. Soc. Japan* **36**, 438.

Yan, B. and Felser, C. (2017). *Ann. Rev. Condens. Mat. Phys.* **8**, 337.

Yang, H., Sun, Y., Zhang, Y., Shi, W.-J., Parkin S.S.P., and Yan, B. (2017). *New J. Phys.* **19**, 015008.

Yao, Y., Kleinman L., MacDonald, A.H., Sinova, J., Jungwirth, T., Wang, D., Wang E., and Niu Q. (2004). *Phys. Rev., Lett.* **92**, 037204.

Yokouchi, T., Kagawa, F., Hirschberger, M., Otani, Y., Nagaosa, N., and Tokura, Y. (2020). *Nature* **586**, 232.

Yosida, K. (1957). *Phys. Rev.* **106**, 893.

Yosida, K. (1996). *Theory of magnetism*, Chap. 8. Springer-Verlag, Berlin.

You, V. and Heine, V. (1982). *J. Phys. F: Metal Phys.* **12**, 177.

Yu, R., Zhang, W., Zhang, H.-J., Zhang, C.-C., Xi, D., and Fang, Z. (2010). *Science* **329**, 61.

Yuasa, S. (2008). *J. Phys. Soc. Japan* **77**, 031001.

Yuasa, S., Fukushima, A., Kubota, H., Suzuki, Y., and Ando, K. (2006). *Appl. Phys. Lett.* **89**, 042505.

Yuasa, S., Miyajima, H., Otani, Y., and Sakuma, A. (1995). *J. Magn. Magn. Mater.* **140**, 79.

Yuasa, S., Nagahama, T., Fukushima, A., Suzuki, Y., and Ando, K. (2004). *Nature Mater.* **3**, 868.

Zahn, P., Mertig, I., Richter, M., and Eschrig, H. (1995). *Phys. Rev. Lett.* **75**, 2996 (1995).

Zahn, P., Binder, J., Mertig, I., Zeller, R., and Dederichs, P.H. (1998). *Phys. Rev. Lett.* **80**, 4309.

Zak, J. (1968). *Phys. Rev.* **168**, 686.

Zak, J. (1989). *Phys. Rev. Lett.* **62**, 2747.

Zener, C. (1951). *Phys. Rev.* **82**, 403.

Zhang, X.-G., Butler, W.H., and Bandyopadhyay, A. (2003). *Phys. Rev. B* **68**, 092402.

Zhang, L. (2016). *New J. Phys.*, **18**, 103039.

Zhang, S. and Levy, P. (1998). *Phys. Rev. Lett.* **81**, 5660.

Zhang, Y., Sun, Y., Yang, H., Zelezny, J., Parkin, S.S.P., Felser C., and Yan, B. (2017). *Phys. Rev. B* **95**, 075128.

Zhang, D., Yan, B., Wu, S-C., Kübler, J., Kreiner, G., Parkin, S.S.P., and Felser, C. (2013). *J. Phys. Cond. Matter* **25**, 206006.

Zhou, X., Hanke, J.-P., Feng, W., Li, F., Guo, G.-Y., Yao, Y., Blügel, S., and Mokrousov, Y. (2019). *Phys. Rev. B* **99**, 104428.

Ziebeck K.R.A. and Neumann, K.-U. (2001). *Alloys and Compounds of d-Elements with Main Group Elements*. In *Landolt-Börnstein, New Series III/32 Pt.C2. pp.64-314*. (ed. H.P.J. Wijn). Springer-Verlag, Heidelberg.

Ziman, J.M. (1964). *Principles of the theory of solids*. Cambridge University Press, Cambridge.

Ziman, J.M. (1967). *Electrons and Phonons*. Clarendon Press, Oxford.

Ziman, J.M. (1969). *Elements of advanced quantum theory*. Cambridge University Press, Cambridge.

Index

E_g (or Γ_{12}) orbitals 152
K-matrix 470
T_{2g} (or $\Gamma_{25'}$) orbitals 152
Δ_1 - transition 324
α-Mn, atomic positions 247
α-Mn, magnetic moments 248
α-Mn, noncollinear order 247
β-Mn 242
γ-Fe, band structure 256
γ-Fe, spiral magnetic structure 255
γ-Mn 242
γ-Fe 250
γ-Fe$_{97}$Co$_3$ 250
γ-Fe, energy-band structure 255

A-phase 451
Abelian group 168
accidental degeneracy 361
accidental magnetic structure 345
accuracy 142
acoustic branch 421
addition theorem 111
adiabatic approximation 5, 31, 390, 403, 473
adiabatic connection 73
adiabatic magnon dispersion relation 394, 397
adiabatic principle 466
adiabatic spin dynamics 389
adiabaticity 394
Adler–Bell–Jackiw anomaly 364
AHE in antiferromagnetic Mn$_3$Sn and Mn$_3$Ge 360
algorithm 34
aliasing 292, 293
allotropic forms 231
allowed **k**-vectors 95
alloy degrees of freedom 471
alloy problem 468
alloying 261
ambient pressure 5, 136
amorphous Al-O 323

amplitude of oscillations 22
analytical expression for GGA 83
and augmentation 136
Anderson condition 434
angle-resolved photoemission 113, 114, 237
angular momentum 109, 133
angular momentum decompositions 145
angular momentum expansion 130
angular momentum index 128
angular momentum operator 3
anisotropy constants 348
anisotropy energy 347
anisotropy ratio 321
anomalous Hall effect 350
anomalous Nernst effect 376
anomalous velocity 353
ansatz 58
anti-Invar 285
anti-unitary transformations 339
antibonding 2p states 273
antibonding hybrids 230
antibonding orbital 139
antibonding orbitals 230
antibonding state 194
antibonding states 205
anticommutator 48
antiferromagnetic δ-Mn 231
antiferromagnetic configurations 220
antiferromagnetic Cr 237
antiferromagnetic CrO$_2$ 273
antiferromagnetic domains 360
antiferromagnetic fcc Mn 424
antiferromagnetic order 160
antiferromagnetic order of type AFI 241
antiferromagnetic phase 252
antiferromagnetic state 282
antiferromagnetic states, AM1, AM2 426

antiferromagnetic superexchange 448
antiferromagnetism 26, 230
antiferromagnets 185
antihedgehog 461
antimonopole 461
antiperovskite 357
antiskyrmion 456
antisymmetric exchange 343, 345
apparent electron mass 98
applied magnetic field 188
APW 108
APW equations 111
arbitrary phase 415
arbitrary spin configurations 159
artificial solid 98
ASA 135
ASW 136
ASW energy parameter 158
ASW envelope function 162
ASW method 145
ASWs 140
asymptote 86
asymptotic form 111
atom averaged moment 259
atomic calculations 149
atomic form factor 249
atomic frame of reference 68
atomic magnetic scattering amplitude 247
atomic orbital moment 172
atomic polyhedron 135
atomic potential 98
atomic scale 466
atomic sphere 67
atomic sphere approximation (ASA) 134, 392
atomic sphere potential 468
atomic units 37
atomization energies 86
atoms 62, 70
attractive metalloid potential 262
augmentation 109, 125, 141

504 *Index*

augmentation conditions 158
augmented Bessel function 257
augmented Bessel spinor 163
augmented functions 139, 171
augmented Hankel function 257
augmented Hankel spinor 162
augmented orbitals 139
augmented plane wave 109
augmented plane wave (APW) 100
augmented plane wave (APW) method 108
augmented spherical wave method (ASW) 136
augmented spinor functions 163
average chemical valence 260
average electronic valence 260
average magnetic valence 264
average moment per particle 176
average saturation magnetization 259
averaged density matrix 68
averaged Fermi velocities 318
Avogadro's number 4
axial vectors 345

back-Fourier transform 295
back-substitution 114
band edge 261
band electrons 173
band energies 45, 349
band gap 88, 241
band index 97
band indices 309
band mass renormalization effects 274
band parameters 194
band picture 173
band structure 99
band structure of bcc Fe 154
band structure of fcc Ni 155
band structure of Li 102
band structure, Cr 236
band structure, Ni 229
band structures, transition metals 196
band-filling ratio 100
band-mass parameter 157
band-structure calculations 221
band-structure problem 170

band-structure properties 184, 193
bands 99
bandwidth 38, 156, 183
bandwidths 390
basic inequality 386
basic ingredients 32
basic translations 91
basis functions 29
bcc coordinated 259
bcc coordinated crystals 282
bcc Fe based alloys 261
bcc lattice 99
behavior of I 191
Berry curvature 350, 351
Berry phase 350, 391
Bessel energy 158
Bessel spinor 163
binary compounds 258, 267
binding energies 223
binding energies, Co 228
bismuth 15
Bloch condition 109
Bloch electrons 7, 24, 27
Bloch function 89, 91, 97, 117, 123, 141, 150
Bloch functions 16, 93, 95, 309
Bloch ket 306
Bloch skyrmion 456
Bloch spinor 289
Bloch states 180
Bloch sum 164
Bloch symmetry 101, 128, 234
Bloch walls 465
Bloch's $T^{3/2}$ law 401
Bloch's law 169
Bloch's magnon contribution 274
Bloch's theorem 89, 91, 391
Bloch-symmetrized core wave function 101
Bloch-transformed expansion coefficients 150
body centered cubic (bcc) 92
body-centered prism 300
body-centered tetragonal 265
Bogoliubov variational free energy 408, 411, 415
Bogoliubov–Peierls inequality 402
Bogoliubov–Peierls variational mean-field theory 473
Bohr magneton 2, 15, 174
Bohr radius 37, 61
Bohr, van Leeuwen 1

Bohr–Sommerfeld quantization rule 16
Boltzmann equation 313, 314
Boltzmann transport equation 313
Boltzmann's constant 1
Boltzmann's equation 16
bonding hybrids 230
bonding orbital 139
bonding orbitals 230
bonding properties 283
bonding states 205
Born approximation 320
Born–Oppenheimer approximation 5, 28, 31, 390
Born–von Kármán boundary conditions, 95
Bose distribution function 429
Bose–Einstein occupation numbers 401
boundary conditions 9, 138, 149
Bravais lattice 90, 123, 125
Bravais vector 141
Brillouin function 175, 176
Brillouin function for small x 177
Brillouin zone 99
Brillouin zone (BZ) 92
Brillouin zone integrals 295
Brillouin zone integration 349
Brillouin zone radius 477
broken SU(2) symmetry 454
brute force 349
bulk bcc Co 420
bulk constituents 299
bulk Fermi surface 306
bulk ferromagnets 347
bulk Green's function 309
bulk moduli 213
bulk modulus 251
bulk properties 95

$C1_b$ structure 267
calculated magnetization 286
canonical ℓ band 157
canonical band concept 114
canonical band picture 216
canonical bands 152, 154
canonical coordinates 1
canonical d bands 153, 200
canonical d state density 195
canonical momenta 1
canonical weighting factor 402
canting 344

Cartesian vectors 92
Cauchy's theorem 310
cavity field 477
cavity-field approximation 478
center of band 156
center of d band 194
center of gravity 156
central sphere 133
central variables 28
change in total energy 303
characteristic energies 147
characteristic potential
 parameters 155
charge density 69, 145, 146,
 159
charge density disturbance 210
charge distribution 33
charge rearrangement 34
charge-density response 188
chemical potential 6, 7, 13, 25,
 313
chemical properties 122
chemical trend 213
chemical valence 263
Chern number 362
chiral anomaly 364
chiral structures 452
chromium 232
CIP 317
CIP orientation 312
classical Hamiltonian 1
classical Heisenberg model 403
classical partition function 1,
 408
classifications of symmetries 93
closed shells 4
closely packed solids 134
clusters 464
Co based alloys 259
Co density of states 262
Co_2-Heusler compounds 276
Co_2CrGa 277
Co_2FeSi 278
Co_2MnAl 277
Co_2MnSi 276
Co_2MnSi electrodes 325
Co_2TiAl 277
Co_2VGa 277
$Co_{1-x}M_x$ with M = B, Si, Sn,
 P, Au 267
Co_2MnAl 267
Co_2MnSn 267
Co_9Cu_7 321
Co_9Cu_7 multilayer 319
Co-Cu interface 318

Co/Cu (001) multilayers 317
Co/Cu/Co 288
coarse magnetic disorder 465
coarse-grained mesh 67, 160
cobalt 173, 225
CoCr alloys 259
coherent t-matrix 472
coherent medium 469, 470
coherent potential 469
coherent potential
 approximation, CPA 468
coherent tunneling 323, 326
collinear antiferromagnetic
 structure 339
collinear spin
 arrangements 160
collinear total energy
 change 421
collinearity 340
collinearly ordered
 antiferromagnet 423
collisions 22
combined correction
 terms 135, 145, 158
commensurate 169
common band 260
common quantization axis 159,
 160
common tangent 239, 282
commutator 48
commute 8
commuting operators 334
CoMnSb 269, 271
compatibility relations 169
compensated magnetic
 state 245
compensated magnetic
 structure 339
compensating magnetic
 moments 270
competing interactions 245
competition 217
complete orthonormal set 34
complete orthonormal set
 (CONS) 32
completeness 97
complex amplitude 302
complex contour
 integration 296
computer compounds 270
concentration dependence 260
conditional probability 72
conduction electrons 289
conductivity 312
conductivity tensor 316, 327

cone structure 337
confined states 299
confinement 302, 318
conservative philosophy of
 approximations 71
constant interstitial
 potential 111
constant minority-electron
 count 267
constants of integration 16
constants of motion 50
constrained calculations 211
constrained search 58
constrained total energy
 calculations 408
constrained total energy
 search 397
constraining the symmetry 202
contact form 289
continuity equation 47, 391
continuous 129
controlled correction 401
convective current 48
convective density 48
convective part 391
convergence 34, 298
convolution 429
coordinate scaling 76
copper 113
core electrons 31, 98
core states 104
corner-sharing triangles 244
correlation 27, 60, 61, 229
correlation contribution 386
correlation energy 62
correlation hole 85
correlation-hole density 86
correspondence principle 16
cosine lattice transform 404
Coulomb interaction 3
Coulomb potential 36, 161
Coulomb potentials 28
Coulomb repulsion 74
Coulombic 33
counting of states 96, 100
couple method 464
coupled equations 30, 50
coupling constant 73, 292
coupling energy 303, 304
coupling term 30
coupling,
 antiferromagnetic 291
coupling, ferromagnetic 291
coupling-constant
 dependence 74

coupling-constant integral 73
covalent bonding 230
CPA condition 470
CPA grand potential 478
CPA integrated density of states 472
CPA results for $Fe_{1-x}Ni_x$ 285
CPA theory 285
CPP 317
CPP orientation 312
Cr^{3+} in $KCr(SO)_4 \cdot 12H_2O$ 177
creation of new states 262
critical behavior 220
critical fluctuations 465
critical vector 305
critical volume 218, 236
critical volumes 218
CrMnSb 270
CrO_2 271
cross-sectional area 20
crystal phase stability 197
crystal Schrödinger equation 100, 150
crystal structure 197
crystal structure sequence 216
crystal structures, sequence 215
crystal-phase stability 199
crystalline solids 116
crystallographic axis 348
Cu 112
Cu_2MnAl 397
Cu_3Au crystal structure 266
CuAuI cell 300
cubic axes 348
cubic system 92
Curie constant 419, 476
Curie temperature 174, 178, 184, 272, 395, 405, 417, 463, 475, 476, 479
Curie temperature in spherical approximation 441
Curie temperature of CoNi 444
Curie temperature, FeRh 428
Curie temperatures 388
Curie temperatures in spherical approximation 447
Curie temperatures mean-field approximation 447
Curie temperatures of Heusler compounds 449
Curie temperatures, Fe, Co, Ni 404

Curie–Weiss behavior 419, 476, 479
Curie–Weiss law 178
Curie–Weiss susceptibility Ni_3Al 437
current 47, 315
current–current correlation function 313
curvature 305
cusps 238
cut-off parameter 428, 429
cyclotron frequency 8
cyclotron mass 18

d state densities 148
Darwin term 53, 333
de Haas–van Alphen (dHvA) oscillations 15
de Haas–van Alphen effect 1, 7, 12, 15, 222
Debye model 197
Debye temperature 198
Debye theory 428
decomposition 309
decomposition of norm 145, 146
decomposition of 1, 163
degeneracies 99
degeneracy 9, 19
degeneracy temperature 7
degree of band filling 200
degrees of freedom 166, 309
delocalized magnetic moments 173, 179
delocalized states 45
densities of states, calculated 192
density matrix 57, 58, 159, 160, 173, 385, 386
density matrix eigenvalues 65
density modulation 79
density of states 9, 22, 24, 95, 96, 180, 191, 200, 232, 261, 302, 307, 471
density of states at Fermi energy 192
density of states formula 472
density of states of sp electrons 263
density of states, Co 261
density of states, Fe 261
density of states, transition metals 193
density operator 58
density parameter r_S 37

density response function 80
density-functional band structure 45
density-functional formalism 35
density-functional theory 27, 28, 39, 385
density-of-states matrix 165
derivative of Fermi function 25
designer solids 287
determinant 97
determinant of an operator 471
determinant of coefficient matrix 152
determinant of single-particle states 32
determination of the Landau coefficients 433
deviation of the magnetic moments 342
dHvA measurements 297
diagonal elements 111
diagonal external potential 67
diagonalize density matrix locally 64
diagonalized density matrix 68
diagonalizing a matrix 98
diamagnetic 7, 15
diamagnetic susceptibility 4
diamagnetic term 15
diatomic molecule 139, 230
dielectric function 81
differentiable 129
differential equation 8
diffraction by crystals 91
diffusion 313
dilute alloys 265
dimensionless density gradient 85
Dirac delta function 83, 307
Dirac formalism 327
Dirac Hamiltonian 54
Dirac single-particle Hamiltonian 46
Dirac theory 2, 46
direct lattice vectors 91
direct structures 347
discontinuity 110
disorder 466
disordered alloys 265
disordered $Fe_{1-x}Ni_x$ 285
disordered local moment 286, 465
disordered local moment picture 473, 476

disordered local moment
  state 466
disordered local moments 389
disordered systems 116
dispersion 299
dispersion relations 45
dissipation constant 432
dissipation constant Ni_3Al 437
dissipative systems 429
distorted oxygen
  octahedra 272
distortion 157
distortion parameter 158
distribution function 313
divergence of the current
  391
documenting numerical
  results 158
domain 92, 465
dominant interaction 26
double counting 35
double exchange 449
double Fourier transform 234
double perovskites 278
double perovskites, Curie
  temperatures 451
double-counting term 198
double-counting terms 43,
  349, 476
double-valued irreducible
  representations 333
down-spin manifold 269
down-spin spectrum gap 270
dynamic approximation 431
dynamic approximation for
  Ni 434
dynamic susceptibility 431,
  433
dynamical spin
  susceptibility 394
dynamical susceptibility 428
dynamical theory 430
Dyson equation 45

early electronic structure
  calculations 39
early transition metal
  atoms 262
easy axes of Fe, Co, and Ni 348
easy axis 347
effective crystal potential 98
effective exchange
  coupling 288
effective exchange potential 39
effective Hamiltonian 30

effective Kohn–Sham
  potential 468
effective magnetic field 55
effective magnetic moment 476
effective mass 7, 45
effective mass
  approximation 15
effective medium 469
effective nuclei Hamiltonian
  30
effective potential 27, 33, 36,
  43, 386, 387, 389
effective potential matrix 59
effective potential, linear
  change 185
effective single-particle
  equation 43
effective vector potential 55
efficiency 124, 142
eigenfunctions 9, 90
eigenkets 306
eigenspinor 289
eigenspinors 168
eigenstate 6
eigenvalue 90
eigenvalues 8
eigenvalues of density
  matrix 60
eigenvalues of spin-density
  matrix 389
elastic constants 217
elastic scattering 315
electric field 262
electric moment 50
electrical resistivity 287, 312
electromagnetic field 47
electron balls 225
electron confinement, partial,
  total 302
electron correlation 272
electron density 471
electron distribution 29
electron gas 6
electron liquid 37
electron magnon
  interaction 271
electron orbit 17
electron sheet 225
electron spin 15, 21, 45, 248,
  387
electron surface 226, 233
electron–electron Coulomb
  interaction 40
electron–electron interaction 5,
  272

electron–electron
  interactions 274
electron–magnon
  interactions 274
electron–nucleon
  interaction 31
electron–phonon coupling 31
electron-density operator 40
electronic configurations 149
electronic function 29
electronic Hamiltonian 31
electronic kinetic energy 31
electronic part 29
electronic structure 27
electronic structure of CrO_2
  272
electronic structure of Ni 228
electronic-structure theory 116
electrostatic screening 98
elementary excitations 401
elementary transition
  metals 122
elements of density matrix 67
Elinvar 281
emergent electromagnetic
  field 456
empirical descriptions 124
empirical potential 112
empirical renormalization 437
empirical tight binding 124
empty lattice 99, 112
empty spheres 272
energetic order of HM, LM
  states 284
energy bands 99
energy bands for Co 227
energy derivative 125
energy derivative of structure
  constant 143
energy differences,
  structural 200
energy functional 42
energy gain mechanism 258
energy moments 147
energy parameter 136
energy range 128
energy–entropy balance 465
energy-band picture 26
energy-band theory 89
energy-dependent
  functions 125
energy-independent
  augmentation 124
energy-independent
  orbital 128, 133

energy-independent orbitals 123
engineered magnetic systems 287
engineering applications 281
enhanced susceptibility 184, 190, 202, 234
enhancement factor 83–86
enthalpy differences 200
entiomorphous pair 452
entropy 386, 387, 425, 467
entropy difference 425
entropy per atom 465
entropy stabilized high-spin–high-temperature ferromagnet 425
entropy-stabilized state 425
envelope function 128, 129, 136, 149, 161
envelope functions 140
environment 149, 265
epitaxy 258
equation of motion 390, 392
equilibrium atomic volumes 217
equilibrium density matrix 386
equivalent atoms 342
equivalent magnetic crystals 334
ergodicity, temporarily broken 473
error integral 13
Euler equation 457
Euler–Lagrange equation 42, 110, 398
Euler–Lagrange equations 32
evanescent states 323
Ewald method 120
Ewald parameter 121
exact exchange 88
exchange 60, 272
exchange and correlation effects 173
exchange constant 191, 244, 245
exchange constants 408, 418
exchange constants of hcp Co 421
exchange contribution 386
exchange contribution per electron 39
exchange coupling 305, 306
exchange function 404
exchange interaction 27
exchange mechanisms 448

exchange potential 33, 34
exchange splitting 160, 223, 229, 273
exchange splitting in antiferromagnetic state 257
exchange splitting, Co 228
exchange term 36, 38
exchange-charge density 33
exchange-correlation energy 42
exchange-correlation energy functional 72
exchange-correlation forces 230
exchange-correlation functional 83
exchange-correlation hole 72
exchange-correlation hole density 74
exchange-correlation integral 190
exchange-correlation kinetic energy 43
exchange-correlation part 72
exchange-correlation potential 60, 161, 388
exchange-correlation sum rule 74
exchange-hole density 75
excitation energies 34, 44
excitation spectra 44
expanded volumes 217
expansion coefficient 122
expansion coefficients 97, 309
expansion of Bloch function 100
expansion theorem 124, 136, 150, 163
expansions 97
expectation value 4
experimental band dispersions, Fe 222
experimental band structure, Fe 222
exponential factor 22
extended states 5, 318
external field 313
external fields 40
external perturbation 185
external potential 40, 161, 385
extremal Fermi surface cross-sections 21
extremum 182
extrinsic effects 319

face centered cubic (fcc) 92
Fano-antiresonance 263
fcc coordinated 259
fcc coordinated crystals 282
fcc lattice 99
fcc Mn (γ-Mn) 424
Fe based alloys 259
Fe films, epitaxial 250
Fe monolayers 250
Fe^{3+} in $NH_4Fe(SO_4)_2 \cdot 12H_2O$ 177
$Fe_{1-x}M_x$ with $M =$ Al, Au, B, Ga, Be, Si, C, Ge, Sn, P 267
Fe_3Cr_n 317, 319
Fe_3Ni calculations 282
Fe_3Pt 281
Fe_3Pt invar 286
$Fe_{65}Ni_{35}$ 281
Fe(001)|MgO(001)|Fe(001) 323
Fe, γ 250
Fe-Co 261
Fe-Cr 261
Fe-Cr alloys 262
Fe-Cr data 260
Fe-Ni alloys 285
Fe-rich bcc Fe-Co 261
Fe-rich Fe-Co alloys 262
Fe-V alloys 262
Fe/Cr/Fe 288
Fe/Cr/Fe multilayers 298
Fe/Cr/Fe/Cu/Fe/Cu multilayers 319
Fe/Cu multilayer 317
Fe/Cu/Fe 288
Fe/Mo/Fe 288
Fe/Mo/Fe(100) trilayers 288
FeCo 67, 68
FeCr 267
FeCr alloys 259
feedback effect 172
FeMnSb 270
FeNi alloys 259
FeNi calculations 282
FeRh 425
Fermi arc 366
Fermi distribution 234
Fermi edge 22
Fermi energy 6, 12, 58, 100, 224, 273, 472
Fermi function 6, 9, 80
Fermi hole 33
Fermi level 45
Fermi radius 6, 36, 84

Fermi surface 1, 173, 224, 274, 305, 316
Fermi surface area normal to magnetic field 22
Fermi surface cross-section, Cr 233
Fermi surface effects 295
Fermi surface nesting 233, 298
Fermi surface topologies 237
Fermi surfaces 9
Fermi temperature 7
Fermi velocity 317
Fermi wave vector 37, 39
Fermi wavelength 84, 291
Fermi's golden rule 316
Fermi–Dirac distribution 303, 388, 391
Fermi–Dirac distribution function 24, 38, 289, 387, 476
Fermi–Dirac function 314
Fermi–Dirac statistics 6, 44, 99
fermions 9
ferrimagnetic insulator Sr_2CrOsO_6 451
ferromagnetic δ-Mn 232
ferromagnetic barriers 304
ferromagnetic double exchange 448
ferromagnetic FeCo 67
ferromagnetic instability 174, 189
ferromagnetic layers 287
ferromagnetic state 464
ferromagnetic transition metals 173
FeV alloys 259
Feynman diagrams 32
field energy 181
field of all other electrons 33
field operators 32
field–particle system 50
filled shell 260
filter function 429
final state 229
fine tuning 232
fine-grained mesh 67, 160
fingerprints of d functions 114
finite temperature density-functional theory 388
finite temperatures 385
finite-temperature exchange-correlation potential 385

first-order perturbation theory 80, 186
first-order phase transition 419
first-order shift of eigenvalues 186
fixed nuclei 29
fluctuating local band picture 389
fluctuation-dissipation theorem 428, 429, 477
fluctuations 245
fluctuations, slow 430
fluid phase 78
folding 302
force theorem 198, 289, 309, 349, 476
form factor 247, 249
form invariance 334, 347
formal properties of density functionals 71
formation of magnetic moments 26
four-component RAPW calculation 55
four-component single-particle wave function 46
four-component wave function 54
Fourier analysis 293
Fourier coefficient 98, 104
Fourier coefficients 81, 407
Fourier component 81
Fourier series 19, 97
Fourier transform 36, 80, 81, 247, 429
fractional filling 216
fractional occupations 44
frame of reference 171
frame of reference, atomic 162
frame of reference, local 162
free electrons 5, 113, 127
free energies, hcp, fcc Co 422
free energy 8, 13, 401, 402, 412, 422, 425, 467, 474
free energy, FeRh 427
free-electron bandwidth 38
free-electron dispersion 99
free-electron eigenvalues 38
free-electron model 98
free-electron value 114
free-particle grand potential 476
free-particle Green's function 116
Freeman–Watson values 249

freeze the positions 29
freezing of nuclear positions 30
Fresnel-type integral 21
Friedel oscillations 82
frustrated Ising model 245
frustrated systems 244
frustration 238, 245
full Hamiltonian 29
full potential 116, 224
full potential method 349
full potential schemes 172
full response 187
functional 58
functional $E[n]$ 41
functional $V[n]$ 41
functional derivative 32, 42, 389
functional integral 439
functional of the density 40
functionals 41
functions describing exchange 61
fundamental translation vectors 91

GaGeLi crystal lattice 339
gap 262
gap, sp density of states 265
gas of free electrons 13
gauge 3, 7
Gauss integral 409
Gaussian orbitals 224
Gaussian statistics 407
Gaussian trial Hamiltonian 428
Gaussian units 437
Gaussian-type orbitals (GTO) 124
$Gd_3SO_4 \cdot 8H_2O$ 177
generalized Bloch function 170
generalized Bloch theorem 168
generalized gradient approximation (GGA) 70, 274
generalized gradient correction (GGA) 250
generalized grand potential 472
generalized Slater–Pauling curve 264
generalized Stoner condition 234
generalized susceptibility 233
generalized translations 167
geometric series 176, 303, 307
GGA 70, 76, 251

giant magnetoresistance (GMR) 287, 312
Gibbs distribution 399
Gibbs grand potential 386, 473
Ginzburg–Landau approach 428
Ginzburg–Landau coefficients 429
global quantization axis 57
global system 333
GMR 312, 317
good quantum number 304, 336
Gordon decomposition 47
gradient approximation (GA) 70
gradient correction 61, 63, 217, 249
gradient term 428
grand canonical density matrices 386
grand canonical ensemble 386
grand potential 303, 308, 387, 467, 473
grand potential change 308
grand potential difference 310
Green's function 120, 137, 307, 471
Green's function formalism 466
Green's function method 116, 302
Green's functions 151, 306
Green's identity 110, 118, 126
ground-state density 39
ground-state energy 34, 40
ground-state magnetic properties 215
ground-state properties 213
group 90
group representation, unitary 90
group velocity 305, 313

Haldane 363, 383
half-filled 3d electron shell 247
half-filled shell 175
half-metallic band structure 271
half-metallic ferrimagnet 280, 447
half-metallic ferromagnet 269
half-metallic ferromagnets 267
Hamiltonian 5, 306
Hamiltonian matrix 98

handedness 347
Hankel energy 158
Hankel functions 124
Hankel spinor 162
Hankel spinor, augmented 170
hard axis 347
harmonic oscillator 8
harmonics 236
Hartree energy 73, 76
Hartree integral 42
Hartree potential 34, 36
Hartree term 33
Hartree–Fock (HF) equations 33
Hartree–Fock approximation 31, 60
Hartree–Fock calculations 112
Hartree–Fock exchange 74
hcp cobalt 421
hcp lattice 418
He atom 5
head 133, 138
head augmentation 138
heavy-fermion systems 45
hedgehog 461
Heine power law 194
Heine's scaling law 213
Heisenberg form 467
Heisenberg Hamiltonian 245
Heisenberg model 389, 392, 394, 464, 477
helical ground state 453
helical spin state 451
helically ordered magnetic metals 26
Helmholtz free energy 2
hematite, α-Fe_2O_3 342
Hermitian 34
Heusler alloys 397
Heusler alloys, C1$_b$ structure 267
Heusler alloys, L2$_1$ structure 267
Heusler compound 444
hexagonal chemical lattice 339
hexagonal lattice 94
HF eigenvalue 36
HF eigenvalues 38
HF Hamiltonian 34
HF level 34
HF wave functions 34
high resistance state 323
high-conductivity channel 318
high-density limit 85, 86

high-moment state (HM) 282, 283
high-spin ferromagnet 425
high-spin ferromagnetic phases 252
high-spin ferromagnetic state 251
high-spin phases 239
high-spin state 238
high-temperature form of Co 420
higher harmonics 22
highest-occupied orbitals 44
highly excited states 34
historical background 89
hole 229
hole arms 226
hole pocket 226
hole sheet 225
hole surface 233
homogeneous DLM effective medium 435
homogeneous interacting electron gas 59
homogeneous magnetic field 7
hopping time 473
host 259, 469
host electronic structure 262
host material 307
hosts 116
Hubbard model 26
Hund's first rule 175
Hund's rule 448
Hund's rules 175
Hund's second rule 172
hybridization 152, 155, 262, 264, 270, 273
hybridized 113
hybridizing bands 240
hydrogen 3
hyperfine field 213
hysteresis loop 288

ideal hexagonal lattice 340
illustrative wave function 103
improve LDA 63
impurities 116, 262
impurity sites 469
incommensurate period 235
incommensurate spin-density 250
incommensurate spiral state 236
independent electrons 173

independent-particle picture 24
indistinguishable electrons 5
inelastic neutron scattering 465
infimum 41, 58
inhomogeneous electron system 60
inhomogeneous system 184
instantaneous position 31
insulators 88
integer multiples 9
integral equation 315
integral equation, susceptibility 187
integrate by parts 11
integrated density matrix 166
integrated density of states 304, 308, 475
integrated fluctuations 418
integrated free-particle density of states 471
integrating by parts 9, 20
integration path 310
interacting electrons 39
interaction potential 1, 28
interatomic electron motion 217
interatomic exchange 177
interatomic self-consistency 149
interface states 318, 320
interference effects 302
interference term 308
interlayer coupling 291, 308
interlayer exchange coupling 304, 311
interlayer exchange coupling strength 293
intermetallic compound 258
intermetallic compounds 116
internal charge current 49
internal charge density 49
internal current 48
internal density 48
internal energy 466
internal magnetic field 177
International Tables 272
interpolation formula 79
interpretative tool 298
interpreting numerical results 158
intersecting bands 257
interstitial region 108, 125, 134
interstitial space 109
interstitial space (IS) 108

intra-atomic exchange 174, 217, 220
intra-atomic noncollinearity 254
intra-band term 187
intrinsic GMR 317
Invar 281
Invar mechanism 285
Invar problem 429
invar region 285
invar region, Curie and Néel temperature 286
invariance of the crystal 345
inverse dynamic susceptibility 432
inverse KKR matrix 469
inverse longitudinal susceptibilities Ni_3Al 436
inverse photoemission 302
inverse structures 347
inverse susceptibility 178
inverse transverse susceptibilities Ni_3Al 436
ion cores 5
ionic charges 35
ionic component 230
ionic motion 31
ionic vibrations 31
ionicity 231
ionization energy 34
iron 173
irreducible representations 333
Ising model 389
Ising spins 244
isomorphic groups 168
iteration 34
itinerant antiferromagnets 423
itinerant electrons 5, 26, 477
itinerant-electron antiferromagnetism 230
itinerant-electron ferromagnet 273
itinerant-electron ferromagnets 258
itinerant-electron magnets 404
itinerant-electron picture 173, 191
itinerant-electron theory 180

joint probability 72

k-space 6
$KCrSe_2$ 271
Kerr angle 329
Kerr effect 327

Kerr rotation 330
Kerr rotation of MnBi 332
kinetic energy 3, 39, 58, 181, 386, 387
kinetic energy price 220
kinetic energy scaling 76
kinetic exchange 449
kink 38
KKR equation 121
KKR matrix 469
KKR method 116
KKR structure constants 150, 152, 257
KKR structure Green's function 468
KKR theory 144
KKR-ASA 149
KKR-ASA equations 152
KKR-ASA matrix 157
KKR-CPA method 265
KKR-CPA results 286
Kohn–Sham equation 43, 59, 74
Kohn–Sham equations 473
Kohn–Sham Hamiltonian 335, 344
Kohn–Sham–Schrödinger equation 169
Koopmans' theorem 34
Korringa–Kohn–Rostoker (KKR) form 120
Korringa–Kohn–Rostoker (KKR) theory 468
Korringa–Kohn–Rostoker Atomic Sphere Approximation 149
Kramers–Kronig relation 430
Kronecker δ-function 411
Kronecker delta 33
Kubo formula 327
Kubo–Greenwood formalism 313
Kubo–Greenwood theory 321

$L2_1$ structure 267
$La_{1-x}Ca_xMnO_3$ 271
labeling convention 93
Lagrange multiplier 42, 210
Lagrange multipliers 398
Lagrange parameter 32, 34
Landé g-factor 175
Landau coefficients 435
Landau expansion 218
Landau free energy 408
Landau gauge 7

Landau level 9, 10
Landau levels 7, 8, 10, 19
Landau susceptibility 12
Landau–Peierls susceptibility 15
Landauer formula 322
Langevin function 403, 405, 474
Langevin temperature 405
Laplace equation 132, 136
LAPW method 129, 132
large component 50
lattice 7
lattice distortions 339, 340
lattice Fourier transform 475, 478
lattice of localized electrons 78
lattice of skyrmions 458
lattice translation vector 89
layer resolved GMR 321
LCAO 123
LDA 61, 83
LDA total energy calculations 418
LDA+U 86
leading κ-dependence 150
leading energy dependence 136
Legendre polynomial 111
level scheme, FeRh 426
Li 102
Lieb–Oxford bound 84
Lifshitz–Kosevich formula 1, 16, 21
lifting degeneracies 100
Lindhard expression 187
Lindhard function 81, 290, 431
Lindhard susceptibility 81
Lindhard susceptibility, interband 187
linear algebra problem 98
linear augmented muffin-tin orbitals (LMTO) 129
linear augmented plane waves (LAPW) 129
linear combination 138
linear density response function 79
linear methods 89, 122
linear muffin-tin orbital method (LMTO) 132
linear response 79, 184
linear response functions 184
linear response theory 84, 429
linearity in energy 123

linearizing potential function 155
lines of high symmetry 152
Liouville equation 314
liquids 116
Lloyd's formula 471
LMTO 134
LMTO method 145
LMTO structure constants 138
LMTO- and KKR-ASA 157
local coordinate system 67, 160
local density of states 318, 319
local direction of magnetization 67
local field interaction 189
local frame of reference 67, 333, 473
local GMR ratios 320
local inhomogeneity parameter 84
local magnetic moments 206
local moment 440
local moment and Curie–Weiss law 441
local system 333
local-density approximation 44
local-density approximation (LDA) 59
local-exchange potential 39
local-moment limit 448
localization 45, 136
localized electron states 88
localized exclusion 268
localized magnetic moment 269
localized magnetic moment picture 282
localized moments 179
localized states 299
locally diagonalized 57
logarithmic derivative 112, 119, 121, 126, 128, 133, 156
logarithmic singularity 38, 86, 291, 295
long period 288, 296
long range Coulomb interaction 39
long wavelength limit 82
long-period oscillations 298
long-time evolution 473
longitudinal fluctuations 410, 412, 416, 423
longitudinal modes 412

longitudinal susceptibilities, nonuniform 205
longitudinal susceptibility 424
Lorentz force equation 16
low resistance state 323
low-energy excitations 388
low-lying excitations 160, 390
low-moment state (LM) 282, 283
low-spin ferromagnetic state 252
low-spin phases 239
low-spin states 238
lowest occupied states 64
LSD 70

macroscopic magnetization 407
Madelung potential 161
magnetic carriers, itinerant 173
magnetic carriers, localized 173
magnetic compounds 258
magnetic dipole interaction 347
magnetic energy gain 411
magnetic field 2, 9, 389
magnetic field, uniform 3
magnetic flux 17
magnetic form factor 248
magnetic induction 50
magnetic instability, condition 182
magnetic layers 307
magnetic moment 174
magnetic moment, Cr 235
magnetic moments 416
magnetic monopole 362
magnetic nanostructures 287
magnetic phase transition 388
magnetic phase transitions 339
magnetic pressure 212, 213, 282
magnetic properties of 174
magnetic quantum numbers 109
magnetic recording material 271
magnetic skyrmion 451
magnetic structure 336
magnetic susceptibilities of... 23
magnetic susceptibility 1, 178
magnetic topological insulator 384

magnetic transition metal 217
magnetic valence 263
magnetic weakness 264
magnetization 181, 386, 391, 405
magnetization at finite temperatures 400
magnetization decrease 400
magnetization vector 166, 428
magneto-crystalline anisotropy 47, 327, 343, 347, 349
magneto-optical Kerr effect 271, 288
magnetoresistance 312
magnetoresistance devices 274
magnitude of magnetic moment 69, 392
magnon densities of states 394, 397
magnon dispersion of Co_2FeSi and Co_2MnSi 399
magnon energies 396
magnon picture 401
magnon spectra 393
magnon spectra in nonprimitive lattices 396
magnon spectrum 394, 398
magnons 271
majority band 222
majority electrons 318
majority-spin electron Fermi surface, bcc Fe 226
majority-spin electrons 154, 224, 232
manganese 231, 237
many-body perturbation theory 78
many-body problem 28, 39
many-body states 29
many-body techniques 85
many-body wave functions 41
mass–velocity 333
mass–velocity term 53
matching conditions 150
matrix elements 114, 309
matrix formalism 56
maximum multiplicity 175
Maxwell's equation 49
Maxwell's equations 47
mean field 403
mean-field approximation 175, 442, 477
mean-field theory 401, 406
metallic binding 103

metallic bonding 149
metallic cohesion 149
metallic magnetism 173
metalloid 259
metalloids 262, 266
metals 5
metamagnetic behaviour 218
metastable states 238
metastable total energy minima 249
method of residues 13, 20
MFA for the multi-sublattice case 442
micromagnetic model 455
microscopic reversibility 315
minimization procedure 32
minority band 222
minority electrons 318
minority Fermi surfaces, Fe 224
minority-spin electron Fermi surface, bcc Fe 225
minority-spin electrons 154, 224, 232, 262
mixed state 57
Mn_3Ga in DO22-structure 448
Mn_3Sn 345
Mn, α 245
Mn, β 242
Mn, δ 231
Mn, γ 240
Mn-Sb covalent interaction 270
$MnAu_4$ 265
MnBi 266
MnMnSb 270
MnSi 430
model grand potential 477
model Hamiltonian 244, 401
molecular beam epitaxy 258, 287, 420
molecular field 177, 180, 385, 403, 474, 477
molecular field constant 181, 182
molecules 70, 116
momentum operator 3
monoclinic distortion 342
monoclinic unit cell 340
monolayers 287, 349
monolayers (ML) 297
monopole 461
monotonic mapping 154
monovalent metals 100

Monte Carlo method 435
Monte Carlo technique 480
Morin transition 345
motion 28
motion of nuclei 30
Mott relation 377
MR in double perovskites 326
muffin-tin 117
muffin-tin approximation 130
muffin-tin APWs 116
muffin-tin potential 468
muffin-tin sphere 115
muffin-tin spheres 108
multi-center integral 124, 136
multi-center integrals 129
multi-sublattice case 442
multiatom effect 464
multilayer 287, 293, 307
multilayer system 299
multilayers 287, 347
multiple reflections 302
multiple scattering 116
multiple scattering theory (MST) 116
multiplet structure 229
multiplicity 175
multipole potential 132

N-electron system 32
Néel skyrmion 456
Néel temperature 250, 424
Néel temperature, calculated 424
nabla operator 3
nearest-neighbor angle 466
nearly free-electron model 100
negative energy case 136
Nernst thermopower 377
nesting 232, 233
nesting Fermi surface 205
nesting translation 234
nesting vector 296
neutral gas 35
neutral quasiparticle 229
neutron diffraction 247
neutron scattering 431, 466
neutron scattering spectra 243
Newton's law 314
Ni based alloys 259
Ni_3Al in dynamic approximation 438
$Ni_{1-x}B_x$ 266
$Ni_{1-x}P_x$ 266
Ni_4Mo crystal structure 266
Ni-Co 261

Ni-Co data 260
Ni-Cu 261
Ni-rich Fe-Ni 261
Ni-Zn 261
nickel 173
NiCo alloys 259
NiCu alloys 259
NiF_2 343
NiMnSb 267
noble gases 5
node 140
non-overlapping spheres 108
noncollinear configurations 220, 415
noncollinear magnetic configuration 336
noncollinear magnetic order 160
noncollinear moment arrangements 282
noncollinear spin arrangements 160
noncollinear spiral moments 415
noncollinear structure 338
nondegenerate state 57
nondiagonal scattering matrix 122
nonequilibrium films 217
nonequilibrium structures 217
noninteracting electrons 5
noninteracting particles 42
nonlocal term 33
nonrelativistic limit 51
nonrelativistic quantum mechanics 46
nonuniformly magnetized systems 63
norm conserving pseudopotential 106
normal mode 414
normalization 54
normalization integral 30
normalization matrix 144, 146
normalized 29
nuclear magnetic resonance 243
nuclear part 29
nuclear positions 29
nuclear spins 289
nuclear system 30
nuclei 28
number of allowed wave vectors 96
number of bands 96

number of magnetic carriers 174, 179, 419, 420, 476
number of modes 410
number of normal modes 428
number of particles 9
number of states 9
number of up and down-spin electrons 24
numerical techniques 96
Nyquist relation 429

occupation number changes 44
occupied states 42
off-diagonal magneto-optical properties 274
off-site total energy 442
on the energy shell t-matrix 471
on-top value 75
one-center integral 124
one-center terms 135
one-dimensional Brillouin zone 295
one-electron reduced density matrix 71, 74
one-particle energy 9
Onsager cavity-field approximation 477
Onsager's formula 18
operator 117
optic branch 421
optical branch 395, 397
optical conductivity 274
optical magnon 395
optical reflectivity 274
optimization condition 412
OPW 101
OPW for Li 103
OPW method 100
orbit 18
orbital contribution 391
orbital magnetic moments 55
orbital polarization 172, 345
orbital susceptibility 15
orbital-polarization corrections 86
orbitals 44, 123
order 466
order on atomic scale 465
ordered compound 265, 286
orientational configurations 474
orientational fluctuations 388
orthogonal 29

orthogonality 100
orthogonality relation 129
orthogonalization 103
orthogonalized plane wave (OPW) 100
orthogonalized plane waves 96
orthonormality 34
orthorhombic distortions 340
orthorhombic lattice distortion 339
orthorhombic unit cell 340
oscillations 114
oscillatory contribution 7
oscillatory correction 14
oscillatory exchange coupling 287, 302
overbinding 191, 213, 251
overlap matrix 146
oxidized interfacial Fe 324

pair exchange interaction 435
parabolic trend 194
parabolic valence band 103
parallel moment arrangement 317
parallelepiped 95
paramagnetic fluctuation 411
paramagnetic fluctuations 413
paramagnetic layer 304
paramagnetic modes 413
paramagnetic moment 479
paramagnetic state 411, 476
paramagnetic susceptibility 15, 424
paramagnetic term 3
parametrization 127
parentage, t_{2g}, e_g 272
partial charge 147
partial Cr d density of states 265
partial reflection 302
partially filled energy bands 149
particle conservation 42
particle density 47, 72, 385
particle number 58
partition function 2, 209, 401, 402, 415
partitioning of space 134
Pauli equation 52, 54
Pauli paramagnetism 22
Pauli principle 46, 315
Pauli spin matrices 46, 65, 333, 386, 389
Pauli spin matrix 160

Pauli spin matrix vector 2
Pauli spin susceptibility 23
Pauli spinor 51
Pauli susceptibility 12, 14, 479
Pauli-paramagnetic susceptibility 7
PdMnSb 267
period of oscillations 15
periodic boundary conditions 6, 9, 35, 95
periodic potential 16
periodic table of the elements 216, 263
periodic zone scheme 295, 296
periodicity 98
permanent magnetic moment 4
permanent magnets 327
perpendicular bisectors 92
perturbation 108, 307
perturbation potential 309
perturbation theory 77, 112, 172
phase diagram, experimental 286
phase diagrams 220
phase shift 296, 302
phase space 473
phase stability 200, 258
phase transition 425, 466
phenomenological facts 27
phonons 30, 271
photoemission 222, 229, 262, 302, 466
photoemission data 224
photoemission process 229
photoemission, Co 225
Planck distribution function 399
plane waves 96, 108, 290
plane-wave ansatz 100
plane-wave expansion 97
plane-wave state 6
point groups 94
point-group operations 343
point-group symmetry 168
Poisson's summation formula 19
polar angles 64, 392
polarization 262
polarization mechanism 262
pole 310
polyatomic system 141
positive background 36
positive energy solutions 50

positron-annihilation radiation 237
potential 5
potential barrier 218, 322
potential energy 30
potential function 152, 154, 156
potential matrix 59
precautions 35
precipitates 250
predictions 271
pressure 285
primitive cells 95
primitive translation 92
principle value 307
prismatic Brillouin zone 296
probability 115
proof 41
pseudopotential theory 104
pseudopotentials 104
PtMnSb 267
pure ℓ band 157
pure host moment 260
PW 101
PW91-GGA 86

quadratic form 417
quantization rule 17
quantized 8
quantized orbital angular momenta 15
quantized orbital motion 24
quantum inequality 386
quantum mechanical justification 29
quantum Monte Carlo calculations 61
quantum Monte Carlo correlation energies 78
quantum Monte Carlo studies 83
quantum numbers 29, 89
quantum well 302
quantum-well picture 298
quantum-well states 302, 318
quasi-gap 272
quasi-particle excitation spectrum 45
quenched angular momentum 175

radial Dirac equation 50
radial Schrödinger equation 50, 109, 171
random configurations 473

random crystal potential 469
random magnetic moment configurations 467
random potential 468
random-phase approximation 79
random-phase approximation (RPA) 430
rank of secular equation 122
rank of the secular 98
rapid quenching 240
rapidly varying densities 85
rare earths 173, 289
Rayleigh–Ritz variational procedure 145, 417
real space 17
real-space notation 32, 100
realistic density of states 25
reciprocal lattice 91
reciprocal lattice vector 97
reciprocal lattice vectors 108
reciprocal space 17, 91
reduced density gradient 84, 86
reduced fluctuations, Ni 418
reduced magnetization 175
reduced magnetization, Ni 418
reduced sublattice magnetization 424
reduced symmetry 347
reduced temperature 175
reduced zone 296
reflection coefficients 304
reflection matrices 310, 311
reflections 94
regular solution 138
regularity 132
relativistic corrections 171, 284
relativistic density-functional 46
relativistic effects 327
relaxation 45
relaxation time 315–317
renormalization 165, 274
representability 41
representation 116, 169
repulsive impurity potential 262
residues 310
resistance 312
resolvent operator 471
response function 82, 187
response functions, static 185
restricted variation 198
restricting relations 340

Rhodes–Wohlfarth plot 179, 180
rhombohedral lattice 343
Riemann zeta function 400
Righi–Leduc effect 379
rigid-band behavior 264
rigid-band formula 261
rigid-band theory 260, 262
rigorous derivation 22
rigorous solution 108
RKKY exchange 288
RKKY exchange constant 289
RKKY-indirect exchange interaction 289
RKKY-type interactions 448
root 11
root tracing 122
roots 97
rotated Pauli spin matrix 65
rotational invariance 347
rotations 93
RPA scaling 62
rutile crystal structure 272
Rydberg 37

saddle point 285
saddle-point approximation 440
sampling 293
sandwich structures 288
saturation magnetization 174, 211, 258, 407
sc-Brillouin zone 92
scalar product 91, 100
scalar relativistic approximation 52, 285
scalar relativistic Hamiltonian 327
scalar relativistic Kohn–Sham Hamiltonian 333
scalar relativistic operator 171
scalar relativistic radial equation 54
scalar relativistic wave equation 52, 171
scalar wave function 160, 333
scale factor for LMTO 156
scaled exchange energy 76
scaled wave function 76
scaling inequalities 77
scaling parameter 158
scaling relations 75, 76
scanning electron microscopy 288
scatterers 471

scattering 116
scattering event 314
scattering of electrons 22
scattering path operator 468, 469
scattering phase shift 119, 121
scattering potential 320
scattering probability 320
scattering processes 313
scattering strength 122
scattering term 314
Schrödinger equation 3, 5, 7, 30, 89, 387, 389
screened Coulomb interaction 39, 84
screening 79, 184, 262, 272
screening length 85
second quantization 32, 57
second-order transition 434
secular determinant 112
secular equation 97, 103, 122, 129, 130, 414
secular matrix 135, 141, 143
Seebeck coefficient 377
self-consistency cycle 422
self-consistency equations 413, 418
self-consistency problem 418
self-consistent 34
self-consistent band theory 222
self-consistent directions 339
self-consistent field 34
self-consistent field problem 43
self-consistent response 202
self-energy 45, 229
self-energy correction 88
self-intermetallic compound 247
semi-empirical band structure, Ni 229
semi-empirical tight-binding models 298
semiconducting gap 269
separation 28
set of coordinates 29
Shandite 356
shape approximation 116, 122
shear distortions 116
short period 296
short periods 288
short-range order 463, 465
short-time description 473
SIC 86
side-jump conductivity 355
simple cubic (sc) 92

simple metals 98
simulated annealing 336
simulated annealing process 345
single-atom scattering t-matrix 316
single-electron norm 145
single-magnon energies 400
single-particle energy 306
single-particle spinor function 387
single-particle spinor functions 166
single-particle states 6
single-particle wave function 35
single-particle wave functions 42
single-site t-matrices 470
single-site t-matrix 468, 469
single-site expansion 141
single-site potential 118
single-site terms 414
single-site wave function 124, 472
singular behavior 219
singularities 116
site decompositions 145
site index 128
site-dependent spinor 162, 170
site-I atom 244
site-I atoms 243
site-II atom 244
site-II atoms 243
size quantization 302
Skyrme 451
skyrmion model 451
skyrmion number 451
Slater determinant 6, 32, 74
Slater exchange 39
Slater orbital 129
Slater–Pauling curve 213, 258, 261, 281
Slater-type orbitals (STO) 124
slowly varying densities 62
small component 50
small wavelength oscillations 100
smeared out charges 5
sobering example 35
solid state 5
solids 62, 70
solubility condition 316
solute 259
Sommerfeld expansion 24, 38

Sommerfeld method 7
sp density of states 262
space group 94, 333, 343
space rotation 167
space translation 167, 334
spacer 287, 291
spacer material 288, 306
spatial distribution 262
special k-points 405
specific heat 38, 400, 466
specific heat
 measurements 274
spectral properties 30
spectral weight 232
spherical approximation 438
spherical average 145
spherical Bessel function 109, 127, 150, 248
spherical Bessel functions 112, 118, 136
spherical Fermi surface 18
spherical Hankel function 150
spherical Hankel functions 136
spherical harmonic 54
spherical harmonic
 expansions 142
spherical harmonics 51, 109, 124
spherical Neumann
 functions 118, 136
spherical potential Dirac
 equation 55
spherical wave 138
spin anisotropy ratio 320
spin as a dynamical variable 46
spin channel 316
spin degeneracy 100, 160, 303
spin degrees of freedom 471
spin density 28, 57, 69, 159
spin density matrix 27
spin fluctuations 271, 401, 406, 407, 422
spin function 162
spin glass 238, 240
spin Hamiltonian 338
spin indices 309
spin observable 50
spin orientation 9
spin polarization 56
spin projection 169, 336
spin quantization axis 347
spin rotation 166, 336, 346
spin rotation matrix 311
spin scaling 85
spin space groups 346

spin spiral 169, 236, 408
spin spirals 404
spin susceptibility 15, 22, 188, 473
spin valve 325
spin waves 465
spin–orbit coupling 55, 171, 249, 273, 327, 336, 343, 344, 347
spin–orbit coupling
 operator 55
spin–orbit coupling term 53
spin–orbit Hamiltonian 333
spin–orbit interaction 47, 50
spin–orbit matrix element 56
spin-1/2 rotation matrices 464
spin-1/2 rotation matrix 64, 159, 161, 333
spin-decomposed d
 occupation 215
spin-degenerate band
 structure 232
spin-density matrix 57, 73, 185, 240
spin-density response 188
spin-density response
 function 185
spin-density wave 235, 236
spin-density wave state 236
spin-density-functional
 theory 56
spin-dependent Hankel
 function 162
spin-dependent structure
 constants 163
spin-down probabilities 472
spin-down resistor 317
spin-electronics 274
spin-flip scattering 315
spin-fluctuation theory 429
spin-orbit coupling 2, 4
spin-parallel states 33
spin-polarized effective
 potentials 55
spin-polarized potentials 154
spin-projected density of
 states 182
spin-reorientation
 transition 462
spin-scaling relation 84
spin-scaling relations 77
spin-space group 336
spin-space group (SSG) 166, 334
spin-space rotations 346

spin-spiral 391
spin-spiral configurations 465
spin-spiral state 232
spin-up d band 261
spin-up probabilities 472
spin-up resistor 317
spin-wave dispersion 394
spin-wave mode 465
spin-wave spectrum 389
spin-wave stiffness
 constant 396, 431
spin-wave stiffness constant of
 Co_2FeSi 399
spin-wave stiffness
 expression 428
spin-wave theory 400
spinor 35
spinor components 390
spinor equation 67
spinor function 160, 333
spinor structure constants 164
spiral energies 443, 466
spiral energies from force
 theorem 446
spiral magnetic order 160
spiral magnetic state 238, 239
spiral magnetic state (SPM) 251
spiral magnetic states 242
spiral magnetic structure 167
spiral state 169
spiral structure 169
spiral structure,
 incommensurate 166
spiral vector 169
spontaneous order 26
sputtering 258, 287
square-root singularity 218
Sr_2CrIrO_6 280
Sr_2CrOsO_6 280
Sr_2CrReO_6 280
Sr_2CrWO_6 280
SSG 167, 334
stability, structure 198
staggered magnetic field 201
standard representation 46
standing wave 235
standing wave free-particle
 solution 117
standing wave Green's
 function 117
state densities of CrMnSb 270
state densities of CrO_2 273
state densities of Fe_3Ni 283, 284

state densities of FeCr 268
state densities of PtMnSb 269
state density of $Cr_{0.2}Co_{0.8}$ 265
state-density peak 264
static multipole potential 132
static nonuniform susceptibilities 430
static positions 40
static susceptibility 4
stationary phase 21
stationary phase approximation 305
stationary points 296
stationary states 116
stationary-phase approximation 296
statistical mechanics 9
step function 161
Stoke's theorem 17
Stoner condition 182, 189, 210, 273
Stoner exchange constant 190
Stoner exchange constants of 191
Stoner excitations 420, 474, 477
Stoner factor 184
Stoner parameter 183, 185, 190, 205, 209, 221
Stoner parameters 203
Stoner products, values 191
Stoner spin-flip excitation continuum 395
Stoner temperatures 388
Stoner theory 173, 180, 385, 388
Stoner–Mermin theory 389
strong ferromagnets 260
structure constants 120, 122, 144
structure Green's function 469
sublattice 230
sublattice magnetization 423
substitutions 11
substrates 217
sum on eigenvalues 43
sum over normal modes 414
sum rule 73, 86
supercell 169
superconductors 31
superexchange 343, 449
superlattice 298
superlattice wave function 320
superparamagnet 465

superposition of oscillatory terms 306
supersaturated CuFe alloys 250
surface integral 111
surface states 114
surface states of Mn_3Sn and Mn_3Ge 373
surfaces 116, 347
susceptibilities of V, Cr, and Pd 203
susceptibility 7, 174, 394, 405, 412
susceptibility calculations 201
susceptibility enhancement 191, 203, 434
susceptibility maximum 205
susceptibility tensor 478
susceptibility tensor, dimensionless 475
susceptibility, $\chi_j(\mathbf{k},\omega)$ 429
susceptibility, molar 4
switching 288
symmetric exchange 343
symmetry 4, 89, 262
symmetry class 348
symmetry group 344
symmetry operation 340, 346
symmetry principle 338
symmetry properties 32, 334, 336
symmetry transformation 166
system of noninteracting electrons 386

t-matrix 307, 308
tail 133, 138
tail augmentation 138, 139
tailored magnets 258
Taylor series 12, 25, 44, 125, 305, 314
technical remarks 114
temperature correction 14
temperature corrections 25
temperature independence 23
temperature-dependent fluctuation components 418
temperature-dependent susceptibility 412
temperature-independent γ 38
ternaries 267
tetragonal crystallographic distortions 241
tetragonal distortion 242, 424
tetragonal lattice distortion 249

tetrahedral order 241
tetrahedron method 96
Th_3P_4 structure 335
theorems 40
theoretical models 281
theory of fluctuations 458
thermal equilibrium 399
thermal excitations 282
thermal expansion 281, 419
thermal gradient 314
thermal properties 389, 399
thermodynamic grand potential 303
thermodynamic quantities 464
thermodynamic relation 2, 14
thermodynamics 387
thin films 347
Thomas–Fermi screened Coulomb potential 82
Thomas–Fermi screening length 82
Thomas–Fermi screening wave number 85
Thomas–Fermi theory 63
three sublattices 447
three-center contributions 143
three-center term 124
three-center terms 135
$Ti Al_3$ crystal structure 266
$TiBe_2$ 430
time reversal 336
time-dependent parameters 392
time-dependent phenomena 390
time-dependent Schrödinger equation 390
time-reversal operation 339
timescales 390, 473
top of d band 194
topological Hall effect 457, 462
topological insulator 383
topological number 451
topological properties 382
topology 224
topology, Fermi surface 295
total angular momentum 51
total energy 181
total energy gain 212, 236
total energy per electron 36, 37
total state of system 29
total-energy calculations 218, 282
total-energy change 406
total-energy difference 298

total-energy differences 205, 408, 464
total-energy differences, FeNi, Fe_3Ni, γ-Fe, 283
total-energy surface 415
trace 57, 58, 308–310, 385, 401
tracing the roots 114
trajectories 16, 116
transition matrix 116
transition metal oxides 272
transition metals, band structure 196
transition probabilities 316
transition probability 322
transition series 192
transition temperature 422
translation 97
translation eigenvalue 92
translation operators 90
translational invariance 35, 93, 304
translational periodicity 89
translational symmetry 133, 144
translations 91
transport properties 271, 317
transverse fluctuations 410, 412, 416, 423
transverse magnetic susceptibility 395
transverse modes 412
transverse susceptibility 424
trial functions 172
trial Hamiltonian 402
trial potential 387
trial wave function 75
triangular antiferromagnet Mn_3Sn 343
triangular configuration 345
triangular shape 260
triangularization 114
tunnel conductance 322
tunnel junction 323
tunnel magnetoresistance 321
two electron-reduced density matrix 71
two sublattice case 443
two-center contributions 142
two-center integral 124
two-center terms 135
two-component functions 50
two-component spinor function 56, 334
two-dimensional Brillouin zone 295

two-particle effects 229
two-site interaction terms 407, 414
two-state model 250
twofold superlattice 233
type-I transition 219
type-II transition 219
type-III transition 219
types of evolution 219

U_3P_4, U_3As_4, U_3Sb_4 335
U_3X_4, (X=P and As) 335
U_3Sb_4 338
unaugmented functions 143
unaugmented spherical waves 142
uncorrelated average 477
unenhanced magnetic susceptibility 290
unenhanced susceptibility 203
ungerade states of Co–Co 277
unhybridized bands 152, 157
unhybridized d bands 195
unified theory 430
unified theory of Slater–Pauling curve 262
unified treatment 180
uniform density scaling 84
uniform positive background 35
uniform static susceptibility 466
uniform susceptibility 190, 413
uniformly scaled wave function 75
unique GGA 71
unit cell 92
unit spinor 162
unit step function 80, 186, 398
unitary transformation 64
unitary transformations 339
units 38, 174
unpaired spins 174
unperturbed medium 469
up-spin d electrons 262
UPdSn 339, 341
uranium compounds 335
useful formula 35

vacuum tunneling measurements 274
valence 37
valence electron 98
valence electron state density 145

valence electrons 99
valence states 5
valence-charge density 147
validity of band structure 222
values of exchange parameters 207
variation 32, 43, 58
variation of the total energy 59
variational estimate 126
variational expression 109
variational parameters 402, 407
variational principle 32, 41, 42, 402, 467
variational treatment 172
VASP 104
VAu_4 265
vector model 474
vector operator 46
vector potential 1, 7, 16
velocity of light 1
velocity of wave packet 16
volume compression 220
volume expansion 220
volume of crystal 95
volume of unit cell 95
von Weizsäcker correction 63
VPd_3 prediction 266

wave function character 224
wave function for spiral state 256
wave function scaling 76
wave packet 16
weak confinement 304
weak crystal potential 100
weak ferromagnet Ni_3Al 436
weak ferromagnet, Sc_3In 428
weak ferromagnetism 342, 346
weakness of Fe 263
wedge-shaped spacer 287
weights 147
Weiss molecular field model 178
Weyl node 362
Weyl points in Co_2MnAl 368
Wiedemann–Franz law 380
Wien2k 132
Wigner lattice 78
Wigner–Seitz approximation 103
Wigner–Seitz atomic spheres 134
Wigner–Seitz boundary conditions 194

Wigner–Seitz radii 213
Wigner–Seitz sphere 152
Wigner–Seitz sphere
  radius 103
Wronskian 106, 119, 151
Wyckoff positions 272

Zak phase 351
Zeeman term 3, 7, 9, 10
zero-point energy 197
zero-point fluctuations
  430
zero-point motion 197

zero-point vibrations 429
zero-slope boundary
  conditions 103
zero-temperature property
  174
$ZrZn_2$ 430